Plant-Microbe Interaction under Xenobiotic Exposure

Swarnendu Roy • Vivekananda Mandal
Editors

Plant-Microbe Interaction under Xenobiotic Exposure

Springer

Editors
Swarnendu Roy
Department of Botany
University of North Bengal
Raja Rammohunpur, West Bengal, India

Vivekananda Mandal
Department of Botany
University of Gour Banga
Malda, West Bengal, India

ISBN 978-981-96-8259-1 ISBN 978-981-96-8260-7 (eBook)
https://doi.org/10.1007/978-981-96-8260-7

This Springer imprint is published by the registered company Springer Nature Singapore Pte Ltd.
The registered company address is: 152 Beach Road, #21-01/04 Gateway East, Singapore 189721, Singapore

Preface

The term 'xenobiotics' refers to a broad class of organic and inorganic compounds which are foreign to the entire biological system and are accidentally introduced in much greater amounts. These include drugs, pesticides, plastics, engineered nanoparticles, industrial chemicals, heavy metals, toxic gases and environmental pollutants. In the coming days, the quantity of such xenobiotic compounds is expected to cross a billion tonnes, in every sphere of human interactions, i.e. soil, water and air. The increasing abundance of xenobiotic compounds in the environment is likely to induce significant impacts on the plant-microbe interaction. In this context, the microbial partners, viz. plant growth-promoting rhizobacteria, endophytes and mycorrhiza, have been known to restrain the impact of environmental pollutants on plant growth and productivity. The microorganisms associated with plants are highly versatile and adaptable to changing environmental conditions, and therefore, the success of these microbial partners under xenobiotic exposure is predicted to define the health status of the plants. In this purview, it is high time to evaluate the extent of xenobiotic exposure in the plants and microbes in a holistic manner. The book thus attempts to present the impact of a wide array of xenobiotic compounds on different plant groups and microbes. It also highlights the recent advances in the strategies and techniques of bioremediation, which include the use of novel microorganisms for rhizosphere engineering, the development of nanosensors, etc. The book will, therefore, cater to the necessity of the research scholars, faculties and scientists working in the field of plant sciences, agriculture and allied disciplines on this burning topic.

Raja Rammohunpur, West Bengal, India — Swarnendu Roy
Malda, West Bengal, India — Vivekananda Mandal

Contents

Editors and Contributors

About the Editors

Swarnendu Roy is currently working as an Assistant Professor at the University of North Bengal, Raja Rammohunpur, West Bengal, India. He has more than 14 years of teaching experience. His doctoral thesis involved understanding the mechanism of salinity stress tolerance in grasses. He has published more than 50 articles in reputed journals with high-impact factors like *Water Research*, *Scientific Reports*, *Plant Physiology and Biochemistry*, *Carbohydrate Polymers*, etc. He has published six edited books with Springer and CRC Press, and also published several book chapters in edited volumes. Under his supervision, two students have been awarded PhD, and he is currently supervising four PhD research scholars who are working on stress management in plants by engineered nanoparticles, synthesis of starch-based biofilms and edible coatings. He received International Travel Support (ITS) from DST, Govt. of India, to present his research paper at the International Conference on Integrative Plant Physiology at Sitges, Spain. He is a Life member of renowned Societies like the Indian Science Congress, the Association of Food Scientists & Technologists (India) and the Society of Plant Research (India).

Vivekananda Mandal is a Professor of Botany at the University of Gour Banga, Malda, India. He has been working in the field of plant-microbe interaction and plant and microbial metabolites bio-prospecting for the last 18 years. He leads the plant-microbe interaction and microbial ecology research group at his institute. He serves as a reviewer for several reputed international journals and research projects. He has published 50 research articles to his credit, including research papers, book chapters and review articles. He has supervised six PhD thesis papers. He is the Principal Investigator of different state and national funded projects. He has also authored ten book chapters in edited books published by Springer-Verlag GmbH, Germany,

Springer Nature Singapore Pte Ltd. and Springer Cham. He has also organized several conferences/workshops as convener/organizing secretary. He has been awarded the 'Best Researcher Award 2022' by the International Research Awards on New Science Inventions, the Science Father, NESIN Awards. He is a Life member of the Association of Microbiologists in India since 2007, a Life member of the Academy for Advancement of Agricultural Sciences, a Life member of the Association of Fungal Biologists since 2024 and a Subject Editor (Botany, Plant Physiology & Biochemistry) in *Vegetos*, An International Journal of Plant Research and Biotechnology, of the Society for Plant Research (SPR), published by Springer Nature.

Contributors

Vahid Alah Jahandideh Mahjen Abadi Soil and Water Research Institute (SWR), Agricultural Research, Education and Extension Organization (AREEO), Karaj, Iran

Malay Kumar Adak Department of Botany, University of Kalyani, Kalyani, Nadia, India

Samarpita Adhikary Microbiology and Computational Biology Laboratory, Department of Botany, Raiganj University, Raiganj, West Bengal, India

Zisan Ahamed Department of Ecological Studies & International Centre for Ecological Engineering (ICEE), University of Kalyani, Kalyani, Nadia, West Bengal, India

Md Masidur Alam Department of Botany, Malda College, Malda, West Bengal, India

Md. Azmalul Alam Department of Botany, Dukhulal Nibaran Chandra College, Murshidabad, India

Kakan Ball Department of Botany, Raiganj University, Raiganj, West Bengal, India

Subhendu Bandyopadhyay Department of Biological Sciences, School of Life Science and Biotechnology, Adamas University, Kolkata, West Bengal, India

Mahadev Bera School of Environment and Disaster Management, Ramakrishna Mission Vivekananda Educational and Research Institute, Kolkata, West Bengal, India

Priya Borah Ecology and Environmental Remediation Laboratory, Department of Botany, Gauhati University, Guwahati, Assam, India

Madhumita G. Chaki Microbiology and Computational Biology Laboratory, Department of Botany, Raiganj University, Raiganj, West Bengal, India

Nilakshi Chakraborty Department of Botany, University of Kalyani, Kalyani, Nadia, India

Pooran Chand Department of Genetics and Plant Breeding, Sardar Vallabhbhai Patel University of Agriculture and Technology, Meerut, Uttar Pradesh, India

Malini Roy Choudhury School of Environment and Disaster Management, Ramakrishna Mission Vivekananda Educational and Research Institute, Kolkata, West Bengal, India

School of Agriculture and Food Sustainability, The University of Queensland, Brisbane, QLD, Australia

Simran Choudhury Department of Life Sciences, Parul Institute of Applied Sciences, Parul University, Vadodara, Gujarat, India

Rudrakshi Chowdhury Department of Biological Sciences, School of Life Science and Biotechnology, Adamas University, Kolkata, West Bengal, India

Chittaranjan Das Department of Botany, Kulti College, Kulti, Paschim Bardhaman, West Bengal, India

Hemanga Kumar Das Department of Biotechnology and Biomedical Engineering, U.V. Patel College of Engineering, Ganpat University, Mehsana, Gujarat, India

Sumanta Das School of Environment and Disaster Management, Ramakrishna Mission Vivekananda Educational and Research Institute, Kolkata, West Bengal, India

School of Agriculture and Food Sustainability, The University of Queensland, Brisbane, QLD, Australia

Tapas Ranjan Das Division of Genetics, ICAR-Indian Agriculture Research Institute, New Delhi, India

Hemen Deka Ecology and Environmental Remediation Laboratory, Department of Botany, Gauhati University, Guwahati, Assam, India

Saikat Dey Division of Agronomy, School of Agriculture and Rural Development, Ramakrishna Mission Vivekananda Educational and Research Institute, Kolkata, West Bengal, India

Anannya Dhar Division of Agronomy, School of Agriculture and Rural Development, Ramakrishna Mission Vivekananda Educational and Research Institute, Kolkata, West Bengal, India

Anwesha Dutta Department of Botany, Dukhulal Nibaran Chandra College, Murshidabad, India

Gargi Department of Bioscience and Biotechnology, Banasthali Vidyapith, Rajasthan, India

Amit Ghosh Department of Biological Sciences, School of Life Science and Biotechnology, Adamas University, Kolkata, West Bengal, India

Santanu Gupta Department of Botany, Malda College, Malda, West Bengal, India

Nishika Jaishee Department of Botany, The University of Burdwan, Bardhaman, India

Soujanya Jana Division of Agronomy, School of Agriculture and Rural Development, Ramakrishna Mission Vivekananda Educational and Research Institute, Kolkata, West Bengal, India

Sahib Kaur Department of Bioscience and Biotechnology, Banasthali Vidyapith, Rajasthan, India

Somanjana Khatua Molecular and Applied Mycology Laboratory, Department of Botany, Faculty of Science, University of Allahabad, Prayagraj, Uttar Pradesh, India

Jinnatunnesa Khatun Department of Botany, The University of Burdwan, Bardhaman, India

Soumita Kumar Department of Microbiology, University of Kalyani, Kalyani, Nadia, West Bengal, India

Vikas Kumar Department of Biotechnology, University Institute of Biotechnology, Chandigarh University, Mohali, Punjab, India

Anupam Kundu Department of Botany, Visva-Bharati (A Central University), Santiniketan, West Bengal, India

Ashutosh Kundu Plant and Microbial Physiology and Biochemistry Laboratory, Department of Botany, University of Gour Banga, Malda, West Bengal, India

Rajat Laller School of Biological and Environmental Sciences, Shoolini University of Biotechnology and Management Sciences, Sultanpur, India

Rohini Lama Department of Botany, Darjeeling Govt. College, Darjeeling, India

Arpita Maity Centre for Life Science and Department of Botany and Forestry, Vidyasagar University, Midnapore, West Bengal, India

Bhaskar Mallick Department of Ecological Studies & International Centre for Ecological Engineering (ICEE), University of Kalyani, Kalyani, Nadia, West Bengal, India

Manab Mandal Department of Botany, Dukhulal Nibaran Chandra College, Murshidabad, India

Narayan Chandra Mandal Mycology and Plant Pathology Laboratory, Department of Botany, Siksha Bhavana (Institute of Science), Visva-Bharati, Santiniketan, West Bengal, India

Supratim Mandal Department of Microbiology, University of Kalyani, Kalyani, Nadia, West Bengal, India

Vivekananda Mandal Plant and Microbial Physiology and Biochemistry Laboratory, Department of Botany, University of Gour Banga, Malda, Mokdumpur, West Bengal, India

Sonal Mishra Department of Bioscience and Biotechnology, Banasthali Vidyapith, Rajasthan, India

Debabrata Misra Plant and Microbial Physiology and Biochemistry Laboratory, Department of Botany, University of Gour Banga, Malda, Mokdumpur, West Bengal, India

Prashanta Kumar Mitra Department of Lifescience, Shri Rawatpura Sarkar University, Raipur, Chattishgarh, India

Amal Kumar Mondal Plant Taxonomy, Biosystematics and Molecular Taxonomy Laboratory, UGC-DRS-SAP & DBT-BOOST-WB Supported Department, Department of Botany and Forestry, Vidyasagar University, Midnapore, West Bengal, India

Arka Mondal Department of Botany, Visva-Bharati (A Central University), Santiniketan, West Bengal, India

Debojyoti Moulick Department of Environmental Science, University of Kalyani, Nadia, West Bengal, India

Seyed Majid Mousavi Soil and Water Research Institute (SWR), Agricultural Research, Education and Extension Organization (AREEO), Karaj, Iran

Chandra Mukherjee Department of Basic Science (Chemistry), The Neotia University, Sarisa, West Bengal, India

Seyed Ali Ghaffari Nejad Soil and Water Research Institute (SWR), Agricultural Research, Education and Extension Organization (AREEO), Karaj, Iran

Ayon Pal Microbiology and Computational Biology Laboratory, Department of Botany, Raiganj University, Raiganj, West Bengal, India

Sudeshna Pal Department of Botany, Dukhulal Nibaran Chandra College, Murshidabad, India

Sumanta Pal Department of Botany, Dukhulal Nibaran Chandra College, Murshidabad, India

Moumita Parveen Mycology and Plant Pathology Laboratory, Department of Botany, Siksha Bhavana (Institute of Science), Visva-Bharati, Santiniketan, West Bengal, India

Nousi Parvin Department of Botany, Visva-Bharati (A Central University), Santiniketan, West Bengal, India

Sarmistha Paul Mycology and Plant Pathology Laboratory, Department of Botany, Siksha Bhavana (Institute of Science), Visva-Bharati, Santiniketan, West Bengal, India

Shouvik Pramanik Department of Botany, Visva-Bharati (A Central University), Santiniketan, West Bengal, India

Sushma Rani School of Life and Basic Sciences, Jaipur National University, Jaipur, Rajasthan, India

Jnanendra Rath Department of Botany, Visva-Bharati (A Central University), Santiniketan, West Bengal, India

Ayan Roy Department of Biotechnology & Bioinformatics, Asian University for Women, Chattogram, Bangladesh

Raj Narayan Roy Department of Botany, Dr. Bhupendra Nath Dutta Smriti Mahavidyalaya, Hatgobindapur, Purba-bardhaman, West Bengal, India

Soumita Roy Department of Botany, University of Kalyani, Kalyani, Nadia, India

Sourav Roy Department of Botany, Visva-Bharati (A Central University), Santiniketan, West Bengal, India

Usha Sabharwal Department of Life Science, Parul Institute of Applied Sciences, Parul University, Waghodia, Vadodara, Gujarat, India

Sanjoy Sadhukhan Department of Botany, Raiganj University, Raiganj, West Bengal, India

Kunal Kumar Saha Mycology and Plant Pathology Laboratory, Department of Botany, Siksha Bhavana (Institute of Science), Visva-Bharati, Santiniketan, West Bengal, India

Subarna Saha Department of Botany, Visva-Bharati (A Central University), Santiniketan, West Bengal, India

Uddhav Sarania Ecology and Environmental Remediation Laboratory, Department of Botany, Gauhati University, Guwahati, Assam, India

Azadeh Sedaghat Soil and Water Research Institute (SWR), Agricultural Research, Education and Extension Organization (AREEO), Karaj, Iran

Sukanta Kumar Sen Department of Botany, Siksha Bhavana, Visva-Bharati, Santiniketan, West Bengal, India

Nitin Sharma Department of Biotechnology, Chandigarh Group of Colleges, Mohali, Punjab, India

Abhishek Singh Department of Genetics and Plant Breeding, Sardar Vallabhbhai Patel University of Agriculture and Technology, Meerut, Uttar Pradesh, India

Somvir Singh Department of Biosciences, University Institute of Biotechnology, Chandigarh University Gharuan, Mohali, Punjab, India

Shivangee Solanki Department of Biotechnology, Gujarat Biotechnology University, Gandhinagar, Gujarat, India

Sudipta Tripathi School of Environment and Disaster Management, Ramakrishna Mission Vivekananda Educational and Research Institute, Kolkata, West Bengal, India

Ecological Impacts of Emerging Pollutants in the Context of Climate Change: Challenges and Solutions

Mahadev Bera, Anannya Dhar, Saikat Dey, Soujanya Jana, Sumanta Das, Malini Roy Choudhury, Sudipta Tripathi, and Debojyoti Moulick

Abstract The interplay between emerging pollutants and climate change presents unprecedented ecological challenges, necessitating a comprehensive understanding and innovative solutions. This chapter provides a detailed overview of the ecological impacts of emerging pollutants and industrial chemicals within the framework of a changing climate. The novelty of this work lies in its integrative strategy, examining not only the direct effects of these pollutants on ecosystems but also their interactions with climate change-induced stressors such as increased temperatures and altered precipitation patterns. This chapter further explores the mechanisms through which emerging pollutants exacerbate climate change impacts on biodiversity, water quality, and ecosystem functioning. By synthesizing recent research, it highlights the synergistic effects that magnify ecological risks, leading to reduced resilience of ecosystems. Furthermore, it focuses on developing adaptive and mitigative strategies tailored to the dynamic nature of both emerging pollutants and climate change. This includes advanced monitoring techniques and the application of nature-based solutions to enhance ecosystem resilience. Additionally, it also addresses policy and regulatory challenges, advocating for integrated environmental management practices that consider the dual threats of pollution and climate change. Moreover, it underscores the value of interdisciplinary collaboration and

M. Bera · S. Tripathi
School of Environment and Disaster Management, Ramakrishna Mission Vivekananda Educational and Research Institute, Kolkata, West Bengal, India

A. Dhar · S. Dey · S. Jana
Division of Agronomy, School of Agriculture and Rural Development, Ramakrishna Mission Vivekananda Educational and Research Institute, Kolkata, West Bengal, India

S. Das · M. R. Choudhury
School of Environment and Disaster Management, Ramakrishna Mission Vivekananda Educational and Research Institute, Kolkata, West Bengal, India

School of Agriculture and Food Sustainability, The University of Queensland, Brisbane, QLD, Australia

D. Moulick (✉)
Department of Environmental Science, University of Kalyani, Nadia, West Bengal, India

S. Roy, V. Mandal (eds.), *Plant-Microbe Interaction under Xenobiotic Exposure*,
https://doi.org/10.1007/978-981-96-8260-7_1

stakeholder engagement in crafting effective solutions. By providing a forward-looking perspective, this chapter's overall objective is to contribute to the global discourse on ecological sustainability, offering insights and practical approaches to mitigate the ecological effects of emerging pollutants in the face of climate change.

Keywords Climate change · Ecological impacts · Emerging pollutants · Environmental challenges · Mitigation strategies · Sustainable solutions

1 Introduction

There are more and more emerging contaminants in environmental matrices due to diverse anthropogenic activities. Many contaminants of emerging concern (examples include pharmaceuticals, microplastics, PFAS (or forever chemicals), BPA, and agricultural chemicals) remain undetected and require identification and quantification across environmental components and biological tissues, although significant advancement has been made in the detection and estimation of trace pollutants in recent decades, primarily due to improved and refined techniques. These pollutants can be mobile and persistent in soil, sediments, water, air, and biological systems, even in low quantities. Comprehensive information on their environmental behavior and the threats they pose to human and ecological health is still lacking, though. Furthermore, because there is frequently a lack of sufficient data for a comprehensive risk assessment, the ecotoxicological impact of some developing micropollutants is poorly understood (Basu et al. 2025; Gavrilescu et al. 2015; Puri et al. 2023).

Significant environmental effects have resulted from the unrestrained expansion of many human activities in recent decades, including urbanization, transportation, agriculture, and industrial operations. Particulate matter, NO_x, SO_2, greenhouse gasses, CO_2, and other air pollutants have increased due to rising living standards and growing consumer demand (Harsh et al. 2024; Choudhury and Moulick 2022; Geissen et al. 2015). While soil pollution is caused by the disposal of hazardous waste, the extensive use of pesticides, sludge, disposable items, and non-biodegradable materials, as well as inadequate waste management facilities, water bodies are also becoming more and more contaminated by a variety of chemicals, nutrients, leachates, and oil spills. Synthetic chemicals including insecticides, beauty products, personal care products, household goods, and medications are examples of emerging pollutants (EPs), which are now widely distributed around the world and are vital to contemporary culture. Water supplies have also been contaminated by biological micropollutants brought about by human activity, such as bacteria and viruses, which are now known to be emerging or reemerging pathogens because of their ability to cause disease. These biological micropollutants, which continue to be a major reason for death worldwide, include enteric bacteria, viruses, mycoplasmas, and protozoa. They are also important causes of waterborne infections. The variety of disease-causing agents is continuously growing, even though many of these illnesses are caused by well-known aquatic pathogens. The resurgence of diseases that were previously believed to be under control is demonstrated by the prevalence of rotavirus, hepatitis A virus, *Legionella*, and *Cryptosporidium* in water sources (Geissen et al. 2015; Gavrilescu et al. 2015).

More than 700 emerging pollutants (EPs) exist in European aquatic habitats, along with their metabolites and transformation products (www.norman-network.net). Synthetic and/or natural chemicals that are not frequently supervised in environmental monitoring but have the potential to enter ecosystems and have known or assumed dangers to human health and the environment are referred to as EPs. Due to improvements in detection technology, some pollutants were only recently discovered, even though they may have been in the environment for a while. In other cases, new sources of EPs have been created by synthesizing new chemicals or alterations in the usage and dumping of existing substances (Vasilachi et al. 2021; Tang et al. 2019). The environmental destiny, behavior, and ecotoxicological impacts of EPs are still mostly unknown, and they are typically not included in standard monitoring programs at the national and international levels. Point sources, such as wastewater treatment plants in cities and industrial areas, or diffuse sources, including air deposition and agricultural practices, can transfer these contaminants into ecosystems (Fig. 1). According to their place of origin, they are divided into more than 20 classifications; some of the most common categories include pesticides (agricultural), disinfection by-products (urban, industry), medications (urban, livestock farming), and industrial chemicals and wood preservation (industry). A coordinated response at several levels is desperately needed, given the possible threats to human health and aquatic ecosystems, the lack of understanding of their environmental dynamics, and the difficulties in sampling and analysis (Geissen et al. 2015).

In addition, with significant effects on life on earth, climate change is becoming a more pressing issue. Numerous chemical, physical, and biological stressors that

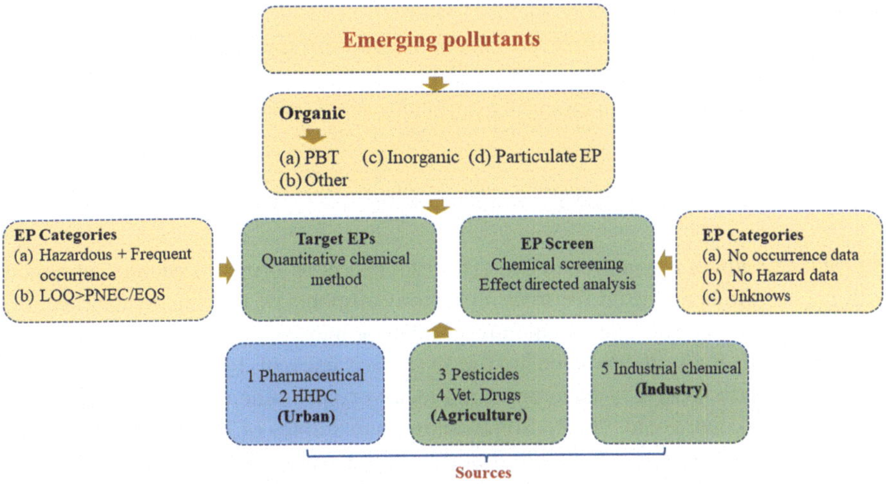

Fig. 1 Different EPs from origins that can be assessed by targeted and screening approaches

are mostly caused by human activity but also arise naturally are present for both people and wildlife. The potential for climate change to alter the distribution of chemical contaminants in the environment and their biological effects is a noteworthy effect that is currently receiving more attention (Noyes et al. 2009). In light of this quickly changing environment, there is an increasing emphasis on proactively evaluating the effects of chemical pollution, with an emphasis on identifying and reducing impacts on the most vulnerable ecosystems and populations. It also emphasizes how crucial stakeholder involvement and interdisciplinary cooperation are to developing successful solutions.

This chapter's general goal is to add to the worldwide conversation on environmental sustainability by offering a forward-looking viewpoint and useful strategies for reducing the ecological effects of new pollutants in the face of climate change.

2 Emerging Pollutants: Types and Source

2.1 Major Categories of Emerging Pollutants

Due to the rapid development of agriculture, industry, and urbanization to satisfy the ever-increasing demand of the growing population, the pollutants are becoming very persistent in the environment, leading to disruption of normal ecological functions. Emerging pollutants or environmental contaminants (EC) are a group of broad-spectrum, either natural or human-made compounds (Table 1) that are not regulated under any existing environmental laws and are considered to have the potential to create negative effects on soil, crops, water, and animal and human health. The common EP includes but is not exhaustive to pharmaceuticals and personal care products, endocrine-disrupting chemicals, pesticides, nanomaterials, micro/nanoplastics, and heavy metals (Moulick et al. 2021; Mishra et al. 2023; Singhal et al. 2023; Naeem et al. 2023). Regulation and management of these emerging pollutants are of utmost need to protect the overall health of the environment. Some of the common emerging pollutants are described below (Table 1).

2.1.1 Pharmaceuticals and Personal Care Products (PPCPs)

Pharmaceuticals is a comprehensive term that refers to drugs/medicines obtained with prescriptions and used for treating various ailments of humans as well as animals. On the other hand, personal care products enhance our daily lives and well-being (Boxall et al. 2012). Due to the extensive uses of PPCPs in human and veterinary medicines, they have appeared as a very prominent group of emerging pollutants (Nikolaou et al. 2007). Some common examples of PPCPs include erythromycin (antibiotics), parabens (antimicrobial preservative), carbamazepine (controls depression and mental ailments), caffeine (stimulant used in beverages and

Table 1 Description of different emerging pollutants and their different sources

Emerging pollutant category	Examples	Primary sources	Typical uses or origins	References
Pharmaceuticals	Antibiotics, analgesics, hormones	Urban wastewater, hospitals, pharmaceutical industries	Human and veterinary medicine	Daughton (2001)
Personal care products (PCPs)	Triclosan, parabens, synthetic musks	Urban wastewater, household products	Antimicrobial agents, cosmetics, fragrances	Montes-Grajales et al. (2017)
Endocrine-disrupting chemicals (EDCs)	Bisphenol A (BPA), phthalates	Plastic manufacturing, urban wastewater, landfill leachate	Plastics, personal care products, food packaging	Lazofsky and Buckley (2022); Khan et al. (2022)
Pesticides	Glyphosate, atrazine, chlorpyrifos	Agricultural runoff, pest control in urban areas	Agriculture, pest management	López-Doval et al. (2017)
Flame retardants	Polybrominated diphenyl ethers (PBDEs), TBBPA	Electronics, furniture, building materials	Fire retardants in electronics, textiles, and construction	Yang et al. (2019)
Per- and polyfluoroalkyl substances (PFAS)	PFOA, PFOS, GenX	Industrial facilities, firefighting foams	Non-stick cookware, water-repellent clothing, firefighting foam	Podder et al. (2021)
Industrial chemicals	Benzene, toluene, xylene, PCBs	Chemical manufacturing, industrial discharges	Solvents, lubricants, coolants, adhesives	Zhong et al. (2022) and Calvo-Flores et al. (2018)
Nanomaterials	Titanium dioxide (TiO_2), silver nanoparticles	Electronics, cosmetics, food packaging	Sunscreens, antimicrobial coatings, food packaging	Bhattu and Singh (2023)
Disinfection by-products (DBPs)	Trihalomethanes, haloacetic acids	Water treatment facilities	By-products from water chlorination and disinfection	Hebert et al. (2010)
Pharmaceutical metabolites	Sulfamethoxazole, carbamazepine metabolites	Human excretion, wastewater treatment plants	Metabolites of pharmaceuticals excreted by humans and animals	Ghosh et al. (2024) and Verlicchi et al. (2010)
Microplastics	Polyethylene (PE), polypropylene (PP)	Plastic waste, personal care products, textiles	Breakdown products of plastic waste, microbeads in cosmetics	Sudarsan et al. (2024)

(continued)

Table 1 (continued)

Emerging pollutant category	Examples	Primary sources	Typical uses or origins	References
Surfactants	Alkylphenols, LAS	Household and industrial cleaners, personal care products	Detergents, emulsifiers, wetting agents	Tyagi et al. (2024)
Herbicides and fungicides	2,4-D, metalaxyl, tebuconazole	Agricultural runoff, urban landscaping	Weed and fungus control in agriculture and landscaping	Mukherjee et al. (2024)
Polychlorinated biphenyls (PCBs)	Various PCB congeners	Industrial sites, electrical equipment, legacy pollution	Insulating fluids and coolants, previously used in electronics	Stadelmann et al. (2024)
Heavy metals	Lead, mercury, cadmium	Industrial discharge, mining, fossil fuel combustion	Batteries, electronics, metal plating, combustion emissions	Yu et al. (2024a, b)
Alkylphenols and bisphenols	Nonylphenol, bisphenol A (BPA)	Urban wastewater, industrial waste	Detergents, plastics, food packaging	Babaei et al. (2024)
Organotin compounds	Tributyltin (TBT), triphenyltin (TPT)	Marine antifouling paints, industrial wastewater	Antifouling agents for ships, wood preservatives	Gomez et al. (2024)
Dispersants	Corexit 9500, DOSS	Oil spill response, industrial discharge	Used in oil spill response to disperse oil in marine environments	Okeke et al. (2022)
Sweeteners	Sucralose, acesulfame	Wastewater effluent, food industry	Artificial sweeteners in food and beverages	Chen et al. (2023)
Stimulants and illicit drugs	Caffeine, cocaine, amphetamines	Urban wastewater, recreational use	Beverage industry, recreational drugs	Ripanda et al. (2022)
Synthetic dyes and pigments	Azo dyes, phthalocyanine	Textile industry, industrial wastewater	Textile and leather dyes, industrial pigments	Millbern et al. (2024)
Plasticizers	Diethyl phthalate (DEP), DBP	Plastic waste, landfill leachate	Additives in plastics, paints, and adhesives	Qadeer et al. (2024)
Antibiotic resistance genes (ARGs)	Various antibiotic resistance genes	Agricultural runoff, hospitals, wastewater treatment	From overuse of antibiotics in humans and animals	Feng et al. (2021)

(continued)

Table 1 (continued)

Emerging pollutant category	Examples	Primary sources	Typical uses or origins	References
Radionuclides	Cesium-137, iodine-131	Nuclear facilities, medical waste	Medical imaging, nuclear power generation	Fan et al. (2023)
Hormones and steroids	Estrone, testosterone	Animal farms, urban wastewater	Growth promoters in agriculture, excreted by humans	Unnikrishan et al. (2024)

food), triclosan (used for manufacturing cosmetics and disinfectants), and diethyl-meta-toluamide (used in developing insect repellents and insecticides) (Osuoha et al. 2023).

Pharmaceutical and personal care products often end up in the ecosystem after their use. They can enter wastewater through bathing, washing, or industrial discharges and then make their way to sewage treatment plants. While these plants can remove some contaminants, many PPCPs remain in the treated wastewater sludge. When this wastewater sludge is applied to agricultural land as fertilizer, it can introduce PPCPs into the soil and contaminate crops. Additionally, PPCPs can leach into groundwater from contaminated soil, potentially affecting drinking water supplies (Daughton and Ternes 1999). The antibiotics fed to the livestock do not get fully used up, and few residues containing the antibiotic resistance genes are excreted in the form of dung of the cattle. Even can be traced in their milk. Even animal manure can contain veterinary drugs (antibiotics, antimicrobials) that can contaminate the ecosystem and enter the food chain, developing resistance to antibiotics in humans. These various pathways expose humans to PPCPs through their consumption of contaminated food and water, posing potential health risks (La Farre et al. 2008).

These PPCPs get bioaccumulated and biomagnified in the food chain and are lethal. A significant concern with the occurrence of PPCPs in the ecosystem is the potential for synergistic toxicity. When combined, even at low concentrations, multiple PPCPs can have a more pronounced toxic effect than any subject compound alone. This is known as synergistic interaction (Cleuvers 2003). Additionally, the widespread use of antibiotics in medical treatment and animal husbandry has contributed to the production of antibiotic-resistant bacteria. These resistant strains can make infections more difficult to treat, posing a serious public health threat (WHO 2015).

2.1.2 Endocrine-Disrupting Chemicals (EDCs)

Endocrine-disrupting compounds (EDCs) are xenobiotic materials found in various products like toys, plastics, PVC pipes, detergents, and cosmetics (Flint et al. 2012). These chemicals can interfere with the body's hormonal system by mimicking, blocking, or altering natural hormone production and function (Zoeller et al. 2012). This can lead to abnormal hormonal signals that disrupt important physiological processes like reproduction, growth, and development (Rodríguez-Hernández et al. 2022). EDCs can enter the food chain through contaminated crops and gather in the fat tissues of animals and humans. Foods from animals, such as meat, milk, and eggs, are particularly susceptible to EDC contamination due to the animals' exposure to contaminated food and water and, sometimes, the illegal use of growth regulators. This widespread contamination poses a significant public health concern (Lu et al. 2020).

They are either natural or synthetic in origin. Estrone, androgens, testosterone, isoflavonoids daidzein, and estriol are the naturally occurring ones, while atrazine, dioxins, polycyclic aromatic hydrocarbons, and bisphenol are regarded as synthetic in origin (Pironti et al. 2021). Endocrine-disrupting compounds enter water systems primarily from industrial and wastewater treatment plant effluents and also from intensively cultivated agricultural soils. Additionally, human activities introduce EDCs into the environment through consumer products like plastic containers, personal care products, detergents, pesticides, and various industrial processes. People can be directly exposed to EDCs by using products containing these chemicals. Furthermore, when these products are discarded in landfills, weathering can cause the release of EDCs into the soil and groundwater, indirectly affecting the environment (Rodríguez-Hernández et al. 2022).

The presence of EDCs in the environment is a global problem because of their harmful effects on wildlife and animals used for human consumption. These chemicals can accumulate in organisms and undergo chemical alteration, leading to further environmental damage (Wang et al. 2021). Some toxic heavy metals like aluminum and manganese, can act as EDCs acidifying aquatic environments and reducing the reproductive success of aquatic animals (Hossain et al. 2023; Correia et al. 2021). Human beings are mainly exposed to EDCs through contaminated environments and by consuming contaminated food and water. Fish consumption is a significant source of EDC exposure. These chemicals can disrupt the endocrine system, resulting in negative health consequences for human populations (Rodríguez-Hernández et al. 2022).

2.1.3 Nanomaterials

Nanomaterials, materials with at least one dimension in the nanometer range (smaller than 100 nm), are increasingly used in various industries (Roy and Hossain 2024). However, their unique properties also make them potential environmental contaminants. Owing to their small size, nanomaterials can enter water bodies with ease and accumulate in sediments, posing risks to a variety of ecosystem (s) (Chowardhara et al. 2024). For example, engineered nanoparticles used in electronics, cosmetics, and pharmaceuticals can contaminate water sources and harm organisms. Furthermore, nanomaterials can interact with other pollutants, creating new and more toxic substances (Pattnaik et al. 2024; El-Kalliny et al. 2023). They can be grouped into four classes: organic-based nanomaterials, inorganic-based nanomaterials, carbon-based nanomaterials, and composite-based nanomaterials (Jeevanandam et al. 2018).

The exposure routes of nanoparticles in humans are either external or internal. External routes include inhalation, ingestion, and dermal absorption via air, water, and food, while internal surgical implant wear results in toxicity, DNA damage, oxidative stress, organ deposition, systemic inflammation, etc. (Moulick et al. 2024; El-Kalliny et al. 2023).

Nanomaterials can cause varying levels of cell damage and oxidative stress depending on their size and charge. They can produce reactive oxygen species (ROS), which can harm cell membranes by oxidizing fatty acid tails in phospholipids. This increases membrane permeability, making cells vulnerable to osmotic stress and reducing uptake of essential nutrients. It also triggers reactions that produce more free radicals, leading to further cell membrane and damage of genetic material (Klaine et al. 2008). High exposure to nanoparticles, such as those found in automobile exhaust in densely populated cities, has been linked to health problems like cardiopulmonary mortality, childhood cancers, myocardial infarction, and inflammatory, blood clotting, and hemolytic responses. Cigarette smoke, which contains nanoparticles, can cause cardiovascular disease, genetic alterations, chronic respiratory illnesses, pancreatic cancer, middle ear disease, and exacerbated asthma (Jeevanandam et al. 2018).

2.1.4 Microplastics

Microplastics, as an emerging pollutant, pose a significant hazard to both the environment and the health of everyone. Microplastics are generally small segments of plastic less than 5 mm of various sizes (Eerkes-Medrano and Thompson 2018). Their ubiquity in the environment, from soil to water bodies, is a result of the breakdown of larger polymer debris (Choi et al. 2020). The ingestion of microplastics through contaminated food products has been associated with various health problems, like respiratory and digestive problems, hormonal disruption, and even neurotoxicity. Furthermore, the existence of microplastics in the blood and breast milk of humans highlights the pervasive nature of this pollution and its potential long-term consequences (Rist et al. 2018). The sources of microplastics in the human food chain include fish and other seafood, drinking water and other beverages, and dietary food products. Microplastics are known to cause respiratory disorders, gastrointestinal disorders, accumulation of fat and liver damage, skin allergies, and irritations. Even it accumulates in blood samples, urine, and breast milk and thus gets transferred to the baby through the placental barrier and breastfeeding (Garai et al. 2024). The use of plastic mulching in agriculture in the form of polyester, polyvinyl chloride, and polyethylene is the main source of microplastic in soil pollution. They degrade the physicochemical and biological attributes of the soil, thereby reducing soil health and productivity (Garai et al. 2024). Upon uptake of microplastic by plant roots grown on contaminated soil, the production of ROS results in oxidative stress, causing a drop in chlorophyll content, photosynthetic efficiency, and leaf area and eventually lowering the yield and quality of the crop (Zantis et al. 2023).

2.1.5 Pesticides

Pesticides, while essential for agricultural productivity, have become a significant concern as emerging pollutants (Roy et al. 2024; Roy et al. 2023). These chemicals, designed to control pests, can contaminate water bodies through runoff, leaching, and aerial drift. Once in the surrounding environment, pesticides can harm organisms, disrupt ecology, and pose risks to the health of the ecosystem (Agrawal et al. 2010). Bioaccumulation, the process where pesticides concentrate in the tissues of organisms over time, further magnifies their impact. For example, DDT is a broad-spectrum pesticide in which, on application, an excess amount reaches water bodies, gets absorbed in fish, and causes biomagnification in the food chain (Brown et al. 2019). The persistent nature of many pesticides can lead to long-term environmental contamination, making it challenging to mitigate their effects (Ghosh et al. 2023). Pesticides, when applied in excess amounts, kill various pollinators and beneficial insects like honeybees, ladybird beetles, and spiders, thus disrupting the natural pest management (Reshi et al. 2025). Even continuous application of the same group of pesticides develops resistance in pests, and they do not get killed upon application. Many pesticides take a longer time to degrade in soil and persist in the environment as contaminants (Whalon et al. 2008). When the vegetables are harvested and consumed before the completion of the waiting period for the pesticide, the pesticide residue gets transferred to humans (Gyawali 2018).

2.1.6 Heavy Metals

Heavy metals, or HMs, are elements that share the common characteristics of having high atomic weights and density. These metals, often cadmium, mercury, arsenic, chromium, lead, and nickel, are naturally found in the Earth's crust (Moulick et al. 2024a, b, 2017). These are beneficial for humans, animals, and plants up to a certain limit, above which it turns into toxicity (Briffa et al. 2020). However, anthropogenic activities can significantly increase their presence in the ecosystem. It is considered an important emerging pollutant of the environment because it accumulates in crops grown on contaminated soils, leading to a threat to plant growth and also to the people consuming those plant products (Moulick et al. 2022, 2023; Majumdar et al. 2024b). Selenium is also considered a potentially toxic element when present in excess amounts in the environment (Moulick et al. 2024). The sources of heavy metals may be natural or anthropogenic. Naturally, they may be produced by the weathering of rocks, volcanic eruptions, and mineral deposits while man-made sources include faulty agricultural practices, industrial activities, construction and urbanization, and waste disposal (Moulick et al. 2019; Saha et al. 2019).

While acute contact with heavy metals can lead to immediate and severe health problems, chronic exposure to even low levels can have subtle yet significant long-term consequences. Over time, these harmful substances can accumulate in various tissues and organs, often without causing noticeable symptoms. The body's detoxification systems may be overwhelmed, allowing heavy metals to persist and interfere with cellular functions. This cumulative effect can lead to a range of health problems, from mild cognitive impairment to severe organ damage (Mahanta et al. 2025). Exposure to these heavy metals causes neurological effects, renal effects, cardiovascular effects, and bone effects and even poses a severe threat due to their carcinogenic potentials (Walter et al. 2005; Orisakwe et al. 2017).

2.2 *Sources and Pathway of Emerging Pollutants*

Emerging pollutants, including pharmaceuticals, pesticides, and microplastics, accumulate in the ecosystem from diverse sources such as industrial discharge, agricultural runoff, wastewater treatment plants, urban runoff, and landfills. These contaminants often persist and bioaccumulate, posing substantial threats to the environment and human health due to their toxicity and resistance to conventional removal processes (Mishra et al. 2023).

2.2.1 Industrial Discharges

Industrial processes generate a significant amount of EPs, containing potentially toxic elements, solvents, and synthetic compounds, which are dumped into the environment through water, soil, and air (Rajaram and Das 2008). Industries such as pharmaceuticals, textiles, and plastics are particularly significant contributors to these pollutants (Kishor et al. 2021). Despite undergoing treatment, effluents from these facilities often contain residues that cannot be entirely removed using traditional wastewater treatment procedures. This is due to the complex nature of these emerging pollutants and the restrictions of traditional treatment processes. As a consequence, these pollutants pose a significant threat to human health and the environment, necessitating the development of more effective and innovative pollution control strategies (Deblonde et al. 2011).

2.2.2 Agricultural Runoff

Pesticides, herbicides, and fertilizers used in agriculture can be washed off fields by rain or irrigation, contaminating nearby water bodies or groundwater (Jacob et al. 2020). Newer pesticides, designed for higher efficacy and sometimes harder to detect, can build up in soils and water, impacting non-target organisms like beneficial insects and aquatic species. Due to the overuse of pesticides like insecticides, acaricides, herbicides, and nematicides, agricultural pollution is increasing at a rapid rate (Sun et al. 2018). Water pollution in the form of algal bloom and eutrophication caused algal bloom and death of various aquatic animals due to the washout of the nitrogenous and phosphatic fertilizers from the fields (Ngatia et al. 2019). Bioaccumulation and biomagnification due to the residual remains of harmful pesticides are a prevalent source of entry of pollutants through contaminated fish (Katagi 2010).

2.2.3 Wastewater Treatment Plants

Although wastewater treatment plants (WWTPs) are designed to eliminate traditional pollutants, they are often inefficient at removing emerging pollutants, especially pharmaceuticals and personal care products (Azaroff et al. 2021). These substances can pass through the treatment procedures and be released into rivers, lakes, and oceans, where they accumulate over time. In some cases, the disintegrated products of these chemicals can be more harmful than the parent compounds. The chemical composition of wastewater can fluctuate significantly throughout the various stages of wastewater treatment, as it undergoes a series of physical, biological, and chemical processes. Additionally, daily variations can occur due to the intermittent delivery of wastewater by slurry trucks or the influx of wastewater from diverse industries (Czarnota et al. 2023).

2.2.4 Urban Runoff and Landfills

Urban areas are significant sources of emerging pollutants due to stormwater runoff and landfill leachate (Müller et al. 2020). Stormwater runoff collects contaminants from roads, buildings, and other surfaces, including microplastics from car tires, personal care products, and household chemicals. These pollutants are then transported into nearby waterways, impacting aquatic ecosystems and potentially contaminating drinking water sources. Landfills, on the other hand, contain a variety of waste materials, including household waste, electronic devices, and industrial refuse. Over time, these wastes degrade and release pollutants into the soil and groundwater. The leaching of contaminants from landfills can pollute drinking water supplies and pose a threat to human and environmental health (Salem et al. 2008). The complex nature of these emerging contaminants and the challenges linked with their removal necessitate the design of sustainable and effective pollution control measures in urban areas (Mishra et al. 2023).

3 Toxicity of Emerging Pollutants to Aquatic and Terrestrial Life—Mechanisms

3.1 Bioaccumulation and Biomagnification

Emerging contaminants can infiltrate and potentially cause suspected or known adverse effects on ecological or human health (La Farre et al. 2008). Bioaccumulation refers to the process where these chemicals accumulate significantly in living organisms, resulting in higher concentrations within the organism compared to their surrounding environment (Nfon et al. 2008). It is the gradual enrichment of toxins within the organism (Chojnacka 2010). Bioaccumulation results when the elimination of pollutants is overcome by their intake. Bioaccumulation can be quantified using the below function (Eq. 1).

$$\text{Bioaccumulation} = \text{Rate of intake} - \text{Rate of elimination} \quad (1)$$

Various factors like the organism's size, genetics, and nutrition influence the degree of bioavailability, absorption, bio-accessibility, and ultimately bioaccumulation (Ivanciuc et al. 2006). Absorption of pollutants occurs in crops through roots or leaves, while for animals, it can happen directly through gills or skin or indirectly through the food chain (Franke et al. 1994). With the help of tropodynamics, the bioaccumulation of emerging pollutants was measured, which describes the tropic transfer of a pollutant in the food chain (Yarsan and Yipel 2013). The ratio of metal content present in a tissue of the organism to its food is called the tropic transfer factor, and the ratio of concentration of the pollutant in organism tissue to the concentration of pollutant in the abiotic medium is defined as the bioaccumulation factor (DeForest et al. 2007). Depending on the ecosystem, whether terrestrial or aquatic, the abiotic medium includes environmental components like water, soil, and sediments (Fang et al. 2016, 2017).

Bioaccumulation occurs through complex food webs, with species higher up the food chain accumulating greater concentrations of chemicals in their bodies, particularly in their tissues (Fig. 2). This accumulation is influenced by food consumption habits, habitat, and the metabolic actions of each species (Daley 2013). Bioaccumulated substances are often broken down into by-products of the parent compound, which may be more toxic; for example, less harmful inorganic mercury is transformed into the highly toxic methylmercury. Bioaccumulation is affected by the binding strength of the chemical within cells and its metabolic half-life. The reduction process generally follows first-order kinetics (Rajput et al. 2023).

In animals, diet is the primary route for bioaccumulation, with minimal contribution from dermal and respiratory exposure, except in soil species. Terrestrial animals generally exhibit lower bioaccumulation levels than marine species. Mammals and birds reach a dynamic equilibrium, where absorption equals removal, resulting in stable chemical concentrations in tissues. The time to reach this balance varies based on the chemical present, type of organism, and food concentration (Gobas et al. 2016; Kelly and Gobas 2003).

Dietary accumulation, or biomagnification, can lead to further bioaccumulation, causing chemical concentrations to increase with rising trophic levels in food chains, a process known as trophic magnification (Sharpe and Mackay 2000; Kelly et al. 2004). This apparent movement of chemicals against the thermodynamic gradient is thought to result from digestion and absorption processes that concentrate ingested chemicals in the gastrointestinal tract (Gobas et al. 1999).

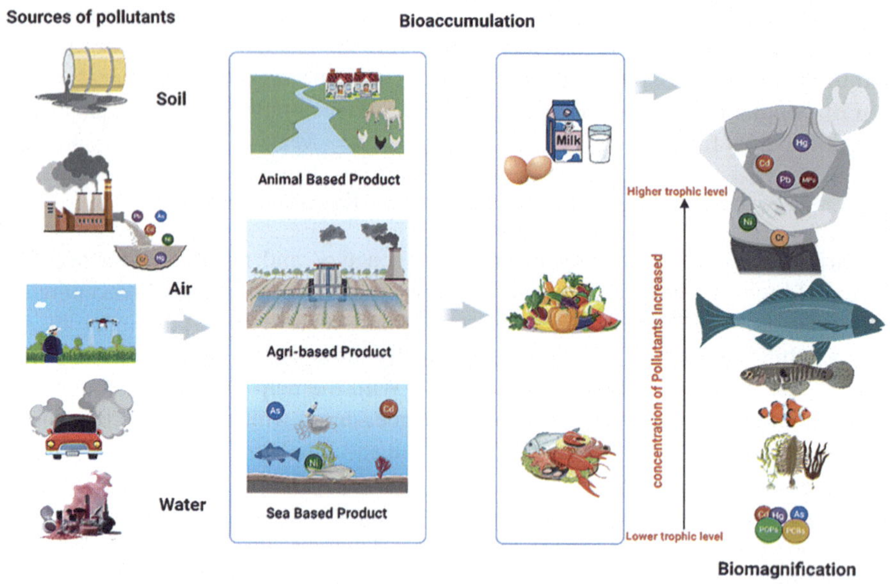

Fig. 2 Sources of various emerging pollutants and their bioaccumulation and biomagnification in soil, plants, animals, and human continuum (Image created with BioRender.com; *license number: QC27GAL2G1)*

3.2 *Toxicity to Aquatic and Terrestrial Life*

Freshwater ecosystems face significant threats from human activities like industrial discharge, agriculture, urban waste, and growing urbanization, leading to increased emerging pollutants (Meijide et al. 2018; Zhu et al. 2018). Pollution has a severe impact on aquatic plant and animal habitats. It disrupts the delicate balance, posing a serious threat to the ecosystem (Schmeller et al. 2018). Van Ginneken et al. (2018) reported that the growth rate, metabolic, feeding, and reproductive actions of freshwater isopod (*Asellus aquaticus*) were determined from exposure to a mixture of heavy metals. Significant Hg levels were traced in the muscles and high Zn levels in the gills and gonads of *Perca fluviatilis* and *Rutilus rutilus* (Łuczyńska et al. 2018). Zhong et al. (2018) examined the contamination effects of eight heavy metals (Cu, Cr, Zn, Mn, Pb, As, Cd, Ni) on fish in 16 freshwater systems across China. It was found that dietary exposure influenced bioaccumulation in zooplankton. Moore and Bringolf (2018) examined nitrate effects on freshwater mussels and reported decreased juvenile production, glochidia attachment, and metamorphosis. Wang et al. (2018) examined the combined influence of fungicides on zebrafish and found synergistic effects during embryonic stages and significant gene expression changes. Demirci et al. (2018) found significant changes in biomarker activities under combined pesticide exposure.

Several pollutants are known to kill soil organisms directly by acute toxicity, or they can contaminate the food supply and influence reproduction. Other indirect influences include effects on growth and development, metabolism, durability, or even genetic effects (Edwards 2002). It was reported that micro- and nanoplastic, a significant emerging pollutant, inhibited growth, reduced chlorophyll content and photosynthesis, induced ultrastructural changes, and affected gene expression in algae. Energy flow impacts aquatic organism productivity, with the type, particle size, and content of micro/nanoparticles affecting their toxic impact on algae (Zhao et al. 2023). Pollutants degrade habitat quality, disrupt environmental functions, and harm organisms through physiological strain and toxicity. They worsen other environmental issues like climate change and habitat loss, affecting biodiversity. Pollution alters species composition, distribution, and richness in terrestrial, marine, and freshwater ecosystems. Chronic air pollution exposure leads to community structure changes, with a reduction in sensitive species and a rise in pollution-resistant ones. This disrupts ecological interactions like pollination and seed dispersal, impacting ecosystem dynamics and stability (Ogwu et al. 2024). Majumdar et al. (2024) reported that toxic metals like mercury, arsenic, lead, and cadmium are causing several ecological turbulences, creating noxiousness to humans if accumulated in high amounts. It has been documented that exposure to polychlorinated biphenyls, a persistent organic pollutant, has been linked to reduced birth size, early sexual maturation, and altered thyroid hormone levels in humans. Different pollutants affect physiological pathways in various ways (Schell et al. 2006). Furthermore, the toxicity level of various pollutants on different aquatic and terrestrial species is illustrated in Table 2.

3.3 *Disruption of Ecosystem Services*

It is very complicated to rectify the influence of emerging contaminants on ecosystem services, as the components of the environment and their interrelations are very complex. These pollutants are found everywhere, are long persistent and mobile, accumulate in most of the components of the ecosystem's services, including soil, water, air, plants, wildlife, and human well-being, and continuously generate environmental footprints that disrupt the integrity of the ecosystems (Yadav et al. 2021a, b). EPs, which comprise pharmaceuticals, personal care products, microplastics, industrial chemicals, and endocrine-disrupting compounds, are increasingly recognized for their ability to interfere with natural processes and ecosystem services. Generally, EPs are present in very minute quantities; still, they can create both chronic and acute toxicity to the living organisms in aquatic and soil ecosystems, resulting in reproductive impairment, endocrine disruption, and interference with immunity and hormonal balances (Shahid et al. 2021). Moreover, EPs can biomagnify and accumulate in the food chain from lower to higher trophic levels (Mofijur et al. 2023). Hence, the transportation, accumulation, and dissemination of the EPs throughout the soil, air, water, plants, animals, and even humans are crucial for assessing their risk. Their presence can disrupt biological and chemical processes in ecosystems, leading to significant ecological and environmental impacts.

Table 2 Toxicity levels of various pollutants on different aquatic and terrestrial species

Pollutant category	Pollutant example	Species affected	Toxicity level (LC50/EC50/LD50)	Exposure pathway	Effects observed	References
Heavy metals	Lead (Pb)	Fish (Rainbow Trout)	0.32 mg/L (LC50, 96 h)	Water	Neurotoxicity, impaired growth, kidney damage	Harshitha et al. (2024)
		Mammals (rat)	500 mg/kg (LD50, oral)	Ingestion	Neurotoxicity, developmental issues	Guševac Stojanović et al. (2024)
	Mercury (Hg)	Crustaceans (Daphnia magna)	0.0002 mg/L (LC50, 48 h)	Water	Reproductive toxicity, growth inhibition	Banaee et al. (2024) and Zaghloul et al. (2020)
		Birds (Mallard Duck)	10 mg/kg (LD50, diet)	Ingestion	Liver and kidney damage, neurotoxicity	Bean et al. (2024)
	Cadmium (Cd)	Fish (Bluegill)	2.2 mg/L (LC50, 96 h)	Water	Oxidative stress, liver damage	Madesh et al. (2024)
		Mammals (Mouse)	88 mg/kg (LD50, oral)	Ingestion	Liver and kidney toxicity	Sheng et al. (2024)
Pesticides	Atrazine	Amphibians (Frog, Xenopus)	3.0 mg/L (LC50, 96 h)	Water	Endocrine disruption, delayed development	Khoshnood (2024)
		Birds (Quail)	940 mg/kg (LD50, oral)	Ingestion	Hormonal imbalances, reproductive toxicity	Arshad et al. (2024)
	Chlorpyrifos	Insects (Honeybee)	0.1 mg/kg (LD50, contact)	Contact/ Ingestion	Neurological effects, mortality	Choi et al. (2024)
		Mammals (Rat)	95 mg/kg (LD50, oral)	Ingestion	Neurotoxicity, respiratory distress	Coppola et al. (2025)

Pharmaceuticals	Ibuprofen	Fish (Fathead Minnow)	1.8 mg/L (LC50, 96 h)	Water	Liver and gill damage, reproductive toxicity	Zhang et al. (2021a, b, c) and Ge et al. (2021)
		Mammals (Rat)	636 mg/kg (LD50, oral)	Ingestion	Gastrointestinal issues, renal impairment	Umoh et al. (2022) and Rakib et al. (2023)
	Carbamazepine	Crustaceans (Daphnia magna)	35 mg/L (EC50, 48 h)	Water	Decreased reproduction, neurotoxicity	Nkoom et al. (2022)
		Mammals (Mouse)	268 mg/kg (LD50, oral)	Ingestion	Liver toxicity, behavioral changes	Boahen et al. (2025)
Endocrine disruptors	Bisphenol A (BPA)	Fish (Medaka)	3 mg/L (LC50, 96 h)	Water	Reproductive disruption, feminization	Chen et al. (2024a, b) and Monchanin et al. (2021)
		Mammals (Rat)	3250 mg/kg (LD50, oral)	Ingestion	Endocrine disruption, reproductive toxicity	
Flame retardants	PBDEs	Fish (Rainbow Trout)	0.6 mg/L (LC50, 96 h)	Water	Developmental toxicity, bioaccumulation	Serafini et al. (2024) and Abbas et al. (2020)
		Birds (Herring Gull)	8 mg/kg (LD50, oral)	Ingestion	Reproductive toxicity, thyroid disruption	
Industrial chemicals	PCBs	Fish (Rainbow Trout)	1 mg/L (LC50, 96 h)	Water	Bioaccumulation, immunotoxicity	Zhang et al. (2024)
		Mammals (Mouse)	4000 mg/kg (LD50, oral)	Ingestion	Liver and skin toxicity, immunotoxicity	
	Benzene	Fish (Fathead Minnow)	28 mg/L (LC50, 96 h)	Water	Neurotoxicity, liver damage	Connell et al. (2023)
		Mammals (Mouse)	930 mg/kg (LD50, oral)	Ingestion	Carcinogenicity, neurotoxicity	

(continued)

Table 2 (continued)

Pollutant category	Pollutant example	Species affected	Toxicity level (LC50/EC50/ LD50)	Exposure pathway	Effects observed	References
Nanomaterials	Silver nanoparticles	Algae (Pseudokirchneriella)	0.1 mg/L (EC50, 72 h)	Water	Inhibition of photosynthesis, oxidative stress	Gusev et al. (2021)
		Mammals (Rat)	2000 mg/kg (LD50, oral)	Ingestion	Oxidative stress, inflammation	
Per- and polyfluoroalkyl substances (PFAS)	PFOA	Fish (Fathead Minnow)	3 mg/L (LC50, 96 h)	Water	Liver damage, developmental issues	Ankley et al. (2021)
		Birds (Chicken)	10 mg/kg (LD50, diet)	Ingestion	Immune system suppression, liver toxicity	

The distribution of pollutants including pesticides, fertilizers, and other plant growth-promoting agrochemicals has been distinguished in global soil ecosystems (Snow et al. 2017). These EPs not only threaten the soil's beneficial microbial profile but also can create a negative impact on their functionality, structure, behavior, and reproductive competencies. Several investigations suggested that pollutants can lead to microbial community structures (Mishra et al. 2023) altering their population involved in organic matter decomposition and nutrient recycling (Ma et al. 2024). Cyanotoxins, micro- and nanoplastics, and other EPs have been identified to disrupt the nutrient availability and their uptake by the plant; thereby, nutrient deficiencies occur, and the plant is often compromised with poor vigor and health (Jiao et al. 2021), and this is why the accumulation of these pollutants is widespread among the soil-dwelling organisms to overall living systems.

Aquatic ecosystems also get contaminated by EPs by involving industrial waste, agrochemicals via surface runoff, municipal solid wastes, and leaching from dump yards (Shahid et al. 2021). According to Carnevali et al. (2018), endocrine-disrupting compounds are the most common pollutants that can be traced in water bodies. These contaminants have been found to interfere with the growth and reproductive systems, thereby disrupting their metabolic and physiological activities (Celino-Brady et al. 2021). Several studies reveal the prevalent presence of endocrine-disrupting chemicals (EDCs) in aquatic ecosystems, highlighting their potential to adversely affect the physiological and reproductive health of fish and amphibian populations (Hagemann et al. 2020; Langston 2020). The consumption of microplastics by aquatic organisms can cause several toxicological consequences, including deteriorating immune balance, reduced disease resistance capability, and disturbed endocrine systems (Garai et al. 2024). Moreover, microplastic particles are also found in the digestive tract of aquatic animals, resulting in bioaccumulation to the higher trophic levels and ultimately in the human food chain (Egbeocha et al. 2018). Other EPs, such as nano-sized copper particles, pesticide residues, liquid crystals, and organic solvents, can compromise water quality and exert detrimental effects on water ecosystems (He et al. 2024).

Several EPs from manufacturing discharges, agricultural runoff, and improper waste handling can infiltrate the soil, water bodies, and the atmosphere, creating complex exposure pathways that affect both wildlife and human populations (Mishra et al. 2023). A substantial body of animal and human research has demonstrated that exposure to air pollutants, including emerging contaminants such as polycyclic aromatic hydrocarbons (PAHs), perfluoroalkyl sulfonates, organophosphorus esters, polybrominated diphenyl ethers (PBDEs), and parabens, can contribute to the development of respiratory, dermatological, cardiovascular, and immune-related diseases, as well as increase mortality risk (Wang et al. 2024). Addressing these pollutants through improved waste management, better water treatment technologies, and stricter regulatory controls is crucial to preserving ecosystem functions and maintaining the services that human society depends on for survival.

4 Climate Change and Pollutant Dynamics

Climate change and environmental pollution are among the most pressing challenges facing both the global ecosystem and human society. These issues are closely interconnected: climate change affects how pollutants behave and are distributed, while pollution exacerbates climate-related changes. Understanding these connections is crucial for measuring risks to ecological health and human safety and for guiding effective policy decisions aimed at minimizing harmful impacts. Climate change refers to long-term changes in temperature, precipitation, wind patterns, and other climatic factors. The production of greenhouse gases (GHGs)—such as carbon dioxide (CO_2), methane (CH_4), and nitrous oxide (N_2O)—has significantly increased due to anthropogenic activities like industrial production, farming practices, and deforestation. The primary cause of global warming is the greenhouse effect, which triggers a series of reactions that impact both human society and natural ecosystems. In addition to altering seasonal cycles and species distributions, rising temperatures are contributing to more regular and severe climatic events, including heatwaves, floods, and droughts. These interconnected deviations highlight the urgency of addressing pollution and climate change together to protect both human health and ecological integrity. Substances that infiltrate the ecosystem and negatively affect living organisms, including humans, are referred to as pollutants. These pollutants can be classified into three categories: chemical, physical, and biological agents. They are often linked to human activities such as waste disposal, industrial manufacture, agriculture, and automobile emissions (Zandalinas et al. 2021).

EPs are a growing concern and include substances like pharmaceuticals, personal care products, flame retardants, microplastics, and per- and polyfluoroalkyl substances (PFAS). The persistence of these pollutants in the environment, their potential to bioaccumulate, and their harmful effects on ecosystems and human health underscore the importance of understanding and managing their presence in our environmental systems. Pollutant dynamics—including their behavior, distribution, bioavailability, and persistence—can be significantly altered by climate change (Mishra et al. 2023; Calvo-Flores et al. 2018; Verlicchi et al. 2010). The important mechanisms involved are as follows:

- *Temperature Variations and Pollutant Behavior:* As global temperatures rise, the mobility and transformation rates of pollutants, particularly volatile chemicals, change. Higher temperatures can increase the volatility and atmospheric dispersion of contaminants like pesticides, heavy metals, and volatile organic compounds (VOCs). This can lead to elevated air concentrations, especially in cities with high levels of industrial and automobile emissions. Additionally, some pollutants may degrade more quickly at higher temperatures, sometimes resulting in the formation of more hazardous metabolites or by-products.
- *Modified Hydrological Cycles and Pollutant Distribution:* Climate change significantly affects hydrological cycles, altering rainfall patterns, evaporation rates, and the frequency of dangerous weather events. Amplified precipitation leads to

greater runoff, which can transport contaminants from industrial, urban, and agricultural sources into rivers, lakes, and oceans. Conversely, drought conditions reduce water levels, concentrating pollutants in surface waters and posing greater risks to aquatic life. Furthermore, storms and floods can cause soil erosion, resulting in sediment migration that carries pollutants to new areas, ultimately affecting the quality of both surface and groundwater.

- *Ocean Acidification and Pollutant Bioavailability:* The absorption of CO_2 into seawater leads to ocean acidification, which alters marine chemistry and affects the behavior of pollutants, particularly heavy metals. Metals such as lead and mercury can become more soluble in acidic environments, increasing their toxicity and bioavailability to marine life. This poses risks to aquatic creatures and to people who consume seafood contaminated with these hazardous metals. Additionally, these changes impact bioaccumulation and biomagnification within marine food chains.
- *Pollutant Persistence and Changes in Soil Properties:* Variations in temperature and rainfall patterns affect the organic matter content, pH, and activity of soil microbes. These changes influence the mobility and persistence of pollutants. For instance, shifts in soil pH can alter the adsorption and desorption processes of heavy metals. Additionally, increased soil temperatures and heightened microbial activity may accelerate the breakdown of certain organic contaminants. Moreover, warming can lead to the thawing of permafrost, releasing contaminants that were previously trapped and creating new contamination risks for these ecosystems.

EPs present significant challenges due to their persistence and often unknown ecological and health impacts. Typical examples include pharmaceuticals, personal care products, and synthetic chemicals that are not completely removed by conventional wastewater treatment methods (Calvo-Flores et al. 2018; Verlicchi et al. 2010). The effects and distribution of these pollutants may be exacerbated by climate change:

- *Increased Contaminant Load in Water Bodies:* Harsh weather events, such as floods and hurricanes, can overwhelm wastewater treatment systems, leading to untreated sewage being released into surface waters. This results in elevated levels of medications, hormones, and personal care products in lakes, rivers, and coastal areas. Additionally, the degradation of these substances into more harmful transformation products can be accelerated by rising temperatures, further complicating their ecological impact.
- *Bioaccumulation and Biomagnification*: Climate change alters species distribution and food webs, which can affect the processes of pollutant bioaccumulation and biomagnification. Higher temperatures may increase the metabolic rates of aquatic species, making them more vulnerable to bioaccumulative toxins like mercury, PCBs, and PFAS. This poses a greater risk to top predators, including humans, who consume contaminated fish and seafood.

5 Monitoring, Assessment, and Mitigation of Emerging Pollutants

Advanced methods are needed for monitoring and evaluating emerging pollutants (EPs) to identify, measure, and comprehend their dynamics within environmental matrices (Table 3). Even at trace quantities, investigative methods such as chromatography and spectrometry (LC-MS, GC-MS, FTIR) enable accurate identification and quantification of EPs such as medications, pesticides, and microplastics. These techniques offer comprehensive insights into pollutant structures and degradation products and are perfect for laboratory examination (Pena-Pereira et al. 2021; Dimpe and Nomngongo 2016). Rapid on-site detection is made possible by sensor-based technologies like biosensors and electrochemical sensors, which are useful for real-time monitoring in situations like wastewater discharge or industrial disasters. Researchers can also identify unknown contaminants and examine their transformation products using non-targeted screening and high-resolution mass spectrometry (HRMS), which improves our comprehension of EPs in challenging environmental conditions. Complementary to these methods, remote sensing technologies allow large-scale, real-time environmental monitoring. Indicators that can indicate the presence and spread of EP include water quality, land use, and temperature anomalies, which are tracked by satellites and drones fitted with hyperspectral and thermal imaging sensors (Dimpe and Nomngongo 2016). Despite being indirect, these methods provide vital information for tracking pollution sources and evaluating environmental effects by revealing pollutant dynamics over large areas, such as agricultural or coastal regions. Integrated monitoring systems can efficiently handle EP concerns and direct pollution management and environmental protection policies by fusing the wide spatial reach of remote sensing with the specificity of analytical procedures (Fuladlu and Altan 2021; Zhang et al. 2021a, b, c) (Table 4). Furthermore, the designing of physiologically based pharmacokinetic (PBPK) models has significantly increased recently. These models describe the biodistribution of chemicals in animals and humans, covering aspects such as intake, distribution, metabolism, and excretion (Deepika et al. 2021; Chou and Lin 2021). Efforts have shifted toward assessing the health effects of pollutant mixtures instead of individual chemicals. Initial comparisons use prioritization schemes based on ranking and scoring systems. Additionally, uncertainty and variability in data are addressed with probabilistic and possibilistic approaches. These prioritization methods are essential for identifying hazards posed by metabolites and transformation compounds compared to their parent compounds. They help quantify dangers to the health of humans and the ecosystem from chemical mixtures and form the basis for assessing cumulative risks from multiple stressors (Fig. 3).

A comprehensive strategy is needed to address emerging pollutants (EPs), integrating legal frameworks, technological advancements, sustainable practices, and public engagement. Governments should establish priority pollutant lists, implement specific regulations for EPs, and promote international cooperation for effective global management of these substances.

Table 3 Different analytical methods for quantifying emerging pollutants

Analytical method	Suitable pollutants	Principle	Limitations	References
Liquid chromatography-mass spectrometry (LC-MS)	Pharmaceuticals, pesticides, personal care products	Separation of compounds via liquid chromatography, followed by mass detection	High cost; requires sample preparation and technical expertise	Almaqdi et al. (2019)
Gas chromatography-mass spectrometry (GC-MS)	Volatile organic compounds (VOCs), polycyclic aromatic hydrocarbons (PAHs), pesticides	Separation of volatile compounds via gas chromatography, followed by mass detection	Limited to volatile compounds; complex sample prep; higher temperature requirements may degrade some analytes	Badea et al. (2020); Pena-Pereira et al. (2021)
Fourier transform infrared spectroscopy (FTIR)	Microplastics, polymer-based contaminants	Measures infrared absorbance to identify functional groups and chemical structures	Limited sensitivity for trace pollutants; requires standards for spectral matching	Mangam et al. (2024) and Dimpe and Nomngongo (2016)
Electrochemical sensors	Heavy metals, nitrates, some pesticides	Pollutants interact with sensor electrodes, altering electrical signals (current or voltage)	Limited to specific ions/metals; often requires calibration; sensitivity may be lower for complex mixtures	Chen et al. (2024a, b)
Biosensors	Pesticides, pharmaceuticals, certain organic pollutants	Utilizes biological components (e.g., enzymes or antibodies) to interact with specific pollutants	Limited to specific pollutants; potential for biological component degradation; higher cost	González-González et al. (2022)
High-resolution mass spectrometry (HRMS)	A wide range of pollutants, including unknown compounds	Provides exact mass measurement, enabling precise molecular identification and screening for unknowns	High cost; complex data analysis; requires skilled operators	Brack et al. (2019)

(continued)

Table 3 (continued)

Analytical method	Suitable pollutants	Principle	Limitations	References
Ultraviolet-visible spectroscopy (UV-Vis)	Organic pollutants, some dyes, and phenolic compounds	Measures absorbance in the UV and visible spectrum, identifying compounds based on their optical properties	Limited to compounds with UV-visible absorbance; lower sensitivity; not ideal for complex mixtures	Qi et al. (2024)
Atomic absorption spectroscopy (AAS)	Heavy metals (e.g., lead, mercury, cadmium)	Measures absorbance of specific metal ions in a sample after atomization	Limited to metal analysis; sample preparation needed; not suitable for organic compounds	Shyamalagowri et al. (2023)
Inductively coupled plasma mass spectrometry (ICP-MS)	Trace metals and metalloids (e.g., arsenic, selenium)	Ionizes samples with plasma and detects metal ions at low concentrations	Expensive equipment; complex operation; requires careful calibration and sample prep	Van Acker et al. (2023)
Ion chromatography (IC)	Inorganic ions, including nitrates, sulfates, and phosphates	Separation of ionic compounds, followed by detection based on conductivity or other parameters	Limited to ionic compounds; high cost for specialized detectors; requires calibration	Mohana Rangan et al. (2021)
Non-targeted screening	Unknown or unidentified pollutants in various matrices	Uses high-resolution analytical techniques (e.g., HRMS) to detect a wide range of compounds	Data-intensive; requires sophisticated software and expertise in data interpretation	Ccanccapa-Cartagena et al. (2019)

- Technological innovations are crucial; strategies such as membrane filtration, activated carbon adsorption, and advanced oxidation processes provide efficient solutions for eliminating EPs from water and wastewater (Gutiérrez et al. 2021).
- Sustainable farming practices, including integrated pest management (IPM) and responsible use of pharmaceuticals in livestock, can significantly reduce the release of pollutants into the ecosystem (Khan et al. 2023).

Table 4 Different remote sensing-based techniques for monitoring emerging pollutants

Techniques	Suitable pollutants	Principle	Limitations	References
Satellite sensing	Air pollutants (e.g., NO_2, SO_2, CO_2, ozone), algal blooms, thermal pollution	Captures spectral data in multiple bands from earth's surface and atmosphere, enabling pollutant mapping and analysis	Limited spatial resolution; some pollutants cannot be directly detected; cloud cover interference	Zhang et al. (2021a, b, c)
Multispectral imaging	Algal blooms, vegetation affected by pollutants, suspended solids in water	Captures data in discrete spectral bands to identify pollutants with unique spectral signatures (e.g., chlorophyll)	Limited spectral resolution; may not capture detailed pollutant types; cloud cover interference	Dunker et al. (2022)
Hyperspectral imaging	Microplastics, algae, heavy metals, oil spills, thermal pollution	Collects data across a wide spectrum, capturing detailed spectral signatures unique to specific pollutants	Expensive equipment; large data sets require advanced processing; limited to surface pollutants	Primpke et al. (2020)
Thermal infrared (TIR) remote sensing	Thermal pollution, oil spills, industrial emissions	Detects infrared radiation, allowing measurement of surface and water temperature anomalies caused by pollutants	Limited to thermal effects; may miss chemical pollutants without thermal signatures	Xu et al. (2021)
Synthetic aperture radar (SAR)	Oil spills, vegetation affected by pollutants, and land cover changes	Uses radar waves to detect surface features, moisture levels, and texture differences, suitable for detecting oil and soil moisture	Lower sensitivity for chemical pollutants; complex data interpretation; limited for certain pollutants	Meng et al. (2024)
Airborne light detection and ranging (LiDAR)	Coastal pollution, sediment distribution, vegetation health	Uses laser pulses to map terrain and vegetation, capturing pollution impacts on land and water bodies	Expensive; limited area coverage per flight; requires specialized processing	Ramzan et al. (2024)
GIS-based spatial analysis	Land-based pollutants, water quality, point and non-point source pollution	Integrates spatial data (land use, hydrology, population) with pollution data to assess pollutant sources and spread	Requires high-quality data inputs; limited by resolution and accuracy of underlying data	Fuladlu and Altan (2021) and Zhang et al. (2021a, b, c)

(continued)

Table 4 (continued)

Techniques	Suitable pollutants	Principle	Limitations	References
Time-series analysis in GIS	Air quality pollutants, water pollution trends, land cover changes	Analyzes temporal data to detect trends, seasonality, and event-driven changes in pollutant levels	Limited by the availability of historical data; requires advanced statistical techniques	Swapna et al. (2024)
Remote sensing of vegetation indices (e.g., NDVI, SAVI)	Pollutants affecting vegetation health, agricultural runoff, and soil degradation	Uses vegetation indices derived from remote sensing data to assess plant health and detect stress due to pollutants	Limited to pollutants affecting vegetation; indirect measurements; sensitive to environmental conditions	Sun et al. (2024) and Soleimany et al. (2021)
Drones (unmanned aerial vehicles—UAVs)	Microplastics, oil spills, localized land and water pollution	Equipped with high-resolution cameras or sensors to capture pollution indicators at fine spatial scales	Limited range and flight time; affected by weather conditions; less effective for large-scale monitoring	Bakirci (2024)

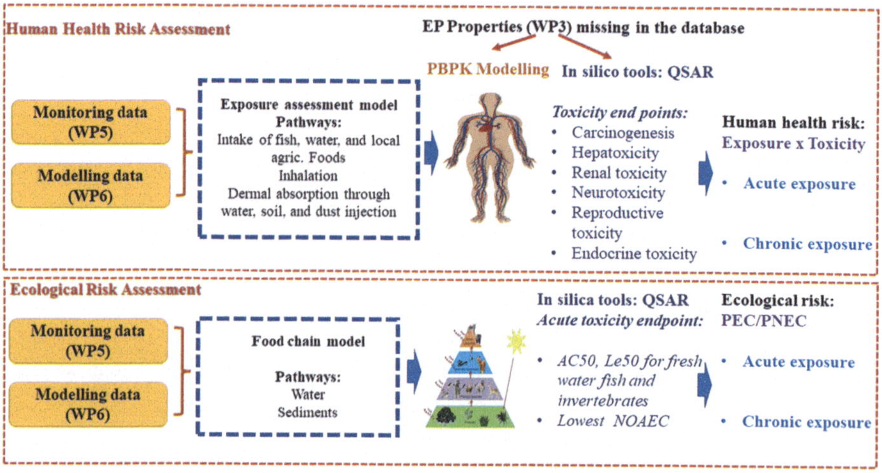

Fig. 3 Multi-scale assessment of ecological and human health risks from acute and chronic exposure to EPs in surface water

- Additionally, pharmaceutical take-back programs and public awareness campaigns perform a crucial role in educating individuals on proper disposal methods, further decreasing environmental contamination.

- Strong monitoring and assessment strategies, utilizing modern analytical techniques and remote sensing technologies, can help track pollution levels and identify hotspots.
- Emphasizing green chemistry principles in product design can lead to the development of biodegradable alternatives (Ncube et al. 2023).
- Proactive management also involves researching innovative degradation techniques and examining contaminant pathways.
- Finally, implementing integrated waste management strategies and promoting cleaner industrial practices can greatly reduce the production of EPs (Fatimah et al. 2020). Together, these approaches can create a comprehensive framework for tackling the challenges posed by emerging pollutant, ultimately protecting environmental health and enhancing human welfare.

6 Future Directions and Research Needs

Emerging pollutants, such as pesticides, PCPs, medications, and microplastics, pose significant threats to ecosystems and human health due to their presence and ability to bioaccumulate. These contaminants often evade traditional wastewater treatment methods, leading to their entry into water systems, which in turn affects biodiversity and contaminates food chains. To mitigate these risks, it is essential to understand their behavior, fate, and long-term effects.

6.1 Area of Interest

- Research on new analytical techniques, such as biosensors and nanotechnology-based detection systems, is essential for identifying and quantifying emerging pollutants at low concentrations (Yadav et al. 2021a, b). Developing real-time monitoring technologies can aid in early detection and enhance response mechanisms.
- Further investigation into the toxicity and bioaccumulation of emerging pollutants across various organisms and ecosystems is crucial (Yu et al. 2024a, b). Understanding these impacts on different species and ecological scales is necessary to inform environmental policies and regulations.
- Improved models are required to predict the behavior and pathways of emerging contaminants in various environmental contexts. Understanding pollutant dynamics can guide enhancements to wastewater treatment and inform environmental management strategies.
- Developing cost-effective and sustainable treatment methods—such as membrane bioreactors, biochar, and advanced oxidation processes—can help reduce the persistence of emerging pollutants (Shah et al. 2020). These technologies

should aim for high efficacy with minimal energy input and low risks of secondary pollution.
- Research into the chronic effects of low-dose exposure to emerging pollutants, particularly through food and water, is urgently needed. Long-term health studies will help establish safe exposure limits and inform public health policies.
- As our understanding of emerging pollutants grows, updated regulations and policies are necessary to limit their release and promote sustainable practices. International cooperation can help establish unified standards and management strategies.

6.2 Research Needs

- *Consistent Risk Assessment Methods:* Regulatory agencies will be better equipped to provide guidelines if they establish uniform procedures for evaluating the risks associated with developing contaminants.
- *Information on Pollutant Mixtures:* There is a deficit of research on the combined influence of multiple contaminants. Since various pollutants coexist in the ecosystem, it is vital to recognize their synergistic effects.
- *Socioeconomic Analysis:* Evaluating the economic impacts of new pollutants, as well as the cost-effectiveness of mitigation and treatment technologies, can aid in resource allocation and inform policy implementation decisions.

7 Conclusion

Different man-made activities, such as industry, urbanization, and agriculture, significantly result in the increasing dominance of emerging pollutants (EPs) in the ecosystem. The persistence, mobility, and potential for bioaccumulation of these pollutants, comprising a range of natural and man-made compounds, make them particularly concerning. While progress in detection and analytical strategies has improved our capability to identify trace contaminants, many EPs remain unidentified, and there is still a lack of comprehensive data on their ecotoxicological impacts. The unknown environmental behavior and interactions of many EPs pose a considerable threat to human health and the ecosystem. Among EPs, pharmaceuticals and personal care products (PPCPs) are of particular concern, as they are frequently found in soils, water bodies, and even crops due to processes like sludge application, wastewater discharge, and agricultural runoff. These contaminants have been associated with adverse effects on ecosystems and human health, including the emergence of antibiotic resistance. Additionally, the complexity of these pollutants can lead to synergistic toxic effects, where combinations of pollutants are more harmful than individual substances. A proactive, multidisciplinary approach is essential for effective EP management, especially given the lack of standardized monitoring and the challenges in assessing these contaminants across various environmental matrices. A coordinated response that incorporates regulatory frameworks, stakeholder involvement, and improved monitoring techniques is crucial to mitigating the ecological impact of these toxins and safeguarding human health. Climate change further complicates these issues, potentially affecting the distribution and toxicity of pollutants. Therefore, addressing new contaminants requires scientific innovation and international collaboration to achieve environmental sustainability.

References

Abbas, Q., Yousaf, B., Ullah, H., Ali, M. U., Ok, Y. S., & Rinklebe, J. (2020). Environmental transformation and nano-toxicity of engineered nano-particles (ENPs) in aquatic and terrestrial organisms. Critical Reviews in Environmental Science and Technology, 50(23), 2523–2581.

Agrawal, A., Pandey, R. S., & Sharma, B. (2010). Water pollution with special reference to pesticide contamination in India. *Journal of water resource and protection*, *2*(5), 432–448.

Almaqdi, K. A., Morsi, R., Alhayuti, B., Alharthi, F., & Ashraf, S. S. (2019). LC-MSMS based screening of emerging pollutant degradation by different peroxidases. BMC biotechnology, 19, 1–16.

Ankley, G. T., Cureton, P., Hoke, R. A., Houde, M., Kumar, A., Kurias, J., … & Valsecchi, S. (2021). Assessing the ecological risks of per-and polyfluoroalkyl substances: Current state-of-the science and a proposed path forward. Environmental toxicology and chemistry, 40(3), 564–605.

Arshad, N., Alam, S., Rafay, M., Jabeen, G., Hussain, K., Hussain, R., … & Ataya, F. S. (2024). Effects of environmental relevant concentrations of acetochlor on growth, hematology, serum biochemistry and histopathology of Japanese quail. Plos one, 19(9), e0306583.

Azaroff, A., Monperrus, M., Miossec, C., Gassie, C., Guyoneaud, R., 2021. Microbial degradation of hydrophobic emerging contaminants from marine sediment slurries (Capbreton Canyon) to pure bacterial strain. J. Hazard. Mater. 402

Babaei, P., Madan, E. N., Güllü, G., Gören, İ. E., Gül, H. K., Dağlıoğlu, N., & Karakuş, P. B. K. (2024). Levels, distribution, sources and human exposure pathways of alkylphenol and alkylphenol ethoxylates in indoor dust in Turkiye. Environmental Pollution, 344, 123447.

Badea, S. L., Geana, E. I., Niculescu, V. C., & Ionete, R. E. (2020). Recent progresses in analytical GC and LC mass spectrometric based-methods for the detection of emerging chlorinated and brominated contaminants and their transformation products in aquatic environment. Science of the Total Environment, 722, 137914.

Bakirci, M. (2024). Evaluating the impact of unmanned aerial vehicles (UAVs) on air quality management in smart cities: A comprehensive analysis of transportation-related pollution. Computers and Electrical Engineering, 119, 109556.

Banaee, M., Zeidi, A., Mikušková, N., & Faggio, C. (2024). Assessing metal toxicity on crustaceans in aquatic ecosystems: a comprehensive review. Biological Trace Element Research, 1–19.

Bean, T. G., Beasley, V. R., Berny, P., Eisenreich, K. M., Elliott, J. E., Eng, M. L., … & Rattner, B. A. (2024). Toxicological effects assessment for wildlife in the 21st century: Review of current methods and recommendations for a path forward. Integrated Environmental Assessment and Management, 20(3), 699–724.

Bhattu, M., & Singh, J. (2023). Recent advances in nanomaterials based sustainable approaches for mitigation of emerging organic pollutants. Chemosphere, 321, 138072.

Boahen, F., Száková, J., Čadková, Z., Štanderová, J., Schwarz, C., Gierus, M., … & Tlustoš, P. (2025). Risk Element Interaction with an Emerging Contaminant in Mammals: The Case of Arsenic and Carbamazepine Orally Taken up by Microtus arvalis L. as a Model Organism. Applied Sciences, 15(3), 1662.

Boxall, A. B., Rudd, M. A., Brooks, B. W., Caldwell, D. J., Choi, K., Hickmann, S., … & Van Der Kraak, G. (2012). Pharmaceuticals and personal care products in the environment: what are the big questions?. *Environmental health perspectives*, *120*(9), 1221–1229.

Basu, A.G., Paul, R.S., Wang, F. and Roy, S. (2025). Impact of microplastics on aquatic flora: Recent status, mechanisms of their toxicity and bioremediation strategies. Chemosphere, 370, 143983.

Brack, W., Hollender, J., de Alda, M. L., Müller, C., Schulze, T., Schymanski, E., … & Krauss, M. (2019). High-resolution mass spectrometry to complement monitoring and track emerging chemicals and pollution trends in European water resources. Environmental Sciences Europe, 31(1), 1–6.

Briffa, J., Sinagra, E., & Blundell, R. (2020). Heavy metal pollution in the environment and their toxicological effects on humans. *Heliyon*, *6*(9).

Brown, D. A., Gossett, R. W., Hershelman, P., Schaefer, H. A., Jenkins, K. D., & Perkins, E. M. (2019). Bioaccumulation and detoxification of contaminants in marine organisms from southern California coastal waters. In *Waste disposal in the oceans* (pp. 171–193). Routledge.

Calvo-Flores, F. G., Isac-García, J., & Dobado, J. A. (2018). Emerging pollutants: origin, structure, and properties. John Wiley & Sons.

Carnevali, O., Santangeli, S., Forner-Piquer, I., Basili, D., & Maradonna, F. (2018). Endocrine-disrupting chemicals in aquatic environment: what are the risks for fish gametes? *Fish physiology and biochemistry*, *44*, 1561–1576.

Ccanccapa-Cartagena, A., Pico, Y., Ortiz, X., & Reiner, E. J. (2019). Suspect, non-target and target screening of emerging pollutants using data independent acquisition: Assessment of a Mediterranean River basin. Science of the total environment, 687, 355–368.

Celino-Brady, F. T., Lerner, D. T., & Seale, A. P. (2021). Experimental approaches for characterizing the endocrine-disrupting effects of environmental chemicals in fish. *Frontiers in Endocrinology*, *11*, 619361.

Chen, P., Wang, J., Xue, Y., Wang, C., Sun, W., Yu, J., & Guo, H. (2024a). From challenge to opportunity: Revolutionizing the monitoring of emerging contaminants in water with advanced sensors. Water Research, 122297.

Chen, Z. W., Shen, Z. W., Hua, Z. L., & Li, X. Q. (2023). Global development and future trends of artificial sweetener research based on bibliometrics. Ecotoxicology and Environmental Safety, 263, 115221.

Chen, Z., Li, X., Gao, J., Liu, Y., Zhang, N., Guo, Y., … & Dong, Z. (2024b). Reproductive toxic effects of chronic exposure to bisphenol A and its analogues in marine medaka (Oryzias melastigma). Aquatic Toxicology, 271, 106927.

Choi, D., Hwang, J., Bang, J., Han, S., Kim, T., Oh, Y., Hwang, Y., Choi, J., Hong, J. (2020). In vitro toxicity from a physical perspective of polyethylene microplastics based on statistical curvature change analysis. *Sci. Total Environ.* 752, 142242 .

Choi, J. Y., Chon, K., Kim, J., Vasamsetti, B. M. K., Kim, B. S., Yoon, C. Y., … & Lee, J. H. (2024). Assessment of Lambda-Cyhalothrin and Spinetoram Toxicity and Their Effects on the Activities of Antioxidant Enzymes and Acetylcholinesterase in Honey Bee (Apis mellifera) Larvae. Insects, 15(8), 587.

Chojnacka, K. (2010). Biosorption and bioaccumulation – the prospects for practical applications. *Environ. Int. 36*, 299e307.

Chou, W. C., & Lin, Z. (2021). Development of a gestational and lactational physiologically based pharmacokinetic (PBPK) model for perfluorooctane sulfonate (PFOS) in rats and humans and its implications in the derivation of health-based toxicity values. Environmental health perspectives, 129(3), 037004.

Choudhury, S. and Moulick, D., 2022. Response of field crops to abiotic stress: Current status and future prospects. CRC Press.

Chowardhara, B., Saha, B., Awasthi, J.P., Deori, B.B., Nath, R., Roy, S., Sarkar, S., Santra, S.C., Hossain, A. and Moulick, D. (2024). An assessment of nanotechnology-based interventions for cleaning up toxic heavy metal/metalloid-contaminated agroecosystems: Potentials and Issues. Chemosphere, p.142178.

Cleuvers, M. (2003). Aquatic ecotoxicity of pharmaceuticals including the assessment of combination effects. *Toxicology letters*, *142*(3), 185–194.

Connell, M. L., Wu, C. C., Blount, J. R., Haimbaugh, A., Kintzele, E. K., Banerjee, D., ... & Baker, T. R. (2023). Adult-Onset Transcriptomic Effects of Developmental Exposure to Benzene in Zebrafish (Danio rerio): Evaluating a Volatile Organic Compound of Concern. International journal of molecular sciences, 24(22), 16212.

Coppola, L., Lori, G., Tait, S., Sogorb, M. A., & Estevan, C. (2025). Evaluation of developmental toxicity of chlorpyrifos through new approach methodologies: a systematic review. Archives of Toxicology, 1–47.

Correia, T. G., Vieira, V. A. R. O., de Moraes Narcizo, A., Zampieri, R. A., Floeter-Winter, L. M., & Moreira, R. G. (2021). Endocrine disruption caused by the aquatic exposure to aluminum and manganese in Astyanax altiparanae (Teleostei: Characidae) females during the final ovarian maturation. *Comparative Biochemistry and Physiology Part C: Toxicology & Pharmacology*, *249*, 109132.

Czarnota, J., Masłoń, A., & Pajura, R. (2023). Wastewater Treatment Plants as a Source of Malodorous Substances Hazardous to Health, Including a Case Study from Poland. *International Journal of Environmental Research and Public Health*, *20*(7), 5379.

Daley, J., 2013. Bioamplification as a Bioaccumulation Mechanism. https://scholar.uwindsor.ca/etd/4758/

Daughton, C. G. (2001). Emerging pollutants, and communicating the science of environmental chemistry and mass spectrometry: pharmaceuticals in the environment. Journal of the American Society for Mass Spectrometry, 12(10), 1067–1076.

Daughton, C. G., & Ternes, T. A. (1999). Pharmaceuticals and personal care products in the environment: agents of subtle change?. *Environmental health perspectives*, *107*(suppl 6), 907–938.

Deblonde, T., Cossu-Leguille, C., & Hartemann, P. (2011). Emerging pollutants in wastewater: a review of the literature. *International journal of hygiene and environmental health*, *214*(6), 442–448.

Deepika, D., Sharma, R. P., Schuhmacher, M., & Kumar, V. (2021). Risk assessment of perfluorooctane sulfonate (PFOS) using dynamic age dependent physiologically based pharmacokinetic model (PBPK) across human lifetime. Environmental Research, 199, 111287.

DeForest, D. K., Brix, K. V., & Adams, W. J. (2007). Assessing metal bioaccumulation in aquatic environments: the inverse relationship between bioaccumulation factors, trophic transfer factors and exposure concentration. *Aquatic toxicology*, *84*(2), 236–246.

Demirci, Ö., Güven, K., Asma, D., Öğüt, S., & Uğurlu, P. (2018). Effects of endosulfan, thiamethoxam, and indoxacarb in combination with atrazine on multi-biomarkers in Gammarus kischineffensis. *Ecotoxicology and environmental safety*, *147*, 749–758.

Dimpe, K. M., & Nomngongo, P. N. (2016). Current sample preparation methodologies for analysis of emerging pollutants in different environmental matrices. TrAC Trends in Analytical Chemistry, 82, 199–207.

Dunker, S., Boyd, M., Durka, W., Erler, S., Harpole, W. S., Henning, S., ... & Wilhelm, C. (2022). The potential of multispectral imaging flow cytometry for environmental monitoring. Cytometry Part A, 101(9), 782–799.

Edwards, C. A. (2002). Assessing the effects of environmental pollutants on soil organisms, communities, processes and ecosystems. *European Journal of Soil Biology*, *38*(3–4), 225–231.

Eerkes-Medrano, D., Thompson, R. (2018). Occurrence, fate, and effect of microplastics in freshwater systems. In: Microplastic contamination in aquatic environments. *Elsevier*, pp. 95–132.

Egbeocha, C. O., Malek, S., Emenike, C. U., & Milow, P. (2018). Feasting on microplastics: ingestion by and effects on marine organisms. *Aquatic Biology*, *27*, 93–106.

El-Kalliny, A. S., Abdel-Wahed, M. S., El-Zahhar, A. A., Hamza, I. A., & Gad-Allah, T. A. (2023). Nanomaterials: a review of emerging contaminants with potential health or environmental impact. *Discover Nano*, *18*(1), 68.

Fan, Y., Pan, D., Yang, M., & Wang, X. (2023). Radiolabelling and in vivo radionuclide imaging tracking of emerging pollutants in environmental toxicology: a review. Science of The Total Environment, 866, 161412.

Fang, H., Huang, L., Wang, J., He, G., & Reible, D. (2016). Environmental assessment of heavy metal transport and transformation in the Hangzhou Bay, China. *Journal of hazardous materials*, *302*, 447–457.

Fatimah, Y. A., Govindan, K., Murniningsih, R., & Setiawan, A. (2020). Industry 4.0 based sustainable circular economy approach for smart waste management system to achieve sustainable development goals: A case study of Indonesia. Journal of cleaner production, 269, 122263.

Feng, G., Huang, H., & Chen, Y. (2021). Effects of emerging pollutants on the occurrence and transfer of antibiotic resistance genes: A review. Journal of Hazardous Materials, 420, 126602.

Flint, S., Markle, T., Thompson, S., Wallace, E. (2012) Bisphenol A exposure, effects, and policy: A wildlife perspective. Journal of Environmental Management. 104, 19–34.

Franke, C., Studinger, G., Berger, G., Böhling, S., Bruckmann, U., Cohors-Fresenborg, D., & Jöhncke, U. (1994). The assessment of bioaccumulation. *Chemosphere*, *29*(7), 1501–1514.

Fuladlu, K., & Altan, H. (2021). Examining land surface temperature and relations with the major air pollutants: A remote sensing research in case of Tehran. Urban Climate, 39, 100958.

Garai, S., Bhattacharjee, C., Sarkar, S., Moulick, D., Dey, S., Jana, S., … & Hossain, A. (2024). Microplastics in the soil–water–food nexus: Inclusive insight into global research findings. *Science of The Total Environment*, 173891.

Gavrilescu, M., Demnerová, K., Aamand, J., Agathos, S., & Fava, F. (2015). Emerging pollutants in the environment: present and future challenges in biomonitoring, ecological risks and bioremediation. *New biotechnology*, *32*(1), 147–156.

Ge, J., Li, H., Liu, P., Zhang, Z., Ouyang, Z., & Guo, X. (2021). Review of the toxic effect of microplastics on terrestrial and aquatic plants. Science of the Total Environment, 791, 148333.

Geissen, V., Mol, H., Klumpp, E., Umlauf, G., Nadal, M., van der Ploeg, M., … & Ritsema, C. J. (2015). Emerging pollutants in the environment: a challenge for water resource management. International soil and water conservation research, 3(1), 57–65.

Ghosh, D., Brahmachari, K., Das, A., Sarkar, S., Dinda, N.K., Moulick, D., Datta, M.K. and Bandyopadhyay, S. (2023). Residual effect of organic manure on weed population and nitrogen uptake in greengram under rice-maize-greengram system. Indian Journal of Weed Science. 55(4), 374–379.

Ghosh, R., Parde, D., Bhaduri, S., Rajpurohit, P., & Behera, M. (2024). Occurrence, fate, transport, and removal technologies of emerging contaminants: A review on recent advances and future perspectives. CLEAN–Soil, Air, Water, 52(12), 2300259.

Gobas, F. A., Burkhard, L. P., Doucette, W. J., Sappington, K. G., Verbruggen, E. M., Hope, B. K., … & Tarazona, J. V. (2016). Review of existing terrestrial bioaccumulation models and terrestrial bioaccumulation modeling needs for organic chemicals. *Integrated environmental assessment and management*, *12*(1), 123–134.

Gobas, F. A., Wilcockson, J. B., Russell, R. W., & Haffner, G. D. (1999). Mechanism of biomagnification in fish under laboratory and field conditions. *Environmental science & technology*, *33*(1), 133–141.

Gomez, N. A., Lompré, J. S., Ferrando, A., Garrido, M., & Domini, C. E. (2024). Update on the status of the contamination by organotin compounds in sediment of Nuevo Gulf, Argentina. Insights from field and experimental studies. Marine Pollution Bulletin, 200, 116087.

González-González, R. B., Flores-Contreras, E. A., Gonzalez-Gonzalez, E., Torres Castillo, N. E., Parra-Saldivar, R., & Iqbal, H. M. (2022). Biosensor constructs for the monitoring of persistent emerging pollutants in environmental matrices. Industrial & Engineering Chemistry Research, 62(11), 4503–4520.

Gusev, A. A., Zakharova, O. V., Vasyukova, I. A., Evtushenko, N. A., Vasilieva, S. G., Lukyanov, A. A., … & Solovchenko, A. E. (2021). Nanoparticles in the Aquatic Environment: The Risks Associated with Them and the Possibilities of Their Mitigation with Microalgae. Moscow University Biological Sciences Bulletin, 76(4), 165–174.

Guševac Stojanović, I., Drakulić, D., Todorović, A., Martinović, J., Filipović, N., & Stojanović, Z. (2024). Acute Toxicity Assessment of Orally Administered Microplastic Particles in Adult Male Wistar Rats. Toxics, 12(3), 167.

Gutiérrez, M., Grillini, V., Pavlović, D. M., & Verlicchi, P. (2021). Activated carbon coupled with advanced biological wastewater treatment: A review of the enhancement in micropollutant removal. Science of the total environment, 790, 148050.

Gyawali, K. (2018). Pesticide uses and its effects on public health and environment. *Journal of Health Promotion*, *6*, 28–36.

Hagemann, L., Buchty-Lemke, M., Maaß, A. L., Schüttrumpf, H., Lehmkuhl, F., & Schwarzbauer, J. (2020). Potential hotspots of persistent organic pollutants in alluvial sediments of the meandering Wurm River, Germany. *Journal of Soils and Sediments*, *20*, 1034–1045.

Harsh, K., Kumar, G., Kumari, P., Kundu, M., Choudhury, S.R., Rakshit, R., Moulick, D. and Das, A., 2024. The altered crop pattern and livelihoods in changing climate. In Waste Management for Sustainable and Restored Agricultural Soil (pp. 345–369). Academic Press.

Harshitha, P., Bose, K., & Dsouza, H. S. (2024). Influence of lead-induced toxicity on the inflammatory cytokines. Toxicology, 503, 153771.

He, W., Cui, Y., Yang, H., Gao, J., Zhao, Y., Hao, N., … & Zhang, M. (2024). Aquatic toxicity, ecological effects, human exposure pathways and health risk assessment of liquid crystal monomers. *Journal of Hazardous Materials*, *461*, 132681.

Hebert, A., Forestier, D., Lenes, D., Benanou, D., Jacob, S., Arfi, C., … & Levi, Y. (2010). Innovative method for prioritizing emerging disinfection by-products (DBPs) in drinking water on the basis of their potential impact on public health. Water research, 44(10), 3147–3165.

Hossain, A., Maitra, S., Sarker, S., Al Mahmud, A., Ahmad, Z., Emon, R.M., Vemuri, H., Malek, M.A., Alam, M.A., Rahman, M.A. and Alam, M.J. (2023). Aluminium stress in plants: consequences and mitigation mechanisms. Beneficial Chemical Elements of Plants: Recent Developments and Future Prospects, pp. 123–168.

Ivanciuc, T., Ivanciuc, O., & Klein, D. J. (2006). Modeling the bioconcentration factors and bioaccumulation factors of polychlorinated biphenyls with posetic quantitative super-structure/activity relationships (QSSAR). *Molecular Diversity*, *10*, 133–145.

Jacob, M. M., Ponnuchamy, M., Kapoor, A., & Sivaraman, P. (2020). Bagasse based biochar for the adsorptive removal of chlorpyrifos from contaminated water. Journal of Environmental Chemical Engineering, 8(4), 103904.

Jeevanandam, J., Barhoum, A., Chan, Y. S., Dufresne, A., & Danquah, M. K. (2018). Review on nanoparticles and nanostructured materials: history, sources, toxicity and regulations. *Beilstein journal of nanotechnology*, *9*(1), 1050–1074.

Jiao, X., Shi, Q., and Gan, J. (2021). Uptake, accumulation and metabolism of PFASs in plants and health perspectives: A critical review. Crit. Rev. Environ. Sci. Technol. 51: 2745–2776.

Katagi, T. (2010). Bioconcentration, bioaccumulation, and metabolism of pesticides in aquatic organisms. *Reviews of environmental contamination and toxicology*, 1–132.

Kelly, B. C., & Gobas, F. A. (2003). An arctic terrestrial food-chain bioaccumulation model for persistent organic pollutants. *Environmental science & technology*, *37*(13), 2966–2974.

Kelly, B. C., Gobas, F. A., & McLachlan, M. S. (2004). Intestinal absorption and biomagnification of organic contaminants in fish, wildlife, and humans. *Environmental Toxicology and Chemistry: An International Journal*, *23*(10), 2324–2336.

Khan, B. A., Nadeem, M. A., Nawaz, H., Amin, M. M., Abbasi, G. H., Nadeem, M., … & Ayub, M. A. (2023). Pesticides: impacts on agriculture productivity, environment, and management strategies. In Emerging contaminants and plants: Interactions, adaptations and remediation technologies (pp. 109–134). Cham: Springer International Publishing.

Khan, S., Naushad, M., Govarthanan, M., Iqbal, J., & Alfadul, S. M. (2022). Emerging contaminants of high concern for the environment: Current trends and future research. Environmental Research, 207, 112609.

Khoshnood, Z. (2024). A review on toxic effects of pesticides in Zebrafish, Danio rerio and common carp, Cyprinus carpio, emphasising Atrazine herbicide. Toxicology Reports, 101694.

Kishor, R., Purchase, D., Saratale, G. D., Saratale, R. G., Ferreira, L. F. R., Bilal, M., … & Bharagava, R. N. (2021). Ecotoxicological and health concerns of persistent coloring pollutants of textile industry wastewater and treatment approaches for environmental safety. *Journal of Environmental Chemical Engineering*, *9*(2), 105012.

Klaine, S. J., Alvarez, P. J., Batley, G. E., Fernandes, T. F., Handy, R. D., Lyon, D. Y., … & Lead, J. R. (2008). Nanomaterials in the environment: behavior, fate, bioavailability, and effects. *Environmental Toxicology and Chemistry: An International Journal*, *27*(9), 1825–1851.

La Farre, M., Pérez, S., Kantiani, L., & Barceló, D. (2008). Fate and toxicity of emerging pollutants, their metabolites and transformation products in the aquatic environment. *TrAC Trends in Analytical Chemistry*, *27*(11), 991–1007.

Langston, W. J. (2020). Endocrine disruption and altered sexual development in aquatic organisms: an invertebrate perspective. *Journal of the Marine Biological Association of the United Kingdom*, *100*(4), 495–515.

Lazofsky, A., & Buckley, B. (2022). Recent trends in multiclass analysis of emerging endocrine disrupting contaminants (EDCs) in drinking water. Molecules, 27(24), 8835.

López-Doval, J. C., Montagner, C. C., de Alburquerque, A. F., Moschini-Carlos, V., Umbuzeiro, G., & Pompêo, M. (2017). Nutrients, emerging pollutants and pesticides in a tropical urban reservoir: Spatial distributions and risk assessment. Science of the Total Environment, 575, 1307–1324.

Lu, X., Sun, J., & Sun, X. (2020). Recent advances in biosensors for the detection of estrogens in the environment and food. *TrAC Trends in Analytical Chemistry*, *127*, 115882.

Łuczyńska, J., Paszczyk, B., & Łuczyński, M. J. (2018). Fish as a bioindicator of heavy metals pollution in aquatic ecosystem of Pluszne Lake, Poland, and risk assessment for consumer's health. *Ecotoxicology and environmental safety*, *153*, 60–67.

Ma, Y., Yang, K., Yu, H., Tan, W., Gao, Y., & Lv, B. (2024). Effects and mechanism of microplastics on organic carbon and nitrogen cycling in agricultural soil: A review. *Soil Use and Management*, *40*(1), e12971.

Madesh, S., Sudhakaran, G., Ramamurthy, K., Kathiravan, M. K., Almutairi, M. H., Almutairi, B. O., … & Arockiaraj, J. (2024). Cadmium and ketoprofen accumulation influences aquatic ecosystem demonstrated using in-vivo zebrafish model. Drug and Chemical Toxicology, 1–16.

Mahanta, S., Kundu, R., Sen, P., Mubarak, K. P. M., Shaw, S., Hossain, A., Dhar, A., Sarkar, S., Santra, S. C., & Moulick, D. (2025). Heavy metal accumulation in vegetable crops: a threat to sustainable agroecosystem. In M. K. Lal, R. K. Tiwari, A. Kumar, R. Kumar, & B. Singh (Eds.), *Abiotic Stress in Underground Vegetables: Response and Mitigation* (1st ed., pp. 1–29). *Academic Press, Elsevier.*

Majumdar, A., Upadhyay, M. K., Ojha, M., Biswas, R., Dey, S., Sarkar, S., … & Roychowdhury, T. (2024). A critical review on the organo-metal (loid) s pollution in the environment: Distribution, remediation and risk assessment. *Science of The Total Environment*, p.175531.

Majumdar, A., Upadhyay, M.K., Giri, B., Yadav, P., Moulick, D., Sarkar, S., Thakur, B.K., Sahu, K., Srivastava, A.K., Buck, M. and Tibbett, M. (2024b). Sustainable water management in rice cultivation reduces arsenic contamination, increases productivity, microbial molecular response, and profitability. Journal of Hazardous Materials, 466, p.133610.175531.

Mangam, V. T., Narla, D., Konda, R. K., & Sarella, P. N. K. (2024). Beyond the spectrum: Exploring unconventional applications of fourier transform infrared (FTIR) spectroscopy. Asian Journal of Pharmaceutical Analysis, 14(2), 86–94.

Meijide, F. J., Da Cuña, R. H., Prieto, J. P., Dorelle, L. S., Babay, P. A., & Nostro, F. L. L. (2018). Effects of waterborne exposure to the antidepressant fluoxetine on swimming, shoaling and anxiety behaviours of the mosquitofish Gambusia holbrooki. *Ecotoxicology and Environmental Safety, 163*, 646–655.

Meng, L., Yan, C., Lv, S., Sun, H., Xue, S., Li, Q., ... & Yan, X. H. (2024). Synthetic aperture radar for geosciences. Reviews of Geophysics, 62(3), e2023RG000821.

Millbern, Z., Trettin, A., Wu, R., Demmler, M., & Vinueza, N. R. (2024). Synthetic dyes: A mass spectrometry approach and applications. Mass Spectrometry Reviews, 43(2), 327–344.

Mishra, R. K., Mentha, S. S., Misra, Y., & Dwivedi, N. (2023). Emerging pollutants of severe environmental concern in water and wastewater: A comprehensive review on current developments and future research. *Water-Energy Nexus*.

Mofijur, M., Hasan, M. M., Ahmed, S. F., Djavanroodi, F., Fattah, I. M. R., Silitonga, A. S., ... & Khan, T. Y. (2023). Advances in identifying and managing emerging contaminants in aquatic ecosystems: Analytical approaches, toxicity assessment, transformation pathways, environmental fate, and remediation strategies. Environmental Pollution, 122889.

Mohana Rangan, S., Krajmalnik-Brown, R., & Delgado, A. G. (2021). An ion chromatography method for simultaneous quantification of chromate, arsenate, selenate, perchlorate, and other inorganic anions in environmental media. Environmental Engineering Science, 38(7), 626–634.

Monchanin, C., Devaud, J. M., Barron, A. B., & Lihoreau, M. (2021). Current permissible levels of metal pollutants harm terrestrial invertebrates. Science of the Total Environment, 779, 146398.

Montes-Grajales, D., Fennix-Agudelo, M., & Miranda-Castro, W. (2017). Occurrence of personal care products as emerging chemicals of concern in water resources: A review. Science of the Total Environment, 595, 601–614.

Moore, A. P., & Bringolf, R. B. (2018). Effects of nitrate on freshwater mussel glochidia attachment and metamorphosis success to the juvenile stage. *Environmental pollution, 242*, 807–813.

Moulick, D. , Santra, S.C. , & Ghosh, D. (2017). Seed priming with Se alleviate As induced phytotoxicity during germination and seedling growth by restricting As translocation in rice (*Oryza sativa* L cv IET -4094). *Ecotoxicology and Environmental Safety*, 145, 449–456

Moulick, D., Chowardhara, B. and Panda, S.K. (2019). Agroecotoxicological aspect of arsenic (As) and cadmium (Cd) on Field crops and its mitigation: Current status and future prospect. Plant-metal interactions, pp.217–246.

Moulick, D., Ghosh, D., Mandal, J., Bhowmick, S., Mondal, D., Choudhury, S., Santra, S.C., Vithanage, M. and Biswas, J.K. (2023). A cumulative assessment of plant growth stages and selenium supplementation on arsenic and micronutrients accumulation in rice grains. Journal of Cleaner Production, 386, p. 135764.

Moulick, D., Ghosh, D., Skalicky, M., Gharde, Y., Mazumder, M.K., Choudhury, S., Biswas, J.K., Santra, S.C., Brestic, M., Vachova, P. and Hossain, A. (2022). Interrelationship among rice grain arsenic, micronutrients content and grain quality attributes: An investigation from genotype× environment perspective. Frontiers in Environmental Science, 10, p. 857629.

Moulick, D., Hossain, A., Barek, V., Chowardhara, B., Mukherjee, A., Pattnaik, B.K., Roy, P., Mahanta, S., Hazra, S., Sarkar, S. and Garai, S. (2024b). Heavy metal stress in the agro-environment: consequences, adaptations and remediation. International Journal of Environmental Science and Technology, pp.1–44.

Moulick, D., Majumdar, A., Choudhury, A., Das, A., Chowardhara, B., Pattnaik, B.K., Dash, G.K., Murmu, K., Bhutia, K.L., Upadhyay, M.K. and Yadav, P. (2024a). Emerging Concern of Nano-pollution in Agro-Ecosystem: Flip Side of Nanotechnology. Plant Physiology and Biochemistry, p.108704.

Moulick, D., Mukherjee, A., Das, A., Roy, A., Majumdar, A., Dhar, A., ... & Hossain, A. (2024). Selenium–An environmentally friendly micronutrient in agroecosystem in the modern era: An overview of 50-year findings. *Ecotoxicology and Environmental Safety, 270*, 115832.

Moulick, D., Samanta, S., Sarkar, S., Mukherjee, A., Pattnaik, B.K., Saha, S., Awasthi, J.P., Bhowmick, S., Ghosh, D., Samal, A.C. and Mahanta, S. (2021). Arsenic contamination, impact and mitigation strategies in rice agro-environment: An inclusive insight. Science of the Total Environment, 800, p.149477.

Mukherjee, P., Banerjee, G., Saha, N., & Mazumdar, A. (2024). Overview on the emergence of pesticide contamination and treatment methodologies. Water, Air, & Soil Pollution, 235(9), 587.

Müller, A., Österlund, H., Marsalek, J., & Viklander, M. (2020). The pollution conveyed by urban runoff: A review of sources. *Science of the Total Environment, 709*, 136125.

Naeem, M., Gill, R., Gill, S. S., Singh, K., Sofo, A., & Tuteja, N. (2023). Emerging contaminants and their effect on agricultural crops. *Frontiers in Plant Science, 14*, 1296252.

Ncube, A., Mtetwa, S., Bukhari, M., Fiorentino, G., & Passaro, R. (2023). Circular economy and green chemistry: the need for radical innovative approaches in the design for new products. Energies, 16(4), 1752.

Nfon, E., Cousins, I. T., & Broman, D. (2008). Biomagnification of organic pollutants in benthic and pelagic marine food chains from the Baltic Sea. *Science of the Total Environment, 397*(1–3), 190–204.

Ngatia, L., Grace III, J. M., Moriasi, D., & Taylor, R. (2019). Nitrogen and phosphorus eutrophication in marine ecosystems. *Monitoring of marine pollution, 1*, 1–17.

Nikolaou, A., Meric, S., & Fatta, D. (2007). Occurrence patterns of pharmaceuticals in water and wastewater environments. *Analytical and bioanalytical chemistry, 387*, 1225–1234.

Nkoom, M., Lu, G., & Liu, J. (2022). Chronic toxicity of diclofenac, carbamazepine and their mixture to Daphnia magna: a comparative two-generational study. Environmental Science and Pollution Research, 29(39), 58963–58979.

Noyes, P. D., McElwee, M. K., Miller, H. D., Clark, B. W., Van Tiem, L. A., Walcott, K. C., … & Levin, E. D. (2009). The toxicology of climate change: environmental contaminants in a warming world. Environment international, 35(6), 971–986.

Ogwu, M. C., Izah, S. C., Imarhiagbe, O., Lori, T., & Aliu, O. O. (2024). Effects of Air Pollutants on Biodiversity. In Air Pollutants in the Context of One Health: Fundamentals, Sources, and Impacts (pp. 341–367). Cham: Springer Nature Switzerland.

Okeke, E. S., Okoye, C. O., Ezeorba, T. P. C., Mao, G., Chen, Y., Xu, H., … & Wu, X. (2022). Emerging bio-dispersant and bioremediation technologies as environmentally friendly management responses toward marine oil spill: A comprehensive review. Journal of Environmental Management, 322, 116123.

Orisakwe, O.E., Dagur, E.A., Mbagwu, H.O.C., & Udowelle, N.A. (2017). Lead Levels in Vegetables from Artisanal Mining Sites of Dilimi River, Bukuru and Barkin Ladi North Central Nigeria, Cancer and Non -Cancer Risk Assessment. Asian Pacific Journal of Cancer Prevention: APJCP, 18 (3), 621–627.

Osuoha, J. O., Anyanwu, B. O., & Ejileugha, C. (2023). Pharmaceuticals and personal care products as emerging contaminants: Need for combined treatment strategy. *Journal of hazardous materials advances, 9*, 100206.

Pattnaik, B.K., Sahu, C. and Moulick, D. (2024). Nanoparticles in soil–water–plant inference. In The nanotechnology driven agriculture (pp. 85–105). CRC Press.

Pena-Pereira, F., Bendicho, C., Pavlović, D. M., Martín-Esteban, A., Díaz-Álvarez, M., Pan, Y., … & Psillakis, E. (2021). Miniaturized analytical methods for determination of environmental contaminants of emerging concern–a review. Analytica chimica acta, 1158, 238108.

Pironti, C., Ricciardi, M., Proto, A., Bianco, P. M., Montano, L., & Motta, O. (2021). Endocrine-disrupting compounds: An overview on their occurrence in the aquatic environment and human exposure. *Water, 13*(10), 1347.

Podder, A., Sadmani, A. A., Reinhart, D., Chang, N. B., & Goel, R. (2021). Per and poly-fluoroalkyl substances (PFAS) as a contaminant of emerging concern in surface water: A transboundary review of their occurrences and toxicity effects. Journal of hazardous materials, 419, 126361.

Primpke, S., Godejohann, M., & Gerdts, G. (2020). Rapid identification and quantification of microplastics in the environment by quantum cascade laser-based hyperspectral infrared chemical imaging. Environmental science & technology, 54(24), 15893–15903.

Puri, M., Gandhi, K., & Kumar, M. S. (2023). Emerging environmental contaminants: A global perspective on policies and regulations. Journal of Environmental Management, 332, 117344.

Qadeer, A., Anis, M., Warner, G. R., Potts, C., Giovanoulis, G., Nasr, S., … & Shuhang, W. (2024). Global environmental and toxicological data of emerging plasticizers: current knowledge, regrettable substitution dilemma, green solution and future perspectives. Green Chemistry.

Qi, X., Lian, Y., Xie, L., Wang, Y., & Lu, Z. (2024). Water quality detection based on UV-Vis and NIR spectroscopy: a review. Applied Spectroscopy Reviews, 59(8), 1036–1060.

Rajaram, T., & Das, A. (2008). Water pollution by industrial effluents in India: discharge scenarios and case for participatory ecosystem specific local regulation. *Futures*, *40*(1), 56–69.

Rajput, A. P., Kulkarni, M., Pingale, P. L., Tekade, M., Shakya, A. K., & Tekade, R. K. (2023). Understanding the bioaccumulation of pharmaceuticals and personal care products. In *Essentials of Pharmatoxicology in Drug Research* (pp. 393–434). Academic Press.

Rakib, M. R. J., Sarker, A., Ram, K., Uddin, M. G., Walker, T. R., Chowdhury, T., … & Idris, A. M. (2023). Microplastic toxicity in aquatic organisms and aquatic ecosystems: a review. Water, Air, & Soil Pollution, 234(1), 52.

Ramzan, F., Ali, M., Raza, M. H., & Zubair, M. (2024). A review: modeling and optimizing diffractive optical elements for enhanced pollutant detection in LiDAR systems. Computational Optics 2024, 13023, 141–149.

Reshi, M. S., Sheikh, T. A., Javaid, D., Ganie, S. Y., Ganie, M. A., & Malik, M. A. (2025). Pesticide Pollution: Potential Risk to Insect Pollinators and Possible Management Strategies. In *Insect Diversity and Ecosystem Services* (pp. 55–73). Apple Academic Press.

Ripanda, A. S., Rwiza, M. J., Nyanza, E. C., Machunda, R. L., & Vuai, S. H. (2022). Contribution of Illicit Drug Use to Pharmaceutical Load in the Environment: A Focus on Sub-Saharan Africa. Journal of Environmental and Public Health, 2022(1), 9056476.

Rist, S., Almroth, B.C., Hartmann, N.B., Karlsson, T.M., 2018. A critical perspective on early communications concerning human health aspects of microplastics. *Sci. Total Environ.* 626, 720–726.

Rodríguez-Hernández, J. A., Araújo, R. G., López-Pacheco, I. Y., Rodas-Zuluaga, L. I., González-González, R. B., Parra-Arroyo, L., … & Parra-Saldívar, R. (2022). Environmental persistence, detection, and mitigation of endocrine disrupting contaminants in wastewater treatment plants–a review with a focus on tertiary treatment technologies. *Environmental Science: Advances*, *1*(5), 680–704.

Roy, D., Ghosh, D., Chander, S., Saha, B. and Moulick, D. (2024). Herbicide Contamination in Ground Water and Their Probable Remediation. In Environmental Contaminants (pp. 111–140). Apple Academic Press.

Roy, D., Ghosh, S., Datta, D., Sreekanth, D., Pawar, D., Mukherjee, P.K., Ghosh, D., Santra, S.C. and Moulick, D. (2023). Climate resilient weed management for crop production. In Climate-Resilient Agriculture, Vol 2: Agro-Biotechnological Advancement for Crop Production (pp. 125–156). Cham: Springer International Publishing.

Roy, S. and Hossain, A. eds., 2024. The Nanotechnology Driven Agriculture: The Future Ahead. CRC Press.

Saha, B., Chowardhara, B., Kar, S., Devi, S.S., Awasthi, J.P., Moulick, D., Tanti, B. and Panda, S.K. (2019). Advances in heavy metal-induced stress alleviation with respect to exogenous amendments in crop plants. Priming and pretreatment of seeds and seedlings: Implication in plant stress tolerance and enhancing productivity in crop plants, pp. 313–332.

Salem, Z., Hamouri, K., Djemaa, R., & Allia, K. (2008). Evaluation of landfill leachate pollution and treatment. *Desalination*, *220*(1–3), 108–114.

Schell, L. M., Gallo, M. V., Denham, M., & Ravenscroft, J. (2006). Effects of pollution on human growth and development: an introduction. *Journal of physiological anthropology*, *25*(1), 103–112.

Schmeller, D. S., Loyau, A., Bao, K., Brack, W., Chatzinotas, A., De Vleeschouwer, F., … & Vredenburg, V. T. (2018). People, pollution and pathogens–Global change impacts in mountain freshwater ecosystems. *Science of the Total Environment*, *622*, 756–763.

Serafini, P. P., Righetti, B. P., Vanstreels, R. E., Bugoni, L., Piazza, C. E., Lima, D., … & Lüchmann, K. H. (2024). Biochemical and molecular biomarkers and their association with

anthropogenic chemicals in wintering Manx shearwaters (Puffinus puffinus). Marine Pollution Bulletin, 203, 116398.
Shah, A. I., Dar, M. U. D., Bhat, R. A., Singh, J. P., Singh, K., & Bhat, S. A. (2020). Prospectives and challenges of wastewater treatment technologies to combat contaminants of emerging concerns. Ecological Engineering, 152, 105882.
Shahid, M. K., Kashif, A., Fuwad, A., & Choi, Y. (2021). Current advances in treatment technologies for removal of emerging contaminants from water–A critical review. *Coordination Chemistry Reviews*, *442*, 213993.
Sharpe, S., & Mackay, D. (2000). A framework for evaluating bioaccumulation in food webs. *Environmental science & technology*, *34*(12), 2373–2379.
Sheng, S., Han, N., Wei, Y., Wang, J., Han, W., Xing, B., … & Zhang, W. (2024). Liver injury induced by exposure to polystyrene microplastics alone or in combination with cadmium in mice is mediated by oxidative stress and apoptosis. Biological Trace Element Research, 202(5), 2170–2183.
Shyamalagowri, S., Shanthi, N., Manjunathan, J., Kamaraj, M., Manikandan, A., & Aravind, J. (2023). Techniques for the detection and quantification of emerging contaminants. Physical Sciences Reviews, 8(9), 2191–2218.
Singhal, R. K., Pandey, S., & Srivastava, S. (2023). Emerging pollutants in the environment: impact and challenges for agriculture and nutritional security. *Frontiers in Ecology and Evolution*, *11*, 1273925.
Snow, D. D., Cassada, D. A., Larsen, M. L., Mware, N. A., Li, X., D'Alessio, M., … & Sallach, J. B. (2017). Detection, occurrence and fate of emerging contaminants in agricultural environments. *Water Environment Research*, *89*(10), 897–920.
Soleimany, A., Grubliauskas, R., & Šerevičienė, V. (2021). Application of satellite data and GIS services for studying air pollutants in Lithuania (case study: Kaunas city). Air Quality, Atmosphere & Health, 14(3), 411–429.
Stadelmann, B., Leonards, P. E., & Brandsma, S. H. (2024). A new class of contaminants of concern? A comprehensive review of liquid crystal monomers. Science of The Total Environment, 174443.
Sudarsan, J. S., Dogra, K., Kumar, R., Raval, N. P., Leifels, M., Mukherjee, S., … & Kumar, M. (2024). Tricks and tracks of prevalence, occurrences, treatment technologies, and challenges of mixtures of emerging contaminants in the environment: With special emphasis on microplastic. Journal of Contaminant Hydrology, 265, 104389.
Sun, S., Sidhu, V., Rong, Y., & Zheng, Y. (2018). Pesticide pollution in agricultural soils and sustainable remediation methods: a review. *Current Pollution Reports*, *4*, 240–250.
Sun, X., Zhou, Y., Jia, S., Shao, H., Liu, M., Tao, S., & Dai, X. (2024). Impacts of mining on vegetation phenology and sensitivity assessment of spectral vegetation indices to mining activities in arid/semi-arid areas. Journal of Environmental Management, 356, 120678.
Swapna, M. S., Korte, D., & Sankararaman, S. (2024). Deciphering air quality index through sample entropy: A nonlinear time series analysis. Gondwana Research, 132, 53–63.
Tang, Y., Yin, M., Yang, W., Li, H., Zhong, Y., Mo, L., … & Sun, X. (2019). Emerging pollutants in water environment: Occurrence, monitoring, fate, and risk assessment. Water Environment Research, 91(10), 984–991.
Tyagi, I., Karri, R. R., Mubarak, N. M., & Dehghani, M. H. (2024). Emerging pollutants in the aqueous solution: sources, classification, and associated health implications. Sustainable Technologies for Remediation of Emerging Pollutants from Aqueous Environment, 1–11.
Umoh, I. U., Umanah, S. J., Obot, G. I., Asukwo, M. O., Effiokette, G. M., & Inam, A. I. (2022). Histomorphological and biochemical alterations on the liver of Wistar rats following co-administration of NSAIDs (Piroxicam, Diclofenac and Ibuprofen). Anatomy Journal of Africa, 11(2), 2175–2184.
Unnikrishan, A., Khalid, N. K., Rayaroth, M. P., Thomas, S., Nazim, A., Aravindakumar, C. T., & Aravind, U. K. (2024). Occurrence and distribution of steroid hormones (estrogen) and other contaminants of emerging concern in a south indian water body. Chemosphere, 351, 141124.

Van Acker, T., Theiner, S., Bolea-Fernandez, E., Vanhaecke, F., & Koellensperger, G. (2023). Inductively coupled plasma mass spectrometry. Nature Reviews Methods Primers, 3(1), 52.

Van Ginneken, M., Blust, R., & Bervoets, L. (2018). Combined effects of metal mixtures and predator stress on the freshwater isopod Asellus aquaticus. *Aquatic Toxicology, 200*, 148–157.

Vasilachi, I. C., Asiminicesei, D. M., Fertu, D. I., & Gavrilescu, M. (2021). Occurrence and fate of emerging pollutants in water environment and options for their removal. Water, 13(2), 181.

Verlicchi, P., Galletti, A., Petrovic, M., & Barceló, D. (2010). Hospital effluents as a source of emerging pollutants: an overview of micropollutants and sustainable treatment options. Journal of hydrology, 389(3–4), 416–428.

Walter, M., Boyd-Wilson, K.S.H., McNaughton, D., & Northcott, G. (2005). Laboratory trials on the bioremediation of aged pentachlorophenol residues. *International Biodeterioration & Biodegradation*, 3, 121–130.

Wang, F., Xiang, L., Leung, K. S. Y., Elsner, M., Zhang, Y., Guo, Y., ... & Tiedje, J. M. (2024). Emerging contaminants: a One Health perspective. *The Innovation*.

Wang, S., Ji, C., Li, F., Zhan, J., Sun, T., Tang, J., & Wu, H. (2021). Tetrabromobisphenol A induced reproductive endocrine-disrupting effects in mussel Mytilus galloprovincialis. *Journal of Hazardous Materials, 416*, 126228.

Wang, Y., Dai, D., Yu, Y., Yang, G., Shen, W., Wang, Q., ... & Zhao, X. (2018). Evaluation of joint effects of cyprodinil and kresoxim-methyl on zebrafish, Danio rerio. *Journal of Hazardous Materials, 352*, 80–91.

Whalon, M. E., Mota-Sanchez, D., & Hollingworth, R. M. (2008). Analysis of global pesticide resistance in arthropods. In *Global pesticide resistance in arthropods* (pp. 5–31). Wallingford UK: CABI.

WHO/UNICEF Joint Water Supply, & Sanitation Monitoring Programme. (2015). *Progress on sanitation and drinking water: 2015 update and MDG assessment*. World Health Organization.

Xu, S., Yang, K., Xu, Y., Zhu, Y., Luo, Y., Shang, C., ... & Wu, C. (2021). Urban Land Surface Temperature Monitoring and Surface Thermal Runoff Pollution Evaluation Using UAV Thermal Remote Sensing Technology. Sustainability, 13(20), 11203.

Yadav, D., Rangabhashiyam, S., Verma, P., et al. (2021a). Environmental and health impacts of contaminants of emerging concerns: Recent treatment challenges and approaches. Chemosphere 272: 129492.

Yadav, N., Garg, V. K., Chhillar, A. K., & Rana, J. S. (2021b). Detection and remediation of pollutants to maintain ecosustainability employing nanotechnology: A review. *Chemosphere, 280*, 130792.

Yang, J., Zhao, Y., Li, M., Du, M., Li, X., & Li, Y. (2019). A review of a class of emerging contaminants: the classification, distribution, intensity of consumption, synthesis routes, environmental effects and expectation of pollution abatement to organophosphate flame retardants (OPFRs). International journal of molecular sciences, 20(12), 2874.

Yarsan, E., & Yipel, M. (2013). The important terms of marine pollution “biomarkers and biomonitoring, bioaccumulation, bioconcentration, biomagnification”. *J Mol Biomark Diagn S, 1*(2).

Yu, Y., Wang, S., Yu, P., Wang, D., Hu, B., Zheng, P., & Zhang, M. (2024a). A bibliometric analysis of emerging contaminants (ECs) (2001–2021): Evolution of hotspots and research trends. Science of The Total Environment, 907, 168116.

Yu, Y., Wang, Z., Yao, B., & Zhou, Y. (2024b). Occurrence, bioaccumulation, fate, and risk assessment of emerging pollutants in aquatic environments: A review. Science of The Total Environment, 923, 171388.

Zaghloul, A., Saber, M., Gadow, S., & Awad, F. (2020). Biological indicators for pollution detection in terrestrial and aquatic ecosystems. Bulletin of the National Research Centre, 44(1), 1–11.

Zandalinas, S. I., Fritschi, F. B., & Mittler, R. (2021). Global warming, climate change, and environmental pollution: recipe for a multifactorial stress combination disaster. Trends in Plant Science, 26(6), 588–599.

Zantis, L. J., Borchi, C., Vijver, M. G., Peijnenburg, W., Di Lonardo, S., & Bosker, T. 2023. Nano- and microplastics commonly cause adverse impacts on plants at environmentally relevant levels: a systematic review. *Sci. Total Environ.*, 867, 161211.

Zhang, J., Zou, T., & Lai, Y. (2021a). Novel method for industrial sewage outfall detection: Water pollution monitoring based on web crawler and remote sensing interpretation techniques. Journal of Cleaner Production, 312, 127640.

Zhang, N., Liu, X., Pan, L., Zhou, X., Zhao, L., Mou, X., … & Wang, X. (2021b). Evaluation of ibuprofen contamination in local urban rivers and its effects on immune parameters of juvenile grass carp. Fish Physiology and Biochemistry, 47(5), 1405–1413.

Zhang, Y., Li, Z., Bai, K., Wei, Y., Xie, Y., Zhang, Y., … & Li, D. (2021c). Satellite remote sensing of atmospheric particulate matter mass concentration: Advances, challenges, and perspectives. Fundamental Research, 1(3), 240–258.

Zhang, Y., Wang, B., Sun, W., Wang, G., Liu, Z., Zhang, X., … & Zhang, H. (2024). Paternal exposures to endocrine-disrupting chemicals induce intergenerational epigenetic influences on offspring: A review. Environment International, 108689.

Zhao, Y., Tao, S., Liu, S., Hu, T., Zheng, K., Shen, M., & Meng, G. (2023). Research advances on impacts micro/nanoplastics and their carried pollutants on algae in aquatic ecosystems: A review. *Aquatic Toxicology*, 106725.

Zhong, M., Wang, T., Zhao, W., Huang, J., Wang, B., Blaney, L., … & Yu, G. (2022). Emerging organic contaminants in Chinese surface water: identification of priority pollutants. *Engineering*, *11*, 111–125.

Zhong, W., Zhang, Y., Wu, Z., Yang, R., Chen, X., Yang, J., & Zhu, L. (2018). Health risk assessment of heavy metals in freshwater fish in the central and eastern North China. *Ecotoxicology and environmental safety*, *157*, 343–349.

Zhu, S., Zhang, Z., & Žagar, D. (2018). Mercury transport and fate models in aquatic systems: A review and synthesis. *Science of the Total Environment*, *639*, 538–549.

Zoeller, R. T., Brown, T. R., Doan, L. L., Gore, A. C., Skakkebaek, N. E., Soto, A. M., … & Vom Saal, F. S. (2012). Endocrine-disrupting chemicals and public health protection: a statement of principles from The Endocrine Society. *Endocrinology*, *153*(9), 4097–4110.

Ecological Impacts of Emerging Pollutants in the Event of Climate Change

Arpita Maity and Amal Kumar Mondal

Abstract Emerging pollutants such as microplastics, nanomaterials, endocrine-disrupting chemicals, personal care products (PCPs), and medications seriously threaten the ecological systems around the world. Climate change can enhance these risks since it is associated with changed precipitation patterns, rising temperatures and more frequent temperate weather events. This study examines the sources of emerging pollutants, their behavior, and how climate change interacts to affect aquatic and terrestrial ecosystems. It also discusses the consequences for ecosystem services, biodiversity, and human health under climate-changing circumstances.

Keywords Climate change · Ecosystem · Endocrine-disrupting chemicals · Pollution

1 Introduction

Emerging pollutants, such as pharmaceuticals, PCPs, and industrial chemicals, are increasingly being detected in aquatic pools (Chiu et al. 2017). Since 2000, these compounds' production has increased from 1 million tons to an estimated 500 million tons a year worldwide (Rasheed et al. 2019). The contaminants are often introduced into the environment via agricultural wastage, improper disposal practices, and wastewater effluents (Srikanth et al. 2019). Emerging pollutants are widely used in modern civilization, including PCPs, fungicides, pesticides, drugs, and

A. Maity
Centre for Life Science and Department of Botany and Forestry, Vidyasagar University, Midnapore, West Bengal, India

A. K. Mondal (✉)
Plant Taxonomy, Biosystematics and Molecular Taxonomy Laboratory, UGC-DRS-SAP & DBT-BOOST-WB Supported Department, Department of Botany and Forestry, Vidyasagar University, Midnapore, West Bengal, India
e-mail: akmondal@mail.vidyasagar.ac.in

S. Roy, V. Mandal (eds.), *Plant-Microbe Interaction under Xenobiotic Exposure*,
https://doi.org/10.1007/978-981-96-8260-7_2

industrial chemicals (Galindo-Miranda et al. 2019). The environment has been exposed to billions of chemical-based materials that have emerged from several anthropogenic activities such as agriculture, building, energy generation, industrialization, mining, and other interference (Lahiri et al. 2023). Additionally, the anthropogenic cast of geogenic chemicals that have produced over 140,000 chemical compositions and different compounds were entirely unknown (Gavrilescu et al. 2015). According to a more recent review of global chemical inventories, it could be more than 350,000 (Kourkoumpas et al. 2015). Recently, several synthetic substances, especially those harmful in very low concentrations, are mixed with other pollutants or discharged into the natural environment as breakdown products. A total of 220 billion tons of chemical wastes are released every year, since only 20% of GHG (greenhouse gas) emissions are attributed to chemical releases (Naidu et al. 2021). The chemicals that enter the environment are agricultural runoff, personal care, and medication pollutants. These pollutants can undergo bioaccumulation and biomagnification via the food chain and pollute in aquatic, terrestrial, and atmospheric environments (Ruhí et al. 2016). Concerns about the prevalence of endocrine-disrupting substances and pharmaceutical compounds in freshwater ecosystems are growing because these substances can harm non-target organisms, causing decreased fertility and fecundity and feminizing and masculinizing females (Ruhí et al. 2016). Moreover, the deteriorating quality of water bodies because of misuse of natural resources is exacerbating the problem of EPs (Emerging pollutants) (Aragão et al. 2020). Worryingly, many of these compounds lack ecotoxicological information or environmental regulation and cause harm to aquatic biota and human health (Aragão et al. 2020; Srikanth et al. 2019). Approximately 9 million people die each year from exposure to contaminated water, food, water, and consumer products (Landrigan et al. 2018). These pollutants include antibiotics, endocrine disruption, microbeads, microplastics, drugs, medicines, hormones, industrial wastage, steroids, and other toxic chemicals. These pollutants contaminate the surface water, soil, and air. Although these pollutants are abiotic and biotic by-products of farming techniques (fertilizers, pesticides, microelements, hazardous waste, and veterinary dung), and PCPs (such as cosmetics, beauty products, antibiotics, and hormones), these pollutants harm the ecosystem (Cheng et al. 2021; Giudice et al. 2021). The escalating emissions of GHGs (e.g., CFC_s, CO_2, CO, CH_4, SO_2, N_2O) and other anthropogenic pollutants have significant environmental concerns due to their substantial contributions to climate change (global warming) and environmental degradation (Patra and Saxena 2010; IPCC 2007). Environmental ecosystems and human health are at risk from these compounds, often not regulated or recently recognized as dangerous. These compounds cause significant risk to the environmental ecosystem and human health. In addition, adverse effects on climate shifts are marked by higher temperatures, rainfall pattern shifts, and more frequently severe weather. In recent decades, Earth's climate has changed significantly, with earth temperatures gradually increasing, precipitation patterns shifting, and towering weather events becoming more legionary and severe (Grant et al. 2016).

Climate change significantly impacts freshwater ecosystems, introducing new contaminants to freshwater ecosystems (Woodward et al. 2010). Temperature changes and CO2 increases are several stresses created by climate variations (Weeb et al. 2008). Other stressors include droughts and extreme floods (Barnett et al. 2005; Milly et al. 2006). Climate change is directly responsible for sea-level rise, severe weather events such as drought and flooding, increased temperatures that cause wildfires, and more variable rainfall (Clarke et al. 2022; Seneviratne et al. 2012). Several reports on climate change predict rapid and extensive changes to global ecosystems with unpredictable impacts on human and ecosystem health (IPCC 2022). A warming climate can exacerbate heat waves that contribute to extremely hot days. Climate change will increase evapotranspiration as the surface temperature rises (Oelbermann et al. 2022; Peng et al. 2023; Trenberth et al. 2014). A warmer climate can also enhance drought-prone environments because of increased land evaporation. Atmospheric temperature changes are linked to heavier precipitation events, such as rainstorms and snowstorms (Trenberth 2012). Shifts in storm and precipitation patterns can affect how pollutants move through water, leading to their concentration or dilution in water bodies. The interaction of climate change and EPs has adversely affected living organisms. The rising water temperature may make aquatic life more susceptible to the detrimental effects of pollutants. For example, in warmer temperatures, some chemicals may evaporate faster, stimulating microbe growth that breaks down pollutants or slows down this breakdown process due to heat stress.

2 Source and Pathways of Emerging Pollutants in the Environment

Emerging pollutants are released through four main routes.

2.1 Domestic and Industrial Wastewater

Discharges from households, personal care products, hospitals, and industries' chemical substances enter the sewage system. Wastewater treatment plants (WWTPs) removed the pollutants (such as organic matter and nitrogen). WWTP waste enters the rivers, lakes, and oceans, carrying residual concentrations of emerging pollutants into natural aquatic systems. In addition, heavy metals, volatile organic compounds released from industrial processes, vehicle emissions, and other chemicals in the atmosphere can contaminate the water and soil.

2.2 Agricultural Runoff

Pesticides, including weed killers, zootoxins (which kill small animals like rodents), fungicides, bactericides, artificial fertilizers, and composite manure, are used in agricultural land. These chemicals lead to contaminating water bodies through leaching, surface runoff, and erosion. It also emits methane gas (flatulent gas) into the atmosphere.

2.3 Atmospheric Deposition

Many industrial and vehicle pollutants drain into the land and water surfaces via rainfall, contaminating the water pools.

2.4 Leaching from Landfills

Improper disposal of personal care products, pharmaceuticals, microplastics, other chemical substances, and other household products in landfills can lead to the leaching of these pollutants into groundwater and surface water. Figure 1 highlights the emerging pollutant source and exposer route in the environment.

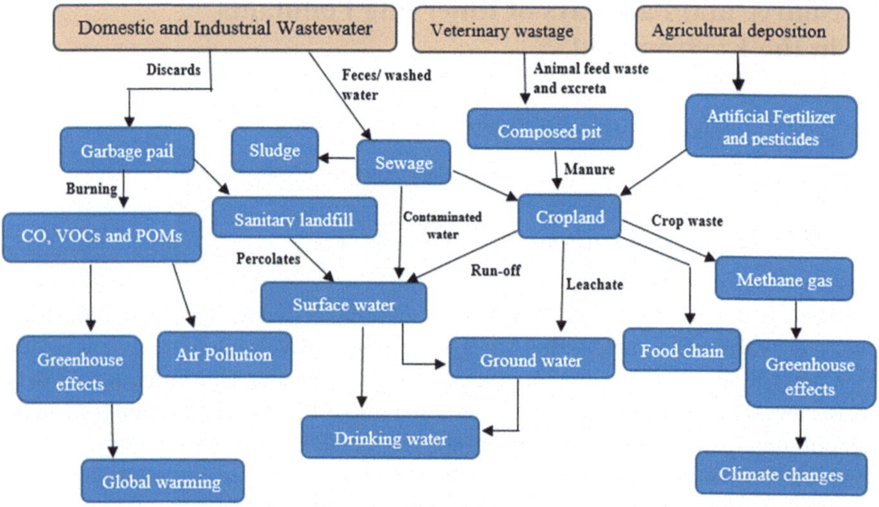

Fig. 1 Route of exposure of EPs to air, water, and soil. VOCs—Volatile organic chemicals, POMS—polycyclic organic matter, CO—carbon monoxide

3 Types, Properties, and Functioning of Emerging Pollutants

EPs have recently been discovered to be responsible for several pollutants: pesticides, EDC medications, PCPs, disinfection by-products, genes, and bacteria resistant to antibiotics. These pollutants were previously considered dangerous for the ecosystem and human health. The leading causes of these EPs are improper disposal of household and PCPs, such as animal wastes, industrial outputs, and agricultural activities. These pollutants can contaminate soil, surface water, and ground water (Ebele et al. 2017). The primary source and the properties of EPs dispersion are highlighted in Fig. 2.

3.1 EPs in Personal Care and Pharmaceutical Products (PCPPs)

This class of substances includes ingredients in nutritional supplements, cosmetic products, and prescribed drugs. These chemicals are used for good agricultural production and personal beauty care. These pollutants leak or leach from agricultural waste and pollute the public water and soil systems (Palli et al. 2019). Hand soaps and toothpaste contain triclosan, an antibiotic that is often detected in the air. Triclosan, an antibacterial substance, was found in clothes, soap, toothpaste, and among other household items. Chlorine and triclosan react with tap water, which is a by-product of chloroform. In the environment, triclosan breaks down into methyl triclosan or dioxins (Gwenzi et al. 2018; Jagini et al. 2021). Triclosan is extremely toxic to different kinds of lower plants (algae) and earthworms. This antibacterial agent causes DNA damage, oxidative stress, and metabolic changes in earthworms (Gillis et al. 2017). Water safety guidelines do not currently include triclosan. Another antibacterial agent, fluoroquinolone, is used to treat animals and humans. This agent has critically harmed the ecosystem. Even at low concentrations,

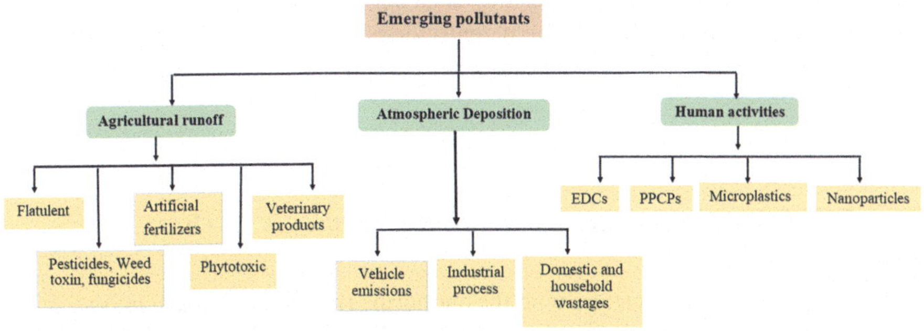

Fig. 2 Types of emerging pollutants from different sources. (EDCs—endocrine-disrupting chemicals, PPCPs—pharmaceuticals and personal care products)

fluoroquinolones will harm soil and groundwater (Chen et al. 2015). Between 2010 and 2021, there was a significant increase in publications on getting rid of new pollutants and antibiotics and a boom in research on the subject (Palacio et al. 2022). Cosmetics contain toxic chemicals and heavy metals, such as M- and O-phenylenediamine, poly- and pre-fluoroalkyl (PFAS) (Xu et al. 2021), isobutyl and isopropyl parabens, mercury, nickel, cadmium, and arsenic. These toxic and heavy metals damage the human reproductive system and biomagnification. According to Tran et al. (2015), pharmaceuticals and medication products are chemicals that are most often utilized in treatment, diagnostics, and diet. Their extensive usage and drainage by cannel and river aquatic ecosystems cause contamination of fresh surface water sources, and damage to human health and biodiversity. Pharmacies were utilized for fisheries, cattle, farm animals, and personal consumption. Livestock is routinely administered extensive medications to lower disease rates and increase the weight of the mammals. Out of the approximately 3000 chemicals included in pharmaceutical items, very little has been field-tested. Researchers have looked at the possibility that hazards to atmospheric ng/L doses might harm humans or wildlife (Richardson and Kimura 2017). The primary sources of pharmaceutical emissions into the atmosphere are medication waste disposal, human urine, feces, and agricultural waste. It indirectly affects the environment by multiplying organic manure and purines in animal farms. They could even infect live things by way of bioaccumulation. Medications used in fisheries are directly spout into surface waters. Pharmaceutical substances that are most commonly mentioned in groundwater and drinking water include antibiotics, antacids, ciprofloxacin, diclofenac, clofibric acid, steroids, antidepressants, analgesics, antipyretics, tranquillizers, salicylic acid, anti-inflammatory drugs, beta-blockers, fluoxetine, nitroglycerin, propranolol, lipid-lowering medications, and stimulants (Tahira et al. 2022). Rather than important effluents, three groups of chemicals were found in sewage water: anxiolytics, antibiotics, and macrolides. According to Gracia-Lor et al. (2012), paroxetine and simvastatin sulfamethazine from the latest class were not found in emerging or wastewater discharge. Benzoylecgonine, cocaine, 2-ethylidene-1, methadone, and its metabolites, 5-dimethyl-3 and 3-diphenyl pyrrolidine, were more commonly detected in groundwater, but at very low (ng/L) amounts. Table 1 provides the list of several EPs properties along with their characteristics. These are only a few of the recently discovered PCPPs poisons that, even at low doses, can be harmful to the environment, and terrestrial and aquatic ecosystems.

3.2 EPs from Endocrine Disruption Chemicals (EDCs)

Certain compounds can interfere with the endocrine (hormone) system in both wildlife and human life. This chemical can mimic, block, and alter the secretory pathway of natural hormones in the body, leading to a range of adverse health effects. EDCs encompass a diverse group of chemicals, including industrial chemicals,

Table 1 Brief overview of the literature regarding EPs as personal care and pharmaceutical products (PCPPs) type and sources

Category of emerging pollutant	Sub-category	Major pollutants	Source	Reference
Personal care products	Antibacterial	Triclosan, Triclocarban	Surface water, wastewater	Bayen et al. (2019); Riva et al. (2018)
	Synthetic musk	Galaxolide, tonalide	Landfill leachate, mangrove sediment	Rathi et al. (2021); Adamson et al. (2017)
	Sunblock agents/UV filters	Zinc oxide, octinoxate, 1,4-dioxane, octocrylene, oxybenzone, benzophenone-4, butyl methoxydibenzoylmethane, benzophenone-3,	Surface water, wastewater, air sample, drinking water	Azaroff et al. (2021)
	UV stabilizers	Benzotriazole	Surface water	Duong et al. (2023)
	Surfactants	Sodium lauryl sulfate, Cocanido propyl betaine	Surface water	Pico et al. (2019)
	Antioxidant	Polyphenol	Surface water	Wang and Wang (2016)
	Antifungal	Phenoxyethanol	Surface water	Teijon et al. (2010)
	Antimicrobial	Paraben, Polyparaben, methylparaben	Surface water	Glassmeyer et al. (2017)
	Insect repellants, mosquito repellant	Diethyltoluamide (DEET)	Wastewater treatment plants (WWTPs), surface water	Tran et al. (2018); Wang and Wang (2016)
	Synthetic fragrance	Phthalates, di-butylphthalate isobytyl and isopropyl, methyl triclosan, musk xylene, galaxolide	Surface water, wastewater, secondary sludge, mangrove sediment	Shahid et al. (2021)

(continued)

Table 1 (continued)

Category of emerging pollutant	Sub-category	Major pollutants	Source	Reference
Pharmaceutical products	Antibacterial	Diazepam, metoprolol, macrolides, sulfonamides, chloramphenicol, sulfamethazine	Hospitals, livestock farms, and domestic wastewater.	Jacob et al. (2021); Shahid et al. (2021); Tewari et al. (2013)
	Antimycotic	Propranolol, diltiazem, terbinafine	Wastewater	Shahid et al. (2021)
	Antibiotic, fluoroquinolone	Levofloxacin, ciprofloxacin moxifloxacin, norfloxacin, sparfloxacin, gatifloxacin, ofloxacin, enrofloxacin	Wastewater	Rathi et al. (2021)
	Antibiotic	Azithromycin, trimethoprim clarithromycin, lincomycin, linezolid, roxithromycin, cefuroxime, doxycycline, sulfanilamide	Wastewater	Sharma et al. (2019); Shahid et al. (2021)
	Cosmetic, psychoactive drug	Caffeine	Water surface, WWTPs	Sharma et al. (2019)
	Antifungal	Fluconazole, voriconazole, voriconazole, sulfamethoxazole thiabendazole		Jacob et al. (2021)
	NSAID (nonsteroidal anti-inflammatory drugs)	Acetaminophen, diclofenac, ketoprofen, naproxen, ibuprofen, terbinafine	Surface water, WWTPs	Sharma et al. (2019)
	Antidepressant	Desipramine, venlafaxine, desmethyl, levamisole	Surface water, ground water	Praveena et al. (2019)
	Antiepileptic	Carbamazepine	Surface water, ground water	Rathi et al. (2021)
	Beta-blocker	Atenolol	Surface water, ground water	Sharma et al. (2019)
	Analgesic	Acetaminophen	Surface water, WWTPs	Sharma et al. (2019)
	Diuretics	Hydrochlorothiazide	Surface water, ground water	Sharma et al. (2019)
	Beta-adrenoreceptor antagonist	Theophylline, metoprolol	Surface water, WWTPs, hospital wastage	Al-Qaim et al. (2018)
	Organic acids	Oxalate, succinate, adipate, maleate, phthalate, and methanesulfonate (MSA)	Surface water, ground water	Stahl et al. (2020)

pesticides, plastics (such as bisphenol A), pharmaceuticals, and PCPs. For example, the androgen receptor's capacity to control male sexual development, that compromised by pesticides, plasticisers, phthalates, and poly-halogenates (Mehrpour et al. 2014). In additionally, a diverse range of endocrine disruptor molecules (EDMs) can be distinguished and classified into two different substance based on their natural source (such as testosterone, ß-sitosterol, estriol, 17-ß-estradiol and estrone) and synthetic source (such as 17a-ethinylestradiol, pesticides such as atrazine, alkylphenol ethoxylate surfactants, dioxins, derivatives of parabens hydrobenzoate, antimicrobial such as triclosan, flame retardant such as polybrominated diphenylethers (PBDE), and plasticizer such as PCBs bisphenol A, perchlorate, Phthalates). It was previously reported that herbicides by-product such as dioxins and pesticides disrupted the functions of thyroids hormone, progesterone receptors, aryl hydrocarbons (Marlatt et al. 2022). Table 2 provides the list of endocrine disruption chemicals (EDCs) properties along with their characteristics. Moreover, estrogen hormones are mostly secreted from human urine, and drainage into wastewater (Tran et al. 2019). As a result, fish exposed to higher level of estrogen chemicals inhibits their reproductive ability (Marlatt et al. 2022). Highest concentration of various plasticizers as phthalates, bisphenol A, PCBs, and perchlorate has been contaminated in Indian river surface (Guo et al. 2011). According to Chakraborty et al. 2016, total PCB concentrations in the Brahmaputra and Hooghly River were determined to be 0.161 μg/L and 0.233 μg/L, respectively. This chemical affected the reproductive system. Perchlorates interfere with iodine transport and reduce thyroid production (Fisher and McLanahan 2008). It causes growth retardation and hemolytic disease and affects the pregnant women, fetuses, and children (Pearce et al. 2010). Raj and Muruganandam (2014) observed that a wide range of perchlorate concentration of 93,500 μg / L in industrial wastewater from Vollore, Tamil Nadu. Das et al. 2014 investigated total 15 phthalates present in drinking water from Okhla industrial region and JNU (Jawaharlal Nehru University) in Delhi.

3.3 Nanoparticles (NPs)

NPs size ranges from 1 to 100 nanometers in diameter. Their size is comparatively related to microbes (viruses and bacteria), protein, and DNA, and characteristically, they have high conductivity, strength, thermal stability, and low permeability. They are also used for different purposes, such as medicine, material science, and electronics (Rauscher et al. 2019). Four distinct types of NPs are present: carbon-based NPs (e.g., carbon black, carbon nanotube, carbon nanofiber, graphene, and fullerenes) (Kumar and Kumbhat 2016), organic based (liposomes, micelles, and dendrimer), inorganic based [aluminum (Au), silver (Ag), titanium dioxide (TiO2), silica dioxide (SiO), zinc oxide (ZnO) etc.], and composed based (metal, ceramic, and polymeric based) (Jeevanandam et al. 2018; Kokkinos et al. 2020). Animals, plants (Buzea et al. 2007; Kumar et al. 2010) and many photochemical reactions, forest fires, and volcanic eruptions (Jeevanandam et al. 2018) naturally produce

Table 2 Brief overview of the literature regarding EPs as of endocrine disruption chemicals (EDCs) properties, type, and sources

Category of emerging pollutant	Sub-category	Major pollutants	Source	Reference
EDCs (endocrine disruption chemicals)	Calcium channel blocker	Diltiazem, verapamil	River water	Williams et al. (2019)
	Beta-blocker	Propanolol	River water	Williams et al. (2019)
	Beta-adrenoreceptor antagonist	Metoprolol	Water surface, river water	Fick et al. (2009); Williams et al. (2019)
	Angiotensin-converting enzyme inhibitor	Enalapril	Water surface	Fick et al. (2009)
	Antimicrobial	Ampicillin, ciproflaxicin erythromycin, lincomycin moxifloxacin, sulfadiazine trimethoprim	Drinking water, surface water, sediments, soil, WWTPs, leachate	Mpatani et al. (2020); Shahid et al. (2021); Yi et al. (2019)
	Hormone	17α ethinyl estradiol, levonorgestrel, norethindrone, cyproterone acetate, testosterone	River water	Al-Odaini et al. (2012)
	Flame retardants	Polychlorinated biphenyls (PCBS), and additive brominated flame retardants (BFRS), n-alkanes, PAHs, phthalates	Indoor dust, road dust	Tue et al. (2013)
	Microplastic and flame retardants	Perfuoroalkyl and polyfuoroalkyl substances, bisphenols, and parabens, PFCs, polychlorinated biphenyls (PCBs)	Tap water	Haron et al. (2023)
	Pesticides and insect repellent	Nonylphenols, DEET (N,N-diethylm-toluamide), DDT, biocides	Sediment, wastewater, and sludge	Dsikowitzky et al. (2020); Juksu et al. (2019); Williams et al. (2019)
	Pesticides and poly aromatic hydrocarbons	Chlorpyrifos, endrin, aldrin, diuron, atrozine, endosulfan, heptachlor, and phenanthrene	Agricultural runoffs, sewage treatment plants, sediments, soil, and surface water	Jacob et al. (2021); Madrid et al. (2020); Shahid et al. (2021)
	Herbicides, fungicides, and molluscicide	Metaldehyde, butachlor, epoxiconazole, sulfamethoxazole, metronidazole	Surface water, aquafarming, and agricultural runoff	De Las et al. (2020); Rathi et al. (2021); Shahid et al. (2021)
	Plasticizers and fire retardants	Dimethyl adipate and tris (1-chloro-2-propyl) phosphate	Domestic and industrial wastewater	Lempart et al. (2020); Shahid et al. (2021)
	Perfluorocarboxylic acids and perfluorosulfonic acids	Perfluorooctanoic acid and perfluorooctanesulfonate	Sediments, groundwater, surface water, and wastewater	Cheng et al. (2021); Shahid et al. (2021); Pitter et al. (2020)

Table 3 Some NPs origin, nature, and source (review in the recent literature -U.S. EPA 2010; Nowack and Bucheli 2007; El-Kalliny et al. 2023)

Nature	Types of NMs	Source	Origin	Example
Natural	Carbon-based	Soot	Geogenic	Fullerenes/buckyballs, black carbon
		Organic colloids	Biogenic	Humic, fulvic acids
		Organisms		Virus
		Aerosols	Atmospheric	Organic acids
		Soot	Pyrogenic	Fullerenes, carbon nanotubes, nanodiamonds, nanowires
	Inorganic	Oxides	Geogenic	Fe-oxide
		Clay		Allophone
		Oxides	Biogenic	Magnetite
		Metals		Silver (Ag), aluminum (Au), platinum (Pt), rhodium (Rh),
		Aerosols	Atmospheric	Sea salt, soil dust
Anthropogenic (engineered, manufactured)	Carbon-based	Soot	Engineered	Fullerene (carbon 20, carbon 60, carbon 70), black carbon, carbon nanotube
		Combustion by-products	By-products	Carbon nanotube, nanodiamonds, nanowires
	Inorganic	Oxides	Engineered	Titanium dioxide (TiO2); zinc oxide (ZnO); cerium oxide (CeO2)
		Metals		Ag, Fe
		Salt		Metal-phosphate, clay, zeolites, ceramics
		Combustion by-products	By-products	Bimetallic nanoscale particles (BNPs). BNPs include elemental iron and a metal catalyst (such as gold, nickel, palladium, or platinum).
		Quantum dots	Engineered	Quantum dots made from cadmium selenide (CdSe), cadmium telluride (CdTe), and zinc selenide (ZnSe)
	Composite	Synthetic polymers	Engineered	Resins, polymeric silver, nanosilver powder

nanoparticles. Otherwise, some NPs are produced by anthropogenic by-products (air craft, vehicles, ships, and trains) (Azarmi et al. 2014; Kumar et al. 2013), and engineered nanoparticles (milling, green chemistry, photocatalysis, and hydrothermal) (Peralta-Videa et al. 2011). Table 3 lists some NPs origin, nature, and sources. Nanomaterials (NM) possess unique characteristics; they can react with biological and environmental systems and alter their toxicity, fate, and dispersal. Nanomaterials (NMs) have direct and accidental discharge into the water surface, whether from wastewater or solid waste effluents. NPs and NMs affect the environments based on size, shape, surface chemistry, and biotic and abiotic matrix (Klaine et al. 2008; Lowry et al. 2010). Petrol-diesel-fueled engines generate 36% of NPs in the atmosphere, and stationary and electronic wastage is burning. In addition, cigarette smoking, building demolition and exhausts, asbestos fibers, and household materials (as lead, glass, platinum, and other toxic elements) emerged anthropogenic nanoparticles (NPs) (Stefani et al. 2005). Several nanoparticles such as organic colloid matter (i.e., humic acids, protein, peptides, fulvic acids) and inorganic matters [such as iron oxide, manganese oxide, titanium dioxide (TiO_2), Ag, zinc oxide (ZnO)] enter into the water and soil surface. However, NPs rapidly aggregate with water contact, which harms aquatic organisms (Lin et al. 2010). Mainly, titanium dioxide (TiO_2), zinc oxide (ZnO), and silver (Ag) contaminate the soil and grow water, and they can impact the food chain.

3.4 *Microplastics (MPs)*

MPs are small polymeric plastics that are less than 5 millimeters in diameter. They originate from different sources such as primary microplastics (personal care, microbeads, cosmetics (i.e., facial polishing, scrubs, and pellets components)) and synthetic cloths. Secondary microplastics are derived from breaking large plastics into small parts, such as plastic bags, bottles, containers, and fishing nets (Syranidou and Kalogerakis 2022). Microplastic is directly released from plastic waste in thermoplastic paint, fisheries, fishing nets, wastewater treatment plants, packaging materials, and industrial waste, contaminating aquatic and terrestrial environments (Elgarahy et al. 2021). Prevalent small polymeric plastic particles are polypropylene (PP), polyethylene tetraphthalate (PET), polyvinyl alcohol (PVA), PVC, PA, PS, and PE; these are wholly found in air, land, and water pool (Elgarahy et al. 2021; Shen et al. 2022; Syranidou and Kalogerakis 2022; Arienzo et al. 2021). MPs are a slow degradation process; they can aggregate in water bodies for a long time. As a result, slowly degraded MPs interact with aquatic organisms (plankton, fish, birds, and invertebrates) (Arienzo et al. 2021; Wu et al. 2019). Day by day, microplastics (MPs) are widely used in human activities, such as inhalation and ingestion. In the long-term persistence, these polymeric microplastics can cause cancer in different organs. Furthermore, MPs are crucially damaging the marine and coastal food web. MPs have the ability to absorb harmful chemicals that aquatic species may ingest and suffer serious health risks (Karthik et al. 2018). The major micropolymeric plastic and source are listed in Table 4.

Table 4 An overview of the literature regarding MPs type and sources

Category of EPS	Major pollutants	Source	References
Microplastic	Polyamide (PA)	Landfills, wastewater	Soto-Salcido et al. (2023)
	Polyethylene (PE)	Landfills can contaminate soil, water, and air burned waste released greenhouse gas	Burelo et al. (2023)
	Polypropylene (PP)	Landfills can contaminate soil, water, and air burned waste released greenhouse gas	Alsabri et al. (2022)
	Polystyrene (PS)	Landfills, sediment, and water	Kik et al. (2020)
	Polyvinylchloride (PVC)	Drinking and bottled water, landfills	Kudzin et al. (2024)
	Polyethylene terephthalate (PET)	Ground water, drinking water, marine water, and sediment	Dhaka et al. (2022)
	Polyvinyl alcohol (PVA)	Wastewater, sludge	Rolsky and Kelkar (2021)
	Polylactic acid (PLA)	Landfills, surface water	Rezvani Ghomi et al. (2021)
	Ethylene vinyl acetate (EVA)	Landfills, surface water	Lopes et al. (2015)
	Nylon	Landfills, surface water	
	Synthetic microfiber	Wastewater, sludge, WWTPs	Arias et al. (2022)

4 Effect of Emerging Pollutants

Emerging pollutants (EPs) are a large variety of natural or synthetic (organic and inorganic) chemicals that have been emitted into the atmosphere but are not yet regulated or monitored effectively. These pollutants include pharmaceuticals, PCPs, EDCs, agrochemicals, nanoparticles, microplastics, pesticides, industrial chemicals, and other organic compounds. As they discharge in natural ecosystems, often through wastewater, agricultural runoff, or industrial discharges, their effects on the environment, wildlife, microbial communities, agriculture, and human health have become a growing concern. Figure 3 highlights the significant negative impact on environment, ecosystem, and climate change by EPs. In addition, Table 3 lists the toxic effects of primary emerging pollutants on the ecosystem, environment, climate change, and human health.

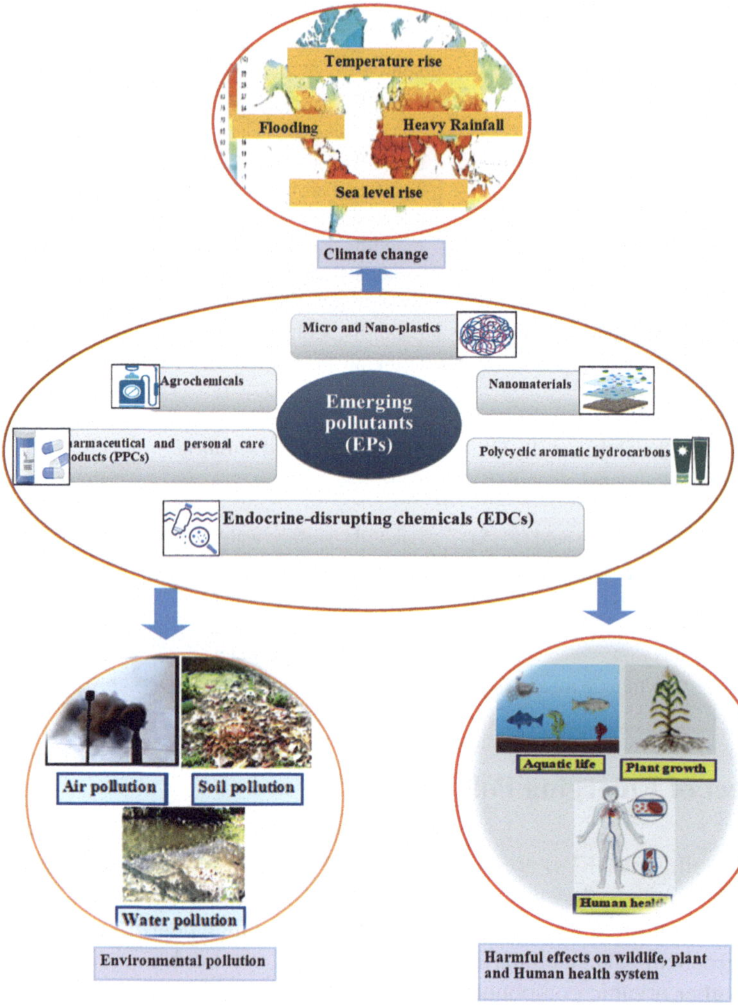

Fig. 3 The significant negative impact of emerging pollutants (EPs) on the environment, ecosystem, and climate change

4.1 Effects of Emerging Pollutants on Climate Changes

Emerging pollutants (such as industrial wastage, domestic and agricultural runoff, personal and pharmaceuticals, and veterinary wastage) contaminate atmospheric environments, surface water, and ecological systems. The EPs are critically harmful to the environment and ecosystems at very low concentrations. The widespread release of synthetic substances, industrial waste and byproducts of anthropogenic activities poses a significant threat to environmental integrity. The global concentration of greenhouse gas (GHG) (CO, CFC, CH_4, etc.) is rising in the environment. The major contributors to the anthropogenic source of GHGs (CH4 and CO_2) are food and agricultural-based industries, such as livestock production (Audsley and Wilkinson 2014). Livestock production industries release 50–60% of methane and 9% of CO_2 gases into the environment (FAO 2013; Mirzaei et al. 2012). The leading greenhouse gases are CH_4, CO_2, and CFC, which are changing the climate and global temperature terribly. GHGs are usually emitted by natural and human activities (Patra and Saxena 2010). Global warming causes climate change, rising the precipitation pattern and temperature, which negatively affects the ecological system. Climate change has terrible impacts on ecosystems, like species distribution change, changes in the timing of biological events (flowering, breeding, and fruiting), increased precipitation and flooding, reduced crop production, and ocean acidification.

4.2 Effects of EPs on the Environment and Ecosystems

The presence of EPs, including flame retardants, pharmaceutical pollutants, personal care and household products, microplastics, nanoparticles, and perfluorinated compounds, has been found in wastewater, on water surfaces, and in sediments. These pollutants contaminate the environment and ecosystem (Ofrydopoulou et al. 2022). Chemicals are critically harmful to living organisms at very low concentrations (Míguez et al. 2021). The major organic pollutants (EPs) such as insect repellent, pesticides, polycyclic aromatic hydrocarbons (PAH), polyamide (PA), PVC, polypropylene (PP), polychlorinated dibenzo-p-dioxins (PCDD), polyethylene tetraphthalate (PET), polyethylene (PE), polyvinyl alcohol (PVA), polychlorinated biphenyls (PCBP), and polystyrene (PS) are frequently found in aquatic environments (Pitarch et al. 2016). Due to industrial manufacturing and the use of large amounts of plastics, synthetic cloths, cosmetics, and textile materials must release massive pollutants into the water surface (Egbuna et al. 2021). Several phthalate congeners have been identified as significant pollutants by the USEPA, such as di-ethyl phthalate, di-methyl phthalate, butyl-benzyl phthalate (BBP), di-n-octyl phthalate (DnOP), di (2-ethylhexyl) phthalate (DEHP), and di-n-butyl phthalate (DnBP) that exposed in atmosphere and have deleterious effects on the body. Bisphenol A (BPA) is a synthetic chemical used in food packaging, bottles, and epoxy resin. BPA does not degrade properly and can accumulate in soil and water. It is very toxic to plants and animals' cell division and inhibits growth, stress response, and cell signaling (Pereira et al. 2015). Alkylphenols are derivatives from alkylphenol ethoxylates (APEs) used in detergents, foods and textiles, paint, pesticides, and active components of spermicides. Alkylphenols can cause testicular damage, reduction of male sexual competence, and reproductive damage in aquatic animals, reptiles, and mammals (White et al. 1994, Legler (1999) Sonnenschein and Soto (1998), Arukwe and Goksoyr (1998). Brominated dioxins, polychlorinated naphthalenes, and furans polychlorinated naphthalenes are mainly produced from forest fires, volcanic eruptions and lightning digesters like natural events and municipal wastage (Safe et al. 2012). They have negative effects on the environment and the economy. They are stored in the biota and human tissue, which causes congenital abnormalities and damages the reproductive and hormonal system (Kanan and Samara 2018; Safe et al. 2012). Perchlorates mainly affect the nervous system and thyroid glands at low concentrations; this effect was more common in mouse pups delivered to infected mothers (Gilbert et al. 2020). Phthalates used in plastic manufacturing industries can enter the environment through leaching and aerosol. It can be toxic to aquatic species, preventing oxidative metabolism and human and animal development. It also caused kidney, liver, and thyroid damage. Accordingly, high-molecular-weight phthalates such as di (2-ethylhexyl) phthalate (DEHP), di-isononyl phthalate (DINP), di-n-octyl phthalate (DnOP), and low-molecular-weight phthalates include di-methyl phthalate, di-ethyl phthalate, di-n-butyl phthalate (DnBP), diisobutyl phthalate (DIBP), butyl-benzyl phthalate (BBP), all are significant pollutants in atmosphere and potentially harmful effects (Kashyap and Agarwal 2018). Polybrominated diphenyl ethers (PBDEs) are a flame retardant and can bioaccumulate and biomagnify in air, soil, and water. Triclosan is a by-product of methyl triclosan, which can reduce growth and reproduction in aquatic species (Rathi et al. 2021) and disturb the soil nitrogen (N2) recycling process.

4.3 *Effects of EP on Human Health*

Most EPs, such as fungicides, pesticides, microplastic, flame-retardants, personal care and pharmaceutical pollutants (antibacterial and antifungal), have been deposited in the water surface, atmosphere, and soil. These pollutants are toxic to humans through drinking water, foodstuff, and air. Among these organic pollutants are phthalate, flame-retardants, organophosphorus, triclosan, polysiloxanes, bisphenol A, synthetic resins, and perchlorates are contaminated with human health (Adeel et al. 2017). Bisphenol A is widely used in plastics industries around the world. It alters the genetic makeup and disrupts the endocrine (thyroid and salivary glands). In addition, it functions as an anti-androgen and negatively affects men's virility, increasing the risk of breast cancer in women (Kawa et al. 2021; Patel 2017). Many pesticides have been found to be carcinogenic and neurotoxic in humans. Anti-inflammatory agents (brominated flame retardant) can cause lung damage, disrupting endocrine and immune cells, abnormalities in female anatomy, neuroprotection, and cancer (Pereira et al. 2015). Tetrabromodiphenyl ether as a flame retardant leads to various microbial diseases, endocrine disruption, and DNA damage. Decabromodiphenyl ether, hexabromodiphenyl ether, and pentabromodiphenyl ether are carcinogens that affect brain development, fertility, thyroid function, and neuronal function (Modesto et al. 2015). Phthalates lead to problematic and abortions pregnancies when exposure is high. A synthetic pesticide can cause increased thyroid sensitivity and ovarian cancer. A synthetic polymer such as polyvinyl alcohol (PVA) can cause skin and eye irritation and changes in vision. Another polymer, polyvinyl chloride (PVC), can have several critical effects, such as liver cancer, liver injury, asthma, dizziness, and fatigue. Dioxin, the by-product of PVC, causes respiratory cancer, Hodgkin lymphoma, prostate cancer, fetal loss, liver damage, and endocrine disruption (Schecter et al. 2010; World Health Organization WHO 2023) (Table 5).

Table 5 Summarized review of recent literature regarding the lethal effects of primary emerging pollutants on the ecosystem, environment, climate change, and human health

EPs category	Pollutant	Effects	Reference
Microplastics	PA	The decomposition of polyamide waste produces harmful liquids and gases like CO, NOx, which pollute the environment. It is very harmful to human health.	Alijagic et al. (2024)
	PE	It reduces the microorganism diversity (as bacteria and fungi) and the growth rate of earthworms.	Slater (2008)
	PP	It can produce greenhouse gases and toxic chemicals that cause water and air pollution and increase global temperature. If it is ingested in the animal body, it can cause digestive blockage and lead to death.	Alsabri et al. (2022)
	PS	It is a dangerous water and air pollutant. It can damage the cell membrane, liver, and nerve tissue. Also, it reduced microbial diversity.	Kik et al. (2020)
	PVC	PVC waste is disposed of in landfills, and SOX and CO_2 are released, contributing to global warming. It is potent for bioaccumulation and more toxic to human health.	Kudzin et al. (2024)
	PET	PET waste burning produces harmful gases such as CO, CFC, sox, and NOx. It reduces soil fertility and affects crop production. It causes cancer and hormonal imbalances in human health. Ingesting PET causes fetal internal injuries in invertebrates and vertebrates.	Khairul Anuar et al. (2022)
	PVA	Easily disposed of, produced greenhouse gases and chemicals cause human health problems.	Rolsky and Kelkar (2021)
	PLA	Decomposed PLA waste releases CH_4 gas that causes global warming. It can affect soil fertility and inhibit the plant growth.	Iwata (2015)
Plasticizer	Phthalates	Phthalates are more potent to accumulate in wildlife and disrupt the food chain through geological processes. It has a very harmful effect on human health and aquatic life. They can also form strong attachments to soil and sediment, so they do not tend to move rapidly into groundwater.	Baloyi et al. (2021)
	Bisphenol A	It can be potent for bioaccumulation in marine life. It can damage the endocrine system of some invertebrates and vertebrates. It also affects the growth and development of fish.	Canesi and Fabbri (2015)
	Dioxin	Dioxins are the by-product of PVC. It can accumulate in animal fat tissues that cause infertility and reproductive impairment.	Kudzin et al. (2024)

(continued)

Table 5 (continued)

EPs category	Pollutant	Effects	Reference
Pesticides	DDT	It can cause eggshells to thin, leading to bird population declines. It accumulates sediments in aquatic species and humans but is not dissolved in water. It is very toxic to aquatic species like fish and invertebrates. It causes eggshell thinning that declines the bird population and causes human infertility.	Kabasenche and Skinner (2014)
	Endrin	It can strongly bind to soil particles and enter into water. It can accumulate in aquatic species (fish and animals). It causes diarrhea, headaches, and damage to the nervous system.	ATSDR. US. (2021)
	Aldrin	It can strongly bind to soil particles and enter into water. It can accumulate in aquatic species. It causes neurological, reproductive, and developmental damage.	ATSDR. US. (2022)
	Diuron	It has the potential to contaminate groundwater. It can affect brain development and be toxic to aquatic life.	Liu (2014)
Insect repellent	DEET	It is toxic to aquatic invertebrates, fish, and birds and more toxic to algae. It can pollute the air and water.	EPA.US. (2014)
PCPs	Atenolol	It is a contaminant to water, and exposure to fish affects the immune system, endocrine system, and metabolism.	Ivantsova and Martyniuk (2023)
	Caffeine	Caffeine easily contaminates air and water. It can affect marine life such as coral reefs, microalgae, and fish. It can potent to accumulate in marine biota, causing growth and reproductive damage.	Vieira et al. (2022)
	Triclosan	It is very toxic to aquatic organisms such as algae and fish. It can be potent in bioaccumulating in the marine food chain. It can affect reproduction, growth, and the endocrine system in marine life. It is found on water surfaces, WWTP, water sediment, and drinking water.	McAvoy et al. (2002); Orvos et al. (2002)
	Triclocarbon	It is degradable and leaching in soil and reduced the soil bacterial colony. Also, it is potent to bioaccumulate in marine organisms like fish, algae, invertebrates, and plants community.	Amigun Taiwo et al. (2022)
	Polyparaben, methylparaben	It contaminates into surface water and sediments. They can damage biofilms on the surface of water and negative effects in marine life. It can also have the potential for biomagnification.	Vale et al. (2022)

	Benzophenone −4	It continuously discharges into the water's surface, contaminating the aquatic environment. It previously observed the neurotoxic effects of BP-3 on aquatic animals, mostly zebrafish.	Moreira et al. (2024)
	Erythromycin	It can contaminate soil and water, causing acute toxicity in aquatic species such as fish and other invertebrates.	Shen et al. (2022)
	Sulfadimidine	It is disposed of in soil and sediment. It interacts with the biogeological cycle that decreases nitrate and nitrite reduction rates. Also, it interacts with the denitrification process and affects diatom diversity.	D'Alessio et al. (2019)
Nanoparticles	TiO_2	NM inhalation could cause inter-alveolar blockage, lung inflammation, lung cancer, and fibrosis. It easily contaminates the soil, air, and water.	Shah et al. (2017)
	CeO_2	NMs strongly interact with soil and water. Lead accumulates in plant tissues and interacts with metabolic processes. It can affect soil microorganisms.	Prakash et al. (2021)
	CdSe	Cd-containing quantum dots leach with the aqueous medium. Cd is a hazardous substance that is very toxic to different organisms and may eventually find its way into the environment.	Kirchner et al. (2005)

PA Polyamide, *PE* Polyethylene, *PP* Polypropylene, *PS* Polystyrene, *PVC* Polyvinyl chloride, *PET* Polyethylene terephthalate, *PLA* Polylactic acid, *PVA* Polyvinyl alcohol, *DEET* N, N-Diethyl-meta-toluamide, *DDT* Dichloro-diphenyl-trichloroethane, *TiO2* Titanium dioxide, *CeO2* Cerium dioxide, *CdSe* Cadmium selenide

5 Climate Change and Its Ecological Consequences

Climate forcing is the term used to describe changes in the net energy level of Earth that results in unanticipated changes in atmospheric temperature that may have cooling or warming effects over time. Changes in aerosol concentration cause global cooling, whereas high concentrations of GHGs cause global warming in environments (Forster et al. 2007). There are a few significant effects of climate change as drastic changes of weather calamities such as drought and flooding, sea level rise, increases in temperature that causes forest fires, and precipitation patterns increase (Clarke et al. 2022; Seneviratne et al. 2012). Recently, it has been reported that climate change will be drastic in the global ecosystem and have unexpected effects on the health system (IPCC 2022). The issue of climate change has become a primary global concern, with significant implications for the delicate balance of global ecosystems. Table 4 mentions the direct and indirect effects of EPs on climate change. As the Earth's climate, warms are being felt throughout the natural world, from the smallest gene to the most expansive biome. Recent studies have shown that the effects of a changing climate are already apparent at every level of ecological organization, including changes in population and life history, alterations in the structure and functioning of entire ecosystems, and shifts in geography (McCarty 2001). These changes can be directly linked to the observed population declines and even global and local extinctions of various species. These impacts have far-reaching consequences for critical economic sectors such as agriculture, forest management, fisheries, and urban development, which are essential for food security and human health. Understanding the causes, consequences, and potential mitigation of these changes will be crucial as we navigate the challenges of a warming world (Scheffers et al. 2016). Figure 4 highlights the climate change influence on extreme weather calamity and pollutant dynamics. Climate change, driven by increased various EPs along with GHG emissions, leads to rising global temperatures, precipitation patterns change, more frequent extreme weather events, biodiversity loss, and sea-level rise. These changes have profound impacts on ecosystems, including:

5.1 Temperature Rise and Habitat Shift

Climate change leads to increased GHG emissions, which has emerged as one of the most pressing environmental issues. High temperature impacts ecosystems, biodiversity, and the intricate balance of natural processes that sustain life on earth. The rise of global temperature is one of the most apparent effects of climate change. This warming environment affects ecosystems, causing species to migrate to cooler regions, often toward the poles or higher altitudes. Consequently, species that are unable to change or relocate risk are going extinct. The Intergovernmental Panel on Climate Change (IPCC) has warned that many species will be at threat if global temperatures increase more than 1.5 °C above pre-industrial levels. Many toxic

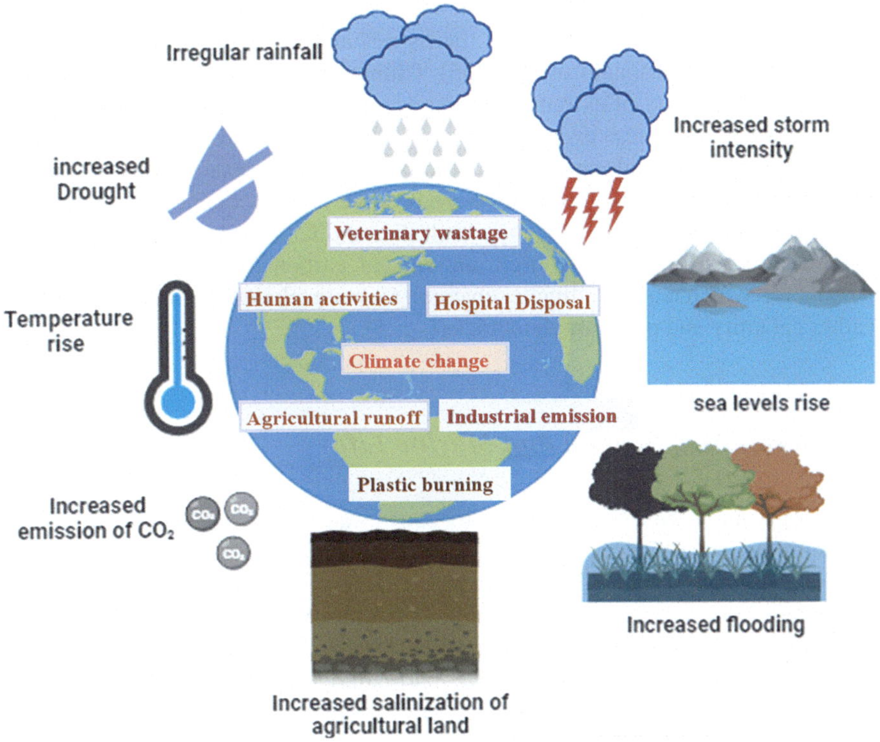

Fig. 4 Climate change influences extreme weather calamity and pollutant dynamics

organic pollutants are rapidly mixed and contaminate the atmosphere at the micro level. For example, winds can disperse polychloride bisphenols (PCBs) into the atmosphere. These organic pollutants are an undeniable indication of the ecological consequences of climate change that have been found in aquatic and terrestrial environments across tropical and polar regions (Lovejoy and Hannah 2005; Penuelas and Filella 2001; Root et al. 2003; Walther et al. 2001). Climate change will have significant effects on specific populations and species. Other ecosystem stresses such as agriculture, chemical pollution, habituated destruction, invasive species, and other microbial diseases are critically declining species and populations in their ecosystem. Bioaccumulation and biomagnification in ecosystems exposed to persistence organic pollutants may be influenced by climate change that can lead to shifts in tropical structure, migratory patterns, feeding behavior, and food source (AMAP 2004; Olafsdottir et al. 1998; Ramsay and Stirling 1982; Stirling et al. 1999).

Biomagnification and bioaccumulation of organic pollutants harm aquatic phytoplankton and zooplankton, which destroy the aquatic food chain (Braune et al. 2005; Macdonald et al. 2005). Global warming melting the ice has a major effect on Hudson Bay polar bears. Hudson Bay polar bears feed on ringed seals (*Pusa hispida* Schreber.), whose populations are reduced due to glacier melting (Furnell and Schweinsburg 1984; Ramsay and Stirling 1982; Stirling et al. 1999). Due to starvation, polar bears will use storage fat burn as an energy source. Some persistent organic pollutants are remobilizing in the tissues and cause chemical toxicity and starvation stress (Macdonald et al. 2005). Climate change is linked to increased frequency and intensity of extreme weather events, such as droughts, floods, hurricanes, and wildfires. These events can destroy ecosystems and habitats, cause species loss, and alter landscapes. For example, wildfires, exacerbated by rising temperatures and prolonged droughts, have become more frequent and severe, particularly in regions like the Amazon, Australia, and California. These fires cause habitat destruction and release high amounts of carbon dioxide (CO2) into the atmosphere, which causes climate change. Climate change influences extreme weather calamity and pollutant dynamics (Fig. 5).

5.2 Ocean Acidification and Marine Life

The absorption of high amounts of CO_2 by ocean water decreases pH levels, leading to an acidification of seawater that may negatively affect marine life. (Raven 2005; Caldeira and Wickett 2003). The high concentration of carbon dioxide (CO_2) in the atmosphere leads to ocean acidification. Previously reported climate change decreased the seawater pH range (Orr et al. 2005). This change in the pH of seawater has detrimental effects on marine life, particularly organisms with skeletons or calcium carbonate shells, such as corals, mollusks, and some plankton species. Coral reefs, often called the "rainforests of the sea," are especially vulnerable. The degradation of these ecosystems affects the marine species and the human communities that depend on coral reefs for food and tourism. In addition to climate change, it increases the global temperature and precipitation pattern and the evaporation rate, which causes high levels of salinity in tropical and subtropical oceans (IPCC 2007). Climate change can melt the ice snow and high precipitation amounts, which cause rising sea water levels (IPCC 2007).

5.3 Sea Level Rise and Disruption of Aquatic Ecosystems

Climate changes affect the species that depend on aquatic habitats, including aquatic plants, fish, and amphibians. Additionally, the loss of glaciers and snowballs, which act as natural reservoirs, threatens water supplies for ecosystems and human populations. The toxic effects of EPS can be enhanced by environmental temperature (Boone and Bridges 1999; Capkin et al. 2006; Gaunt and Barker 2000; Silbergeld 1973). The most common herbicide, atrazine toxicity toward catfish (*Ictalurus punctatus* Rafinesque.), was found with temperature increase or dissolved oxygen (DO) level decrease in water (Gaunt and Barker 2000). Global warming can increase the persistent organic pollutants (POPs) dieldrin's lethal effects on freshwater (Silbergeld 1973). In addition, insecticide carbaryl becomes more toxic to green

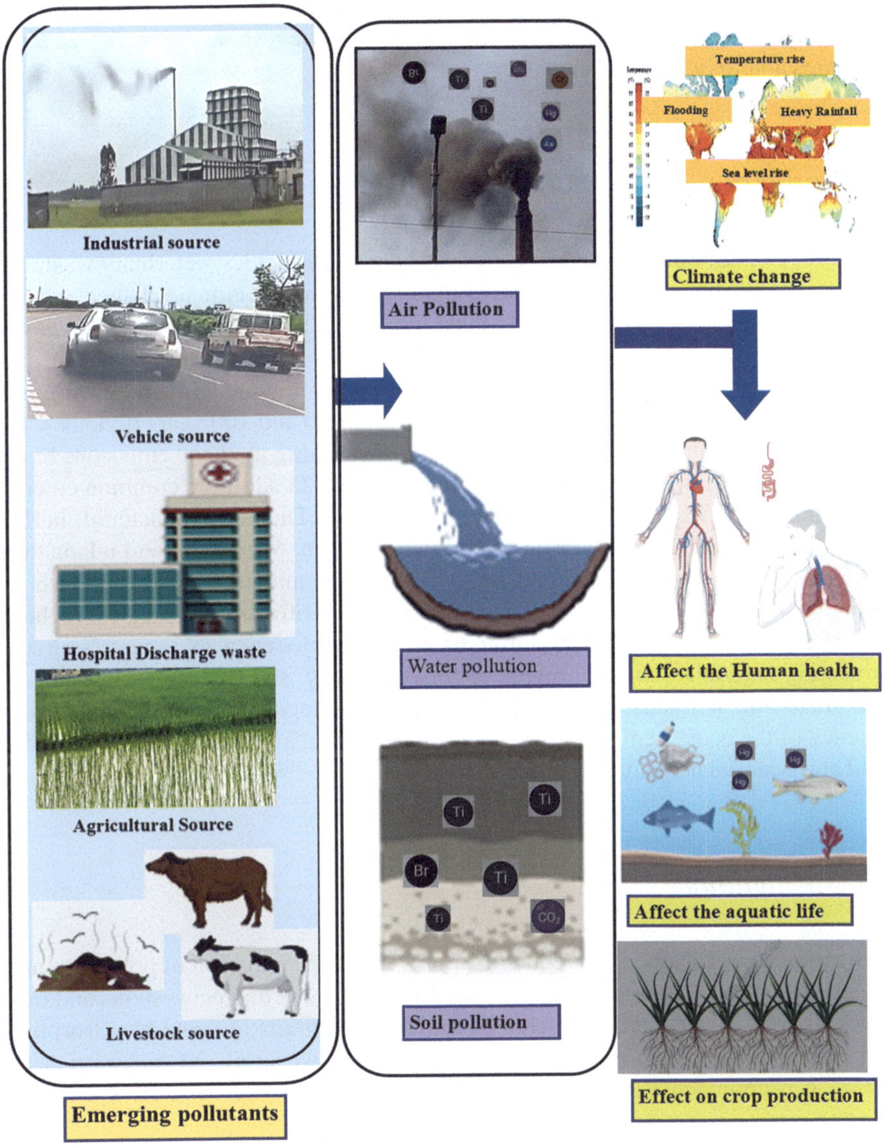

Fig. 5 The interaction of climate change and EPs has adversely affected living organisms and the ecosystem

frogs (*Lithobates clanitans* Latreille.) with temperature increases from 17 °C to 27 °C (Boone and Bridges 1999). According to Capkin et al. (2006), the insecticide endosulfan caused an increase in the mortality rates of rainbow trout (*Oncorhynchus mykiss* Walbaum.). Other herbicides, DDT, are more toxic at low temperatures and increase the toxic effects of pyrethroids at warming temperatures in leopard frogs

(*Rana* spp.) (Materna et al. 1995) and water fleas (*Daphnia magna* Straus.) (Ratushnyak et al. 2005), illustrating the species-specific response to increased temperatures and toxicant exposures.

6 Mitigation and Adaptation Strategies

Different EPs such as polymeric plastics, herbicides, pesticides, veterinary wastage, and greenhouse gases in the atmosphere significantly impact climate change and the ecosystem. GHGs have the potential to increase global temperature, which is known as global warming. One of the potent greenhouse gases is carbon dioxide (CO_2), and the other is methane (CH_4). IPCC proposed that only 1 ton of CH_4 will be warming around 25 to 72 tons of carbon dioxide (CO_2) above 20 and 100 years periods. CH_4 is more potent than carbon dioxide. Therefore, decreasing methane emissions has a more significant impact on mitigating climate change. In addition, common emerging pollutants (such as PVC, phthalate, bisphenol A, Dioxin, antibacterial, herbicides) directly and indirectly influence climate change. Mitigation and adaptation strategies are crucial to the global response to environmental challenges such as pollution, biodiversity loss, and climate change. Mitigation strategies include decreasing the emission of GHGs through transitioning to renewable energy sources, protecting natural carbon, improving energy efficiency, and protecting natural carbon sinks like wetlands and forests. Adaptation strategies involve enhancing the potency of the ecosystem, restoring habitat destruction, and developing sustainable land and water management practices.

6.1 Mitigation

Mitigation strategies focus on reducing or preventing the emission of pollutants and greenhouse gases, improving natural resources, and protecting ecosystems and biodiversity loss. Other methods such as filtration, biodegradation, and bioadsorption have been reducing the emerging pollutants (Okonkwo et al. 2006; Vieira et al. 2022). Key approaches include the following.

6.1.1 Reducing Greenhouse Gas Emissions

Increasing renewable energy sources like hydroelectric power, solar, and wind reduces dependence on fossil fuels, major contributors to greenhouse gas emissions. Improving energy efficiency in transportation, buildings, and industry can significantly lower energy consumption and emissions. This includes enhancing insulation, upgrading infrastructure, and adopting energy-efficient technologies. Carbon capture and storage (CCS) technology captures and stores the industrial emerging carbon dioxide, preventing direct emission into the atmosphere. A high amount of greenhouse gas emissions (methane) originates from rumen fermentation, composted manure and the use of agricultural livestock, which accelerate atmospheric pollution and climate change. These pollutants are reduced by replacing cornstarch with grass-fed and mineral fertilizers that reduce methane and CO_2 emissions in environments (Hristov et al. 2013).

6.1.2 Removal of Synthetic Substances from Water Surface

Filtration and biodegradation mainly remove emerging pollutants like alkylphenols, phthalates, parabens, bisphenol A, pesticides, antibiotics, and natural and synthetic estrogen (17β-estradiol, estrone, 17α-ethynylestradiol). Biological processes like fungal bioreactors (Espinosa-Ortiz et al. 2016), anaerobic digester systems (Yang et al. 2016), and activated sludge treatment (Krah et al. 2016) can also remove emerging pollutants. Microalgae, bacteria, and fungi may digest emerging pollutants (EPs) through the enzymatic (Xiong et al. 2018; Roccuzzo et al. 2021). For example, marine algal species such as *Dunaliella salina* (Dunal.) Teodoresco., *Nannochloropsis oculata* (Droop.) Hibberd, *P. subcordiformis* G.S.West, and *P. globosa* Lagerheim can remove 43% to 91% nonylphenol from polluted water through biodegradation and bioabsorption (Wang et al. 2019). White-rot fungi can degrade 60 to 100% of polysteric plastic substances such as nonylphenol and bisphenol A from the water surface by extracellular enzyme secretion (Grelska and Noszczyńska 2020). Many literatures reported that the UV/peroxide, ozonation (Alvarez et al. 2009), photocatalysis (Varma et al. 2020), and Fenton, known as advanced oxidation processes, which are frequently used in the removal of EPs (BPA, estrol, diethylstibesterol and many pesticides, herbicides) from the water surface (Irmak et al. 2005).

6.1.3 Sustainable Agriculture and Forestry

Implementing practices such as organic farming, crop rotation, reduced application of fertilizer and pesticides and applying natural pest control management, reduces the environmental impact of agriculture, conserves soil health, and enhances crop production and biodiversity. Planting and restoring forests enhance carbon sequestration, provide habitat for wildlife, and protect soil erosion and watersheds. Forest conservation and sustainable management also prevent deforestation and habitat destruction.

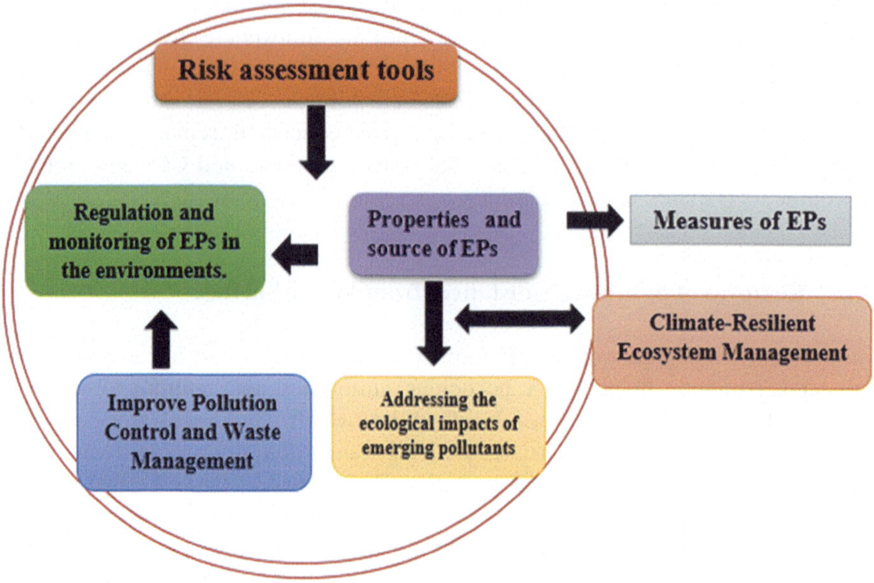

Fig. 6 Concept for mitigation and adaptation strategies of climate change, pollution, and biodiversity loss

6.1.4 Pollution Control and Waste Management

Implementing stricter regulations on industrial emissions, controlling the use of harmful chemicals and plastics, and promoting cleaner production technologies can reduce pollution. Encouraging waste recycling and reduction and the circular economy helps minimize the environmental footprint of products and reduces the amount of waste that is disposed of in landfills and oceans.

6.2 Adaptation Strategies

Adaptation strategies aim to increase the resilience of natural and human systems to the impacts of climate change and other environmental stressors. These strategies involve preparing for and managing changes in weather patterns, sea levels, and ecosystem dynamics (Fig. 6).

6.2.1 Ecosystem-Based Adaptation

Protecting the natural ecosystems, such as wetlands, mangroves, and coral reefs, provides natural barriers against extreme weather events, flooding, drought, and food webs. Conservation and restoration efforts enhance ecosystem persistency and climate impact. Also, increasing the species and genetic diversity helps species adapt to changing conditions and enhances the ecosystem's stability and function.

6.2.2 Climate-Resilient Infrastructure and Urban Planning

Designing and constructing infrastructure that can resist extreme weather events, such as heatwaves, floods, hurricanes, and drought conditions, is crucial for protecting communities. This includes building flood defense, reinforcing buildings, and upgrading transportation networks. Urban planning that integrates green spaces improves water and waste management, and promotes public transportation can reduce urban heat waves, vector-borne disease, and greenhouse gas emissions.

6.2.3 Water Resource Management

Implementing water-saving technologies, improving irrigation practices, and promoting water recycling can help manage water resources more sustainably, especially in drought-prone areas. Integrated water resource management (IWRM) involves managing water, land, and resources to optimize social and economic welfare while maintaining ecosystem sustainability.

6.2.4 Public Health and Community Resilience

Improving the healthcare infrastructure, monitoring and controlling diseases, and improving emergency response systems are critical for protecting public health from climate-related impacts, such as heatwaves, food- and water-borne diseases, vector-borne diseases, and air quality issues.

6.2.5 Public Awareness and Education

Educating the public and community about the sources and risks of emerging pollutants, as well as their impacts on climate change and environmental issues, promoting sustainable practice, and encouraging community-based adaptation can increase resilience and empower individuals to take action.

7 Conclusion

The ecological impacts of emerging pollutants associated with climate change present important and multifaceted concerns. The pollutants, including agricultural runoff, microplastic, pharmaceuticals, personal care products, pesticides, and industrial chemicals, continue to infiltrate ecosystems, and the concurrent stresses of a changing climate often aggravate their effects. Climate change is linked to ocean acidification, rising sea levels, extreme weather calamities as temperatures rise, flooding, forest fires, drought, and precipitation patterns shift. These extreme calamities affect the atmosphere and biodiversity, causing an increase in greenhouse gases and their contaminant substance. This extensive assessment investigates the emerging pollutant sources and the properties that affect climate change and ecology. The interactions between pollutants and extreme weather events, such as rising temperatures, sea level rise, shifts in precipitation patterns, and ocean acidification, can lead to synergistic effects that amplify harm to the atmosphere, biodiversity, ecosystem functions, and human health. The bioaccumulation, persistence, and potential toxicity of emerging pollutants pose substantial risks to wildlife, particularly when combined with climate-induced stressors like habitat destruction and extreme weather calamity. In addition, higher temperatures enhance the chemical toxicity of emerging pollutants, which harm aquatic organisms (Brian et al. 2008; Heugens et al. 2001; Schiedek et al. 2007). Several literatures reported that membrane filtration, bioabsorption, and biological treatment transmit minimal EC removal and advanced oxidation processes as UV/peroxide, ozonation, and photocatalytic can effectively remove most emerging pollutants. In addition, ecological conservation, restoration, and adaptive management techniques can reduce pollutant climate change's impact, which enhances ecosystem resilience.

Author Contribution A. Maity played a crucial role in bibliometric analysis, data collection, and manuscript writing. She also contributed to the methodology and research design. AK Mondal guided manuscript preparation and assisted in data analysis. He also contributed to the final revision of the manuscript.

References

Adamson, D.T., Piña, E.A., Cartwright, A.E., Rauch, S.R., Hunter Anderson, R., Mohr, T., Connor, J.A., 2017. 1, 4-Dioxane drinking water occurrence data from the third unregulated contaminant-monitoring rule. *Sci. Total Environ.* 596–597: 236–245.

Adeel, M., Song, X., Wang, Y., Francis, D., Yang, Y., (2017). Environmental impact of estrogens on human, animal and plant life: A critical review. *Environ. Int.* 99: 107–119.

Agency for Toxic Substances and Disease Registry (US); (2022). Toxicological Profile for Aldrin and Dieldrin.

Agency for Toxic Substances and Disease Registry (US); (2021). Toxicological Profile for Endrin.

Alijagic, A., Kotlyar, O., Larsson, M., Salihovic, S., Hedbrant, A., Eriksson, U., Karlsson, P., Persson, A., Scherbak, N., Färnlund, K., Engwall, M., Särndahl, E. (2024). Immunotoxic, genotoxic, and endocrine disrupting impacts of polyamide microplastic particles and chemicals. *Environment International*. 183:108412. https://doi.org/10.1016/j.envint.2023.108412.

Al-Odaini, N.A., Zakaria, M.P., Yaziz, M.I., Surif, S., Kannan, N. (2012). Occurrence of synthetic hormones in sewage effluents and Langat River and its tributaries, Malaysia, International Journal of Environmental Analytical Chemistry.1-13. https://doi.org/10.1080/03067319.2012.727810

Alsabri, A., Tahir, F., Al-Ghamdi, S.G. (2022). Environmental impacts of polypropylene (PP) production and prospects of its recycling in the GCC region. *Materials Today: Proceedings*. 56(4): 2245–2251. https://doi.org/10.1016/j.matpr.2021.11.574.

Alvarez, P., Beltrán, F., Masa, F., Pocostales, J. (2009). A comparison between catalytic ozonation and activated carbon adsorption/ozoneregeneration processes for wastewater treatment. *Appl Catal B Environ*. 92(3):393–400.

Al-Qaim, F. F., Mussa, Z. H., Yuzir, A., Tahrim, N. A., Hashim, N., and Azman, S. J. W. (2018). Transportation of different therapeutic classes of pharmaceuticals to the surface water, sewage treatment plant, and hospital samples, Malaysia. Malaysia. 10:916. https://doi.org/10.3390/w10070916.

Amigun Taiwo, A., Mustapha, S., Jimoh Oladejo, T., Folahan Amoo, A., Elabor, R. (2022). Occurrence, Effects, Detection, and Photodegradation of Triclosan and Triclocarban in the Environment: A Review. *International Journal of Environmental Analytical Chemistry*. 1–19. https://doi.org/10.1080/03067319.2022.2106860.

AMAP (Arctic Monitoring and Assessment Program). AMAP Assessment 2002: persistent organic pollutants in the Arctic. Oslo, Norway; 2004. Available at: http://www.amap.no/

Aragão, R. B.D. A., Semensatto, D., Calixto, L.A., Labuto, G. (2020). Pharmaceutical market, environmental public policies and water quality: the case of the São Paulo Metropolitan Region, Brazil, *Cadernos de Saúde Pública*. 36:11.

Arienzo, M., Ferrara, L., Trifuoggi, M. (2021).The dual role of microplastics in marine environment: sink and vectors of pollutants. *J. Mar. Sci. Eng*. 9(6):642

Arias, A.H., Alfonso, M.B., Girones, L., Piccolo, C.M., Marcovecchio, J.E. (2022). Synthetic microfibers and tyre wear particles pollution in aquatic systems: Relevance and mitigation strategies. *Environmental Pollution*. 295:118607. https://doi.org/10.1016/j.envpol.2021.118607.

Arukwe, A, Goksoyr, A. (1998). Xenobiotics, xenoestrogens and reproduction disturbances in fish. *Sarsia*. 83: 225–241.

Audsley, E and Wilkinson, M. (2014). What is the potential for reducing national greenhouse gas emissions from crop and livestock production systems? *Journal of Cleaner Production*. 73:263–268 https://doi.org/10.1016/j.jclepro.2014.01.066

Azarmi, F., Kumar P., Mulheron, M. (2014). The exposure to coarse, fine and ultrafine particle emissions from concrete mixing, drilling and cutting activities. *J Hazard Mater*. 279:268–79.

Azaroff, A., Monperrus, M., Miossec, C., Gassie, C., Guyoneaud, R. (2021). Microbial degradation of hydrophobic emerging contaminants from marine sediment slurries (Capbreton Canyon) to pure bacterial strain. *J. Hazard. Mater*. 402:123477.

Baloyi, N. D., Tekere, M., Maphangwa, K. W., Masindi, V. (2021). Insights Into the Prevalence and Impacts of Phthalate Esters in Aquatic Ecosystems. *Frontiers in Environmental Science*. 9. https://doi.org/10.3389/fenvs.2021.684190.

Barnett, T. P., Adam, J. C. & Lettenmaier, D. P. (2005). Potential impacts of a warming climate on water availability in snow-dominated regions. *Nature*. 438: 303–309. https://doi.org/10.1038/nature04141

Bayen, S., Estrada, E. S., Zhang, H., Lee, W. K., Juhel, G., Smedes, F., et al. (2019). Partitioning and bioaccumulation of legacy and emerging hydrophobic organic chemicals in mangrove ecosystems. *Environ. Sci. Technol*. 53:2549–2558. https://doi.org/10.1021/acs.est.8b06122

Boone, MD., Bridges, CM. (1999). The effect of temperature on the potency of carbaryl for survival of tadpoles of the green frog (Rana clamitans). *Environ Toxicol Chem*.18 (7):1482–4

Braune BM, Outridge PM, Fisk AT, Muir DCG, Helm PA, Hobbs K, et al. (2005). Persistent organic pollutants and mercury in marine biota of the Canadian Arctic: an overview of spatial and temporal trends. *Sci Total Environ.* 351:4–56.

Brian, J. V., Harris, C. A., Runnalls, T. J., Fantinati, A., Pojana, G., Marcomini, A., Booy, P., Lamoree, M., Kortenkamp, A., Sumpter, J. P. (2008). Evidence of temperature-dependent effects on the estrogenic response of fish: Implications with regard to climate change. *Science of The Total Environment.* 397, (1–3):72–81. https://doi.org/10.1016/j.scitotenv.2008.02.036.

Burelo, M., Hernández-Varela, J.D., Medina, I.D., Treviño-Quintanilla, C.D. (2023). Recent developments in bio-based polyethylene: Degradation studies, waste management and recycling. *Heliyon.* 9;11. https://doi.org/10.1016/j.heliyon.2023.e21374.

Buzea, C., Pacheco, I.I., Robbie, K. (2007). Nanomaterials and nanoparticles: sources and toxicity. Biointerphases. 2:MR17–MR71. 1934-8630/2007/2,,4…/MR17/55/$23.00

Caldeira, K., Wickett, M.E. (2003). Anthropogenic carbon and ocean pH. *Nature.* 425:365–365. doi: https://doi.org/10.1038/425365a.

Canesi L, Fabbri E. (2015). Environmental Effects of BPA: Focus on Aquatic Species. Dose Response. 13(3):1559325815598304. doi: https://doi.org/10.1177/1559325815598304.

Capkin, E., Altinok, I., Karahan, S.(2006). Water quality and fish size affect toxicity of endosulfan, an organochlorine pesticide, to rainbow trout. *Chemosphere. 64*(10): 1793–800.

Chakraborty, P., Zhang, G., Li, J., Sivakumar, A. Jones, K. C. (2016). Occurrence and sources of selected organochlorine pesticides in the soil of seven major Indian cities: Assessment of air-soil exchange. *Environmental Pollution.* 204: 74–80.

Chen, G., Li, M., Liu, X. (2015). Fluoroquinolone antibacterial agent contaminants in soil/groundwater: a literature review of sources, fate, and occurrence. *Water Air Soil Pollut.* 226, 418.

Cheng, N., Wang, B., Wu, P., Lee, X., Xing, Y., Chen, M., Gao, B. (2021). Adsorption of emerging contaminants from water and wastewater by modified biochar: a review. *Environ. Pollut.* 273:116448.

Chiu, M. C., Hunt, L., Resh, V. H. (2017). Climate-change influences on the response of macroinvertebrate communities to pesticide contamination in the Sacramento River, California watershed. *Science of the total environment.* 581–582: 741–749.

Clarke, B., Otto, F., Stuart-Smith, R., Harrington, L., (2022). Extreme weather impacts of climate change: an attribution perspective. *Environ. Res.: Clim.* 1 (1), 012001 https://doi.org/10.1088/2752-5295/ac6e7d.

D'Alessio M., Durso L. M., Miller D. N., Woodbury B., Ray C., Snow D. D. (2019). Environmental fate and microbial effects of monensin, lincomycin, and sulfamethazine residues in soil. *Environ. Pollut.* 246:60–68. https://doi.org/10.1016/j.envpol.2018.11.093

Das M. T., Ghosh P. and Thakur I. S. (2014). Intake estimates of phthalate esters for South Delhi population based on exposure media assessment. *Environ Pollut.* 189:118–25.

De Las, H.I., Molina, R., Segura, Y., Hülsen, T., Molina, M., Gonzalez-Benítez, N., Melero, J., Mohedano, A., Martínez, F., Puyol, D. (2020). Contamination of N-poor wastewater with emerging pollutants does not affect the performance of purple phototrophic bacteria and the subsequent resource recovery potential. *J. Hazard. Mater.* 385:121617.

Dhaka, V., Singh, S., Anil, A.G., Naik, T. S. SK., Garg, S., Samuel, J., Kumar, M., Ramamurthy, C. P., Singh, J. (2022). Occurrence, toxicity and remediation of polyethylene terephthalate plastics. A review. *Environ Chem Lett.* 20:1777–1800. https://doi.org/10.1007/s10311-021-01384-8.

Dsikowitzky, L., Crawford, S. E., Nordhaus, I., Lindner, F., Dwiyitno, D., Irianto, H. E., et al. (2020). Analysis and environmental risk assessment of priority and emerging organic pollutants in sediments from the tropical coastal megacity Jakarta, Indonesia, *Regional Stud. Mar. Sci.* 34: 101021. https://doi.org/10.1016/j.rsma.2019.101021

Duong, H. T., Kadokami, K., Nguyen, D. T., Trinh, H. T., Doan, N. H., Mizukawa, H., et al. (2023). Occurrence, potential sources, and risk assessment of pharmaceuticals and personal care products in atmospheric particulate matter in Hanoi. *Vietnam Environ Sci Pollut Res Int.* 30(12):34814–34826. doi: https://doi.org/10.1007/s11356-022-24630-0.

Ebele, A.J., Abdallah, M.-A.-E., Harrad, S. (2017). Pharmaceuticals and personal care products (PPCPs) in the freshwater aquatic environment. *Emerging Contaminants*. 3: 1–16.

Egbuna, C., Amadi, CN., Patrick-Iwuanyanwu, KC., Ezzat, SM., Awuchi, CG., Ugonwa, PO., Orisakwe, OE. (2021). Emerging pollutants in Nigeria: A systematic review. *Environmental Toxicology and Pharmacology*. 85:103638.

EPA, U.S.E.P.A. (2010). Emerging contaminants nanomaterials. https://archive.epa.gov/region9/mediacenter/web/pdf/emerging_contaminant_ nanomaterials.pdf.

EPA.US. (2014). DEET (N,N-Diethyl-meta-toluamide) Interim Registration Review Decision Case Number 0002.

El-Kalliny, A.S., Abdel-Wahed, M.S., El-Zahhar, A.A., Hamza, IA., Gad-Allah, TA. (2023). Nanomaterials: a review of emerging contaminants with potential health or environmental impact. *Discover Nano* 18 (1):68. https://doi.org/10.1186/s11671-023-03787-8

Elgarahy, A.M., Akhdhar, A., Elwakeel, K.Z. (2021). Microplastics prevalence, interactions, and remediation in the aquatic environment: a Critical Review. *J. Environ. Chem. Eng.* 9 (5): 106224.

Espinosa-Ortiz, E.J., Rene, E.R., Pakshirajan, K., van Hullebusch, E.D., Lens, P.N. (2016). Fungal pelleted reactors in wastewater treatment: Applications and perspectives. *Chem. Eng. J.* 283: 553–571.

FAO (Food and Agricultural Organization). (2013). Tackling climate change through livestock: A global assessment of emissions and mitigation opportunities. Food and Agricultural Organization of the United Nations (FAO) Rome.

Forster P, Ramaswamy V, Artaxo P, et al. (2007). Changes in atmospheric constituents and in radiative forcing. In: Solomon S, Qin D, Manning M, et al (eds.), Climate change 2007: the physical science basis. Contribution of Working Group I to the Fourth Assessment Report of the Intergovernmental Panel on Climate Change (Cambridge, United Kingdom and New York, NY, USA: Cambridge University Press. 135–136.

Fick, J., Soderstrom, H., Lindberg, R. H., Phan, C., Tysklind, M., Larsson D.G.J. (2009). Contamination of surface, ground, and drinking water from pharmaceutical production. *Environmental Toxicology and Chemistry*. 28(12): 2522–2527.

Fisher, J. McLanahan, E. (2008). Evidence for perchlorate altering thyroid function. *Toxicol. Lett.*, 180(182).

Furnell DJ, Schweinsburg RE. (1984). Population dynamics of central Canadian Arctic island polar bears. *J Wildlife Manage*. 48:72228.

Galindo-Miranda, J.M., Guízar-González, C., Becerril-Bravo, E.J., Moeller-Chávez, G., León-Becerril, E., Vallejo-Rodríguez, R. (2019). Occurrence of emerging contaminants in environmental surface waters and their analytical methodology – a review. *Water Supply*. 19:1871–1884.

Gaunt, P., Barker, SA. (2000). Matrix solid phase dispersion extraction of triazines from catfish tissues; examination of the effects of temperature and dissolved oxygen on the toxicity of atrazine. *Int J Environ Pollut*. 13 (1–6):284–312.

Gavrilescu, M., Demnerová, K., Aamand, J., Agathos, S., Fava, F. (2015). Emerging pollutants in the environment: present and future challenges in biomonitoring, ecological risks and bioremediation. *N. Biotechnol.* 32:147–156.

Giudice, L.C., Llamas-Clark, E.F., DeNicola, N., Pandipati, S., Zlatnik, M.G., Decena, D. C.D., Woodruff, T.J., Conry, J.A. (2021). Climate change, women's health, and the role of obstetricians and gynecologists in leadership. *Int. J. Gynecology and Obstetrics*. 155: 345–356.

Gilbert, ME., O'Shaughnessy, KL., Axelstad, M. (2020). Regulation of thyroid-disrupting chemicals to protect the developing brain. *Endocrinology*. 161(10).

Gillis, J.D., Price, G.W., Prasher, S., (2017). Lethal and sub-lethal effects of triclosan toxicity to the earthworm Eisenia fetida assessed through GC–MS metabolomics. *J. Hazard. Mater.* 323, 203–211.

Glassmeyer, S. T., Furlong, E. T., Kolpin, D. W., Batt, A. L., Benson, R., Boone, J. S., Conerly, O., Donohue, M. J., King, D. N., Kostich, M. S., Mash, H. E., Pfaller, S. L., Schenck, K. M., Simmons, J. E., Varughese, E. A., Vesper, S. J., Villegas, E. N., Wilson, V. S. (2017). Nationwide reconnaissance of contaminants of emerging concern in source and treated drinking waters of the United States. *Science of The Total Environment.* 581–582:909–922. https://doi.org/10.1016/j.scitotenv.2016.12.004.

Grant, P.R., Grant, B. R., Huey, R. B., Johnson, M. T. J., Knoll, A. H., Schmitt, J. (2016). Evolution caused by extreme events. *Philosophical transactions – Royal Society. Biological sciences.* 372: 20160146.

Gracia-Lor, E., Sancho, J.V., Serrano, R., Hernández, F. (2012). Occurrence and removal of pharmaceuticals in wastewater treatment plants at the Spanish Mediterranean area of Valencia. Chemosphere 87, 453–462. Gruber, K., 2018. Cleaning up our future health. *Nature.* 555. S20–S22.

Grelska, A., Noszczyńska, M. (2020).White Rot Fungi Can Be a Promising Tool for Removal of Bisphenol A, Bisphenol S, and Nonylphenol from Wastewater. *Environ. Sci. Pollut. Res.* 27: 39958–39976.

Guo, Y., Alomirah, H., Cho, H. S., Minh, T.B., Mohd, M.A., Nakata, H., Kannan, K. (2011). Occurrence of phthalate metabolites in human urine from several Asian countries. Environmental Science & Technology. 45(7): 3138-3144. PMID: 21395215. https://doi.org/10.1021/es103879m

Gwenzi, W., Mangori, L., Danha, C., Chaukura, N., Dunjana, N., Sanganyado, E., (2018). Sources, behaviour, and environmental and human health risks of hightechnology rare earth elements as emerging contaminants. *Sci. Total Environ.* 636, 299–313.

Haron, D. E. M., Yoneda, M., Hod, R., Ramli, M. R., Aziz, M. Y. (2023). Assessment of 18 endocrine disrupting chemicals in tap water samples from Klang Valley, Malaysia. *Environ. Sci. Pollut. Res. Int.* 30, 111062–111075. https://doi.org/10.1007/s11356-023-30022-9.

Heugens, E.H., Hendriks, A.J., Dekker, T., van Straalen, N.M., Admiraal, W. (2001). A review of the effects of multiple stressors on aquatic organisms and analysis of uncertainty factors for use in risk assessment. *Crit. Rev. Toxicol.* 31: 247–284.

Hristov, A.N., Oh, J., Lee, C., Meinen, R., Montes, F., Ott, T., Firkins, J., Rotz, A., Dell, C., Adesogan, A., Yang, W., Tricarico, J., Kebreab, E., Waghorn, G., Dijkstra, J. & Oosting, S. (2013). Mitigation of greenhouse gas emissions in livestock production – A review of technical options for non-CO2 emissions. FAO Animal Production and Health .177.

Irmak, S., Erbatur, O., Akgerman, A. (2005). Degradation of 17β-estradiol and bisphenol A in aqueous medium by using ozone and ozone/UV techniques. *J. Hazard. Mater.* 126:54–62.

IPCC (2007). Summary for policy makers. Climate Change 2007: Synthesis Report. Fourth Assessment Report of the Intergovernmental Panel for Climate Change. Available online at http://www.ipcc.ch/pdf/assessment-report/ar4/syr/ar4_syr_spm.pdf

IPCC (2022). Climate Change 2022: Impacts, Adaptation and Vulnerability. Contribution of Working Group II to the Sixth Assessment Report of the Intergovernmental Panel on Climate Change. https://www.ipcc.ch/2022/02/28/pr-wgii-ar6/

Ivantsova, E. Martyniuk, C.J. (2023). A synthesis on the sub-lethal toxicity of atenolol, a beta-blocker, in teleost fish. *Environmental Toxicology and Pharmacology.* 102. https://doi.org/10.1016/j.etap.2023.104236.

Iwata T. (2015). Biodegradable and bio-based polymers: future prospects of eco-friendly plastics. *Angew. Chem. Int. Ed.* 54 (11):3210–3215.

Jacob, R.S., Araújo, C.V., de Souza Santos, L.V., Moreira, V.R., Lebron, Y.A.R., Lange, L. C. (2021). The environmental risks of pharmaceuticals beyond traditional toxic effects: Chemical differences that can repel or entrap aquatic organisms. *Environ. Pollut.* 268:115902.

Jagini, S., Thaduri, S., Konda, S., Saranga, V.K., Dheeravath, B., Vurimindi, H. (2021). Emerging contaminant (Triclosan) removal by adsorption and oxidation process: comparative study. *Modeling Earth Systems and Environment.* 7:2431–2438

Jeevanandam J, et al. (2018). Review on nanoparticles and nanostructured materials: history, sources, toxicity and regulations. *Beilstein J Nanotechnol.* 9(1):1050–74.

Juksu, K., Zhao, J.-L., Liu, Y.-S., Yao, L., Sarin, C., Sreesai, S., et al. (2019). Occurrence, fate and risk assessment of biocides in wastewater treatment plants and aquatic environments in Thailand. *Sci. Total Environ.* 690: 1110–1119. https://doi.org/10.1016/j.scitotenv.2019.07.097.

Kabasenche, W.P., Skinner, M.K. (2014). DDT, epigenetic harm, and transgenerational environmental justice. *Environ Health.* 13: 62. https://doi.org/10.1186/1476-069X-13-62

Kanan S, Samara F (2018). Dioxins and furans: a review from chemical and environmental perspectives. *Trends Environ Anal Chem* 17:1–13. https://doi.org/10.1016/j.teac.2017.12.001

Kashyap, D., Agarwal, T. (2018). Concentration and factors affecting the distribution of phthalates in the air and dust: A global scenario. *Science of The Total Environment.* 635:817–827. https://doi.org/10.1016/j.scitotenv.2018.04.158.

Kawa, I.A., Fatima, Q., Mir, S.A., Jeelani, H., Manzoor, S., Rashid, F., (2021). Endocrine disrupting chemical Bisphenol A and its potential effects on female health. *Diabetes Metab. Syndr.* 15, 803–811.

Karthik, R., Robin, R., Purvaja, R., Ganguly,D., Anandavelu, I., Raghuraman, R., Hariharan, G., Ramakrishna, A., Ramesh, R. (2018). Microplastics along the beaches of southeast coast of India. *Sci. Total Environ.* 645:1388–1399

Khairul Anuar NFS, Huyop F, Ur-Rehman G, Abdullah F, Normi YM, Sabullah MK, Abdul Wahab R. (2022). An Overview into Polyethylene Terephthalate (PET) Hydrolases and Efforts in Tailoring Enzymes for Improved Plastic Degradation. *Int J Mol Sci.* 20;23 (20):12644. https://doi.org/10.3390/ijms232012644

Klaine SJ, et al. (2008). Nanomaterials in the environment: behavior, fate, bioavailability, and effects. *Environ Toxicol Chem.* 27(9):1825–51.

Kokkinos P, Mantzavinos D, Venieri D. (2020). Current trends in the application of nanomaterials for the removal of emerging micropollutants and pathogens from water. *Molecules.* ;25(9):2016.

Kourkoumpas, D.-S., Karellas, S., Kouloumoundras, S., Koufodimos, G., Grammelis, P., Kakaras, E. (2015). Comparison of waste-to-energy processes by means of life cycle analysis principles regarding the global warming potential impact: applied case studies in Greece, France and Germany. *Waste Biomass Valoriz.* 6: 605–621.

Krah, D., Ghattas, A.-K., Wick, A., Bröder, K., Ternes, T.A. (2016). Micropollutant degradation via extracted native enzymes from activated sludge. *Water Res.* 95: 348–360.

Kudzin, M.H., Piwowarska, D., Festinger, N., Chruściel, J.J. (2024). Risks Associated with the Presence of Polyvinyl Chloride in the Environment and Methods for Its Disposal and Utilization. *Materials. 17(1):*173. https://doi.org/10.3390/ma17010173

Kumar, P., Fennell P, Robins A. (2010). Comparison of the behaviour of manufactured and other airborne nanoparticles and the consequences for prioritising research and regulation activities. *J Nanopart Res.* 12(5):1523–30

Kumar, P., Pirjola, L., Ketzel, M., Harrison, R.M. (2013). Nanoparticle emissions from 11 non-vehicle exhaust sources–a review. *Atmos Environ.* 67:252–77

Kumar, N., Kumbhat, S. (2016). Carbon-based nanomaterials in Essentials in Nanoscience and Nanotechnology. 5:189–236

Kik, K., Bukowska, B., Sicińska, P. (2020). Polystyrene nanoparticles: Sources, occurrence in the environment, distribution in tissues, accumulation and toxicity to various organisms. *Environmental Pollution.* 262:114297. https://doi.org/10.1016/j.envpol.2020.114297

Kirchner, C., Liedl, T., Kudera, S., Pellegrino, T., Muñoz Javier, A., Gaub, H. E., Stölzle, S., Fertig, N., Parak, W. J. (2005). Cytotoxicity of colloidal CdSe and CdSe/ZnS nanoparticles. *Nano Letters.* 5 (2): 331–338, https://doi.org/10.1021/nl047996m, 2-s2.0-14644405573.

Lahiri, A., Daniel, S., Kanthapazham, R., Vanaraj, R., Thambidurai, A., Peter, L.S. (2023). A critical review on food waste management for the production of materials and biofuel. *J. Hazardous Mater. Adv.* 100266.

Landrigan, P.J., Fuller, R., Acosta, N.J., Adeyi, O., Arnold, R., Baldé, A.B., Bertollini, R., Bose-O'Reilly, S., Boufford, J.I., Breysse, P.N. (2018). The Lancet Commission on pollution and health. *Lancet.* 391: 462–512.

Legler J. (1999). Development of a stably transfected estrogen receptor-mediated luciferase reporter gene assaying the human T47D breast cancer cell line. *Toxicol Sci,* 48(1): 55–56.

Lempart, A., Kudlek, E., Dudziak, M. (2020). The potential of the organic micropollutants emission from swimming accessories into pool water. *Environ. Int.* 136, 105442.

Lopes, D., Ferreira, M.J., Russo, R., Dias, MJ. (2015). Natural and synthetic rubber/waste – Ethylene-Vinyl Acetate composites for sustainable application in the footwear industry. *Journal of Cleaner Production*. 92. 230–236. https://doi.org/10.1016/j.jclepro.2014.12.063.

Lin D, et al. (2010). Fate and transport of engineered nanomaterials in the environment. *J Environ Qual*.39 (6):1896–908.

Liu, J. (2014). Diuron. *Encyclopedia of Toxicology (Third Edition)*. 215–216. https://doi.org/10.1016/B978-0-12-386454-3.00140-8.

Lovejoy TE, Hannah L. (2005). Climate change and biodiversity. New Haven, CT: Yale University Press; 2005.

Lowry GV, et al. (2010). Environmental occurrences, behavior, fate, and ecological effects of nanomaterials: an introduction to the special series. *J Environ Qual*. 39(6):1867–74.

Macdonald RW, Harner T, Fyfe J. (2005). Recent climate change in the Arctic and its impact on contaminant pathways and interpretation of temporal trend data. *Sci Total Environ*. 342 (1–3):5–86.

Madrid, F., Rubio-Bellido, M., Morillo, E. (2020). Extraction of nonylphenol, pyrene and phenanthrene from sewage sludge and composted biosolids by cyclodextrins and rhamnolipids. *Sci. Total Environ*. 715:136986.

Marlatt, V.L., Bayen, S., Castaneda-Cortès, D., Delbès, G., Grigorova, P., Langlois, V.S., Martyniuk, C.J., Metcalfe, C.D., Parent, L., Rwigemera, A., Thomson, P., Van Der Kraak, G. (2022). Impacts of endocrine disrupting chemicals on reproduction in wildlife and humans. *Environ. Res*. 208:112584.

Materna EJ, Rabeni CF, Lapoint TW. (1995). Effects of the synthetic pyrethroid insecticide, esfenvalerate, on larval leopard frogs (*Rana* spp.). *Environ Toxicol Chem*. 14(4): 613–22.

McAvoy, DC., Schatowitz, B., Jacob, M., Hauk, A., Eckhoff, WS. (2002). Measurement of triclosan in wastewater treatment systems. *Environ Toxicol Chem*. 21:1323–29.

McCarty, J.P. (2001). Ecological Consequences of Recent Climate Change. *Conservation Biology*.15 (1): 320–331.

Milly, P. C D., Dunne, K. A. Vecchia, A. V. (2006). Global pattern of trends in streamflow and water availability in a changing climate. *Nature*. 438:347–350. https://doi.org/10.1038/nature04312

Mirzaei, AG., Alireza, S, Hasan, F., Sohrab, R., Mohammad, S., Mohamad T. (2012). Garlic in Ruminants Feeding. *Asian Journal of Biological Sciences*. 5(7):328–340. https://doi.org/10.3923/ajbs.2012.328.340

Mehrpour, O., Karrari, P., Zamani, N., Tsatsakis, A.M., Abdollahi, M. (2014). Occupational exposure to pesticides and consequences on male semen and fertility: a review. *Toxicol. Lett*. 230:146–156.

Míguez, L., Esperanza, M., Seoane, M., Cid, A. (2021). Assessment of cytotoxicity biomarkers on the microalga *Chlamydomonas reinhardtii* exposed to emerging and priority pollutants. *Ecotoxicology and Environmental Safety*. 208:111646. https://doi.org/10.1016/j.ecoenv.2020.111646.

Modesto, T., Tiemeier, H., Peeters, R.P., Jaddoe, V.W., Hofman, A., Verhulst, F.C., Ghassabian, A., (2015). Maternal mild thyroid hormone insufficiency in early pregnancy and attention-deficit/hyperactivity disorder symptoms in children. *JAMA Pediatr*. 169: 838–845

Moreira, ALP., Souza, JACR., De Souza, JF., Mamede, JPM., Farias, D., Luchiari, AC. (2024). Long-term effects of embryonic exposure to benzophenone-3 on neurotoxicity and behavior of adult zebrafish. *Sci Total Environ*. 15:908.168403. doi: https://doi.org/10.1016/j.scitotenv.2023.168403.

Mpatani, F.M., Aryee, A.A., Kani, A.N., Guo, Q., Dovi, E., Qu, L., Li, Z., Han, R. (2020). Uptake of micropollutant-bisphenol A, methylene blue and neutral red onto a novel bagasse-b-cyclodextrin polymer by adsorption process. *Chemosphere*. 259: 127439.

Naidu, R., Biswas, B., Willett, I.R., Cribb, J., Kumar Singh, B., Paul Nathanail, C., Coulon, F., Semple, K.T., Jones, K.C., Barclay, A., Aitken, R.J. (2021). Chemical pollution: A growing peril and potential catastrophic risk to humanity. *Environ. Int*. 156: 106616.

Nowack B. and Bucheli T.D. (2007). Occurrence, behavior and effects of nanoparticles in the environment, Review. *Environmental Pollution.* 150: 5–22.

Oelbermann, M., Morgan, S., Echarte, L., (2022). Elevated carbon dioxide and temperature effects on soil properties from sole crops and intercrops. *Soil Use Manag.* 38 (1), 435–447. https://doi.org/10.1111/sum.12752.

Olafsdottir K, Skirnisson K, Gylfadottir G, Johannesson T. (1998). Seasonal fluctuations of organochlorine levels in the common elder (Somateria mollissima) in Iceland. *Environ Pollut.*103 (2–3):153–8.

Okonkwo, J.O., Sibali, L.L., McCrindle, R., Senwo, Z.N. (2006). An improved activated carbon method to quantify dichlorodiphenyltrichloroethane (DDT) in surface water. *Environ. Chem. Lett.* 5: 121–123.

Orr, JC., Fabry, VJ., Aumont, O., Bopp, L., Doney, SC., Feely, RA., Gnanadesikan, A., Gruber, N., Ishida, A., Joos, F., Key, RM., Lindsay, K,, Maier-Reimer, E., Matear, R., Monfray, P., Mouchet A, Najjar RG, Plattner GK, Rodgers KB, Sabine CL, Sarmiento JL, Schlitzer R, Slater RD, Totterdell IJ, Weirig MF, Yamanaka Y, Yool A. (2005). Anthropogenic ocean acidification over the twenty-first century and its impact on calcifying organisms. *Nature.* ;437:681–686. https://doi.org/10.1038/nature04095.

Ofrydopoulou, A., Nannou, C., Evgenidou, E., Christodoulou, A., Lambropoulou, D. (2022). Assessment of a wide array of organic micropollutants of emerging concern in wastewater treatment plants in Greece: Occurrence, removals, mass loading and potential risks. Science of The Total Environment. 802:149860. https://doi.org/10.1016/j.scitotenv.2021.149860

Orvos, D.R., Versteeg, D.J., Inauen, J., Capdevielle, M., Rothenstein, A., Cunningham, V. (2002). Aquatic toxicity of triclosan. Environ Toxicol Chem. 21(7), 1338–49. PMID: 12109732.

Palacio, D.A., Aranda, F.L., Rivas, B.L., (2022). Removal of antibiotic emerging pollutants: an overview. *J. Chil. Chem. Soc.* 67:5547–5561.

Palli, L., Spina, F., Varese, G.C., Vincenzi, M., Aragno, M., Arcangeli, G., Mucci, N., Santianni, D., Caffaz, S., Gori, R., (2019). Occurrence of selected pharmaceuticals in wastewater treatment plants of Tuscany: An effect-based approach to evaluate the potential environmental impact. Int. J. Hyg. Environ. Health. 222: 717–725.

Patra, A., Saxena, J (2010). ChemInform Abstract: A New Perspective on the Use of Plant Secondary Metabolites to Inhibit Methanogenesis in the Rumen. *Phytochemistry.* 71. (11–12):1198–222. https://doi.org/10.1016/j.phytochem.2010.05.010

Patel, S. (2017). Disruption of aromatase homeostasis as the cause of a multiplicity of ailments: a comprehensive review. *J. Steroid Biochem. Mol. Biol.* 168:19–25.

Pearce, E. N., Lazarus, J. H., Smyth, P. P., He, X., Dall'amico, D., Parkes, A. B., Burns, R., Smith, D. F., Maina, A., Bestwick, J. P., Jooman, M., Leung, A. M., Braverman, L. E. (2010). Perchlorate and thiocyanate exposure and thyroid function in first-trimester pregnant women. *J Clin Endocrinol Metab.* 95(7):3207–15. doi: https://doi.org/10.1210/jc.2010-0014.

Peralta-Videa JR, et al. (2011). Nanomaterials and the environment: a review for the biennium 2008–2010. *J Hazard Mater.* 186(1):1–15.

Peng, Y., Xu, H., Wang, Z., Li, L., Shang, J., Li, B., et al., (2023). Effects of intercropping and drought on soil aggregation and associated organic carbon and nitrogen. *Soil Use Manag.* 39 (1): 316–328. https://doi.org/10.1111/sum.12866.

Pereira, L.C., de Souza, A.O., Bernardes, M.F.F., Pazin, M., Tasso, M.J., Pereira, P.H., Dorta, D.J., (2015). A perspective on the potential risks of emerging contaminants to human and environmental health. *Environ. Sci. Pollut. Res.* 22: 13800–13823.

Penuelas J, Filella L. (2001). Responses to a warming world. *Science.* 294:793–5

Pico, Y., Alfarhan, A., Barcelo, D. (2019). Nano- and microplastic analysis: Focus on their occurrence in freshwater ecosystems and remediation technologies. *TrAC Trends in Analytical Chemistry.* 113: 409–425. https://doi.org/10.1016/j.trac.2018.08.022.

Pitarch, E., Cervera, M. I., Portoles, T., Ibanez, M., Barreda, M., Renau-Pruñonosa, A., Ignacio Morell, I., López, J. F., Albarrán, F., Hernandez, F. (2016). Comprehensive monitoring of organic micro-pollutants in surface and groundwater in the surrounding of a solid-waste treatment plant of Castellón, Spain. *The Science of The Total Environment.* 548–549:211–220. https://doi.org/10.1016/j.scitotenv.2015.12.166

Pitter, G., Da Re, F., Canova, C., Barbieri, G., Zare Jeddi, M., Daprà, F., Manea, F., Zolin, R., Bettega, A.M., Stopazzolo, G. (2020). Serum levels of perfluoroalkyl substances (PFAS) in adolescents and young adults exposed to contaminated drinking water in the Veneto region, Italy: a cross-sectional study based on a health surveillance program. *Environ. Health Perspect.* 128. 027007.

Prakash, V., Peralta-Videa, J., Tripathi, D.K. Ma, X., Sharma, S. (2021). Recent insights into the impact, fate and transport of cerium oxide nanoparticles in the plant-soil continuum. *Ecotoxicology and Environmental Safety.* 221:112403. https://doi.org/10.1016/j.ecoenv.2021.112403.

Praveena, S. M., Rashid, M. Z. M., Nasir, F. A. M., Yee, W. S., Aris, A. Z. (2019). Occurrence and potential human health risk of pharmaceutical residues in drinking water from Putrajaya (Malaysia). *Ecotoxicol. Environ. Saf.* 180: 549–556. https://doi.org/10.1016/j.ecoenv.2019.05.051

Raj, J.R.A. Muruganandam, L. (2014). Occurrence of perchlorate in drinking, surface, ground and effluent water from various parts of South India. *International Journal of Innovations in Engineering and Technology.* 4(4): 1–7.

Ramsay MA, Stirling I. (1982). Reproductive biology and ecology of female polar bears in western Hudson Bay. *Nat Can.* 109.(94146).

Rasheed, T., Bilal, M., Nabeel, F., Adeel, M., Iqbal, H.M.N. (2019). Environmentally-related contaminants of high concern: Potential sources and analytical modalities for detection, quantification, and treatment. *Environment International.* 52–66: 0160–4120. https://doi.org/10.1016/j.envint.2018.11.038.

Rathi, B.S., Kumar, P.S., Show, P.-L. (2021). A review on effective removal of emerging contaminants from aquatic systems: current trends and scope for further research. *J. Hazard. Mater.* 409: 124413.

Rauscher, H., Mech, A., Gibson, P., Gilliland, D., Held, A., Kestens, V., Koeber, R., Linsinger, T. and Stefaniak, E., (2019). Identification of nanomaterials through measurements. *Publications Office of the European Union.* ISBN 978-92-76-10371-4, https://doi.org/10.2760/053982

Raven J. 2005. Ocean acidification due to increasing atmospheric carbon dioxide. London, *The Royal Society;* 2005.

Ratushnyak, A. A., Andreeva, M. G., Trushin, M.V. (2005). Effects of type II pyrethroids on Daphnia magna: dose and temperature dependences. *Riv Biol-Biol Forum.* 98 (2): 349–57.

Rezvani Ghomi, E., Khosravi, F., Saedi Ardahaei, A., Dai, Y., Neisiany, RE., Foroughi, F., Wu, M., Das, O., Ramakrishna, S. (2021). The Life Cycle Assessment for Polylactic Acid (PLA) to Make It a Low-Carbon Material. *Polymers (Basel).* 13(11):1854. https://doi.org/10.3390/polym13111854.

Riva, F., Castiglioni, S., Fattore, E., Manenti, A., Davoli, E., Zuccato, E., (2018). Monitoring emerging contaminants in the drinking water of Milan and assessment of the human risk. *Int. J. Hyg. Environ. Health.* 221: 451–457.

Richardson, S.D., Kimura, S.Y., (2017). Emerging environmental contaminants: challenges facing our next generation and potential engineering solutions. *Environ. Technol. Innov.* 8: 40–56.

Rolsky C, Kelkar V. (2021). Degradation of Polyvinyl Alcohol in US Wastewater Treatment Plants and Subsequent Nationwide Emission Estimate. *Int J Environ Res Public Health.* 18(11):6027. https://doi.org/10.3390/ijerph18116027.

Root TL, Price JT, Hall KR, Schneider SH, Rosenzweig C, Pounds JA.(2003). Fingerprints of global warming on wild animals and plants. *Nature.* 421(6918):57–60.

Roccuzzo, S., Beckerman, A.P., Trögl, J. (2021). New perspectives on the bioremediation of endocrine disrupting compounds from wastewater using algae-, bacteria- and fungi-based technologies. *Int. J. Environ. Sci. Technol.* 18: 89–106.

Ruhí, A., Acuña, V., Barceló, D., Huerta, B., Mor, J.R., Rodríguez-Mozaz, S., Sabater, S. (2016). Bioaccumulation and trophic magnification of pharmaceuticals and endocrine disruptors in a Mediterranean river food web. *Science of the total environment.* 540:250–259.

Safe, S., Hutzinger, O., Hill, TA. (2012). Polychlorinated dibenzop-dioxins and-furans (PCDDs/PCDFs): Sources and environmental impact, epidemiology, mechanisms of action, health risks. Springer, New York.

Schecter, A., Haffner, D., Colacino, J., et al. (2010). Polybrominated diphenyl ethers (PBDEs) and hexabromocyclodecane (HBCD) in composite U.S. food samples. *Environ. Health Perspect.* 118 (3):357–362, https://doi.org/10.1289/ehp.0901345.

Scheffers, B.R., De Meester, L., Bridge, T. C., Hoffmann, A. A., Pandolfi, J. M., Corlett, R. T., Butchart, S. H., Pearce-Kelly, P., Kovacs, K. M., Dudgeon, D., Pacifici, M., Rondinini, C., Foden, W. B., Martin, T. G., Mora, C., Bickford, D., Watson, J. E. (2016). The broad footprint of climate change from genes to biomes to people. *Science.* 11: 354 (6313):aaf7671. https://doi.org/10.1126/science.aaf7671.

Schiedek, D., Sundelin, B., Readman, J.W., Macdonald, R.W. (2007). Interactions between climate change and contaminants. *Mar. Pollut. Bull.* 54:1845–1856.

Seneviratne, S.I., Nicholls, N., Easterling, D., Goodess, C.M., Kanae, S., Kossin, J., et al., (2012). Changes in climate extremes and their impacts on the natural physical environment. In: Field, C.B., Barros, V., Stocker, T.F., Qin, D., Dokken, D.J., Ebi, K.L., Mastrandrea, M.D., Mach, K.J., Plattner, G.-K., Allen, S.K., Tignor, M., Midgley, P.M. (Eds.), Managing the Risks of Extreme Events and Disasters to Advance Climate Change Adaptation, A Special Report of Working Groups I and II of the Intergovernmental Panel on Climate Change (IPCC). Cambridge University Press, Cambridge, UK, and New York, NY, USA, pp. 109–230. https://doi.org/10.7916/d8-6nbt-s431.

Shahid, M.K., Kashif, A., Fuwad, A., Choi, Y. (2021). Current advances in treatment technologies for removal of emerging contaminants from water–A critical review. *Coord. Chem. Rev.* 442. 213993.

Sharma, B.M., Becˇanová, J., Scheringer, M., Sharma, A., Bharat, G.K., Whitehead, P.G., Klánová, J., Nizzetto, L. (2019). Health and ecological risk assessment of emerging contaminants (pharmaceuticals, personal care products, and artificial sweeteners) in surface and groundwater (drinking water) in the Ganges River Basin, India, *Sci. Total Environ.* 646, 1459–1467.

Shah, SN., Shah, Z., Hussain, M., Khan, M. (2017). Hazardous Effects of Titanium Dioxide Nanoparticles in Ecosystem. *Bioinorg Chem Appl.*4101735. https://doi.org/10.1155/2017/4101735.

Shen, D., Gu, X., Zheng, Y., Delgado-Moreno, L., Jia, W., Ye, Q., Wang, W. (2022). The fate of erythromycin in soils and its effect on soil microbial community structure. *Science of The Total Environment.* 820. 153373. https://doi.org/10.1016/j.scitotenv.2022.153373.

Slater, K. (2008). 6-Environmental impact of polyester and polyamide textiles. Polyesters and Polyamides. *Woodhead Publishing Series in Textiles.* 171–199.

Silbergeld EK. (1973). Dieldrin—effects of chronic sublethal exposure on adaptation to thermal-stress in freshwater fish. Environ Sci Technol. 7(9):846–9.

Sonnenschein, C., Soto, AM. (1998). An updated review of environmental estrogen and androgen mimics and antagonists. *J Steroid Biochem Mol Biol.* 165: 143–150.

Soto-Salcido L.A., Pihlajamäki A., Mänttäri M. (2023). Reuse of end—of—Life membranes through accelerated polyamide degradation. *Waste Manag.* 171:124–133. https://doi.org/10.1016/j.wasman.2023.08.025.

Srikanth, K., Sukesh, K., Rao, AR., Pavan, G., Ravishankar, GA. (2019). Emerging Contaminants Effect on Aquatic Ecosystem: Human Health Risks. *Agricultural research & technology.* 19(4): 556104.

Stahl, C., Cruz, M. T., Bañaga, P. A., Betito, G., Braun, R. A., Aghdam, M. A., Cambaliza, M. O., Lorenzo, G. R., MacDonald, A. B., Hilario, M. R. A., Pabroa, P. C., Yee, J. R., Simpas, J. B., and Sorooshian, A. (2020). Sources and characteristics of size-resolved particulate organic acids and methanesulfonate in a coastal megacity: Manila, Philippines, Atmos. *Chem. Phys.* 20: 15907–15935. https://doi.org/10.5194/acp-20-15907-2020

Stefani D, Wardman D, Lambert T. (2005). The implosion of the Calgary General Hospital: ambient air quality issues. *J Air Waste Manag Assoc.* 55(1):52–9.

Stirling I, Lunn NJ, Iacozza J. (1999). Long-term trends in the population ecology of polar bears in western Hudson Bay in relation to climatic change. *Arctic.* 52(3):294–306.

Syranidou, E., Kalogerakis, N. (2022). Interactions of microplastics, antibiotics and antibiotic resistant genes within WWTPs. *Sci. Total Environ.* 804: 150141.

Tahira, M., Saima, M., Rahmat, A., Abdul, N., Afsar, K., (2022). Technologies for Removal of Emerging Contaminants from Wastewater. In: Muharrem, I., Olcay Kaplan, I. (Eds.), Wastewater Treatment. *IntechOpen, Rijeka.* 9.

Teijon, G., Candela, L., Tamoh, K., Molina-Díaz, A., Fernández-Alba, A. (2010). Occurrence of emerging contaminants, priority substances (2008/105/CE) and heavy metals in treated wastewater and groundwater at Depurbaix facility (Barcelona, Spain). *Sci. Total Environ.* 408: 3584–3595.

Tewari, S., Jindal, R., Kho, Y. L., Eo, S., Choi, K. J. C. (2013). Major pharmaceutical residues in wastewater treatment plants and receiving waters in Bangkok, Thailand, and associated ecological risks. *Chemosphere.* 91. 697–704. doi:https://doi.org/10.1016/j.chemosphere.2012.12.042

Tran, N., Drogui, P., Brar, S.K. (2015). Sonochemical techniques to degrade pharmaceutical organic pollutants. *Environ. Chem. Lett.* 13:251–268.

Tran, N.H., Reinhard, M., Gin, K.-Y.-H. (2018). Occurrence and fate of emerging contaminants in municipal wastewater treatment plants from different geographical regions-a review. *Water Res.* 133: 182–207.

Tran, N.H., Reinhard, M., Khan, E., Chen, H., Nguyen, V.T., Li, Y., Goh, S.G., Nguyen, Q. B., Saeidi, N., Gin, K.-Y.-H. (2019). Emerging contaminants in wastewater, stormwater runoff, and surface water: Application as chemical markers for diffuse sources. *Sci. Total Environ.* 676: 252–267.

Trenberth, K.E., (2012). Changes in precipitation with climate change. Clim. Res. 47 (1–2), 123–138. https://doi.org/10.3354/cr00953.

Trenberth, K.E., Dai, A., van der Schrier, G., Jones, P.D., Barichivich, J., Briffa, K.R., et al., (2014). Global warming and changes in drought. Nat. Clim. Chang. 4 (1), 17–22. https://doi.org/10.1038/nclimate2067.

Tue, N. M., Takahashi, S., Suzuki, G., Isobe, T., Viet, P. H., Kobara, Y., et al. (2013). Contamination of indoor dust and air by polychlorinated biphenyls and brominated flame retardants and relevance of non-dietary exposure in Vietnamese informal e-waste recycling sites. *Environ. Int.* 51:160–167. https://doi.org/10.1016/j.envint.2012.11.006

Vale, F., Sousa, CA., Sousa, H., Santos, L., Simões, M. (2022). Parabens as emerging contaminants: Environmental persistence, current practices and treatment processes. *Journal of Cleaner Production.* 347: 131244. https://doi.org/10.1016/j.jclepro.2022.131244.

Varma, K.S., Tayade, R.J., Shah, K.J., Joshi, P.A., Shukla, A.D., Gandhi, V.G. (2020). Photocatalytic degradation of pharmaceutical and pesticide compounds (PPCs) using doped TiO2 nanomaterials: A review. *Water-Energy Nexus.* 3:46–61.

Vieira, L.R., Soares, A.M.V.M., Freitas, R. (2022). Caffeine as a contaminant of concern: A review on concentrations and impacts in marine coastal systems. *Chemosphere.* 286 (2):131675. https://doi.org/10.1016/j.chemosphere.2021.131675.

Walther GR, Post E, Convey P, Menzel A, Parmesan C, Beebee TJC, et al. (2001). Ecological responses to recent climate change. *Nature.* 294:389–95.

Wang, J., Wang, S. (2016). Removal of pharmaceuticals and personal care products (PPCPs) from wastewater: a review. *J. Environ. Manage.* 182: 620–640.

Wang, L., Xiao, H., He, N., Sun, D., Duan, S. (2019). Biosorption and Biodegradation of the Environmental Hormone Nonylphenol By Four Marine Microalgae. *Sci. Rep.* 9: 5277.

Weeb, B., Hannah, D. M., Moore, R. D., Brown, L. E., Nobilis, F. (2008). Recent Advances in Stream and River Temperature Research. Hydrological Processes. 22 (7): 902–918. https://doi.org/10.1002/hyp.6994.

Woodward, G., Perkins, D.M., Brown, L.E. (2010). Climate change and freshwater ecosystems: impacts across multiple levels of organization. *Philosophical transactions - Royal Society. Biological sciences* 365: 1549.

White, R., Jobling, S., Hoare, SA., Sumpter, JP., Parker, MG. (1994). Environmentally persistent alkylphenolic compounds are estrogenic. *Endocrinology* 135: 175–182.

Williams, M., Kookana, R.S., Mehta, A., Yadav, S.K., Tailor, B.L. and Maheshwari, B. (2019). Emerging contaminants in a river receiving untreated wastewater from an Indian urban centre. *Science of the Total Environment.* 647: 1256–1265.

World Health Organization, (2023). Dioxins. World Health Organization. https://www.who.int/news-room/fact-sheets/detail/dioxins-and-their-effects-on-human-health

Wu, P., Huang, J., Zheng, Y., Yang, Y., Zhang, Y., He, F., Chen, H., Quan, G., Yan, J., Li, T., Gao, B. (2019). Environmental occurrences, fate, and impacts of microplastics, *Ecotoxicol. Environ. Saf.* 184: 109612.

Xiong, J.-Q., Kurade, M.B., Jeon, B.-H. (2018). Can Microalgae Remove Pharmaceutical Contaminants from Water? Trends Biotechnol. 36:30–44.

Xu, B., Liu, S., Zhou, J.L., Zheng, C., Weifeng, J., Chen, B., Zhang, T., Qiu, W. (2021). PFAS and their substitutes in groundwater: Occurrence, transformation and remediation. *J. Hazard. Mater.* 412: 125159.

Yang, S., Hai, F.I., Price, W.E., McDonald, J., Khan, S.J., Nghiem, L.D. (2016). Occurrence of trace organic contaminants in wastewater sludge and their removals by anaerobic digestion. *Bioresour. Technol.* 210: 153–159.

Yi, X., Lin, C., Ong, E. J. L., Wang, M., Zhou, Z. J. C. (2019). Occurrence and distribution of trace levels of antibiotics in surface waters and soils driven by non-point source pollution and anthropogenic pressure. *Chemosphere.* 216:213–223. https://doi.org/10.1016/j.chemosphere.2018.10.087.

Challenges of Emerging Pollutants for the Ecological Receptors in the Twenty-First Century

Chittaranjan Das

Abstract Emerging pollutants (EPs) are present in both mobile and persistent forms across various environmental matrices, affecting ecological receptors such as, soil, water, sediments, and air, even at low concentrations originating from various human activities. Micro contaminants enter the environment from various human-made sources, including pesticides, cosmetics, personal and household care items, pharmaceuticals, industrial chemicals, surfactants, plastic products, and nanomaterials. These substances are widely distributed globally and are essential for contemporary society. The precise aims of this review include forecasting the assessment of risks associated with pollutants in the terrestrial, aquatic, and atmospheric environments while considering the impacts of severe weather events driven by climate change. Recent progress has been made in investigating emerging pollutants, their sources, behavior, and movement through the environment, as well as the effects of climate change, biodiversity, and human responses on the biosphere. Certain micropollutants, although not regularly monitored, hold the potential to infiltrate aquatic ecosystems and can negatively impact both ecological systems and human health. The ecotoxicological implications of specific emerging micropollutants are still largely unclear, as there is often inadequate data to assess their associated risks. The article explores the ecological risks associated with the presence and movement of emerging pollutants within ecosystems. Therefore, to address the ecological risks associated with EPs, more research is needed to identify science-based solutions that can mitigate chemical hazards to both living beings and humans.

Keywords Aquatic pollutants · Chemical compounds · Emerging contaminants · Ecosystem health · Microplastics pollution · Personal care products

---

C. Das (✉)
Department of Botany, Kulti College, Kulti, Paschim Bardhaman, West Bengal, India

S. Roy, V. Mandal (eds.), *Plant-Microbe Interaction under Xenobiotic Exposure*,
https://doi.org/10.1007/978-981-96-8260-7_3

1 Introduction

Emerging pollutants contaminants include both natural and synthetic chemicals, as well as microorganisms of anthropogenic origin, which are not commonly monitored and pose detrimental effects on the environment and human health (Philip et al. 2018). Emerging pollutants (EPs) are diverse synthetic chemicals, including pesticides, cosmetics, personal care products, and pharmaceuticals, which are utilized globally and essential for contemporary society (Thomaidis et al. 2012). These pollutants are introduced into the environment through various human activities and distributed across multiple environmental matrices. However, they have the potential to enter ecosystems and may lead to recognized or suspected negative impacts on both ecological systems and the health of human beings (Adeel et al. 2017; Rudd et al. 2014). The origins and routes of these contaminants are increasingly linked with wastewater and waste generated by industrial, agricultural, or municipal operations (Ramírez-Malule et al. 2020). Chemical micropollutants frequently arise from the breakdown of organic substances, resulting in the accumulation of persistent metabolites, or they may originate from the discharge of products, such as pharmaceuticals into the ecosystem (Sørensen et al. 2007). Emerging contaminants (ECs) comprise a diverse range of chemical substances from various human activities, including domestic, healthcare, agricultural, and industrial practices within the environment (Martín-Pozo et al. 2019). These contaminants include pharmaceuticals (both animal and people medications), hormones and steroids, disinfection byproducts, self-care and sanitation products, household items, surfactants, flame retardants, agrochemicals (such as growth enhancers, fertilizers, and pesticides), industrial chemicals, trace elements, microplastics, and nanoparticles (NPs), among various others (Taheran et al. 2018; Wilkinson et al. 2017; Lodeiro et al. 2019). Each of these substances is a potentially hazardous material that can threaten environmental resources, interfere with biochemical processes in the ecosystem, and adversely affect the health of living organisms. This includes endocrine disruption, mutagenesis, congenital disorders, and carcinogenesis (Barroso et al. 2019). Recently, a variety of micropollutants, referred to as ECs, has been detected globally in various types of water, such as surface, drinking, ground, and wastewater (Teijon et al. 2010; Valcarcel et al. 2011; Cabeza et al. 2012; Meffe et al. 2014; Gao et al. 2014). Additionally, microbial contaminants, including antibiotic-resistant bacteria and their associated genes, are recognized as emerging contaminants (Czekalski et al. 2014; Sanderson et al. 2016; Glassmeyer et al. 2017). Previous research has highlighted that primary risks associated with pharmaceuticals within the environment include their potential for toxicity of acute and chronic to living organisms, development of antibiotic-resistant microorganisms, and disruption of endocrine functions in both humans as well as animals (Felis et al. 2019; Li et al. 2020). An important concern is the presence of metabolites from pharmaceuticals, pesticides, and veterinary medications that an reach higher levels in water compare to their original compounds. These metabolites exist consistently identified across various aquatic environments (la Farre et al. 2008). Specific metabolites exhibit more excellent stability and chemical reactivity than their parent substances; consequently, they may increase toxicity and lead to lethal consequences for living organisms. For example, some steroidal medications are linked to structural changes in various aquatic species (Bottoni et al. 2010). Therefore, this article aims to provide a comprehensive review on the occurrence of ECs in various sources, along with their potential hazards related to toxic loads affecting aquatic ecosystems and the environment.

2 Methods

The present study accessed databases, such as Science Direct, PubMed, and Google Scholar, to find pertinent articles published in English. The keywords are associated with emerging contaminants and pollution in different environmental metrics. The approach used to find and choose relevant manuscripts for this review included pulling together 202 manuscripts from three designated databases. A comprehensive review was carried out, analyzing 114 articles after excluding 125 articles due to irrelevant titles and abstracts, along with 9 duplicate articles. Subsequently, 12 unrelated articles were eliminated, leading to a total 102 articles deemed relevant to the focus of this study, which were ultimately included in the review. The article is based on previously published literature related to issues concerning emerging contaminants/pollution. A number of studies were part of the review that explored the impacts of emerging contaminants on various ecological receptors.

3 Pathways for Environmental Release of ECs

Emerging contaminants are present in both rural and urban areas, originating from two main sources such as, point and non-point soutces. Point source releases originate from treatment plants of wastewater, which include effluents related to homes, industries, and healthcare facilities. On the other hand, non-point sources arise from stormwater runoff related to agricultural practices, such as livestock and aquaculture, as well as from urban environments (Parida et al. 2021; du Plessis 2022; Morin-Crini et al. 2022; Aguilar-Aguilar et al. 2023). Conventional methods for treating wastewater in facilities were not specifically engineered to eliminate these contaminants efficiently. Consequently, numerous studies have documented the presence of these substances in treated effluents, with concentrations generally falling within ng/L and µg/L (Tang et al. 2018; Subedi et al. 2015). Effectively handling wastewater that contains various chemical pollutants and harmful microorganisms poses a significant challenge in environmental engineering, particularly in countries with lower and middle incomes. Besides the discharge of effluents from wastewater treatment facilities, agricultural practices and landfills leachate, where household waste is kept, serve as a substantial source of new pollutants impacting land ecosystems (Qi et al. 2018; Rogers et al. 2021; Qian et al. 2024). The construction sector poses considerable environmental challenges due to the production of numerous pollutants, such as building and tearing down debris, dust, fly ash, and plastic waste, throughout the building process (Hong et al. 2021; Zhong et al. 2022). These contaminants may adversely impact the quality of life of local residents and construction personnel (Kang et al. 2021). Moreover, tiny pollutants, such as ultrafine particles, micro and nano-size plastics, and engineered nanomaterials, can enter the air through various processes, such as volatilization, aerosol creation, and diffusion. Airborne pollutants have the ability to move both nearby and distant locations through processes like wet or dry precipitation, as well as during windy weather (Barroso et al. 2019; Yang et al. 2021). These pollutants are significantly influenced by various weather conditions, including rainfall, temperature, wind speed, and the mixing of air in the lower atmosphere (Sridharan et al. 2021). The environmental matrices, mainly consisting of outdoor air, atmospheric deposits, and dust from urban areas, have become a pathway for airborne pollutants to disperse and source of new contaminants impacting both terrestrial and aquatic ecosystems (Yang et al. 2021; Mbachu et al. 2020). Environmental contaminants can infiltrate the ecosystem via numerous channels, such as mining, agriculture, construction, industrial operations, chemical spills, and oil leaks. They also stem from diffuse sources, such as stormwater drainage systems, roadways, parking lots, and wastewater treatment plants (Pal et al. 2010; Tong et al. 2022). Contaminants present in soil or landfills have the potential to seep into adjacent underground water sources (Gogoi et al. 2018; Pradhan et al. 2023). Furthermore, river systems and wind can carry these pollutants from urban, agricultural, and industrial areas to far-off regions, eventually impacting marine ecosystems (Tong et al. 2022). Fig. 1 illustrates the origins of various anthropogenic contaminants and pollutants within environmental matrices

4 Wastewater Contaminants

The environment and quality of water resources are negatively impacted by the release of industrial flow due to a lack of more efficient wastewater treatment methods. Refinery effluents exhibit a diverse range of phenolic compounds, organic solvents, and various hydrocarbons, which are influenced by specific configurations and plant operational practices. These substances are highly toxic, chemically stable, and typically recognized as carcinogenic (Diya'uddeen et al. 2011). Electroplating wastewater is a major contributor to heavy metal pollution, containing substances such as cadmium, copper, nickel, and chromium, all of which are toxic and detrimental to living organisms (Wei et al. 2013). Additionally, pulp and paper industries produce effluent that contains chlorophenols and acidic compounds (Khan et al. 2019). Most endocrine disruptors and their byproducts and breakdown products are recognized as non-biodegradable, persisting in water even after

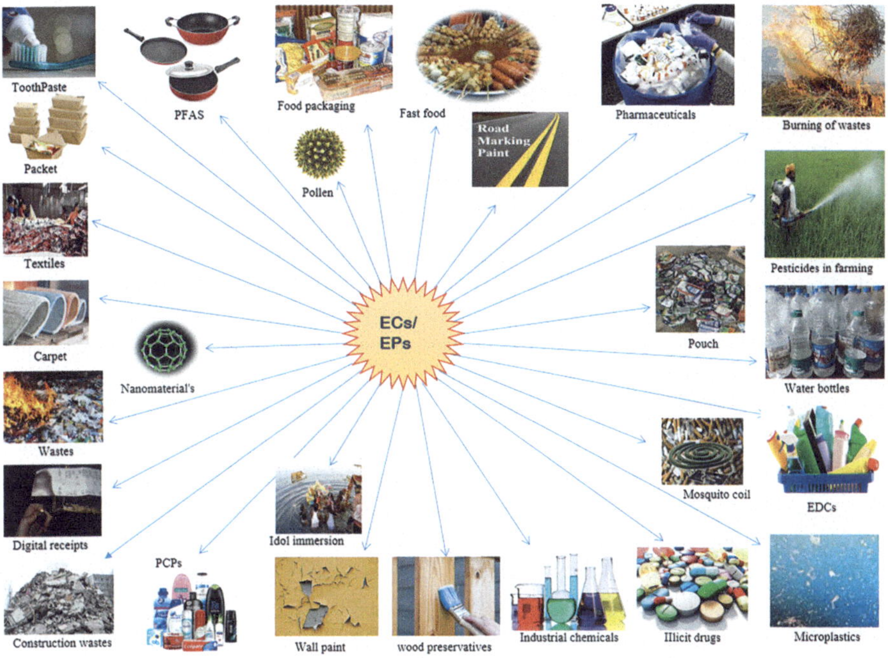

Fig. 1 Sources of emerging contaminants (ECs)/pollutants (EPs) from different receptors (Li et al. 2013; Lou et al. 2014; Gavrilescu et al. 2015; Montes-Grajales et al. 2017; Philip et al. 2018; Naeem et al. 2023)

standard treatment methods are applied. This persistence leads to their bioaccumulation within the ambience, ultimately allowing them to enter the food web (Rodriguez-Narvaez et al. 2017). Moreover, tannery industry is one of the most environmentally damaging sectors, primarily due to the diverse range of chemicals utilized to transform animal hides into leather. Substances, such as chromium salts, phenolics, tannins, and organic matter, are routinely discharged into the ambience through wastewater from the industry tannery. These contaminants pose significant ambience hazards to both water-based ecosystems and human beings of human (Chandra et al. 2011). Regarding the harmful impacts of surfactants on living organisms, several references highlighted issues, such as the gradual breakdown of membrane systems, which disrupts the metabolism of energy and oxygen and nutrient transport in various species. Additionally, these substances can weaken protective structures in organisms and lead to various physiological changes. Furthermore, surfactants have the potential to decrease photosynthesis and chlorophyll levels in plants, as well as affecting various bacterial species (Rebello et al. 2014).

5 Pharmaceutical Compounds

Pharmaceutical compounds are important for maintaining human health, but their presence in the environment can lead to substantial environmental and human health risks. The levels of persistent organic contaminants, including pharmaceuticals and their metabolites, steadily increase within the natural ambience due to human actions (Stroski et al. 2020). The most commonly used pharmaceuticals consist of pain relievers, anti-inflammatory drugs, and synthetic antibiotics. Notable examples include β-lactams like penicillin and amoxicillin, heart medications, such as β-blockers and diuretics, various estrogens and hormonal agents such as estrone, estradiol, and estriol, 17α-ethinyl estradiol, as well as antiepileptic medications such as carbamazepine (Li 2014). Pharmaceutical compounds, including antibiotics, analgesics, and anti-inflammatory medications, are typically administered orally through feed or injection. These substances often enter aquatic environments through household waste, solid waste disposal, or toilets (Stefanakis and Becker 2015). Additionally, biological and toxicological impacts of these emerging pollutants can contribute to increased bacterial resistance to antibiotics and reduced fertility, reproductive issues, and declines in marine populations (Samuelsen et al. 2014). Furthermore, regarding chronic toxicity, the majority of specific pharmaceuticals, including antibiotics, antibacterial agents, anti-inflammatory, and non-steroidal anti-inflammatory medications, can harm the reproductive health of fish and crustaceans even at low or environmentally relevant amounts (ng L^{-1}/μg L^{-1}) (Overturf et al. 2015; Santos et al. 2010). For example, erythromycin is an antibiotic that is part of a class of medications used to combat infections in people. It works by attaching to the 50S subunit of bacteria, which prevents them from producing proteins (Liu et al. 2014).

6 Pesticides

Pestcide polution affects the environment and human health by contaminating soil, water and air which contributes to biodiversity loss and health issues ranging from acute to chronic diseases. The chemicals used to make pesticides typically fall into categories, such as organophosphates, carbamates, organochlorides, pyrethroids, and others (Rasheed et al. 2021). Some of these, such as parathion, atrazine, chlordane, dichlorodiphenyldichloroethylene, malathion, and DDT, are recognized as being detrimental to both people and the environment. Among the various types of pesticides, organochlorides pose the most significant risk and directly harm people's well-being (Rani et al. 2017). These substances contribute to the decline of biodiversity, destroy bird habitats, harm pollinating insects, and represent a significant danger to species at risk of extinction (Intisar et al. 2022). Pesticide transfer to aquatic environments occurs through runoff from the application site. The solubility of these compounds determines their capacity to bioaccumulate in soil, sediments, and living organisms (Gonzalez et al. 2012). Numerous pesticides have been linked to various health and environmental concerns (Zheng et al. 2016; Goulson 2014). Chemical pesticides are associated with numerous harmful health issues, which can affect the skin, digestive system, and nervous system, and even increase the risk of cancer. They can also impact respiratory health, reproductive functions, and hormonal balance (Séralini et al. 2014; Thakur et al. 2014; Mostafalou and Abdollahi 2013). Moreover, significant exposure to pesticides through occupational, accidental, or deliberate means, including hospitalization and even fatalities (Gunnell et al. 2007). Agricultural practices today largely depend on the widespread use of chemical pesticides, which have been associated with negative impacts on people's well-being, wildlife, and the environment (Goulson 2014). Some pesticides that are limited in use are super toxic and can release cancer-causing substances that accumulate in non-target species, such as humans, aquatic environments, algae, and microbes (Bilal et al. 2019).

7 Microplastics

Microplastic are associated with considerable risks to the environment and human health by potentially altering ecosystems and affecting the quality of soil and water. These substances enter the human body through the consumption of food and water as well as inhalation from the air and may accumulate in various tissues and organs. Microplastics are vehicles for various pollutants, including PAHs, PCBs, and heavy metals (Barletta et al. 2019). These microplastics can release these harmful substances from their surfaces, leading to chemical pollution in both water and sediment (Ahmed et al. 2021). Additionally, microplastics can negatively affect the physiological functions of aquatic plants, interfering with vital processes, such as photosynthesis and nutrient uptake. This disruption may lead to slower growth rates, altered reproductive behaviors, and a decline in the overall health of aquatic vegetation (Ge et al. 2021; Ceschin et al. 2023). Moreover, aquatic organisms often mistake microplastics for food, leading to their ingestion throughout various levels of the food chain (Egbeocha et al. 2018). This ingestion can cause physical harm, including internal injuries, obstructions, and disturbances in digestive processes. The accumulation of microplastics in the digestive tracts of aquatic organisms can lead to malnutrition and reduced energy levels (Rakib et al. 2023; Alak et al. 2022; Harmon et al. 2024). The harmful effects of pollutants linked to microplastics encompass hormonal disruptions, weakened immune responses, and an increased vulnerability to numerous diseases. Such effects can profoundly affect the health and lifespan of aquatic life (Sreelakshmi and Chitra 2021). Previous research has shown that exposure to microplastics may negatively impact male fertility and diminish sperm quality, which could complicate the chances of conception (D'Angelo and Meccariello 2021). Furthermore, microplastics can trigger inflammation within the body, which could result in various health complications, such as cardiovascular diseases, cancer, and autoimmune disorders (Sharma et al. 2020;

Amato-Lourenço et al. 2021; Ghosh et al. 2023). Microplastics can also cause oxidative stress, which is harmful and may lead to damage to cells and DNA. This kind of damage is linked to a range of health problems, including neurodegenerative disorders and reproduction issues (Wang et al. 2023). Hence, microplastics are emerging major threat to public health and the environment, posing risks to both human well-being and ecosystems.

8 Nanomaterials

In recent years, nanoparticles have growing interest for their diverse applications across various sectors, including consumer products, healthcare, soil management, and aquatic ecosystems. They have proven useful in areas such as textiles (Som et al. 2011), water purification (Bottero et al. 2006), environmental remediation (Pak et al. 2020), cancer therapies (Lammers et al. 2008), radiological applications (Gupta and Gupta 2006), and cosmetic products (Botta et al. 2011). Nanoparticles are increasingly recognized and utilized due to their distinctive properties arising from their small size and extensive surface area (Auffan et al. 2010). While these remarkable traits provide various advantages, they also pose particular risks to living organisms. Their tiny dimensions allow nanoparticles to easily traverse biological barriers, potentially jeopardizing health. These particles can enter the human body via the lungs, digestive system, or skin, potentially resulting in issues, such as brain toxicity, lung inflammation, and cardiovascular complications (Oberdorster et al. 2005). Research indicates that certain nanoparticles can lead to lasting cellular damage due to organ impairment and oxidative stress associated with their size and composition. Magrez et al. (2006) discovered that the size of carbon-based nanoparticles significantly influences their toxic effects on lung cancer cells. Various factors, including composition, surface characteristics, size, crystallinity, and aggregation, are believed to impact the toxicity of nanoparticles (Mancuso and Cao 2014). Additionally, nanoparticles (NPs) can be produced from numerous human activities, including cooking, vehicle emissions, power generation, aircraft engines, chemical manufacturing, ore processing, smelting, and welding (Jeevanandam et al. 2018). Nanoparticles can enter the environment through three primary pathways: (1) during the production of raw materials and nano-enhanced products, (2) while these products are in use, and (3) after the disposal of products containing nanoparticles during waste management (Gottschalk et al. 2013).

9 Fast Food Additives

The consumption of fast food has adverse effects on human health. Recent studies indicate that the consumption of processed and fast foods could lead to more premature deaths compared to cigarette smoking (Murray et al. 2013). Individuals from economically disadvantaged backgrounds in these regions face a twice-fold rise in heart attacks, a similar doubling in the risk of diabetes, and a four-fold increase in the likelihood of renal failure (Fuchs 2011; Mozaffarian et al. 2016). Additionally, refined carbohydrates are linked not just to obesity and diabetes; they might also contribute to the onset of dementia, mental health issues, and some forms of cancer (Sánchez-Villegas et al. 2011; Gangwisch et al. 2015). Regularly eating fast food has been linked to weight gain and an increase in abdominal fat. It may also cause problems with insulin and glucose regulation, disrupt lipid and lipoprotein levels, and trigger systemic inflammation and oxidative stress (Bahadoran et al. 2016). Additionally, consuming more fast food prior to pregnancy is connected to a higher risk of developing diabetes during pregnancy and an increased likelihood of having a high birth weight for first-time mothers (Wen et al. 2013). Fuhrman (2018) indicated that the consumption of unhealthy fast and processed foods can increase the risk of experiencing a stroke in early life by seven times (Fuhrman 2018). Furthermore, the increased consumption of fast food or junk food is associated with depressive symptoms, heighten the risk of psychiatric problems, and aggressive behaviors among children and adolescents (Zahedi et al. 2014; Hemmingsson 2018). Conversely, nitrates and nitrites, which are prevalent in processed meats, are considered potential carcinogens associated with various cancers, including breast, prostate, pancreatic, and colorectal cancers. Additionally, fast food are associated with medical conditions, like non-alcoholic fatty liver disease and insulin resistance (Quist et al. 2018; Perdomo et al. 2019; Ma et al. 2018). Therefore, regular consumption of fast food may result in a lack of essential nutrients whch could lead to nutrient deficiences.

10 Self-Care Consumer Products

Consumer products often contain chemicals that pose unique challenges in contrast to pollutants emerge from industrial sources. The presence of chemicals in consumer products has the potential to impact a vast number of individuals within a brief timeframe. The chemicals found in everyday products are constantly in contact with consumers (Csiszar et al. 2016). These substances can enter the human body through phthalates (Philippat et al. 2015), benzophenone-3 (Krause et al. 2012), bisphenol-A (Bemrah et al. 2014), and polybrominated diphenyl ethers (Gump et al. 2014). The variety of chemicals in consumer products raises concerns regarding the simultaneous exposure to multiple substances found in different items (Kortenkamp and Faust 2018). Self-care products enter our water systems through sewage from showers, baths, and laundry. Additionally, these substances can also be released straight into surface waters while people are enjoying activities, such as sunbathing and swimming. Worries about their harmful impacts could involve possible estrogen-like effects (such as ultraviolet filters and parabens), developmental toxicity linked to UV blockers, and a notable buildup in living organisms (Houtman 2010; Richardson 2009). The same chemical can present varying levels of risk based on its application in a product and how people engage with it (Tao et al. 2018). For instance, Phthalates can be present in shampoos, food storage containers, and cell phones. Furthermore, exposure to phthalates and the associated health risks can differ on the specific product (Hartle et al. 2016; Bao et al. 2015).

10.1 Endocrine-Disrupting Chemicals (EDCs) in Cosmetics

Endocrine-disrupting chemicals (EDCs), such as bisphenol-A and phthalates present in plastics, are compounds that can imitate or interfere with the action of hormones in the endocrine system, acting as either stimulators or blockers. These substances primarily impact the women's reproductive system, which may increase the likelihood of various reproductive issues, including difficulties with fertility, growth anomalies, and hormone-related cancers, such as breast carcinoma (Matuszczak et al. 2019; Laws et al. 2021). In addition, EDCs interfere with the hormonal systems of water-dwelling creatures, leading to issues in their reproductive, developmental, and physiological processes. Recent research has highlighted the widespread occurrence of endocrine-disrupting chemicals (EDCs) in aquatic environments, indicating their possible negative impact on the health of fish and amphibians (Carnevali et al. 2018; You and Song 2021; Celino-Brady et al. 2020; Langston 2020). Hence, EDCs have the potential to affect human health by interfering with the body's hormonal system which is present in numerous consumer products and may enter the environment through air and water.

10.2 Triclosan (TCS)

The widespread use of Triclosan and other antiseptic substances contributes to pollution in aquatic ecosystems. This compound is frequently present in numerous consumer items (Weatherly and Gosse 2017), such as toothpaste, hand soaps, shampoos, and cosmetics. Moreover, it is also used in healthcare environments, where it is included in antiseptics, disinfectants, and various medical devices (Montagnini et al. 2021). Extensive studies show that Triclosan (TCS), a widely used phenolic compound, exhibits endocrine-disrupting effects and traits associated with carcinogenic potential (Carey et al. 2016; Liu et al. 2022). Recent studies indicate that exposure to TCS could be associated with a range of negative health effects, such as liver and kidney toxicity (Ena et al. 2018; Tang et al. 2018) and disruption of thyroid function (Milanović et al. 2023). Moreover, exposure to low concentrations of TCS negatively impacts the DNA and increases the incidence of morphological disorders in sperm (Jurewicz et al. 2018).

10.3 Paraben

Parabens are commonly found in pharmaceuticals, cosmetics, and food as preservatives. However, they can disrupt the immune system and have been linked to reproductive issues and neurotoxicity. Additionally, they may mimic estrogen, which raises concerns about their potential connection to breast cancer. Their widespread use in cosmetics is attributed to their neutral impact on various cosmetic attributes, including taste, scent, color, and texture (Petric et al. 2021; Fransway et al. 2019). Handa et al. (2006) explored how parabens affect the skin and their possible connection to skin cancer. Their research showed that when keratinocytes are exposed to ultraviolet B (UVB) radiation along with methylparaben, it leads to oxidative stress and lipid oxidation. Handa et al. (2006) explored how parabens affect the skin and their possible link to skin cancer. Their research showed that skin exposure to ultraviolet B (UVB) rays and methylparaben causes lipid oxidation and oxidative stress in keratinocytes. This chain reaction can damage the skin, including the DNA of skin cells, potentially increasing cancer risk. There are worries about these factors impacting endocrine function, cancer development, infertility, sperm production, fat formation, exposure during pregnancy, and non-allergic skin issues, along with their psychological and environmental effects (Fransway et al. 2019).

10.4 Phthalate

Phthalate acid esters (PAEs) are extensively utilized in workplace environments, primarily serving as plasticizers across various applications (Dutta et al. 2020). Certain phthalates are incorporated into polyvinyl chloride (PVC) plastics, which are used to produce various items, including medical tubes, plastic wraps, and garden pipes (Przybylińska and Wyszkowski 2016). Clinical research has identified a variety of abnormalities, including obesity, endometriosis, early onset puberty, asthma, and overweight conditions (Durmaz et al. 2018). Moreover, phthalates can affect testicular cells, which might result in alterations to the external genitalia and a decrease in the organ's weight (Hlisníková et al. 2020). Numerous studies investigating the impact of phthalate toxicity on children have mainly focused on several factors, such as pregnancy outcomes, genital development, semen quality, early onset of puberty, thyroid function, respiratory problems, and neurodevelopment (Eales et al. 2022).

10.5 Black Carbon(BC)

Black carbon (BC), commonly referred to as called charcoal, is a dark material made from wood using dry distillation. It is present as a contaminant in cosmetics, which can adversely affect both the environment and human well-being. This material is processed into a fine, lightweight powder that is a pigment in cosmetic products, providing a rich dark or black hue (IARC 2010). Breathing in black carbon regularly can reduce lung function in people (Wellmann et al. 2006). Research in epidemiology focusing on people exposed to black carbon through inhalation and skin contact has shown an increased prevalence of serious skin keratosis and leukoplakia. Certain conditions are early warning signs that could raise the chances of developing cancerous lesions (Boland et al. 2014; Yong et al. 2019).

10.6 Silica (SiO_2)

Silica is used in cosmetics or for making colloidal substances and gels and can be found in both amorphous and crystalline forms. The crystalline, non-hydrated form often gets much attention regarding possible health risks. Research has demonstrated that prolonged contaminant dust particles of crystalline silica suspended in the air can lead to blood vessel diseases, lung diseases, such as silicosis, and pneumoconiosis due to extensive fibrosis. These findings were derived from an epidemiological study focusing on workers who were occupationally exposed to silica dust (Brown and Rushton 2005). Furthermore, the extended existence of silica within the lungs leads to the stimulation of macrophages. This extended exposure leads to the secretion of pro-inflammatory cytokines, resulting in inflammation. Additionally, it stimulates neutrophils to generate free radicals, ultimately resulting in genotoxicity and the development of neoplastic lesions (Lin et al. 2006).

11 Heavy Metals

Heavy metals potentially risks to the environment by contaminating soil, water and ecosystem, resulting in health issues and ecological problems. These metals can interfere with biological systems, create oxidative stress and potentially cause cancer. There have been recorded cases of allergic dermatitis and systemic reactions attributed to components present in cosmetic products (Abalaka et al. 2020). Studies indicate that beauty products may be contaminated with heavy metals, such as arsenic, cadmium, lead, and mercury (Arshad et al. 2020; Saadatzadeh et al. 2019; Lim et al. 2018). The most detrimental and undesirable metal contaminants identified include arsenic, cadmium, lead, and mercury (Chen et al. 2019; Borowska and Brzóska 2015). Exposure to these substances poses significant risks to children's health, potentially resulting in problems, such as intellectual disabilities, neurocognitive disorders, behavioral issues, respiratory complications, cancer, and heart disease (Al Osman et al. 2019). Lipsticks are widely used cosmetics worldwide, raising concerns regarding the possible presence of heavy metal contaminants due to skin contact and unintentional ingestion (Attard and Attard 2022; Khodabakhshi and Bagherzadeh 2024). Prolonged exposure to these substances may result in significant health risks, including nerve damage, an increased likelihood of cancer, and potential harm to various organs (Borowska and Brzóska 2015).

11.1 Arsenic (As)

Arsenic pollution prevalent, stemming from natural and human-made sources that contribute to its occurence in water, soil and food. Its toxicity interfere with cellular functions, resulting in a range of health problems. Previous studies have identified a link between arsenic exposure and the development of various health problems. Studies indicated that arsenic trends to accumulating in keratin-rich tissues, such as the skin, hair, and nails. Earlier studies has identified a link between arsenic contaminants and the onset of diseases that may lead to cancer, including skin cancers, such as Bowen's disease as well as carcinoma of basal cells (Sawada 2018). The carcinogenic mechanism of arsenic is likely linked to the production of reactive oxygen species (ROS) inside the cell, which results in oxidative stress. Furthermore, oxidative stress can damage the DNA, leading to genetic mutations that could eventually result in cancer (Martinez et al. 2011). In addition, arsenic is currently identified as a cancer-causing substance linked to breast, lung, and several other forms of cancer (Khanjani et al. 2017; Gamboa-Loira et al. 2017).

11.2 Cadmium (Cd)

Cadmium containants presents substantial risks to the environment, as it is toxic heavy metal capable of accumulating in the food chain, which may result in multiple health issues. Cadmium and its derivatives mainly get into the body through breathing and are absorbed via the skin. Once these substances get into the body, they are absorbed into the bloodstream and predoninantly transporated to the organs of the liver and kidneys. Symptoms associated with poisoning may encompass fever, pneumonia, breathing failure, yellowing of teeth, as well as dysfunction of the liver and kidneys (Borowska and Brzóska 2015). Additionally, there is a documented association between cadmium contaminant and an increased incidence of cancers, such as lung, kidney, and prostate (Il'yasova and Schwartz 2005). Cadmium is associated with the development of cancers in the stomach, breast, and other internal organs (O'Reilly and Buchanan 2019; So et al. 2019). Moreover, cadmium-induced hypomethylation of genes leads to an overproduction of factors that promote cell growth, which can play a role in developing cancerous tumors (Venza et al. 2014).

11.3 Lead(Pb)

Lead pollution present serious threat to the environment and human well-being. It is persistant contaminants in the environment and has the potential to buildup in soil, water and living beings. Moreover, lead is non-biodegrable and can persist in the environment for long durations, creating cntinuous risks. Additionally, lead is a harmful material are present in some cosmetic products, capable of infiltrate the skin. Once it enters the bloodstream, it has the potential to accumulate in different organs and tissues over time. Lead poisoning presents as a neurological disorder characterized by challenges in concentration, slower reaction times, and headaches. It is a fundamental ingredient in numerous lipsticks, resulting in significant absorption caused by inadequate consumer practices and associated with anemia and abdominal discomfort (Borowska and Brzóska 2015). For instance, exposure to lead has been associated with cognitive difficulties, particularly affecting both kids and pregnant women (Karri et al. 2016).

11.4 Mercury (Hg)

Mercury pollution present serious risks to the environment and human health. It is a long lasting, bioaccumulative and hazardous pollutant that can buildup in the food chain. Mercury can enter the environment from multiple sources, such as industrial activities, mining operation, and the cobustion of fossil fuels, leading to contamination of water and soil. Mercury and its compounds can be absorbed through the skin surface and its related components, such as hair follicles as well as sweat glands and sebaceous glands. Epidemiological research indicates a higher mortality rate among individuals who have been exposed to mercury vapor. Mercury's pro-oxidant properties are associated with causing DNA damage and increasing the production of pro-inflammatory cytokines (Skalny et al. 2022). The precise mechanism through which mercury may potentially induce carcinogenic effects is still not fully understood.

12 Industrial Chemicals

Industrial chemical contaminants are substantial concerns for environmental as wel as human health pespectives. These substances emitted from different industrial activities, can result in extensive contamination of air, water and soil, leadind to various health issues. Different industrial chemicals, such as alkylphenols, alkylphenol ethoxylates, and bisphenol-A are used as surfactants, plasticizers, and to make polycarbonate plastic. These compounds can lead to oxidative stress in marine organisms at different levels. Research shows that these chemicals have structures similar to hormones, which may lead to similar effects in the body, although they usually require much higher amounts to have an impact (Houtman 2010). Many new compounds, such as insect repellent and DEET, could potentially threaten both living organisms and human well-being (Corbel et al. 2009). These substances are usually present in minimal quantities in the ambience, but they could potentially harm ecosystems as well as the food web (Eggen et al. 2010). Moreover, many ECs are derived from synthetic chemicals that play a key role in today's society. Some examples are perfluorinated compounds, byproducts from water disinfection, gasoline additives, pharmaceuticals, synthetic nanomaterials, and UV filters (Gavrilescu et al. 2015). Disinfection chemicals utilized in swimming pools and treatment of consuming water are essential in protecting people from waterborne illnesses. The frequent and widespread application of the chemical substances leads to the creation of byproducts of disinfection, chlorinated DBPs (CDBPs), in treated water. Almost everyone in developed areas comes into contact with these substances through drinking water and swimming pools (Richardson and Ternes 2014).

13 Particular Matter (PM)

Particulate matter (PM) has adverse effects on the environment and human human. It may cause airway irritation, resulting in coughing, wheezing and difficulty breathing. Prolong exposure to PM is associated with higher risks of developing chronic respiratory diseases. Earlier studies reported an inverse relationship between particulate matter (PM) inhalation and life expectancy (Dziubanek et al. 2017). PM exerts detrimental effects on living organisms and various vital organs within the body, particularly the lungs (Al-Hamdan et al. 2017). Furthermore, PM2.5 particles, which have a diameter smaller than 2.5 μm, are a well-known air contaminant that can seriously harm Community health (Brauer et al. 2016). Higher levels of PM2.5 exposure have been linked to a greater chance of progressing with atherosclerosis (Woo et al. 2021), various respiratory illnesses (Pun et al. 2017), and cancer of the lung (Yang et al. 2020). The detrimental impacts of PM2.5 are largely due to its small size, which allows it to evade the body's defense mechanisms, penetrate the bronchial passages, and reach the alveoli (Al-Hamdan et al. 2017). Furthermore, its complex makeup often attracts harmful agents, such as polycyclic aromatic hydrocarbons, endotoxins, sulfates, and heavy metals, resulting in toxicity (Dong et al. 2020; Faraji Ghasemi et al. 2020). It is important to highlight that substances, such as black carbon (Bové et al. 2019) and cadmium (Tinkov et al. 2018), are identified as hazardous and significantly affect environmental health.

14 Pollen Grains

Pollen is vital for plant reproduction but can adversely affect the environment and human health if it becomes a contaminant. According to the World Allergy Organization (WAO) pollen is identified as a major contributor to allergic rhinitis and asthma on the global scale (Pawankar et al. 2013). It is estimated that approximately 20–30% of the global population experiences allergic rhinitis, which may manifest through symptoms, such as sneezing, nasal congestion, a runny nose, and irritation of the nasal passages (Pawankar et al. 2013; Ravindra et al. 2021). Pollen grains typically range in size from 10 to 80 micrometers, making them too large to be inhaled and unable to penetrate the deeper areas of the breathing system. In contrast, particles smaller than pollen are significantly reduced in size and carry pollen allergens, allowing access to the lower respiratory pathways, potentially resulting in respiratory illnesses (Jones et al. 2021; Ravindra et al. 2022). Breathing in pollen can cause some people to have an asthma attack, leading to inflammation in the airways and a quick tightening of the bronchial muscles. To manage this airway inflammation, mast cells and basophils release histamines, leading to a secondary asthmatic response that occurs 2 to 24 hours after the initial reaction. This secondary response counteracts the developing pro-inflammatory mediators (Kailaivasan and Davies 2018). Research in aerobiology suggests that an increase in pollinosis cases in the future, attributed to factors, such as climate alteration, urban expansion, industrial growth, and the emission of pollutants (D'Amato et al. 2015; Camacho et al. 2020). Pollen phenology is closely connected to seasons and climatic conditions (Rojo et al. 2015).

15 Biological Contaminants

Biological contaminants such as, microorganisms and toxins can present considerable risks to the environment and human health. Contaminants may enter the food chain which can affect human health. Agricultural soils are frequently overlooked as sources of pathogens that can impact both humans and animals, potentially leading to various diseases that affect our air, water, and food supply (Singh et al. 2023). Some bacteria, such as *Bacillus anthracis*, *Burkholderia pseudomallei*, and *Vibrio cholera*, can lead to serious infections and, in certain situations, even death when there is direct contact (Steffan et al. 2020; Limmathurotsakul et al. 2016). Moreover, pathogens transmitted through food, including *Salmonella enterica* and *Escherichia coli* O157:H7, can infiltrate the food supply, leading to outbreaks that pose significant health risks (Gonzalez-Martin et al. 2014; Scott et al. 2017). Resistance to antimicrobial agents is often described as a quiet pandemic and become a significant issue within living organism-related contaminants (Mah 2021). In addition, different substances besides antibiotics, such as heavy metals, disinfectants, biocides, and non-antibiotic drugs, can change how bacteria act and contribute to the

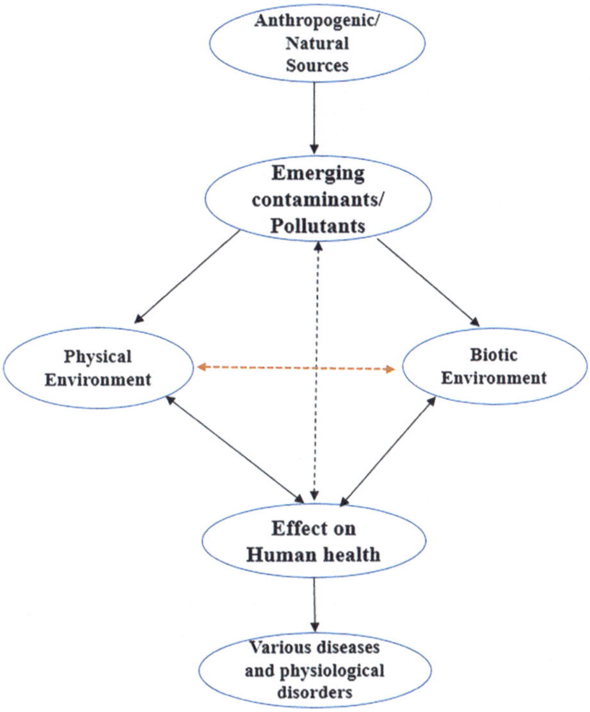

Fig. 2 Schematic pathway of ECs/EPs from source to receptor within the environment and their effect on human health (Lei et al. 2015; Gavrilescu et al. 2015; Stuart et al. 2012)

development of antibiotic resistance (Lu et al. 2018; Pál et al. 2015). The relationship of ECs/EPs from various sources to receptor within the environmental components and their effect on human health are represented by Fig. 2.

16 Conclusion

The article provides a comprehensive overview of recent research regarding the impact of emerging contaminants on various ecological receptors, especially human health and the environment. Environmental contaminants continue to pose onging and substantial challenges to water quality, air purity, soil health, natural resources, ecosystems, and human well-being. The increasing prevalence of ECs generates concerns about their lasting and unpredictable effects on ecosystems, water quality, and environmental health. The present extensive review meticulously analyzed the origins, behavior, transport mechanisms, and ultimate fate of ECs from multiple viewpoints within the environment. The impact of pollutants on the global well-being of our planet encompasses humans and animals, as well as their interconnected ecosystems, all viewed through the perspective of holistic health. Moreover, nanomaterials are a rapidly growing sector for society and potential source of new emerging pollutants. Hence, global attention is needed toward the source, pathways of entrance in the environment, and health-related toxic xenobiotic effects of the ECs. Further investigation is necessary to explore methods for preventiong the production and exposure of ECs compounds. The review identified a significant gap in the in the scientific literature regarding research on the environmental impacts of various nanoparticles. Moreover, future more study should be investigate the impact of emerging contaminants on aquatic, terrestrial, and soil systems. Further research is necessary to develop more effective tehnologies for the degradation and removal of ECs from aquatic systems.

References

Aguilar-Aguilar A, de León-Martínez LD, Forgionny A et al (2023) A systematic review on the current situation of emerging pollutants in Mexico: A perspective on policies, regulation, detection, and elimination in water and wastewater. Sci Total Environ 905:167426. doi:https://doi.org/10.1016/j.scitotenv.2023.167426

Adeel M, Song X, Wang Y, Francis D, Yang Y (2017) Environmental impact of estrogens on human, animal and plant life: A critical review. Environ Int 99:107–119. doi:https://doi.org/10.1016/j.envint.2016.12.010

Ahmed MB, Rahman MS, Alom J et al. (2021) Microplastic particles in the aquatic environment: A systematic review. Sci Total Environ 775:145793. doi:https://doi.org/10.1016/j.scitotenv.2021.145793

Alak G, Uçar A, Parlak V et al. (2022) Identification, characterization of microplastic and their effects on aquatic organisms. Chem Ecol 38:967–987. doi:https://doi.org/10.1080/02757540.2022.2126461

Abalaka SE, Enem SI, Idoko IS, Sani NA, Tenuche OZ, Ejeh SA, Sambo WK (2020) Heavy metals bioaccumulation and health risks with associated histopathological changes in *Clarias gariepinus* from the kado fish market, Abuja, Nigeria. J Health Pollut 10:1–12. doi:https://doi.org/10.5696/2156-9614-10.26.200602

Amato-Lourenço LF et al. (2021) Presence of airborne microplastics in human lung tissue. J Hazard Mater. 416:126124. doi:https://doi.org/10.1016/j.jhazmat.2021.126124

Arshad H, Mehmood MZ, Shah MH, Abbasi AM (2020) Evaluation of heavy metals in cosmetic products and their health risk assessment. Saudi Pharm. J. 28:779–790. doi:https://doi.org/10.1016/j.jsps.2020.05.006

Al Osman M, Yang F, Massey IY (2019) Exposure routes and health effects of heavy metals on children. Biometals 32:563–573. doi:https://doi.org/10.1016/j.jsps.2020.05.006

Auffan M, Bottero JY, Chaneac C, Rose J (2010) Inorganic manufactured nanoparticles: How their physicochemical properties influence their biological effects in aqueous environments. Nanomed Nanotechnol Biol Med 5:999–1007. doi:https://doi.org/10.2217/nnm.10.61

Al-Hamdan AZ, Albashaireh RN, Al-Hamdan MZ, Crosson WL (2017) The association of remotely sensed outdoor fine particulate matter with cancer incidence of respiratory system in the USA. J. Environ Sci Health A Tox Hazard Subst Environ Eng 52(6):547–554. doi:https://doi.org/10.1080/10934529.2017.1284432

Attard T, Attard E (2022) Heavy metals in cosmetics Environ. Impact Remediat, Heavy Met. pp. 1632–1635. doi:https://doi.org/10.5772/intechopen.102406

Bao J, Wang M, Ning X, Zhou Y, He Y, Yang J et al (2015) Phthalate concentrations in personal care products and the cumulative exposure to female adults and infants in Shanghai. J Toxicol Environ Health A 78:325–341. doi:https://doi.org/10.1080/15287394.2014.968696

Brauer M et al (2016) Ambient air pollution exposure estimation for the global burden of disease 2013. Environ Sci Technol 50(1):79–88. doi:https://doi.org/10.1021/acs.est.5b03709

Bové H et al (2019) Ambient black carbon particles reach the fetal side of human placenta. Nat Commun 10(1):3866. doi:https://doi.org/10.1038/s41467-019-11654-3

Bahadoran Z, Mirmiran P, Azizi F (2016) Fast food pattern and cardiometabolic disorders: a review of current studies. Health Promot Perspect 5:231–240. doi:https://doi.org/10.15171/hpp.2015.028

Botta C, Labille J, Auffan M, Borschneck D, Miche H, Cabie M et al. (2011) TiO_2-based nanoparticles released in water from commercialized sunscreens in a life-cycle perspective: structures and quantities. Environ Pollut159:1543–1550. doi:https://doi.org/10.1016/j.envpol.2011.03.003

Bottero JY, Rose J, Wiesner MR. (2006) Nanotechnologies: Tools for sustainability in a new wave of water treatment processes. Integr Environ Assess Manag 2:391–5. doi:https://doi.org/10.1002/ieam.5630020411

Barletta M, Lima ARA, Costa MF (2019) Distribution, sources and consequences of nutrients, persistent organic pollutants, metals and microplastics in South American estuaries. Sci Total Environ 651:1199–1218. doi:https://doi.org/10.1016/j.scitotenv.2018.09.276

Barroso PJ, Santos JL, Martín J, Aparicio I, Alonso E (2019) Emerging contaminants in the atmosphere: analysis, occurrence and future challenges. Crit Rev Environ Sci Technol 49:104–171. doi:https://doi.org/10.1080/10643389.2018.1540761.

Bilal M et al (2019) Persistence of pesticides-based contaminants in the environment and their effective degradation using laccase-assisted biocatalytic systems. Sci. Total Environ 695:133896. doi:https://doi.org/10.1016/j.scitotenv.2019.133896

Bottoni P, Caroli S, Caracciolo AB (2010) Pharmaceuticals as priority water contaminants, Toxicol Environ Chem 92:549–565. doi:https://doi.org/10.1080/02772241003614320

Borowska S, Brzóska MM (2015) Metals in Cosmetics: Implications for Human Health. J Appl Toxicol 35:551–572. doi:https://doi.org/10.1002/jat.3129

Brown TP, Rushton L (2005) Mortality in the UK industrial silica sand industry: 1. assessment of exposure to respirable crystalline silica. Occup Environ Med 62:442–445. doi:https://doi.org/10.1136/oem.2004.017715

Boland S, Hussain S, Baeza-Squiban A (2014) Carbon black and titanium dioxide nanoparticles induce distinct molecular mechanisms of toxicity Wiley Interdiscip. Rev. Nanomed Nanobiotechnol. 6:641–652. doi:https://doi.org/10.1002/wnan.1302

Bemrah N, Jean J, Rivière G, Sanaa M, Leconte S, Bachelot M et al (2014) Assessment of dietary exposure to bisphenol A in the French population with a special focus on risk characterization for pregnant French women. Food Chem Toxicol 72:90–97. doi:https://doi.org/10.1016/j.fct.2014.07.005.

Csiszar SA, Meyer DE, Dionisio KL, Egeghy P, Isaacs KK, Price PS et al (2016) Conceptual framework to extend life cycle assessment using near-field human exposure modeling and high-throughput tools for chemicals. Environ Sci Technol 50:11922–11934. doi:https://doi.org/10.1021/acs.est.6b02277

Carey DE, Zitomer DH, Hristova KR, Kappell AD, McNamara PJ (2016) Triclocarban influences antibiotic resistance and alters anaerobic digester microbial community structure. Environ Sci Technol 50:126–134. doi:https://doi.org/10.1021/acs.est.5b03080

Chen QY, DesMarais T, Costa M (2019) Metals and Mechanisms of Carcinogenesis. Annu. Rev. Pharmacol. Toxicol 59:537–554. doi:https://doi.org/10.1146/annurev-pharmtox-010818-021031

Ceschin S, Mariani F, Di Lernia D et al. (2023) Effects of microplastic contamination on the aquatic plant *Lemna minuta* (least duckweed). Plants 12:207. doi:https://doi.org/10.3390/plants12010207

Chandra R, Bharagava RN, Kapley A, Purohit HJ (2011) Bacterial diversity, organic pollutants and their metabolites in two aeration lagoons of common effluent treatment plant (CETP) during the degradation and detoxification of tannery wastewater. Bioresour Technol 102:2333–23. doi:https://doi.org/10.1016/j.biortech.2010.10.087

Cabeza Y, Candela L, Ronen D, Teijon G (2012) Monitoring the occurrence of emerging contaminants in treated wastewater and groundwater between 2008 and 2010. The Baix

Llobregat (Barcelona, Spain). J Hazard Mater 239:32–39. doi:https://doi.org/10.1016/j.jhazmat.2012.07.032

Corbel V, Stankiewicz M, Pennetier C, Fournier C, Stojan J, Girard E et al. (2009) Evidence for inhibition of cholinesterases in insect and mammalian nervous systems by the insect repellent DEET. BMC Biol 7. doi:https://doi.org/10.1186/1741-7007-7-47.

Czekalski N, Gascón Díez E, Bürgmann H (2014) Wastewater as a point source of antibiotic-1031 resistance genes in the sediment of a freshwater lake. ISME J 8:1381–1390. doi:https://doi.org/10.1038/ismej.2014.8

Celino-Brady FT, Lerner DT, Seale AP (2020) Experimental approaches for characterizing the endocrine-disrupting effects of environmental chemicals in fish. Front Endocrinol 11:619361. doi:https://doi.org/10.3389/fendo.2020.619361

Carnevali O, Santangeli S, Forner-Piquer I et al (2018) Endocrine-disrupting chemicals in aquatic environment: What are the risks for fish gametes? Fish Physiol Biochem 44:1561–1576. doi:https://doi.org/10.1007/s10695-018-0507-z

Camacho I, Caeiro E, Nunes C, Morais-Almeida M (2020) Airborne pollen calendar of Portugal: a 15-year survey (2002–2017) Allergol. Immunopathol 48:94–201. doi:https://doi.org/10.1016/j.aller.2019.06.012

D'Amato G et al. (2015) Meteorological conditions, climate change, new emerging factors, and asthma and related allergic disorders. A statement of the World Allergy Organization. World Allergy Org J 8(1):25. doi:https://doi.org/10.1186/s40413-015-0073-0

D'Angelo S, Meccariello R (2021) Microplastics: a threat for male fertility. Int J Environ Res Publ Health 18(5):2392. doi:https://doi.org/10.3390/ijerph18052392

Diya'uddeen BH, Daud WMAW, Abdul Aziz AR (2011) Treatment technologies for petroleum refinery effluents: a review, Process Saf. Environ Protect 89:95–105. doi:https://doi.org/10.1016/j.psep.2010.11.003

Dong Z, Su F, Zhang Z, Wang S (2020) Observation of chemical components of PM2.5 and secondary inorganic aerosol formation during haze and sandy haze days in Zhengzhou, China. Environ Sci 88:316–325. doi:https://doi.org/10.1016/j.jes.2019.09.016

Dziubanek G, Spychała A, Marchwińska-Wyrwał E, Rusin M, Hajok I, Ćwieląg-Drabek M, Piekut A (2017) Long-term exposure to urban air pollution and the relationship with life expectancy in cohort of 3.5 million people in Silesia. Sci Total Environ 580:1–8. doi:https://doi.org/10.1016/j.scitotenv.2016.11.217

Dutta DK, Haggerty DA, Rappolee DM, Ruden (2020) Phthalate exposure and long-term epigenomic consequences: a review. Front Genet 11:405. doi:https://doi.org/10.3389/fgene.2020.00405

Durmaz E, Erkekoglu P, Asci A, Akçurin S, Bircan İ, Kocer-Gumusel B (2018) Urinary phthalate metabolite concentrations in girls with premature thelarche. Environ Toxicol Pharmacol 59:172–181. doi:https://doi.org/10.1016/j.etap.2018.03.010

du Plessis A (2022) Persistent degradation: Global water quality challenges and required actions. One Earth 5: 129–131. doi:https://doi.org/10.1016/j.oneear.2022.01.005

Eales J, Bethel A, Galloway T, Hopkinson P, Morrissey K, Short RE, Garside R (2022) Human health impacts of exposure to phthalate plasticizers: An overview of reviews. Environ Int 158:106903. doi:https://doi.org/10.1016/j.envint.2021.106903

Eggen T, Monika Moeder M, Arukwe A (2010) Municipal landfill leachates: A significant source for new and emerging pollutants. Sci Total Environ. 408:5147–5157. doi:https://doi.org/10.1016/j.scitotenv.2010.07.049

Egbeocha CO, Malek S, Emenike CU et al (2018) Feasting on microplastics: ingestion by and effects on marine organisms. Aquat Biol. 27:93–106. doi:https://doi.org/10.3354/ab00701

Ena L, Lim JS, Son JY, Park YJ, Lee YH et al (2018) Evaluation of subchronic exposure to Triclosan on hepatorenal and reproductive toxicities in prepubertal male rats. J Toxic Environ Health A 81:421–431. doi:https://doi.org/10.1080/15287394.2018.1451188

Faraji Ghasemi F et al. (2020) Levels and ecological and health risk assessment of PM2.5-bound heavy metals in the northern part of the Persian Gulf. Environ Sci Pollut Res Int. 27(5):5305–5313. doi:https://doi.org/10.1007/s11356-019-07272-7

Felis E, Kalka J, Sochacki A, Kowalska K, Bajkacz S, Harnisz M, Korzeniewska E (2019) Antimicrobial pharmaceuticals in the aquatic environment – occurrence and environmental implications. Eur J Pharmacol 172813, doi:https://doi.org/10.1016/j.ejphar.2019.172813.

Fuchs FD (2011) Why do black Americans have higher prevalence of hypertension? An enigma still unsolved. Hypertension 57:379–380. doi:https://doi.org/10.1161/HYPERTENSIONAHA.110.16319

Fuhrman J (2018) The hidden dangers of fast and processed food. Am J Lifestyle Med 12: 375–381. doi:https://doi.org/10.1177/1559827618766483

Fransway AF, Fransway PJ, Belsito DV, Yiannias JA (2019) Paraben Toxicology. Dermatitis 30:32–45. doi:https://doi.org/10.1097/DER.0000000000000428

Gamboa-Loira B, Cebrián ME, Franco-Marina F, López-Carrillo L (2017) Arsenic Metabolism and Cancer Risk: A Meta-Analysis. Environ Res 156:551–558. doi:https://doi.org/10.1016/j.envres.2017.04.016

Gao D, Li Z, Wen Z, Ren N (2014) Occurrence and fate of phthalate esters in full-scale domestic wastewater treatment plants and their impact on receiving waters along the Songhua River in China. Chemosphere 95:24–32. doi:https://doi.org/10.1016/j.chemosphere.2013.08.009

Gavrilescu M et al (2015) Emerging pollutants in the environment: present and future challenges in biomonitoring, ecological risks and bioremediation, New Biotechnol 32(1):147–156. doi:https://doi.org/10.1016/j.nbt.2014.01.001

Ge J, Li H, Liu P et al (2021) Review of the toxic effect of microplastics on terrestrial and aquatic plants. Sci Total Environ 791:148333. doi:https://doi.org/10.1016/j.scitotenv.2021.148333

Gonzalez S, Lopez-Roldan R, Cortina JL (2012) Presence and biological effects of emerging contaminants in Llobregat River basin: A review. Environ Pollut 161:83–92. doi:https://doi.org/10.1016/j.envpol.2011.10.002

Ghosh S, Sinha JK, Ghosh S, Vashisth K, Han S, Bhaskar R (2023) Microplastics as an emerging threat to the global environment and human health, Sustainability 15(14):10821. doi:https://doi.org/10.3390/su151410821

Glassmeyer ST et al (2017) Nationwide reconnaissance of contaminants of emerging concern in source and treated drinking waters of the United States. Sci Total Environ 581–582:909–922. doi:https://doi.org/10.1016/j.scitotenv.2016.12.004

Gottschalk F, Sun T, Nowack B (2013) Environmental concentrations of engineered nanomaterials: Review of modeling and analytical studies. Environ Pollut 181:287–300. doi:https://doi.org/10.1016/j.envpol.2013.06.003

Gogoi A, Mazumder P, Tyagi VK et al (2018) Occurrence and fate of emerging contaminants in water environment: A review. Groundw Sustain Dev 6:169–180. doi:https://doi.org/10.1016/j.gsd.2017.12.009

Gupta AK, Gupta M. (2006) Synthesis and surface engineering of iron oxide nanoparticles for biomedical applications. Biomaterials 26:3995–4021. doi:https://doi.org/10.1016/j.biomaterials.2004.10.012

Gangwisch JE, Hale L, Garcia L et al. (2015) High glycemic index diet as a risk factor for depression: analyses from the Women's Health Initiative. Am J Clin Nutr 102:454–463. doi:https://doi.org/10.3945/ajcn.114.103846

Gonzalez-Martin C, Teigell-Perez N, Valladares B et al (2014) The global dispersion of pathogenic microorganisms by dust storms and its relevance to agriculture. Adv Agron 127:1–41. doi:https://doi.org/10.1016/B978-0-12-800131-8.00001-7

Gunnell D, Eddleston M, Phillips MR, Konradsen F (2007) The global distribution of fatal pesticide self-poisoning: systematic review. BMC Public Health 7:357. doi:https://doi.org/10.1186/1471-2458-7-357

Goulson D (2014) Ecology: pesticides linked to bird declines. Nature. 511:295–6. doi:https://doi.org/10.1038/nature13642

Gump BB, Yun S, Kannan K (2014) Polybrominated diphenyl ether (PBDE) exposure in children: possible associations with cardiovascular and psychological functions. Environ Res 132:244–250. doi:https://doi.org/10.1016/j.envres.2014.04.009

Harmon SM, Chen Q, Ma C et al. (2024) The effects of microplastic pollution on aquatic organisms. In Microplastic Contamination in Aquatic Environments (Elsevier), pp. 355–379. doi:https://doi.org/10.1016/B978-0-12-813747-5.00008-4

Hartle JC, Fox MA, Lawrence RS (2016) Probabilistic modeling of school meals for potential bisphenol A (BPA) exposure. J Expo Sci Environ Epidemiol 26:315–323. doi:https://doi.org/10.1038/jes.2015.58

Houtman CJ (2010) Emerging contaminants in surface waters and their relevance for the production of drinking water in Europe. J Integrative Environ Sci 7(4):271–295. doi:https://doi.org/10.1080/1943815X.2010.511648

Hong J, Kang H, An J et al (2021) Towards environmental sustainability in the local community: Future insights for managing the hazardous pollutants at construction sites. J Hazard Mater 403:123804. doi:https://doi.org/10.1016/j.jhazmat.2020.123804

Hemmingsson E (2018) Early childhood obesity risk factors: socioeconomic adversity, family dysfunction, offspring distress, and junk food self-medication. Curr Obes Rep 7:204–209. doi:https://doi.org/10.1007/s13679-018-0310-2

Handa O, Kokura S, Adachi S, Takagi T, Naito Y, Tanigawa T, Yoshida N, Yoshikawa T (2006) Methylparaben Potentiates UV-Induced Damage of Skin Keratinocytes. Toxicology 227:62–72. doi:https://doi.org/10.1016/j.tox.2006.07.018

Hlisníková H, Petrovičová I, Kolena B, Šidlovská M, Sirotkin A (2020) Effects and mechanisms of phthalates' action on reproductive processes and reproductive health: a literature review. Int J Environ Res Public Health 17:1–37. doi:https://doi.org/10.3390/ijerph17186811

la Farre M, Perez S, Kantiani L, Barcelo D (2008) Fate and toxicity of emerging pollutants, their metabolites and transformation products in the aquatic environment. TrAC Trends Anal Chem 27:991–1007. doi:https://doi.org/10.1016/j.trac.2008.09.010

IARC (2010) Monographs on the evaluation of carcinogenic risks to humans Volume 93 Carbon Black, Titanium Dioxide, and Talc; IARC: Lyon, France.

Il'yasova D, Schwartz GG (2005) Cadmium and renal cancer. Toxicol Appl Pharmacol 207:179–186. doi:https://doi.org/10.1016/j.taap.2004.12.005

Intisar A et al. (2022) Occurrence, toxic effects, and mitigation of pesticides as emerging environmental pollutants using robust nanomaterials – A review. Chemosphere. 293:133538. doi:https://doi.org/10.1016/j.chemosphere.2022.133538

Jeevanandam J, Barhoum A, Chan YS et al. (2018) Review on nanoparticles and nanostructured materials: history, sources, toxicity and regulations. Beilstein J Nanotechnol 9:1050–1074. doi:https://doi.org/10.3762/bjnano.9.98

Jones PJ, Koolhof IS, Wheeler AJ, Williamson GJ, Lucani C, Campbell SL, Bowman DJMS, Cooling Gasparrini NA, Johnston FH (2021) Characterizing non-linear associations between airborne pollen counts and respiratory symptoms from the Air Rater smartphone app in Tasmania, Australia: A case time series approach. Environ Res 200:111484. doi:https://doi.org/10.1016/j.envres.2021.111484

Jurewicz J, Radwan M, Wielgomas B, Kałużny P, Klimowska A et al (2018) Environmental levels of Triclosan and male fertility. Environ Sci Pollut Res Int 25:5484–5490. doi:https://doi.org/10.1007/s11356-017-0866-5

Kailaivasan T, Davies JM (2018) The molecular allergology of subtropical grass pollen. Mol Immunol 100:126–135. doi:https://doi.org/10.1016/j.molimm.2018.03.012

Karri V, Schuhmacher M, Kumar V (2016) Heavy metals (Pb, Cd, as and MeHg) as risk factors for cognitive dysfunction: a general review of metal mixture mechanism in brain. Environ Toxicol Pharmacol 48:203–213. doi:https://doi.org/10.1016/j.etap.2016.09.016

Khanjani N, Jafarnejad AB, Tavakkoli L (2017) Arsenic and Breast Cancer: A Systematic review of epidemiologic studies. Rev Environ Health 32:267–277. doi:https://doi.org/10.1515/reveh-2016-0068

Khodabakhshi A, Bagherzadeh F (2024) Risk assessment of heavy metals in brands of lipsticks commonly used in Shahrekord, Iran. J Shahrekord Univ Med Sci 26:42–45. doi:https://doi.org/10.34172/jsums.771

Kortenkamp A, Faust M (2018) Regulate to reduce chemical mixture risk. Science 361:224–226. doi:https://doi.org/10.1126/science.aat9219

Krause M, Klit A, Blomberg Jensen M, Søeborg T, Frederiksen H, Schlumpf M et al. (2012) Sunscreens: are they beneficial for health? An overview of endocrine disrupting properties of UV-filters. Int J Androl 35:424–436. doi:https://doi.org/10.1111/j.1365-2605.2012.01280.x

Khan NA, Khan SU, Islam DT, Ahmed S, Farooqi IH, Isa MH, Hussain A, Changani F, Dhingra A (2019) Performance evaluation of column-SBR in paper and pulp wastewater treatment: optimization and bio-kinetics. Desalin Water Treat 156:204–219. doi:https://doi.org/10.5004/dwt.2019.23775

Kang H, Sung S, Hong J et al (2021) Development of a real-time automated monitoring system for managing the hazardous environmental pollutants at the construction site. J Hazard Mater 402:123483. doi:https://doi.org/10.1016/j.jhazmat.2020.123483

Langston WJ (2020) Endocrine disruption and altered sexual development in aquatic organisms: an invertebrate perspective. J Mar Biol Assoc U. K. 100:495–515. doi:https://doi.org/10.1017/S0025315420000533

Laws MJ, Neff AM, Brehm E et al (2021) Endocrine disrupting chemicals and reproductive disorders in women, men, and animal models. Adv Pharmacol 92:151–190. doi:https://doi.org/10.1016/bs.apha.2021.03.008

Lodeiro C, Capelo JL, Oliveira E, Lodeiro JF (2019) New toxic emerging contaminants: beyond the toxicological effects. Environ Sci Pollut Res 26:1–4. doi:https://doi.org/10.1007/s11356-018-3003-1

Li Y, Zhang L, Ding J, Liu X (2020) Prioritization of pharmaceuticals in water environment in China based on environmental criteria and risk analysis of top-priority pharmaceuticals. J Environ Manag 253:109732. doi:https://doi.org/10.1016/j.jenvman.2019.109732.

Li X, Zheng W, Kelly WR (2013) Occurrence and removal of pharmaceutical and hormone contaminants in rural wastewater treatment lagoons. Sci Total Environ 445–446:22–28. doi:https://doi.org/10.1016/j.scitotenv.2012.12.035

Li WC (2014) Occurrence, sources, and fate of pharmaceuticals in aquatic environment and soil. Environ Pollut 187:193–201. doi:https://doi.org/10.1016/j.envpol.2014.01.015

Lammers T, Hennink WE, Storm G (2008) Tumour-targeted nanomedicines: Principles and practice. Br J Cancer 99:392–7. doi:https://doi.org/10.1038/sj.bjc.6604483

Lei M, Zhang L, Lei J, Zong L, Li J, Wu Z, Wang Z (2015) Overview of emerging contaminants and associated human health effects. Biomed Res Int 404796. doi:https://doi.org/10.1155/2015/404796

Limmathurotsakul D, Golding N, Dance DAB et al (2016) Predicted global distribution of Burkholderia pseudomallei and burden of melioidosis. Nat Microbiol 1:15008. doi:https://doi.org/10.1038/nmicrobiol.2015.8

Lim DS, Roh TH, Kim MK, Kwon YC, Choi SM, Kwack SJ, Kim KB, Yoon S, Kim HS, Lee BM (2018) Non-Cancer, cancer, and dermal sensitization risk assessment of heavy metals in cosmetics. J Toxicol Environ Health A 81:432–452. doi:https://doi.org/10.1080/15287394.2018.1451191

Lin W, Huang YW, Zhou XD, Ma Y (2006) In vitro toxicity of silica nanoparticles in human lung cancer cells. Toxicol Appl Pharmacol 217:252–259. doi:https://doi.org/10.1016/j.taap.2006.10.004

Lu J, Jin M, Nguyen SH et al (2018) Non-antibiotic antimicrobial Triclosan induces multiple antibiotic resistance through genetic mutation. Environ Int 118:257–265. doi:https://doi.org/10.1016/j.envint.2018.06.004

Liu F, Zhang Y, Wang F (2022) Environmental relevant concentrations of Triclosan affected developmental toxicity, oxidative stress, and apoptosis in zebrafish embryos. Environ Toxicol 37:848–857. doi:https://doi.org/10.1002/tox.23448

Liu J, Lu G, Ding J, Zhang Z, Wang Y (2014) Science of the Total Environment Tissue distribution, bioconcentration, metabolism, and effects of Erythromycin in crucian carp (*Carassius auratus*) Sci Total Environ 490:914–920. doi:https://doi.org/10.1016/j.scitotenv.2014.05.055
Ma L, Hu L, Feng X, Wang S (2018) Nitrate and Nitrite in Health and Disease. Aging Dis 9:938–945. doi:https://doi.org/10.14336/AD.2017.1207
Mah TF (2021) Giving antibiotics an assist. Science 372:1153. doi:https://doi.org/10.1126/science.abj3062
Martín-Pozo L, B. de Alarcon-Gomez, Rodríguez-Gomez R., García-Corcoles MT, Çipa M, Zafra-Gomez A (2019) Analytical methods for the determination of emerging contaminants in sewage sludge samples. A review. Talanta 192:508–533. doi:https://doi.org/10.1016/j.talanta.2018.09.056.
Meffe R et al (2014) Fate of para-toluenesulfonamide (p-TSA) in groundwater under anoxic conditions: Modelling results from a field site in Berlin (Germany). Environ Sci Pollut Res 21(1):568–583. doi:https://doi.org/10.1007/s11356-013-1902-8
Morin-Crini N, Lichtfouse E, Liu G et al (2022) Worldwide cases of water pollution by emerging contaminants: a review. Environ Chem Lett 20:2311–2338. doi:https://doi.org/10.1007/s10311-022-01447-4
Matuszczak E, Komarowska MD, Debek W et al (2019) The impact of Bisphenol A on fertility, reproductive system, and development: a review of the literature. Internet J Endocrinol 2019:4068717. doi:https://doi.org/10.1155/2019/4068717
Mbachu O, Jenkins G, Pratt C et al (2020) A new contaminant superhighway? A review of sources, measurement techniques and fate of atmospheric microplastics. Water Air Soil Pollut 231:85. doi:https://doi.org/10.1007/s11270-020-4459-4
Magrez A, Kasas S, Salicio V, Pasquier N, Seo JW, Celio M et al (2006) Cellular toxicity of carbon-based nanomaterials. Nano Lett 6(6):1121–1125. doi:https://doi.org/10.1021/nl060162e
Mancuso L, Cao G (2014) Acute toxicity test of CuO nanoparticles using human mesenchymal stem cells. Toxicol Mech Methods 24(7):449–454. doi:https://doi.org/10.3109/15376516.2014.928920
Mozaffarian D, Benjamin EJ et al (2016) Heart disease and stroke statistics – 2016 update: a report from the American Heart Association. Circulation 133:e38-e360. doi:https://doi.org/10.1161/CIR.0000000000000350
Martinez VD, Vucic EA, Becker-Santos DD, Gil L, Lam WL (2011) Arsenic exposure and the induction of human cancers. J Toxicol 431287. doi:https://doi.org/10.1155/2011/431287
Mostafalou S, Abdollahi M (2013) Pesticides and human chronic diseases: evidences, mechanisms, and perspectives. Toxicol Appl Pharmacol 268:157–177. doi:https://doi.org/10.1016/j.taap.2013.01.025
Montagnini BG, Forcato S, Pernoncine KV, Monteiro MC, Pereira MRF et al (2021) Developmental and reproductive outcomes in male rats exposed to Triclosan: two-generation study. Front Endocrinol. 12:738980. doi:https://doi.org/10.3389/fendo.2021.738980
Montes-Grajales D, Fennix-Agudelo M, Miranda-Castro W (2017) Occurrence of personal care products as emerging chemicals of concern in water resources: A review. Sci Total Environ 595:601–614. doi:https://doi.org/10.1016/j.scitotenv.2017.03.286
Milanović M, Đurić L, Milošević N, Milić N (2023) Comprehensive insight into triclosan-from widespread occurrence to health outcomes. Environ Sci Pollut Res Int 30(10):25119–25140. doi:https://doi.org/10.1007/s11356-021-17273-0
Murray CJ, Atkinson C, Bhalla K et al (2013) US Burden of Disease Collaborators. The state of US health, 1990–2010: burden of diseases, injuries, and risk factors. JAMA 310:591–608. doi:https://doi.org/10.1001/jama.2013.13805.
Naeem M, Gill R, Gill SS, Singh K, Sofo A, Tuteja N (2023) Editorial: Emerging contaminants and their effect on agricultural crops. Front Plant Sci 14:1296252. doi:https://doi.org/10.3389/fpls.2023.1296252.

Overturf MD, Anderson JC, Pandelides Z, Beyger L, Holdway DA (2015) Pharmaceuticals and personal care products: a critical review of the impacts on fish reproduction. Crit Rev Toxicol 45:469–491. doi:https://doi.org/10.3109/10408444.2015.1038499

Oberdorster G, Oberdorster E, Oberdorster J (2005) Nanotoxicology: an emerging discipline evolving from studies of ultrafine particles. Environ Health Perspect 113:823–839. doi:https://doi.org/10.1289/ehp.7339

O'Reilly D, Buchanan P (2019) Calcium channels and cancer stem cells. Cell Calcium 81:21–28. doi:https://doi.org/10.1016/j.ceca.2019.05.006

Parida VK, Saidulu D, Majumder A et al (2021) Emerging contaminants in wastewater: A critical review on occurrence, existing legislations, risk assessment, and sustainable treatment alternatives. J Environ Chem Eng 9:105966. doi:https://doi.org/10.1016/j.jece.2021.105966

Philip JM, Aravind UK, Aravindakuma CT (2018) Emerging contaminants in Indian environmental matrices – A review. Chemosphere 190:307–326. doi:https://doi.org/10.1016/j.chemosphere.2017.09.120

Pradhan B, Chand S, Chand S et al (2023) Emerging groundwater contaminants: A comprehensive review on their health hazards and remediation technologies. Groundw Sustain Dev 20:100868. doi:https://doi.org/10.1016/j.gsd.2022.100868/

Pal A, Gin KYH, Lin AYC et al (2010) Impacts of emerging organic contaminants on freshwater resources: Review of recent occurrences, sources, fate and effects. Sci Total Environ 408:6062–6069. doi:https://doi.org/10.1016/j.scitotenv.2010.09.026

Pál C, Papp B, Lázár V (2015) Collateral sensitivity of antibiotic-resistant microbes. Trends Microbiol 23:401–407. doi:https://doi.org/10.1016/j.tim.2015.02.009

Pak T, Luz LFDL, Tosco T, Costa GSR, Rosa PRR, Archilha NL (2020) Pore-scale investigation of the use of reactive nanoparticles for in situ remediation of contaminated groundwater source. Proc Natl Acad Sci 117(24):13366–73. doi:https://doi.org/10.1073/pnas.1918683117

Perdomo CM, Frühbeck G, Escalada J (2019) Impact of nutritional changes on non-alcoholic fatty liver disease. Nutrients 11:677. doi:https://doi.org/10.3390/nu11030677

Pawankar R, Holgate ST, Walter Canonica G, Lockey RF, Blaiss MS (2013) WAO White Book on Allergy – 2013 Update, World Allergy Organisation, Milwaukee, Winsconsin

Petric Z, Ruzi'c J, Zuntar I (2021) The Controversies of Parabens – An Overview Nowadays. Acta Pharm 71:17–32. doi:https://doi.org/10.2478/acph-2021-0001

Philippat C, Bennett D, Calafat AM, Picciotto IH (2015) Exposure to select phthalates and phenols through use of personal care products among Californian adults and their children. Environ Res 140:369–376. doi:https://doi.org/10.1016/j.envres.2015.04.009

Przybylińska PA, Wyszkowski M (2016) Environmental contamination with phthalates and its impact on living organisms. Ecol Chem Eng 23:347–356. doi:https://doi.org/10.1515/eces-2016-0024

Pun VC, Kazemiparkouhi F, Manjourides J, Suh HH (2017) Long-term PM2.5 exposure and respiratory, cancer, and cardiovascular mortality in older US adults. Am J Epidemiol 186(8):961–969. doi:https://doi.org/10.1093/aje/kwx166

Qi C, Huang J, Wang B et al (2018) Contaminants of emerging concern in landfill leachate in China: A review. Emerging Contam 4:1–10. doi:https://doi.org/10.1016/j.emcon.2018.06.001

Qian Y, Hu P, Lang-Yona N et al (2024) Global landfill leachate characteristics: Occurrences and abundances of environmental contaminants and the microbiome. J Hazard Mater 461:132446. doi:https://doi.org/10.1016/j.jhazmat.2023.132446

Quist AJL, Inoue-Choi M, Weyer PJ, Anderson KE, Cantor KP et al (2018) Ingested nitrate and nitrite, disinfection byproducts, and pancreatic cancer risk in postmenopausal women. Int J Cancer 142:251–261. doi:https://doi.org/10.1002/ijc.31055

Rakib MRJ, Sarker A, Ram K et al (2023) Microplastic toxicity in aquatic organisms and aquatic ecosystems: a review. Water Air Soil Pollut 234:52. doi:https://doi.org/10.1007/s11270-023-06062-9

Ramírez-Malule H, Quiñones-Murillo DH, Manotas-Duque B (2020) Emerging contaminants as global environmental hazards. A bibliometric analysis. Emerg Contam 6:179–193. doi:https://doi.org/10.1016/j.emcon.2020.05.001

Rani M, Shanker U, Jassal V (2017) Recent strategies for removal and degradation of persistent & toxic organochlorine pesticides using nanoparticles: A review. J Environ Manage 190:208–222. doi:https://doi.org/10.1016/j.jenvman.2016.12.068

Rasheed T et al (2021) Tailored functional materials as robust candidates to mitigate pesticides in aqueous matrices – a review. Chemosphere.282:131056. doi:https://doi.org/10.1016/j.chemosphere.2021.131056

Richardson SD (2009) Water analysis: emerging contaminants and current issues. Anal Chem 81:4645–4677. doi:https://doi.org/10.1021/ac9008012

Richardson SD, Ternes TA (2014) Water analysis: emerging contaminants and current issues. Anal Chem 86(6):2813–2848. doi:https://doi.org/10.1021/ac500508t

Rodriguez-Narvaez OM, Peralta-Hernandez JM, Goonetilleke A, Bandala ER (2017) Treatment technologies for emerging contaminants in water: a review. Chem Eng J 323:361–380. doi:https://doi.org/10.1016/j.cej.2017.04.106

Rogers ER, Zalesny RS, Lin CH (2021) A systematic approach for prioritizing landfill pollutants based on toxicity: Applications and opportunities. J Environ Manag 284:112031. doi:https://doi.org/10.1016/j.jenvman.2021.112031

Rebello S, Asok AJ, Mundayoor S, Jisha MS (2014) Surfactants: toxicity, remediation and green surfactants. Environ Chem Lett 12:275–287. doi:https://doi.org/10.1007/s10311-014-0466-2

Ravindra K, Goyal A, Mor S (2022) Pollen allergy: Developing multi-sectorial strategies for its prevention and control in lower and middle-income countries. Int J Hygiene Environ Health 242:113951. doi:https://doi.org/10.1016/j.ijheh.2022.113951

Ravindra K.et al (2021) Pollen Calendar to depict seasonal periodicities of airborne pollen species in a city situated in Indo-Gangetic plain, India. Atmos. Environ 262:118649. doi:https://doi.org/10.1016/j.atmosenv.2021.118649

Rojo J. et al (2015) Effect of land uses and wind direction on the contribution of local sources to airborne pollen. Sci Total Environ 538:672–682. doi:https://doi.org/10.1016/j.scitotenv.2015.08.074

Rudd MA, Ankley GT, Boxall AB, Brooks BW (2014) International scientists' research priorities for pharmaceuticals and personal care products in the environment. Integr Environ Assess Manage 10:576–587. doi:https://doi.org/10.1002/ieam.1551

Santos LHMLM, Araújo AN, Fachini A, Pena A, Delerue-Matos C, Montenegro MCBSM (2010) Ecotoxicological aspects related to the presence of pharmaceuticals in the aquatic environment. J Hazard Mater 175:45–95. doi:https://doi.org/10.1016/j.jhazmat.2009.10.100

Sánchez-Villegas A, Toledo E, de Irala J, Ruiz-Canela M, Pla-Vidal J, Martínez-González MA (2011) Fast-food and commercial baked goods consumption and the risk of depression. Public Health Nutr 15:424–432. doi:https://doi.org/10.1017/S1368980011001856

Sharma MD, Elanjickal AI, Mankar JS, Krupadam RJ (2020) Assessment of cancer risk of microplastics enriched with polycyclic aromatic hydrocarbons. J Hazard Mater 398:122994. doi:https://doi.org/10.1016/j.jhazmat.2020.122994

Sørensen SR, Holtze MS, Simonsen A, Aamand J (2007) Degradation and mineralization of nanomolar concentrations of the herbicide dichlobenil and its persistent metabolite 2,6-dichlorobenzamide by Aminobacter spp. isolated from Dichlobenil-treated soils. Appl Environ Microbiol 73:399–406. doi:https://doi.org/10.1128/AEM.01498-06

Samuelsen OB, Lunestad BT, Farestveit E, Grefsrud ES, Hannisdal R, Holmelid B, Tjensvoll T, Agnalt AL (2014) Mortality and deformities in European lobster (*Homarus gammarus*) juveniles exposed to the anti-parasitic drug teflubenzuron. Aquat Toxicol 149:8–15. doi:https://doi.org/10.1016/j.aquatox.2014.01.019

Sanderson H, Fricker C, Brown RS, Majury A, Liss SN (2016) Antibiotic resistance genes as an emerging environmental contaminant. Environ Rev 24:205–218. doi:https://doi.org/10.1139/er-2015-0069

Som C, Wick P, Krug H, Nowack B (2011) Environmental and health effects of nanomaterials in nanotextiles and facade coatings. Environ Int 37:1131–1142. doi:https://doi.org/10.1016/j.envint.2011.02.013

So CL, Saunus JM, Roberts-Thomson SJ, Monteith GR (2019) Calcium Signaling and Breast Cancer. Semin. Cell Dev Biol 94:74–83. doi:https://doi.org/10.1016/j.semcdb.2018.11.001

Stefanakis A, Becker JA (2015) A review of emerging contaminants in water: Classification, sources and potential risks (Impact of Water Pollution on Human Health and Environmental Sustainability, Edition: 1st) Ed E McKeown and G. Bugyi (Publisher: IGI Global) Chapter 3 pp 55–80. doi:https://doi.org/10.4018/978-1-4666-9559-7.ch003

Subedi B, Balakrishna K, Sinha RK et al (2015) Mass loading and removal of pharmaceuticals and personal care products, including psychoactive and illicit drugs and artificial sweeteners, in five sewage treatment plants in India. J Environ Chem Eng 3:2882–2891. doi:https://doi.org/10.1016/j.jece.2015.09.031

Sridharan S, Kumar M, Singh L et al (2021) Microplastics as an emerging source of particulate air pollution: A critical review. J Hazard Mater 418:126245. doi:https://doi.org/10.1016/j.jhazmat.2021.126245

Sreelakshmi T, Chitra K (2021) Microplastics contamination in the environment: An ecotoxicological concern. Int J Zool Invest 7:230–258. doi:https://doi.org/10.33745/ijzi.2021.v07i01.019

Singh BK, Yan ZZ, Whittaker M et al (2023) Soil microbiomes must be explicitly included in One Health policy. Nat Microbiol 1–6. doi:https://doi.org/10.1038/s41564-023-01386-y

Scott RA, Thilmony R, Harden LA et al (2017) Escherichia coli O157: H7 converts plant-derived choline to glycine betaine for osmoprotection during pre-and post-harvest colonization of injured lettuce leaves. Front Microbiol 8:2436. doi:https://doi.org/10.3389/fmicb.2017.02436

Steffan JJ, Derby JA, Brevik EC (2020) Soil pathogens that may potentially cause pandemics, including severe acute respiratory syndrome (SARS) coronaviruses. Curr Opin Environ Sci Health 17:35–40. doi:https://doi.org/10.1016/j.coesh.2020.08.005

Skalny AV, Aschner M, Sekacheva MI, Santamaria A, Barbosa F, Ferrer B, Aaseth J, Paoliello MMB, Rocha JBT, Tinkov AA (2022) Mercury and Cancer: where are we now after two decades of research? Food Chem Toxicol 164:113001. doi:https://doi.org/10.1016/j.fct.2022.113001

Sawada N (2018) Association between arsenic intake and cancer-from the viewpoint of epidemiological study. Nihon Eiseigaku Zasshi 73:265–268.doi:https://doi.org/10.1265/jjh.73.265

Saadatzadeh A, Afzalan S, Zadehdabagh R, Tishezan L, Najafi N, Seyedtabib M, Noori SMA (2019) Determination of Heavy Metals (Lead, Cadmium, Arsenic, and mercury) in authorized and unauthorized cosmetics. Cutan Ocul Toxicol 38:207–211. doi:https://doi.org/10.1080/15569527.2019.1590389

Séralini GE, Clair E, Mesnage R, Gress S, Defarge N, Malatesta M et al (2014) Republished study: long-term toxicity of a Roundup herbicide and a Roundup-tolerant genetically modified maize. Environ Sci Eur 26:14. doi:https://doi.org/10.1186/s12302-014-0014-5

Stroski KM, Luong KH, Challis JK, Chaves-Barquero LG, Hanson ML, Wong CS (2020) Wastewater sources of per-and polyfluorinated alkyl substances (PFAS) and pharmaceuticals in four Canadian Arctic communities. Sci Total Environ 708:134494. doi:https://doi.org/10.1016/j.scitotenv.2019.134494

Stuart M, Lapworth D, Crane E, Hart A. (2012) Review of risk from potential emerging contaminants in UK groundwater. Sci Total Environ 446:1–21. doi:https://doi.org/10.1016/j.scitotenv.2011.11.072

Tao M, Li D, Song R, Suh S, Keller AA (2018) OrganoRelease – a framework for modeling the release of organic chemicals from the use and post-use of consumer products. Environ Pollut 234:751–761. doi:https://doi.org/10.1016/j.envpol.2017.11.058.

Tang Y, Wu MVMY, Beland FA, Olson GR (2018) Role of peroxisome proliferator-activated receptor alpha (PPARα) and PPARα-mediated species differences in triclosan-induced liver toxicity. Arch Toxicol 92:3391–3402. doi:https://doi.org/10.1007/s00204-018-2308-7

Thakur DS, Khot R, Joshi PP, Pandharipande M, Nagpure K. (2014) Glyphosate poisoning with acute pulmonary edema. Toxicol Int 21:328–30. doi:https://doi.org/10.4103/0971-6580.155389

Tong X, Mohapatra S, Zhang J et al (2022) Source, fate, transport and modelling of selected emerging contaminants in the aquatic environment: Current status and future perspectives. Water Res 217: 118418. doi:https://doi.org/10.1016/j.watres.2022.118418

Taheran M, Naghdi M, Brar SK, Verma M, Surampalli RY (2018) Emerging contaminants: here today, there tomorrow. Environ Nanotechnol Monit Manage 10:122–126. doi:https://doi.org/10.1016/j.enmm.2018.05.010.

Thomaidis NS, Asimakopoulos AG, Bletsou AA (2012) Emerging contaminants: a tutorial mini-review. Global NEST J14:72–9. doi:https://doi.org/10.30955/gnj.000823

Teijon G, Candela L, Tamoh K, Molina-Díaz A, Fernández-Alba AR (2010) Occurrence of emerging contaminants, priority substances (2008/105/CE) and heavy metals in treated wastewater and groundwater at Depurbaix facility (Barcelona, Spain). Sci Total Environ 408(17):3584–3595. doi:https://doi.org/10.1016/j.scitotenv.2010.04.041

Tinkov AA et al (2018) Cadmium and atherosclerosis: a review of toxicological mechanisms and a meta-analysis of epidemiologic studies. Environ Res 162:240–260. doi:https://doi.org/10.1016/j.envres.2018.01.008

Yang X, Gong J, Zhang X et al (2021) Evaluation of the combined toxicity of multiwalled carbon nanotubes and cadmium on earthworms in soil using multi-level biomarkers. Ecotoxicol Environ Saf 221:112441. doi:https://doi.org/10.1016/j.ecoenv.2021.112441

Yang D, Liu Y, Bai C, Wang X, Powell CA (2020) Epidemiology of lung cancer and lung cancer screening programs in China and the United States. Cancer Lett 468:82–87. doi:https://doi.org/10.1016/j.canlet.2019.10.009

Yong M, Anderle L, Levy L, McCunney RJ (2019) Carbon Black and Lung Cancer Mortality-A Meta-regression analysis based on three occupational cohort studies. J Occup Environ Med 61:949–954. doi:https://doi.org/10.1097/JOM.0000000000001713

You HH, Song G (2021) Review of endocrine disruptors on male and female reproductive systems. Comp Biochem Physiol C Toxicol Pharmacol 244:109002. doi:https://doi.org/10.1016/j.cbpc.2021.109002

Valcarcel Y, Alonso SG, Rodríguez-Gil JL, Gil A, Catala M (2011) Detection of pharmaceutically active compounds in the rivers and tap water of the Madrid region (Spain) and potential ecotoxicological risk. Chemosphere 84(10):1336–1348. doi:https://doi.org/10.1016/j.chemosphere.2011.05.014

Venza M, Visalli M, Biondo C, Oteri R, Agliano F, Morabito S, Caruso G, Caffo M, Teti D, Venza I (2014) Epigenetic effects of cadmium in cancer: focus on melanoma. Curr Genomics 15(6):420–35. doi:https://doi.org/10.2174/138920291506150106145932

Wang F et al (2023) Polystyrene microplastics induce endoplasmic reticulum stress, apoptosis and inflammation by disrupting the gut microbiota in carp intestines, Environ Pollut 323:121233. doi:https://doi.org/10.1016/j.envpol.2023.121233

Weatherly LM, Gosse JA (2017) Triclosan exposure, transformation, and human health effects. J Toxicol Environ Health B Crit Rev 20:447–469. doi:https://doi.org/10.1080/10937404.2017.1399306

Wilkinson J, Hooda PS, Barker J, Barton S, Swinden J (2017) Occurrence, fate and transformation of emerging contaminants in water: an overarching review of the field. Environ Pollut 231:954–970. doi:https://doi.org/10.1016/j.envpol.2017.08.032.

Wei X, Kong X, Wang S, Xiang H, Wang J, Chen J (2013) Removal of heavy metals from electroplating wastewater by thin-film composite nanofiltration hollow-fiber membranes. Ind Eng Chem Res 52:17583–17590. doi:https://doi.org/10.1021/ie402387u

Wen LM, Simpson JM, Rissel C, Baur LA (2013) Maternal "junk food" diet during pregnancy as a predictor of high birth weight: findings from the healthy beginnings trial. Birth 40:46–51. doi:https://doi.org/10.1111/birt.12028

Wellmann J, Weiland SK, Neiteler G, Klein G, Straif K (2006) Cancer mortality in German carbon black workers 1976–98. Occup Environ Med 63:513–521. doi:https://doi.org/10.1136/oem.2006.026526

Woo KS et al (2021) The impact of particulate matter air pollution (PM2.5) on atherosclerosis in modernizing China: a report from the CATHAY study. Int J Epidemiol 50(2):578–588. doi:https://doi.org/10.1093/ije/dyaa235

Zhong B, Guo J, Zhang L et al (2022) A block chain-based framework for on-site construction environmental monitoring: Proof of concept. Build Environ 217:109064. doi:https://doi.org/10.1016/j.buildenv.2022.109064

Zahedi H, Kelishadi R, Heshmat R, Motlagh ME, Ranjbar SH et al (2014) Association between junk food consumption and mental health in a national sample of Iranian children and adolescents: the CASPIAN-IV study. Nutrition 30:1391–1397. doi:https://doi.org/10.1016/j.nut.2014.04.014

Zheng S, Chen B, Qiu X, Chen M, Ma Z, Yu X (2016) Distribution and risk assessment of 82 pesticides in Jiulong River and estuary. Chemosphere 144:1177–1192. doi:https://doi.org/10.1016/j.chemosphere.2015.09.050

Effect of Xenobiotic Exposure to the AMF Community: A Concern on Their Diversity, Viability, and Host Physio-biochemical Impacts

Ashutosh Kundu and Vivekananda Mandal

Abstract **Xenobiotic exposure** refers to introducing foreign substances (drugs, environmental pollutants, food additives, pesticides, and industrial chemicals) into an organism's body that are not naturally produced or expected to be found in living organisms. Xenobiotics can cause metabolic changes, cellular damage and chromosomal aberration, disrupt cell signaling pathways, and long-term exposure exerts toxicity and increase cancer risk in animal systems. Xenobiotic exposure disrupts microbial community structure and function, reduce microbial diversity, and alter the abundance of beneficial microbes, leading to shifts in species composition. Arbuscular mycorrhizal fungi (AMF) are obligate biotrophs associated symbiotically with almost 70–90% of plant roots. The AMFs colonize in the root regions of the host plant and transport mineral nutrients from soil to plant. In return, the fungus uptakes the required carbon from plants. This nutrient exchange affects the carbon (C) cycle, key soil processes, and plant fitness, significantly influencing microbial and plant ecosystems. AMFs are present in various settings, such as highly disrupted ecosystems with heavily contaminated soils from heavy metals, excessive drought, salt, and xenobiotics. Some AMF species are known to mitigate such stresses and boost plant growth and development. In this book chapter the effect of xenobiotics on the AMF, their diversity, viability in natural soil, and host physio-biochemical impacts will be explored, and to find out the strategies of the application of AMF inoculum for sustainable crop production and maintaining soil biodiversity and improving nutrient cycling under xenobiotic stress.

Keywords Biodiversity · Mycorrhizal fungi · Nutrient cycling · Soil fertility

A. Kundu · V. Mandal (✉)
Plant and Microbial Physiology and Biochemistry Laboratory, Department of Botany, University of Gour Banga, Malda, West Bengal, India
e-mail: vivek.bot@ugb.ac.in

S. Roy, V. Mandal (eds.), *Plant-Microbe Interaction under Xenobiotic Exposure*,
https://doi.org/10.1007/978-981-96-8260-7_4

1 Introduction

The term "xenobiotic exposure" describes the introduction of foreign compounds into an organism's body that are neither generated nor anticipated to be present there naturally. These xenobiotic substances can include various chemicals, such as drugs, environmental pollutants, synthetic food additives, pesticides, and industrial chemicals (Štefanac et al. 2021). Xenobiotics are very harmful to all living entities, including humankind. Some xenobiotics are not easily broken down and can accumulate in tissues, leading to long-term toxic effects. Xenobiotics can produce reactive oxygen species (ROS), which assist in cellular injury (Farag and Alagawany 2018; Klotz and Steinbrenner 2017). Prolonged exposure can trigger inflammatory responses and alter immune function. Certain xenobiotics can cause mutations or disrupt cell signaling pathways, increasing cancer risk (Lu et al. 2020).

In soil ecosystems, exposure to xenobiotics, including herbicides and pesticides, can reduce microbial diversity and alter the abundance of beneficial microbes like nitrogen-fixing bacteria and AMF, impacting soil quality and plant growth (Meena et al. 2020). Some microorganisms can adapt to xenobiotic stress through genetic mutations or horizontal gene transfer, leading to the development of resistance (Top and Springael 2003). This is particularly evident in antibiotic resistance, where exposure to antimicrobial agents selects resistant strains. Microorganisms exposed to certain xenobiotics, such as pollutants, can evolve the ability to metabolize and detoxify these compounds, enabling them to survive in contaminated environments (Chakraborty et al. 2017).

AMFs are obligate biotrophs and cannot live without their association with plants. AMF is the most common underground symbiosis, associated with almost 70–90% of plants studied (Shi et al. 2023). In colonizing AMF with the host plant, the AMF transfers to the plant by trapping and transporting mineral nutrition from soil to plant (Berruti et al. 2016). In return, the fungus uptakes the required carbon from plants as photosynthetically derived food. This nutrient interchange significantly impacts microbial and plant ecosystems by affecting the C cycle, important soil procedures, and plant fitness (Giovannini et al. 2020). AMF assists the host plant's roots in obtaining water and nourishment from faraway soils through their vast hyphal matrix, reducing the need for synthetic fertilizers and water management (Kakouridis et al. 2022). AMFs are present in various settings, such as highly disrupted ecosystems with heavily contaminated soils from heavy metals, excessive drought, salt, and xenobiotics (Zou et al. 2015; Singh et al. 2019; Branco et al. 2022). AMF supports plant development and ameliorates several stresses, as well as resistance to salt (Branco et al. 2022), nitrogen shortage, drought stress (Zou et al. 2015), and heavy metal stress (Singh et al. 2019), by promoting nutrient absorption. Many experiments have been accomplished on the physiological, biochemical, microbiological, and molecular techniques by which AMF supports plants' ability to forage for nutrients and withstand stress (Salvioli di Fossalunga and Novero 2019). As biofertilizers and protectors, AMF can be used in forest management, agronomy, ecological restoration, and phytoremediation (Phour et al. 2020; Zhu et al. 2022). Beneficial associations of AMF with roots contribute to sustainable horticulture and agriculture. AMF colonization will improve the performance and productivity of the majority of crops (Emmanuel and Babalola 2020). However, due to the extensive fertilizer and pesticide use in modern agriculture, such beneficial microorganisms are unavailable naturally or inactive in all agroecosystems. Hence, assessment of AMF diversity and their physio-biochemical impact under xenobiotic exposure could draw critical attention for sustainable crop production and eco-restoration in agricultural settings.

2 Effect of Xenobiotic Exposure on AMF Diversity and Viability

2.1 Chemical Fertilizer

Chemical fertilizers are substances synthesized or extracted to supply vital nutrients for plant growth, yield, biomass, and all-round development. These fertilizers play a critical role in modern agriculture by increasing crop yield, improving soil enrichment, and meeting the global food demand. They are typically categorized based on the nutrients they provide. The primary macronutrients needed by plants are potassium (K), phosphorus (P), and nitrogen (N). In 1965, the total fertilizer consumption was approximately 46.3 million metric tons. By 2022, this figure had risen to over 200 million metric tons, marking a substantial increase (Fig. 1). Continuous application of chemical fertilizers, especially nitrogen-based ones, can lower the soil's pH, increasing acidity. Acidic soils can negatively affect plant development and reduce the emergence of vital nutrients (Goulding 2016). Excessive reliance on chemical fertilizers often neglects organic inputs like compost and manure. This reduces soil organic matter, vital for maintaining soil fertility, water retention, and microbial activity (Brady and Weil 2008). Chemical fertilizers can break up the balance of soil microorganisms. Beneficial organisms like bacteria and fungi, including mycorrhizal fungi, may decline, while harmful microbes may proliferate. This disruption can reduce soil biodiversity and impair nutrient cycling (Mäder et al. 2002; Geisseler and Scow 2014). High levels of chemical fertilizers, especially phosphorus, can reduce AMF colonization. When plants have abundant access to readily available nutrients, especially phosphorus, the need for symbiosis with AMF diminishes, leading to reduced colonization rates. Excessive nitrogen can also negatively affect AMF by altering the soil pH or causing nutrient availability imbalances (Smith and Read 2008; Błaszkowski 2012). Long-time use of chemical fertilizers can lead to a deterioration in AMF diversity. According to Antunes et al. (2012), the impact of AMF symbiosis on fungal and plant development was changed by the long-time application of specific soil fertilization treatments. AMF obtained from nitrogen- or phosphate-insufficient soils have higher populations but impede plant development more than as per the optimum foraging hypothesis; AMF was extracted

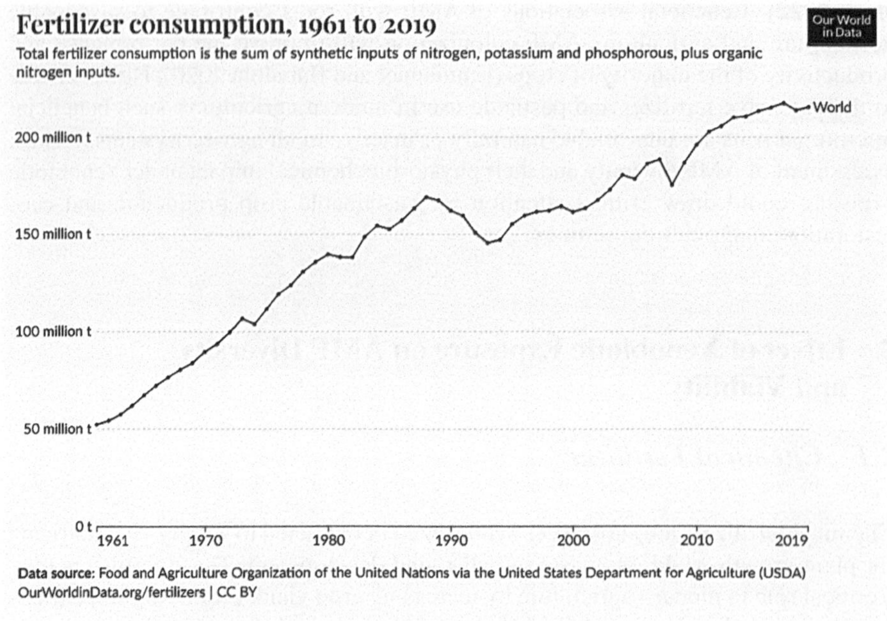

Fig. 1 Fertilizer consumption in the World (1961 to 2019)

from non-deficient soils. A similar result was observed by Ma et al. (2021), which was that extended use of NPK considerably reduced the alpha diversity of rhizosphere AMF. Also, it caused a possible decline in the mutualistic relationship between AMF and plants.

Certain AMF species are particularly vulnerable to elevated nutrition levels and might be outcompeted or suppressed. This leads to a less diverse AMF community, which can affect overall soil health and plant resilience (Treseder 2004). In some cases, moderate synthetic fertilizers can enhance AMF colonization by promoting overall plant growth, leading to more extensive root systems that AMF can colonize. However, this is usually only seen at low or moderate fertilization levels and depends on the balance between nitrogen, phosphorus, and other soil conditions. Fall et al. (2023) found that applying AMF with a 50% NPK treatment significantly boosted the plant's capacity to grow, colonize roots, produce chlorophyll on leaves, and contain ambient amounts of N, P, and K in tissues. Ziane et al. (2017) experimented with *Lycopersicon esculentum* plants with two AMF species, *Funneliformis mosseae* and *Septoglomus constrictum,* and used three rates of chemical fertilizer—mono-ammonium phosphate, ammonium sulfate, and NPK—applied at 0, 50%, and 100% of the required dose—mixed with the AMF inoculant for the production of industrial tomatoes. Compared to native AMF alone, introduced AMF dramatically raised mycorrhizal roots' growth, yield, and colonization levels.

Additionally, the yield from a 50% application of the recommended fertilizer dose was identical to that from the complete fertilizer dose without AMF inoculation. Another observation of Trejo et al. (2021) shows that AMF, when paired with 25% fertilization, mycorrhizal inoculation already equaled 100% chemical fertilization in pineapple crops, while in conjunction with 50% fertilization, it exceeded it in the majority of the yield variables that were assessed. In mycorrhizal plants with 50% fertilization, the fruit mass and organoleptic features were markedly more significant than in the non-infected controls and the treatments infected with AMF and combined with 0%, 25%, 75%, and 100% of a dosage of synthetic fertilizer. Mycorrhizal fungus inoculation in the field can increase fruit quality and prevent yield loss while reducing the need for chemical fertilizer treatment by 50%. However, the P fertilizer dosage was unaffected by the diversity, colonization, and community structure of tomato plants' AMF roots. Long-term fertilization with triple superphosphate and ammonium sulfate reduced the overall quantity of AMF spores and varied the species diversity (Bhadalung et al. 2005). They noted that the total amounts of *Entrophospora schenckii*, *Glomus mosseae*, *Glomus* sp.1, *G. geosporum*-like, and *Scutellospora fulgida* spores reduced in response to fertilization. These AMF species responded to fertilization by declining in absolute numbers, while *Acaulospora* sp.1 was insensitive to fertilization. However, soil nitrate-nitrogen and pH were the only factors influencing AMF diversity in roots (Zhang et al. 2021).

2.2 *Pesticides*

Pesticides are biological or chemical compounds that control, deter, or eradicate pests that damage crops and productivity. Pesticides are a group of chemicals that include insecticides, fungicides, herbicides, garden chemicals, home disinfectants, and rodenticides that are used to both eliminate and guard against pests (El Nemr et al. 2012; Hassaan and El Nemr 2020). These pests include insects, weeds, fungi, rodents, and other organisms that can damage plants, spread diseases, or pose health risks to humans and animals. Pesticides can be classified based on (i) the chemical structure, (ii) the entry mechanism, and (iii) the activity on the organisms they target (Yadav and Devi 2017). Chemical pesticides are divided into four categories based on their chemical composition: carbamate, organophosphate, organochlorine, or pyrethroid (Hassaan and El Nemr 2020). Global pesticide use has seen a significant increase over the past few decades. In 2022, agricultural pesticide consumption reached over 3.7 million metric tons of active chemicals, reflecting a 4.1% increase from 2021, a 13.5% increase over the last decade, and a doubling since 1999 (Fig. 2). Pesticides can disrupt soil microbial communities, essential for nutrient cycling and breakdown of organic substances, and plant health. Some pesticides, especially broad-spectrum ones, can kill not only target pests but also beneficial microorganisms, leading to reduced soil microbial diversity (Zhang and Wang 2020). Changes in microbial diversity can lead to imbalanced nutrient cycling and a decline in soil fertility over time. Using pesticides can reduce soil organic matter as microbial populations that help decompose organic materials decline. A reduction in organic

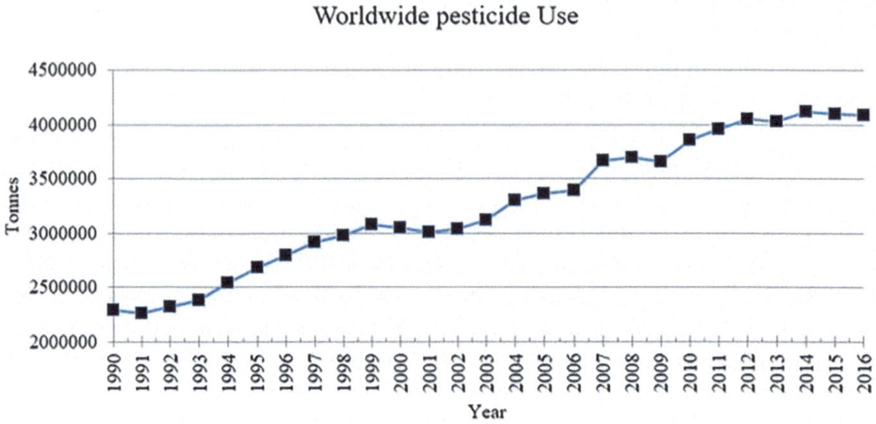

Fig. 2 Worldwide pesticide consumption (1990 to 2016). (Source: http://www.fao.org/faostat/en/#data/RP/visualize)

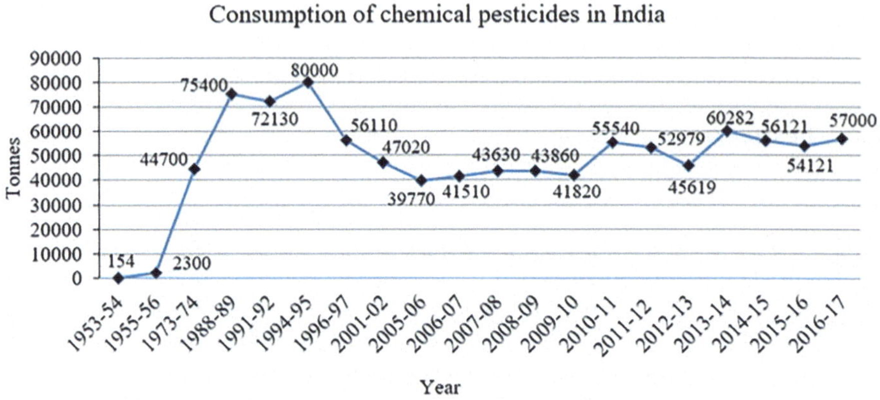

Fig. 3 Pesticide consumption in India (1954 to 2017). (Source: http://www.fao.org/faostat/en/#data/RP, Chand and Birthal (1997))

matter can degrade soil structure, leading to compaction and poor aeration (Rasmussen et al. 1998; Zhang and Wang 2020). The microbial enzymes play critical roles in processes like nitrogen cycling, organic matter decomposition, and phosphorus availability (Aktar et al. 2009). Reduced enzyme activity is often an indicator of declining soil health and biological activity. Pesticides can negatively affect beneficial symbiotic relationships in the soil, such as those between plants and AMF or nitrogen-fixing bacteria. This can reduce plant growth and nutrient uptake efficiency (Fig. 3).

Pesticides, especially fungicides, can directly target soil fungi, reducing the diversity of AMF. This reduced diversity can impact the ecosystem functions provided by different AMF species, including nutrient cycling and soil structure improvement (Mahmood et al. 2016). They can inhibit the growth and reproduction of AMF, leading to lower spore counts in the soil. Since AMF relies on spore germination for spreading and colonization, a reduced spore count can have long-term consequences on their ability to establish symbiosis with plants (Choudhary and Kaushik 2020). Pesticides can affect AMF's ability to colonize plant roots by directly inhibiting fungal growth or altering root exudates that AMF relies on for colonization. Lower colonization rates reduce plants' benefits from AMF, such as improved nutrient and water uptake (Choudhary and Kaushik 2020). As mentioned below, each category of pesticides has a different impact on the AMF diversity and their host physiological influences. The effects of different categories of pesticides that affect the AMF diversity and functionality are elaborated below.

2.2.1 Herbicides

Herbicides are a type of pesticide especially designed to destroy or impede the growth of undesired vegetation. They are used to manage or eradicate unwanted plants (weeds) that engaged with different crops for resources like sunlight, H_2O, and nutrients. They are widely used in agriculture, landscaping, forestry, and gardens. Herbicides, including glyphosate, oxyfluorfen, and paraquat, are often utilized. According to research, chemical weed management before seeding with glyphosate and paraquat improved wheat roots' early AMF colonization. The first research on glyphosate's influence on AMF symbiosis in GM soybeans in Brazil was done by Malty et al. (2006). They noted that glyphosate had an adverse effect on the germination of spores in three AMF species, the most susceptible of which was *Claradioglomus etunicatum*, followed by *Dentiscutata. heterogama* and *Gigaspora margarita*. Vieira et al. (2007) showed that sulfentrazone significantly reduced the mycorrhizal colonization of soybean (*Glycine max*) by approximately 50%. Additionally, the use of dicamba resulted in a reduction in spore number (deNovais et al. 2019). The study by Sheng et al. (2012) observed that in continuous durum wheat, glyphosate changed the species makeup of AMF, causing *F. mosseae* to increase and another *Glomus* species to decrease. Wilkes et al. (2020) and Druille et al. (2013) analyzed the impact of glyphosate on AMF spore viability and the colonization of grassland plant roots in the region of flooding Pampa in Argentina. When glyphosate was applied at dosages equivalent to or less than the recommended dosage, spore viability was seven to eight times poorer in the glyphosate-treated soils than in the glyphosate-untreated control soils. Compared to the control soils that were not treated with glyphosate, the spore viability in the glyphosate-treated soils was 7–8 times lower. Following glyphosate treatments, there was a decrease in *Lolium multiflorum* root colonization due to decreased spore viability.

2.2.2 Insecticides and Nematicides

Insecticides are chemical substances designed to kill or manage the insect population threatening crops, human health, and domestic environments (Ansari et al. 2014). Nematicides are chemical substances created especially to control or eliminate plant-parasitic nematodes, which are tiny worms that may seriously harm crops. These pests attack plant roots, reducing nutrient uptake, stunting growth, and lowering crop yields. According to earlier assessments, pesticides and nematodes positively or neutrally impacted AMF (Trappe et al. 1984; Hamel and Strullu 2006). In contrast, the shifting effects of systemic and contact insecticides were documented with fungicides (Sarr et al. 2013; Deliopoulos et al. 2008). It has been reported that the insecticide/acaricide phoxim stops AM from colonizing carrots but not green onions (Wang et al. 2011a, b). However, the nematicide aldicarb did not affect the potatoes' AMF colonization (Goulson 2013).

Nowadays, the most popular class of pesticides is neonicotinoids used globally because of their broad range of effectiveness against several pests of many crops. Imidacloprid and thiamethoxam, two neonicotinoids, were investigated by Malfatti et al. (2023) on the spore viability and germination in two different AMF species: *R. clarus* and *Gi. albida*. Both neonicotinoids remarkably and dose-dependently decreased spore germination. Therefore, the reduced spore germination may eventually affect how the AM symbiosis functions.

2.2.3 Fungicides

Fungicides are chemical or biological agents designed to prevent, eradicate, or stop the growth of fungus and their spores. These products are essential for horticulture, agriculture, and plant protection since fungal infections may seriously harm crops, lowering yields and causing financial losses. Fungi are responsible for various plant diseases, including rusts, blights, mildews, and root rots, which can severely affect the health and productivity of plants. Fungicides target fungal pathogens but often have broad-spectrum effects, impacting non-target beneficial soil fungi and microorganisms. This can reduce soil biodiversity and disrupt critical microbial interactions necessary for soil health, such as mycorrhizal fungi, decomposer fungi, and beneficial bacteria (Kobayashi and Croucher 2009). Reduced microbial diversity can lead to imbalances in nutrient cycling and a decline in soil fertility. According to Carrenho et al. (1998), fungicide treatments on citrus trees impact the variety of AMFs. Their investigation found that *Gi. ramisporophora, Cetraspora gilmore, and Glomus macrocarpum* were less susceptible to Fosetyl-AI and metalaxyl than other species. The effects of fungicides (benomyl, pentachloronitrobenzene, and captan) on particular fungal species were seen in pea roots colonized by three distinct AMF species. *Gi. rosea, F. mosseae,* and *C. etunicatum*. Modest dosages of mancozeb and carbendazim sprayed on *Pisum sativum* (pea) plants showed unimodal response curves for hyphal phosphorus (P) absorption and fractional root colonization, whereas high doses inhibited and low dosages encouraged. Another observation by Bharat (2011) shows that the use of four fungicides (carbendazim, mancozeb, copper sulfate, and aureofungin) reduced the AMF colonization significantly with time, and mancozeb shows the highest reduction, followed by aureofungin, copper sulfate, and carbendazim. Okiobe et al. (2022) studied three different groups of fungicides on *R. irregularis* and found that Signam (Pyraclostrobin and boscalid) fungicides lowered the appearance of arbuscules. Topas (Penconazole) reduced the hyphal growth of AMF, and Cuprozin (Copper oxychloride 35%) reduced the vesicle structure but did not decrease the root colonization. A similar observation was found by Jin et al. (2013), where colonization, host growth, and P uptake were reduced in pea and chickpea plants in the absence of disease pressure by the systemic fungicides Allegiance™ FL, Apron Maxx® RTA®, Vitaflo® 280, Crown® and Trilex® AL. Still, fungicides had no discernible effect on glomalin-related protein synthesis or AMF sporulation. Half-life, or the time it takes for 50% of the applied pesticide to dissipate from the soil, is a common way to represent pesticide persistence. However, because of the impacts of soil and climate, DT50 is intrinsically unpredictable. A list of pesticides used in agricultural settings is given below (Table 1). It is to be noted that these pesticides are used in high preferences; however, no risk assessment on the AMF has been done.

3 Effect of Xenobiotic Exposure on the Host Physio-biochemical Impacts

As stated by the Food and Agriculture Organization (FAO), the demand for fertilizer nutrients has been rising consistently since the 1960s and is anticipated to reach over 200 million tons by 2020, up from 184.02 million tons in 2015. It has long been known that meiotic processes are influenced by nutritional status. Zeyad et al. (2019) studied the genotoxic effect of wastewater on *Allium cepa* root and observed that the mitotic index (MI) remarkably declined as the concentration of sewage rose. The decline in MI showed that the mitotic phase was affected by genotoxic factors; hence, the prophase cell percentage increased, revealing distinct forms of chromosomal aberrations. Atrazine-treated sorghum likewise had a significant degree of meiotic aberrations (Lee et al. 1974). Subsequent research revealed even more subtle alterations to the meiotic nuclear environment. During diplotene and early diakinesis, more nucleoli were commonly seen in experimental plants (Currie and Liang 1996). *Helianthus annuus* seeds were treated with the herbicide maleic hydrazine at doses ranging from 10^{-5} M to 10^{-2} M to assess its meiotic impact (Kaymak 2005). Even at the lowest concentration examined, this study shows a considerable rise in anomalies in the first and second meiotic divisions. Additionally, it discovers a startling degree of anomalies in the following generation. Table 2 lists xenobiotic effects on different plants and their influence on meiosis.

Table 1 List of different types of xenobiotics and their half-life with residues in soil

Xenobiotic sources	Properties of pesticides	Half-life (days)	Residues in soil (μg/kg)	Risk for plants	References
Pyriproxyfen	Insecticides	918	**ND**	*Capsicum annuum*	Fantke and Juraske (2013)
Pyrethrins	Insecticides	0.04	**ND**	*Solanum lycopersicum*	Fantke and Juraske (2013)
Dimethachlon	Fungicides	8.26	**ND**	*Panax notoginseng*	Fu et al. (2020)
Boscalid	Fungicides	10.03	**0–52.7**	*Panax notoginseng*	Fu et al. (2020)
Iprodione	Fungicides	6.60	**20.19–51.33**	*Panax notoginseng*	Fu et al. (2020)
Dimethomorph	Fungicides	6.88	**78.61–118.9**	*Panax notoginseng*	Fu et al. (2020)
Bifenthrin	Termiticides	1262	**ND**	ND	Manzoor and Pervez (2017)
Fipronil	Termiticides	555	**ND**	ND	Manzoor and Pervez (2017)
Fenbuconazole	Fungicides	7.52	**96.76–473.4**	*Panax notoginseng*	Fu et al. (2020)
Chlorothalonil	BS fungicides	3.43	**14.94–84.54**	*Panax notoginseng*	Fu et al. (2020)
Tebuconazole	Fungicides	17.90	**255.4–642.3**	*Panax notoginseng*	Fu et al. (2020)
Propiconazole	Fungicides	22.17	**66.28–107.7**	*Panax notoginseng*	Fu et al. (2020)
Triadimenol	Fungicides	23.93	**68.43–107.7**	*Panax notoginseng*	Fu et al. (2020)
Myclobutanil	Fungicides	24.34	**5.68–8.68**	*Panax notoginseng*	Fu et al. (2020)
Diniconazole	Fungicides	26.57	**12.22–22.94**	*Panax notoginseng*	Fu et al. (2020)
Chlorsulfuron	Herbicides	0.3	**ND**	**ND**	Fantke and Juraske (2013)
Dalapon	Herbicides	35	**ND**	**ND**	Fantke and Juraske (2013)
Bifenthrin	Insecticides	2.2	**ND**	**ND**	Fantke and Juraske (2013)
Pyrimethanil	Fungicides	41.37	**6.31–13.69**	*Panax notoginseng*	Fu et al. (2020)
Procymidone	Fungicides	42.46	**55.35–166.3**	*Panax notoginseng*	Fu et al. (2020)
Chlorpyrifos	Insecticides	58.60	**19.96–44.74**	*Panax notoginseng*	Fu et al. (2020)
Triadimefon	Fungicides	58.76	**5.78–8.24**	*Panax notoginseng*	Fu et al. (2020)

(continued)

Table 1 (continued)

Xenobiotic sources	Properties of pesticides	Half-life (days)	Residues in soil (μg/kg)	Risk for plants	References
Cyhalothrin	Insecticides	21.90	**ND**	*Panax notoginseng*	Fu et al. (2020)
Cypermethrin	Insecticides	17.24	**12.53–14.67**	*Panax notoginseng*	Fu et al. (2020)

Xenobiotics also damage plants' pollen mother cells (PMCs), so pollen viability and the reproductive rate of plants also decrease. The maximum amounts of sterile pollen grains were produced by adhesion, lagging chromosomes, and chromosomal breakage during meiosis after 35 days of treatment with 2, 4, 5-T (2, 4, 5-Trichlorophenoxyacetic acid). The effects of glyphosate and pendimethalin sprayed on *Vigna mungo* leaves showed abnormalities, such as binucleate cells, chromatin bridges, anaphase I laggards, and disturbed spindle development after 21 days. In another experiment, the seed sprayed with 0.1, 0.5, and 1.0% dosages of the insecticides (monocrotophos, phosphamidon, dimethoate, and DDVP (2, 2-dichlorovinyl dimethyl phosphate) or dichlorvos) for 24 h showed reduced pollen viability in each treatment, and the level of sterility increased as the concentration increased. Additionally, the authors emphasized that tested pesticides are just as disruptive as ionizing radiation. Similar outcomes were seen when combining cadmium and chromium to treat *H. vulgare*.

Additionally, the number of pollen grains in each anther remarkably decreased, and the sterility of pollen significantly increased due to this combined heavy metal treatment (Mittal and Srivastava 2014). The impact of the insecticide Profex Super (profenofos 40% + cypermethrin 4% EC) on tomato (*Lycopersicon esculentum*) pollen germination in vitro was investigated. The insecticide was administered at recommended doses of 30 ml/15 L and 60 ml/15 L, and it was observed that germination continued to decline at the prescribed dosages in comparison to the untreated control (Meshram and Chaturvedi 2017). Aberrant PMCs with premature mobility, stickiness, or lagging chromosomes were much more prevalent in all plant species gathered from the polluted areas. The frequency of aberrant PMCs varied significantly between species. *Cyperus arenarius* had the highest effect, with 99% of PMCs having abnormalities, as opposed to 13% for the control, while *Ochtochloa compressa* had only 7% of PMCs with defects, as opposed to the control's 2%. Additionally, the study found a higher quantity of sterile pollen grains and unreduced dyads. Therefore, numerous xenobiotic stressors might affect the plant reproductive efficiency and the ecosystem as a whole.

Table 2 Effect of xenobiotics in different plants and how they affect meiosis

Xenobiotics	Species	Chromosomal changes	References
Insecticide	*Vicia faba*	Stickiness, bridges, fragment	Amer and Ali 1983
	Capsicum annuum	Stickiness, laggards, bridges, micronuclei, spindle aberrations	Devadas et al. (1986)
	Allium cepa	Stickiness, laggards, bridges, precocious movements	Kuchy et al. (2016)
Herbicide	*Triticum durum*	Laggards, bridges	Razu et al. (2014)
	Hordeum vulgare	Stickiness, bridges, fragments, micronuclei	Wuu (1967)
	Sorghum vulgare	Laggards, bridges, aneuploidy	Soriano (1984)
	Vicia faba	Stickiness, laggards, bridges, aneuploidy, micronuclei	Amer and Ali (1974)
Fungicide	*Hordeum vulgare*	Stickiness, bridges, fragments, micronuclei	Wuu (1967)
	Allium cepa	Stickiness, laggards, bridges, precocious movements	Kuchy et al. (2016)
	Capsicum annuum	Stickiness, laggards, bridges, univalents, multivalents	Prakash et al. (1988)
	Lathyrus sativus	Univalents	Kumar and Sinha (1991)
Heavy metals	*Glycine max*	Stickiness, laggards, bridges, univalent, precocious movements	Kumar (2007)
	Zea mays	Laggards, bridges, fragments, Univalents, micronuclei, precocious movements	Kumar and Rai (2010)
	Vicia faba	Stickiness, laggards, bridges	George (2000)
	Hordeum vulgare	Laggards, bridges, micronuclei, precocious movements	Mittal and Srivastava (2014)
	Lathyrus sativus	Univalents	Kumar and Ritambhara (2006)
	Capsicum annuum	Stickiness, laggards, bridges, fragments, micronuclei, precocious movements	Aslam et al. (2017)
	Vicia faba	Stickiness, laggards, bridges	George (2000)

4 Effect of Xenobiotic Exposure on the AMF-Host Physio-biochemical Impacts

AMF has a significant role in influencing the host physio-biochemical parameters. They are also known to reduce toxicity and xenobiotic stress tolerance to host plants. AMF may collect a significant amount of xenobiotics, such as heavy metals (HMs), via fungal structures and reduce the absorption of HMs inside the plant host. AMF uses intraradical and extraradical mycelium to accumulate HMs (Wu et al. 2016a, 2019). They are considered excellent biological tools for preserving the environmental balance of HMs. Adsorption of HMs on the surface of the fungal structure happens when HMs are reduced from one form to another by charge transfer from the thiol group amino acids like cysteine and glutathione residues (Park et al. 2006; Joutey et al. 2015). Thus, through the buildup of HMs on the surface of hyphae, AMF decreases the translocation of HMs to the host plant's cytoplasm (Gonzalez-Guerrero et al. 2008; Wu et al. 2015). Glomalin, a glue-like material that mimics extracellular polymeric substances (EPS), is secreted by AMF on the fungal surface. EPS has a strong affinity for metal adsorption by ion exchange, surface precipitation, and chelation because of its carboxyl, amine, phosphoric, and hydroxyl functional groups (More et al. 2014; Li and Yu 2014). The fungal structure absorbs a portion of chromium (Cr), and it is adsorbed to the fungal surface. This structure can attach to histidine analogs or carboxylic ligands before adhering to cell walls (Wu et al. 2015, 2016b). Furthermore, certain AMFs have been shown to reduce the toxicity of HMs by promoting metal compartmentalization via intraradical fungal structures (Degenhardt 2000; Macfie and Welbourn 2000; Krzesłowska 2011; Ghasemi Siani et al. 2017). Glomalin is a temperature- and water-resistant glycoprotein that AMF produces through the mycelium and spore walls (Wang et al. 2019). Various research suggests that glomalin can effectively chelate heavy metals (HMs) and reduce their bioavailability in the rhizosphere by means of metal speciation (Malekzadeh et al. 2016; Gujre et al. 2021). According to some research, 1 g of glomalin secreted by specific AMF strains may effectively chelate up to 1.12 mg of lead, 0.08 mg of cadmium, and 4.3 mg of copper from HM-contaminated soils. This suggests that glomalin has varying capacities depending on the kind and concentration of metal (Gonzalez-Chavez et al. 2004). A 140-day experimental treatment revealed that glomalin absorbed 0.21–1.78% of lead and 0.38–0.98% of nickel from soil polluted with HMs (Wu et al. 2014). It was verified that mycorrhizal inoculation, specifically with *Glomus aggregatum*, demonstrated a significant reduction in the concentrations of Cd, Pb, Zn, and Cu in the shoot by the production of glomalin, which chelated HMs in the soil (Nafady and Elgharably 2018). Earlier research by Gonzalez-Guerrero et al. (2008) shows that HMs like Cd, Cu, and Zn were stored in spore walls and vacuoles of AM fungi as observed through electron dispersive X-ray synchrotron.

5 Xenobiotics Ameliorating Effects of AMF

Heavy metal contamination is a global issue that has raised severe concerns for human health and the environment (Cai et al. 2019). Agriculturally contaminated soil has garnered much recognition in the last decade. One technology that successfully reduces HM toxicity is phytoremediation; nevertheless, the nature of metal toxicities and poor plant growth rate are significant limitations of this strategy. AMFs are globally used to support plant establishment against heavy metal-contaminated soil by raising growth and yield development. Various health diseases caused by HMs, like Cd and Pb, may be due to the accumulation of these HMs in food crops, fruits, and vegetables (Chen et al. 2018; Qin et al. 2021). AMF moderately decreased Cd transport and accumulation in maize plants (Zhuo et al. 2020). Transport of as in various lentil genotypes colonized by AMF has been reported, and it was found in reducing arsenic phytotoxicity (Alam et al. 2019). Table 3 lists AMF species colonized with different host plants and their mitigation against HM-contaminated soil.

Chen et al. (2018), through SR-μXRF imaging, found a high Cd assemblage in fungal arbuscules and intercellular hyphae. Chen et al. (2018) state that fungal structures like arbuscules were where Cd was primarily stored, and its distribution to plant cells was restricted. According to Wu et al. (2016a), both intraradical and extraradical hyphae of AMF can reduce metal toxicity in plants by chelating with other substances in the cytoplasm or by removing heavy metals from the cell wall and storing them in the vacuole (Fig. 4). Specific genes have been identified in AM fungi that withstand metal stress and retain cellular homeostasis. For instance, the *Gigaspora rosea* GrosMT1 gene, the *Gigaspora margarita* GmarMT1 gene, and the zinc transporter GinZnT1 in *R. intraradices* can all encode MTs and control the ability of fungi to reduce oxidative stress, shielding plants from metal-induced oxidative damage. According to a study by Adeyemi et al. (2021), AMF, like *F. mosseae,* was the highest tolerant in easing metal toxicity by the retention of HMs in the roots, consequently lowering translocation of Cu, Pb, and Zn by 21.8%, 57.6%, and 67.3% correspondingly in the upper region of soybean (*Glycine max* L.) and enhancing total plant production by 59.1%.

6 Conclusion

Diverse groups of xenobiotics are encrypting in every ecosystem, including the agricultural field. The impact of such xenobiotics on plants, microbial physiology, and biochemical and molecular domains is enormous. With the increase of those chemicals, future plant health and productivity vis-a-vis animal food supply are at extreme risk. In this junction, the AMF, one of the critical soil dwellers and plant colonizers, can profoundly mitigate several xenobiotic factors through its unique adaptive biochemical and molecular mechanism. These subterranean symbionts form symbiotic relationships with the plant roots of most vascular plants, greatly influencing nutrient dynamics and soil structure. The partnership primarily benefits plants by enhancing the soaking of vital nutrients, particularly nitrogen and phosphorus, often limited in soils. The plant and AMF relations are fundamental to the functioning of many ecosystems, as they influence plant diversity, productivity, and resilience. Without these symbiotic associations, many plant species would struggle to thrive, which would, in turn, impact the entire ecosystem. AMF also mitigates xenobiotic stress like environmental pollutants, chemical fertilizers, pesticides, and industrial chemicals. The mechanisms by which AMF contributes to the degradation of xenobiotics are still not fully understood, which is why more biochemical, physiological, and molecular studies are necessary. In contrast, they also have a

Table 3 AMF's ability to assist plants in simulating HM stress

AMF species	Host plant	AMF response to host	Heavy metals	Reference
Funneliformis mosseae, Rhizophagus intraradices, Claroideoglomus etunicatum	*Solanum lycopersicum*	By mitigating oxidative stress, AMF decreases the production of malonaldehyde and H_2O_2 and is also effective against Cd stress	Cd	Hashem et al. (2016)
F. mosseae, Glomus versiforme, R. intraradices	*Zea mays*	AMF moderately decreased Cd concentration and increased soil pH	Cd	Zhuo et al. (2020)
G. versiforme, R. intraradices	*Lonicera japonica*	Roots had higher Cd amounts, while shoots had lower values	Cd	Jiang et al. (2016)
G. aggregatum, G. intraradices, G. elunicatum, G. versiforme	*Medicago sativa*	Gradually reduce Cd concentrations in treated plants	Cd	Zhang et al. (2019)
G. intraradices	*Zea mays*	Pb Phytostabilization	Pb	Sudová and Vosátka (2007)
G. mosseae	*Glycine max, Lens culinaris*	Zn, Ni Phytoextraction	Zn, Ni	Jamal et al. (2002)
C. claroideum , C. etunicatum, F. mosseae, R. intraradices	*Cajanus cajan*	By boosting the production of thiols, AMF supplements enhance growth, nutrient intake, rhizobial symbiosis, decreased oxidative stress, and Cd absorption	Cd	Bisht et al. (2023)
R. fasciculatus, R. intraradices, F. mosseae, G. aggregatum	*Zea mays*	Phytoextraction	Cd, Cr, Ni, Pb	Singh et al. (2019)
G. mosseae, G. intraradices	*Helianthus annuus*	*G. mosseae* and sludge amendments produced the highest glomalin concentrations and metal uptakes	Cr, Mn, Ni, Cu, Zn, Al, Pb, Co, Mo, Fe, and Si	Sayın et al. (2019)
F. mosseae	*Lavandula angustifolia*	Phytoremediation	Pb, Ni	Pirsarandib et al. (2022)
G. monosporum, G. clarum, Gigaspora nigra	*Trigonella foenum-graecum*	Phytostabilization	Cd	Abdelhameed and Metwally (2019)
F. mosseae, R. intraradices	*Robinia pseudoacacia*	Pb concentrations in the leaves were reduced by AMF symbiosis, although biomass accumulation and photosynthetic pigment contents were enhanced	Pb	Yang et al. (2015)

(continued)

Table 3 (continued)

AMF species	Host plant	AMF response to host	Heavy metals	Reference
G. versiforme, R. intraradices	*Lonicera japonica*	Due to the enhancement of P feeding and the decrease in shoot Cd contents, the Gv and Ri symbiosis reduced the Cd toxicity of *L. japonica*	Cd	Jiang et al. (2016)
G. intraradices	*Sorghum bicolor*	Phytostabilization	Cu	Toler et al. (2005)
F. mosseae	*Phragmites australis*	*F. mosseae* enhanced the buildup of Ti in roots and altered the arrangement of Ti in reeds	TiO2NPs	Xu et al. (2019)

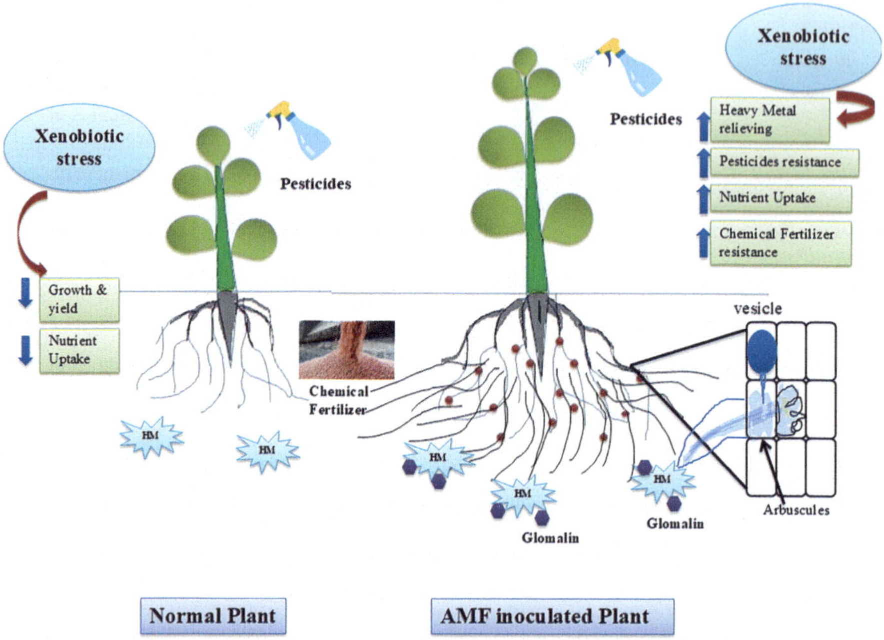

Fig. 4 AMF-mediated direct and indirect remediation of xenobiotic stress, such as HMs, pesticides, and fertilizer, on plants. The host plant's morphology and physiology change due to AM association, aiding in the plant's ability to withstand HM stress. Glomalin is a glycoprotein produced by AM-associated plants that have the ability to bind metals. Consequently, chelating HM into fungal structures and using glomalin to immobilize the metal in the mycorrhizosphere leads to AM metal tolerance

tolerance limit to these xenobiotic factors. Therefore, more vivid research is required to investigate their selective exploitation in mitigating their xenobiotic stress in the host and AMF diversity in the soil. Such research could shed light on how AMF influences the breakdown of chemical compounds, their interactions with soil microbes involved in degradation, and how they might alter xenobiotic metabolism within the plant-soil system. Excessive use of xenobiotics can create a toxic environment in the soil that disrupts the natural community structure of AMF, leading to reduced diversity and shifts in species composition, altered community composition, and decreased ecosystem functions. So, we have to minimize xenobiotics and adopt biofertilizers such as AMF, a sustainable approach to improving soil health, enhancing crop productivity, and protecting ecosystems for sustainable agriculture.

References

Abdelhameed, R. E., and Metwally, R. A. 2019. Alleviation of cadmium stress by arbuscular mycorrhizal symbiosis. *International Journal of Phytoremediation*, *21*(7), 663–671. https://doi.org/10.1080/15226514.2018.1556584

Adeyemi, N. O., Atayese, M. O., Sakariyawo, O. S., Azeez, J. O., Sobowale, S. P. A., Olubode, A., ... and Adeoye, S. 2021. Alleviation of heavy metal stress by arbuscular mycorrhizal symbiosis in *Glycine max* (L.) grown in copper, lead and zinc contaminated soils. *Rhizosphere*, *18*, 100325. https://doi.org/10.1016/j.rhisph.2021.100325

Aktar, M. W., Sengupta, D., and Chowdhury, A. 2009. Impact of pesticides use in agriculture: their benefits and hazards. *Interdisciplinary Toxicology, 2(1)*, 1–12.

Alam, M. Z., Hoque, M. A., Ahammed, G. J., and Carpenter-Boggs, L. 2019. Arbuscular mycorrhizal fungi reduce arsenic uptake and improve plant growth in *Lens culinaris*. *PLoS One*, *14*(5), e0211441. https://doi.org/10.1371/journal.pone.0211441

Amer, S. M., and Ali, E. M. 1974. Cytological effects of pesticides V. Effects of some herbicides on *Viciafaba*. *Cytologia* 39, 633–643. https://doi.org/10.1508/cytologia.39.633

Amer, S.M. and Ali, E.M., 1983. Cytological Effects of Pesticides XIV. Effect of the insecticide Dipterex "trichlorphon" on Vicia faba plant. Cytologia, 48(4), pp.761–770

Ansari, M. S., Moraiet, M. A., and Ahmad, S. 2014. Insecticides: impact on the environment and human health. *Environmental deterioration and human health: Natural and Anthropogenic Determinants*, 99–123.

Antunes, P. M., Lehmann, A., Hart, M. M., Baumecker, M., and Rillig, M. C. 2012. Long-term effects of soil nutrient deficiency on arbuscular mycorrhizal communities. *Functional Ecology*, *26*(2), 532–540. https://doi.org/10.1111/j.1365-2435.2011.01953.x

Aslam, R., Bhat, T. M., Choudhary, S., and Ansari, M. Y. K. 2017. An overview of the genotoxicity of heavy metal in a spice crop (*Capsicum annuum* L.) in respect to cytomorphological behaviour. *Caryologia* 70, 42–47. https://doi.org/10.1080/00087114.2016.1258884

Berruti, A., Lumini, E., Balestrini, R., and Bianciotto, V. 2016. Arbuscular mycorrhizal fungi as natural biofertilizers: let's benefit from past successes. *Frontiers in Microbiology*, *6*, 1559. https://doi.org/10.3389/fmicb.2015.01559

Bhadalung, N. N., Suwanarit, A., Dell, B., Nopamornbodi, O., Thamchaipenet, A., and Rungchuang, J. 2005. Effects of long-term Np-fertilization on abundance and diversity of arbuscular mycorrhizal fungi under a maize cropping system. *Plant and Soil*, *270*, 371–382. https://doi.org/10.1007/s11104-004-1829-4

Bharat, N. K. 2011. Effect of commonly used fungicides on arbuscular-mycorrhizal association with apple roots. *World Journal of Science and Technology*, *1*(6), 61–64

Bisht, A., Bhalla, S., Kumar, A., Kaur, J., and Garg, N. 2023. Arbuscular mycorrhizae impart Cd tolerance in *Cajanscajan* (L.) Millsp. by upregulating the expression of metallothionein (CcMT1) and phytochelatin synthase (CcPCS1) genes. *Journal of Plant Growth Regulation*, *42*(6), 3947–3966. https://doi.org/10.1007/s00344-022-10864-2

Błaszkowski, J. 2012. Arbuscular Mycorrhizal Fungi (Glomeromycota). *Springer*. ISBN (Hardback): 978-83-89648-82-2

Brady, N.C., and Weil, R.R. 2008. The Nature and Properties of Soils (14th Edition). *Prentice Hall*. ISBN (Hardback): 978-0-02-946030-6

Branco, S., Schauster, A., Liao, H.L., Ruytinx, J. 2022. Mechanisms of stress tolerance and their effects on the ecology and evolution of mycorrhizal fungi. *New Phytologist 235*(6): 2158–2175. https://doi.org/10.1111/nph.18308

Brito, I., Carvalho, M., and Goss, M. J. 2013. Soil and weed management for enhancing arbuscular mycorrhiza colonization of wheat. *Soil Use and Management*, *29*(4), 540–546. https://doi.org/10.1111/sum.12069

Cai, L. M., Wang, Q. S., Luo, J., Chen, L. G., Zhu, R. L., Wang, S., and Tang, C. H. 2019. Heavy metal contamination and health risk assessment for children near a large Cu-smelter in central China. *Science of the Total Environment*, *650*, 725–733. https://doi.org/10.1016/j.scitotenv.2018.09.081

Carrenho, R., Trufem, S. F., and Bononi, V. L. 1998. Arbuscularmycorrhizal fungi in *Citrus sinensis*/*C. limon* treated with Fosetyl-Al and Metalaxyl. *Mycological Research*, *102*(6), 677–682. https://doi.org/10.1017/S0953756297005649

Chakraborty, J., Das, S., and Roy, S. 2017. Bioremediation of xenobiotics. In P. Kumar & M. Khodadoust (Eds.), *Microbial Ecology: Current Advances from Genomics, Metagenomics and Other Omics Approaches* (pp. 243–270). Springer.

Chen, B., Nayuki, K., Kuga, Y., Zhang, X., Wu, S., and Ohtomo, R. 2018. Uptake and intraradical immobilization of Cadmium by arbuscular mycorrhizal fungi as revealed by a stable isotope tracer and synchrotron radiation μX-ray fluorescence analysis. *Microbes and Environments*, *33*(3), 257–263. https://doi.org/10.1264/jsme2.ME18010

Choudhary, A., and Kaushik, N. 2020. Impact of pesticides on the diversity and population dynamics of AMF in agricultural soils. *International Journal of Agriculture and Biology, 22(2)*, 451–458.

Currie, R. S., and Liang, G. H. 1996. Cytological and morphological effects of atrazine and propazine application on grain sorghum. *Cytologia* 61, 359–363. https://doi.org/10.1508/cytologia.61.359

deNovais, C. B., Avio, L., Giovannetti, M., de Faria, S. M., Siqueira, J. O., and Sbrana, C. 2019. Interconnectedness, length and viability of arbuscular mycorrhizal mycelium as affected by selected herbicides and fungicides. *Applied Soil Ecology*, *143*, 144–152. https://doi.org/10.1016/j.apsoil.2019.06.013

Degenhardt, B. 2000. Cell wall adaptations to multiple environmental stresses in maize roots. *Journal of Experimental Botany. 51, 595–603*. https://doi.org/10.1093/jexbot/51.344.595.

Deliopoulos, T., Haydock, P. P., and Jones, P. W. 2008. Interaction between arbuscular mycorrhizal fungi and the nematicide aldicarb on hatch and development of the potato cyst nematode, *Globodera pallida*, and yield of potatoes. *Nematology*, *10*(6), 783–799..

Devadas, N., Rajam, M. V., and Subhash, K. 1986. Comparative mutagenicity of four organophosphorus insecticides in meiotic system of red pepper. *Cytologia* 51, 645–653. https://doi.org/10.1508/cytologia.51.645

Druille, M., Cabello, M. N., Omacini, M., and Golluscio, R. A. 2013. Glyphosate reduces spore viability and root colonization of arbuscular mycorrhizal fungi. *Applied Soil Ecology*, *64*, 99–103. https://doi.org/10.1016/j.apsoil.2012.10.007

El Nemr, A., Moneer, A. A., El-Sikaily, A., and Khaled, A. 2012. PCBs and pesticides in the soft tissue of bivalves along the coast of north Egypt. *Blue Biotechnology Journal*, *1*(2), 301.

Emmanuel, O. C., and Babalola, O. O. 2020. Productivity and quality of horticultural crops through co-inoculation of arbuscular mycorrhizal fungi and plant growth-promoting bacteria. *Microbiological Research*, *239*, 126569. https://doi.org/10.1016/j.micres.2020.126569

Fall, A. F., Nakabonge, G., Ssekandi, J., Founoune-Mboup, H., Badji, A., Ndiaye, A., ... and Ekwangu, J. 2023. Combined effects of indigenous arbuscular mycorrhizal fungi (AMF) and NPK fertilizer on growth and yields of maize and soil nutrient availability. *Sustainability*, *15*(3), 2243. https://doi.org/10.3390/su15032243

Fantke, P., and Juraske, R. 2013. Variability of pesticide dissipation half-lives in plants. *Environmental Science & Technology*, *47*(8), 3548–3562. https://doi.org/10.1021/es303525x

Farag, M. R., and Alagawany, M. 2018. Erythrocytes as a biological model for screening of xenobiotics toxicity. *Chemico-biological Interactions*, *279*, 73–83.

Fu, Y., Dou, X., Lu, Q., Qin, J., Luo, J., and Yang, M. 2020. Comprehensive assessment for the residual characteristics and degradation kinetics of pesticides in *Panax notoginseng* and planting soil. *Science of the Total Environment*, *714*, 136718. https://doi.org/10.1016/j.scitotenv.2020.136718

Geisseler, D., and Scow, K.M. 2014. Long-term effects of mineral fertilizers on soil microorganisms—A review. *Soil Biology and Biochemistry*, *75*, 54–63. https://doi.org/10.1016/j.soilbio.2014.03.023

George, N. M. 2000. Evaluation on mutagenic effects of the three heavy metals on *Vicia faba* plants. *Cytologia 65,* 75–82. https://doi.org/10.1508/cytologia.65.75

Ghasemi Siani, N., Fallah, S., Pokhrel, L. R., and Rostamnejadi, A. 2017. Natural amelioration of Zinc oxide nanoparticle toxicity in fenugreek (*Trigonella foenum-gracum*) by arbuscular mycorrhizal (*Glomus intraradices*) secretion of glomalin. *Plant Physiology and Biochemistry*, *112*, 227–238. https://doi.org/10.1016/j.plaphy.2017.01.001

Giovannini, L., Palla, M., Agnolucci, M., Avio, L., Sbrana, C., Turrini, A., Giovannettim, M. 2020. Arbuscular mycorrhizal fungi and associated microbiota as plant biostimulants: research strategies for the selection of the best-performing inocula. *Agronomy* *10*(1):106. https://doi.org/10.3390/agronomy10010106

Gonzalez-Chavez, M. C., Carrillo-Gonzalez, R., Wright, S. F., and Nichols, K. A. 2004. The role of glomalin, a protein produced by arbuscular mycorrhizal fungi, in sequestering potentially toxic elements. *Environmental Pollution*, *130*(3), 317–323. https://doi.org/10.1016/j.envpol.2004.01.004

Gonzalez-Guerrero, M., Melville, L. H., Ferrol, N., Lott, J. N., Azcon-Aguilar, C., and Peterson, R. L. 2008. Ultrastructural localization of heavy metals in the extraradical mycelium and spores of the arbuscularmycorrhizal fungus *Glomus intraradices*. *Canadian Journal of Microbiology*, *54*(2), 103–110. https://doi.org/10.1139/W07-119

Goulding, K.W.T. 2016. Soil acidification and the importance of liming agricultural soils with particular reference to the United Kingdom. *Soil Use and Management, 32*(3), 390–399.

Goulson, D. 2013. An overview of the environmental risks posed by neonicotinoid insecticides. *Journal of Applied Ecology*, *50*(4), 977–987. https://doi.org/10.1111/1365-2664.12111

Gujre, N., Agnihotri, R., Rangan, L., Sharma, M. P., and Mitra, S. 2021. Deciphering the dynamics of glomalin and heavy metals in soils contaminated with hazardous municipal solid wastes. *Journal of Hazardous Materials*, *416*, 125869. https://doi.org/10.1016/j.jhazmat.2021.125869

Hamel, C., and Strullu, D. G. 2006. Arbuscular mycorrhizal fungi in field crop production: Potential and new direction. *Canadian Journal of Plant Science*, *86*(4), 941–950. https://doi.org/10.4141/P05-099

Hashem, A., Abd Allah, E. F., Alqarawi, A. A., Al Huqail, A. A., Egamberdieva, D., and Wirth, S. 2016. Alleviation of cadmium stress in *Solanum lycopersicum* L. by arbuscular mycorrhizal fungi via induction of acquired systemic tolerance. *Saudi Journal of Biological Sciences*, *23*(2), 272–281. https://doi.org/10.1016/j.sjbs.2015.11.002

Hassaan, M. A., and El Nemr, A. 2020. Pesticides pollution: Classifications, human health impact, extraction and treatment techniques. *The Egyptian Journal of Aquatic Research*, *46*(3), 207–220.

Jakobsen, I., Murmann, L. M., and Rosendahl, S. 2021. Hormetic responses in arbuscular mycorrhizal fungi. *Soil Biology and Biochemistry, 159*, 108299. https://doi.org/10.1016/j.soilbio.2021.108299

Jamal, A., Ayub, N., Usman, M., and Khan, A. G. 2002. Arbuscular mycorrhizal fungi enhance zinc and nickel uptake from contaminated soil by soybean and lentil. *International Journal of Phytoremediation, 4*(3), 205–221. https://doi.org/10.1080/15226510208500083

Jiang, Q. Y., Zhuo, F., Long, S. H., Zhao, H. D., Yang, D. J., Ye, Z. H., … and Jing, Y. X. 2016. Can arbuscular mycorrhizal fungi reduce Cd uptake and alleviate Cd toxicity of *Lonicera japonica* grown in Cd-added soils? *Scientific Reports, 6*(1), 21805. https://doi.org/10.1038/srep21805

Jin, H., Germida, J. J., and Walley, F. L. 2013. Suppressive effects of seed-applied fungicides on arbuscular mycorrhizal fungi (AMF) differ with fungicide mode of action and AMF species. *Applied Soil Ecology, 72*, 22–30. https://doi.org/10.1016/j.apsoil.2013.05.013

Joutey, N.T., Sayel, H., Bahafid, W., El Ghachtouli, N. 2015. Mechanisms of hexavalent chromium resistance and removal by microorganisms. *Reviews of Environmental Contamination and Toxicology. 233*, 45–69. https://doi.org/10.1007/978-3-319-10479-9_2.

Kakouridis, A., Hagen, J. A., Kan, M. P., Mambelli, S., Feldman, L. J., Herman, D. J., … and Firestone, M. K. 2022. Routes to roots: direct evidence of water transport by arbuscular mycorrhizal fungi to host plants. *New Phytologist, 236*(1), 210–221. https://doi.org/10.1111/nph.18281

Karpouzas, D. G., Papadopoulou, E., Ipsilantis, I., Friedel, I., Petric, I., Udikovic-Kolic, N., … and Martin-Laurent, F. 2014. Effects of nicosulfuron on the abundance and diversity of arbuscular mycorrhizal fungi used as indicators of pesticide soil microbial toxicity. *Ecological Indicators, 39*, 44–53. https://doi.org/10.1016/j.ecolind.2013.12.004

Kaymak, F. 2005. Cytogenetic effects of maleic hydrazide on *Helianthus annuus* L. *Pakisthan Journal of Biological Sciences 8,* 104–108. https://doi.org/10.3923/pjbs.2005.104.108

Klotz, L. O., and Steinbrenner, H. 2017. Cellular adaptation to xenobiotics: Interplay between xenosensors, reactive oxygen species and FOXO transcription factors. *Redox Biology, 13*, 646–654. https://doi.org/10.1016/j.redox.2017.07.015

Kobayashi, K., and Croucher, R. 2009. Effects of fungicides on soil microorganisms. *Microbes and Environments, 24(1),* 1–9.

Krzesłowska, M. 2011. The cell wall in plant cell response to trace metals: Polysaccharide remodeling and its role in defense strategy. *Acta Physiologie Plantarum 33, 35–51.* https://doi.org/10.1007/s11738-010-0581-z.

Kuchy, A., Wani, A., and Kamili, A. 2016. Cytogenetic effects of three commercially formulated pesticides on somatic and germ cells of *Allium cepa. Environmental Science and Pollution Research* 23, 6895–6906. https://doi.org/10.1007/s11356-015-5912-6

Kumar, G. 2007. Comparative genotoxic potential of mercury and Cadmium in soybean. *Turkish Journal of Biology 31,* 13–18.

Kumar, G., and Rai, P. K. 2010. The genotoxic potential of two heavy metals in inbred lines of maize (*Zea mays*). *Turkish Journal of Botany 34,* 39–46.

Kumar, G., and Ritambhara, T. 2006. Individual and combined effects of lead and gamma irradiations on genetic recombination in *Lathyrus sativus. Journal of Phytological Research 19,* 215–220.

Kumar, U., and Sinha, S. S. N. 1991. Genotoxic effects of two pesticides (Rogor and Bavistin) and an antibiotic (Streptomycin) in meiotic cells of grass pea (*Lathyrus sativus* L.). *Cytologia* 56, 209–214. https://doi.org/10.1508/cytologia.56.209

Law, C. 1963. An effect of potassium on chiasma frequency and recombination. *Genetica 33,* 313–329. https://doi.org/10.1007/BF01725768

Lee, K., Rao, G., Barnett, F., and Liang, G. 1974. Further evidence of meiotic instability induced by Atrazine in grain sorghum. *Cytologia 39,* 697–702. https://doi.org/10.1508/cytologia.39.697

Li, W.W., Yu, H.Q. 2014. Insight into the roles of microbial extracellular polymer substances in metal biosorption. *Bioresource Technology 160,* 15–23. https://doi.org/10.1016/j.biortech.2013.11.074.

Lu, P. C., Shahbaz, S., and Winn, L. M. 2020. Benzene and its effects on cell signaling pathways related to hematopoiesis and leukemia. *Journal of Applied Toxicology*, *40*(8), 1018–1032.

Ma, Y., Zhang, H., Wang, D., Guo, X., Yang, T., Xiang, X., … and Chu, H. 2021. Differential responses of arbuscular mycorrhizal fungal communities to long-term fertilization in the wheat rhizosphere and root endosphere. *Applied and Environmental Microbiology*, *87*(17), e00349–21. https://doi.org/10.1128/AEM.00349-21

Macfie, S.M.,Welbourn, P.M. 2000. The cell wall as a barrier to uptake of metal ions in the unicellular green alga *Chlamydomonas reinhardtii* (Chlorophyceae). *Archives of Environmental Contamination and Toxicology. 39*, 413–419. https://doi.org/10.1007/s002440010122.

Mäder, P., Fließbach, A., Dubois, D., Gunst, L., Fried, P., and Niggli, U. 2002. Soil fertility and biodiversity in organic farming. *Science, 296*(5573), 1694–1697. https://doi.org/10.1126/science.1071148

Mahmood, I., Imadi, S. R., Shazadi, K., Gul, A., and Hakeem, K. R. 2016. "Effects of pesticides on environment." In: *Plant, Soil and Microbes* (pp. 253–269). Springer, Cham. https://doi.org/10.1007/978-3-319-27455-3_13

Malekzadeh, E., Aliasgharzad, N., Majidi, J., Abdolalizadeh, J., and Aghebati-Maleki, L. 2016. Contribution of glomalin to Pb sequestration by arbuscular mycorrhizal fungus in a sand culture system with clover plant. *European Journal of Soil Biology*, *74*, 45–51. https://doi.org/10.1016/j.ejsobi.2016.03.003

Malfatti, A. D. L. R., Oliveira Filho, L. C. I., Carniel, L. S. C., Mallmann, G. C., Da Cruz, S. P., and Klauberg-Filho, O. 2023. Risk assessment tests of neonicotinoids on spore germination of arbuscular mycorrhizal fungi *Gigaspora albida* and *Rhizophagus clarus*. *Journal of Soils and Sediments*, *23*(3), 1295–1303. https://doi.org/10.1007/s11368-022-03419-1

Malty, J. D. S., Siqueira, J. O., and Moreira, F. M. D. S. 2006. Effects of glyphosate on soybean symbiotic microorganisms, in culture media and in greenhouse. *Pesquisa Agropecuária Brasileira*, *41*, 285–291.

Manzoor, F., and Pervez, M. 2017. HPLC analysis to determine the half-life and bioavailability of the termiticides bifenthrin and fipronil in soil. *Journal of Economic Entomology*, *110*(6), 2527–2533. https://doi.org/10.1093/jee/tox249

Meena, R. S., Kumar, S., Datta, R., Lal, R., Vijayakumar, V., Brtnicky, M., … and Marfo, T. D. 2020. Impact of agrochemicals on soil microbiota and management: A review. *Land*, *9*(2), 34.

Meshram, R. E. E. N. A., and Chaturvedi, A. L. K. A. 2017. Effect of insecticide on in vitro pollen germination of *Lycopersicon esculentum* (Mill.) of F1 hybrid variety Laxmi. *International Journal of Applied and Natural Sciences*, *6*(4), 1–10.

Mittal, N., and Srivastava, A. K. 2014. Cadmium and chromium induced aberrations in the reproductive biology of *Hordeum vulgare*. *Cytologia* 79, 207–214. https://doi.org/10.1508/cytologia.79.207

More, T.T., Yadav, J.S.S., Yan, S., Tyagi, R.D., Surampalli, R.Y. 2014. Extracellular polymeric substances of bacteria and their potential environmental applications. *Journal of Environmental Management, 144*, 1–25. https://doi.org/10.1016/j.jenvman.2014.05.010.

Nafady, N. A., and Elgharably, A. 2018. Mycorrhizal symbiosis and phosphorus fertilization effects on *Zea mays* growth and heavy metals uptake. *International Journal of Phytoremediation*, *20*(9), 869–875.

Okiobe, S.T., Meidl, P., Koths, T., Olschewsky, D., Rillig, M.C. and Lammel, D.R. 2022. Root colonization by arbuscular mycorrhizal fungi is reduced in tomato plants sprayed with fungicides. *Frontiers in Agronomy 4*:1028195. https://doi.org/10.3389/fagro.2022.1028195

Park, D., Yun, Y.S., Park, J.M. 2006. Comment on the removal mechanism of hexavalent chromium by biomaterials or biomaterial-based activated carbons. *Industrial & Engineering Chemistry Research 45*, 2405–2407. https://doi.org/10.1021/ie0509387.

Phour, M., Sehrawat, A., Sindhu, S.S., Glick, B.R. 2020. Interkingdom signaling in plant rhizomicrobiome interactions for sustainable agriculture. *Microbiological Research 241*:126589. https://doi.org/10.1016/j.micres.2020.126589

Pirsarandib, Y., Hassanpouraghdam, M. B., Rasouli, F., Aazami, M. A., Puglisi, I., and Baglieri, A. 2022. Phytoremediation of soil contaminated with heavy metals via arbuscular mycorrhiza (*Funneliformis mosseae*) inoculation ameliorates the growth responses and essential oil content in lavender (*Lavandula angustifolia* L.). *Agronomy*, *12*(5), 1221.

Prakash, N., Lakshmi, N., and Harini, I. 1988. Cytological effects of agricultural chemicals. II. Effects of fungicides "Bavistin" and "Deltan" on chilli (*Capsicum annuum* L.). *Cytologia 53, 709–715.* https://doi.org/10.1508/cytologia.53.709

Qin, G., Niu, Z., Yu, J., Li, Z., Ma, J., and Xiang, P. 2021. Soil heavy metal pollution and food safety in China: Effects, sources and removing technology. *Chemosphere*, *267*, 129205. https://doi.org/10.1016/j.chemosphere.2020.129205

Rasmussen, P. E., Goulding, K. W., Brown, J. R., Grace, P. R., Janzen, H. H., and Körschens, M. 1998. Long-term agroecosystem experiments: assessing agricultural sustainability and global change. *Science, 282(5390)*, 893–896. https://doi.org/10.1126/science.282.5390.893

Razu, M., Zaman, S., Akhter, R., Rahman, M., Rahaman, M. H., Mazid, M., et al. 2014. Morphological and cytological effects of two herbicides on tetraploid wheat (*Triticum durum* L.). *Journal of Biosciences* 20, 143–151

Salvioli di Fossalunga, A., Novero, M. 2019. To trade in the field: the molecular determinants of arbuscular mycorrhiza nutrient exchange. *Chemical and Biological Technologies in Agriculture 6*(1):12. https://doi.org/10.1186/s40538-019-0150-7

Sarr, B., Ndiaye, F., and Diop, T. A. 2013. Effect of two types of insecticides on mycorrhization and development of two potato varieties (*Solanum tuberosum*). *International Journal of Biological and Chemical Sciences*, *7*(5), 1902–1909.

Sayın, F. E., Khalvati, M. A., and Erdinçler, A. 2019. Effects of sewage sludge application and arbuscularmycorrhizal fungi (*G. mosseae* and *G. intraradices*) interactions on the heavy metal phytoremediation in chrome mine tailings. *Frontiers in Plant Science*, *112*, 217–224.

Schreiner, R. P., and Bethlenfalvay, G. J. 1997. Plant and soil response to single and mixed species of arbuscular mycorrhizal fungi under fungicide stress. *Applied Soil Ecology*, *7*(1), 93–102. https://doi.org/10.1016/S0929-1393(96)00141-2

Sheng, M., Hamel, C., and Fernandez, M. R. 2012. Cropping practices modulate the impact of glyphosate on arbuscular mycorrhizal fungi and rhizosphere bacteria in agroecosystems of the semiarid prairie. *Canadian Journal of Microbiology*, *58*(8), 990–1001. https://doi.org/10.1139/w2012-080

Shi, J., Wang, X., Wang, E. 2023. Mycorrhizal symbiosis in plant growth and stress adaptation: from genes to ecosystems. *Annual Review of Plant Biology 74*(1): 569–607. https://doi.org/10.1146/annurevarplant-061722-090342

Singh, G., Pankaj, U., Chand, S., and Verma, R. K. 2019. Arbuscular mycorrhizal fungi-assisted phytoextraction of toxic metals by *Zea mays* L. from tannery sludge. *Soil and Sediment Contamination: An International Journal*, *28*(8), 729–746. https://doi.org/10.1080/15320383.2019.1657381

Smith, S.E., and Read, D.J. 2008. Mycorrhizal Symbiosis (3rd Ed.). Academic Press. ISBN: 9780080559346, 0080559344

Soriano, J. D. 1984. Herbicide-induced chromosomal aberrations and inheritance of a digenic seedling mutation in sorghum. *Cytologia 49,* 201–207. https://doi.org/10.1508/cytologia.49.201

Štefanac, T., Grgas, D., and Landeka Dragičević, T. 2021. Xenobiotics—division and methods of detection: a review. *Journal of Xenobiotics*, *11*(4), 130–141. https://doi.org/10.3390/jox11040009

Sudová, R., and Vosátka, M. 2007. Differences in the effects of three arbuscular mycorrhizal fungal strains on P and Pb accumulation by maize plants. *Plant and Soil*, *296*, 77–83. https://doi.org/10.1007/s11104-007-9291-8

Toler, H.D., Morton, J.B., and Cumming, J.R., 2005. Growth and metal accumulation of mycorrhizal sorghum exposed to elevated copper and zinc. *Water, Air, & Soil Pollution* 164, 155–172. https://doi.org/10.1007/s11270-005-2718-z.

Top, E. M., and Springael, D. 2003. The role of mobile genetic elements in bacterial adaptation to xenobiotic organic compounds. *Current Opinion in Biotechnology*, 14(3), 262–269. https://doi.org/10.1016/S0958-1669(03)00066-1

Trappe, J. M., Molina, R., and Castellano, M. 1984. Reactions of mycorrhizal fungi and mycorrhiza formation to pesticides. *Annual Review of Phytopathology*, *22*(1), 331–359.

Trejo, D., Sangabriel-Conde, W., Gavito-Pardo, M. E., and Banuelos, J. 2021. Mycorrhizal inoculation and chemical fertilizer interactions in pineapple under field conditions. *Agriculture*, *11*(10), 934.

Treseder, K.K. 2004. A meta-analysis of mycorrhizal responses to nitrogen, phosphorus, and carbon additions. *Ecology Letters,* *7*(11), 1091–1102. https://doi.org/10.1111/j.1469-8137.2004.01159.x

Vieira, R. F., Silva, C. M. M., and Silveira, A. P. 2007. Soil microbial biomass C and symbiotic processes associated with soybean after sulfentrazone herbicide application. *Plant and Soil*, *300*, 95–103. https://doi.org/10.1007/s11104-007-9392-4

Wang, F. Y., Shi, Z. Y., Tong, R. J., and Xu, X. F. 2011a. Dynamics of phoxim residues in green onion and soil as influenced by arbuscular mycorrhizal fungi. *Journal of Hazardous Materials*, *185*(1), 112–116. https://doi.org/10.1016/j.jhazmat.2010.09.004

Wang, F. Y., Tong, R. J., Shi, Z. Y., Xu, X. F., and He, X. H. 2011b. Inoculations with arbuscular mycorrhizal fungi increase vegetable yields and decrease phoxim concentrations in carrot and green onion and their soils. *PLoS One*, *6*(2), e16949. https://doi.org/10.1371/journal.pone.0016949

Wang, L., Cho, D. W., Tsang, D. C., Cao, X., Hou, D., Shen, Z., ... and Poon, C. S. 2019. Green remediation of As and Pb contaminated soil using cement-free clay-based stabilization/solidification. *Environment International*, *126*, 336–345. https://doi.org/10.1016/j.envint.2019.02.057

Wilkes, T. I., Warner, D. J., Davies, K. G., and Edmonds-Brown, V. 2020. Tillage, glyphosate and beneficial arbuscular mycorrhizal fungi: optimizing crop management for plant–fungal symbiosis. *Agriculture*, *10*(11), 520.

Wu, S., Zhang, X., Chen, B., Wu, Z., Li, T., Hu, Y., Sun, Y., Wang, Y. 2016a. Chromium immobilization by extraradical mycelium of arbuscular mycorrhiza contributes to plant chromium tolerance. *Environmental and Experimental Botany* *122*, 10–18. https://doi.org/10.1016/j.envexpbot.2015.08.006.

Wu, S., Zhang, X., Huang, L., Chen, B. 2019. Arbuscular mycorrhiza and plant chromium tolerance. *Soil Ecology Letters* *1*, 94–104. https://doi.org/10.1007/s42832-019-0015-9

Wu, S., Zhang, X., Sun, Y., Wu, Z., Li, T., Hu, Y., ... and Chen, B. 2016b. Chromium immobilization by extra-and intraradical fungal structures of arbuscular mycorrhizal symbioses. *Journal of Hazardous Materials*, *316*, 34–42. https://doi.org/10.1016/j.jhazmat.2016.05.017

Wu, S., Zhang, X., Sun, Y., Wu, Z., Li, T., Hu, Y., Su, D., Lv, J., Li, G., Zhang, Z., Zheng, L., Zhang, J., Chen, B. 2015. Transformation and immobilization of Chromium by Arbuscular Mycorrhizal Fungi as revealed by SEM-EDS, TEM-EDS, and XAFS. *Environmental Science & Technology. 49*, 14036–14047. https://doi.org/10.1021/acs.est.5b03659.

Wu, Z., McGrouther, K., Huang, J., Wu, P., Wu, W., and Wang, H. 2014. Decomposition and the contribution of glomalin-related soil protein (GRSP) in heavy metal sequestration: field experiment. *Soil Biology and Biochemistry*, *68*, 283–290. https://doi.org/10.1016/j.soilbio.2013.10.010

Wuu, K. 1967. Chromosomal aberrations induced by pesticides in meiotic cells of barley. *Cytologia* 32, 31–41. https://doi.org/10.1508/cytologia.32.31

Xu, Z., Wu, Y., Xiao, Z., Ban, Y., and Belvett, N. 2019. Positive effects of *Funneliformis mosseae* inoculation on reed seedlings under water and TiO 2 nanoparticles stresses. *World Journal of Microbiology and Biotechnology*, *35*, 1–13.https://doi.org/10.1007/s11274-019-2656-3.

Yadav, I. C., and Devi, N. L. 2017. Pesticides classification and its impact on human and environment. *Environmental Science and Engineering*, *6*(7), 140–158.

Yang, Y., Han, X., Liang, Y., Ghosh, A., Chen, J., and Tang, M. 2015. The combined effects of arbuscular mycorrhizal fungi (AMF) and lead (Pb) stress on Pb accumulation, plant growth

parameters, photosynthesis, and antioxidant enzymes in *Robinia pseudoacacia* L. *PloS One*, *10*(12), e0145726.https://doi.org/10.1371/journal.pone.0145726.
Zhang, C., and Wang, S. 2020. "Pesticides and soil health: impact on soil microbial communities and ecosystem functions." *Current Opinion in Environmental Science & Health, 15*, 100–107.
Zhang, S., Luo, P., Yang, J., Irfan, M., Dai, J., An, N., … and Han, X. 2021. Responses of arbuscular mycorrhizal fungi diversity and community to 41-year rotation fertilization in brown soil region of northeast China. *Frontiers in Microbiology*, *12*, 742651. https://doi.org/10.3389/fmicb.2021.742651
Zhang, X. F., Hu, Z. H., Yan, T. X., Lu, R. R., Peng, C. L., Li, S. S., and Jing, Y. X. 2019. Arbuscular mycorrhizal fungi alleviate Cd phytotoxicity by altering Cd subcellular distribution and chemical forms in *Zea mays*. *Ecotoxicology and Environmental Safety*, *171*, 352–360. https://doi.org/10.1016/j.ecoenv.2018.12.097
Zeyad, M.T., Kumar, M. and Malik, A., 2019. Mutagenicity, genotoxicity and oxidative stress induced by pesticide industry wastewater using bacterial and plant bioassays. Biotechnology Reports, 24, p.e00389.https://doi.org/10.1016/j.btre.2019.e00389
Zhu, B., Gao, T., Zhang, D., Ding, K., Li, C., Ma, F. 2022. Functions of arbuscular mycorrhizal fungi in horticultural crops. *Scientia Horticulturae 303*:111219. https://doi.org/10.1016/j.scienta.2022.111219
Zhuo, F., Zhang, X. F., Lei, L. L., Yan, T. X., Lu, R. R., Hu, Z. H., and Jing, Y. X. 2020. The effect of arbuscular mycorrhizal fungi and biochar on the growth and Cd/Pb accumulation in *Zea mays*. *International Journal of Phytoremediation*, *22*(10), 1009–1018. https://doi.org/10.1080/15226514.2020.1725867
Ziane, H., Meddad-Hamza, A., Beddiar, A., and Gianinazzi, S. 2017. Effects of arbuscular mycorrhizal fungi and fertilization levels on industrial tomato growth and production. *International Journal of Agriculture and Biology*, *19*(2), 341–347.
Zou, Y. N., Srivastava, A. K., Ni, Q. D., and Wu, Q. S. 2015. Disruption of mycorrhizal extraradical mycelium and changes in leaf water status and soil aggregate stability in root box-grown trifoliate orange. *Frontiers in Microbiology*, *6*, 203. https://doi.org/10.3389/fmicb.2015.00203

Xenobiotics in the Hydrosphere: Natural Degradation Pathways and Future Concerns

Usha Sabharwal, Shivangee Solanki, and Hemanga Kumar Das

Abstract Xenobiotics are synthetic chemicals that are alien to an ecosystem. Regrettably, they have become ubiquitous in the hydrosphere due to anthropogenic activities. They invade aquatic ecosystems via diverse means and persist due to their exceptional chemical stability. The degradation pathways of these chemicals in the hydrosphere involve a complex interplay of biological, chemical, and physical processes. This review explores and elucidates the diverse mechanisms by which xenobiotics undergo degradation in aquatic environments. Microbial biodegradation, influenced by diverse microbial communities, plays a pivotal role in transforming xenobiotics into metabolites that may exhibit altered toxicity profiles. Additionally, photolysis and abiotic degradation processes driven by factors such as solar radiation and water chemistry contribute to the breakdown of xenobiotics over time. Additionally, factors influencing degradation rates, such as chemical structure, environmental conditions, and microbial communities, are discussed in detail. Despite these natural processes, concerns regarding the persistence and potential accumulation of xenobiotics in the hydrosphere remain significant. Future research directions include enhancing our comprehension of the role of the aquatic microbiota, investigating emerging contaminants, and formulating sustainable management practices to mitigate the long-term consequences of xenobiotic contamination in aquatic ecosystems. Addressing these challenges will be pivotal in safeguarding water quality and ecosystem health in the face of increasing xenobiotic contamination.

Keywords Abiotic degradation · Aquatic contamination · Hydrosphere · Microbial degradation · Xenobiotics

U. Sabharwal
Department of Life Science, Parul Institute of Applied Sciences, Parul University, Waghodia, Vadodara, Gujarat, India

S. Solanki
Department of Biotechnology, Gujarat Biotechnology University, Gandhinagar, Gujarat, India

H. K. Das (✉)
Department of Biotechnology and Biomedical Engineering, U.V. Patel College of Engineering, Ganpat University, Mehsana, Gujarat, India

S. Roy, V. Mandal (eds.), *Plant-Microbe Interaction under Xenobiotic Exposure*,
https://doi.org/10.1007/978-981-96-8260-7_5

1 Introduction

Degradation is the ultimate destiny of every chemical released in the environment. Due to the chemical nature, physical stability, structural chemistry, and biological activity, an extreme degree of heterogeneity exists in the period of breakdown of different kinds of polymers to their respective monomers. Benzenes, biphenyls, anilines, halogenated aliphatic, and pesticides are common examples of slow-degrading compounds (Johnson et al. 2004; Terrazas-Salgado et al. 2022; Shi et al. 2022; Lee et al. 2023; Sandoval et al. 2023).

Microorganism-mediated degradation or biodegradation of any substance is dependent on the physicochemical environment, bioavailability of substrate and/or nutrients, predation, and metabolic efficiency of the microbial population responsible for degrading them (Ayilara et al. 2023; Zhu et al. 2024; Yu et al. 2023). Microorganisms are the major force responsible for biodegradation in any ecosystem. However, substances introduced into the environment as a result of human activities are novel to the native microbial populations. The indigenous enzymatic processes do not identify the rare chemical structures and/or properties, therefore, cannot metabolize them into harmless monomers. These compounds remain in nature for a very long time, therefore, seriously undermining the safety and quality of biotic and abiotic components exposed to them (Othman et al. 2021; Ciuffi et al. 2024).

There have been numerous field experiments and studies documenting a number of compounds that cannot be broken down through natural processes and are primary sources of ecotoxicity and pollution. These compounds are known as xenobiotic pollutants or xenobiotics (Palmer & Herat, 2021). Azo dyes, phenolics, polycyclic aromatic hydrocarbons (PAHs), halogenated compounds, personal care products (PCPs), pharmaceutical active compounds (hACs), pesticides, nitroaromatic compounds, triazines, and chlorinated substances are just a few examples of notoriously xenobiotic compounds. A recent study showed the potential of some microbial populations to make xenobiotics adaptive in nature in an attempt to use them as a prospective growth-stimulating substrate. Regrettably, so much is left unexplored with regard to these adaptation processes (Wang et al. 2024a, b; Li et al. 2024; De Filippis et al. 2024).

This chapter provides review of xenobiotics in the hydrosphere, with an emphasis on their natural degradation processes and future concerns. We review current knowledge about how these long-lived pollutants interact with aquatic biota, the microorganisms that have coevolved to degrade them, and the environmental factors that regulate rates of degradation. This chapter also discusses the huge knowledge gaps in our current understanding of adaptation mechanisms and points out how advanced technologies can be utilized to gain deeper insights into the molecular processes that result in adaptation, thereby providing the foundation for revolutionary breakthroughs in environmental remediation strategies (Wang et al. 2024a, b; Guerra et al. 2018). By uncovering these processes in detail, we hope to offer a thorough framework for confronting the increasing threat of xenobiotic contamination of water environments and creating more efficient bioremediation methods.

Natural organic matter (NOM) plays an important role in controlling sorption behavior of ionizable compounds. Observed that dissolved humic substances promoted sorption of cationic drugs, such as propranolol and metoprolol to suspended particles in river water through the formation of ternary complexes but inhibited sorption of anionic compounds, such as ibuprofen and naproxen through competitive interactions (Paniagua-López et al. 2024; Brown et al. 2016).

Aging processes influence desorption kinetics as well. According to some research, newly contaminated sediments desorbed pesticides (chlorpyrifos and pendimethalin) more easily compared to aged samples, and desorption rates were reduced by 65–80% after 3 months of aging. This "aging effect" is credited with the diffusion of compounds into micropores and increased bonding with the passage of time (Zhang et al. 2025; Pan et al. 2022; Chang et al. 2022).

Temperature changes within aquatic settings may have a substantial effect on sorption-desorption equilibria. Some of the studies indicate that the sorption of xenobiotics that are associated with microplastics, such as bisphenol A and nonylphenol, is reduced by about 30% when water temperature goes from 10 °C to 25 °C, which could lead to a greater bioavailability in warmer periods or in hot discharge zones (Nguyen et al. 2023; Martín et al. 2022; Pannetier et al. 2020).

Knowing some of these sorption-desorption processes is critical for the formulation of efficient strategies for the control of xenobiotic pollutants and their degradation by biotic or abiotic means (Haider et al. 2022).

2 Aquatic Ecotoxicity and Ecotoxicokinetics

Geological aspect of hydrosphere pollution provides essential insights into the transport, persistence, and level of risk of xenobiotics. In-depth ecotoxicological studies are a blend of geography, geophysics, meteorology, and oceanography (Akhtar et al. 2021). Findings from such studies assist in determining the occurrence and fate of xenobiotics at a global level. Over the course of decades, a large number of studies and investigations have highlighted the significance of water as a transport medium and accumulation repository for xenobiotics. While surface water is critically polluted due to manufacturing industries dumping their waste effluent into rivers, the groundwater is at a high risk of pollution due to the seeping of rainwater containing xenobiotic pollutants (Adeola et al. 2022). However, only water-soluble xenobiotics are able to transverse through soil strata while the remaining are trapped within interparticle spaces or soil particle pores. Trichloroethene and tetrachloroethene have been identified as common groundwater pollutants (Ball et al. 1997; Parsons et al. 1984).

Rainfall is a paramount driver responsible for the transport of the majority of xenobiotic substances from the troposphere. "Rain-out" and "wash-out" are terms dedicated to the movement of xenobiotics in the hydro-ecosystem (Kim et al. 2021; Štefanac et al. 2021). Immobilized water bodies, such as lakes, which lack water influx or efflux sources, accumulate xenobiotics solely from rainfall. Hence, they serve as catchment basins for xenobiotic fall-out. Large water bodies, such as oceans, serve not only as a "sink" by receiving rainfall but also as a "source" by re-sequestering chemicals into the troposphere by reverse diffusion, evaporation, and aerosol formation (sea spray) (Butler et al. 2024; Lin et al. 2023). The hydrosphere includes both the water layer system and oceanic surface currents. These circulating systems are responsible for the global transport of xenobiotics and, hence, must be taken into consideration. As far as surface water is concerned, water as well as wind movements influence the interoceanic surface water exchange and ultimately, xenobiotic transport (Hartman et al. 2021; Wang et al. 2022a, b).

Xenobiotics predominantly originate from industrialization, urbanization, transportation, and construction, and find their way into water bodies via airborne particle deposition (from the troposphere), surface water from roads and land surfaces, commercial effluent, sewage disposal, waste from combustion of fossil fuels and solid waste (Roy et al. 2022; Kolawole and Iyiola 2023), and leaching (Kolawole and Iyiola 2023). Prolonged exposure to high concentrations of xenobiotics is not only fatal to the ecological integrity but also cytotoxic to aquatic life (Curpan et al. 2022). Xenobiotics have been recorded to inflict oxidative injury to the cell. A fairly recent study evaluated the fish fauna of a man-made lake in Nigeria. The fish exhibited elevated oxidative stress responses in the face of xenobiotic stress due to altered homeostasis, overproduction of ROS (reactive oxygen species) and RNS (reactive nitrogen species), and impairment of the antioxidant system. Furthermore, several life-sustaining metabolic processes in aquatic life forms are severely compromised (Iyiola et al. 2024; Muluye et al. 2023).

Dyes, paints, pesticides, herbicides, organophosphorus, nitrophenols, morpholine, synthetic pyrethroids, carbamates, and β-Cypermethrin (an insecticide) are regarded as significant sources of hydrosphere pollution. They are categorized as xenobiotic pollutants as they degrade very slowly and exhibit exceptional persistence. Even in trace amounts, they block sunlight from penetrating through the surface and hinder gaseous exchange among marine animals (Wei et al. 2023; Štefanac et al. 2021) .

3 Fate of Xenobiotics in the Hydrosphere: Natural Degradation Pathways

Poor water quality is a critical threat to the human population, global ecosphere, and the world economy. The agricultural and industrial sectors are greatly impacted by the reduction in water quality. Such adversities eventually translate to diseases and pandemic outbreaks (Zhu et al. 2023; Bukowska et al. 2022). Rivers are an important element of the hydrosphere as they sustain an array of global ecosystem services, such as drinking water needs, irrigation, navigation, transportation, recreation, and waste disposal. Regrettably, unwarranted littering and disposal of untreated or poorly treated effluent into freshwater bodies has led to a significant fall in the world's water quality (Mindubaev et al. 2021; Xu et al. 2024). The following sections present a brief discussion of various degradation pathways of xenobiotics identified in an aquatic ecosystem (Summarized in Table 1).

3.1 Photoinitiated Pathways

Xenobiotic degradation pathways that involve the absorption of radiation are identified as photoinitiated reactions (Table 1). This mechanism is significant in surface waters where sunlight penetration is high. The UV radiation can cause the cleavage of chemical bonds in xenobiotics, leading to their transformation into less harmful compounds. They can be further classified on the basis of the wavelength of radiation absorbed. The photolytic pathway is a first-order reaction wherein a xenobiotic substance absorbs high energy UV radiation (typically, 254 nm), absorption of which leads to the cleavage of a covalent bond in the substrate (He et al. 2024; Fairbanks et al. 2021). Photosensitized reactions are those wherein a sensitizer molecule first absorbs radiation and transfers it onto another molecule (potentially, a xenobiotic compound) (Calza et al. 2012; Beijersbergen Van Henegouwen 1991; Kowalska et al. 2021). Humic substances (humic acids and fulvic acids) have been identified as the most common and abundantly occurring natural sensitizers. In the hydrosphere, a myriad of natural sensitizers namely, heme/porphyrin-based molecules (chlorophyll, bilirubin, hemocyanin, and hemoglobin), carotenoid, flavonoids, etc., are observed albeit in minute concentrations (Zhao et al. 2022). The photocatalytic reaction is yet another pathway through which xenobiotics are naturally degraded in the hydrosphere. They can occur either homogeneously or heterogeneously (Baveye et al. 1999).

A wide assortment of studies has highlighted the environmental relevance of photodegradation pathways. Xenobiotic pesticides, such as carbaryl and napropamide, photodegrade readily (Da Silva et al. 2008; Yuan et al. 2025). Antibacterial drugs, such as fluoroquinolones, have also exhibited susceptibility to photodegradation (Rusu et al. 2023; Chen et al. 2024). The persistence and degradability of antidepressants have been excessively studied by researchers. It has been hypothesized that photosensitized reactions involving humic substances are largely responsible for their natural degradation (Ait Chait et al. 2020; Bojanowska-Czajka et al. 2021). All pharmaceuticals were reduced via direct or indirect photoreactions. Chemically, hydroxyl radicals were identified as prime drivers of the reaction (Zhou et al. 2024; Garg and Chandra 2022). Additionally, dinitrotoluene and trinitrotoluene are effectively degraded when exposed to ozone and UV radiations, as are carbamate insecticides and trihalomethanes (Valkova et al. 2001; Fernandes et al. 2023). The aforementioned photochemical reactions play a crucial role in determining the fate of xenobiotics in aquatic ecosystems. Nonetheless, a dearth of experimental investigations retards the development of mitigative strategies to curb xenobiotic pollution. Extensive comprehension of these oxidation techniques can facilitate the safe, feasible, and rapid elimination of xenobiotic contaminants (Hartman et al. 2021; Adamovsky et al. 2024).

Table 1 Natural degradation pathways of xenobiotics in aquatic ecosystems with examples and effectiveness

Degradation pathway	Description	Examples	Effectiveness	Ref.
Photoinitiated degradation	Degradation triggered by absorption of solar radiation (direct photolysis) or through sensitizers/catalysts (indirect photolysis)	Carbaryl pesticide degradation in surface waters ($t_1/_2$ = 5–7 days); fluoroquinolone antibiotics degradation via hydroxyl radical formation	High in clear surface waters; limited in turbid waters or at depths >2 m	MacCallum (1989) and Yousif and Haddad (2013)
Microbial degradation	Transformation of xenobiotics by microorganisms through enzymatic processes; can occur aerobically or anaerobically	Degradation of PCBs by *Pseudomonas* sp. (60–85% removal in 30 days); dehalogenation of trichloroethene by *Desulfitobacterium* sp. (>90% removal)	Very high under optimal conditions; can be limited by bioavailability and toxicity	Amanna et al. (2021) and Kakirde et al. (2010)
Sorption processes	Binding of xenobiotics to suspended particles, sediments, and organic matter, and reducing bioavailability	PCB sorption to organic-rich sediments (Kd 10^3–10^5 L/kg); tetracycline sorption to clay minerals (>95% removal)	Moderate to high for hydrophobic compounds; reversible process; not true degradation	Knezovich et al. (1987) and Skic et al. (2023)
Plant-mediated degradation	Uptake and transformation of xenobiotics by aquatic plants through enzymatic processes and/or rhizosphere effects	*E. crassipes* removal of phenolics (85%); *L. minor* degradation of pharmaceuticals (60–75%); *P. australis* removal of PCBs (40–55%)	Moderate to high; dependent on plant species, pollutant properties, and environmental conditions	Abhilash et al. (2009) and Ur Rahman et al. (2024)
Hydrolysis	Chemical reaction of xenobiotics with water, often pH-dependent	Organophosphate pesticide degradation in alkaline waters ($t_1/_2$ = 2–14 days); carbamate hydrolysis (variable rates)	Low to moderate; highly dependent on chemical structure and pH	Koppel et al. (2017) and Stanley (2017)
Redox reactions	Chemical oxidation or reduction of xenobiotics by naturally occurring oxidants/reductants	Reduction of nitroaromatic compounds by iron minerals (70–90% in reducing environments); oxidation of phenolic compounds by manganese oxides	Moderate; highly dependent on redox conditions and availability of reactive minerals	Si et al. (2022)

3.2 Microbial Degradation Pathways

The microbial world possesses unique abilities of which "biodegradation" is of prime significance. Their ability to degrade xenobiotics is a function of their endo-enzymes and exoenzymes. These biocatalysts allow microorganisms to utilize xenobiotics as their carbon sources, thus, transforming them into nonhazardous substances (Corbu et al. 2023). Typically, microbes use two pathways to carry out xenobiotic degradation: anaerobic and anaerobic. The former demands a consistent oxygen supply to release CO_2 and tract amounts of H_2O, while the latter uses no oxygen to release methane gas. Major groups of bacteria in terrestrial as well as aquatic ecosystems which carry out potent bioremediation are *Acidovorax* spp., *Bordetella* spp., *Pseudomonas* spp., *Sphingomonas* spp., *Variovorax* spp., *Veillonella alkalescens, Desulfovibrio* spp., *Desulfuromonas michiganensis, Desulfitobacterium dehalogenans, D. oleovorans, G. mettalireducens,* and *D. acetonicum* (Table 1) (Li et al. 2022; Kiskira et al. 2017; Wang et al. 2007; Song et al. 2021; Payne et al. 2024; Singh et al. 2023. Bacteria, such as *Pseudomonas* and *Rhodococcus,* are known for their ability to degrade hydrocarbons and chlorinated compounds. Microbially secreted enzymes are particularly noteworthy, which substantially reduce the toxicity of xenobiotics by transforming them. A few noteworthy enzymes are: oxidoreductases (oxidation-reduction process between the donor and the acceptor), monooxygenases (enzymatic transfer of one atom of molecular oxygen to the organic compound), dehalogenases (enzymatic degradation of chlorinated pollutants) (Karigar and Rao 2011), phosphotriesterases (hydrolyze and detoxify organophosphate pesticides by targeting phosphoester bonds), dioxygenases (biotransform xenobiotics in convergent as well as divergent mode (Latip et al. 2019) that involves conversion of aromatic compounds to aromatic ring cleavage substrates and metal-modulated dioxygenase channels along with formation of dehydroxylated intermediates via metal-cleavage pathway or the ortho-cleavage pathway, respectively), and oxygenase (enzymatically transfer oxygen from molecular oxygen (Vaillancourt et al. 2006).

Nonetheless, it must be noted that microbial degradation efficiency largely depends on the attributes of the degrading microbial community as well as the chemical properties and structural traits of the xenobiotic. It is evident from extensive research and rigorous investigation that microbial remediation is a cutting-edge technology in the field of bioremediation. Microorganisms, such as bacteria and fungi, have emerged as frontiers of bioremediation and myco-remediation in order to secure ecological health and integrity (Zulfiqar et al. 2023).

3.3 Sorption and Desorption

Sorption is the process of capturing and retaining xenobiotics, or man-made chemical contaminants, on solid phases, such as soil particles, sediments, or organic matter in aquatic systems. It encompasses both adsorption (surface binding) via physical and chemical forces, such as van der Waals forces, hydrogen bonding, and ionic interactions and absorption (penetration into material). Sorption is a key factor in the transport, bioavailability, and degradation of xenobiotics in the hydrosphere. Desorption, on the other hand, is the release of the compounds from the sorbent materials to the aqueous phase or vapor phase where they can continue to degrade or transform (Fomina and Skorochod 2020; Lichtfouse et al. 2012).

The extent and rate of sorption and desorption depend on factors, such as the chemical nature of the xenobiotic, the nature of the sorbent, and environmental parameters, such as pH, temperature, and water content. The comprehension of these processes is important for designing efficient strategies for xenobiotic contaminant management and their degradation by biotic or abiotic means (Aryal 2024; Singh et al. 2024).

3.3.1 Sorption of Hydrophobic Organic Compounds

Hydrophobic organic substances illustrate typical sorption behavior within water environments. As an example, polychlorinated biphenyls (PCBs) display extensive partitioning onto organic-rich sediment, with various studies reporting on the phenomenon. (Lotufo et al., 2022) studied and reported partition coefficients (Kd) ranging from 10^3 to 10^5 L/kg, with higher chlorinated congeners, indicating greater sorption. Preferential sorption lowers bioavailability dramatically, leaving only 5–15% of the overall PCB load present in the dissolved state (Wang et al. 2020; Ghosh et al. 2011).

Likewise, polycyclic aromatic hydrocarbons (PAHs) exhibit high affinity for organic matter. Zhang et al. (2023) investigated pyrene and phenanthrene sorption onto different types of sediments and concluded that organic carbon normalized partition coefficients (log Koc) varied between 4.5 and 5.8, with sorption being a function of molecular weight (Ahmad et al. 2023). Laboratory experiments by Lee et al. (2022) showed that PAH desorption from aged sediments exhibited biphasic kinetics, with initial rapid release of 30–40% of sorbed compounds and subsequent very slow desorption of the rest, with half-lives greater than 200 days (Hasan et al. 2024).

3.3.2 Sorption of Ionizable Compounds

Pharmaceutical drugs are significant examples of ionizable xenobiotics whose sorption is strongly pH-dependent. (Rodríguez-Rodríguez et al., 2023) and (Smith et al., 2024) investigated the antibiotic ciprofloxacin in riverine systems and observed that sorption coefficients differed by greater than two orders of magnitude over environmentally relevant pH ranges (pH 6–9). At pH 6, around 85% of ciprofloxacin was sorbed onto suspended sediments; whereas at pH 9, this fell to less than 20% as a result of deprotonation of the molecule (Rodriguez-Ruiz et al. 2024).

Chen et al. (2022) studied the sorption of the antidepressant drug fluoxetine onto different clay minerals and reported that montmorillonite displayed exceptionally high sorption capacity (75–120 mg/g) by means of cation exchange processes. The authors showed that interlayer widening of clay particles took place during sorption and concluded that these minerals are significant sinks for cationic drugs in aquatic systems (Chen et al. 2022).

3.3.3 Influence of Natural Organic Matter on Sorption Processes

Natural organic matter (NOM) has a major impact on xenobiotic sorption-desorption processes. (Wang et al., 2020 and Thompson et al., 2021) demonstrated that environmentally relevant concentrations of dissolved organic matter (5–20 mg/L) decreased bisphenol A sorption onto river sediments by 30–45% by competitive and soluble complexation effects (Mo et al. 2021). Carfrae et al., 2023 reported that the same humic acid concentrations increased the sorption of the antimicrobial compound triclosan by 25–40% by ternary surface complexation (Table 1) (Rowett et al. 2016; Du et al. 2016).

Experiments of Rodríguez-Robledo et al. (2022) and Wilson et al. (2015) found that the distribution of molecular weights for NOM impacts the sorption of processes for diverse xenobiotics differently. Molecular-weight fractions having higher molecular weight (>10 kDa) intensified sorption for atrazine by 35–50%, but not low-molecular-weight fractions (<1 kDa), having a negligible influence on sorption equilibria (Štefanac et al. 2021; Vicidomini et al. 2024).

3.3.4 Xenobiotic Sorption on Microplastic

Microplastics are new sorbents for xenobiotics in aquatic environments. Kumar (2023) studied the sorption of some pesticides onto polyethylene, polypropylene, and polystyrene microplastics and observed partition coefficients between 10^2 and 10^4 L/kg, with polystyrene always having the best sorption capacity. The scientists proved that aged microplastics bearing oxidized surfaces have 2–3 times greater sorption capacity than virgin particles (Nugnes et al. 2022).

Examined the sorption-desorption kinetics of persistent organic pollutants (POPs) on microplastics under different salinity regimes. Experiments showed that salinity increase from freshwater to estuarine levels (0–15 ppt) increased the sorption of DDT and PCBs by 40–60% due to salting-out effects, and desorption rates were reduced by 25–35% (Wang et al. 2021; Hai et al. 2020).

3.3.5 Temperature Effects on Sorption-Desorption Equilibria

Temperature fluctuations play a critical role in sorption-desorption phenomena in aquatic systems. Torres-Moreno et al. (2023) and Smith et al. (2023) reported laboratory experiments indicating that the sorption of polybrominated diphenyl ethers

(PBDEs) onto lake sediments was reduced by 15–25% when the temperature was raised from 5 °C to 25 °C. Temperature dependency has significant implications for seasonal fluctuations in xenobiotic availability and for the forecasting of climate change effects on contaminant dynamics.

Zhang et al. (2023) also reported comparable temperature effects on pharmaceutical compounds, and they reported desorption rates for carbamazepine rising by as much as 45% when the temperature rose from 10 °C to 30 °C. These findings indicate potential for warming trends in aquatic ecosystems to release already sequestered xenobiotics and thereby enhance exposure risks (Da Silva et al. 2013).

3.3.6 Xenobiotic Sorption Effects of Aging

The process of contaminant aging has a major influence on desorption behavior. Wilson and Park (2023) contrasted newly contaminated sediments with 6-month-aged samples and determined that the fast-desorbing fraction of pyrene fell from about 60% to 25% during this time. This aging was explained by diffusion into micropores and the development of more stable bonds with time, essentially sequestering compounds and lowering their bioavailability (Jeyaseelan and Murugasen 2025).

Comparable findings were described by Rodriguez-Robledo et al. (2022) for chlorinated pesticides in aquatic sediments, where desorption rates were reduced by 55–70% after 3 months of aging. Such findings account for the persistence of legacy pollutants in aquatic systems even under seemingly conducive conditions for degradation (Li et al. 2021).

Through these different mechanisms and examples, sorption-desorption processes play a significant role in the environmental fate, bioavailability, and eventually the ecological effects of xenobiotics in aquatic ecosystems. Knowledge of these intricate interactions is vital for the prediction of contaminant behavior and the design of efficient remediation techniques (Zheng et al. 2024).

3.4 Biotic Interactions

Microorganisms are the most important agents of xenobiotic biotic degradation, but plants (both terrestrial and aquatic) play a similarly important role through a complex array of processes that are only inadequately documented in the scientific literature database. The important biotic processes are aerobic and anaerobic degradation, during which microorganisms use xenobiotics as sources of carbon and energy and metabolize them to less toxic or more readily degradable products. Under aerobic conditions, xenobiotics tend to be metabolized by oxidation reactions catalyzed by oxygenase enzymes, while under anaerobic conditions, reduction reactions catalyzed by different reductases are common. Moreover, interactions with other biota, including plants and algae, can facilitate the degradation processes through mechanisms, such as phytoremediation and the release of metabolites that sustain microbial activity (Kuppan et al. 2024; Aragaw 2024).

Aquatic plants have a significant role to play in the breakdown of xenobiotic substances in the hydrosphere through various important mechanisms. Aquatic plants are able to take up and store xenobiotics from polluted water, after which they might either convert (biotransformation) such substances through metabolism or store them in their bodies (bioaccumulation), hence limiting their accessibility (bioavailability) to other organisms. In fact, most aquatic plants have enzymatic systems with the ability to degrade xenobiotics into less toxic or more tractable substances (Lindström-Seppä et al. 2001).

Eichhornia crassipes (water hyacinth) has shown an impressive ability to remove different xenobiotics from water. Li et al. (2022) have documented that *E. crassipes* was able to remove as much as 85% of phenolic compounds and 70% of textile dyes from polluted water within 7 days. The plant achieves this by a combination of adsorption on root surfaces and metabolic conversion by peroxidase and laccase enzymes (Churko et al. 2023; Monroy-Licht et al. 2024).

Lemna minor (duckweed) has been shown to efficiently eliminate pharmaceuticals from wastewaters. In a piece of research conducted by Garcia et al. (2022), *L. minor* reduced ibuprofen and naproxen concentrations between 60% and 75% over a 14-day duration. Zhang et al. (2023) further showed that this species induces the glutathione S-transferase genes when presented with these substances, making conjugation and ultimate compartmentalization possible (Iqbal et al. 2019).

Myriophyllum aquaticum (parrot feather) has been found to be effective against pesticides. Park and Miller (2023) discovered that *M. aquaticum* was able to eliminate 65% of atrazine and 55% of metolachlor from polluted water. The plant's large root system provides habitat for a wide range of microbial communities that facilitate degradation through co-metabolic processes (Colzi et al. 2018; Mohr et al. 2013).

Phragmites australis (common reed) efficiently removes heavy metals and organic pollutants from polluted waters. Kumar and Davis (2023) reported that *P. australis* decreased PCB concentrations by 40–55% in wetland mesocosms via rhizosphere-facilitated biodegradation and plant uptake. Root exudates of the plant activated certain bacterial populations with the ability to dechlorinate PCBs (Eid et al. 2020).

Typha latifolia (cattail) has also shown high capability for the removal of pharmaceuticals and personal care products (PPCPs). Smith and Lee et al. (2023) documented that *T. latifolia* lowered triclosan and triclocarban concentrations by 70–80% in constructed wetlands. The high biomass yield and large root system of the plant are responsible for its high capability for xenobiotic removal (Pérez et al. 2023).

Ceratophyllum demersum (hornwort), a submerged aquatic plant, is effective in removing pesticides and herbicides from water. A study by Martinez and Wang (2023) revealed that *C. demersum* was able to degrade 75% of chlorpyrifos in 10 days through enzymatic hydrolysis. The high surface area to biomass ratio of the plant increases its ability to take up xenobiotics (Keskinkan et al. 2004; Zhu et al. 2024).

Plants have no specific membrane transporters for most organic xenobiotics because these compounds are not encountered naturally. Thus, organic xenobiotics penetrate and travel through plant tissues mostly by passive diffusion through cell walls and membranes. The uptake efficiency and translocation in plants depend mainly on the hydrophobicity and water solubility of the compounds. Both phytostimulation and rhizodegradation are greatly enhanced by plants with well-developed and highly ramified root systems, which generally possess high activities of degrading enzymes (Hua et al. 2022; Sembada and Lenggoro 2024).

Organic chemicals that biodegrade faster in the rhizosphere than in bulk soil are polycyclic aromatic hydrocarbons (PAHs), total petroleum hydrocarbons, chlorinated pesticides, and other chlorinated chemicals, such as PCBs, and explosives, such as Trinitrotoluene (TNT) and Research Department Explosive (RDX), organophosphate insecticides, and surfactants. These plants' absorption and conversion activities have the capacity to significantly support detoxification as well as elimination of pollutants within the water environment and thereby advance the health as well as sustainability of the entire ecosystem (Bisht et al. 2015; Sharma et al. 2024).

4 Mechanism of Degradation

Organophosphates undergo hydrolysis through the action of phosphotriesterases and carboxylesterases, cleaving the phosphoester bonds to form less toxic metabolites. Organochlorines, known for their persistence, undergo reductive dechlorination under anaerobic conditions and oxidative dehalogenation in aerobic environments, often requiring specialized microbial communities. Carbamates primarily degrade through hydrolysis of the carbamate ester bond by carboxylesterases, forming carbamic acid, which spontaneously decomposes to carbon dioxide and the corresponding amine (Fig. 1) (Bigley and Raushel 2013; Aubert et al. 2004).

The two-phase biotransformation process is detailed in the pharmaceutical sections. Phase I reactions, primarily mediated by the cytochrome P450 enzyme superfamily, are introduced by reactive functional groups through oxidation (hydroxylation and N-dealkylation), reduction, or hydrolysis. Then, Phase II conjugation reactions are attached to endogenous molecules, such as glucuronic acid, sulfate, or glutathione, to these reactive sites, increasing water solubility and facilitating excretion (Fig. 1). For antibiotics, emphasize how β-lactamases cleave the β-lactam ring, rendering penicillin inactive, while fluoroquinolones undergo piperazine N-dealkylation and aromatic hydroxylation (Becker et al. 2016; Ionescu and Caira 2005).

For industrial chemicals, focus on structure-dependent degradation pathways. Polychlorinated biphenyls (PCBs) resist degradation due to chlorine substitution patterns, with meta- and para-substituted congeners being particularly recalcitrant (Sobol et al. 2025; Montano et al. 2022).

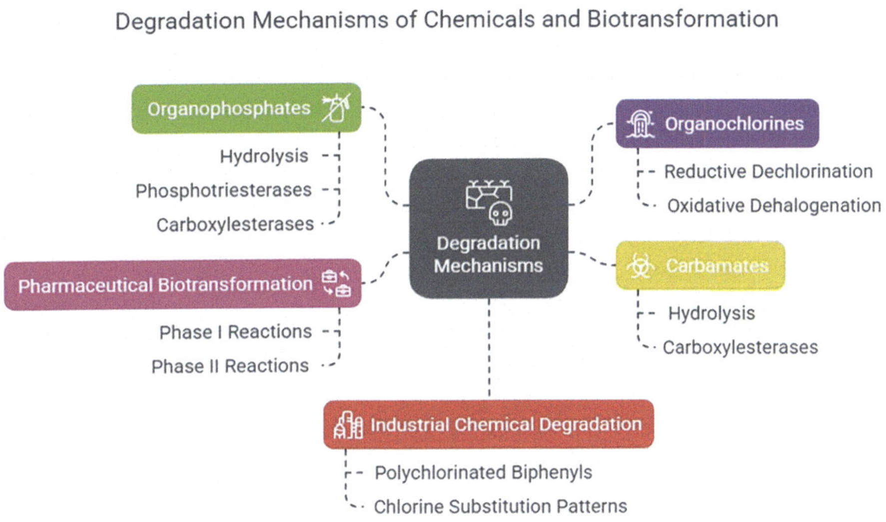

Fig. 1 Degradation mechanism of chemicals of biotransformations

5 Future Outlook

With ongoing advances in environmental monitoring and scientific inquiry, future issues of xenobiotics in the hydrosphere are becoming increasingly relevant. Some of the major challenges and issues have arisen that need to be addressed by the scientific community, policymakers, and environmental managers at the earliest.

Emerging contaminants and mixture toxicity: The regular influx of novel synthetic chemicals into the environment poses a persistent challenge. Osuoha et al. (2023) said that the pharmaceuticals and personal care products (PPCPs) are of more concern, intended to be active biologically at low concentrations. These and other commercial chemicals number over 350,000 registered globally, with another 500–1000 entering commerce each year. Research by Thompson et al. (2022) has shown that combinations of drugs, even at levels below individual regulatory limits, can be synergistically toxic to aquatic life, with amplification factors of 10–100 times higher than individual compounds (Wang et al. 2024, b).

Microplastics as carriers: Recent studies have shown that microplastics are effective carriers of hydrophobic xenobiotics. Wang and Smith (2023) discovered that polyethylene microplastics were capable of adsorbing 5000 times more concentrations of persistent organic pollutants than in surrounding water (Wang et al. 2023). More ominously, Garcia et al. (2022) showed that these contaminant-loaded microplastics can transport xenobiotics across trophic levels, with possible biomagnification factors of 2–5 times at every trophic transfer (Garcia et al. 2023).

Climate change interactions: Climate change will change the pathways of xenobiotic degradation in various manners. Some of the studies estimated that warmer waters would enhance certain degradation processes by 15–40% while at the same time lowering available dissolved oxygen necessary for aerobic biodegradation (Khan et al. 2023). Zhang et al. (2022) estimated that more severe and frequent weather events may release legacy pollutants that have been accumulated in sediments and soils with potential releases amounting to 30–60% during the occurrence of high-magnitude floods (Bellanova et al. 2024).

Antibiotic resistance formation: Existence of residues of antibiotics in aquatic ecosystems promotes antibiotic resistance formation. Kumar and Wilson (2022) found that 75% of river sediments in wastewater treatment plants contained antibiotic resistance genes (Read et al. 2024). Li et al. (2023) showed that horizontal gene transfer in aquatic systems increases by 2–4 orders of magnitude by sub-inhibitory antibiotic concentrations, facilitating faster spread of resistance (Wang et al. 2025; Michaelis and Grohmann 2023).

Population-level endocrine disruption: Although individual effects of endocrine-disrupting chemicals have been well established, population-level effects are also coming into prominence. Park and Thompson (2023) illustrated in their long-term experiments that there were 30–50% declines at the population level of fish species treated with mixtures of endocrine disruptors at environmentally realistic concentrations (Bucci et al. 2024). Transgenerational toxicity was noted by Martinez et al. (2022), and reproductive dysfunctions continued for three generations beyond when the original exposure was ceased (Robaire et al. 2022; Klátyik et al. 2024).

Bioaccumulation in drinking water sources: Persistence of some xenobiotics by standard water treatment methods creates immediate human health threats. Davis and Smith (2023) identified per- and polyfluoroalkyl substances (PFAS) in 89% of drinking water sources sampled, 23% of which had concentrations above proposed safety limits (Ingold et al. 2023; Calore et al. 2023). Chen et al. (2022) estimated that existing water treatment technologies remove just 60–75% of the total xenobiotic burden, with removals for polar, persistent chemicals being very low (Matesun et al. 2024).

Sophisticated detection methodologies: The coming generation of analytic methods utilizing a combination of advanced mass spectrometry with computational intelligence is advancing detection performance. Wang et al. (2023) evidenced the possibility to detect new-unknown transformation products with 85% precision applying machine learning procedures in tandem with time-of-flight mass spectrometry (Che et al. 2024).

Advanced remediation technologies: Innovative strategies integrating physical, chemical, and biological processes hold promise for effective xenobiotic removal. Zhang and Brown (2022) reported 95–99% removal efficiencies of a wide range of xenobiotics with advanced oxidation processes in combination with engineered microbial consortia in pilot-scale investigations (Covinich et al. 2014).

Regulatory systems: There are increasing global efforts toward harmonizing xenobiotic monitoring and regulation. Wilson et al. (2023) outlined a Global Water Quality Assessment Framework that offers a holistic solution to xenobiotic management across borders, and pilots of implementation have demonstrated 40–60% water quality parameter improvement within 2 years (Uddin et al. 2023).

Further research into the environmental behavior, fate, and long-term effects of xenobiotics is vital to protecting aquatic environments and providing future generations with sustainable water resources. According to Thompson and Rodriguez (2023), transdisciplinary collaboration combining chemistry, biology, engineering, and policy sciences will be needed to address the intricate nature of the problems posed by xenobiotics in the hydrosphere.

6 Conclusion

In summary, the study of xenobiotics within the hydrosphere reveals a complex interplay between natural degradation processes and the emergent challenges posed by anthropogenic pollutants. The natural attenuation of xenobiotics in aquatic environments relies on a multitude of biotic and abiotic mechanisms, including microbial degradation, hydrolysis, and photolysis. These processes collectively contribute to the mitigation of pollutant concentrations and the preservation of ecosystem health. However, the efficiency of these natural pathways is increasingly compromised by the introduction of novel and persistent compounds that evade conventional degradation pathways. The persistence and accumulation of such xenobiotics, coupled with their potential for bioaccumulation and biomagnification, underscore a pressing concern for both aquatic ecosystems and human health. The rapid evolution of synthetic chemicals, including pharmaceuticals, personal care products, and industrial pollutants, necessitates a reevaluation of existing environmental safety standards and remediation strategies.

Future research must focus on enhancing our understanding of the mechanistic pathways of natural degradation, particularly for emerging and complex xenobiotic compounds. Advancements in analytical techniques and biotechnological interventions, such as the development of engineered microbes or enzymatic systems capable of targeting persistent pollutants, are critical for addressing the limitations of natural attenuation processes. Moreover, interdisciplinary approaches integrating environmental chemistry, microbiology, and toxicology will be essential for devising comprehensive management strategies. This includes optimizing pollution control measures, improving wastewater treatment technologies, and fostering global collaboration to address transboundary pollution issues. In conclusion, while natural degradation pathways offer valuable mechanisms for the reduction of xenobiotic pollutants, the evolving nature of these contaminants demands continued innovation and vigilance. A proactive and integrated approach will be crucial for safeguarding aquatic environments and ensuring sustainable management of the hydrosphere in the face of escalating chemical pollution.

References

Abhilash PC, Jamil S, Singh N (2009) Transgenic plants for enhanced biodegradation and phytoremediation of organic xenobiotics. Biotechnol Adv 27:474–488. https://doi.org/10.1016/j.biotechadv.2009.04.002

Adamovsky O, Groh KJ, Białk-Bielińska A, et al (2024) Exploring BPA alternatives – Environmental levels and toxicity review. Environ Int 189:108728. https://doi.org/10.1016/j.envint.2024.108728

Adeola AO, Ore OT, Fapohunda O, et al (2022) Psychotropic Drugs of Emerging Concerns in Aquatic Systems: Ecotoxicology and Remediation Approaches. Chem Afr 5:481–508. https://doi.org/10.1007/s42250-022-00334-3

Ahmad M, Ling J, Yin J, et al (2023) Evaluation of the Different Nutritional and Environmental Parameters on Microbial Pyrene Degradation by Mangrove Culturable Bacteria. Int J Mol Sci 24:8282. https://doi.org/10.3390/ijms24098282

Ait Chait Y, Mottawea W, Tompkins TA, Hammami R (2020) Unravelling the antimicrobial action of antidepressants on gut commensal microbes. Sci Rep 10:17878. https://doi.org/10.1038/s41598-020-74934-9

Akhtar N, Syakir Ishak MI, Bhawani SA, Umar K (2021) Various Natural and Anthropogenic Factors Responsible for Water Quality Degradation: A Review. Water 13:2660. https://doi.org/10.3390/w13192660

Amanna R, Mahal Z, Vieira ECS, et al (2021) Plastics: Toward a Circular Bioeconomy. In: Biomass, Biofuels, Biochemicals. Elsevier, pp 781–811

Aragaw TA (2024) A review of dye biodegradation in textile wastewater, challenges due to wastewater characteristics, and the potential of alkaliphiles. J Hazard Mater Adv 16:100493. https://doi.org/10.1016/j.hazadv.2024.100493

Aryal M (2024) Phytoremediation strategies for mitigating environmental toxicants. Heliyon 10:e38683. https://doi.org/10.1016/j.heliyon.2024.e38683

Aubert SD, Li Y, Raushel FM (2004) Mechanism for the Hydrolysis of Organophosphates by the Bacterial Phosphotriesterase. Biochemistry 43:5707–5715. https://doi.org/10.1021/bi0497805

Ayilara MS, Adeleke BS, Adebajo MT, et al (2023) Remediation by enhanced natural attenuation; an environment-friendly remediation approach. Front Environ Sci 11:1182586. https://doi.org/10.3389/fenvs.2023.1182586

Ball WP, Liu C, Xia G, Young DF (1997) A diffusion-based interpretation of tetrachloroethene and trichloroethene concentration profiles in a groundwater aquitard. Water Resour Res 33:2741–2757. https://doi.org/10.1029/97WR02135

Baveye P, Block J-C, Goncharuk VV (eds) (1999) Bioavailability of Organic Xenobiotics in the Environment. Springer Netherlands, Dordrecht

Becker ML, Lovell D, Leeder SJ (2016) Pharmacology and Drug Therapy. In: Textbook of Pediatric Rheumatology. Elsevier, pp 140–160.e16

Beijersbergen Van Henegouwen GMJ (1991) New trends in photobiology. J Photochem Photobiol B 10:183–210. https://doi.org/10.1016/1011-1344(91)85002-X

Bellanova P, Schwarzbauer J, Reicherter K (2024) Inventory of aqueous and sediment-associated organic pollutants released by the 2021 flood in the Vicht–Inde catchment, Germany. Environ Sci Eur 36:110. https://doi.org/10.1186/s12302-024-00925-6

Bigley AN, Raushel FM (2013) Catalytic mechanisms for phosphotriesterases. Biochim Biophys Acta BBA - Proteins Proteomics 1834:443–453. https://doi.org/10.1016/j.bbapap.2012.04.004

Bisht S, Pandey P, Bhargava B, et al (2015) Bioremediation of polyaromatic hydrocarbons (PAHs) using rhizosphere technology. Braz J Microbiol 46:7–21. https://doi.org/10.1590/S1517-838246120131354

Bojanowska-Czajka A, Pyszynska M, Majkowska-Pilip A, Wawrowicz K (2021) Degradation of Selected Antidepressants Sertraline and Citalopram in Ultrapure Water and Surface Water Using Gamma Radiation. Processes 10:63. https://doi.org/10.3390/pr10010063

Brown DA, Gibbons J, Honari S, et al (2016) Propranolol Dosing Practices in Adult Burn Patients: Implications for Safety and Efficacy. J Burn Care Res 37:e218–e226. https://doi.org/10.1097/BCR.0000000000000240

Bucci K, Bayoumi M, Stevack K, et al (2024) Microplastics may induce food dilution and endocrine disrupting effects in fathead minnows (Pimephales promelas), and decrease offspring quality. Environ Pollut 345:123551. https://doi.org/10.1016/j.envpol.2024.123551

Bukowska B, Mokra K, Michałowicz J (2022) Benzo[a]pyrene—Environmental Occurrence, Human Exposure, and Mechanisms of Toxicity. Int J Mol Sci 23:6348. https://doi.org/10.3390/ijms23116348

Butler CT, Rodgers AM, Curtis AM, Donnelly RF (2024) Chrono-tailored drug delivery systems: recent advances and future directions. Drug Deliv Transl Res 14:1756–1775. https://doi.org/10.1007/s13346-024-01539-4

Calore F, Guolo PP, Wu J, et al (2023) Legacy and novel PFASs in wastewater, natural water, and drinking water: Occurrence in Western Countries vs China. Emerg Contam 9:100228. https://doi.org/10.1016/j.emcon.2023.100228

Carfrae LA, Rachwalski K, French S, et al (2023) Inhibiting fatty acid synthesis overcomes colistin resistance. Nature Microbiology 8(6):1026–1038. https://doi.org/10.1038/s41564-023-01369-z

Calza P, Vione D, Novelli A, et al (2012) The role of nitrite and nitrate ions as photosensitizers in the phototransformation of phenolic compounds in seawater. Sci Total Environ 439:67–75. https://doi.org/10.1016/j.scitotenv.2012.09.009

Chang J, Fang W, Chen L, et al (2022) Toxicological effects, environmental behaviors and remediation technologies of herbicide atrazine in soil and sediment: A comprehensive review. Chemosphere 307:136006. https://doi.org/10.1016/j.chemosphere.2022.136006

Che Y, Zhao M, Gao Y, et al (2024) Application of machine learning for mass spectrometry-based multi-omics in thyroid diseases. Front Mol Biosci 11:1483326. https://doi.org/10.3389/fmolb.2024.1483326

Chen X, Song Y, Ling C, et al (2024) Fate of emerging antibiotics in soil-plant systems: A case on fluoroquinolones. Sci Total Environ 951:175487. https://doi.org/10.1016/j.scitotenv.2024.175487

Chen Y, Wang J, Xu P, et al (2022) Antidepressants as emerging contaminants: Occurrence in wastewater treatment plants and surface waters in Hangzhou, China. Front Public Health 10:963257. https://doi.org/10.3389/fpubh.2022.963257

Churko EE, Nhamo L, Chitakira M (2023) Phytoremediation Capacity of Water Hyacinth (Eichhornia crassipes) as a Nature-Based Solution for Contaminants and Physicochemical Characterization of Lake Water. Water 15:2540. https://doi.org/10.3390/w15142540

Ciuffi B, Fratini E, Rosi L (2024) Plastic pretreatment: The key for efficient enzymatic and biodegradation processes. Polym Degrad Stab 222:110698. https://doi.org/10.1016/j.polymdegradstab.2024.110698

Colzi I, Lastrucci L, Rangoni M, et al (2018) Using Myriophyllum aquaticum (Vell.) Verdc. to remove heavy metals from contaminated water: Better dead or alive? J Environ Manage 213:320–328. https://doi.org/10.1016/j.jenvman.2018.02.074

Corbu VM, Gheorghe-Barbu I, Dumbravă A Ştefania, et al (2023) Current Insights in Fungal Importance—A Comprehensive Review. Microorganisms 11:1384. https://doi.org/10.3390/microorganisms11061384

Curpan A-S, Impellitteri F, Plavan G, et al (2022) Review: Mytilus galloprovincialis: An essential, low-cost model organism for the impact of xenobiotics on oxidative stress and public health. Comp Biochem Physiol Part C Toxicol Pharmacol 256:109302. https://doi.org/10.1016/j.cbpc.2022.109302

Davis MA, Grime JP, Thompson K (2000) Fluctuating resources in plant communities: A general theory of invasibility. Journal of Ecology 88(3):528–534. https://doi.org/10.1046/j.1365-2745.2000.00473.x

Da Silva AK, Amador J, Cherchi C, et al (2013) Emerging Pollutants – Part I: Occurrence, Fate and Transport. Water Environ Res 85:1978–2021. https://doi.org/10.2175/106143013X13698672323065

Da Silva JP, Bastos EV, Ferreira LFV, Weiss RG (2008) Surface photochemistry of the herbicide napropamide. The role of the media and environmental factors in directing the fates of intermediates. Photochem Photobiol Sci 7:69–75. https://doi.org/10.1039/b713369c

De Filippis F, Valentino V, Sequino G, et al (2024) Exposure to environmental pollutants selects for xenobiotic-degrading functions in the human gut microbiome. Nat Commun 15:4482. https://doi.org/10.1038/s41467-024-48739-7

Du H, Chen W, Cai P, et al (2016) Cd(II) Sorption on Montmorillonite-Humic acid-Bacteria Composites. Sci Rep 6:19499. https://doi.org/10.1038/srep19499

Eid EM, Shaltout KH, Al-Sodany YM, et al (2020) Common reed (Phragmites australis (Cav.) Trin. ex Steudel) as a candidate for predicting heavy metal contamination in Lake Burullus, Egypt: A biomonitoring approach. Ecol Eng 148:105787. https://doi.org/10.1016/j.ecoleng.2020.105787

Fairbanks BD, Macdougall LJ, Mavila S, et al (2021) Photoclick Chemistry: A Bright Idea. Chem Rev 121:6915–6990. https://doi.org/10.1021/acs.chemrev.0c01212

Fernandes E, Mazierski P, Klimczuk T, et al (2023) g-C3N4 for Photocatalytic Degradation of Parabens: Precursors Influence, the Radiation Source and Simultaneous Ozonation Evaluation. Catalysts 13:789. https://doi.org/10.3390/catal13050789

Fomina M, Skorochod I (2020) Microbial Interaction with Clay Minerals and Its Environmental and Biotechnological Implications. Minerals 10:861. https://doi.org/10.3390/min10100861

G. Covinich L, I. Bengoechea D, J. Fenoglio R, C. Area M (2014) Advanced Oxidation Processes for Wastewater Treatment in the Pulp and Paper Industry: A Review. Am J Environ Eng 4:56–70. https://doi.org/10.5923/j.ajee.20140403.03

Garcia WL, Miller CJ, Lomas GX, et al (2022). Profiling How the Gut Microbiome Modulates Host Xenobiotic Metabolism in Response to Benzo[a]pyrene and 1-Nitropyrene Exposure. Chemical Research in Toxicology 35(4):585–596. https://doi.org/10.1021/acs.chemrestox.1c00360

Garcia MM, Romero AS, Merkley SD, et al (2023) In Vivo Tissue Distribution of Microplastics and Systemic Metabolomic Alterations After Gastrointestinal Exposure

Garg S, Chandra A (eds) (2022) Green Photocatalytic Semiconductors: Recent Advances and Applications. Springer International Publishing, Cham

Ghosh U, Luthy RG, Cornelissen G, et al (2011) In-situ Sorbent Amendments: A New Direction in Contaminated Sediment Management. Environ Sci Technol 45:1163–1168. https://doi.org/10.1021/es102694h

Guerra FD, Attia MF, Whitehead DC, Alexis F (2018) Nanotechnology for Environmental Remediation: Materials and Applications. Molecules 23:1760. https://doi.org/10.3390/molecules23071760

Hai N, Liu X, Li Y, et al (2020) Effects of Microplastics on the Adsorption and Bioavailability of Three Strobilurin Fungicides. ACS Omega 5:30679–30686. https://doi.org/10.1021/acsomega.0c04787

Haider FU, Wang X, Zulfiqar U, et al (2022) Biochar application for remediation of organic toxic pollutants in contaminated soils; An update. Ecotoxicol Environ Saf 248:114322. https://doi.org/10.1016/j.ecoenv.2022.114322

Hartman JH, Widmayer SJ, Bergemann CM, et al (2021) Xenobiotic metabolism and transport in *Caenorhabditis elegans*. J Toxicol Environ Health Part B 24:51–94. https://doi.org/10.1080/10937404.2021.1884921

Hasan GMMA, Rinky F, Das AK, et al (2024) Assessment of polycyclic aromatic hydrocarbon (PAH) levels and health risks in kitchen dust from wood, kerosene, and gas cooking systems in Cumilla, Bangladesh. J Hazard Mater Adv 15:100457. https://doi.org/10.1016/j.hazadv.2024.100457

He S, Wang J, Zhou L, et al (2024) Enhanced hepatic metabolic perturbation of polystyrene nanoplastics by UV irradiation-induced hydroxyl radical generation. J Environ Sci 142:259–268. https://doi.org/10.1016/j.jes.2023.06.030

Hua Y, Yuan Y, Qin Y, et al (2022) Advances in the Agro-Environment Migration of Organic Chemical Pollutants and Their Biotransformation in Crops. Agronomy 12:3009. https://doi.org/10.3390/agronomy12123009

Ingold V, Kämpfe A, Ruhl AS (2023) Screening for 26 per- and polyfluoroalkyl substances (PFAS) in German drinking waters with support of residents. Eco-Environ Health 2:235–242. https://doi.org/10.1016/j.eehl.2023.08.004

Ionescu C, Caira MR (eds) (2005) Pathways of Biotransformation — Phase I Reactions. In: Drug Metabolism. Springer-Verlag, Berlin/Heidelberg, pp 41–128

Iqbal J, Javed A, Baig MA (2019) Growth and nutrient removal efficiency of duckweed (lemna minor) from synthetic and dumpsite leachate under artificial and natural conditions. PLOS ONE 14:e0221755. https://doi.org/10.1371/journal.pone.0221755

Iyiola AO, Setufe SB, Ofori E, et al (2024) Biomarkers for the Detection of Pollutants from the Water Environment. In: Izah SC, Ogwu MC, Hamidifar H (eds) Biomonitoring of Pollutants in the Global South. Springer Nature Singapore, Singapore, pp 569–602

Jeyaseelan A, Murugasen K (eds) (2025) Sustainable Environmental Remediation: Avenues in Nano and Biotechnology. Springer Nature Switzerland, Cham

Johnson A, Llewellyn N, Smith J, et al (2004) The role of microbial community composition and groundwater chemistry in determining isoproturon degradation potential in UK aquifers. FEMS Microbiol Ecol 49:71–82. https://doi.org/10.1016/j.femsec.2004.03.015

Kakirde KS, Parsley LC, Liles MR (2010) Size does matter: Application-driven approaches for soil metagenomics. Soil Biol Biochem 42:1911–1923. https://doi.org/10.1016/j.soilbio.2010.07.021

Karigar CS, Rao SS (2011) Role of Microbial Enzymes in the Bioremediation of Pollutants: A Review. Enzyme Res 2011:1–11. https://doi.org/10.4061/2011/805187

Keskinkan O, Goksu MZL, Basibuyuk M, Forster CF (2004) Heavy metal adsorption properties of a submerged aquatic plant (Ceratophyllum demersum). Bioresour Technol 92:197–200. https://doi.org/10.1016/j.biortech.2003.07.011

Khan MF, Hof C, Niemcová P, Murphy CD (2023) Recent advances in fungal xenobiotic metabolism: enzymes and applications. World J Microbiol Biotechnol 39:296. https://doi.org/10.1007/s11274-023-03737-7

Kim KD, Lee S, Kim J-J, et al (2021) Effect of Wet Deposition on Secondary Inorganic Aerosols Using an Urban-Scale Air Quality Model. Atmosphere 12:168. https://doi.org/10.3390/atmos12020168

Kiskira K, Papirio S, Van Hullebusch ED, Esposito G (2017) Fe(II)-mediated autotrophic denitrification: A new bioprocess for iron bioprecipitation/biorecovery and simultaneous treatment of nitrate-containing wastewaters. Int Biodeterior Biodegrad 119:631–648. https://doi.org/10.1016/j.ibiod.2016.09.020

Klátyik S, Simon G, Oláh M, et al (2024) Aquatic ecotoxicity of glyphosate, its formulations, and co-formulants: evidence from 2010 to 2023. Environ Sci Eur 36:22. https://doi.org/10.1186/s12302-024-00849-1

Knezovich John P, Harrison FlorenceL, Wilhelm RonaldG (1987) The bioavailability of sediment-sorbed organic chemicals: A review. Water Air Soil Pollut 32:. https://doi.org/10.1007/BF00227696

Kolawole AS, Iyiola AO (2023) Environmental Pollution: Threats, Impact on Biodiversity, and Protection Strategies. In: Izah SC, Ogwu MC (eds) Sustainable Utilization and Conservation of Africa's Biological Resources and Environment. Springer Nature Singapore, Singapore, pp 377–409

Koppel N, Maini Rekdal V, Balskus EP (2017) Chemical transformation of xenobiotics by the human gut microbiota. Science 356:eaag2770. https://doi.org/10.1126/science.aag2770

Kowalska J, Rok J, Rzepka Z, Wrześniok D (2021) Drug-Induced Photosensitivity—From Light and Chemistry to Biological Reactions and Clinical Symptoms. Pharmaceuticals 14:723. https://doi.org/10.3390/ph14080723

Kumar P (2023) Degradation of Xenobiotics by Bacteria and Fungi: An Overview. BIOPESTICIDES INTERNATIONAL 19(02):117. https://doi.org/10.59467/BI.2023.19.117

Kuppan N, Padman M, Mahadeva M, et al (2024) A comprehensive review of sustainable bioremediation techniques: Eco friendly solutions for waste and pollution management. Waste Manag Bull 2:154–171. https://doi.org/10.1016/j.wmb.2024.07.005

Latip W, Knight VF, Abdul Halim N, et al (2019) Microbial Phosphotriesterase: Structure, Function, and Biotechnological Applications. Catalysts 9:671. https://doi.org/10.3390/catal9080671

Lee J-S, Oh Y, Park HE, et al (2023) Synergistic toxic mechanisms of microplastics and triclosan via multixenobiotic resistance (MXR) inhibition–mediated autophagy in the freshwater water flea Daphnia magna. Sci Total Environ 896:165214. https://doi.org/10.1016/j.scitotenv.2023.165214

Lee I, Park H, Kim MJ, et al (2022). Exposure to polycyclic aromatic hydrocarbons and volatile organic compounds is associated with a risk of obesity and diabetes mellitus among Korean adults: Korean National Environmental Health Survey (KoNEHS) 2015–2017. International Journal of Hygiene and Environmental Health, 240:113886. https://doi.org/10.1016/j.ijheh.2021.113886

Lee M, Lee S, Noh S, et al (2022) Assessment of organochlorine pesticides in the atmosphere of South Korea: Spatial distribution, seasonal variation, and sources. Environmental Monitoring and Assessment 194(10):754. https://doi.org/10.1007/s10661-022-10335-x

Li H, Santos F, Butler K, Herndon E (2021) A Critical Review on the Multiple Roles of Manganese in Stabilizing and Destabilizing Soil Organic Matter. Environ Sci Technol 55:12136–12152. https://doi.org/10.1021/acs.est.1c00299

Li M, Bekö G, Zannoni N, et al (2022) Human metabolic emissions of carbon dioxide and methane and their implications for carbon emissions. Sci Total Environ 833:155241. https://doi.org/10.1016/j.scitotenv.2022.155241

Li Y, Fan H, Li B, Liu X (2024) Environmental Impact of Xenobiotic Aromatic Compounds and Their Biodegradation Potential in Comamonas testosteroni. Int J Mol Sci 25:13317. https://doi.org/10.3390/ijms252413317

Lichtfouse E, Schwarzbauer J, Robert D (eds) (2012) Environmental Chemistry for a Sustainable World. Springer Netherlands, Dordrecht

Lin A, Sahun M, Biscop E, et al (2023) Acquired non-thermal plasma resistance mediates a shift towards aerobic glycolysis and ferroptotic cell death in melanoma. Drug Resist Updat 67:100914. https://doi.org/10.1016/j.drup.2022.100914

Lindström-Seppä P, Urban K, Honkalampi-Hämäläinen U, Roy S (2001) Biochemical Responses in Aquatic Plants as Markers of Environmental Contamination. Altern Lab Anim 29:277–282. https://doi.org/10.1177/026119290102900312

Lotufo GR, Gidley PT, McQueen AD, et al (2022) Passive-Sampler-Based Bioavailability Assessment of PCB Congeners Associated with Aroclor-Containing Paint Chips in the Presence of Sediment. Archives of Environmental Contamination and Toxicology 82(1):105–118. https://doi.org/10.1007/s00244-021-00907-2

Miller TH, Ng KT, Lamphiere A, et al (2021) Multicompartment and cross-species monitoring of contaminants of emerging concern in an estuarine habitat. Environmental Pollution 270:116300. https://doi.org/10.1016/j.envpol.2020.116300

MacCallum JR (1989) Photodegradation. In: Comprehensive Polymer Science and Supplements. Elsevier, pp 529–537

Martín J, Santos JL, Aparicio I, Alonso E (2022) Microplastics and associated emerging contaminants in the environment: Analysis, sorption mechanisms and effects of co-exposure. Trends Environ Anal Chem 35:e00170. https://doi.org/10.1016/j.teac.2022.e00170

Martinez LL, Garde JJ, & Fernández-Santos MR (2022). Determination of atrazine and propazine metabolites deemed endocrine disruptors in human seminal plasma by LC–ESI-MS/MS. Chemical and Biological Technologies in Agriculture, 9(1), 22. https://doi.org/10.1186/s40538-022-00285-4

Matesun J, Petrik L, Musvoto E, et al (2024) Limitations of wastewater treatment plants in removing trace anthropogenic biomarkers and future directions: A review. Ecotoxicol Environ Saf 281:116610. https://doi.org/10.1016/j.ecoenv.2024.116610

Michaelis C, Grohmann E (2023) Horizontal Gene Transfer of Antibiotic Resistance Genes in Biofilms. Antibiotics 12:328. https://doi.org/10.3390/antibiotics12020328

Mindubaev AZ, Babynin EV, Bedeeva EK, et al (2021) Biological Degradation of Yellow (White) Phosphorus, a Compound of First Class Hazard. Russ J Inorg Chem 66:1239–1244. https://doi.org/10.1134/S0036023621080155

Mo L, Wang Q, Bi E (2021) Effects of endogenous and exogenous dissolved organic matter on sorption behaviors of bisphenol A onto soils. J Environ Manage 287:112312. https://doi.org/10.1016/j.jenvman.2021.112312

Mohr S, Schott J, Maletzki D, Hünken A (2013) Effects of toxicants with different modes of action on Myriophyllum spicatum in test systems with varying complexity. Ecotoxicol Environ Saf 97:32–39. https://doi.org/10.1016/j.ecoenv.2013.06.030

Monroy-Licht A, Carranza-Lopez L, De La Parra-Guerra AC, Acevedo-Barrios R (2024) Unlocking the potential of Eichhornia crassipes for wastewater treatment: phytoremediation of aquatic pollutants, a strategy for advancing Sustainable Development Goal-06 clean water. Environ Sci Pollut Res 31:43561–43582. https://doi.org/10.1007/s11356-024-33698-9

Montano L, Pironti C, Pinto G, et al (2022) Polychlorinated Biphenyls (PCBs) in the Environment: Occupational and Exposure Events, Effects on Human Health and Fertility. Toxics 10:365. https://doi.org/10.3390/toxics10070365

Muluye T, Fetahi T, Engdaw F, Mohammed A (2023) Cyanotoxins in African waterbodies: occurrence, adverse effects, and potential risk to animal and human health. Environ Geochem Health 45:7519–7542. https://doi.org/10.1007/s10653-023-01724-3

Nguyen TL, Ayub A, Anam A, et al (2023) Fabrication of ZnO decorated porous chitosan beads for the sustainable bioremediation of Cr(VI) contaminated water. J Environ Chem Eng 11:110445. https://doi.org/10.1016/j.jece.2023.110445

Nugnes R, Russo C, Lavorgna M, et al (2022) Polystyrene microplastic particles in combination with pesticides and antiviral drugs: Toxicity and genotoxicity in Ceriodaphnia dubia. Environ Pollut 313:120088. https://doi.org/10.1016/j.envpol.2022.120088

Osuoha JO, Anyanwu BO, Ejileugha C (2023) Pharmaceuticals and personal care products as emerging contaminants: Need for combined treatment strategy. J Hazard Mater Adv 9:100206. https://doi.org/10.1016/j.hazadv.2022.100206

Othman AR, Hasan HA, Muhamad MH, et al (2021) Microbial degradation of microplastics by enzymatic processes: a review. Environ Chem Lett 19:3057–3073. https://doi.org/10.1007/s10311-021-01197-9

Pan L, Mao L, Zhang H, et al (2022) Modified Biochar as a More Promising Amendment Agent for Remediation of Pesticide-Contaminated Soils: Modification Methods, Mechanisms, Applications, and Future Perspectives. Appl Sci 12:11544. https://doi.org/10.3390/app122211544

Park M, Kang D, Yoon W, et al (2018). Programmed Hierarchical Hybrid Nanostructures from Fullerene-Dendrons and Pyrene-Dendrons. Small 14(49):1803291. https://doi.org/10.1002/smll.201803291

Paniagua-López M, García-Robles H, Aguilar-Garrido A, et al (2024) Vegetation establishment in soils polluted by heavy metal(loid)s after assisted natural remediation. Plant Soil 497:257–275. https://doi.org/10.1007/s11104-024-06521-0

Pannetier P, Morin B, Le Bihanic F, et al (2020) Environmental samples of microplastics induce significant toxic effects in fish larvae. Environ Int 134:105047. https://doi.org/10.1016/j.envint.2019.105047

Parsons F, Wood PR, DeMarco J (1984) Transformations of Tetrachloroethene and Trichloroethene in Microcosms and Groundwater. J AWWA 76:56–59. https://doi.org/10.1002/j.1551-8833.1984.tb05282.x

Payne L, Alcolea-Medina A, Snell LB, et al (2024) The first reported case of human infection with Variovorax durovernensis; a novel Variovorax sp. Isolated from the prosthetic aortic graft of a shepherd. Clin Infect Pract 21:100332. https://doi.org/10.1016/j.clinpr.2023.100332

Pérez DJ, Lombardero LR, Doucette WJ (2023) Influence of exposure time, physicochemical properties, and plant transpiration on the uptake dynamics and translocation of pharmaceutical and personal care products in the aquatic macrophyte Typha latifolia. Sci Total Environ 896:165107. https://doi.org/10.1016/j.scitotenv.2023.165107

Read DS, Gweon HS, Bowes MJ, et al (2024) Dissemination and persistence of antimicrobial resistance (AMR) along the wastewater-river continuum. Water Res 264:122204. https://doi.org/10.1016/j.watres.2024.122204

Robaire B, Delbes G, Head JA, et al (2022) A cross-species comparative approach to assessing multi- and transgenerational effects of endocrine disrupting chemicals. Environ Res 204:112063. https://doi.org/10.1016/j.envres.2021.112063

Rodriguez-Ruiz JP, Lin Q, Van Heirstraeten L, et al (2024) Long-term effects of ciprofloxacin treatment on the gastrointestinal and oropharyngeal microbiome are more pronounced after longer antibiotic courses. Int J Antimicrob Agents 64:107259. https://doi.org/10.1016/j.ijantimicag.2024.107259

Rodríguez-Cruz MS, Sánchez-Martín MJ (Eds.) (2022) Pesticides in Soils: Occurrence, Fate, Control and Remediation (Vol. 113). Springer International Publishing https://doi.org/10.1007/978-3-030-90546-0.

Rodríguez-Robledo V, Vendrell A, García-Cifuentes A, et al (2022) Determination of atrazine and propazine metabolites deemed endocrine disruptors in human seminal plasma by LC–ESI-MS/MS. Chemical and Biological Technologies in Agriculture 9(1):22. https://doi.org/10.1186/s40538-022-00285-4

Rodríguez-Rodríguez CE, Ramírez-Morales D, Masis-Mora M, et al (2023). Occurrence and risk assessment of pharmaceuticals in hospital wastewater in Costa Rica. Chemosphere 339:139746. https://doi.org/10.1016/j.chemosphere.2023.139746

Rowett CJ, Hutchinson TH, Comber SDW (2016) The impact of natural and anthropogenic Dissolved Organic Carbon (DOC), and pH on the toxicity of triclosan to the crustacean Gammarus pulex (L.). Sci Total Environ 565:222–231. https://doi.org/10.1016/j.scitotenv.2016.04.170

Roy H, Alam SR, Bin-Masud R, et al (2022) A Review on Characteristics, Techniques, and Waste-to-Energy Aspects of Municipal Solid Waste Management: Bangladesh Perspective. Sustainability 14:10265. https://doi.org/10.3390/su141610265

Rusu A, Munteanu A-C, Arbănaşi E-M, Uivarosi V (2023) Overview of Side-Effects of Antibacterial Fluoroquinolones: New Drugs versus Old Drugs, a Step Forward in the Safety Profile? Pharmaceutics 15:804. https://doi.org/10.3390/pharmaceutics15030804

Sandoval C, Calle Y, Godoy K, Farías J (2023) An Updated Overview of the Role of CYP450 during Xenobiotic Metabolization in Regulating the Acute Myeloid Leukemia Microenvironment. Int J Mol Sci 24:6031. https://doi.org/10.3390/ijms24076031

Sembada AA, Lenggoro IW (2024) Transport of Nanoparticles into Plants and Their Detection Methods. Nanomaterials 14:131. https://doi.org/10.3390/nano14020131

Sharma S, Pathania S, Bhagta S, et al (2024) Microbial remediation of polluted environment by using recombinant E. coli: a review. Biotechnol Environ 1:8. https://doi.org/10.1186/s44314-024-00008-z

Shi F, Almerick T Boncan D, Wan HT, et al (2022) Hepatic metabolism gene expression and gut microbes in offspring, subjected to in-utero PFOS exposure and postnatal diet challenges. Chemosphere 308:136196. https://doi.org/10.1016/j.chemosphere.2022.136196

Si L, Branfireun BA, Fierro J (2022) Chemical Oxidation and Reduction Pathways of Mercury Relevant to Natural Waters: A Review. Water 14:1891. https://doi.org/10.3390/w14121891

Singh GB, Vinayak A, Mudgal G, Kesari KK (2024) Azo dye bioremediation: An interdisciplinary path to sustainable fashion. Environ Technol Innov 36:103832. https://doi.org/10.1016/j.eti.2024.103832

Singh SB, Carroll-Portillo A, Lin HC (2023) Desulfovibrio in the Gut: The Enemy within? Microorganisms 11:1772. https://doi.org/10.3390/microorganisms11071772

Skic K, Boguta P, Klimkowicz-Pawlas A, et al (2023) Effect of sorption properties on the content, ecotoxicity, and bioaccumulation of polycyclic aromatic hydrocarbons (PAHs) in bottom sediments. J Hazard Mater 442:130073. https://doi.org/10.1016/j.jhazmat.2022.130073

Sobol Ł, Dyjakon A, Dlugogorski BZ (2025) Dioxin-like polychlorinated biphenyls (dl-PCB) in hydrochars and biochars: Review of recent evidence, pollution levels, critical gaps, formation mechanisms and regulations. J Hazard Mater 486:136615. https://doi.org/10.1016/j.jhazmat.2024.136615

Song L, Niu X, Zhang N, Li T (2021) Effect of biochar-immobilized Sphingomonas sp. PJ2 on bioremediation of PAHs and bacterial community composition in saline soil. Chemosphere 279:130427. https://doi.org/10.1016/j.chemosphere.2021.130427

Stanley LA (2017) Drug Metabolism. In: Pharmacognosy. Elsevier, pp 527–545

Štefanac T, Grgas D, Landeka Dragičević T (2021) Xenobiotics—Division and Methods of Detection: A Review. J Xenobiotics 11:130–141. https://doi.org/10.3390/jox11040009

Smith CD, Payne SE, Morace JL, Nilsen EB (2023) Organohalogenated contaminants in multiple life stages of the Pacific lamprey (Entosphenus tridentatus), Oregon, USA. Environmental Pollution 335:122363. https://doi.org/10.1016/j.envpol.2023.122363

Smith K, Stone W, Botha A, et al (2024) Riverine mycobiome dynamics: From South African tributaries to laboratory bioreactors. Mycology 15(4):631–650. https://doi.org/10.1080/21501203.2023.2278309

Smith M, Love DC, Rochman CM, Neff RA (2018) Microplastics in Seafood and the Implications for Human Health. Current Environmental Health Reports 5(3):375–386. https://doi.org/10.1007/s40572-018-0206-z

Terrazas-Salgado L, García-Gasca A, Betancourt-Lozano M, et al (2022) Epigenetic Transgenerational Modifications Induced by Xenobiotic Exposure in Zebrafish. Front Cell Dev Biol 10:832982. https://doi.org/10.3389/fcell.2022.832982

Thompson JR, García-Arenal F (1998) The Bundle Sheath-Phloem Interface of Cucumis sativus Is a Boundary to Systemic Infection by Tomato Aspermy Virus. Molecular Plant-Microbe Interactions® 11(2):109–114. https://doi.org/10.1094/MPMI.1998.11.2.109

Thompson TJ, Briggs MA, et al (2021) Groundwater discharges as a source of phytoestrogens and other agriculturally derived contaminants to streams. Science of The Total Environment 755:142873. https://doi.org/10.1016/j.scitotenv.2020.142873

Torres-Moreno AC, Mejia-Grau K, Puente-DelaCruz L, et al (2023) Polybrominated diphenyl ethers (PBDEs), polychlorinated biphenyls (PCBs), organochlorine pesticides (OCPs) in human breast milk from Colombia: A probabilistic risk assessment approach. Chemosphere 339:139597. https://doi.org/10.1016/j.chemosphere.2023.139597

Uddin MG, Rahman A, Nash S, et al (2023) Marine waters assessment using improved water quality model incorporating machine learning approaches. J Environ Manage 344:118368. https://doi.org/10.1016/j.jenvman.2023.118368

Ur Rahman S, Qin A, Zain M, et al (2024) Pb uptake, accumulation, and translocation in plants: Plant physiological, biochemical, and molecular response: A review. Heliyon 10:e27724. https://doi.org/10.1016/j.heliyon.2024.e27724

Vaillancourt FH, Bolin JT, Eltis LD (2006) The Ins and Outs of Ring-Cleaving Dioxygenases. Crit Rev Biochem Mol Biol 41:241–267. https://doi.org/10.1080/10409230600817422

Valkova N, Lépine F, Valeanu L, et al (2001) Hydrolysis of 4-Hydroxybenzoic Acid Esters (Parabens) and Their Aerobic Transformation into Phenol by the Resistant *Enterobacter cloacae* Strain EM. Appl Environ Microbiol 67:2404–2409. https://doi.org/10.1128/AEM.67.6.2404-2409.2001

Vicidomini C, Palumbo R, Moccia M, Roviello GN (2024) Oxidative Processes and Xenobiotic Metabolism in Plants: Mechanisms of Defense and Potential Therapeutic Implications. J Xenobiotics 14:1541–1569. https://doi.org/10.3390/jox14040084

Wang B, Smith M, Liu Y, et al (2023) Microplastic isolation method for wastewater and sludge samples by removal of excess organic and inorganic interferences. Chemosphere 329:138625. https://doi.org/10.1016/j.chemosphere.2023.138625

Wang F, Grundmann S, Schmid M, et al (2007) Isolation and characterization of 1,2,4-trichlorobenzene mineralizing Bordetella sp. and its bioremediation potential in soil. Chemosphere 67:896–902. https://doi.org/10.1016/j.chemosphere.2006.11.019

Wang H, Li X, Zhao X, et al (2022a) A review on heterogeneous photocatalysis for environmental remediation: From semiconductors to modification strategies. Chin J Catal 43:178–214. https://doi.org/10.1016/S1872-2067(21)63910-4

Wang X, Sial MU, Bashir MA, et al (2022b) Pesticides Xenobiotics in Soil Ecosystem and Their Remediation Approaches. Sustainability 14:3353. https://doi.org/10.3390/su14063353

Wang K, Yan D, Chen X, et al (2024a) New insight to the enriched microorganisms driven by pollutant concentrations and types for industrial and domestic wastewater via distinguishing the municipal wastewater treatment plants. Environ Pollut 361:124789. https://doi.org/10.1016/j.envpol.2024.124789

Wang Y, Li X, Yang H, et al (2024b) A review of tire wear particles: Occurrence, adverse effects, and control strategies. Ecotoxicol Environ Saf 283:116782. https://doi.org/10.1016/j.ecoenv.2024.116782

Wang J, Tan Z, Peng J, Qiu Q, Li M (2016) The behaviors of microplastics in the marine environment. Marine Environmental Research 113:7–17. https://doi.org/10.1016/j.marenvres.2015.10.014

Wang X, Chen A, Chen B, Wang L (2020) Adsorption of phenol and bisphenol A on river sediments: Effects of particle size, humic acid, pH and temperature. Ecotoxicology and Environmental Safety 204:111093. https://doi.org/10.1016/j.ecoenv.2020.111093

Wang L-C, Lin JC-T, Dong C-D, et al (2021) The sorption of persistent organic pollutants in microplastics from the coastal environment. J Hazard Mater 420:126658. https://doi.org/10.1016/j.jhazmat.2021.126658

Wang Y, Song Y, Zhang D, et al (2025) Effects of nitrogen-driven eutrophication on the horizontal transfer of extracellular antibiotic resistance genes in water-sediment environments. Environ Res 274:121317. https://doi.org/10.1016/j.envres.2025.121317

Wei X, Peng H, Li Y, et al (2023) Pyrethroids exposure alters the community and function of the internal microbiota in Aedes albopictus. Ecotoxicol Environ Saf 252:114579. https://doi.org/10.1016/j.ecoenv.2023.114579

Wilson BA, Addo-Mensah AK, Mendez MO (2015) In situ impacts of a flooding event on contaminant deposition and fate in a riparian ecosystem. Journal of Soils and Sediments 15(11):2244–2256. https://doi.org/10.1007/s11368-015-1145-z

Wilson ID, Nicholson JK (2003) Topics in Xenobiochemistry: Do metabolic pathways exist for xenobiotics? The micro-metabolism hypothesis. Xenobiotica 33(9):887–901. https://doi.org/10.1080/00498250310001598221

Wilson NM, Christou M, Turner CR, et al (1984). Binding and metabolism of benzo[a]pyrene and 7,12-dimethylbenz[a]anthracene by seven purified forms of cytochrome P-450. Carcinogenesis, 5(11), 1475–1483. https://doi.org/10.1093/carcin/5.11.1475

Xu C, Hu C, Lu J, et al (2024) Lake plastisphere as a new biotope in the Anthropocene: Potential pathogen colonization and distinct microbial functionality. J Hazard Mater 461:132693. https://doi.org/10.1016/j.jhazmat.2023.132693

Yousif E, Haddad R (2013) Photodegradation and photostabilization of polymers, especially polystyrene: review. SpringerPlus 2:398. https://doi.org/10.1186/2193-1801-2-398

Yu Y, Huang W, Yu W, et al (2023) Metagenomic insights into the mechanisms of triphenyl phosphate degradation by bioaugmentation with Sphingopyxis sp. GY. Ecotoxicol Environ Saf 263:115261. https://doi.org/10.1016/j.ecoenv.2023.115261

Yuan Y, Li D, Huang H, et al (2025) Direct photodegradation of aromatic carbamate pesticides: Kinetics and mechanisms in aqueous vs. non-aqueous media. J Hazard Mater 489:137648. https://doi.org/10.1016/j.jhazmat.2025.137648

Zhang Y, Chen Z, Shi Y et al (2023) Revealing the sorption mechanisms of carbamazepine on pristine and aged microplastics with extended DLVO theory. Science of The Total Environment, 874, 162480. https://doi.org/10.1016/j.scitotenv.2023.162480

Zhang G, Lan T, Yang G, Li J, Zhang K (2022) Contamination, spatial distribution, and source contribution of persistent organic pollutants in the soil of Guiyang city, China: A case study. Environmental Geochemistry and Health 44(10):3265–3278. https://doi.org/10.1007/s10653-021-01089-5

Zhang Z, Liu S, Wang X, Huang S, Sun K, Xia X (2023). Differences in structure and composition of soil humic substances and their binding for polycyclic aromatic hydrocarbons in different climatic zones. Environmental Pollution 322:121121. https://doi.org/10.1016/j.envpol.2023.121121

Zhang B, Mu H, Li H, et al (2025) Nationwide prediction of pesticide residual levels in soil: Implications on the resulting risk and prioritization framework. Environ Int 197:109355. https://doi.org/10.1016/j.envint.2025.109355

Zhang J, Ren S, Xu W et al (2022) Effects of plastic residues and microplastics on soil ecosystems: A global meta-analysis. J Hazard Mater 435:129065. https://doi.org/10.1016/j.jhazmat.2022.129065

Zhao X, Zhang Y, Long T, et al (2022) Regulation Mechanism of Plant Pigments Biosynthesis: Anthocyanins, Carotenoids, and Betalains. Metabolites 12:871. https://doi.org/10.3390/metabo12090871

Zheng X, Lin H, Du D, et al (2024) Remediation of heavy metals polluted soil environment: A critical review on biological approaches. Ecotoxicol Environ Saf 284:116883. https://doi.org/10.1016/j.ecoenv.2024.116883

Zhou H, Wang H, Yue C, et al (2024) Photocatalytic degradation by TiO2-conjugated/coordination polymer heterojunction: Preparation, mechanisms, and prospects. Appl Catal B Environ 344:123605. https://doi.org/10.1016/j.apcatb.2023.123605

Zhu M, Qi X, Yuan Y, et al (2023) Deciphering the distinct successional patterns and potential roles of abundant and rare microbial taxa of urban riverine plastisphere. J Hazard Mater 450:131080. https://doi.org/10.1016/j.jhazmat.2023.131080

Zhu Q-M, Wang C, Liu J-W, et al (2024) Degradation profile of environmental pollutant 17β-estradiol by human intestinal fungus Aspergillus niger RG13B1 and characterization of genes involved in its degradation. J Hazard Mater 461:132617. https://doi.org/10.1016/j.jhazmat.2023.132617

Zulfiqar U, Haider FU, Maqsood MF, et al (2023) Recent Advances in Microbial-Assisted Remediation of Cadmium-Contaminated Soil. Plants 12:3147. https://doi.org/10.3390/plants12173147

Xenobiotic Interaction with Algal Community: Cutting-Edge Insights and Future Horizons

Shouvik Pramanik, Anupam Kundu, Nousi Parvin, Arka Mondal, Subarna Saha, Sourav Roy, and Jnanendra Rath

Abstract Exposure to xenobiotic chemicals through food and the environment is a growing concern for both humans and animals. Pesticides, used since 2500 BC to control pests and improve crop yields, persist in the environment, accumulating in soil, water, and living organisms. These chemicals pose significant threats to aquatic ecosystems, particularly impacting algal communities essential for ecological balance and primary productivity. While xenobiotics are toxic to algae, these organisms can accumulate and bioremediate these compounds, making them crucial in studies of environmental detoxification and ecosystem resilience. Understanding the interactions between xenobiotics and algae is key to assessing ecological risks and mitigating the impact of these pollutants. This chapter provides a comprehensive overview of the interactions between xenobiotics and algae, exploring the types of xenobiotics that affect algal communities, their uptake and toxicity mechanisms, and the key factors influencing these processes. By examining the adverse and adaptive responses of algae to these pollutants, this chapter offers critical insights into the complex dynamics at play. The goal is to enhance our understanding of xenobiotic-algal interactions and to inform the development of strategies for protecting and restoring aquatic ecosystems in the face of increasing environmental challenges.

Keywords Algae · Cyanobacteria · Phycoremediation · Toxicity · Xenobiotic

S. Pramanik · A. Kundu · N. Parvin · A. Mondal · S. Saha · S. Roy · J. Rath (✉)
Department of Botany, Visva-Bharati (A Central University), Santiniketan, West Bengal, India
e-mail: jnanendra.rath@visva-bharati.ac.in

S. Roy, V. Mandal (eds.), *Plant-Microbe Interaction under Xenobiotic Exposure*,
https://doi.org/10.1007/978-981-96-8260-7_6

1 Introduction

Urbanization, industrialization, and globalization drive economic growth but also lead to environmental pollution, particularly through the release of pharmaceuticals, personal care products, and pesticides. Weak government policies on drug disposal and corporate irresponsibility exacerbate groundwater contamination with xenobiotics. This increasing discharge of harmful substances into aquatic environments poses significant threats to ecological balance (Aroh 2018; Desai et al. 2022; Miglani et al. 2022). Pollutants, particularly microplastics (MP), are prevalent in aquatic ecosystems, especially near urban areas. Smaller microplastics (20–150 μm) are more abundant, with 531–3485 MP/kg in the Brahmaputra River and 525–1752 MP/kg in the Indus River. Larger microplastics (150 μm–5 mm) are found in lower concentrations, ranging from 20 to 240 MP/kg in the Brahmaputra and 60–340 MP/kg in the Indus River (Tsering et al. 2021). Pharmaceuticals, particularly antibiotics, in freshwater systems are also raising concerns about resistance. In various aquatic ecosystems across India, ofloxacin was found at the highest concentration (145.3794 ng/mL) in river water samples, followed by amoxicillin (3.033 ng/mL) and erythromycin (2.171 ng/mL) (Akhter et al. 2023). Xenobiotics exert toxic effects on algae, yet these organisms can accumulate and bioremediate such compounds. This dual capacity makes algae a fascinating subject for further study, offering potential insights into environmental detoxification and the resilience of aquatic ecosystems (Lu et al. 2021; Abdelfattah et al. 2023). Understanding complex interactions between xenobiotics and algal communities is crucial for determining the ecological risks of these pollutants and developing ways to reduce their impact (Fleeger et al. 2003; Mathew et al. 2017). Xenobiotics affect algal communities through various mechanisms, including direct toxicity, interference with photosynthetic processes, and disruption of cellular functions (Petsas and Vagi 2017). The types of xenobiotics that impact algae are diverse, ranging from heavy metals that can bind to cellular components and inhibit enzyme activity or that may be taken up and metabolized (Patterson et al. 2010; Nowicka 2022). The mechanisms of uptake, which include adsorption, passive diffusion, and active transport, are influenced by both the xenobiotics' physicochemical properties and the algae's physiological state. Once inside the cells, xenobiotics can induce oxidative stress, disrupt membrane integrity, and interfere with essential metabolic pathways, leading to reduced growth, altered pigmentation, and even cell death (Nowicka 2022; Abdelfattah et al. 2023). Various environmental factors also shape the extent and nature of xenobiotic-algal interactions. Light intensity, temperature, pH, and nutrient availability can all influence the susceptibility of algae to xenobiotics and modulate the severity of their effects (Sharma et al. 2021). Additionally, the community structure and diversity of algal populations play a crucial role in determining the overall resilience of the ecosystem. Diverse communities may be better equipped to cope with xenobiotic stress due to functional redundancy and the presence of tolerant species that can maintain ecosystem function in the face of chemical perturbation (Nogales et al. 2011; Morrissey et al. 2021). Recent advances in analytical techniques, including omics approaches, have improved our understanding of algae's molecular and biochemical responses to xenobiotics. These technologies identify genes, proteins, and metabolites involved in detoxification and stress responses, offering insights into algae's adaptive strategies and revealing previously inaccessible xenobiotic-algal interaction dynamics (Pandey et al. 2019; Sharma et al. 2022). The implications of these interactions extend beyond the immediate effects on algal physiology and community structure. Xenobiotics can influence algal community dynamics, alter food web interactions, and accumulate toxic compounds in higher trophic levels. As such, there is a growing interest in leveraging the natural detoxification abilities of algae for bioremediation purposes (Perhar and Arhonditsis 2014; Borgå et al. 2022). Algal-based bioremediation strategies, which involve algae to remove or neutralize xenobiotics from contaminated environments, hold promise for sustainable and cost-effective environmental management (Abdelfattah et al. 2023). This chapter provides a comprehensive overview of xenobiotic interactions with algal communities, addressing the types of xenobiotics, mechanisms of uptake and toxicity, and the environmental factors that shape these interactions. It highlights recent technological advances, the importance of algal diversity in mitigating xenobiotic impacts, and the potential of algae-based bioremediation strategies. By integrating recent research, this chapter offers key insights and outlines future research directions to enhance environmental management practices.

2 Current Understanding of Xenobiotic-Algal Interactions

2.1 Types of Xenobiotics Affecting Algal Communities

Xenobiotics are foreign environmental pollutants, such as pesticides, heavy metals, pharmaceuticals, and industrial chemicals (PCBs, PAHs). Algae can absorb or adsorb these substances when introduced into aquatic systems, leading to harmful effects on cellular processes, growth inhibition, and changes in algal community structure. All of these xenobiotics are illustrated in Fig. 1 and discussed below.

2.1.1 Pesticides and Herbicides

Pesticides and herbicides are xenobiotic synthetic substances that can majorly impact algal communities in aquatic ecosystems. These compounds can be hazardous to algae, resulting in decreased growth, lower photosynthetic efficiency, and even cell death. The toxicity depends on various factors, including concentration, exposure time, and specific species of algae involved (Staley et al. 2015; Li and Mcintyre 2024). Organochlorines, like dichlorodiphenyltrichloroethane (DDT), are persistent insecticides that bioaccumulate in the food chain. In aquatic systems, DDT can adsorb onto algal surfaces or be absorbed into cells via passive diffusion

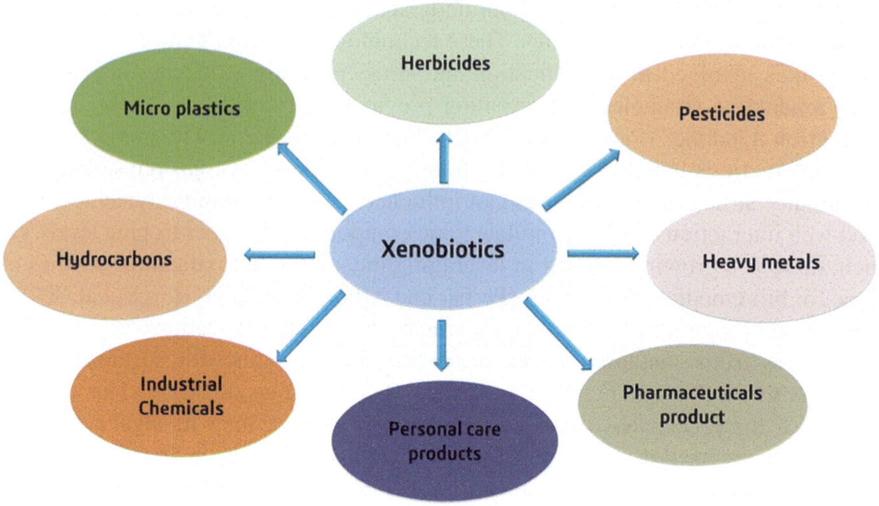

Fig. 1 Types of xenobiotics

due to its lipophilicity (Jayaraj et al. 2016; Verasoundarapandian et al. 2022). DDT inhibits photosynthesis by disrupting electron transport in chloroplasts, reducing algal growth. Its persistence in sediments poses long-term exposure risks, affecting species diversity and ecosystem dynamics. Decades after its ban, DDT residues continue to threaten aquatic ecosystems, highlighting its lasting ecological impact (Sudharshan et al. 2012; Baghour 2019; Chen et al. 2020a). Glyphosate, a widely used herbicide, disrupts algal metabolic pathways by inhibiting photosynthesis, leading to reduced growth or death. This disturbance alters community composition, with some species more resistant and others more vulnerable, ultimately changing species dominance (Iummato et al. 2019; Klátyik et al. 2024). A recent study found that glyphosate significantly inhibited photosynthesis in freshwater algae by interfering with the photosynthetic electron transport chain. This disruption led to increased reactive oxygen species (ROS) production, lipid peroxidation, and superoxide dismutase (SOD) activity, ultimately reducing the growth and biomass accumulation of *Chlorella vulgaris* in Tibetan water bodies (Zhang et al. 2023). Atrazine and diuron, both photosystem II inhibitors, disrupt photosynthesis in algae. Atrazine reduces chlorophyll production and growth, while diuron, used in agriculture and antifouling coatings, decreases photosynthetic ability. Their presence can alter species composition in algal communities (Haynes et al. 2000; Zhu et al. 2009; Castro et al. 2022). A study found that atrazine and diuron impair the photosynthetic efficiency of benthic diatoms in intertidal mudflats, with diuron being more toxic. At higher concentrations, photosynthesis was fully inhibited, while at lower concentrations, the herbicides exhibited a synergistic effect (Legrand et al. 2006). Azoxystrobin, a fungicide that disrupts mitochondrial respiration in fungi, can also affect algae by impairing mitochondrial electron transport. This disruption can harm aquatic ecosystems and promote freshwater cyanobacterial dominance by altering species competition (Lu et al. 2019).

2.1.2 Heavy Metals and Metalloids

Heavy metals and metalloids, as xenobiotics, significantly influence algal communities in aquatic environments, causing severe ecological disruption. Lead (Pb), mercury (Hg), cadmium (Cd), and arsenic (As) are not naturally prevalent in ecosystems, but they become harmful in increased amounts due to anthropogenic activities such as industrial discharges, mining, agricultural runoff, and inappropriate waste disposal (Singh et al. 2023, 2024). In aquatic ecosystems, pollutants like lead can be concentrated in algae through adsorption or absorption. Lead inhibits chlorophyll synthesis, damages photosystem II, reduces photosynthetic efficiency and growth rates, and causes oxidative stress by generating reactive oxygen species (ROS), leading to cellular damage (Pasricha et al. 2021; Nowicka 2022). A recent study indicates that *Chlorella sorokiniana* exhibited significant ultrastructural and physiological disruptions when exposed to lead, which caused severe effects. Photosynthesis was heavily inhibited, while respiration increased dramatically (Carfagna et al. 2013). Mercury, a highly toxic metal, affects algae by inhibiting photosynthesis and inducing oxidative stress, reducing chlorophyll content and increasing ROS generation (Nowicka 2022). Mercury affects algal growth and morphology, reducing cell density and altering cell structure, with potential biomagnification risks. A study found that mercury exposure at 3 mg/L significantly decreased cell density in *Spirulina maxima*, *Nannochloropsis oculata*, *Chlorella vulgaris*, and *Porphyridium cruentum*, highlighting its toxicity (Arsad et al. 2020). Cadmium severely affects aquatic algae by altering cell growth, morphology, and metabolism. A study found that cadmium exposure in *Chlorella sorokiniana* reduced photosynthesis and energy production but increased antioxidant defenses and nutrient recycling. It also impacted carotenoid and lipid metabolism (León-Vaz et al. 2021). Arsenic, a toxic metalloid, inhibits growth and photosynthesis in algae and cyanobacteria by reducing chlorophyll content and inducing oxidative stress through reactive oxygen species (ROS). It disrupts metabolism and nutrient uptake, affecting growth, health, and species composition (Ferrari et al. 2013; Mao et al. 2022; Zaidi et al. 2024). A study found that *Tolypothrix tenuis* and *Nostoc minutum* respond differently to arsenic, with As(III) being more toxic than As(V). While *T. tenuis* was sensitive to As(V), *Nostoc minutum* showed enhanced growth and pigment production at high As(V) concentrations, accumulating significant arsenic (Ferrari et al. 2013). A study showed that *Chlorella thermophila* and *Leptolyngbya* sp. can tolerate and accumulate arsenic, indicating their potential for bioremediation. Both algae initially boosted antioxidant defenses but differed in toxicity responses and bioaccumulation abilities, highlighting their role in arsenic bioremediation in polluted waters (Mao et al. 2022).

2.1.3 Pharmaceuticals and Personal Care Products (PPCPs)

Pharmaceuticals and personal care products (PPCPs) comprise diverse substances with potential adverse effects on aquatic environments. Pharmaceuticals, such as antibiotics (e.g., tetracycline, amoxicillin), antidepressants (e.g., fluoxetine, sertraline), anti-inflammatory drugs (e.g., ibuprofen, naproxen), hormones (e.g., ethinyl estradiol, testosterone), and anticonvulsants (e.g., carbamazepine, valproic acid), can disrupt algal communities by affecting photosynthesis, growth, and reproductive processes, as well as by inducing physiological stress and altering nutrient dynamics (Ebele et al. 2017; Xin et al. 2021). Personal care products, including cosmetics (e.g., sunscreens with oxybenzone), fragrances, antibacterial agents (e.g., triclosan), and shampoos, can similarly impact algae by causing toxicity, modifying nutrient availability, and altering growth patterns. Substances like steroids (e.g., cortisol) and antihistamines (e.g., cetirizine) can further disrupt hormonal balance and physiological responses in aquatic organisms (Ebele et al. 2017; Bilal et al. 2020). Different types of PPCPs, categorized according to their chemical nature, have varying effects on different algal species. These effects manifest differently depending on the species of algae exposed to the PPCPs.

2.1.4 Industrial Chemicals

Industrial chemicals like dioxins, furans, PAHs, PCBs, xylene, BPA, phthalates, benzene, toluene, and detergents such as alkylphenol ethoxylates have significant toxic effects on algal communities (Xia et al. 2016; Koukoulakis et al. 2020; Abootalebi Jahromi et al. 2020). PCBs accumulate in algal tissues, with *Chlamydomonas reinhardtii* showing ultra-hydrophobic PCBs (Kow >6) enter cells with minimal restriction. Surface adsorption to the cell wall contributes less than 10% to total bioaccumulation, emphasizing the importance of internal uptake in the bioaccumulation process (Jabusch and Swackhamer 2004). Pennate diatoms and euglenoids absorb dissolved PCBs in rivers, with increased absorption during bloom seasons. PCBs accumulate more in lipid-rich samples, but no strong link was found between lipid content and PCB association, suggesting other algal components are involved. During blooms, PCBs partition more to live algae, while detritus is more significant in non-bloom conditions (Fitzgerald and Steuer 2006). A study on the marine alga *Phaeodactylum tricornutum* showed that exposure to PAHs, especially from oil spills, increased sensitivity under UV-B exposure, with toxicity 2–3 times higher compared to PAHs alone. This confirms a phototoxic effect and highlights the compounded impact of UV-B and PAHs in aquatic environments (Okay and Karacik 2007). BTEX compounds, common industrial solvents, have been found to affect cellular activities in algae such as *Euglena gracilis*, which exhibit dose-dependent toxic reactions to BTEX, with solvent xylenes being the most toxic, inhibiting growth, followed by ethylbenzene, toluene, and benzene (Penga et al.

2014). BPA, an endocrine disruptor, reduces growth and reproduction in *Chlorella pyrenoidosa*, showing maximum inhibition on day six. Both BPA and BPS exhibit a "low promotion and high inhibition" effect, with combined exposure increasing oxidative stress and antioxidant enzyme activity, indicating a synergistic toxic effect (Li et al. 2022). Detergents such as nonylphenol and octylphenol have been shown to be acutely toxic to algae at concentrations of 17–3000 μg/L, and chronic toxicity has been shown in the range of 3.7–6 μg/L (Acir and Guenther 2018). These case studies underscore industrial chemicals' pervasive and detrimental impact on algal communities, leading to ecological imbalances and potential disruptions in aquatic ecosystems.

2.2 *Mechanisms of Xenobiotic Uptake and Cellular Response in Algae*

Xenobiotics enter algae through adsorption onto extracellular polymeric substances (EPS) and absorption into cells via passive diffusion or active transport. They may be metabolized or induce oxidative stress by generating reactive oxygen species (ROS), damaging lipids, proteins, and DNA. They can also inhibit photosynthesis by disrupting the photosynthetic apparatus and electron transport chain and interfere with cellular functions, including enzyme activity (Petsas and Vagi 2017; Abdelfattah et al. 2023). An overview of the mechanisms of xenobiotic uptake and cellular response is illustrated in Fig. 2.

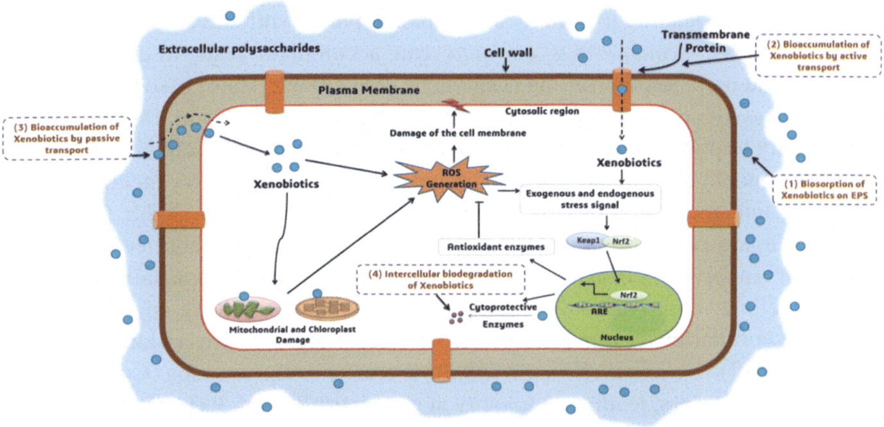

Fig. 2 Illustration of xenobiotic uptake and cellular reactions in algal cell

2.2.1 Cellular Uptake and Transport

Xenobiotics can enter cells by various mechanisms and may have adverse effects. The type of xenobiotic and its physicochemical features significantly impact the cellular intake process in algae (Štefanac et al. 2021; Miglani et al. 2022). Size, polarity, solubility, and structural complexity are all key factors for xenobiotics. These features influence how easily a xenobiotic crosses the cell membrane and the mechanisms involved in its uptake. Understanding these pathways is vital for evaluating the environmental impact of xenobiotics and developing bioremediation strategies (Dordio and Carvalho 2013; Hejna et al. 2022). Xenobiotics enter algal cells via passive diffusion, active transport, adsorption, and absorption. Each mechanism varies by algal species, making the study of xenobiotic uptake complex and species-specific (Sutherland and Ralph 2019; Xiong et al. 2021; Hejna et al. 2022). Passive diffusion allows xenobiotics, including heavy metals like cadmium, lead, and mercury, to enter algae and cyanobacteria driven by concentration gradients. This energy-free process accumulates these substances, which can disrupt cellular functions and photosynthesis (Nowicka 2022; Abdelfattah et al. 2023). Xenobiotics move from high to low concentration through passive diffusion. Nonpolar, lipid-soluble compounds can pass through the hydrophobic cell membrane, while polar compounds, large molecules, and ions cannot. Pollutants like sulfamethoxazole, carbamazepine, trimethoprim, and florfenicol enter microalgae cells via this process (Song et al. 2019; Abdelfattah et al. 2023). Exposure to pollutants can alter cell membrane permeability, affecting passive diffusion and toxin sensitivity. In *Anabaena*, perfluorinated alkyl acids increased the toxicity of most herbicides, except atrazine. Conversely, perfluorooctane sulfonate heightened the toxicity of paraquat and diuron, reduced atrazine's toxicity, and had no significant effect on 2,4-D, impacting microorganism growth and metabolism (Rodea-Palomares et al. 2015). For instance, studies have shown that *Chlorella vulgaris* can accumulate cadmium rapidly, leading to toxicity and impaired photosynthetic efficiency (Bajguz 2011). Similarly, *Scenedesmus obliquus* can accumulate lead, altering cellular structures and metabolite profiling (Danouche et al. 2022). These case studies highlight how passive diffusion facilitates xenobiotic uptake in algae, revealing their vulnerability to pollutants and broader ecological implications. Active transport, an energy-dependent mechanism, moves foreign compounds like heavy metals, organic pollutants, and metalloids across cell membranes (Ankit et al. 2022; Dey et al. 2024). Comparative genomic analysis in *Chlamydomonas* sp. identified new metal transporters involved in iron and copper homeostasis, revealing specialized roles in different cellular compartments. The study highlights the diverse and specialized metal transport pathways that evolved across various algal species (Blaby-Haas and Merchant 2012). *Chlamydomonas reinhardtii* uptake cadmium ions (Cd^{2+}) through metal ion transporters, leading to toxicity at concentrations of 300 μM. Metal transport transporters demonstrate the organisms' potential in bioremediation and environmental cleanup through their ability to handle and detoxify diverse xenobiotics (Mosulén et al. 2003). Algae and cyanobacteria interact with xenobiotics through biosorption. Their cell walls, with a fibril matrix and amorphous fraction, enhance biosorption due to mechanical strength, flexibility, and intercellular spaces. Surface porosity and roughness affect heavy metal adsorption. Smaller cell sizes increase surface area for adsorption, but excessive biomass can reduce efficiency due to active site repulsion (Abdelfattah et al. 2023). Extracellular polymeric substances (EPS) derived from algae can accelerate metal biosorption. EPS, consisting of polysaccharides, proteins, and lipids, enhances the adsorption of pollutants like pesticides and heavy metals by binding them to the surface of the cells (Yadav et al. 2021; Kalita and Baruah 2023). Exopolysaccharides, polymeric substances, effectively adsorb and absorb positively charged metal ions, transporting them into the cytoplasm. In contrast, heavy metals are directly adsorbed from the substratum when using nonliving biomass (Kalita and Baruah 2023). Algal cell walls, with functional groups like −COOH, −OH, $-HPO_4^{2-}$, SO_4^{2-}, $-NH_2$, and −SH, facilitate xenobiotic adsorption. Chelation, hydrogen bonding, and ion exchange are key mechanisms, with varying bioadsorption capacities for pharmaceuticals. For instance, *Chlorella* sp., *Chlamydomonas* sp., and *Mychonastes* sp. adsorb 4.74 to 2.95 mg/g of 7-amino cephalosporanic acid, while *Scenedesmus obliquus* and *Chlorella vulgaris* show 7.3% and 100% adsorption efficiency for sulfamethazine and metronidazole, respectively. Tetracycline adsorption can reach up to 295.34 mg/g in *Scenedesmus quadricauda* and 56.25 mg/g in *Tetraselmis suecica* (Hejna et al. 2022). Cyanobacteria like *Nostoc* sp. effectively remove heavy metals, particularly nickel (Ni II), with maximal biosorption of 188.8 mg/g of cell dry weight. This is linked to increased EPS, protein, and carbohydrate content. FT-IR and GC-MS analyses show nickel removal is associated with changes in functional groups and silicon-containing heterocyclic and aromatic diacid compounds (Ghorbani et al. 2022). The chromium adsorption potential of polysaccharides from *Lyngbya putealis* is optimal at an initial chromium concentration of 30–40 mg/L, pH 2, and 45 °C. The adsorption capacity of EPS increased from 45 to 157 mg/g as Cr(VI) concentration rose from 10 to 30 mg/L. These interactions are crucial for bioremediation, reducing pollutant bioavailability and toxicity (Kiran and Kaushik 2008).

2.2.2 Biochemical and Molecular Responses

Algae encounter xenobiotics such as pollutants, heavy metals, and synthetic chemicals, triggering biochemical and molecular responses to detoxify and mitigate damage. Understanding these responses is crucial for assessing pollutant impacts on aquatic ecosystems and exploring algae-based bioremediation (Mathew et al. 2017). Algae interact with xenobiotics through various phases and pathways. While biochemical and molecular responses to xenobiotics are well studied in higher organisms, they are less explored in algae. Understanding these processes is crucial for environmental monitoring and bioremediation (Touliabah et al. 2022; Malchi et al. 2022; Bastolla et al. 2023). After entering an algal cell, the fate of xenobiotics depends on their nature and mode of action, leading to different cellular responses (Patterson et al. 2010; Wang et al. 2019a). Some xenobiotics may be directly metabolized or sometimes form more toxic compounds, and others are detoxified through biochemical pathways (Štefanac et al. 2021; Touliabah et al. 2022). In contrast, others interact with specific receptor proteins, triggering a cascade of molecular events that alter the expression of genes involved in stress response, detoxification, and repair mechanisms (Baldwin 2019; Nowicka 2022; Zhu et al. 2023). Exposure to xenobiotics like heavy metals, pesticides, or pharmaceuticals induces oxidative stress in algae, leading to the overproduction of reactive oxygen species (ROS). To counteract this, algae activate a sophisticated antioxidant defense system at both cellular and molecular levels (Sule et al. 2022). At the cellular level, key antioxidant enzymes such as superoxide dismutase (SOD), catalase (CAT), and peroxidases (POD) are upregulated (Rajput et al. 2021). SOD catalyzes the conversion of superoxide anions into hydrogen peroxide, which is then broken down by CAT into water and oxygen, thereby preventing harmful accumulation of ROS. Peroxidases further detoxify hydrogen peroxide and organic peroxides, protecting cellular structures from oxidative damage (Wang et al. 2018; Heo et al. 2020). Heavy metal contamination has spurred interest in remediation techniques. Microalgae remove excess transition metal ions from the cytoplasm and redirect them either outside the cell or to other subcellular regions, involving transporter genes in their metal detoxification systems (Saravanakumar et al. 2024). Algal cells have the affinity to bind metals up to 10% of their total biomass (Sarma et al. 2024). The cell wall polysaccharides of some algal species are reported to affect ion exchange reactions with polyvalent metals (Redalje et al. 2018). *Pseudochlorococcum typicum* and *Scenedesmus quadricauda* var. *quadrispina* demonstrated greater tolerance to heavy metals, with *P. typicum* achieving notable heavy metal accumulation, accumulating 97% of Hg^{2+}, 86% of Cd^{2+}, and 70% of Pb^{2+} within the first 30 min (Shanab et al. 2012). Additionally, heavy metals may bind to and be sequestered by metal-binding proteins such as phytochelatins or metallothioneins. Small, metal-binding proteins play crucial roles in managing metal toxicity and homeostasis in plants, algae, and other organisms (Chaudhary et al. 2018). Phytochelatins are small, cysteine-rich peptides that bind heavy metal ions, forming stable complexes that reduce metal bioavailability and toxicity by sequestering the metals in vacuoles. This binding detoxifies the metals and protects cellular components from oxidative damage (Yadav 2010). Metallothioneins, on the other hand, are small proteins that also bind various metal ions, including essential and toxic metals like cadmium (Cd) and copper (Cu). They play a key role in maintaining metal homeostasis, neutralizing reactive oxygen species (ROS) to prevent oxidative stress, and managing intracellular metal concentrations (Florianczyk 2007). Together, these proteins enable algae to adapt to and survive in metal-contaminated environments by reducing metal-induced stress and ensuring cellular protection and balance (Tripathi and Poluri 2021). Marine alga *Ulva compressa* responds to increasing copper concentrations by initially elevating glutathione (GSH) and phytochelatins (PCs), then upregulating metallothioneins (MTs) to aid in copper accumulation and detoxification. When transferred to copper-free seawater, the alga releases copper along with GSH, indicating an extracellular detoxification mechanism (Navarrete et al. 2019). Other xenobiotics may interact with receptor proteins, such as transcription factors or membrane receptors, leading to the activation of signal transduction pathways. This can result in the upregulation or downregulation of specific genes (Zheng et al. 2016). Xenobiotics upregulate genes for specific antioxidant enzymes by activating stress-responsive transcription factors, such as nuclear factor erythroid 2-related factor 2 (Nrf2) analogs, which enhance their transcription. Antioxidant enzymes may also undergo post-translational modifications like phosphorylation to boost activity and stability. This response integrates with other stress pathways, creating a coordinated defense mechanism against oxidative stress (Veal et al. 2007; He et al. 2020; Kanwugu and Glukhareva 2023). Cadmium (Cd) exposure in *Dunaliella salina* induces distinct physiological and genetic responses based on duration. Long-term exposure increases lipid peroxidation, reduces pigment levels, and causes cell damage. Activities of antioxidant enzymes (APX, SOD, GST, GPX, GR) and ROS scavenging gene expression vary with exposure duration (Zhu et al. 2019). In the case of organic compound nonylphenol exposure, antioxidant responses are triggered in different microalgae, with effects varying by its concentration and exposure time. Nonylphenol-tolerant species like *Chlorella vulgaris* and *Chlorella* sp. showed significant increases in antioxidant enzyme activities (SOD, CAT, POD) at high nonylphenol levels, while nonylphenol-sensitive species like *Selenastrum capricornutum* and *Scenedesmus quadricauda* exhibited weaker or insignificant responses (Gao et al. 2017). Sometimes, xenobiotics get metabolized in the biotransformation process by changing their chemical structure and toxicity (Hejna et al. 2022). Biotransformation in algal cytoplasm involves converting hazardous hydrophobic xenobiotic compounds into non-toxic hydrophilic forms through methylation, oxidation, hydroxylation, and acylation. These processes occur via two enzyme-dependent processes: metabolic (employing organic pollutants as a carbon source and energy) and co-metabolic mechanisms (degrading micropollutants through non-specific enzymes in the catabolism of other substrates) (Kumar et al. 2023). *Microcystis aeruginosa*, *Synechococcus* sp., *Chlamydomonas reinhardtii*, *Desmodesmus subspicatus*, *Scenedesmus dimorphus*, *Scenedesmus obliquus*, and *Chlorella pyrenoidosa* can transform various xenobiotics and hormones through enzymatic processes such as hydrolysis, methylation, CYP450 oxidation, and conjugation. They break down compounds like strobilurin, sulfamethoxazole, atenolol, estradiols, progesterone, and norgestrel. For example, estradiols are converted to estrone and estriol, while progesterone and norgestrel undergo hydroxylation, reduction, and oxidation (Hejna et al. 2022). This complex interaction of cellular and molecular responses allows algae to adapt and thrive in environments contaminated with xenobiotics.

2.3 Factors Influencing Xenobiotic-Algal Interaction

Xenobiotics often have no natural counterpart in these environments, making them particularly disruptive to ecological processes. Algal communities, which play a fundamental role in aquatic ecosystems as primary producers, are especially vulnerable to xenobiotics (Liu et al. 2021). Understanding how these pollutants interact with algae is crucial for assessing the broader ecological impacts and developing strategies to mitigate pollution effects.

2.3.1 Chemical Properties of Xenobiotics

The chemical structure of a xenobiotic affects its reactivity, solubility, and ability to penetrate algal cells. Water solubility is crucial; algae absorb more soluble compounds more readily, while less soluble ones may accumulate in sediments and pose long-term risks through resuspension or food chain transfer (Luthe et al. 2008). Persistent organic pollutants (POPs) such as polychlorinated biphenyls (PCBs) and dioxins are highly hydrophobic and remain in the environment for long periods, leading to prolonged exposure and potential bioaccumulation in algae (Barceló and Petrovic 2007). Hydrophobic (lipophilic) compounds tend to accumulate in cell membranes, potentially causing more damage. Higher concentrations of xenobiotics typically result in more significant toxicity, but even low concentrations can cause chronic stress in algal communities.

2.3.2 Environmental Conditions

The pH of the water affects the speciation and solubility of many xenobiotics, which in turn influences their bioavailability to algae. For example, metals like copper are more toxic in acidic conditions (Luthe et al. 2008). Temperature affects the metabolic rates of algae and the chemical reactivity of xenobiotics. Higher temperatures can increase the uptake and toxicity of some xenobiotics (Solé et al. 2015). Light impacts photosynthesis and can interact with xenobiotics to produce reactive oxygen species (ROS), leading to oxidative stress in algae. Salinity affects the bioavailability and toxicity of xenobiotics, particularly metals and organic pollutants, in marine environments (Muñoz-García et al. 2019). Xenobiotics can disrupt nutrient cycling by affecting algal growth and metabolism, which is crucial for cycling nitrogen and phosphorus in aquatic ecosystems. Their impact on algal photosynthesis and primary production can influence carbon sequestration, affecting global carbon cycles (Petsas and Vagi 2017).

2.3.3 Algal Species Characteristics

Different algal species have varying tolerance levels to xenobiotics, depending on their evolutionary history, physiological adaptations, and ecological niche (Kurade et al. 2016). The duration of xenobiotic exposure affects algal communities: short-term, high-dose exposure causes acute effects, while long-term, low-dose exposure leads to chronic stress and adaptation. The timing of exposure relative to the algal life cycle (e.g., during growth or reproduction) significantly influences the outcome

(Ding et al. 2020). The composition of the algal community affects its response to xenobiotics, with diverse communities potentially being more resilient due to tolerant species. Algal cell wall structure influences xenobiotic uptake; species with thicker or more complex walls may resist certain chemicals better. Active metabolic processes can either detoxify or activate xenobiotics, with some algae metabolizing and detoxifying them while others may convert them into more toxic forms (Kurade et al. 2016). A recent study found that faster-growing algae may be more susceptible to xenobiotics because they have higher nutrient and energy demands, making them more vulnerable to disruptions (Gilbert et al. 2003). Some algal species can acclimate to xenobiotics by adjusting their physiological processes, such as upregulating detoxification enzymes. Over time, algal populations may develop genetic resistance to xenobiotics, leading to shifts in community composition and potentially reducing overall biodiversity (Sánchez-Fortún et al. 2009).

3 Cutting-Edge Insights

3.1 Recent Advances in Understanding Xenobiotic-Algal Interactions

The potential of algae for environmental bioremediation and detoxification has spurred interest in studying xenobiotic-algal interactions. Recent advances in omics, microscopy, bioimaging, and bioanalytical methods have expanded our understanding of these interactions, revealing new insights into algal responses and mechanisms for bioremediation.

3.1.1 Omics Approaches (Genomics, Transcriptomics, Proteomics)

Omics technologies, including genomics, transcriptomics, and proteomics, have advanced the study of xenobiotic-algal interactions by offering detailed molecular insights. Genomics has identified genes involved in algae xenobiotic uptake, resistance, and metabolism. For instance, *Chlamydomonas reinhardtii* expressing the 1-pyrroline-5-carboxylate synthetase (P5CS) gene shows enhanced growth and metal binding in toxic environments. Genetic modifications in *Chlorella* sp. DT with the mercuric reductase (merA) gene from *Bacillus megaterium* doubled Hg (II) bioremediation rates compared to unmodified strains (Saravanakumar et al. 2024). Pyro-tag sequencing is an excellent genomic technique to analyze bacterial-microalgal consortia in fixed-bed photobioreactors (Rathour et al. 2024). Exposure to spiramycin and amoxicillin enhances photosynthesis-related genes (psbA, psaB, mycB, and rbcL) in *Microcystis aeruginosa*. The biodegradation process occurs in two phases: Phase I converts pharmaceuticals to more hydrophilic compounds, while Phase II involves various intracellular enzymes like soluble inorganic pyrophosphatases, malate/pyruvate dehydrogenase, catalase, and laccases. Microalgae *Chlamydomonas reinhardtii* also removes PO_4^{3-} (68–94%) and NO_3^- (70–97%) through Na^+/Pi co-transporters, H^+/Pi symporters, nitrate reductase (NR), and nitrite reductase (NiR) (Goswami et al. 2022). *Chlorella sorokiniana* and *Chlorella fusca* have a higher affinity for the xenobiotic substrate metflurazon, which is biotransformed by N-demethylation by a cytochrome P450 monooxygenase (Thies et al. 1996). Transcriptomic analyses, including high-throughput RNA sequencing (RNA-Seq), are crucial for understanding algal responses to xenobiotics. RNA-Seq profiling of *Saccharina japonica* exposed to copper (Cu^{2+}) reveals Cu^{2+} toxicity impacts on biosynthesis, metabolism, and photosynthesis. Transcriptomic studies also identify key regulatory networks involved in oxidative stress response, energy metabolism, and cellular repair. For example, 70% of expressed genes in *Ectocarpus siliculosus* are regulated in response to oxidative stress, and in *Sargassum vulgare*, transcriptomic studies reveal changes in antioxidants, amino acids, fatty acids, and phenolic compounds due to natural acidification (Beaulieu 2019). In a consortia development, the diatom *Thalassiosira pseudonana* transports and catabolizes 2,3-dihydroxypropane-1-sulfonate (DHPS), which is utilized by *Ruegeria pomeroyi* as carbon source, expressing several metabolism genes according to transcriptomic studies (Rathour et al. 2024). Proteomics, the large-scale study of proteins, complements genomics and transcriptomics by detailing the functional output of genes. It helps identify and quantify proteins involved in xenobiotic stress, degradation, and resistance in algae. For example, in brown algae *Sargassum fusiforme*, cadmium stress induces antioxidative defense and metal detoxification proteins, while downregulating key metabolic enzymes, like Rubisco. In *Gracilaria lemaneiformis*, lead stress exposure results in the expression of fourteen Pb stress-regulated proteins related to antioxidation, signal transduction, and metabolism (Beaulieu 2019). Proteomic data analysis has led to a better understanding of the mechanisms adopted by algae to detoxify and sequester xenobiotics at the protein level, providing potential targets to enhance algal bioremediation capabilities.

3.1.2 Bioimaging and Microscopy Techniques

Bioimaging and microscopy offer valuable insights into xenobiotic-algal interactions by visualizing the localization and effects of xenobiotics within algal cells. Fluorescence microscopy, including confocal laser scanning microscopy (CLSM), is used to study the intracellular localization of xenobiotics, such as BaP. Light microscopy documents morphological changes, like cell shrinkage and chloroplast structure alterations, in response to heavy metal stress (Rathour et al. 2024). Electron microscopy, including scanning electron microscopy (SEM) and transmission electron microscopy (TEM), offers high-resolution images of ultrastructural modifications in algae exposed to xenobiotics. TEM is used to study the disruption of thylakoid membranes in chloroplasts due to heavy metals, while SEM reveals surface deformities and protective mechanisms against pollutants (Singh and Nagaraj 2006; Rathour et al. 2024).

3.1.3 Novel Analytical Methods for Xenobiotic Detection

The advancement of novel analytical methods for improved recognition, quantification, and characterization of xenobiotics in algal systems has enabled more precise monitoring of xenobiotic levels and their interactions in algal cells. LC-MS and GC-MS have contributed to the identification and characterization of the molecular structures of xenobiotics and their products in algae. These methods have provided detailed mechanisms of absorption, transformation, and degradation of xenobiotics by algae (Gohain et al. 2023; Saravanakumar et al. 2024). Mass spectrometric studies facilitate the study of the metabolic pathways involved in xenobiotic degradation and transformation. Surface-enhanced laser-desorption-ionization MS (SELDI-TOF-MS), an advanced method of MALDI-TOF-MS, can analyze differential expression of signature proteins in blue mussels, *Mytilus edulis*, exposed to heavy metals and polynuclear aromatic hydrocarbons (PAHs) (Singh and Nagaraj 2006).

3.2 Predicting Xenobiotic Effects on Algal Communities Using Machine Learning and Modeling

Deep learning techniques, including convolutional neural networks (CNNs) and recurrent neural networks (RNNs), improve the accuracy of forecasting algal blooms by analyzing complex correlations between phytoplankton density and environmental factors. These models outperform traditional approaches by effectively processing large datasets from marine monitoring networks (Park et al. 2024). Using patterns found in water quality data, support vector machines (SVMs) are used to categorize and forecast the presence of xenobiotic impacts on algal communities. They are beneficial when dealing with non-linear relationships in datasets (Cruz et al. 2021). The random forests learning technique analyzes different decision trees to predict the effects of numerous xenobiotic variables on algae, and majority voting is employed to increase prediction accuracy (Malwe and Sharma 2023). Dynamic simulation models incorporate physiological, chemical, and biological characteristics to mimic xenobiotics' effects on algal growth and interactions within aquatic ecosystems. They offer perceptions of the possible long-term effects of contaminants (Gianfreda and Rao 2008). Based on a wealth of environmental monitoring data, data-driven models forecast the dynamics of algal blooms and community changes due to exposure to xenobiotics. Data-driven approach models aid in environmental management and mitigation efforts by predicting future circumstances using historical data (Connon et al. 2019). Assessments of environmental threats can now be made with more excellent reliability because of machine learning models' higher predictive accuracy compared to older techniques. Environmental threat assessments can now be more reliable because of machine learning models' higher predictive accuracy than older techniques (Anifowose and Anifowose 2024). These models offer versatility in addressing diverse ecological problems since they may be tailored to varied aquatic environments and scales. By integrating with real-time monitoring systems, machine learning models offer the potential for proactive environmental management and rapid response to pollution events (Imam et al. 2024). Bridging the gap between predictive modeling and policy implementation is vital for effective environmental management. Machine learning and modeling tools are essential for predicting xenobiotic impacts on algal communities, aiding in sustainable management and environmental preservation. Ongoing development aims to enhance their predictive accuracy and integration with ecological management strategies.

4 Effects of Xenobiotics on Algal Physiology and Ecology

4.1 Impacts on Algal Growth, Photosynthesis, and Nutrient Cycling

Xenobiotics disrupt algal cellular processes, reducing chlorophyll production and photosynthesis, stunting growth, or killing algal populations. This exacerbates eutrophication and affects aquatic ecosystems by altering nutrient availability and shifting algal communities. These disruptions lead to imbalances, harmful algal blooms, and compromised ecosystem health, emphasizing the need for effective environmental management. The effects of xenobiotics on algal physiology and ecology are discussed below.

4.1.1 Impacts on Algal Growth

Xenobiotics can impact algal growth by inducing toxicity and disrupting nutrient availability. Heavy metals can interfere with essential enzymatic processes necessary for cellular growth and division, leading to reduced growth rates or even death of algae (You et al. 2021). The uptake and catabolism of acetyl CoA derivatives through beta-oxidation pathways stimulate algal blooms in *Chlamydomonas* sp., illustrating the complex interactions between xenobiotics and algal physiology in aquatic ecosystems (Ashwath et al. 2023). Certain xenobiotics can bind to essential nutrients like nitrogen and phosphorus, making them unavailable to algae and impeding their growth. For example, studies have shown that *Desmodesmus armatus* and *Microcystis aeruginosa* exposed to polyphenols like tannic acid and gallic acid displayed reduced efficacy at higher pH levels, likely due to polymerization that decreases nutrient bioavailability (Laue et al. 2014). Conversely, regardless of pH, the alkaloid gramine exhibited increasing toxicity over time. Pharmaceuticals and xenobiotics in wastewater, which often contain high nutrient loads, have been used to cultivate microalgae while simultaneously treating pollutants. Nie et al. (2009) demonstrated that the growth of *Scenedesmus obliquus* was inhibited by 60.6% at 16 mg/L of butylated hydroxyanisole and by 82.4% at 60 mg/L of norfloxacin. Similarly, *Chlorella vulgaris* effectively removed 94% of diazinon at 20 mg/L but exhibited over 30% growth inhibition at higher concentrations (Ianelly and Elsa 2024). The microalga *Chlamydomonas reinhardtii* was found to biodegrade phenol under stress, though this process was energy-intensive and affected growth (Nazos et al. 2017). Studies on compounds like atrazine, copper sulfate, and methyl parathion reveal significant growth inhibition and alterations in algal biochemical properties. *Scenedesmus intermedius* showed adaptation to chloramphenicol exposure, with resistant cells emerging due to spontaneous mutations (Sánchez-Fortún et al. 2009). *Cymbella* sp. and *Scenedesmus quadricauda* exhibited complete growth inhibition at high naproxen concentrations, further illustrating the diverse impacts of xenobiotics on algae (Xin et al. 2021). Research on xenobiotic-algal interactions has shown that using glucose as a carbon source enhances the growth of *Scenedesmus obliquus* and increases phenolic compound biodegradation rates compared to inorganic or limited carbon sources. A dynamic energy budget model for aerobic degradation revealed that while glucose boosts growth rates, it may also inhibit biodegradation due to increased competition for oxygen. This understanding can improve wastewater treatment and bioremediation strategies (Lika and Papadakis 2009). More recently, it was discovered that *Chlamydomonas mexicana* can degrade bisphenol-A, with elevated concentrations of this compound leading to higher carbohydrate levels in the cells as a stress response (Ji et al. 2014).

4.1.2 Impacts on Photosynthesis

Xenobiotics impact algal photosynthesis by damaging chlorophyll and altering light absorption. For instance, mercury disrupts chlorophyll synthesis, reducing pigment levels and impairing the algae's ability to capture and convert light energy, thereby decreasing photosynthetic efficiency (Singh et al. 2022). Herbicides like atrazine disrupt the structure of photosynthetic pigments, such as chlorophyll a and b, reducing light absorption and photosynthetic rates, which hampers algal growth and ecosystem health. Similarly, high phenol concentrations inhibit the photosynthetic rate in *Chlamydomonas reinhardtii* (Nazos et al. 2017). Effects of mesotrione on algae, focusing on photosynthetic responses and mesotrione accumulation in *Microcystis* sp. and *Scenedesmus quadricauda* (Ni Yan et al. 2014). Antibiotic residues have been shown to decrease the photosynthetic rate in the freshwater green algae *Raphidocelis subcapitata* (Sharma et al. 2021). Kumar et al. (2014) highlighted using chlorophyll fluorescence to assess photochemical responses in algae exposed to environmental xenobiotics. The green alga *Scenedesmus opoliensis* demonstrated significant changes in photosynthesis rates under long-term exposure to micromolar amounts of three common non-selective herbicides, providing molecular indicators of environmental stress due to water pollution with these herbicides (Fodorpataki et al. 2009). Gao and Tam (2011) compared the effects of nonylphenol (NP) on growth, photochemistry, and biochemistry in *Chlorella vulgaris* and *Selenastrum capricornutum*, assessing their NP degradation abilities. The ecotoxicological impacts of sulfonamides and fluoroquinolones on photosynthesis were investigated in *Chlorella vulgaris* and *Chrysosporum ovalisporum* (Chen et al. 2020b). Xenobiotics, such as herbicides like atrazine, disrupt photosynthetic pathways by inhibiting key enzymes in the electron transport chain, affecting processes like plastoquinone reduction and reducing oxygen production (Zhu et al. 2009). Inhibition of photosynthetic pathways by xenobiotics leads to decreased photosynthetic efficiency and energy production in algae, reducing oxygen output. This decrease can adversely affect aquatic ecosystems by lowering oxygen levels, disrupting ecological balance, and impacting aquatic organisms (Sachdev et al. 2021). Exposing *Chlamydomonas reinhardtii* (strain CC125), *Pseudokirchneriella subcapitata* (strain CPCC37), and two strains of *Microcystis aeruginosa* (CPCC299 and CPCC632) to 4-octylphenol, 4-nonylphenol, and β-estradiol at concentrations ranging from 0.1 to 5 μg/mL revealed significant alterations in PSII energy fluxes, demonstrating the impact of endocrine disruptors on photosynthetic activity in non-target organisms lacking endocrine systems (Perron and Juneau 2011). Another example is the impact of the heavy metal cadmium, which can interfere with the electron transport chain by binding to thiol groups in proteins, leading to reduced photosynthetic activity and lower oxygen production (Pietrini et al. 2006; Branca et al. 2020). *Chlamydomonas reinhardtii* can degrade phenol, with its efficiency affected by conditions influencing photosynthesis. This underscores the potential for using xenobiotic-modulated algae in bioremediation of phenol-contaminated environments (Nazos et al. 2017). A recent study reveals that sulfamethoxazole (SMX) significantly inhibits *Chlorella pyrenoidosa* growth and carbon fixation, inducing oxidative stress and altering enzyme activities. It disrupts carbonic anhydrase, affects Rubisco, and triggers CO_2 compensation mechanisms, highlighting the ecological risks of long-term, low-concentration SMX exposure on algal populations and carbon fixation (Zhou et al. 2024). These disruptions highlight xenobiotics' significant role in impairing photosynthetic pathways and influencing aquatic environments.

4.1.3 Impacts on Nutrient Cycling

Xenobiotics, like triphenyl phosphate, can significantly impair nutrient uptake and overall health in freshwater microalgae such as *Chlorella vulgaris* and *Scenedesmus obliquus*, affecting their growth and ecosystem health (Nikkilä 2001; Wang et al. 2019b). These toxic effects underscore the vulnerability of microalgae to pollutants, highlighting the need for careful management of xenobiotic contaminants. Pollutants like heavy metals and synthetic chemicals can alter algal membrane permeability, impairing the uptake of essential nutrients such as nitrogen and phosphorus (Rai et al. 2016). Heavy metals, like chromium, cadmium, and lead, can bind to membrane proteins and disrupt the function of nutrient transporters, thereby reducing the efficiency with which algae absorb these vital nutrients from their environment (Xiao et al. 2023). This impairment can lead to nutrient deficiencies and hinder algal growth. Xenobiotics that disrupt algal metabolism, such as herbicides like atrazine and glyphosate, inhibit key enzymes like nitrate reductase and ammonium assimilation, impeding the conversion of inorganic nutrients into organic forms and causing nutrient imbalances (Cobb 2022). Consequently, these disruptions can cause nutrient imbalances, leading to problems like algal blooms or eutrophication, and highlight the broader ecological impacts of xenobiotics by altering nutrient dynamics in aquatic environments.

4.2 Trophic Cascades and Ecosystem-Level Consequences

Trophic cascades occur when changes at one trophic level impact others, often with widespread ecological effects. For example, reducing apex predators can lead to increased consumption of primary producers like algae. Biomagnification exacerbates this, as toxins increase in concentration at higher trophic levels. This can affect human health, water quality, biodiversity, and ecosystem stability (Heath et al. 2014). Algae, as primary producers in aquatic food webs, support higher trophic levels. However, toxins accumulated by algae, such as pesticides and heavy metals, are passed up the food chain through biomagnification. These toxins concentrate in the tissues of herbivores and increase in concentration as they move up to apex predators (Fig. 3) (Saidon et al. 2024). Trophic cascades can arise when the removal of apex predators leads to increased herbivore populations, resulting in

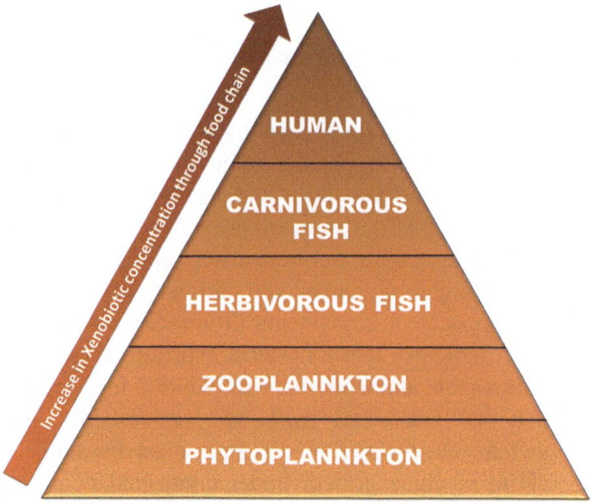

Fig. 3 Biomagnification of xenobiotics through trophic levels

excessive grazing on algae. This can cause algal blooms, further concentrating toxins and exacerbating biomagnification through the food chain (Best and Stachowicz 2012).

Algae are the primary producers at the base of aquatic food webs. Higher trophic levels, including fish, zooplankton, and, finally, apex predators, depend on them for survival. Certain algae can accumulate hazardous substances through biomagnification, including pesticides, heavy metals, and other contaminants. The food chain then passes these elements through. The quantity of toxins rises as a species moves up the food chain because they are concentrated in the tissues of herbivorous animals that consume these tainted algae (Ramanan et al. 2016). This can trigger a trophic cascade: reduced apex predators lead to increased herbivorous fish and overgrazing algae, causing blooms that further concentrate toxins and exacerbate biomagnification (Patiño et al. 2023). Algal blooms that concentrate toxins can severely impact ecosystems by causing biodiversity loss. Sensitive species may face extinction or population declines, leading to unstable ecosystems that are less resilient to environmental changes and new disturbances (Carpenter et al. 2010). Toxic algal accumulation can degrade water quality, creating "dead zones" with low oxygen levels that harm aquatic life. Nutrient contamination from agricultural runoff often exacerbates these conditions by fueling algal blooms and toxin buildup (Bashir et al. 2020). Algal blooms that biomagnify toxins pose severe health risks to humans, including fish and shellfish poisoning, cancer, and liver damage from toxins like microcystins. They also have significant economic impacts, leading to losses in the fishing, tourism, and water treatment industries due to damaged aquatic resources and declining water quality (Oh et al. 2023).

5 Future Horizons

5.1 Developing Novel Bioremediation Strategies Using Algae

Xenobiotics are toxic to algae, but these organisms can accumulate and bioremediate them. This dual capacity makes algae a compelling subject for study, offering insights into environmental detoxification and aquatic ecosystem resilience (Lu et al. 2021; Abdelfattah et al. 2023). Bioremediation of xenobiotics by algae, known as phycoremediation, leverages their ability to absorb and detoxify pollutants. Their widespread presence and high pollutant accumulation capacity make them ideal for mitigating xenobiotic impacts in ecosystems (Touliabah et al. 2022; Ankit et al. 2022). Recent research indicates that long-term exposure to pharmaceutical compounds like sulfamethoxazole and trimethoprim (7.5–8.5 ng/L) reduces the growth of microalgae such as *Isochrysis galbana*, *Chaetoceros neogracile*, and *Nannochloropsis oculata* (Goswami et al. 2022). In closed systems, such as photobioreactors, they optimize algae growth under controlled conditions, enhancing their ability to remove xenobiotics from wastewater. These systems can be tailored to target specific pollutants (Chojnacka et al. 2022). Some xenobiotics can be degraded through co-metabolic processes, transforming them into non-toxic by-products while the microorganisms metabolize other primary carbon sources. Some microalgae produce reactive oxygen species (ROS) under light conditions, which can help break down xenobiotics through photodegradation (Solovchenko et al. 2022). Light-dependent processes in photobioreactors, coupled with photosynthetic activity, can enhance xenobiotic degradation. Bacterial and algal-bacterial airlift reactors outperformed traditional biofilters in removing toluene at an empty bed residence time of 1.1 min (Lebrero et al. 2016). The photobioreactor environment, particularly for wastewater treatment, can remove nutrients like nitrogen and phosphorus, which often accompany xenobiotics. Removing these nutrients can create an environment that is less conducive to the persistence of certain xenobiotics (Solovchenko et al. 2022). Optimizing light, temperature, pH, and nutrients in photobioreactors can enhance the growth of microorganisms with high xenobiotic degradation potential. However, scaling up faces challenges like uniform light distribution and contamination control. Recent strategies for immobilizing microalgal biomass have improved the efficiency of wastewater treatment processes in removing nutrients (Goswami et al. 2022). Genetic modification of algae can enhance their ability to degrade specific xenobiotics (Joutey et al. 2013). Genetic modification can enhance algae's ability to target xenobiotics by introducing enzymes like cytochrome P450s or laccases, which break down complex pollutants. CRISPR technology has been used to knock out genes that limit algae's tolerance or degradation capabilities, improving their bioremediation potential (Stein et al. 2018). Genetic engineering can also make algae more resistant to the toxic effects of xenobiotics, allowing them to survive and remain effective in highly contaminated environments (Kumar et al. 2013). Another novel approach is co-culturing algae with bacteria, which can create synergistic systems where bacteria degrade complex organic pollutants into simpler compounds that algae can then uptake and further detoxify (Khanpour-Alikelayeh and Partovinia 2021). The interaction between the two organisms enhances the degradation of xenobiotics, especially in wastewater treatment applications. Chlorella vulgaris consortia with native wastewater-grown bacteria effectively treated municipal wastewater, removing over 95% of nitrogen and phosphorus. Additionally, consortia of *Scenedesmus* sp. and *Paenibacillus* sp. ISTP10 demonstrated complete (100%) lindane degradation (Ryu et al. 2014). Thus, synergistic algal-bacterial systems are innovative in removing xenobiotics from the environment. Incorporating nanoparticles into algal systems can improve xenobiotic degradation capacity (Priya et al. 2024). For instance, magnetic nanoparticles can be attached to algae, allowing for easy separation and recovery of the algae after treatment. Similarly, photocatalytic nanoparticles (like titanium dioxide) can be combined with algae to enhance the photodegradation of pollutants (Chaudhary et al. 2023). Algae can be incorporated into biomimetic membranes with nano-scale pores that selectively filter and degrade xenobiotics as they pass through (Priya et al. 2024). Microalgae have been investigated for their potential in bioremediating plastic pollution. While no single microorganism is universally effective, co-cultures of different microorganisms have shown improved plastic biodegradation (Katare et al. 2021; Scopetani et al. 2022). Studies have shown that microalgae can interact with plastic particles, aiding in adsorption and degradation. For example, the brown algae *Fucus vesiculosus* effectively adsorbs microplastics, largely due to its alginic acid content, which has a strong affinity for plastic particles (Dey et al. 2021; Cui and Liu 2021). Microalgae like *Chlorella* sp., *Scenedesmus* sp., and *Raphidocelis subcapitata* have been tested for plastic bioremediation, showing varying levels of success. Their ability to interact with and remove microplastics highlights their potential role in mitigating plastic pollution in aquatic ecosystems (Nolte et al. 2017; Veerasingam et al. 2020). Table 1 summarizes the efficiency of different algae in removing various xenobiotics. These strategies showcase algae's versatility in tackling xenobiotic pollutants and emphasize their potential for innovative environmental remediation applications.

Table 1 Different xenobiotics and their removal efficiency by different algae

Sl. no.	Xenobiotic	Microalgae	Removal efficiency (%)	Reference
1	Methylene blue	*Chlorella vulgaris*	83.04	Chin et al. (2020)
2	Bisphenol-A	*Desmodesmus* sp.	78	Wang et al. (2023)
3	Atrazine	*Anabaena variabilis*	85	Liu et al. (2021)
4	Synazol	*Spirogyra* sp.	85	Khalaf (2008)
5	Lindane	*Lyngbya majuscula*	88	Ughy et al. (2015)
6	Ibuprofen	*Nannochloris* sp.	40	Bai and Acharya (2017)
7	Carbamazepine	*Chlorella sorokiniana*	30	Bai and Acharya (2017)
8	Triclosan	*Nannochloris* sp.	100	Bai and Acharya (2017)
9	Norgestrel	*Chlorella pyrenoidosa*	68	Maes et al. (2014)
10	17a-ethinylestradiol	*Desmodesmus subspicatus*	65	Peng et al. (2014)
11	Enrofloxacin	*Chlamydomonas mexicana*	25	Xiong et al. (2017)
13	Lead	*Pseudochlorococcum typicum*	75	Shanab et al. (2012)
14	Cadmium	*Spirulina platensis*	67	Gunasundari (2017)
15	Toluene	*Oscillatoria* sp.	75	Ei-Sheekh et al. (2012)
16	Endosulfan	*Microcystis aeruginosa*	80	Mudhoo et al. (2019)
17	Naphthalene	*Spirulina platensis*	67	Touliabah et al. (2022)
18	Arsenic	*Maugeotia genuflexa*	50	Sarı et al. (2011)

5.2 *Integrating Xenobiotic-Algal Interaction Research into Environmental Policy and Management*

Integrating xenobiotic-algal interaction research into environmental risk management is crucial for mitigating pollutant impacts on aquatic ecosystems and human health. Risk assessment, including hazard identification, dose-response evaluation, exposure assessment, and risk characterization, is vital. Heavy metals like cadmium and mercury disrupt algal photosynthesis and nutrient uptake, underscoring the need for strict industrial discharge limits and effective waste management practices (Vardhan et al. 2019). The impact of herbicides like atrazine on algal growth highlights the need for stricter pesticide use guidelines to prevent water runoff. Research on pharmaceutical bioaccumulation and trophic transfer underscores the importance of enhancing waste treatment processes to reduce pharmaceutical residues in aquatic environments (Puckowski et al. 2016). The role of xenobiotics in harmful algal blooms and eutrophication highlights the need to manage nutrient pollution from agricultural runoff and sewage. Improved nutrient regulation and monitoring can help prevent ecological degradation. Despite bans, persistent organic pollutants (POPs) like PCBs and DDT continue to cause reproductive dysfunction and long-term population declines in aquatic organisms (Johnson et al. 2013). The persistence of pollutants like POPs, PCBs, and DDT necessitates continuous monitoring, strict regulation, and the adoption of cleaner technologies. Developed nations have improved control measures, but developing countries face significant challenges due to less advanced technologies. Research on xenobiotic interactions in India is crucial for shaping environmental policies, guided by regulations such as the Environmental Protection Act of 1986, the Water and Air Pollution Acts, and the Hazardous Waste Management Rules of 2016 (Sahu 2021). Indian research institutions, such as NEERI and IITR, are at the forefront of studying the toxicological impacts of xenobiotics on various ecosystems (Kumar et al. 2022). Research on bioremediation using microorganisms like algae and cyanobacteria is crucial for developing sustainable methods to remove xenobiotics. The Central Pollution Control Board (CPCB) monitors pollutant levels and impacts, integrating findings into national action plans and Sustainable Development Goals. Collaboration between the CPCB and the National River Conservatory Directorate (NRCD) enhances remediation strategies, addressing the environmental challenges posed by xenobiotics (Patel et al. 2020). The Indian Institute of Toxicological Research highlights the importance of assessing the toxicological consequences of drug and chemical exposure, with high-throughput genomic screenings expanding our capacity to evaluate xenobiotic effects (Flora 2008). Risk assessment (RA), a science-based process integral to food safety, includes hazard identification, characterization, exposure assessment, and risk characterization (Durham et al. 2023). However, challenges arise in extrapolating data across species and populations due to genetic variability and the complexities of microbe-drug-chemical interactions. The National Institutes of Health (NIH) in the USA has also highlighted xenobiotics' risks to human health and other organisms (Chiu et al. 2013). Limited toxicological data for many compounds can skew risk assessments, often favoring well-studied chemicals like bisphenol-A and triclosan. Food additives, including flavoring agents, preservatives, and sweeteners, are only considered safe if certified by the Joint Expert Committee on Food Additives (JECFA). Effective management of xenobiotics depends on various organizations' collaborative efforts in their regulation and oversight (Boobis et al. 2017). Translational research is crucial for translating laboratory findings into practical xenobiotic risk management strategies, bridging the gap between science and clinical application. It is essential for protecting human health and the environment. However, challenges remain in enforcing regulations and addressing research gaps. Future efforts should focus on innovative, cost-effective solutions adapted to India's environmental conditions to ensure that research informs policy effectively and safeguards ecosystems and public health.

6 Conclusion

The persistent nature of xenobiotics in aquatic ecosystems necessitates careful monitoring and management to protect ecosystem health. Future research should focus on understanding the molecular mechanisms of xenobiotic uptake and toxicity in algae, mainly using advanced omics technologies. It is also crucial to examine the effects of pollutant mixtures on the influence of algal community structure and to develop innovative bioremediation approaches, including novel algal strains. Long-term monitoring is essential to assess the lasting impacts of xenobiotics. Additionally, researchers should inform policy and regulatory frameworks for effective management and restoration of aquatic environments. Protecting algae from xenobiotic impacts is critical for maintaining overall ecosystem health.

References

Abdelfattah A, Ali SS, Ramadan H, El-Aswar EI, Eltawab R, Ho S-H, Elsamahy T, Li S, El-Sheekh MM, Schagerl M, Kornaros M, Sun J (2023) Microalgae-based wastewater treatment: Mechanisms, challenges, recent advances, and future prospects. Environmental Science and Ecotechnology 13:100205. https://doi.org/10.1016/j.ese.2022.100205

Abootalebi Jahromi F, Moore F, Keshavarzi B, Mohebbi-Nozar SL, Mohammadi Z, Sorooshian A, Abbasi S (2020) Bisphenol A (BPA) and polycyclic aromatic hydrocarbons (PAHs) in the surface sediment and bivalves from Hormozgan Province coastline in the Northern Persian Gulf: A focus on source apportionment. Mar Pollut Bull 152:110941. https://doi.org/10.1016/j.marpolbul.2020.110941

Acir I-H, Guenther K (2018) Endocrine-disrupting metabolites of alkylphenol ethoxylates – A critical review of analytical methods, environmental occurrences, toxicity, and regulation. Science of The Total Environment 635:1530–1546. https://doi.org/10.1016/j.scitotenv.2018.04.079

Akhter S, Bhat MA, Ahmed S, Siddiqi WA, Ahmad S, Shrimal H (2023) Profiling of Antibiotic Residues in Surface Water of River Yamuna Stretch Passing through Delhi, India. Water 15(3):527. https://doi.org/10.3390/w15030527

Anifowose B, Anifowose F (2024) Artificial intelligence and machine learning in environmental impact prediction for soil pollution management – case for EIA process. Environmental Advances 17:100554. https://doi.org/10.1016/j.envadv.2024.100554

Ankit, Bauddh K, Korstad J (2022) Phycoremediation: Use of Algae to Sequester Heavy Metals. Hydrobiology 1(3):288–303. https://doi.org/10.3390/hydrobiology1030021

Aroh K (2018) Globalization, Industralization and Population Growth the Fundamental Drivers of Pollution in a Given Society (India). SSRN Electronic Journal. https://doi.org/10.2139/ssrn.3453740

Arsad S, Kholifah S, Prabawati E, Kadim MK, Sari L, Kilawati Y (2020) Effect of Mercury on Growth of Several Microalgae. Asian Journal of Water 17:13–17. https://doi.org/10.3233/AJW200029

Baghour M (2019) Algal Degradation of Organic Pollutants. In: Martínez LMT, Kharissova OV, Kharisov BI (eds) Handbook of Ecomaterials. Springer International Publishing, Cham, pp 565–586

Bai X, Acharya K (2017) Algae-mediated removal of selected pharmaceutical and personal care products (PPCPs) from Lake Mead water. Science of the Total Environment 581:734–740

Bajguz A (2011) Suppression of Chlorella vulgaris Growth by Cadmium, Lead, and Copper Stress and Its Restoration by Endogenous Brassinolide. Arch Environ Contam Toxicol 60(3):406–416. https://doi.org/10.1007/s00244-010-9551-0

Baldwin WS (2019) Phase 0 of the Xenobiotic Response: Nuclear Receptors and Other Transcription Factors as a First Step in Protection from Xenobiotics. Nucl Receptor Res 6:101447. https://doi.org/10.32527/2019/101447

Barceló D, Petrovic M (2007) Pharmaceuticals and personal care products (PPCPs) in the environment. Anal Bioanal Chem 387(4):1141–1142. https://doi.org/10.1007/s00216-006-1012-2

Bashir I, Lone FA, Bhat RA, Mir SA, Dar ZA, Dar SA (2020) Concerns and threats of contamination on aquatic ecosystems. Bioremediation and biotechnology: sustainable approaches to pollution degradation :1–26

Bastolla CLV, Lima D, Mattos JJ, Dias VHV, Righetti BPH, Gomes CHAM, Cella H, Reis IMM, Saldaña-Serrano M, Ferreira CP, Bícego MC, Taniguchi S, Zacchi FL, Bainy ACD (2023) Comparative biochemical and molecular responses of biotransformation and antioxidant systems in three species of *Crassostrea* (Sacco, 1897) oysters exposed to chrysene. Comparative Biochemistry and Physiology Part C: Toxicology & Pharmacology 270:109642. https://doi.org/10.1016/j.cbpc.2023.109642

Beaulieu L (2019) Insights into the regulation of algal proteins and bioactive peptides using proteomic and transcriptomic approaches. Molecules 24(9):1708

Best RJ, Stachowicz JJ (2012) Trophic cascades in seagrass meadows depend on mesograzer variation in feeding rates, predation susceptibility, and abundance. Marine Ecology Progress Series 456:29–42

Bilal M, Mehmood S, Iqbal HMN (2020) The Beast of Beauty: Environmental and Health Concerns of Toxic Components in Cosmetics. Cosmetics 7(1):13. https://doi.org/10.3390/cosmetics7010013

Blaby-Haas CE, Merchant SS (2012) The ins and outs of algal metal transport. Biochimica et Biophysica Acta (BBA) - Molecular Cell Research 1823(9):1531–1552. https://doi.org/10.1016/j.bbamcr.2012.04.010

Boobis A, Cerniglia C, Chicoine A, Fattori V, Lipp M, Reuss R, Verger P, Tritscher A (2017) Characterizing chronic and acute health risks of residues of veterinary drugs in food: latest methodological developments by the joint FAO/WHO expert committee on food additives. Critical Reviews in Toxicology 47(10):889–903. https://doi.org/10.1080/10408444.2017.1340259

Borgå K, McKinney MA, Routti H, Fernie KJ, Giebichenstein J, Hallanger I, Muir DC (2022) The influence of global climate change on accumulation and toxicity of persistent organic pollutants and chemicals of emerging concern in Arctic food webs. Environmental Science: Processes & Impacts 24(10):1544–1576

Branca JJV, Pacini A, Gulisano M, Taddei N, Fiorillo C, Becatti M (2020) Cadmium-induced cytotoxicity: effects on mitochondrial electron transport chain. Frontiers in cell and developmental biology 8:604377

Carfagna S, Lanza N, Salbitani G, Basile A, Sorbo S, Vona V (2013) Physiological and morphological responses of Lead or Cadmium exposed Chlorella sorokiniana 211-8K (Chlorophyceae). SpringerPlus 2(1):147. https://doi.org/10.1186/2193-1801-2-147

Carpenter SR, Cole JJ, Kitchell JF, Pace ML (2010) Trophic cascades in lakes: lessons and prospects. Trophic cascades: predators, prey and the changing dynamics of nature :55–69

Castro M, Silva J, Machado B, Guimarães P, Lombardi A, Martins C, Zanette J (2022) Differential Effects of Atrazine to Chlorophyceae Species and Association to Morphology, Photosynthesis, Chlorophyll Content and Glutathione S -Transferase Activity. Environmental Toxicology and Chemistry 41. https://doi.org/10.1002/etc.5339

Chaudhary K, Agarwal S, Khan S (2018) Role of Phytochelatins (PCs), Metallothioneins (MTs), and Heavy Metal ATPase (HMA) Genes in Heavy Metal Tolerance. pp 39–60

Chaudhary P, Ahamad L, Chaudhary A, Kumar G, Chen W-J, Chen S (2023) Nanoparticle-mediated bioremediation as a powerful weapon in the removal of environmental pollutants. Journal of Environmental Chemical Engineering 11(2):109591

Chen C, Zou W, Chen S, Zhang K, Ma L (2020a) Ecological and health risk assessment of organochlorine pesticides in an urbanized river network of Shanghai, China. Environmental Sciences Europe 32(1):42. https://doi.org/10.1186/s12302-020-00322-9

Chen S, Zhang W, Li J, Yuan M, Zhang J, Xu F, Xu H, Zheng X, Wang L (2020b) Ecotoxicological effects of sulfonamides and fluoroquinolones and their removal by a green alga (Chlorella vulgaris) and a cyanobacterium (Chrysosporum ovalisporum). Environmental pollution 263:114554

Chin JY, Chng LM, Leong SS, Yeap SP, Yasin NHM, Toh PY (2020) Removal of synthetic Dye by Chlorella vulgaris microalgae as natural adsorbent. Arabian Journal for Science and Engineering 45:7385–7395

Chiu WA, Euling SY, Scott CS, Subramaniam RP (2013) Approaches to advancing quantitative human health risk assessment of environmental chemicals in the post-genomic era. Toxicology and applied pharmacology 271(3):309–323

Chojnacka K, Skrzypczak D, Izydorczyk G, Mikula K, Szopa D, Moustakas K, Witek-Krowiak A (2022) Biodegradation of pharmaceuticals in photobioreactors–a systematic literature review. Bioengineered 13(2):4537–4556

Cobb AH (2022) Herbicides and plant physiology. John Wiley & Sons

Connon RE, Hasenbein S, Brander SM, Poynton HC, Holland EB, Schlenk D, Orlando JL, Hladik ML, Collier TK, Scholz NL, others (2019) Review of and recommendations for monitoring contaminants and their effects in the San Francisco Bay- Delta. San Francisco Estuary and Watershed Science 17(4)

Cruz RC, Reis Costa P, Vinga S, Krippahl L, Lopes MB (2021) A review of recent machine learning advances for forecasting harmful algal blooms and shellfish contamination. Journal of Marine Science and Engineering 9(3):283

Cui C, Liu W (2021) Recent advances in wet adhesives: Adhesion mechanism, design principle and applications. Progress in Polymer Science 116:101388. https://doi.org/10.1016/j.progpolymsci.2021.101388

Danouche M, Ghachtouli NE, Aasfar A, Bennis I, Arroussi HE (2022) Pb(II)-phycoremediation mechanism using Scenedesmus obliquus: cells physicochemical properties and metabolomic profiling. Heliyon 8(2). https://doi.org/10.1016/j.heliyon.2022.e08967

Desai M, Njoku A, Nimo-Sefah L (2022) Comparing Environmental Policies to Reduce Pharmaceutical Pollution and Address Disparities. Int J Environ Res Public Health 19(14):8292. https://doi.org/10.3390/ijerph19148292

Dey S, Samanta P, Ghosh AR, Banerjee S, Sen K (2024) State-of-the-art microalgae-based bioreactor wastewater treatment for the elimination of emerging contaminants: A mechanistic review. Cleaner Water 2:100027. https://doi.org/10.1016/j.clwat.2024.100027

Dey TK, Uddin MdE, Jamal M (2021) Detection and removal of microplastics in wastewater: evolution and impact. Environ Sci Pollut Res 28(14):16925–16947. https://doi.org/10.1007/s11356-021-12943-5

Ding T, Li W, Cai M, Jia X, Yang M, Yang B, Li J (2020) Algal toxicity, accumulation and metabolic pathways of galaxolide. Journal of hazardous materials 384:121360

Dordio AV, Carvalho AJP (2013) Organic xenobiotics removal in constructed wetlands, with emphasis on the importance of the support matrix. Journal of Hazardous Materials 252–253:272–292. https://doi.org/10.1016/j.jhazmat.2013.03.008

Durham SK, Rudmann DG, Rudmann KC, Swenberg JA (2023) Chapter 16 - Risk Assessment. In: Haschek WM, Rousseaux CG, Wallig MA, Bolon B, Bolon B, Heinz-Taheny KM, Rudmann DG, Mahler BW (eds) Haschek and Rousseaux' s Handbook of Toxicologic Pathology (Fourth Edition). Academic Press, pp 617–628

Ebele AJ, Abou-Elwafa Abdallah M, Harrad S (2017) Pharmaceuticals and personal care products (PPCPs) in the freshwater aquatic environment. Emerging Contaminants 3(1):1–16. https://doi.org/10.1016/j.emcon.2016.12.004

Ei-Sheekh MM, Ghareib MM, El-Souod GW (2012) A (2012) Biodegradation of Phenolic and Polycyclic Aromatic Compounds by Some Algae and Cyanobacteria. J Bioremed Biodegrad 3(133):2

Ferrari SG, Silva PG, González DM, Navoni JA, Silva HJ (2013) Arsenic tolerance of cyanobacterial strains with potential use in biotechnology. Revista Argentina de Microbiología 45(3):174–179. https://doi.org/10.1016/S0325-7541(13)70021-X

Fitzgerald SA, Steuer JJ (2006) Association of polychlorinated biphenyls (PCBs) with live algae and total lipids in rivers—a field-based approach. Science of The Total Environment 354(1):60–74. https://doi.org/10.1016/j.scitotenv.2004.11.025

Fleeger JW, Carman KR, Nisbet RM (2003) Indirect effects of contaminants in aquatic ecosystems. Sci Total Environ 317(1–3):207–233. https://doi.org/10.1016/S0048-9697(03)00141-4

Flora SJS (2008) Status of Toxicological Research in India. Chem Res Toxicol 21(7):1317–1319. https://doi.org/10.1021/tx8001803

Florianczyk B (2007) Metallothioneins and its role in metal regulation, binding of reactive oxygen species, apoptosis and cell differentiation. Journal of Pre-clinical and clinical research 1(1)

Fodorpataki L, Bartha C, Keresztes ZG (2009) Stress-physiological reactions of the green alga Scenedesmus opoliensis to water pollution with herbicides. aquatic 17:24

Gao QT, Tam NFY (2011) Growth, photosynthesis and antioxidant responses of two microalgal species, Chlorella vulgaris and Selenastrum capricornutum, to nonylphenol stress. Chemosphere 82(3):346–354

Gao QT, Wong YS, Tam NFY (2017) Antioxidant responses of different microalgal species to nonylphenol-induced oxidative stress. J Appl Phycol 29(3):1317–1329. https://doi.org/10.1007/s10811-017-1065-y

Ghorbani E, Nowruzi B, Nezhadali M, Hekmat A (2022) Metal removal capability of two cyanobacterial species in autotrophic and mixotrophic mode of nutrition. BMC Microbiol 22:58. https://doi.org/10.1186/s12866-022-02471-8

Gianfreda L, Rao MA (2008) Interactions between xenobiotics and microbial and enzymatic soil activity. Critical Reviews in Environmental Science and Technology 38(4):269–310

Gilbert P, McBain AJ, Rickard AH (2003) Formation of microbial biofilm in hygienic situations: a problem of control. International biodeterioration & biodegradation 51(4):245–248

Gohain A, Sarma R, Pardeep S (2023) Recent Advancements in Bioremediation of Xenobiotics Using Microbes. In: Singh R, Singh P, Tripathi S, Chandra KK, Bhadouria R (eds) Xenobiotics in Urban Ecosystems. Springer International Publishing, Cham, pp 355–374

Goswami RK, Agrawal K, Verma P (2022) An exploration of natural synergy using microalgae for the remediation of pharmaceuticals and xenobiotics in wastewater. Algal Research 64:102703

Gunasundari E (2017) Adsorption isotherm, kinetics and thermodynamic analysis of Cu (II) ions onto the dried algal biomass (Spirulina platensis). Journal of Industrial and Engineering Chemistry 56:129–144

Haynes D, Ralph P, Prange J, Dennison B (2000) The Impact of the Herbicide Diuron on Photosynthesis in Three Species of Tropical Seagrass. Marine Pollution Bulletin - MAR POLLUT BULL 41:288–293. https://doi.org/10.1016/S0025-326X(00)00127-2

He F, Ru X, Wen T (2020) NRF2, a Transcription Factor for Stress Response and Beyond. Int J Mol Sci 21(13):4777. https://doi.org/10.3390/ijms21134777

Heath MR, Speirs DC, Steele JH (2014) Understanding patterns and processes in models of trophic cascades. Ecology letters 17(1):101–114

Hejna M, Kapuścińska D, Aksmann A (2022) Pharmaceuticals in the Aquatic Environment: A Review on Eco-Toxicology and the Remediation Potential of Algae. International Journal of Environmental Research and Public Health 19(13):7717. https://doi.org/10.3390/ijerph19137717

Heo S, Kim S, Kang D (2020) The Role of Hydrogen Peroxide and Peroxiredoxins throughout the Cell Cycle. Antioxidants (Basel) 9(4):280. https://doi.org/10.3390/antiox9040280

Ianelly T-C, Elsa C-G (2024) Potential degradation efficiency of Chlorella vulgaris towards methamidophos and diazinon. Algal Research :103566

Imam M, Adam S, Dev S, Nesa N (2024) Air quality monitoring using statistical learning models for sustainable environment. Intelligent Systems with Applications 22:200333

Iummato M, Fassiano A, Graziano M, Dos Santos AFONSO M, Rios M, Juarez A (2019) Effect of glyphosate on the growth, morphology, ultrastructure and metabolism of Scenedesmus vacuolatus. Ecotoxicology and Environmental Safety 172:471–479. https://doi.org/10.1016/j.ecoenv.2019.01.083

Jabusch TW, Swackhamer DL (2004) Subcellular accumulation of polychlorinated biphenyls in the green alga Chlamydomonas reinhardtii. Environmental Toxicology and Chemistry 23(12):2823–2830. https://doi.org/10.1897/03-431.1

Jayaraj R, Megha P, Sreedev P (2016) Organochlorine pesticides, their toxic effects on living organisms and their fate in the environment. Interdiscip Toxicol 9(3–4):90–100. https://doi.org/10.1515/intox-2016-0012

Ji M-K, Kabra AN, Choi J, Hwang J-H, Kim JR, Abou-Shanab RA, Oh Y-K, Jeon B-H (2014) Biodegradation of bisphenol A by the freshwater microalgae Chlamydomonas mexicana and Chlorella vulgaris. Ecological engineering 73:260–269

Johnson LL, Anulacion BF, Arkoosh MR, Burrows DG, da Silva DA, Dietrich JP, Myers MS, Spromberg J, Ylitalo GM (2013) Effects of legacy persistent organic pollutants (POPs) in fish—current and future challenges. Fish physiology 33:53–140

Joutey NT, Bahafid W, Sayel H, El Ghachtouli N (2013) Biodegradation: involved microorganisms and genetically engineered microorganisms. Biodegradation-life of science 1:289–320

Kalita N, Baruah PP (2023) Cyanobacteria as a potent platform for heavy metals biosorption: Uptake, responses and removal mechanisms. Journal of Hazardous Materials Advances 11:100349. https://doi.org/10.1016/j.hazadv.2023.100349

Kanwugu ON, Glukhareva TV (2023) Activation of Nrf2 pathway as a protective mechanism against oxidative stress-induced diseases: Potential of astaxanthin. Archives of Biochemistry and Biophysics 741:109601. https://doi.org/10.1016/j.abb.2023.109601

Katare Y, Singh P, Sankhla MS, Singhal M, Jadhav EB, Parihar K, Nikalje BT, Trpathi A, Bhardwaj L (2021) Microplastics in aquatic environments: sources, ecotoxicity, detection & remediation. Biointerface Res Appl Chem 12:3407–3428

Khalaf MA (2008) Biosorption of reactive dye from textile wastewater by non-viable biomass of Aspergillus niger and Spirogyra sp. Bioresource Technology 99(14):6631–6634

Khanpour-Alikelayeh E, Partovinia A (2021) Synergistic and antagonistic effects of microbial co-culture on bioremediation of polluted environments. Microbial Rejuvenation of Polluted Environment: Volume 2 :229–265

Kiran B, Kaushik A (2008) Chromium binding capacity of *Lyngbya putealis* exopolysaccharides. Biochemical Engineering Journal 38(1):47–54. https://doi.org/10.1016/j.bej.2007.06.007

Klátyik S, Simon G, Oláh M, Takács E, Mesnage R, Antoniou MN, Zaller JG, Székács A (2024) Aquatic ecotoxicity of glyphosate, its formulations, and co-formulants: evidence from 2010 to 2023. Environmental Sciences Europe 36(1):22. https://doi.org/10.1186/s12302-024-00849-1

Koukoulakis KG, Kanellopoulos PG, Chrysochou E, Costopoulou D, Vassiliadou I, Leondiadis L, Bakeas E (2020) Atmospheric Concentrations and Health Implications of PAHs, PCBs and PCDD/Fs in the Vicinity of a Heavily Industrialized Site in Greece. Applied Sciences 10(24):9023. https://doi.org/10.3390/app10249023

Kumar A, Nighojkar A, Varma P, Prakash NJ, Kandasubramanian B, Zimmermann K, Dixit F (2023) Algal mediated intervention for the retrieval of emerging pollutants from aqueous media. Journal of Hazardous Materials 455:131568

Kumar KS, Dahms H-U, Lee J-S, Kim HC, Lee WC, Shin K-H (2014) Algal photosynthetic responses to toxic metals and herbicides assessed by chlorophyll a fluorescence. Ecotoxicology and environmental safety 104:51–71

Kumar S, Dagar VK, Khasa YP, Kuhad RC (2013) Genetically modified microorganisms (GMOs) for bioremediation. Biotechnology for environmental management and resource recovery :191–218

Kumar V, Garg VK, Kumar S, Biswas JK (2022) Omics for environmental engineering and microbiology systems. CRC Press

Kurade MB, Kim JR, Govindwar SP, Jeon B-H (2016) Insights into microalgae mediated biodegradation of diazinon by Chlorella vulgaris: microalgal tolerance to xenobiotic pollutants and metabolism. Algal research 20:126–134

Laue P, Bährs H, Chakrabarti S, Steinberg CE (2014) Natural xenobiotics to prevent cyanobacterial and algal growth in freshwater: contrasting efficacy of tannic acid, gallic acid, and gramine. Chemosphere 104:212–220

Lebrero R, Ángeles R, Pérez R, Muñoz R (2016) Toluene biodegradation in an algal-bacterial airlift photobioreactor: Influence of the biomass concentration and of the presence of an organic phase. Journal of environmental management 183:585–593

Legrand H, Herlory O, Guarini J-M, Blanchard G, Richard P (2006) Inhibition of microphytobenthic photosynthesis by the herbicides atrazine and diuron. Cahiers de biologie marine 47(1):39–45

León-Vaz A, Romero LC, Gotor C, León R, Vigara J (2021) Effect of cadmium in the microalga *Chlorella sorokiniana*: A proteomic study. Ecotoxicology and Environmental Safety 207:111301. https://doi.org/10.1016/j.ecoenv.2020.111301

Li J, Wang Y, Li N, He Y, Xiao H, Fang D, Chen C (2022) Toxic Effects of Bisphenol A and Bisphenol S on Chlorella Pyrenoidosa under Single and Combined Action. Int J Environ Res Public Health 19(7):4245. https://doi.org/10.3390/ijerph19074245

Li S, Mcintyre H (2024) Toxicity Assessment of 36 Herbicides to Green Algae: Effects of Mode of Action and Chemical Family. Agrochemicals 3(2):164–180. https://doi.org/10.3390/agrochemicals3020012

Lika K, Papadakis IA (2009) Modeling the biodegradation of phenolic compounds by microalgae. Journal of Sea Research 62(2–3):135–146

Liu R, Li S, Tu Y, Hao X (2021) Capabilities and mechanisms of microalgae on removing micropollutants from wastewater: A review. Journal of Environmental Management 285:112149

Lu T, Zhang Q, Lavoie M, Zhu Y, Ye Y, Yang J, Paerl HW, Qian H, Zhu Y-G (2019) The fungicide azoxystrobin promotes freshwater cyanobacterial dominance through altering competition. Microbiome 7:128. https://doi.org/10.1186/s40168-019-0744-0

Lu T, Zhang Q, Zhang Z, Hu B, Chen J, Chen J, Qian H (2021) Pollutant toxicology with respect to microalgae and cyanobacteria. Journal of Environmental Sciences 99:175–186. https://doi.org/10.1016/j.jes.2020.06.033

Luthe G, Garcia Boy R, Jacobus J, Smith BJ, Rahaman A, Robertson LW, Ludewig G (2008) Xenobiotic Geometry and Media pH Determine Cytotoxicity through Solubility. Chem Res Toxicol 21(5):1017–1027. https://doi.org/10.1021/tx700214p

Maes HM, Maletz SX, Ratte HT, Hollender J, Schaeffer A (2014) Uptake, Elimination, and Biotransformation of 17α-Ethinylestradiol by the Freshwater Alga *Desmodesmus subspicatus*. Environ Sci Technol 48(20):12354–12361. https://doi.org/10.1021/es503574z

Malchi T, Eyal S, Czosnek H, Shenker M, Chefetz B (2022) Plant pharmacology: Insights into in-planta kinetic and dynamic processes of xenobiotics. Critical Reviews in Environmental Science and Technology 52(19):3525–3546. https://doi.org/10.1080/10643389.2021.1946360

Malwe AS, Sharma VK (2023) Application of artificial intelligence approaches to predict the metabolism of xenobiotic molecules by human gut microbiome. Frontiers in Microbiology 14:1254073

Mao Q, Xie Z, Irshad S, Zhong Z, Liu T, Pei F, Gao B, Li L (2022) Effect of arsenic accumulation on growth and antioxidant defense system of *Chlorella thermophila* SM01 and *Leptolyngbya* sp. XZMQ. Algal Research 66:102762. https://doi.org/10.1016/j.algal.2022.102762

Mathew BB, Singh H, Biju VG, Krishnamurthy NB (2017) Classification, Source, and Effect of Environmental Pollutants and Their Biodegradation. J Environ Pathol Toxicol Oncol 36(1):55–71. https://doi.org/10.1615/JEnvironPatholToxicolOncol.2017015804

Miglani R, Parveen N, Kumar A, Ansari MohdA, Khanna S, Rawat G, Panda AK, Bisht SS, Upadhyay J, Ansari MN (2022) Degradation of Xenobiotic Pollutants: An Environmentally Sustainable Approach. Metabolites 12(9):818. https://doi.org/10.3390/metabo12090818

Morrissey KL, Iveša L, Delva S, D'Hondt S, Willems A, De Clerck O (2021) Impacts of environmental stress on resistance and resilience of algal-associated bacterial communities. Ecol Evol 11(21):15004–15019. https://doi.org/10.1002/ece3.8184

Mosulén S, Domínguez MJ, Vigara J, Vílchez C, Guiraum A, Vega JM (2003) Metal toxicity in Chlamydomonas reinhardtii. Effect on sulfate and nitrate assimilation. Biomol Eng 20(4–6):199–203. https://doi.org/10.1016/s1389-0344(03)00053-4

Mudhoo A, Bhatnagar A, Rantalankila M, Srivastava V, Sillanpää M (2019) Endosulfan removal through bioremediation, photocatalytic degradation, adsorption and membrane separation processes: a review. Chemical Engineering Journal 360:912–928

Muñoz-García A, Mestanza O, Isaza JP, Figueroa-Galvis I, Vanegas J (2019) Influence of salinity on the degradation of xenobiotic compounds in rhizospheric mangrove soil. Environmental Pollution 249:750–757

Navarrete A, González A, Gómez M, Contreras RA, Díaz P, Lobos G, Brown MT, Sáez CA, Moenne A (2019) Copper excess detoxification is mediated by a coordinated and complementary induction of glutathione, phytochelatins and metallothioneins in the green seaweed *Ulva compressa*. Plant Physiology and Biochemistry 135:423–431. https://doi.org/10.1016/j.plaphy.2018.11.019

Nazos TT, Kokarakis EJ, Ghanotakis DF (2017) Metabolism of xenobiotics by Chlamydomonas reinhardtii: Phenol degradation under conditions affecting photosynthesis. Photosynth Res 131(1):31–40. https://doi.org/10.1007/s11120-016-0294-2

Ni Yan NY, Lai JinHu LJ, Wan JinBao WJ, Chen LianShui CL (2014) Photosynthetic responses and accumulation of mesotrione in two freshwater algae.

Nie X, Gu J, Lu J, Pan W, Yang Y (2009) Effects of norfloxacin and butylated hydroxyanisole on the freshwater microalga Scenedesmus obliquus. Ecotoxicology 18:677–684

Nikkilä A (2001) Effects of organic material on the bioavailability, toxicokinetics and toxicity of xenobiotics in freshwater organisms. University of Jyväskylä

Nogales B, Lanfranconi MP, Piña-Villalonga JM, Bosch R (2011) Anthropogenic perturbations in marine microbial communities. FEMS Microbiology Reviews 35(2):275–298. https://doi.org/10.1111/j.1574-6976.2010.00248.x

Nolte TM, Hartmann NB, Kleijn JM, Garnæs J, van de Meent D, Jan Hendriks A, Baun A (2017) The toxicity of plastic nanoparticles to green algae as influenced by surface modification, medium hardness and cellular adsorption. Aquatic Toxicology 183:11–20. https://doi.org/10.1016/j.aquatox.2016.12.005

Nowicka B (2022) Heavy metal–induced stress in eukaryotic algae—mechanisms of heavy metal toxicity and tolerance with particular emphasis on oxidative stress in exposed cells and the role of antioxidant response. Environ Sci Pollut Res Int 29(12):16860–16911. https://doi.org/10.1007/s11356-021-18419-w

Oh J-W, Pushparaj SSC, Muthu M, Gopal J (2023) Review of Harmful Algal Blooms (HABs) Causing Marine Fish Kills: Toxicity and Mitigation. Plants 12(23):3936

Okay OS, Karacik B (2007) Photoinduced toxicity of selected PAHs to the marine microalga Phaeodactylum tricornutum. J Environ Sci Health A Tox Hazard Subst Environ Eng 42(6):707–714. https://doi.org/10.1080/10934520701304344

Pandey A, Tripathi PH, Tripathi AH, Pandey SC, Gangola S (2019) Chapter 2 - Omics technology to study bioremediation and respective enzymes. In: Bhatt P (ed) Smart Bioremediation Technologies. Academic Press, pp 23–43

Park J, Patel K, Lee WH (2024) Recent advances in algal bloom detection and prediction technology using machine learning. Science of The Total Environment 938:173546. https://doi.org/10.1016/j.scitotenv.2024.173546

Pasricha S, Mathur V, Garg A, Lenka S, Verma K, Agarwal S (2021) Molecular mechanisms underlying heavy metal uptake, translocation and tolerance in hyperaccumulators-an analysis: Heavy metal tolerance in hyperaccumulators. Environmental Challenges 4:100197. https://doi.org/10.1016/j.envc.2021.100197

Patel N, Khan MD, Shahane S, Rai D, Chauhan D, Kant C, Chaudhary VK (2020) Emerging pollutants in aquatic environment: source, effect, and challenges in biomonitoring and bioremediation-a review. Pollution 6(1):99–113

Patiño R, Christensen VG, Graham JL, Rogosch JS, Rosen BH (2023) Toxic algae in inland waters of the conterminous United States—A review and synthesis. Water 15(15):2808

Patterson AD, Gonzalez FJ, Idle JR (2010) XENOBIOTIC METABOLISM – A VIEW THROUGH THE METABOLOMETER. Chem Res Toxicol 23(5):851–860. https://doi.org/10.1021/tx100020p

Peng F-Q, Ying G-G, Yang B, Liu S, Lai H-J, Liu Y-S, Chen Z-F, Zhou G-J (2014) Biotransformation of progesterone and norgestrel by two freshwater microalgae (Scenedesmus obliquus and Chlorella pyrenoidosa): transformation kinetics and products identification. Chemosphere 95:581–588

Penga C, Leec J-W, Sichania H, Ng J (2014) Toxic effects of individual and combined effects of BTEX on Euglena gracilis. Journal of Hazardous Materials 284:10–18. https://doi.org/10.1016/j.jhazmat.2014.10.024

Perhar G, Arhonditsis GB (2014) Aquatic ecosystem dynamics following petroleum hydrocarbon perturbations: A review of the current state of knowledge. Journal of Great Lakes Research 40:56–72. https://doi.org/10.1016/j.jglr.2014.05.013

Perron M-C, Juneau P (2011) Effect of endocrine disrupters on photosystem II energy fluxes of green algae and cyanobacteria. Environmental Research 111(4):520–529. https://doi.org/10.1016/j.envres.2011.02.013

Petsas AS, Vagi MC (2017) Effects on the photosynthetic activity of algae after exposure to various organic and inorganic pollutants. Chlorophyll 37

Pietrini F, Iannelli RM, Bianconi D, Massacci A (2006) Cadmiumi interaction with thiols and photo-synthesis in higher plants. Advances in plant physiol 8:313–26

Priya AK, Muruganandam M, Kumar A, Senthilkumar N, Shkir M, Pandit B, Imran M, Prakash C, Ubaidullah M (2024) Recent advances in microbial-assisted degradation and remediation of xenobiotic contaminants; challenges and future prospects. Journal of Water Process Engineering 60:105106

Puckowski A, Mioduszewska K, Łukaszewicz P, Borecka M, Caban M, Maszkowska J, Stepnowski P (2016) Bioaccumulation and analytics of pharmaceutical residues in the environment: A review. Journal of pharmaceutical and biomedical analysis 127:232–255

Rai R, Agrawal M, Agrawal SB (2016) Impact of Heavy Metals on Physiological Processes of Plants: With Special Reference to Photosynthetic System. In: Singh A, Prasad SM, Singh RP (eds) Plant Responses to Xenobiotics. Springer Singapore, Singapore, pp 127–140

Rajput VD, Harish, Singh RK, Verma KK, Sharma L, Quiroz-Figueroa FR, Meena M, Gour VS, Minkina T, Sushkova S, Mandzhieva S (2021) Recent Developments in Enzymatic Antioxidant Defence Mechanism in Plants with Special Reference to Abiotic Stress. Biology (Basel) 10(4):267. https://doi.org/10.3390/biology10040267

Ramanan R, Kim B-H, Cho D-H, Oh H-M, Kim H-S (2016) Algae–bacteria interactions: evolution, ecology and emerging applications. Biotechnology advances 34(1):14–29

Rathour RK, Sharma D, Ullah S, Mahmoud E-HM, Sharma N, Kumar P, Bhatt AK, Ahmad I, Bhatia RK (2024) Bacterial–microalgal consortia for bioremediation of textile industry wastewater and resource recovery for circular economy. Biotechnol Environ 1(1):6. https://doi.org/10.1186/s44314-024-00005-2

Redalje DG, Duerr EO, de la Noüe J, Mayzaud P, Nonomura AM, Cassin R (2018) Algae as ideal waste removers: biochemical pathways. In: Biotreatment of agricultural wastewater. CRC press, pp 91–110

Rodea-Palomares I, Makowski M, Gonzalo S, González-Pleiter M, Leganés F, Fernández-Piñas F (2015) Effect of PFOA/PFOS pre-exposure on the toxicity of the herbicides 2,4-D, Atrazine, Diuron and Paraquat to a model aquatic photosynthetic microorganism. Chemosphere 139:65–72. https://doi.org/10.1016/j.chemosphere.2015.05.078

Ryu B-G, Kim EJ, Kim H-S, Kim J, Choi Y-E, Yang J-W (2014) Simultaneous treatment of municipal wastewater and biodiesel production by cultivation of Chlorella vulgaris with indigenous wastewater bacteria. Biotechnology and bioprocess engineering 19:201–210

S AD, Ashwath P, Manjunatha KG, Aksahy SD, Rao R, Reddy D, Vittal R (2023) Pesticide and Xenobiotic Metabolism in Aquatic Organisms. In: Rather MA, Amin A, Hajam YA, Jamwal A, Ahmad I (eds) Xenobiotics in Aquatic Animals. Springer Nature Singapore, Singapore, pp 1–66

Sachdev S, Ansari SA, Ansari MI, Fujita M, Hasanuzzaman M (2021) Abiotic stress and reactive oxygen species: Generation, signaling, and defense mechanisms. Antioxidants 10(2):277

Sahu P (2021) Policies and Legal Aspects of Environmental Management. Environmental Management: Issues and Concerns in Developing Countries :343–367

Saidon NB, Szabó R, Budai P, Lehel J (2024) Trophic transfer and biomagnification potential of environmental contaminants (heavy metals) in aquatic ecosystems. Environmental pollution 340:122815

Sánchez-Fortún S, Marvá F, Rouco M, Costas E, López-Rodas V (2009) Toxic effect and adaptation in Scenedesmus intermedius to anthropogenic chloramphenicol contamination: genetic versus physiological mechanisms to rapid acquisition of xenobiotic resistance. Ecotoxicology 18(5):481–487. https://doi.org/10.1007/s10646-009-0303-8

Saravanakumar K, Sathyanarayanan K, Khan MS, Park SJ, Yoo HM, Cho SH, Ahn G, Ahamed MAA, Padmanabhan S, MubarakAli D, Cho N (2024) Recent progress in biotechnological approaches for diverse applications of algae: an overview. Int J Environ Sci Technol 21(3):3453–3474. https://doi.org/10.1007/s13762-023-05321-w

Sarı A, Uluozlü ÖD, Tüzen M (2011) Equilibrium, thermodynamic and kinetic investigations on biosorption of arsenic from aqueous solution by algae (Maugeotia genuflexa) biomass. Chemical Engineering Journal 167(1):155–161

Sarma U, Hoque ME, Aswani T, Nethravathy V, Rajagopal S (2024) Microalgae in removing heavy metals from wastewater–An advanced green technology for urban wastewater treatment. Journal of Hazardous Materials Advances :100444

Scopetani C, Chelazzi D, Cincinelli A, Martellini T, Leiniö V, Pellinen J (2022) Hazardous contaminants in plastics contained in compost and agricultural soil. Chemosphere 293:133645. https://doi.org/10.1016/j.chemosphere.2022.133645

Shanab S, Essa A, Shalaby E (2012) Bioremoval capacity of three heavy metals by some microalgae species (Egyptian Isolates). Plant Signal Behav 7(3):392–399. https://doi.org/10.4161/psb.19173

Sharma L, Siedlewicz G, Pazdro K (2021) The toxic effects of antibiotics on freshwater and marine photosynthetic microorganisms: State of the art. Plants 10(3):591

Sharma P, Singh SP, Iqbal HMN, Tong YW (2022) Omics approaches in bioremediation of environmental contaminants: An integrated approach for environmental safety and sustainability. Environmental Research 211:113102. https://doi.org/10.1016/j.envres.2022.113102

Singh N, Afzal S, Singh NK, Sohrab S, Mishra SK, Agrawal SC (2022) Physiological, morphological, and biochemical responses of metals and metalloids on algae. In: Metals Metalloids Soil Plant Water Systems. Elsevier, pp 271–286

Singh OV, Nagaraj NS (2006) Transcriptomics, proteomics and interactomics: unique approaches to track the insights of bioremediation. Briefings in Functional Genomics 4(4):355–362

Singh V, Ahmed G, Vedika S, Kumar P, Chaturvedi SK, Rai SN, Vamanu E, Kumar A (2024) Toxic heavy metal ions contamination in water and their sustainable reduction by eco-friendly methods: isotherms, thermodynamics and kinetics study. Sci Rep 14(1):7595. https://doi.org/10.1038/s41598-024-58061-3

Singh V, Singh N, Rai SN, Kumar A, Singh AK, Singh MP, Sahoo A, Shekhar S, Vamanu E, Mishra V (2023) Heavy Metal Contamination in the Aquatic Ecosystem: Toxicity and Its Remediation Using Eco-Friendly Approaches. Toxics 11(2):147. https://doi.org/10.3390/toxics11020147

Solé M, Varó I, González-Mira A, Torreblanca A (2015) Xenobiotic metabolism modulation after long-term temperature acclimation in juveniles of Solea senegalensis. Mar Biol 162(2):401–412. https://doi.org/10.1007/s00227-014-2588-2

Solovchenko AE, Vasilieva SG, Zaitsev P, Lukyanov AA, Skripnikova EV, Antal TK (2022) Approaches to rapid screening of pharmaceutical xenobiotic effects on microalgae via monitoring of photosynthetic apparatus condition. Journal of Applied Phycology :1–9

Song C, Wei Y, Qiu Y, Qi Y, Li Y, Kitamura Y (2019) Biodegradability and mechanism of florfenicol via Chlorella sp. UTEX1602 and L38: Experimental study. Bioresour Technol 272:529–534. https://doi.org/10.1016/j.biortech.2018.10.080

Staley ZR, Harwood VJ, Rohr JR (2015) A Synthesis of the Effects of Pesticides on Microbial Persistence in Aquatic Ecosystems. Crit Rev Toxicol 45(10):813–836. https://doi.org/10.3109/10408444.2015.1065471

Štefanac T, Grgas D, Landeka Dragičević T (2021) Xenobiotics—Division and Methods of Detection: A Review. J Xenobiot 11(4):130–141. https://doi.org/10.3390/jox11040009

Stein HP, Navajas-Pérez R, Aranda E (2018) Potential for CRISPR genetic engineering to increase xenobiotic degradation capacities in model fungi. Approaches in Bioremediation: The New Era of Environmental Microbiology and Nanobiotechnology :61–78

Sudharshan S, Naidu R, Mallavarapu M, Bolan N (2012) DDT remediation in contaminated soils: a review of recent studies. Biodegradation 23(6):851–863. https://doi.org/10.1007/s10532-012-9575-4

Sule RO, Condon L, Gomes AV (2022) A Common Feature of Pesticides: Oxidative Stress—The Role of Oxidative Stress in Pesticide-Induced Toxicity. Oxid Med Cell Longev 2022:5563759. https://doi.org/10.1155/2022/5563759

Sutherland DL, Ralph PJ (2019) Microalgal bioremediation of emerging contaminants - Opportunities and challenges. Water Research 164:114921. https://doi.org/10.1016/j.watres.2019.114921

Thies F, Backhaus T, Bossmann B, Grimme LH (1996) Xenobiotic biotransformation in unicellular green algae (involvement of cytochrome P450 in the activation and selectivity of the pyridazinone pro-herbicide metflurazon). Plant physiology 112(1):361–370

Touliabah HE-S, El-Sheekh MM, Ismail MM, El-Kassas H (2022) A Review of Microalgae- and Cyanobacteria-Based Biodegradation of Organic Pollutants. Molecules 27(3):1141. https://doi.org/10.3390/molecules27031141

Tripathi S, Poluri KM (2021) Metallothionein-and Phytochelatin-Assisted Mechanism of Heavy Metal Detoxification in Microalgae. In: Hasanuzzaman M (ed) Approaches to the Remediation of Inorganic Pollutants. Springer, Singapore, pp 323–344

Tsering T, Sillanpää M, Sillanpää M, Viitala M, Reinikainen S-P (2021) Microplastics pollution in the Brahmaputra River and the Indus River of the Indian Himalaya. Science of The Total Environment 789:147968. https://doi.org/10.1016/j.scitotenv.2021.147968

Ughy B, Nagy CI, Kós PB (2015) Biomedical potential of cyanobacteria and algae. Acta Biologica Szegediensis 59(suppl. 2.):203–224

Vardhan KH, Kumar PS, Panda RC (2019) A review on heavy metal pollution, toxicity and remedial measures: Current trends and future perspectives. Journal of Molecular Liquids 290:111197

Veal EA, Day AM, Morgan BA (2007) Hydrogen Peroxide Sensing and Signaling. Molecular Cell 26(1):1–14. https://doi.org/10.1016/j.molcel.2007.03.016

Veerasingam S, Ranjani M, Venkatachalapathy R, Bagaev A, Mukhanov V, Litvinyuk D, Verzhevskaia L, Guganathan L, Vethamony P (2020) Microplastics in different environmental compartments in India: Analytical methods, distribution, associated contaminants and research needs. TrAC Trends in Analytical Chemistry 133:116071. https://doi.org/10.1016/j.trac.2020.116071

Verasoundarapandian G, Lim ZS, Radziff SBM, Taufik SH, Puasa NA, Shaharuddin NA, Merican F, Wong C-Y, Lalung J, Ahmad SA (2022) Remediation of Pesticides by Microalgae as Feasible Approach in Agriculture: Bibliometric Strategies. Agronomy 12(1):117. https://doi.org/10.3390/agronomy12010117

Wang F, Guan W, Xu L, Ding Z, Ma H, Ma A, Terry N (2019a) Effects of Nanoparticles on Algae: Adsorption, Distribution, Ecotoxicity and Fate. Applied Sciences 9(8):1534. https://doi.org/10.3390/app9081534

Wang L, Huang X, Lim DJ, Laserna AKC, Li SFY (2019b) Uptake and toxic effects of triphenyl phosphate on freshwater microalgae Chlorella vulgaris and Scenedesmus obliquus: Insights from untargeted metabolomics. Science of the Total Environment 650:1239–1249

Wang M, Lian Y, Wang Y, Zhu L (2023) The role and mechanism of quorum sensing on environmental antimicrobial resistance. Environmental Pollution :121238

Wang Y, Branicky R, Noë A, Hekimi S (2018) Superoxide dismutases: Dual roles in controlling ROS damage and regulating ROS signaling. J Cell Biol 217(6):1915–1928. https://doi.org/10.1083/jcb.201708007

Xia X, Carroll-Haddad A, Brown N, Utell MJ, Mallon T, Hopke PK (2016) Polycyclic Aromatic Hydrocarbons and Polychlorinated Dibenzo-p-Dioxins/Dibenzofurans in Microliter Samples of Human Serum as Exposure Indicators. J Occup Environ Med 58(8):S72–S79. https://doi.org/10.1097/JOM.0000000000000743

Xiao X, Li W, Jin M, Zhang L, Qin L, Geng W (2023) Responses and tolerance mechanisms of microalgae to heavy metal stress: A review. Marine environmental research 183:105805

Xin X, Huang G, Zhang B (2021) Review of aquatic toxicity of pharmaceuticals and personal care products to algae. Journal of Hazardous Materials 410:124619. https://doi.org/10.1016/j.jhazmat.2020.124619

Xiong J-Q, Kurade MB, Jeon B-H (2017) Biodegradation of levofloxacin by an acclimated freshwater microalga, Chlorella vulgaris. Chemical Engineering Journal 313:1251–1257

Xiong Q, Hu L-X, Liu Y-S, Zhao J-L, He L-Y, Ying G-G (2021) Microalgae-based technology for antibiotics removal: From mechanisms to application of innovational hybrid systems. Environment International 155:106594. https://doi.org/10.1016/j.envint.2021.106594

Yadav APS, Dwivedi V, Kumar S, Kushwaha A, Goswami L, Reddy BS (2021) Cyanobacterial Extracellular Polymeric Substances for Heavy Metal Removal: A Mini Review. Journal of Composites Science 5(1):1. https://doi.org/10.3390/jcs5010001

Yadav SK (2010) Heavy metals toxicity in plants: An overview on the role of glutathione and phytochelatins in heavy metal stress tolerance of plants. South African Journal of Botany 76(2):167–179. https://doi.org/10.1016/j.sajb.2009.10.007

You X, Xu N, Yang X, Sun W (2021) Pollutants affect algae-bacteria interactions: a critical review. Environmental Pollution 276:116723

Zaidi S, Hayat S, Pichtel J (2024) Arsenic-induced plant stress: Mitigation strategies and omics approaches to alleviate toxicity. Plant Physiology and Biochemistry 213:108811. https://doi.org/10.1016/j.plaphy.2024.108811

Zhang Y, Chen Z, Li X, Wu X, Chen L, Wang G (2023) Photosynthesis Responses of Tibetan Freshwater Algae Chlorella vulgaris to Herbicide Glyphosate. International Journal of Environmental Research and Public Health 20(1). https://doi.org/10.3390/ijerph20010386

Zheng J-W, Li D-W, Lu Y, Chen J, Liang J-J, Zhang L, Yang W-D, Liu J-S, Lu S-H, Li H-Y (2016) Molecular exploration of algal interaction between the diatom *Phaeodactylum tricornutum* and the dinoflagellate *Alexandrium tamarense*. Algal Research 17:132–141. https://doi.org/10.1016/j.algal.2016.04.019

Zhou Y, Yue Y, Chen X, Wu F, Li W, Li P, Han J (2024) Physiological-biochemical responses and transcriptomic analysis reveal the effects and mechanisms of sulfamethoxazole on the carbon fixation function of *Chlorella pyrenoidosa*. Science of The Total Environment 917:170460. https://doi.org/10.1016/j.scitotenv.2024.170460

Zhu J, Cai Y, Wakisaka M, Yang Z, Yin Y, Fang W, Xu Y, Omura T, Yu R, Zheng ALT (2023) Mitigation of oxidative stress damage caused by abiotic stress to improve biomass yield of microalgae: A review. Science of The Total Environment 896:165200. https://doi.org/10.1016/j.scitotenv.2023.165200

Zhu J, Patzoldt WL, Radwan O, Tranel PJ, Clough SJ (2009) Effects of Photosystem-II-Interfering Herbicides Atrazine and Bentazon on the Soybean Transcriptome. The Plant Genome 2(2). https://doi.org/10.3835/plantgenome2009.02.0010

Zhu Q-L, Guo S-N, Wen F, Zhang X-L, Wang C-C, Si L-F, Zheng J-L, Liu J (2019) Transcriptional and physiological responses of *Dunaliella salina* to cadmium reveals time-dependent turnover of ribosome, photosystem, and ROS-scavenging pathways. Aquatic Toxicology 207:153–162. https://doi.org/10.1016/j.aquatox.2018.12.007

Impacts of Industrial Effluents on Plants and Their Effects on the Food Chain

Md. Azmalul Alam, Sudeshna Pal, Sumanta Pal, Anwesha Dutta, and Manab Mandal

Abstract Industrial effluents, wastewater from manufacturing operations, pose serious risks to human health and the environment due to contaminants such as heavy metals, organic materials, suspended solids, and excessive nutrients. These pollutants degrade water bodies, soil, and biodiversity, leading to toxic accumulation in the food chain and harmful ecological effects. Under the Environmental Quality Act 1974 (EQA), industrial effluents are classified as major contributors to water pollution, with key pollutants including suspended solids (SS), ammoniacal nitrogen (NH_3-N), and biochemical oxygen demand (BOD). Hazardous substances like arsenic, cadmium, lead, and mercury can cause poisoning and organ damage through bioaccumulation. Aquatic ecosystems suffer from oxygen depletion and algal blooms, while heavy metals in soil reduce agricultural productivity and harm plant health. Industrial wastewater also disrupts plant-pollinator relationships, leading to cascading ecological consequences. The rapid industrialization driving economic growth has worsened pollution, making effective wastewater management crucial. Addressing these challenges requires stringent regulations, advanced treatment technologies, and sustainable industrial practices. Techniques like filtration, chemical treatment, and bioremediation can minimize pollution before discharge. Industries should adopt green technologies and resource-efficient methods to reduce environmental impact. Collaborative efforts among regulators, industries, and communities are essential for protecting ecosystems, maintaining biodiversity, and ensuring long-term sustainability. By prioritizing proactive pollution management and stringent wastewater treatment, societies can mitigate the harmful effects of industrial effluents and promote a healthier environment for future generations.

Keywords Biomagnification · Crop production · Food chain · Human health · Toxicity to plants Industrial effluents

---

M. A. Alam · S. Pal · S. Pal · A. Dutta · M. Mandal (✉)
Department of Botany, Dukhulal Nibaran Chandra College, Murshidabad, India

S. Roy, V. Mandal (eds.), *Plant-Microbe Interaction under Xenobiotic Exposure*,
https://doi.org/10.1007/978-981-96-8260-7_7

1 Introduction

Industrial effluents, the wastewater produced by manufacturing processes, pose severe threats to human-environment interaction and health outcomes due to their diverse range of pollutants. Heavy metals, organic substances, suspended particulates, pathogenic microbes, and an excessive amount of nutrients are frequently found in these effluents, each contributing to a range of negative impacts on water bodies, soil, and public health (Guerra 2005). Industrial growth and urban expansion have substantially higher production of waste and environmental degradation, impacting ecosystems and natural systems (Voumik and Sultana 2022). Environmental pollution happens when the environment cannot manage and neutralize harmful substances or by-products from industrial effluents. In Malaysia, water contamination is a major challenge. It typically leads aquatic ecosystems to become more fragile, whereas air pollution negatively impacts human respiratory health. Identifying contaminants in water poses an immense barrier, particularly those arising from industrial waste, complicating efforts to manage pollution (Afroz and Rahman 2017).

Industrial effluents, defined in Regulation 2 of the Environmental Quality Act 1974 (EQA) as any liquid waste or wastewater produced by industrial manufacturing processes, including water treatment for supply or activities at industrial sites, are a significant source of water pollution. According to Regulation 4 of the EQA, industrial premises must provide an advance written notice to the Director General of the Department of Environment (DOE) before starting work, construction, or upgrades. These industrial effluents often contain hazardous compounds which are toxic, ignitable, corrosive, or reactive. Whenever these wastes are not appropriately managed, they may eventually find their way to rivers, ultimately affecting water systems (Musarat et al. 2023). The effluents often contain microorganisms, many of which are critical in causing waterborne disease outbreaks (Okereke et al. 2016).

In Malaysia, key pollutants include suspended solids (SS), ammoniacal nitrogen (NH_3-N), and BOD. A high BOD level, indicating bacterial oxygen depletion from untreated or partially treated sewage, is often accompanied by elevated NH_3-N levels from household organic waste and suspended solids from several activities like clearing lands and earthquakes (Huang et al. 2015). To evaluate and monitor water quality, the DOE uses the Water Quality Index (WQI) method to assess the condition of river water (Uddin et al. 2020). Stream, lake, and coastal water pollution has become a serious issue, rendering these natural resources unsuitable for human consumption and leisure activities such as swimming and boating (Bashir et al. 2020). Municipal sewage, home waste materials, and industrial emissions are major causes of pollution in rivers (Muralikrishna and Manickam 2017). Furthermore, agricultural practices can lead to water pollution by introducing excess silt and chemicals into water sources through runoff and deterioration of soil. Most manufacturing activities inevitably result in the production of industrial effluent, involving the use of large amounts of water (Mishra et al. 2023). While merely a tiny portion is consumed in the manufacturing process, the greater part is expelled, often contaminated with various materials used in the plant. These contaminants consist of raw materials, undesired chemical compounds, finished and intermediate goods, by-products, and other processing substances, all of which contribute to the pollution of receiving water bodies (Kumar et al. 2022). Extremely hazardous contaminants such as arsenic, cadmium, lead, and mercury can accumulate throughout the food chain of the natural environment, posing a significant threat to organisms at all levels. This bioaccumulation poses risks to wildlife as well as to humans, potentially leading to significant health issues (Mitra et al. 2022b). In addition to hindering seed germination, heavy metals can also impede development, reduce biomass, interfere with the formation of chlorophyll, cause nutritional deficits in plants, and at high concentrations, induce cell damage and plant death (Sethy and Ghosh 2013; Haider et al. 2021). Pesticides, volatile organic compounds (VOCs), and polycyclic aromatic hydrocarbons (PAHs) can contaminate water sources, endangering aquatic life, lowering water quality, disrupting ecosystems, and challenging water treatment efforts (Honda and Suzuki 2020). Drinking water sources can be contaminated by microbes like bacteria and viruses; they can cause diseases and also be able to spread them. Nitrogen and phosphorus together lead to eutrophication, causing oxygen deprivation and harmful algal blooms in water bodies (Bashir et al. 2020; Coffey et al. 2018). Rivers, groundwater, and lakes are the key sources of water used for irrigation, industrial purposes, and consumption by humans and animals (Matthews 2023). However, due to climatic changes, floods and droughts are occurring more often worldwide. Moreover, water pollution rises due to industrial, agricultural, and domestic waste, which has significantly deteriorated both the quality and quantity of potable water (Du Plessis 2022; Evans et al. 2019). Therefore, effective wastewater treatment is crucial to improving the quality and availability of drinking water. Polluted water is characterized by excessive hazardous contaminants that render it unsuitable for domestic use. Human activities are often the cause of water contamination, with contaminants generated from different sources like heavy metals, sewage leaks, industrial discharges, oil spills, animal waste, chemical residues, eroded sediments, agricultural practices, littering, and deforestation involving fertilizers, herbicides, and pesticides (Akhtar et al. 2021). These sectors collectively consume about a third of the earth's replenishable freshwater reserves, introducing a range of natural and synthetic chemical contaminants (Ahmed et al. 2021a, b). Mining operations, smelters, foundries, and other metal-based industries are point source areas that can cause particularly severe environmental pollution (Chaney 1991). Despite water covering 75% of the surface of the earth, freshwater is often inadequately distributed in terms of location, timing, and quality (Srivastava and Yadav 2023). Further factors affecting freshwater availability include pollution, urbanization, population growth, and changes in the global climate. The United Nations projects that by 2050, almost 50% of the world's population will reside in water-scarce regions. Pollution, urbanization, population increase, alterations in the worldwide climate, leaching of heavy metals, sediment re-suspension, and metal evaporation from water resources to soil and groundwater are other variables influencing freshwater availability (Nriagu 1989). Improper disposal of industrial effluents can lead to water pollution, soil contamination, aquatic harm, and health issues. Exposure to pollutants can cause skin diseases and respiratory issues and may even cause cancer. Specifically, biomagnification can be defined as the process by which heavy metals from contaminated plants move into the food chain and accumulate in species higher up the chain may have an impact on human health (Okereke et al. 2016).

A wide range of treatment technologies are employed in order to mitigate the harmful impacts of industrial effluents, among them physical treatment (separation, flotation, and screening to remove suspended particles and particulate debris), chemical treatment (eliminate the surface charges of particles), biological treatment (by utilizing microorganisms to break down organic matter), membrane filtration (heavy metals and organic compounds are removed the contaminants by reverse osmosis and nanofiltration) and advanced oxidation processes (AOPs) (techniques tend to involve ozonation, UV irradiation, and Fenton's reagent break down organic pollutants and eliminate microorganisms (Rawat et al. 2022; Behera et al. 2021; Kamali et al. 2019a, b). For the effective management of industrial effluents, extremely strict regulations, including emission standards, compliance monitoring, and evaluations of environmental impacts, must be adopted (Magat and Viscusi 1990; Ziolko et al. 2011). The sickness known as Minamata was caused by the discharge of wastewater tainted with mercury into Japan's Minamata Bay, a serious neurological disorder that impacts both humans and wildlife. This case underscores the devastating health impacts of mercury pollution (Harada 1995). Industrial effluents containing cadmium polluted rice paddies, resulting in cadmium accumulation in rice grains. This contamination caused health problems in individuals consuming the affected rice (Genchi et al. 2020). Industrial effluents carry heavy metals and other pollutants that pose serious risks to plants, aquatic ecosystems, and human health (Aziz et al. 2023). These pollutants can inhibit plant growth, disrupt nutrient uptake, and cause toxicity that multiplies through aquatic life (e.g., fish, etc.) to the food chain, ultimately impacting human health (Alengebawy et al. 2021; Sonone et al. 2020).

2 Characteristics and Components of Industrial Effluents

Industrial effluents present major environmental issues, especially concerning their effects on plant life and the broader food chain. The expansion of industrial activities has led to a notable increase in the discharge of these effluents into natural environments (Singh et al. 2023). Ecosystems may be adversely affected by these effluents, a complex mixture of heavy metals, including both inorganic and organic substances, and other potentially hazardous substances. When industrial effluents are discharged into water bodies, their impact can be extensive. Aquatic life frequently bears the brunt of pollution first, as pollutants have the ability to disturb the delicate balance of undersea ecosystems (Bashir et al. 2020; Ali et al. 2023). Fish and other marine creatures may be potentially hazardous to pollutants comparable to heavy metals that can have an adverse effect on their survival, growth and development, and ability to reproduce. Additionally, the presence of organic contaminants can cause eutrophication, where an overabundance of nutrients stimulates the excessive growth of algae, reducing oxygen levels in the water and creating "dead zones" where aquatic life faces challenges for their survival (Bhagowati and Ahamad 2018; Mallin 2009). Soil quality is another major concern. The contamination from industrial waste has the potential to change the chemical and physical structure of soil, lowering its fertility and negatively impacting plant life. For instance, heavy metals can build up in the soil and promote long-term contamination that may persist long after the pollution source has been eradicated (Alengebawy et al. 2021; Angon et al. 2024). This degradation of soil quality can deteriorate crop yields, retard plant development, and alter natural vegetation patterns. The impacts on plant life have a cascading effect on the entire food chain (Blum 2013). Plants are the primary producers in most ecosystems, forming the initial point of the food chain. When water or soil contamination affects plants, it can endanger each and every level of the food web (Gavrilescu 2021). Herbivores that consume contaminated plants may experience health problems, not only these herbivores but also the higher consumers that feed on them. This disruption can result in a reduction in biodiversity and alter the structural and functional changes in ecosystems (Maron and Crone 2006; Kozlov 2022).

2.1 *Classification of Industrial Effluents Based on Their Characteristics*

Based on their comparable parameters, the characteristics of industrial effluents are often classified into three main classes.

2.1.1 Physical Characteristics

The fundamental physical characteristics of wastewater quality are its temperature, turbidity or clarity, suspended particle quantity, colors, and odor. An emerging concern is microplastics—tiny plastic particles resulting from the breakdown of larger plastic waste—which present significant environmental threats (Ghosh et al. 2023; Ziani et al. 2023).

2.1.2 Chemical Characteristics

Solids that are suspended, colloidal, and dissolved (both organic and mineral) can be found in industrial wastewaters. They might also have high or low amounts of colorful materials and be too acidic or alkaline.

Chemical characteristics of effluents often involve:

I. Heavy metals [e.g., As, Cd, Hg, Pb, and Zn, etc.]

II. Chemical Oxygen Demand (COD) and Total Organic Carbon (TOC).
III. pH levels,
IV. Toxic substances, including phenol and cyanide.
V. Dissolved Oxygen (DO).
VI. Nutrients like nitrogen, phosphorus, and potassium (Chalaris et al. 2023; Singare et al. 2010)
VII. The leaking of oil, especially from the marine and petroleum sectors, adds organic contaminants and hydrocarbons—these compounds can have detrimental effects on the ecosystems of land as well as water because they are difficult to break down (Das and Chandran 2010).

2.1.3 Biological Characteristics

A large portion of the food supply for organic bacteria that feed on dead stuff comes from domestic wastewater. This makes household wastewater very putrescible, biodegradable, and unstable.

Biological parameters include:

I. Biological Oxygen Demand (BOD) (Lokhande et al. 2012).
II. Microbial populations, pathogens (e.g., bacteria, viruses, and protozoans like *Giardia* and *Cryptosporidium)* (Ariyadasa et al. 2023).
III. Radioactive isotopes introduced by radioactive waste, mostly from the mining and nuclear industries, can retard biological processes in plants and animals and induce mutations (Folkers and Gunter 2022), (Fazal et al. 2015) (Fig. 1).

2.2 Industry-Specific Effluents

2.2.1 Iron and Steel Industry

Organic and inorganic debris, nutrients, and hazardous substances, including phenol, ammonia, and cyanide, are frequently found in the wastewater from this industry (nitrogen, phosphorus, and potassium) and are potential hazards (Sathya et al. 2022).

2.2.2 Leather Industry

Elevated concentrations of heavy metals, like cadmium (Cd), chromium (Cr), lead (Pb), manganese (Mn), nickel (Ni), and zinc (Zn), are a fundamental characteristic of wastewaters (Niamat et al. 2023).

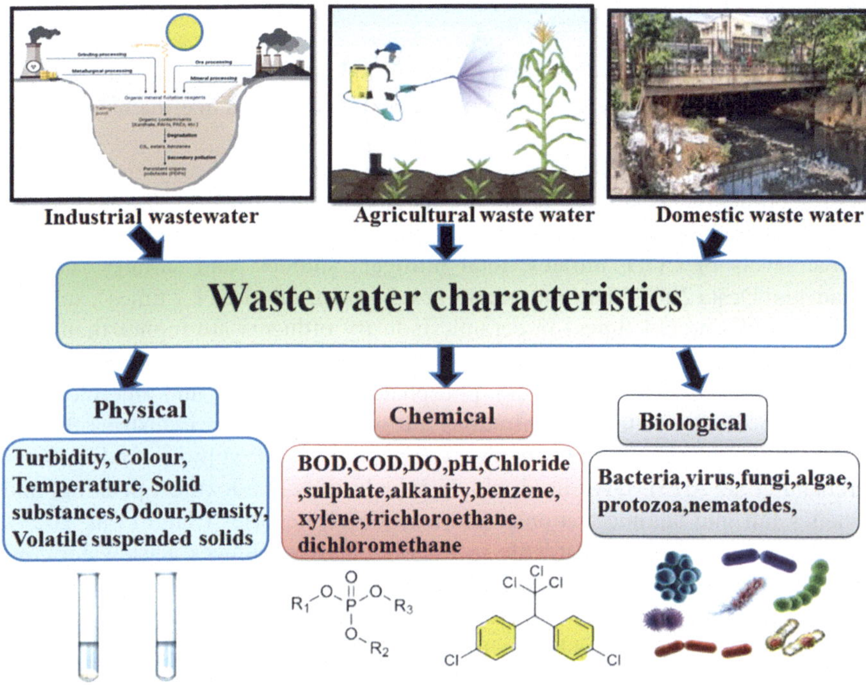

Fig. 1 Major Characteristics of wastewater (industrial effluents)

2.2.3 Microplastics

Having a high specific surface area, the particles of microplastics can absorb various pollutants, including PAHs and heavy metals (Talukdar et al. 2024).

2.2.4 Oily Wastewater

This type contains elements like PAHs, phenols, and petroleum hydrocarbons, which are harmful (De et al. 2022).

2.2.5 Pulp and Paper Industry

Wastes generated from this industry often have a high Biological Oxygen Demand (BOD) and COD ratio due to significant lignin content. High concentrations of resins, fatty acids, dispersed particulate matter, and complex chlorinated by-products, including benzofuran, dibenzo-p-dioxin, dichlorophenol, trichlorophenol, and pentachlorophenol, are also present (Jahan & Singh 2023).

2.2.6 Textile Industry

Sulfate (S), calcium (Ca), magnesium (Mg), chloride (Cl), and magnesium (Cl) are the most prevalent inorganic and organic mixed states observed in effluents (Kothari et al. 2020).

In the Ethiopian Rift Valley, Lake Hawassa is polluted with heavy metals due to industrial effluents. The textile factory discharges, notably from the Hawassa textile factory, have high levels of arsenic, copper, chromium, mercury, and lead, but have minimal levels of COD, nitrates, total nitrogen, sulfides, and sulfates (Gebre-Mariam and Desta 2002). The phosphates (24.5 times), sulfides (1.7 times), sulfates (2 times), and zinc (14 times) in ceramic industry effluents are higher than EPA guidelines. Additionally, the effluents from soft drink factories exceed regulations, exhibiting elevated COD (1.6 times), phosphates (20.5 times), and zinc elements (20.4 times). Heavy metal concentrations are greater in textile industries than in ceramic effluents, which have higher concentrations of sulfides and sulfates (Berehanu et al. 2015). Dyes used widely in industries like textiles, printing, and cosmetics fall into anionic, cationic, and non-ionic categories (Saigl et al. 2023). The major user of dyes is the textile sector, which also produces wastewater containing dangerous substances, including acids, solvents made from organic matter, and softeners that are harmful to the environment and human health (Tamjidi et al. 2024). The battery industry is highly toxic, producing wastewater with high levels of cadmium, nickel, and silver (Mashhadikhan et al. 2022). Radioactive wastes are released by nuclear power facilities; pressurized water reactors discharge more liquid tritium than boiler water reactors (Kónya and Nagy 2018). With wastewater having high levels of COD and Biological Oxygen Demand (BOD), the food industry is one of the major contributors to pollution due to the massive usage of water and chemicals. Chocolate, one of the food industries, requires a lot of water because it contaminates most frequently (Sathya et al. 2022). To mitigate the environmental impacts of different kinds of industrial effluents, an individualized management system and adjusted treatment are needed.

3 Direct Toxicity on Plants Due to Industrial Effluents

There are a lot of dangerous substances in wastewater that might hurt plants and animals on land. This happens because of elevated ammonia and chlorine levels, a heavy concentration of materials that require oxygen, and elevated amounts of toxic metals and organic contaminants (Tchounwou et al. 2012 ; Jaishankar et al. 2014). The last several decades have seen rapid industrialization that has greatly increased the amount of contaminants in the environment. When hazardous industrial wastes are improperly treated and released into water bodies, they can have detrimental effects on a range of living things, both directly and indirectly. Pollutants such as heavy metals should be taken very seriously because they are persistent and non-biodegradable in nature (Singh et al. 2024). When aquatic organisms ingest these toxic metals, they can cause severe health issues in other animals and, ultimately, through the food chain, in humans. Heavy metals can be teratogenic and carcinogenic and may induce oxidative stress, organ damage, neurological impairments, and stunted growth (Bilal et al. 2013). Industrial contaminants like phenolic compounds are harmful because they prevent typical microbial activity, which makes biological treatment procedures more difficult (Rajkumar and Palanivelu 2004). Phenolic exposure can cause symptoms like cyanosis, low body temperature, sweating, reflex loss, impaired breathing, and respiratory distress. The companies that produce pulp and paper have a substantial impact on water pollution because their effluents contain organic compounds that are chlorinated, tannins, and resins, which can induce genotoxicity and mutagenicity (Lindholm-Lehto et al. 2015). Paper pulp effluents frequently include lignin and its derivatives, which are weakly degradable and might change into hazardous chemicals during biological treatment, perhaps upsetting aquatic species' hormonal balance (Oikari et al. 1983). For example, some wood sterols like β-sitosterol and stigmasterol can bind to estrogen receptors in fish, causing reproductive disruption (Tremblay and Van Der Kraak 1999) (Fig. 2 and Table 1).

Textile industry effluents are also highly toxic due to their complex mixture of hazardous constituents. Oily scum and chromium compounds may interact to create a colloidal substance that blocks sunlight and lowers the amount of dissolved oxygen in water bodies. Furthermore, chlorine-bound organic additives used in many

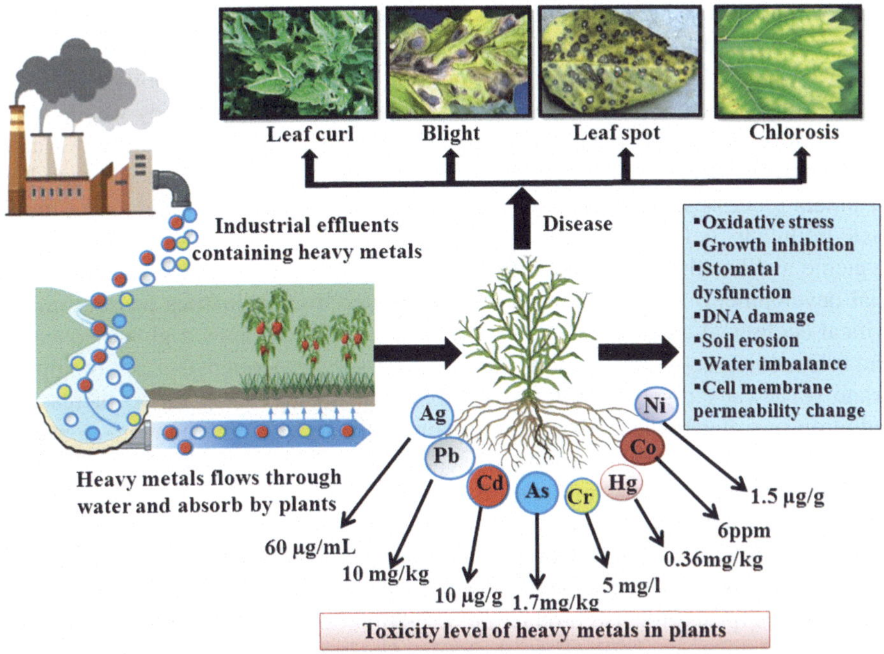

Fig. 2 Toxicity to plants due to industrial effluents

Table 1 Heavy metal toxicity and their harmful effects on various crop plants

Heavy metals	Toxic form	Plant	Effect	Reference
Zn	Zn^{2+}	Soybean	Chlorosis and inhibits, generation of biomass, cell division, absorption of nutrients, and enzymes, leading to a notable reduction in production.	Hassan et al. (2022)
AS	As^{3+}	Maize, soybean, Rice	Reduced stomatal conductance, poor growth, photosynthesis, aberrations and reductions of plant height, leaf area, and biomass, germination percentage.	Rahman et al. (2023)
Cd	Cd^{2+}	Wheat, Rice, mung bean	Shoot and root length, plant development, soluble protein, reduced chlorophyll levels, oxidative stress genotoxicity.	Ahmed et al. (2020)
Cu	Cu salts	Tomato	Reduced growth, closed stomata photosynthesis, protein content	Nazir et al. (2019)
Ni	Ni^{2+}	Basil	Cellular damage, oxidative stress, reduced chlorophyll content.	Georgiadou et al. (2018)

textile sectors are carcinogenic. According to the case study report published in Tribune on April 7, 2009, farmers in Bhatinda, Punjab, India, suffered from cancer because they irrigated their farmland with hazardous sludge from industries because there was not enough canal water (Kant 2012). Toxicological studies on PFOA have demonstrated adverse effects in CD-1 mice, including pregnancy loss, increased liver weight, and reduced weight gain in mothers, in addition to reduced odds of postnatal survival, delayed eye opening, and stunted newborn development. In a similar vein, studies on cynomolgus monkeys exposed to PFOS showed reduced levels of triiodothyronine (T3), increased liver weight, decreased body weight, and lowered levels of estrogen and cholesterol (Lau et al. 2006). High salinity in water can negatively affect non-salt-adapted microorganisms and interfere with aerobic treatment processes. Therefore, implementing effective removal strategies for these toxic pollutants before their release into water bodies is crucial for maintaining a hygienic and healthy environment (Ahmed et al. 2021a, b).With the rise in industrial development, the safe disposal of wastewater from industries has become a critical ecological challenge. Environmental degradation is now a global concern, and maintaining ecosystem health is a pressing issue for ecologists. Lack of appropriate wastewater disposal systems and treatment facilities has led to increasing pollution in water bodies, adversely affecting soil, aquatic environments, agriculture, flora, and fauna due to the presence of toxic and persistent chemicals. Discharging industrial effluents into freshwater bodies deteriorates water quality, which is vital for sustaining aquatic life, primary productivity, and the food chain (Gurunadha Rao et al. 2001). However, if industries and the safety of treated industrial effluent for the environment are guaranteed by pollution control boards, it could be repurposed for fish production, irrigation of non-edible cash crops, aquaculture, and other uses (Wong et al. 2001). Consequently, water-living vascular plants are being increasingly recognized for their potential in screening, conducting phytotoxicity studies of chemicals, and serving as effective bioindicators (USEPA 1996).

The floating, vascular cryptogam *Lemna minor* is especially remarkable because of its quick growth, straightforward structure (single flat, oval, leaf-like frond with a single root), simplicity of maintenance, vegetative reproduction, cost-effectiveness, and widespread distribution in freshwater bodies (Ziegler et al. 2023). These attributes make it an ideal candidate for laboratory toxicity testing compared to other aquatic plants. Toxicity testing and toxicological evaluation of sodium chloride using *L. minor* bioassays from tannery effluents have been conducted by Verma et al. (1999, 2001). Their studies found that when sodium chloride concentrations rise relative to controls, plant growth declines (Polla et al. 2022). The ecological impact of cadmium and linear alkyl benzene sulfonate, both individually and in combination, resulted in a significant reduction in the incorporation of 14C into proteins, phospholipids, DNA, and RNA (Singh et al. 1994). Furthermore, *L. minor* has high sensitivity to boron, causing noticeable drops in growth rates and an increase in the percentage of aberrant fronds, such as chlorotic, necrotic, and dead ones. Future toxicological problems are anticipated as a result of increased mining activity and pesticide use increasing the quantity of boron accessible biologically (Butterwick et al. 1989).

Additionally, exposure to 5 μg/l of triphenyltin, a widely used antifouling agent in paints, caused the amount of chlorophyll content to decrease, chloroplast activity, and reduced peroxidase and nitrate reductase activity (Song and Huang 2001). The toxicity of heavy metal cadmium to *L. minor* varies with environmental factors like temperature and pH (Chawla et al. 1991; Bergmann et al. 2000; Chowdhury et al. 2000). Key pollutants in the nitrogenous fertilizer business are lubricating oils, sulfuric acid, caustic soda, ammonia, urea, arsenic, phosphate, and fluoride, according to V.K. Singh and J. Singh. A recent study used *L. minor* bioassays to assess the impact of fertilizer factory wastewater and sewage wastewater from the Phoolpur industrial area of Allahabad, U.P., India (Singh and Singh 2006).

Heavy metals are defined as compounds with high atomic weight and density greater than 5 g/cm^3 (Zhang et al. 2019). While these metals are essential in trace amounts for different biological processes, they become toxic when their concentrations exceed safe limits. Cadmium (Cd), copper (Cu), lead (Pb), and zinc (Zn) are particularly concerning due to their environmental and health impacts. The main sources of heavy metals are mining, industrial processes, and agriculture; among the latter are fertilizers, pesticides, wastewater, and livestock dung (Li et al. 2013). Recently, the rapid increase in heavy metal pollution, particularly in agriculture, resulted in soil contamination and plant uptake, posing urgent environmental and health challenges. Plant root systems are the main site of contact for the metal ions that collect as heavy metals from the soil throughout ecosystems. One of the several transmission mechanisms for phytostabilization—which immobilizes metals in soil—and rhizofiltration—which uses plant roots to absorb hazardous substances—is phytoextraction. These metals can damage plants and subsequently enter the food chain, creating health risks to humans (Cho-Ruk et al. 2005; Tangahu et al. 2011). Since pesticides are artificial substances mostly employed in agriculture to control pests, integrated pest management (IPM) systems depend heavily on them (Damalas and Eleftherohorinos 2011). However, overuse of these leads to bioaccumulation in food chains, endangering non-target species, including mammals. Pesticide residues can persist in soil, air, and water and are considered prime environmental threats as they are carcinogenic in nature and have long-term persistence (Yadav et al. 2014). The health risks linked to toxicants are increasing because of the accumulation in the food chain and persistent environmental presence (Dórea 2008). Acute and chronic illnesses, including lung cancer, kidney failure, osteoporosis, and heart failure, can be brought on by toxins (Alissa and Ferns 2011). The study conducted by Tong et al. between 2003 and 2009 highlighted the dangers that heavy metal exposure brings to human health in China, emphasizing that the primary mode of exposure is ingestion (Osman et al 2019). Analogously, pesticides present noteworthy hazards via tainted food, water, and atmosphere, resulting in asthma, cancer, hypersensitivity, hormonal imbalances, congenital abnormalities, low birth weight, and fatality (Wickerham et al. 2012). It covers their classification, the ways they contaminate agricultural soils, and their effects on plant species, from infection to death. Along with examining the negative effects that these chemical pollutants have on human health individually as well as collectively, the research also examines the connections between pesticides and heavy metals in plants and soil. The methods by which these pollutants affect ecosystems and public health are clarified by this comprehensive review (Alengebawy et al. 2021).

4 Reduction in Crop Productivity

Many industries generate waste that contains trace metals, which can appear in both liquid and solid forms. Liquid wastes are frequently dumped straight into sewers, which causes metals to accumulate in sewage sludge that may then be utilized as compost or soil conditioners (Webber 1981). Since water generates the essential aquatic environment and supplies moisture, its quality matters greatly to seed germination. Seed germination is highly susceptible to pollutants, and significantly contaminated industrial effluents can also hinder the process. Despite this, wastewater from municipalities and industrial effluents like paper mills, breweries, distilleries, sugar mills, and dairies are frequently used to irrigate agricultural fields (Kushwaha 2013). Sludge from the tannery business, for example, has a high chromium content despite giving crops low-acting nitrogen (Guo et al. 2022). According to Jerath and Sahai (1982), the effluent from fertilizer factories, with high levels of concentrations in nitrogenous forms, negatively affected seed germination and seedling growth of maize. The increasing pollution of soils from persistent use of industrial effluents tainted with heavy metals represents a significant threat to global crop production and the environment. These metals accumulated in soils and plants can find their way into the food chain and cause various health and ecological problems (Kashem and Singh 1999; Stolt et al. 2006; Jamali et al. 2007). Among the most common toxic metals are arsenic, mercury, cadmium, lead, and uranium, with cadmium (Cd) being particularly harmful. One of the main pollutants in soil is cadmium, and excessive discharges from industrial processes have made soil pollution worse (Lawlor 2004). It is smoothly absorbed by plant roots and translocated throughout the plant (Benavides et al., 2005), leading to reduced growth and biomass yield due to altered physiological processes. Photosynthesis, leaf hydration status, and leaf rotting are all impacted by cadmium, which affects ATPase activity in the plasma membrane and alters lipid composition by increasing reactive oxygen species (ROS) production (Demirevska-Kepova et al. 2006). It also decreases antioxidant activity in various plant species (Sethy and Ghosh 2013), which can impact the expression of tolerance to phytotoxicity (Mobin and Khan 2006).

Studies have indicated that plants react more favorably to treated wastewater (TW) than to groundwater. No appreciable changes are made to the soil's chemical or physical characteristics by wastewater irrigation (Sou et al. 2013). Despite the fact that heavy metals are accumulated in the soil, plants that absorb wastewater usually only show very low levels of lead (Pb), nickel (Ni), and zinc (Zn) (Vongdala et al. 2018). Hayat et al. (2002) found that, as compared to groundwater, the effluent from oil refineries had greater concentrations of phosphate, calcium, magnesium, potassium, and sulfate, and it generated more wheat and mustard seeds than groundwater (Hayat et al. 2002). Narasimha Rao et al. (1992) assessed the quality of effluent from the paperboard industry and its impact on alluvial soil crops. They demonstrated that irrigation of rice and cotton on loamy to sandy loam soils with this wastewater poses no risks. However, irrigating tobacco and chilies with this wastewater led to low-quality products and lower crop yields (Murali et al. 2022). Baruah et al. (1993) investigated the effects of paper mill effluent on the germination and growth of rice (*Oryza sativa* L. var. Masuri) seedlings (Kamalakar et al. 1991; Vishnuvardhan and Thoti 2018). Their study revealed that higher concentrations of effluent inhibit both germination and seedling growth. Additionally, rice seeds from effluent-affected areas were less viable, and viable seeds showed delayed germination compared to seeds from uncontaminated areas (Wang 1991). Singh et al. (2002) evaluated the agro-potential of effluent from Century Pulp and Paper Mill on *Triticum aestivum* var. UP-2329 (wheat) crops grown in two different soil types with varying effluent concentrations. They found that applying diluted effluent enhanced the total biomass that plants produce (height, shoots, and roots), the amount of protein, carbs, and lipids in wheat grains, and the levels of chlorophyll. In contrast, undiluted effluent inhibited plant growth and sharply reduced yield (Table 2).

According to research by Sahai et al. (1983), rice (*Oryza sativa*) seedling development and germination were negatively impacted by a significant amount of distillery waste (over 25%), with roots being more vulnerable than shoots (Rahman et al. 2019). They attributed these effects to the high levels of various cations and anions in the effluent. Similar findings were made by Chaudhary (1983), who discovered that high pH levels and chromium were possible causes of the suppression of wheat (*Triticum aestivum*) and rice (*Oryza sativa*) germination, seedling development, and biochemical content at concentrations over 5% of the effluent from Bansbari Leather and Shoe Factory (Arshad et al. 2017; Gao et al. 2022).

Table 2 Effect of Industrial effluents on seedlings development

Species	Control		Treated		References
	Root length (in cm)	Shoot Length (in cm)	Root length (in cm)	Shoot Length (in cm)	
Oryza sativa	14.09 ± 0.56	17.20 ± 0.66	6.16 ± 0.303	5.451 ± 0.113	Vishnuvardhan and Thoti (2018)
Triticum vulgare	12.631 ± 0.31	9.86 ± 0.12	6.026 ± 0.23	3.191 ± 0.11	Vishnuvardhan and Thoti (2018)
Zea mays	21.84 ± 2.50	16.38 ± 2.74	13.00 ± 6.91	13.44 ± 2.80	Odutayo et al. (2016)
Allium cepa	4.70 ± 1.55	–	1.94 ± 0.52	–	Odutayo et al. (2016)

Average of Six replications, Mean ± S.D

The combined effects of wastewater from distilleries and mixed fertilizer factories on mung beans (*Phaseolus radiatus* Linn) were studied (Sahai & Neelam 1987). They discovered that plant growth was enhanced by effluents up to a 25% concentration, indicating that such combinations may be used for irrigation. However, Chaudhary et al. (1987) and Shrivastava and Mathur (1987) reported that high-level concentrations of industrial effluents suppressed growth, while lower concentrations promoted it in agricultural and vegetable crops (Zaidi et al. 2015).

Kumar et al. (1990) studied the effects of Saurashtra chemicals effluent, containing oil, grease, phenols, and various ions, on guar (*Cymopsis tetragonaloba* L.) and found that concentrations of 25% or higher of the effluent inhibited plant growth, while undiluted effluent completely prevented germination of the seeds (Mandal et al. 2022). As reported by Bahadur and Sharma (1990), industrial effluents significantly decreased the length of the shoot, seed germination proportions, and length of the root of *Triticum aestivum* L., with shoots considerably more affected than roots (Kanwal et al. 2020).

Maize seeds were more susceptible to changes in the quality of water than cowpea seeds, as determined by Biney (1991) in germination tests (Salian et al. 2018). Swaminathan and Vaidheswaran (1991) found that dyeing factory effluents favored seed germination, but higher concentrations (above 50%) hindered groundnut seedling growth. Mishra and Behera (1991) reported that increasing levels of effluents from the paper industry noticeably decreased water imbibing capacity, germination, and growth pigment content in rice seedlings (Biswas et al. 2017). Sharma and Naik (1991) studied the effects of steel mill effluent on *Abelmoschus esculentus* (L.) Moench, observing reductions in the percentage of germination and parameters for seedling growth. Patel and Kumar (1991) noted that increasing concentrations of pharmaceutical factory effluent led to decreased germination, seedling growth, and pod production in mustard (*Brassica juncea* L.) (Chowardhara et al. 2019). The effects of sewage pollution and industrial discharge on *Phaseolus mungo (=Vigna mungo)* germination and seedling development were studied by Ramanujan (1991). Chapagain (1991) found that polluted water from Dhobikhola, containing tannery and sewage waste, inhibited seed germination rate and growth of seedlings of vegetables, with toxicity varying by species. According to Goel and Kulkarni (1994), higher levels of industrial effluents hampered crop development and germination rates, whereas lower concentrations promoted these activities. According to Sharma (2000) and Pokhrel (2000), the heavy metals effect on Brassicaceae crops, finding significant impacts on germination and growth. Pokharel et al. (2000) informed that root growth *of Allium cepa* was negatively impacted by industrial effluents with 100% concentrated pollutants (Dutta et al. 2018). Pokhrel and Lekhak (2001) noted that in *Brassica juncea, Brassica oleracea var. botrytis, and Raphanus sativus*, lead (Pb) affected the rate of germination, seedling growth, and moisture content. Lead accumulation is higher in roots than in shoots (Soares et al. 2020). Acharya (2001) studied the effects of brewery effluents on crops, finding that higher concentrations suppressed germination of seed and seedling development (Ramana et al. 2002).

4.1 Effect of Effluents on Soil

Soil quality is crucial for vegetation, and the practice of discharging industrial effluents into aquatic systems and using contaminated water for irrigation can lower soil quality. According to Sharma et al. (1990), it has been estimated that effluent from the steel mill and its effects on the land, noting changes in the levels of soil nutrients as well as raised, high amounts of NO_2, NH_4, iron, and phenol, including decreases in calcium and magnesium and increases in total nitrogen, sulfur, and phosphate (Rajbala 2012) (Tables 3 and 4).

The steel mill effluent irrigation increased readily oxidized organic matter and sulfur in soil but significantly decreased total nitrogen (Chaudhary et al. 2023). Sah et al. (2000) found that effluents from paper mills and Bhrikuti Pulp, Gorkha Brewery Limited, and Shree Distillery are beneficial for irrigation because they contain high organic matter or phosphorus content and do not significantly affect soil salinity or alkalinity. With the exception of convertible potassium, which was decreased in irrigated soil, when comparing irrigated soil to non-irrigated soil,

Table 3 Recommended maximum concentration (mg/L) of heavy metals in agricultural soil

Heavy metal	Recommended maximum concentration (mg/L)
Fe	5.0
Mn	0.2
Zn	2.0
Cu	0.2
B	2.0
Cd	0.01
Cr	0.1
Ni	1.2

Source: FAO, 1985: Indian Standard, 1982

Table 4 Mean available concentration of heavy metals in contaminated and uncontaminated (Concentration: mg kg-1)

Metals in soil samples	Contaminated soil	Uncontaminated soil	References
Fe	22.80	4.02	Rahmani and Rezaei (2007)
Cu	2.67	1.07	Rahmani and Rezaei (2007)
Ni	40	28.11	Parizanganeh et al. (2012)
Zn	5.10	1.83	Rahmani and Rezaei (2007)
Cd	0.42	0.12	Rahmani and Rezaei (2007)
Co	8	18.53	Parizanganeh et al. (2012)
Pb	4.48	1.76	Rahmani and Rezaei (2007)
B	9.2	2	Lucho-Constantino et al. (2005)

Acharya (2001) discovered that brewery effluents increased most of the chemical factors in irrigated soil (Gorfie et al. 2022).

5 Biomagnification and Bioaccumulation

Naturally occurring heavy metal substances with densities greater than 5 g/cm^3 and atomic numbers higher than 20 pose significant environmental and health risks (Ahmed et al. 2021a, b). These metals are released into soils and surface waters through natural processes such as physical and chemical weathering as well as volcanic eruptions (Dixit et al. 2015). However, human activity significantly exacerbates their existence. Leachates from mining operations, landfills, urban runoff, and municipal wastewater and industrial effluent from the electroplating, metal-finishing, and electronics industries substantially increase the levels and alter the circulation in the environment of these heavy metals (Briffa et al. 2020).

The biota can quickly and widely collect heavy metals as a result of a variety of types of pollution (Zhang et al. 2023). For instance, extremely toxic metals that break down slowly are introduced by industrial effluent and pose serious risks to ecosystems and human health (Razzak et al. 2022). Metals like selenium (Se) and mercury (Hg) are particularly concerning because of their high mobility and propensity to be transformed and concentrated in the aquatic food chain (Córdoba-Tovar et al. 2021). Carcinogenic heavy metals like trivalent chromium [Cr(III)], divalent lead [Pb(II)], and pentavalent arsenic [As(V)] are notable inorganic pollutants that bind to sediments and act as secondary contaminants in aquatic ecosystems (Córdoba-Tovar et al. 2021; Dey et al. 2024).

In aquatic environments, metals are taken up by amphipod cuticles, permeable membranes of fish gills, and other sensitive tissues (Saidon et al. 2023). They accumulate in microscopic organisms like phytoplankton and zooplankton, which have high surface-to-volume ratios (Botterell et al. 2018). Heavy metals, which are neurotoxins, can adversely affect the neurological functions of fish, leading to malformations and biochemical alterations that impact general health and survival. These outcomes are frequently employed as gauges of the degree of environmental pollution (Kumar et al. 2024; Shahjahan et al. 2022; Yousif et al. 2021). Heavy metal distribution in the food chain is influenced by the trophic levels of organisms, with higher trophic levels accumulating elevated concentrations of contaminants (Zheng et al. 2023; Gao et al. 2021).

The build-up of potentially toxic elements (PTEs) in aquatic environments depends on their mobility in the water column and their absorption by suspended particles. This leads to increased concentrations of contaminants at higher trophic levels due to biomagnification, where pollutants biomagnify in the food chains (Suedel et al. 1994). Aquatic organisms are typically classified into four trophic levels: Level 0 (producers, such as aquatic plants), Level 1 (primary consumers, such as zooplankton), Level 2 (secondary consumers, e.g., forage fish), and Level 3 (tertiary consumers, e.g., predatory fish). Because of the pollutant concentration, biomagnification occurs and is conserved through the food chain, but the biomass decreases, leading to higher concentrations at higher levels (Saidon et al. 2023) (Fig. 3).

The rate of metal accumulation and removal varies by species, affecting their half-lives and influencing how these metals are transferred through the food chain. Persistent metals enter organisms directly or indirectly from the environment or through their diet, leading to biomagnification in top predators (Khan et al. 2023).

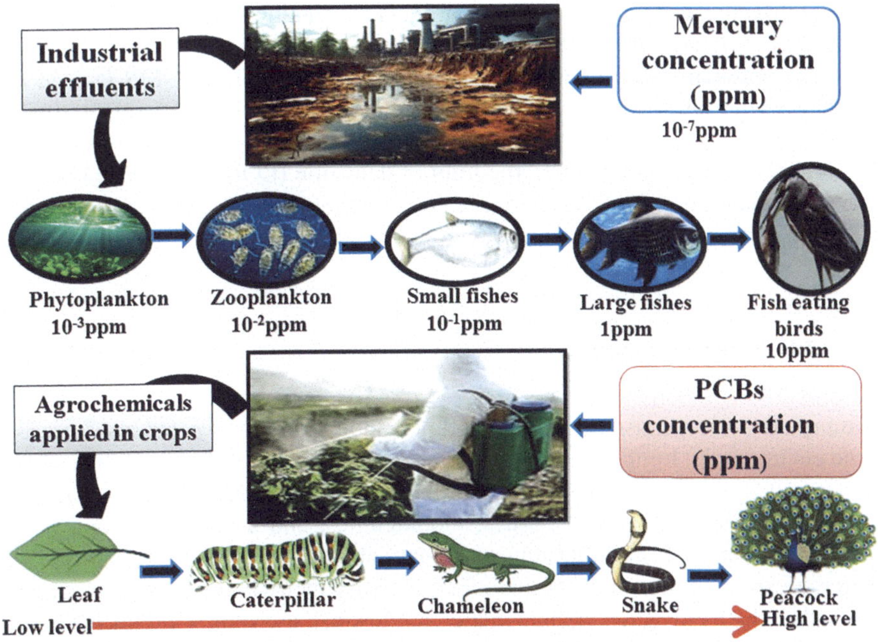

Fig. 3 Biomagnification and bioaccumulation due to industrial effluents

Despite numerous studies on the metal trophic magnification potential and metalloids, there is still little knowledge of precise mechanisms and variables determining the trophodynamics in complex ecosystems (Vijver and Peijnenburg 2011).

5.1 Quantifying Trophic Transfer of Heavy Metals

As a chemical goes up through the food chain, its concentration increases and reaches greater concentrations within predators than in their surroundings. This phenomenon is known as biomagnification (Drouillard 2008). To assess biomagnification and trophic transfer, researchers first identify the trophic levels of organisms in an ecosystem using methods like gut content analysis or stable isotope analysis. The trophic transfer factor (TTF) or trophic transfer coefficient (TTC) is then calculated to determine the trophic magnification factors of contaminants from lower to higher trophic levels (Suedel et al. 1994).

5.2 Elemental Analysis

Typically, analytical methods like inductively coupled plasma-mass spectrometry (ICP-MS) or inductively coupled plasma-optical emission spectroscopy (ICP-OES) are used to evaluate the amounts of contaminants in aquatic samples, which include sediments, fish, and arthropods (Djedjibegovic et al. 2011). ICP-OES is suited for analyzing pollutants with high levels of concentration, while ICP-MS is used for substances with lower concentrations and regulatory limits (Rabieh et al. 2020). To guarantee data accuracy, both approaches need validation using variables including the limit of quantification (LOQ), the limit of detection (LOD), and the calibration range. Atomic Fluorescence Spectrometry (AFS) is also used for analyzing mercury (Hg) and arsenic (As) concentrations (Gómez-Ariza et al. 2005).

5.3 Mercury Dynamics

Mercury (Hg) is a persistent, mobile, and poisonous metal that poses a severe risk to the environment. Long-distance transport of inorganic mercury is possible, and in aquatic settings, it may be methylated into methylmercury (Me-Hg), which is very poisonous, bioaccumulates in tissues, biomagnifies, and ascends the food chain (Al-Sulaiti et al. 2022). Sulfur concentrations, pH, dissolved oxygen, and organic matter are some of the variables that affect mercury methylation (Luo et al. 2022). Wetlands, which have high organic matter and low pH, are significant for Hg methylation (Liem-Nguyen et al. 2020).

The Cape Fear River (CFR) in North Carolina is a notable example of a region with high potential for Hg production and biomagnification. Historical sources of Hg contamination include coal-burning power plants and waste incineration, which have been reduced in recent years but continue to impact Hg levels in aquatic species (Kim et al. 2015). Due to their ideal biogeochemical circumstances, the wetland regions of the CFR contribute to significant rates of mercury methylation and bioaccumulation (Hardy et al. 2024; Missimer et al. 2024).

6 Impacts on Biodiversity

Around 850 million people lived on Earth during the start of the industrial period, coexisting with an astonishingly diverse array of species. By the twentieth century, it was evident that natural resources have a finite amount and that activity by humans was pushing these boundaries, which led to a decline in biodiversity (Western 2001). The rate of species extinction has increased, climate change has accelerated, and pollution is degrading the quality of our ecosystem and its water bodies (Eissa and Zaki 2011).

Biodiversity encompasses the variety of living organisms—including plants, animals, fungi, and microbes—on Earth. It encompasses species, genetic, and ecological diversity. Although in the past people believed that humans and nature were distinct, modern viewpoints acknowledge that humans are essential to biodiversity (Cepic et al. 2021). The population growth and technological innovation drive have resulted in increased biodiversity exploitation, threatening sustainable livelihoods and potentially leading to mass extinctions (Shivanna 2020).

In Nigeria, the biodiversity of aquatic ecosystems is under severe threat from pollution caused by various human activities. These include the excessive applications of agrochemicals like fertilizers and pesticides in agriculture, industrialization, urbanization, and the construction of infrastructure such as dams and roads. These operations release pollutants into water bodies, which degrade the water's quality and endanger aquatic life (Isukuru et al. 2024). The water crisis is a pressing issue, with only a small percentage of the Earth's water being fresh and available for use. Most freshwater is locked in ice or deep groundwater, making surface water from lakes, rivers, and rainfall crucial yet limited. Industrial activities and poor management have further exacerbated water scarcity, especially in arid regions (Musie and Gonfa 2023). Loss of biodiversity is a significant environmental challenge worldwide. The rise in biodiversity loss rate is largely driven by climate change and anthropogenic pressures, including chemical pollution (Muluneh 2021). Pollutants released into the environment pose a major threat to the health of living organisms and ecosystems. The process of biomagnification in the food chain can impair human health and disrupt ecosystem services, such as food and water supply, climate regulation, and disease control (Mustafa et al. 2024).

To address these issues, holistic approaches like One Health and Eco-Health have emerged. These frameworks acknowledge the interconnectedness of humans, animals, and ecosystems, suggesting that protecting ecosystems is essential for safeguarding human health (Marselle et al. 2021). In high-income countries, regulations like the European Union's Water Framework Directive aim to improve water quality and monitor contaminants. Tropical areas, on the other hand, suffer many difficulties since they have the highest biodiversity but also the fewest resources. These areas have extensive chemical pollution from mining and other businesses like agriculture, which exacerbates the consequences of climate change (Bolan et al. 2023).

It is imperative to do further study on the consequences of chemical contamination on tropical ecosystems. Most deaths from pollution occur in middle-class and lower-class nations, which cannot frequently address these problems. Creating successful mitigation measures requires an understanding of how chemical pollutants impact ecosystem services and biodiversity in these areas (Tuomisto et al. 2017).

To sum up, the relationship between pollution, biodiversity loss, and human activity emphasizes the need for ecological emergency management and study. Protecting ecosystems and understanding the full impacts of environmental pollution are crucial steps toward ensuring a sustainable future for both biodiversity and human health.

7 Indirect Effect on Food Availability Due to Industrial Effluents

Rapid urbanization and industrialization are essential for both economic expansion and human well-being. However, they have also emerged as major sources of water pollution that pose serious risks to environmental stability and the health of ecosystems (Acheampong and Opoku 2023). The increasing rate of resource loss, especially from industrial effluents, emphasizes how urgently this problem must be solved. Industrial effluents are the major contributor to water pollution, with their impact evidenced by the degradation of water quality and the depletion of drinkable water sources (Jahan and Singh 2023; Saravanakumar et al. 2022). Water resources are under extreme stress due to the improper treatment and release of such effluents into rivers, as well as the decreased quantity and quality of groundwater. To protect water availability and quality, it is imperative to apply sustainable treatment strategies for industrial effluents (Singh et al. 2023; Al-Tohamy et al. 2022).

Effluents from industrial activities can be classified into three main types based on the manufacturing processes involved:

(i) *Inorganic Process Waste*: Generated from chemical industries.
(ii) *Organic Process Waste*: Originated from industries like textiles, food processing, dairies, breweries, and certain chemical industries.
(iii) *Chemical Waste*: Include leftovers from the production of fertilizer, pesticides, dyes, and raw materials.

Of them, the complicated and problematic character of organic effluents poses major remediation issues (Sathya et al. 2022; Kamali et al. 2019a, b). Certain industrial effluents are hazardous and are not biodegradable, having permanent physical-chemical characteristics. These effluents often exhibit a diverse array of concentrations for standard parameters. For instance, their pH can vary from 5 to 12, and they can show high levels of BOD (100–3000 mgO_2/L) and chemical oxygen demand (COD, 10–2250 mgO_2/L) (Ayeche 2012). The diverse composition of these effluents makes them difficult to treat and poses serious risks to both aquatic and terrestrial ecosystems (Bashir et al. 2020; Mishra et al. 2023; Al-Tohamy et al. 2022; Naidoo and Olaniran 2013). The creation of long-term strategies for the pre-treatment of industrial effluents has been the focus of national and international governmental and non-governmental organizations. A great deal of research and analysis has focused on the harmful properties of polluted water and how they affect all living things and have global consequences (Saravanan et al. 2021).

This study highlights the sources and impacts of toxic industrial effluents, their role in spreading diseases, and their detrimental effects on air and water ecosystems. Furthermore, it investigates the possibility that cleaning industrial effluent using micropore membrane filters based on MOFs might be a helpful way to lessen the threats that industrial pollution poses to the environment and human health (Dey et al. 2023).

8 Direct and Indirect Impact on Ecosystem and Food Chain

The effects of industrial effluents on the food chain can be both direct and indirect, influencing various organisms and trophic levels within ecosystems (Guerra 2005; Saaristo et al. 2018). Here's a breakdown of how industrial effluents affect the food chain:

8.1 Direct Impacts on the Food Chain

8.1.1 Direct Toxicity to Primary Producers

Industrial effluents can directly impact primary producers, i.e., phytoplankton in aquatic ecosystems or plants in terrestrial ecosystems. Pollutants like heavy metals, pesticides, and nutrients [such as nitrogen (N) and phosphorus (P)] can suppress photosynthesis, lowered development, and disrupt the reproductive cycles of these organisms. This directly affects the availability of food for herbivores and other organisms higher up in the food chain (Alengebawy et al. 2021).

8.1.2 Bioaccumulation in Herbivores

Herbivores that feed on contaminated plants or algae can accumulate pollutants in their tissues through bioaccumulation. These pollutants are like heavy metals, persistent organic pollutants (POPs), and other toxic chemicals. Over time, when herbivores ingest polluted food sources, the amount of contaminants in their systems may rise (Latorre-Padilla et al. 2021).

8.1.3 Impact on Predators and Top Predators

Industrial effluents have the potential to directly impact predators and dominant predators in the food chain through processes of bioaccumulation and biomagnification. Due to a process known as biomagnification, predators that eat contaminated herbivores or other smaller prey species may absorb more pollutants into their systems. This can lead to health issues, reduced reproductive success, and even population declines in sensitive species (Miller et al. 2020).

8.2 *Indirect Impacts on the Food Chain*

8.2.1 Disruption of Trophic Dynamics

Industrial effluents can disrupt trophic dynamics within ecosystems. For example, pollution-induced declines in primary producers can lead to food shortages for herbivores and subsequently impact predators that consume them. This can disrupt the balance of predator-prey relationships and affect the overall stability of food webs (Ali et al. 2023; Afrad et al. 2020).

8.2.2 Loss of Biodiversity

Pollution from industrial wastewater can lead to habitat deterioration and biodiversity loss. Decreases in sensitive species cause changes in the composition of the food web. This can reduce the resilience of ecosystems, and they are capable of adapting to environmental changes (Feckler et al. 2023; Sigmund et al. 2023).

8.2.3 Impacts on Human Health

Industrial pollutants that enter the food chain can pose risks to human health when consumed. For example, human consumers may be in danger from fish that has been tainted by industrial sources of mercury or other heavy metals. In a similar vein, crops raised in contaminated soils may collect toxic materials that compromise the security and safety of food (Mitra et al. 2022; Rai et al. 2019).

8.2.4 Economic Implications

Pollution's effects on the food chain can have economic consequences, particularly for industries dependent on fisheries, agriculture, and tourism. Food source contamination can damage customer confidence, lower a product's market value, and cause financial losses for the communities that rely on these resources.

9 Long-Term Consequences

Industrial effluents can have long-term environmental consequences that persist well beyond the initial discharge occurrences. These repercussions can affect ecosystems, biodiversity, human health, and socio-economic stability. Several notable long-term environmental consequences of industrial effluents are as follows:

9.1 Soil Contamination

Persistent discharges of industrial effluents have heavy metals, pesticides, solvents, and other chemicals that can lead to soil contamination. These pollutants may persist in soils for years, impacting soil fertility, microbial communities, and plant health. Ingesting contaminated foods or grazing on contaminated grounds can put human and animal health at risk, and contaminated soils can reduce agricultural production (Alengebawy et al. 2021; Neethu et al. 2020).

9.2 Water Pollution and Aquatic Ecosystems

Continuous discharge of effluents into water bodies can lead to long-term water pollution. Pollutants such as heavy metals, nutrients, organic chemicals, and thermal pollution can degrade water quality, disrupt aquatic ecosystems, and harm aquatic organisms. Fish and other aquatic creatures can collect persistent pollutants, which can be harmful to human health and have an impact on fisheries and leisure activities (Bashir et al. 2020; Mishra et al. 2023).

9.3 Eutrophication

Industrial effluents containing excessive nutrients like nitrogen and phosphorus can contribute to the eutrophication of aquatic environments. This process leads to excessive algae growth, depletes oxygen levels in water, and creates "dead zones" where the survival of aquatic life becomes a major challenge. Eutrophication alters ecosystem dynamics, reduces biodiversity, and impairs the ecological and economic value of affected water bodies (Bhagowati and Ahamad 2018).

9.4 Loss of Biodiversity

Loss of biodiversity, fragmentation of habitats, and habitat degradation can all be caused by industrial pollution. Direct impacts such as habitat degradation from contamination and indirect effects like disruptions to food chains and ecosystem services can lead to declines in species populations and loss of biodiversity at local and regional scales. Reduced biodiversity affects ecosystem resilience and the capacity of ecosystems to adapt to environmental changes (Rafferty 2024; Adla et al. 2022).

9.5 Human Health Risks

Long-term exposure to contaminants from industrial effluents can pose significant risks to human health. Bioaccumulation and biomagnification of toxic substances in the ecosystem food chain can lead to chronic health effects such as cancer, immune system suppression, neurological disorders, and reproductive issues. Communities living near industrial sites or downstream from discharge points may be particularly vulnerable to health impacts (Jaishankar et al. 2014; Alijagic et al. 2024).

9.6 Climate Change Impacts

Some industrial activities contribute to greenhouse gas emissions and climate change. For example, combustion processes in industries release carbon dioxide (CO2), hydrofluorocarbons (HFCs), and other strong greenhouse gases released by industrial refrigerants and solvents. Climate change can exacerbate environmental stresses on ecosystems already impacted by industrial pollution, leading to further declines in biodiversity and ecosystem health (Afifa et al. 2024; Ofremu et al. 2024).

9.7 Legacy Pollution

Even after industrial activities cease or regulations improve, legacy pollution from historical discharges can persist in the environment. Contaminants can remain in soils, sediments, and groundwater for decades or longer, continuing to pose risks to ecosystems and human health. Legacy contamination site remediation is an expensive and difficult process that calls for long-term management and monitoring plans (Padhye et al. 2023).

Comprehensive strategies that place a high priority on pollution prevention, the adoption of environmentally friendly manufacturing technologies, strict enforcement of regulations, the rehabilitation of degraded natural environments, and involvement of the public in environmental management are needed to address the long-term environmental effects of industrial effluents (Awewomom et al. 2024). In order to reduce the current and potential effects of industrial activity on the environment and on people, sustainable development strategies that limit negative effects on the environment and encourage resource efficiency are crucial (Mensah 2019).

10 Remedies

Heavy metal pollution from industrial activities, including mining, metal processing, and nuclear operations, poses significant environmental and health risks due to their toxicity and potential for bioaccumulation (Tchounwou et al. 2012; Briffa et al. 2020). For low concentrations, conventional removal techniques like ion exchange and evaporation are expensive and ineffective, producing troublesome sludge (Rajasulochana and Preethy 2016; Vidu et al. 2020).

10.1 Method for the Removal of Heavy Metals

There are various cost-effective and eco-friendly technologies for the removal of heavy metals from polluted water.

10.1.1 Biosorption

Biosorption utilizes biological materials to remove heavy metal ions from wastewater. It is an energy-independent process where metals bind to the surfaces or active sites of biosorbents (Xie 2024), (Karnwal 2024). Biosorption takes advantage of the inherent qualities of biosorbents such as bacteria, algae, fungi, and plant materials and is frequently more cost-effective than traditional techniques (Abdi and Kazemi 2015).

10.1.2 Biosorbents

High biosorption capability, advantageous physical characteristics, and simplicity in solution separation are necessary for effective biosorbents (Okoro et al. 2022; Elgarahy et al. 2021). They include various biological materials such as bacteria, algae, fungi, and agricultural residues. Several factors influence biosorption efficiency:

- *pH*: Affects ionization of both the metals and biosorbent functional groups.
- *Temperature*: Affects the resilience and efficiency of biosorption.
- *Biosorbent Dosage*: Greater doses often result in more efficiency but may also cost more.
- *Retention Time*: Typically, longer times enhance removal efficiency.
- *Functional Groups*: Essential for metal ion binding (Vijayaraghavan and Yun 2008).

10.1.3 Adsorption Isotherms

Adsorption isotherms, which show the correlation between the concentration of heavy metals in the solution and the quantity adsorbed on the biosorbent, characterize equilibrium in biosorption (Febrianto et al. 2008). These isotherms help understand the biosorbent's capacity and surface properties (Pathirana et al. 2019).

10.1.4 Biosorption Mechanism

The biosorption mechanism involves complex interactions between metal ions and biosorbent surfaces. The process is influenced by the structure of the biosorbent's cell wall and the chemistry of metal complexes (Javanbakht et al. 2013). The majority of biosorption research is done in lab settings. Scaling up is necessary for industrial uses. For example, *Pleurotus* waste mushroom compost that has been treated under certain circumstances (Chia-Chay et al. 2010). A significant cause of contamination with heavy metals that affects soil, water, and the air, industrial effluents have detrimental effects on aquatic life and ecosystems (Jaiswal et al. 2018). Heavy metals, particularly non-essential ones like Cd (cadmium), Hg (mercury), and Pb (lead), are highly toxic and persist in the environment due to their non-biodegradable nature (Tchounwou et al. 2012). In contrast, metals like zinc (Zn) and copper (Cu) are essential in trace amounts for metabolic processes but can be harmful at elevated concentrations (Jomova et al. 2022).

10.1.5 Role of Endophytes

Microorganisms known as endophytes reside within plant tissues without posing a threat. They are diverse and have the potential for bioremediation due to their ability to produce enzymes and secondary metabolites that can detoxify heavy metals (Chaudhary et al. 2022; Burragoni and Jeon 2021). Endophytic bacteria are beneficial for the restoration and bioremediation of the environment because they can increase the growth and development of plants and enhance stress tolerance (De-Bashan et al. 2011).

10.1.6 Bioremediation Techniques

Utilizing biological processes to purify contaminated surroundings is known as bioremediation. It makes use of plants' and microbes' capacities to change or eliminate toxins (Imam et al. 2021; Saeed et al. 2023). Techniques include:

- *Microbial Fuel Cells (MFCs)*: Reduce metal resistance.
- *Biofilm-mediated Bioremediation*: Detoxify heavy metal-polluted environments.
- *Endophyte-mediated remediation*: Increases metal ion mobilization and pollution uptake by plants (Sharma and Kumar 2021).

10.1.7 Traditional Methods

Traditional physical and chemical methods for treating industrial effluents are costly and often ineffective at low metal concentrations. Bioremediation offers a more sustainable and eco-friendly alternative (Saravanan et al. 2022). However, issues like leachate from landfills and soil contamination require new initiatives for waste management and pollution control (Dagwar and Dutta 2024).

10.1.8 Endophyte Benefits

Endophytes improve plant resistance to environmental stressors and enhance nutrient uptake and metal absorption (García-Latorre et al. 2021). They produce growth-promoting hormones and bioactive compounds that can help remediate metal contamination (Gond et al. 2021). Additionally, endophytes help to lessen the negative effects of heavy metals on soil health and plant development (Mukherjee et al. 2021).

10.1.9 Metabolites of Endophytes

Endophytes produce different beneficial metabolic compounds and enzymes that can cleanse pollutants and improve plant health. These include compounds with medicinal potential, contributing to their value in bioremediation and ecological restoration (Sharma and Kumar 2021; Rana et al. 2019).

10.2 Strategies for the Removal of Heavy Metals

Overcoming the detrimental effects of industrial effluents necessitates a multidimensional approach involving prevention, treatment, regulation, and restoration. Here are some key remedies and strategies to mitigate and overcome the impact of industrial effluents:

10.2.1 Pollution Prevention at the Source

- *Cleaner Production Practices*: Promote the use of cleaner production techniques and technology by industries to reduce pollution and waste output (Santos et al. 2021).
- *Substitution of Hazardous Substances*: To lessen the impact of industrial activities on the environment, swap out hazardous chemicals for safer substitutes (Syeda et al. 2021).
- *Water Conservation*: Implement water-saving technologies and practices to reduce the volume of wastewater generated (Kulkarni 2011).

10.2.2 Effluent Treatment and Control

- *Wastewater Treatment*: Companies must detoxify their effluents to remove contaminants before releasing them into the environment. Treatment technologies may include physical, chemical, and biological methods tailored to specific contaminants (Sonune and Ghate 2004; Osman 2014).
- *Advanced Treatment Technologies*: Invest in advanced treatment technologies like membrane filtration, activated carbon adsorption, and oxidation processes to effectively remove pollutants (Sathya et al. 2022; Adane et al. 2021).
- *Effluent Quality Monitoring*: Establish routine monitoring procedures to spot developing contaminants and make sure discharge regulations are being followed (Ahmad et al. 2008).

10.2.3 Regulatory Measures and Compliance

- *Stringent Regulations*: Implement stringent environmental laws and standards for industrial discharges, including limits on pollutant concentrations and discharge volumes (Earnhart and Rassier 2016).
- *Permitting and Monitoring*: Require companies to get permits before releasing wastewater and to conduct routine evaluations of the quality of their effluent and its impact on the environment (Morris et al. 2016).
- *Penalties for Non-compliance*: Impose penalties and fines for industries that violate environmental regulations, incentivizing compliance with discharge limits (Mokhtar et al. 2024).

10.2.4 Environmental Management and Restoration

- *Ecosystem Restoration*: Restore damaged ecosystems impacted by industrial pollution by implementing programs to restore wetlands, reforest, and habitat (Mitsch and Mander 1997).
- *Contaminated Site Remediation*: Utilize methods including soil cleaning, bioremediation, and phytoremediation to eliminate or reduce pollutants from polluted areas (Sinha et al. 2007).
- *Long-term Monitoring*: Create long-term monitoring procedures to evaluate cleanup operations' efficacy and guarantee ecological recovery (Borja et al. 2010; Lohner and Dixon 2013).

10.2.5 Public Awareness and Stakeholder Engagement

- *Education and Outreach*: Encourage responsible environmental stewardship by educating the public on the effects of industrial effluents on the environment and human health (Ellerbusch et al. 2006).
- *Community Engagement*: Involve local communities, NGOs, and stakeholders in decision-making processes related to industrial pollution control and remediation efforts (Goven et al. 2012).
- *Citizen Science*: Encourage citizen science initiatives to track water quality and environmental health, fostering public participation in environmental monitoring and protection (McKinley et al. 2016).

10.2.6 International Cooperation and Technology Transfer

- *Technology Transfer*: To enhance industrial pollution control and management, facilitate the transfer of clean technology and best practices from developed to economically developing nations (Zhang et al. 2014).

- *International Agreements*: Support and comply with international agreements and conventions aimed at reducing industrial pollution and promoting sustainable development goals (Letchumanan and Kodama 2000).

11 Future Challenges

In the future, addressing industrial wastewater cleanup will require negotiating several changing obstacles:

- *Emerging Contaminants*: The emergence of new materials, including microplastics, complicates wastewater treatment, etc. Treatment techniques that are novel or cutting-edge are often required for these contaminants.
- *Regulatory Changes*: As environmental regulations tighten, industries will need to adapt their treatment processes to meet new standards, potentially requiring significant updates to existing technologies.
- *Resource Efficiency*: Future difficulties include finding equilibrium between the costs of treatment and resource efficiency. This includes creating technologies that not only successfully remove pollutants but also reclaim precious assets from effluents.
- *Climate Change*: Extreme weather events increase the volume and unpredictability of industrial effluents and have an impact on the efficacy of current treatment solutions.
- *Technological Innovation*: Keeping up with rapidly evolving treatment technologies, such as nanotechnology and bioremediation, will be crucial. The challenge is integrating these innovations effectively and economically.
- *Public and Investor Pressure*: Investors and the public are calling for more open and sustainable practices. Businesses must ensure that their wastewater treatment is efficient and compliant with regulations pertaining to social and environmental responsibility.
- *Complex Waste Streams*: The growing complexity of waste streams with diverse contaminants will require more sophisticated and adaptable treatment processes.
- *Aging Infrastructure*: A lot of institutions are still using antiquated infrastructure, which raises the possibility of system failures and makes it more difficult to implement new treatment methods.
- *Data Management*: Large-scale data management and analysis pertaining to wastewater treatment procedures is becoming more and more important. Optimizing these processes through data use is essential for ensuring regulatory compliance.
- *Global Standards*: For multinational companies, aligning treatment practices with varying international standards and regulations presents a challenge, as they must navigate diverse requirements across different regions. Addressing these challenges will necessitate continuous research, technological advancement, and flexible management strategies.

12 Conclusion

The impacts on plants and ecosystems due to industrial effluents are both profound and complex, highlighting the critical need for effective remediation and management strategies. Many different types of contaminants, including heavy metals, dangerous compounds, and excess nutrients, are frequently seen in industrial wastewaters. These pollutants have the ability to seriously impair natural systems. Exposure to these contaminants can have detrimental effects on plants. Contaminants can stifle growth, damage root systems, impair photosynthesis, and even lead to plant mortality. This not only affects the health of individual plants but can also reduce agricultural productivity and alter the composition of plant communities. Such perturbations may impact the relationships between pollinators and plants as well as the larger food chain, with cascading impacts across the ecosystem. The overall state of ecosystems is also affected. Contaminants from industrial effluents can degrade soil quality, harm aquatic environments, and disrupt the balance of flora and fauna. Pollutants have the ability to leak into soil and rivers, harming these vital resources over time. Aquatic ecosystems are particularly vulnerable, as pollutants can reduce oxygen levels in water and alter nutrient balances. This can lead to detrimental phenomena such as algal blooms, which further deplete oxygen and create dead zones where most aquatic life cannot survive. Furthermore, the buildup of poisons in the food chain can have an impact on a variety of animals, including big predators and microbes, and eventually pose health hazards to humans. To address these impacts effectively, a diversified strategy is needed. Stringent regulations must be enforced to control and limit the types and quantities of pollutants that can be released. Advanced treatment technologies need to be adopted to ensure that industrial effluents are adequately cleaned before they are discharged into the environment. This includes innovations in filtration, chemical treatment, and bioremediation techniques that can target specific contaminants more effectively. Moreover, proactive pollution management is essential. This involves not only the implementation of effective treatment solutions but also the adoption of sustainable industrial practices that minimize waste production and reduce environmental impact. Companies must be encouraged to invest in green technologies and practices that support environmental stewardship. Ultimately, industries, regulators, and communities must work together to protect plant health and maintain the integrity of ecosystems. We can mitigate the negative impacts of industrial pollutants, safeguard natural systems, and advance ecological balance by placing a high priority on efficient effluent treatment and sustainable practices. Ensuring the resilience of these systems is crucial for maintaining the health and well-being of all living organisms and securing a sustainable future for our planet.

References

Abdi, O. & Kazemi, Mosstafa. (2015). A review study of biosorption of heavy metals and comparison between different biosorbents. Journal of Materials and Environmental Science 6. 1386-1399.

Acheampong, A. O., & Opoku, E. E. O. (2023). Environmental degradation and economic growth: Investigating linkages and potential pathways. Energy Economics, 123, 106734. https://doi.org/10.1016/j.eneco.2023.106734

Adane, T., Adugna, A. T., & Alemayehu, E. (2021). Textile industry effluent treatment techniques. Journal of Chemistry, 2021, 1–14. https://doi.org/10.1155/2021/5314404

Adla, K., Dejan, K., Neira, D., & Dragana, Š. (2022). Degradation of ecosystems and loss of ecosystem services. In Elsevier eBooks (pp. 281–327). https://doi.org/10.1016/b978-0-12-822794-7.00008-3

Afifa, N., Arshad, K., Hussain, N., Ashraf, M. H., & Saleem, M. Z. (2024). Air pollution and climate change as grand challenges to sustainability. The Science of the Total Environment, 928, 172370. https://doi.org/10.1016/j.scitotenv.2024.172370

Afrad, M. S. I., Monir, M. B., Haque, M. E., Barau, A. A., & Haque, M. M. (2020). Impact of industrial effluent on water, soil and Rice production in Bangladesh: a case of Turag River Bank. Journal of Environmental Health Science and Engineering, 18(2), 825–834. https://doi.org/10.1007/s40201-020-00506-8

Afroz, R., & Rahman, A. (2017). Health impact of river water pollution in Malaysia. International Journal of ADVANCED AND APPLIED SCIENCES, 4(5), 78–85. https://doi.org/10.21833/ijaas.2017.05.014

Ahmad, M., Bajahlan, A. S., & Hammad, W. S. (2008). Industrial effluent quality, pollution monitoring and environmental management. Environmental Monitoring and Assessment, 147(1–3), 297–306. https://doi.org/10.1007/s10661-007-0121-5

Ahmed, J., Thakur, A., & Goyal, A. (2021a). Industrial wastewater and its toxic effects. In The Royal Society of Chemistry eBooks (pp. 1–14). https://doi.org/10.1039/9781839165399-00001

Ahmed, S. F., Kumar, P. S., Rozbu, M. R., Chowdhury, A. T., Nuzhat, S., Rafa, N., Mahlia, T., Ong, H. C., & Mofijur, M. (2021b). Heavy metal toxicity, sources, and remediation techniques for contaminated water and soil. Environmental Technology & Innovation, 25, 102114. https://doi.org/10.1016/j.eti.2021.102114

Ahmed, T., Noman, M., Manzoor, N., Shahid, M., Abdullah, M., Ali, L., Wang, G., Hashem, A., Al-Arjani, A. F., Alqarawi, A. A., Abd_Allah, E. F., & Li, B. (2020). Nanoparticle-based amelioration of drought stress and cadmium toxicity in rice via triggering the stress responsive genetic mechanisms and nutrient acquisition. Ecotoxicology and Environmental Safety, 209, 111829. https://doi.org/10.1016/j.ecoenv.2020.111829

Akhtar, N., Ishak, M. I. S., Bhawani, S. A., & Umar, K. (2021). Various natural and anthropogenic factors responsible for water quality degradation: a review. Water, 13(19), 2660. https://doi.org/10.3390/w13192660

Alengebawy, A., Abdelkhalek, S. T., Qureshi, S. R., & Wang, M. (2021). Heavy Metals and pesticides Toxicity in agricultural soil and plants: ecological risks and human health implications. Toxics, 9(3), 42. https://doi.org/10.3390/toxics9030042

Ali, N., Khan, M. H., Ali, M., Sidra, N., Ahmad, S., Khan, A., Nabi, G., Ali, F., Bououdina, M., & Kyzas, G. Z. (2023). Insight into microplastics in the aquatic ecosystem: Properties, sources, threats and mitigation strategies. The Science of the Total Environment, 913, 169489. https://doi.org/10.1016/j.scitotenv.2023.169489

Alijagic, A., Suljević, D., Fočak, M., Sulejmanović, J., Šehović, E., Särndahl, E., & Engwall, M. (2024). The triple exposure nexus of microplastic particles, plastic-associated chemicals, and environmental pollutants from a human health perspective. Environment International, 188, 108736. https://doi.org/10.1016/j.envint.2024.108736

Alissa, E. M., & Ferns, G. A. (2011). Heavy metal poisoning and cardiovascular disease. Journal of Toxicology, 2011, 1–21. https://doi.org/10.1155/2011/870125

Al-Sulaiti, M. M., Soubra, L., & Al-Ghouti, M. A. (2022). The Causes and effects of mercury and methylmercury contamination in the marine environment: a review. Current Pollution Reports, 8(3), 249–272. https://doi.org/10.1007/s40726-022-00226-7

Al-Tohamy, R., Ali, S. S., Li, F., Okasha, K. M., Mahmoud, Y. A., Elsamahy, T., Jiao, H., Fu, Y., & Sun, J. (2022). A critical review on the treatment of dye-containing wastewater: Ecotoxicological and health concerns of textile dyes and possible remediation approaches for environmental safety. Ecotoxicology and Environmental Safety, 231, 113160. https://doi.org/10.1016/j.ecoenv.2021.113160

Angon, P. B., Islam, M. S., Kc, S., Das, A., Anjum, N., Poudel, A., & Suchi, S. A. (2024). Sources, effects and present perspectives of heavy metals contamination: Soil, plants and human food chain. Heliyon, 10(7), e28357. https://doi.org/10.1016/j.heliyon.2024.e28357

Ariyadasa, S., Taylor, W., Weaver, L., McGill, E., Billington, C., & Pattis, I. (2023). Nonbacterial Microflora in Wastewater Treatment Plants: an Underappreciated Potential Source of Pathogens. Microbiology Spectrum, 11(3). https://doi.org/10.1128/spectrum.00481-23

Arshad, M., Khan, A. H. A., Hussain, I., Badar-Uz-Zaman, N., Anees, M., Iqbal, M., Soja, G., Linde, C., & Yousaf, S. (2017). The reduction of chromium (VI) phytotoxicity and phytoavailability to wheat (Triticum aestivum L.) using biochar and bacteria. Applied Soil Ecology, 114, 90–98. https://doi.org/10.1016/j.apsoil.2017.02.021

Awewomom, J., Dzeble, F., Takyi, Y. D., Ashie, W. B., Ettey, E. N. Y. O., Afua, P. E., Sackey, L. N. A., Opoku, F., & Akoto, O. (2024). Addressing global environmental pollution using environmental control techniques: a focus on environmental policy and preventive environmental management. Discover Environment, 2(1). https://doi.org/10.1007/s44274-024-00033-5

Ayeche, R. (2012). Treatment by Coagulation-Flocculation of Dairy Wastewater with the Residual Lime of National Algerian Industrial Gases Company (NIGC-Annaba). Energy Procedia, 18, 147–156. https://doi.org/10.1016/j.egypro.2012.05.026

Aziz, K. H. H., Mustafa, F. S., Omer, K. M., Hama, S., Hamarawf, R. F., & Rahman, K. O. (2023). Heavy metal pollution in the aquatic environment: efficient and low-cost removal approaches to eliminate their toxicity: a review. RSC Advances, 13(26), 17595–17610. https://doi.org/10.1039/d3ra00723e

Bashir, I., Lone, F. A., Bhat, R. A., Mir, S. A., Dar, Z. A., & Dar, S. A. (2020). Concerns and threats of contamination on aquatic ecosystems. In Springer eBooks (pp. 1–26). https://doi.org/10.1007/978-3-030-35691-0_1

Behera, M., Nayak, J., Banerjee, S., Chakrabortty, S., & Tripathy, S. K. (2021). A review on the treatment of textile industry waste effluents towards the development of efficient mitigation strategy: An integrated system design approach. Journal of Environmental Chemical Engineering, 9(4), 105277. https://doi.org/10.1016/j.jece.2021.105277

Berehanu, B., Lemma, B., & Tekle-Giorgis, Y. (2015). Chemical composition of industrial effluents and their effect on the survival of fish and eutrophication of Lake Hawassa, Southern Ethiopia. Journal of Environmental Protection, 06(08), 792–803. https://doi.org/10.4236/jep.2015.68072

Bergmann, B., Cheng, J., Classen, J., & Stomp, A. (2000). In vitro selection of duckweed geographical isolates for potential use in swine lagoon effluent renovation. Bioresource Technology, 73(1), 13–20. https://doi.org/10.1016/s0960-8524(99)00137-6

Bhagowati, B., & Ahamad, K. U. (2018). A review on lake eutrophication dynamics and recent developments in lake modeling. Ecohydrology & Hydrobiology, 19(1), 155–166. https://doi.org/10.1016/j.ecohyd.2018.03.002

Bilal, M., Shah, J. A., Ashfaq, T., Gardazi, S. M. H., Tahir, A. A., Pervez, A., Haroon, H., & Mahmood, Q. (2013). Waste biomass adsorbents for copper removal from industrial wastewater—A review. Journal of Hazardous Materials, 263, 322–333. https://doi.org/10.1016/j.jhazmat.2013.07.071

Biswas, S., Bhattacharya, A., & Basak, P. (2017). Effect of Tannery Effluent on Germination and Early Seedling Growth of Oryza sativa Var. IET-4786. In Springer eBooks (pp. 243–251). https://doi.org/10.1007/978-981-10-5349-8_23

Blum, W. E. (2013). Soil and Land Resources for Agricultural Production: General Trends and Future Scenarios-A Worldwide Perspective. International Soil and Water Conservation Research, 1(3), 1–14. https://doi.org/10.1016/s2095-6339(15)30026-5

Bolan, S., Padhye, L. P., Jasemizad, T., Govarthanan, M., Karmegam, N., Wijesekara, H., Amarasiri, D., Hou, D., Zhou, P., Biswal, B. K., Balasubramanian, R., Wang, H., Siddique, K. H., Rinklebe, J., Kirkham, M., & Bolan, N. (2023). Impacts of climate change on the fate of contaminants through extreme weather events. The Science of the Total Environment, 909, 168388. https://doi.org/10.1016/j.scitotenv.2023.168388

Borja, Á., Dauer, D. M., Elliott, M., & Simenstad, C. A. (2010). Medium- and long-term recovery of estuarine and coastal ecosystems: patterns, rates and restoration effectiveness. Estuaries and Coasts, 33(6), 1249–1260. https://doi.org/10.1007/s12237-010-9347-5

Botterell, Z. L., Beaumont, N., Dorrington, T., Steinke, M., Thompson, R. C., & Lindeque, P. K. (2018). Bioavailability and effects of microplastics on marine zooplankton: A review. Environmental Pollution, 245, 98–110. https://doi.org/10.1016/j.envpol.2018.10.065

Briffa, J., Sinagra, E., & Blundell, R. (2020). Heavy metal pollution in the environment and their toxicological effects on humans. Heliyon, 6(9), e04691. https://doi.org/10.1016/j.heliyon.2020.e04691

Burragoni, S. G., & Jeon, J. (2021). Applications of endophytic microbes in agriculture, biotechnology, medicine, and beyond. Microbiological Research, 245, 126691. https://doi.org/10.1016/j.micres.2020.126691

Butterwick, L., De Oude, N., & Raymond, K. (1989). Safety assessment of boron in aquatic and terrestrial environments. Ecotoxicology and Environmental Safety, 17(3), 339–371. https://doi.org/10.1016/0147-6513(89)90055-9

Cepic, M., Bechtold, U., & Wilfing, H. (2021). Modelling human influences on biodiversity at a global scale–A human ecology perspective. Ecological Modelling, 465, 109854. https://doi.org/10.1016/j.ecolmodel.2021.109854

Chalaris, M., Gkika, D. A., Tolkou, A. K., & Kyzas, G. Z. (2023). Advancements and sustainable strategies for the treatment and management of wastewaters from metallurgical industries: an overview. Environmental Science and Pollution Research, 30(57), 119627–119653. https://doi.org/10.1007/s11356-023-30891-0

Chaney, R. (1991). The heavy elements: chemistry, environmental impact, and health effects. Journal of Environmental Quality, 20(4), 876. https://doi.org/10.2134/jeq1991.00472425002000040028x

Chaudhary, P., Agri, U., Chaudhary, A., Kumar, A., & Kumar, G. (2022). Endophytes and their potential in biotic stress management and crop production. Frontiers in Microbiology, 13. https://doi.org/10.3389/fmicb.2022.933017

Chaudhary, S., Sindhu, S. S., Dhanker, R., & Kumari, A. (2023). Microbes-mediated sulphur cycling in soil: Impact on soil fertility, crop production and environmental sustainability. Microbiological Research, 271, 127340. https://doi.org/10.1016/j.micres.2023.127340

Chawla, G., Singh, J., & Viswanathan, P. N. (1991). Effect of pH and temperature on the uptake of cadmium by Lemna minor L. Bulletin of Environmental Contamination and Toxicology, 47(1), 84–90. https://doi.org/10.1007/bf01689457

Chia-Chay, T., Redzwan, G., Hong-Hooi, L., Soon-Kong, Y., Surif, S., & Abdul-Talib, S. (2010). Copper (II) biosorption characteristic of pleurotus spent mushroom compost. International Conference on Science and Social Research (CSSR 2010), Kuala Lumpur, Malaysia, 2010, Pp. 6-10. https://doi.org/10.1109/cssr.2010.5773873

Cho-Ruk, K., Kurukote, J., Supprung, P., & Vetayasupo, S. (2005). Perennial plants in the phytoremediation of lead-contaminated soils. Biotechnology (Faisalabad), 5(1), 1–4. https://doi.org/10.3923/biotech.2006.1.4

Chowardhara, B., Borgohain, P., Saha, B., Awasthi, J. P., Moulick, D., & Panda, S. K. (2019). Phytotoxicity of Cd and Zn on three popular Indian mustard varieties during germination and early seedling growth. Biocatalysis and Agricultural Biotechnology, 21, 101349. https://doi.org/10.1016/j.bcab.2019.101349

Chowdhury, S. A., Sultana, N., Huque, K. S., & Huque, Q. M. E. (2000). Manure Based Duckweed Production in Shallow Sink : Effect of Genera on Biomass and Nutrient Yield of Duckweed under the Same Nutritional and Management Conditions. Asian-Australasian Journal of Animal Sciences, 13(5), 686–693. https://doi.org/10.5713/ajas.2000.686

Coffey, R., Paul, M. J., Stamp, J., Hamilton, A., & Johnson, T. (2018). A review of water quality responses to air temperature and precipitation changes 2: Nutrients, algal blooms, sediment, pathogens. JAWRA Journal of the American Water Resources Association, 55(4), 844–868. https://doi.org/10.1111/1752-1688.12711

Córdoba-Tovar, L., Marrugo-Negrete, J., Barón, P. R., & Díez, S. (2021). Drivers of biomagnification of Hg, As and Se in aquatic food webs: A review. Environmental Research, 204, 112226. https://doi.org/10.1016/j.envres.2021.112226

Dagwar, P. P., & Dutta, D. (2024). Landfill leachate a potential challenge towards sustainable environmental management. The Science of the Total Environment, 926, 171668. https://doi.org/10.1016/j.scitotenv.2024.171668

Damalas, C. A., & Eleftherohorinos, I. G. (2011). Pesticide exposure, safety issues, and risk assessment indicators. International Journal of Environmental Research and Public Health, 8(5), 1402–1419. https://doi.org/10.3390/ijerph8051402

Das, N., & Chandran, P. (2010). Microbial Degradation of Petroleum Hydrocarbon Contaminants: An Overview. Biotechnology Research International, 2011, 1–13. https://doi.org/10.4061/2011/941810

De Medeiros, A. D. M., Da Silva, C. J. G., Junior, De Amorim, J. D. P., Durval, I. J. B., De Santana Costa, A. F., & Sarubbo, L. A. (2022). Oily wastewater treatment: methods, challenges, and trends. Processes, 10(4), 743. https://doi.org/10.3390/pr10040743

De-Bashan, L. E., Hernandez, J., & Bashan, Y. (2011). The potential contribution of plant growth-promoting bacteria to reduce environmental degradation—A comprehensive evaluation. Applied Soil Ecology, 61, 171–189. https://doi.org/10.1016/j.apsoil.2011.09.003

Demirevska-Kepova, K., Simova-Stoilova, L., Stoyanova, Z. P., & Feller, U. (2006). Cadmium stress in Barley: growth, leaf pigment, and protein composition and detoxification of reactive oxygen species. Journal of Plant Nutrition, 29(3), 451–468. https://doi.org/10.1080/01904160500524951

Dey, S., Rajak, P., & Sen, K. (2024). Bioaccumulation of metals and metalloids in seafood: A comprehensive overview of mobilization, interactive effects in eutrophic environments, and implications for public health risks. Journal of Trace Elements and Minerals, 100141. https://doi.org/10.1016/j.jtemin.2024.100141

Dey, T. K., Hou, J., Sillanpää, M., & Pramanik, B. K. (2023). Metal-organic framework membrane for waterborne micro/nanoplastics treatment. Chemical Engineering Journal, 474, 145715. https://doi.org/10.1016/j.cej.2023.145715

Dixit, R., Wasiullah, N., Malaviya, D., Pandiyan, K., Singh, U., Sahu, A., Shukla, R., Singh, B., Rai, J., Sharma, P., Lade, H., & Paul, D. (2015). Bioremediation of Heavy Metals from Soil and Aquatic Environment: An Overview of Principles and Criteria of Fundamental Processes. Sustainability, 7(2), 2189–2212. https://doi.org/10.3390/su7022189

Djedjibegovic, J., Larssen, T., Skrbo, A., Marjanović, A., & Sober, M. (2011). Contents of cadmium, copper, mercury and lead in fish from the Neretva river (Bosnia and Herzegovina) determined by inductively coupled plasma mass spectrometry (ICP-MS). Food Chemistry, 131(2), 469–476. https://doi.org/10.1016/j.foodchem.2011.09.009

Dórea, J. G. (2008). Persistent, bioaccumulative and toxic substances in fish: Human health considerations. The Science of the Total Environment, 400(1–3), 93–114. https://doi.org/10.1016/j.scitotenv.2008.06.017

Drouillard, K. (2008). Biomagnification. In Elsevier eBooks (pp. 441–448). https://doi.org/10.1016/b978-008045405-4.00377-3

Du Plessis, A. (2022). Persistent degradation: Global water quality challenges and required actions. One Earth, 5(2), 129–131. https://doi.org/10.1016/j.oneear.2022.01.005

Dutta, J., Ahmad, A., & Singh, J. (2018). Study of industrial effluents induced genotoxicity on Allium cepa L. Caryologia, 71(2), 139–145. https://doi.org/10.1080/00087114.2018.1447631

Earnhart, D., & Rassier, D. G. (2016). "Effective regulatory stringency" and firms' profitability: the effects of effluent limits and government monitoring. Journal of Regulatory Economics, 50(2), 111–145. https://doi.org/10.1007/s11149-016-9304-8

Eissa, A. E., & Zaki, M. M. (2011). The impact of global climatic changes on the aquatic environment. Procedia Environmental Sciences, 4, 251–259. https://doi.org/10.1016/j.proenv.2011.03.030

Elgarahy, A., Elwakeel, K., Mohammad, S., & Elshoubaky, G. (2021). A critical review of biosorption of dyes, heavy metals and metalloids from wastewater as an efficient and green process. Cleaner Engineering and Technology, 4, 100209. https://doi.org/10.1016/j.clet.2021.100209

Ellerbusch, F., Gute, D. M., Desmarais, A. M., & Woodin, M. (2006). Community engagement as a component of revitalization: Lessons learned from the Technical Outreach Services to Communities programme. Local Environment, 11(5), 515–535. https://doi.org/10.1080/13549830600853437

Evans, A. E., Mateo-Sagasta, J., Qadir, M., Boelee, E., & Ippolito, A. (2019). Agricultural water pollution: key knowledge gaps and research needs. Current Opinion in Environmental Sustainability, 36, 20–27. https://doi.org/10.1016/j.cosust.2018.10.003

Fazal, S., Zhang, B., & Mehmood, Q. (2015). Biological treatment of combined industrial wastewater. Ecological Engineering, 84, 551–558. https://doi.org/10.1016/j.ecoleng.2015.09.014

Febrianto, J., Kosasih, A. N., Sunarso, J., Ju, Y., Indraswati, N., & Ismadji, S. (2008). Equilibrium and kinetic studies in adsorption of heavy metals using biosorbent: A summary of recent studies. Journal of Hazardous Materials, 162(2–3), 616–645. https://doi.org/10.1016/j.jhazmat.2008.06.042

Feckler, A., Wolfram, J., Schulz, R., & Bundschuh, M. (2023). Reducing pollution to levels not harming biodiversity and ecosystem functions: A perspective on the post-2020 Global Biodiversity Framework. Current Opinion in Environmental Science & Health, 35, 100495. https://doi.org/10.1016/j.coesh.2023.100495

Folkers, C., & Gunter, L. P. (2022). Radioactive releases from the nuclear power sector and implications for child health. BMJ Paediatrics Open, 6(1), e001326. https://doi.org/10.1136/bmjpo-2021-001326

Gao, Y., Duan, Z., Zhang, L., Sun, D., & Li, X. (2022). The Status and Research Progress of Cadmium Pollution in Rice- (Oryza sativa L.) and Wheat- (Triticum aestivum L.) Cropping Systems in China: A Critical Review. Toxics, 10(12), 794. https://doi.org/10.3390/toxics10120794

Gao, Y., Wang, R., Li, Y., Ding, X., Jiang, Y., Feng, J., & Zhu, L. (2021). Trophic transfer of heavy metals in the marine food web based on tissue residuals. The Science of the Total Environment, 772, 145064. https://doi.org/10.1016/j.scitotenv.2021.145064

García-Latorre, C., Rodrigo, S., & Santamaría, O. (2021). Endophytes as plant nutrient Uptake-Promoter in plants. In Sustainable development and biodiversity (pp. 247–265). https://doi.org/10.1007/978-3-030-65447-4_11

Gavrilescu, M. (2021). Water, soil, and plants interactions in a threatened environment. Water, 13(19), 2746. https://doi.org/10.3390/w13192746

Gebre-Mariam, Z., & Desta, Z. (2002). The chemical composition of the effluent from Awassa Textile Factory and its effects on aquatic biota. SINET Ethiopian Journal of Science, 25(2). https://doi.org/10.4314/sinet.v25i2.18084

Genchi, G., Sinicropi, M. S., Lauria, G., Carocci, A., & Catalano, A. (2020). The effects of cadmium toxicity. International Journal of Environmental Research and Public Health, 17(11), 3782. https://doi.org/10.3390/ijerph17113782

Georgiadou, E. C., Kowalska, E., Patla, K., Kulbat, K., Smolińska, B., Leszczyńska, J., & Fotopoulos, V. (2018). Influence of Heavy Metals (Ni, Cu, and Zn) on Nitro-Oxidative Stress Responses, Proteome Regulation and Allergen Production in Basil (Ocimum basilicum L.) Plants. Frontiers in Plant Science, 9. https://doi.org/10.3389/fpls.2018.00862

Ghosh, S., Sinha, J. K., Ghosh, S., Vashisth, K., Han, S., & Bhaskar, R. (2023). Microplastics as an emerging threat to the global environment and human health. Sustainability, 15(14), 10821. https://doi.org/10.3390/su151410821

Gómez-Ariza, J. L., Lorenzo, F., & García-Barrera, T. (2005). Comparative study of atomic fluorescence spectroscopy and inductively coupled plasma mass spectrometry for mercury and arsenic multispeciation. Analytical and Bioanalytical Chemistry, 382(2), 485–492. https://doi.org/10.1007/s00216-005-3094-7

Gond, D. P., Jha, S. S., Kumar, A., & Singh, S. K. (2021). Plant growth promoting bacteria and its role in green remediation. In Elsevier eBooks (pp. 149–163). https://doi.org/10.1016/b978-0-12-823828-8.00007-4

Gorfie, B. N., Tuhar, A. W., Keraga, A. S., & Woldeyohannes, A. B. (2022). Effect of brewery wastewater irrigation on soil characteristics and lettuce (Lactuca sativa) crop in Ethiopia. Agricultural Water Management, 269, 107633. https://doi.org/10.1016/j.agwat.2022.107633

Goven, J., Langer, E., Baker, V., Ataria, J., & Leckie, A. (2012). Community engagement in the management of biosolids: Lessons from four New Zealand studies. Journal of Environmental Management, 103, 154–164. https://doi.org/10.1016/j.jenvman.2012.02.007

Guerra, R. (2005). WATER ANALYSIS | Industrial Effluents. In Elsevier eBooks (pp. 289–299). https://doi.org/10.1016/b0-12-369397-7/00111-4

Guo, S., Yu, C., Zhao, X., Chen, Y., Wang, J., Su, M., Yang, X., & Yang, J. (2022). The chromium migration risk from tannery sludge into shallow soil and groundwater: Influence factors, modeling, and microbial response. Journal of Cleaner Production, 374, 133776. https://doi.org/10.1016/j.jclepro.2022.133776

Gurunadha Rao, V.V.S., Dhar, R.L. & Subrahmanyam, K. Assessment of Contaminant Migration in Groundwater from an Industrial Development Area, Medak District, Andhra Pradesh, India. Water, Air, & Soil Pollution 128, 369–389 (2001). https://doi.org/10.1023/A:1010307026457

Haider, F. U., Liqun, C., Coulter, J. A., Cheema, S. A., Wu, J., Zhang, R., Wenjun, M., & Farooq, M. (2021). Cadmium toxicity in plants: Impacts and remediation strategies. Ecotoxicology and Environmental Safety, 211, 111887. https://doi.org/10.1016/j.ecoenv.2020.111887

Hamid Reza Rahmani and Mosleheddin Rezaei, 2007. Effect of Industrial Effluents of Zob-Ahan on Soil, Water and Vegetable Plants. Journal of Applied Sciences, 7: 2519-2522. https://doi.org/10.3923/jas.2007.2519.2522

Harada, M. (1995). Minamata Disease: Methylmercury poisoning in Japan caused by environmental pollution. Critical Reviews in Toxicology, 25(1), 1–24. https://doi.org/10.3109/10408449509089885

Hardy, A., Skrabal, S. A., Addison, L., & Emslie, S. D. (2024). Biomagnification of mercury in an estuarine food web. Marine Pollution Bulletin, 205, 116604. https://doi.org/10.1016/j.marpolbul.2024.116604

Hassan, M. U., Nawaz, M., Mahmood, A., Shah, A. A., Shah, A. N., Muhammad, F., Batool, M., Rasheed, A., Jaremko, M., Abdelsalam, N. R., Hasan, M. E., & Qari, S. H. (2022). The role of zinc to mitigate heavy metals toxicity in crops. Frontiers in Environmental Science, 10. https://doi.org/10.3389/fenvs.2022.990223

Hayat, S., Ahmad, I., Azam, Z., & Samiullah. (2002). Longterm effect of oil refinery wastewater on crops yield, heavy metal accumulation in soil and crop produce. ResearchGate. https://www.researchgate.net/publication/279648847

Honda, M., & Suzuki, N. (2020). Toxicities of polycyclic aromatic hydrocarbons for aquatic animals. International Journal of Environmental Research and Public Health, 17(4), 1363. https://doi.org/10.3390/ijerph17041363; https://enviro2.doe.gov.my/ekmc/wp-content/uploads/2020/04/INDUSTRIAL-EFFLUENT

Huang, Y. F., Ang, S. Y., Lee, K. M., & Lee, T. S. (2015). Quality of water resources in Malaysia. In InTech eBooks. https://doi.org/10.5772/58969

Imam, A., Kanaujia, P. K., Ray, A., & Suman, S. K. (2021). Removal of Petroleum Contaminants Through Bioremediation with Integrated Concepts of Resource Recovery: A Review. Indian Journal of Microbiology, 61(3), 250–261. https://doi.org/10.1007/s12088-021-00928-4

Isukuru, E. J., Opha, J. O., Isaiah, O. W., Orovwighose, B., & Emmanuel, S. S. (2024). Nigeria's water crisis: Abundant water, polluted reality. Cleaner Water, 2, 100026. https://doi.org/10.1016/j.clwat.2024.100026

Jahan, N. S., & Singh, A. (2023). Causes and impact of industrial effluents on receiving water bodies: a review. Malaysian Journal of Science and Advanced Technology, 111–121. https://doi.org/10.56532/mjsat.v3i2.144

Jaishankar, M., Tseten, T., Anbalagan, N., Mathew, B. B., & Beeregowda, K. N. (2014). Toxicity, mechanism and health effects of some heavy metals. Interdisciplinary Toxicology, 7(2), 60–72. https://doi.org/10.2478/intox-2014-0009

Jaiswal, A., Verma, A., & Jaiswal, P. (2018). Detrimental effects of heavy metals in soil, plants, and aquatic ecosystems and in humans. Journal of Environmental Pathology Toxicology and Oncology, 37(3), 183–197. https://doi.org/10.1615/jenvironpatholtoxicoloncol.2018025348

Jamali, M. K., Kazi, T. G., Arain, M. B., Afridi, H. I., Jalbani, N., & Memon, A. R. (2007). Heavy Metal Contents of Vegetables Grown in Soil, Irrigated with Mixtures of Wastewater and Sewage Sludge in Pakistan, using Ultrasonic-Assisted Pseudo-digestion. Journal of Agronomy and Crop Science, 193(3), 218–228. https://doi.org/10.1111/j.1439-037x.2007.00261.x

Javanbakht, V., Alavi, S. A., & Zilouei, H. (2013). Mechanisms of heavy metal removal using microorganisms as biosorbent. Water Science & Technology, 69(9), 1775–1787. https://doi.org/10.2166/wst.2013.718

Jomova, K., Makova, M., Alomar, S. Y., Alwasel, S. H., Nepovimova, E., Kuca, K., Rhodes, C. J., & Valko, M. (2022). Essential metals in health and disease. Chemico-Biological Interactions, 367, 110173. https://doi.org/10.1016/j.cbi.2022.110173

Kamalakar, J. A., Sharma, D., & Melkania, U. (1991). Effect of paper and pulp factory effluents on the growth and development of maize and sunflower. International Journal of Environmental Studies, 39(1–2), 55–63. https://doi.org/10.1080/00207239108710681

Kamali, M., Costa, M. E., Aminabhavi, T. M., & Capela, I. (2019a). Sustainability of treatment technologies for industrial biowastes effluents. Chemical Engineering Journal, 370, 1511–1521. https://doi.org/10.1016/j.cej.2019.04.010

Kamali, M., Suhas, D., Costa, M. E., Capela, I., & Aminabhavi, T. M. (2019b). Sustainability considerations in membrane-based technologies for industrial effluents treatment. Chemical Engineering Journal, 368, 474–494. https://doi.org/10.1016/j.cej.2019.02.075

Kant, R. (2012). Textile dyeing industry an environmental hazard. Natural Science, 04(01), 22–26. https://doi.org/10.4236/ns.2012.41004

Kanwal, A., Farhan, M., Sharif, F., Hayyat, M. U., Shahzad, L., & Ghafoor, G. Z. (2020). Effect of industrial wastewater on wheat germination, growth, yield, nutrients and bioaccumulation of lead. Scientific Reports, 10(1). https://doi.org/10.1038/s41598-020-68208-7

Karnwal, A. (2024). Unveiling the promise of biosorption for heavy metal removal from water sources. Desalination and Water Treatment, 319, 100523. https://doi.org/10.1016/j.dwt.2024.100523

Kashem, M.A., Singh, B.R. Heavy Metal Contamination of Soil and Vegetation in the Vicinity of Industries in Bangladesh. Water, Air, & Soil Pollution 115, 347–361 (1999). https://doi.org/10.1023/A:1005193207319

Khan, B. N., Ullah, H., Ashfaq, Y., Hussain, N., Atique, U., Aziz, T., Alharbi, M., Albekairi, T. H., & Alasmari, A. F. (2023). Elucidating the effects of heavy metals contamination on vital organ of fish and migratory birds found at fresh water ecosystem. Heliyon, 9(11), e20968. https://doi.org/10.1016/j.heliyon.2023.e20968

Kim, K., Kabir, E., & Jahan, S. A. (2015). A review on the distribution of Hg in the environment and its human health impacts. Journal of Hazardous Materials, 306, 376–385. https://doi.org/10.1016/j.jhazmat.2015.11.031

Kónya, J., & Nagy, N. M. (2018). Nuclear energy production. In Elsevier eBooks (pp. 159–185). https://doi.org/10.1016/b978-0-12-813643-0.00007-x

Kothari, V., Vij, S., Sharma, S., & Gupta, N. (2020). Correlation of various water quality parameters and water quality index of districts of Uttarakhand. Environmental and Sustainability Indicators, 9, 100093. https://doi.org/10.1016/j.indic.2020.100093

Kozlov, M. V. (2022). Population dynamics of herbivorous insects in polluted landscapes. Current Opinion in Insect Science, 54, 100987. https://doi.org/10.1016/j.cois.2022.100987

Kulkarni, S.. (2011). Innovative Technologies for Water Saving in Irrigated Agriculture. International Journal of Water Resources and Arid Environments. 1. 226-231. https://www.researchgate.net/publication/229050067

Kumar, M., Singh, S., Jain, A., Yadav, S., Dubey, A., & Trivedi, S. P. (2024). A review on heavy metal-induced toxicity in fishes: Bioaccumulation, antioxidant defense system, histopathological manifestations, and transcriptional profiling of genes. Journal of Trace Elements in Medicine and Biology, 83, 127377. https://doi.org/10.1016/j.jtemb.2023.127377

Kumar, R., Qureshi, M., Vishwakarma, D. K., Al-Ansari, N., Kuriqi, A., Elbeltagi, A., & Saraswat, A. (2022). A review on emerging water contaminants and the application of sustainable removal technologies. Case Studies in Chemical and Environmental Engineering, 6, 100219. https://doi.org/10.1016/j.cscee.2022.100219

Kushwaha, J. P. (2013). A review on sugar industry wastewater: sources, treatment technologies, and reuse. Desalination and Water Treatment, 53(2), 309–318. https://doi.org/10.1080/19443994.2013.838526

Latorre-Padilla, N., Meynard, A., Oyarzun, F. X., & Contreras-Porcia, L. (2021). Ingestion of contaminated kelps by the herbivore Tetrapygus Niger: Negative effects on food intake, growth, fertility, and early development. Marine Pollution Bulletin, 167, 112365. https://doi.org/10.1016/j.marpolbul.2021.112365

Lau, C., Thibodeaux, J. R., Hanson, R. G., Narotsky, M. G., Rogers, J. M., Lindstrom, A. B., & Strynar, M. J. (2006). Effects of Perfluorooctanoic Acid Exposure during Pregnancy in the Mouse. Toxicological Sciences, 90(2), 510–518. https://doi.org/10.1093/toxsci/kfj105

Lawlor, D. W. (2004). Mengel, K. and Kirkby, E. A. Principles of plant nutrition. Annals of Botany, 93(4), 479–480. https://doi.org/10.1093/aob/mch063

Letchumanan, R., & Kodama, F. (2000). Reconciling the conflict between the `pollution-haven' hypothesis and an emerging trajectory of international technology transfer. Research Policy, 29(1), 59–79. https://doi.org/10.1016/s0048-7333(99)00033-5.

Li, Z., Ma, Z., Van Der Kuijp, T. J., Yuan, Z., & Huang, L. (2013). A review of soil heavy metal pollution from mines in China: Pollution and health risk assessment. The Science of the Total Environment, 468–469, 843–853. https://doi.org/10.1016/j.scitotenv.2013.08.090

Liem-Nguyen, V., Skyllberg, U., & Björn, E. (2020). Methylmercury formation in boreal wetlands in relation to chemical speciation of mercury(II) and concentration of low molecular mass thiols. The Science of the Total Environment, 755, 142666. https://doi.org/10.1016/j.scitotenv.2020.142666

Lindholm-Lehto, P. C., Knuutinen, J. S., Ahkola, H. S. J., & Herve, S. H. (2015). Refractory organic pollutants and toxicity in pulp and paper mill wastewaters. Environmental Science and Pollution Research, 22(9), 6473–6499. https://doi.org/10.1007/s11356-015-4163-x

Lohner, T. W., & Dixon, D. A. (2013). The value of long-term environmental monitoring programs: an Ohio River case study. Environmental Monitoring and Assessment, 185(11), 9385–9396. https://doi.org/10.1007/s10661-013-3258-4

Lokhande, R. S., Singare, P. U., & Pimple, D. S. (2012). Study on Physico-Chemical Parameters of Waste Water Effluents from Taloja Industrial Area of Mumbai, India. International Journal of Ecosystem, 1(1), 1–9. https://doi.org/10.5923/j.ije.20110101.01

Lucho-Constantino, C. A., Prieto-García, F., Del Razo, L. M., Rodríguez-Vázquez, R., & Poggi-Varaldo, H. M. (2005). Chemical fractionation of boron and heavy metals in soils irrigated with wastewater in central Mexico. Agriculture Ecosystems & Environment, 108(1), 57–71. https://doi.org/10.1016/j.agee.2004.12.013

Luo, H., Cheng, Q., He, D., Sun, J., Li, J., & Pan, X. (2022). Recent advances in microbial mercury methylation: A review on methylation habitat, methylator, mechanism, and influencing factor. Process Safety and Environmental Protection, 170, 286–296. https://doi.org/10.1016/j.psep.2022.12.007

Magat, W. A., & Viscusi, W. K. (1990). Effectiveness of the EPA's regulatory enforcement: the case of industrial effluent standards. The Journal of Law and Economics, 33(2), 331–360. https://doi.org/10.1086/467208

Mallin, M. A. (2009). Effect of human land development on water quality. In Elsevier eBooks (pp. 67–94). https://doi.org/10.1016/b978-0-12-374192-9.00004-2

Mandal, S., Hwang, S., & Shi, S. Q. (2022). Guar gum, a low-cost sustainable biopolymer, for wastewater treatment: A review. International Journal of Biological Macromolecules, 226, 368–382. https://doi.org/10.1016/j.ijbiomac.2022.12.039

Maron, J. L., & Crone, E. (2006). Herbivory: effects on plant abundance, distribution and population growth. Proceedings of the Royal Society B Biological Sciences, 273(1601), 2575–2584. https://doi.org/10.1098/rspb.2006.3587

Marselle, M. R., Hartig, T., Cox, D. T., De Bell, S., Knapp, S., Lindley, S., Triguero-Mas, M., Böhning-Gaese, K., Braubach, M., Cook, P. A., De Vries, S., Heintz-Buschart, A., Hofmann, M., Irvine, K. N., Kabisch, N., Kolek, F., Kraemer, R., Markevych, I., Martens, D., ... Bonn, A. (2021). Pathways linking biodiversity to human health: A conceptual framework. Environment International, 150, 106420. https://doi.org/10.1016/j.envint.2021.106420

Mashhadikhan, S., Amooghin, A. E., Sanaeepur, H., & Shirazi, M. M. A. (2022). A critical review on cadmium recovery from wastewater towards environmental sustainability. Desalination, 535, 115815. https://doi.org/10.1016/j.desal.2022.115815

Matthews, K. R. (2023). Agricultural water. In Elsevier eBooks (pp. 67–77). https://doi.org/10.1016/b978-0-12-819524-6.00006-9

McKinley, D. C., Miller-Rushing, A. J., Ballard, H. L., Bonney, R., Brown, H., Cook-Patton, S. C., Evans, D. M., French, R. A., Parrish, J. K., Phillips, T. B., Ryan, S. F., Shanley, L. A., Shirk, J. L., Stepenuck, K. F., Weltzin, J. F., Wiggins, A., Boyle, O. D., Briggs, R. D., Chapin, S. F., ... Soukup, M. A. (2016). Citizen science can improve conservation science, natural resource management, and environmental protection. Biological Conservation, 208, 15–28. https://doi.org/10.1016/j.biocon.2016.05.015

Mensah, J. (2019). Sustainable development: Meaning, history, principles, pillars, and implications for human action: Literature review. Cogent Social Sciences, 5(1). https://doi.org/10.1080/23311886.2019.1653531

Miller, M. E., Hamann, M., & Kroon, F. J. (2020). Bioaccumulation and biomagnification of microplastics in marine organisms: A review and meta-analysis of current data. PLoS ONE, 15(10), e0240792. https://doi.org/10.1371/journal.pone.0240792

Mishra, R. K., Mentha, S. S., Misra, Y., & Dwivedi, N. (2023). Emerging pollutants of severe environmental concern in water and wastewater: A comprehensive review on current developments and future research. Water-Energy Nexus, 6, 74–95. https://doi.org/10.1016/j.wen.2023.08.002

Missimer, T. M., MacDonald, J. H., Tsegaye, S., Thomas, S., Teaf, C. M., Covert, D., & Kassis, Z. R. (2024). Natural Background and the Anthropogenic Enrichment of Mercury in the Southern Florida Environment: A Review with a Discussion on Public Health. International Journal of Environmental Research and Public Health, 21(1), 118. https://doi.org/10.3390/ijerph21010118

Mitra, S., Chakraborty, A. J., Tareq, A. M., Emran, T. B., Nainu, F., Khusro, A., Idris, A. M., Khandaker, M. U., Osman, H., Alhumaydhi, F. A., & Simal-Gandara, J. (2022). Impact of heavy metals on the environment and human health: Novel therapeutic insights to counter the toxicity. Journal of King Saud University - Science, 34(3), 101865. https://doi.org/10.1016/j.jksus.2022.101865

Mitsch, W., & Mander, Ü. (1997). Remediation of ecosystems damaged by environmental contamination: applications of ecological engineering and ecosystem restoration in central and eastern Europe. Ecological Engineering, 8(4), 247–254. https://doi.org/10.1016/s0925-8574(97)00021-9

Mobin, M., & Khan, N. A. (2006). Photosynthetic activity, pigment composition and antioxidative response of two mustard (Brassica juncea) cultivars differing in photosynthetic capacity subjected to cadmium stress. Journal of Plant Physiology, 164(5), 601–610. https://doi.org/10.1016/j.jplph.2006.03.003

Mokhtar, Z., Kenway, S., Pikaar, I., & Mat Nashir, I. (2024). Challenges for Compliance with Industrial Effluent Regulations—the Industry Perspective. Preprints. https://doi.org/10.20944/preprints202404.1793.v1

Morris, L., Colombo, V., Hassell, K., Kellar, C., Leahy, P., Long, S. M., Myers, J. H., & Pettigrove, V. (2016). Municipal wastewater effluent licensing: A global perspective and recommendations for best practice. The Science of the Total Environment, 580, 1327–1339. https://doi.org/10.1016/j.scitotenv.2016.12.096

Mukherjee, A., Bhowmick, S., Yadav, S., Rashid, M. M., Chouhan, G. K., Vaishya, J. K., & Verma, J. P. (2021). Re-vitalizing of endophytic microbes for soil health management and plant protection. 3 Biotech, 11(9). https://doi.org/10.1007/s13205-021-02931-4

Muluneh, M. G. (2021). Impact of climate change on biodiversity and food security: a global perspective—a review article. Agriculture & Food Security, 10(1). https://doi.org/10.1186/s40066-021-00318-5

Murali, S., Jawahar, D., Balasubramaniam, P., Maheswari, M., Ejilane, J., & Pazhanivelan, S. (2022). Impact of paper board industry effluent irrigation to agroforestry system on changes in soil quality. International Journal of Plant & Soil Science, 24–34. https://doi.org/10.9734/ijpss/2022/v34i1931085

Muralikrishna, I. V., & Manickam, V. (2017). Industrial wastewater treatment technologies, recycling, and reuse. In Elsevier eBooks (pp. 295–336). https://doi.org/10.1016/b978-0-12-811989-1.00013-0

Musarat, M. A., Alaloul, W. S., Qureshi, A. H., & Ghufran, M. (2023). Construction waste to energy, technologies, economics, and challenges. In Elsevier eBooks (pp. 51–60). https://doi.org/10.1016/b978-0-323-93940-9.00027-x

Musie, W., & Gonfa, G. (2023). Fresh water resource, scarcity, water salinity challenges and possible remedies: A review. Heliyon, 9(8), e18685. https://doi.org/10.1016/j.heliyon.2023.e18685

Mustafa, S. A., Al-Rudainy, A. J., & Salman, N. M. (2024). Effect of environmental pollutants on fish health: An overview. The Egyptian Journal of Aquatic Research. https://doi.org/10.1016/j.ejar.2024.02.006

Naidoo, S., & Olaniran, A. (2013). Treated wastewater effluent as a source of microbial pollution of surface water resources. International Journal of Environmental Research and Public Health, 11(1), 249–270. https://doi.org/10.3390/ijerph110100249

Nazir, F., Hussain, A., & Fariduddin, Q. (2019). Hydrogen peroxide modulate photosynthesis and antioxidant systems in tomato (Solanum lycopersicum L.) plants under copper stress. Chemosphere, 230, 544–558. https://doi.org/10.1016/j.chemosphere.2019.05.001

Neethu, S., Midhun, S. J., Radhakrishnan, E., & Jyothis, M. (2020). Microbially synthesized nanomaterials for remediation of contaminated soil and water environment. In Elsevier eBooks (pp. 157–176). https://doi.org/10.1016/b978-0-12-821199-1.00015-8

Niamat, H., Yasmeen, R., Mustafa, M. D., Zahid, M. A., Noor, Z., & Abbas, J. (2023). Mitigating the harmful impacts of industrial effluents; the potential of biological treatment techniques. MARKHOR (the Journal of Zoology), 12–17. https://doi.org/10.54393/mjz.v4i02.77

Nriagu, J. O. (1989). A global assessment of natural sources of atmospheric trace metals. Nature, 338(6210), 47–49. https://doi.org/10.1038/338047a0

Odutayo, F. O. I., Oyetade, O., Esan, E. B., & Adaramola, F. B. (2016). Effects of Industrial Effluent on The Environment Using Allium cepa and Zea mays as Bioindicators. ResearchGate. https://www.researchgate.net/profile/Feyisara-Adaramola/publication/320235292

Ofremu, G. O., Raimi, B. Y., Yusuf, S. O., Dziwornu, B. A., Nnabuife, S. G., Eze, A. M., & Nnajiofor, C. A. (2024). Exploring the Relationship between Climate Change, Air Pollutants and Human Health: Impacts, Adaptation, and Mitigation Strategies. Green Energy and Resources, 100074. https://doi.org/10.1016/j.gerr.2024.100074

Oikari, A., Lönn, B., Castrén, M., Nakari, T., Snickars-Nikinmaa, B., Bister, H., & Virtanen, E. (1983). Toxicological effects of dehydroabietic acid (DHAA) on the trout, Salmo gairdneri Richardson, in fresh water. Water Research, 17(1), 81–89. https://doi.org/10.1016/0043-1354(83)90288-9

Okereke, J. N., Ogidi, O. I., Obasi, K. O., Department of Biotechnology, Federal University of Technology, Owerri, Department of Science Laboratory Technology, Federal Polytechnic, Ekowe, & Department of Toxicology and Environmental Management, University of Agriculture, Umudike. (2016). Environmental and health impact of industrial wastewater effluents in Nigeria – a review. In International Journal of Advanced Research in Biological Sciences (Vol. 3, Issue 6, pp. 55–67). https://ijarbs.com/pdfcopy/june2016/ijarbs8.pdf

Okoro, H. K., Pandey, S., Ogunkunle, C. O., Ngila, C. J., Zvinowanda, C., Jimoh, I., Lawal, I. A., Orosun, M. M., & Adeniyi, A. G. (2022). Nanomaterial-based biosorbents: Adsorbent for efficient removal of selected organic pollutants from industrial wastewater. Emerging Contaminants, 8, 46–58. https://doi.org/10.1016/j.emcon.2021.12.005

Osman, M. (2014). Waste water Treatment in Chemical Industries: The concept and current technologies. Hydrology Current Research, 05(01). https://doi.org/10.4172/2157-7587.1000164

Osman, M. A., Yang, F., & Massey, I. Y. (2019). Exposure routes and health effects of heavy metals on children. BioMetals, 32(4), 563–573. https://doi.org/10.1007/s10534-019-00193-5

Padhye, L. P., Srivastava, P., Jasemizad, T., Bolan, S., Hou, D., Shaheen, S. M., Rinklebe, J., O'Connor, D., Lamb, D., Wang, H., Siddique, K. H., & Bolan, N. (2023). Contaminant containment for sustainable remediation of persistent contaminants in soil and groundwater. Journal of Hazardous Materials, 455, 131575. https://doi.org/10.1016/j.jhazmat.2023.131575

Parizanganeh, A. H., Bijnavand, V., Zamani, A. A., & Hajabolfath, A. (2012). Concentration, distribution and comparison of total and bioavailable heavy metals in top soils of Bonab district in Zanjan province. Open Journal of Soil Science, 02(02), 123–132. https://doi.org/10.4236/ojss.2012.22018

Pathirana, C., Ziyath, A. M., Jinadasa, K., Egodawatta, P., Sarina, S., & Goonetilleke, A. (2019). Quantifying the influence of surface physico-chemical properties of biosorbents on heavy metal adsorption. Chemosphere, 234, 488–495. https://doi.org/10.1016/j.chemosphere.2019.06.074

Polla, W., Regaldo, L., Reno, U., Popielarz, A., Gervasio, S., Fernández, V., & Gagneten, A. (2022). Effect of a glyphosate formulation on freshwater plankton: A community combined metric approach. Journal of Environmental Biology, 43(1), 35–42. https://doi.org/10.22438/jeb/43/1/mrn-1471

Rabieh, S., Bayaraa, O., Romeo, E., Amosa, P., Calnek, K., Idaghdour, Y., Ochsenkühn, M. A., Amin, S. A., Goldstein, G., & Bromage, T. G. (2020). MH-ICP-MS analysis of the freshwater and saltwater Environmental Resources of Upolu Island, Samoa. Molecules, 25(21), 4871. https://doi.org/10.3390/molecules25214871

Rafferty, J. P. (2024, September 3). Biodiversity loss | Causes, Effects, & Facts. Encyclopedia Britannica. https://www.britannica.com/science/biodiversity-loss/Ecological-effects

Rahman, M. A., Rahman, M. S., Mohiuddin, K., Chowdhury, M. a. H., & Chowdhury, M. a. K. (2019). Germination and seedling growth of rice (Oryza sativa L.) as affected by varying concentrations of loom-dye effluent. Journal of the Bangladesh Agricultural University, 17(2), 153–160. https://doi.org/10.3329/jbau.v17i2.41938

Rahman, M. M., Das, A. K., Sultana, S., Ghosh, P. K., Islam, M. R., Keya, S. S., Ahmed, M., Nihad, S. a. I., Khan, M. a. R., Lovell, M. C., Rahman, M. A., Ahsan, S. M., Anik, T. R., Fnu, P., Tran, L. P., & Mostofa, M. G. (2023). Biochar potentially enhances maize tolerance to arsenic toxicity by improving physiological and biochemical responses to excessive arsenate. Biochar, 5(1). https://doi.org/10.1007/s42773-023-00270-6

Rai, P. K., Lee, S. S., Zhang, M., Tsang, Y. F., & Kim, K. (2019). Heavy metals in food crops: Health risks, fate, mechanisms, and management. Environment International, 125, 365–385. https://doi.org/10.1016/j.envint.2019.01.067

Rajasulochana, P., & Preethy, V. (2016). Comparison on efficiency of various techniques in treatment of waste and sewage water—A comprehensive review. Resource-Efficient Technologies, 2(4), 175–184. https://doi.org/10.1016/j.reffit.2016.09.004

Rajbala, N. (2012). Impact of steel industry waste on Physico-Chemical property of soil. International Journal of Environmental Sciences, 2(3). https://doi.org/10.6088/ijes.00202030003

Rajkumar, D., & Palanivelu, K. (2004). Electrochemical treatment of industrial wastewater. Journal of Hazardous Materials, 113(1–3), 123–129. https://doi.org/10.1016/j.jhazmat.2004.05.039

Ramana, S., Biswas, A., Kundu, S., Saha, J., & Yadava, R. (2002). Effect of distillery effluent on seed germination in some vegetable crops. Bioresource Technology, 82(3), 273–275. https://doi.org/10.1016/s0960-8524(01)00184-5

Rana, K. L., Kour, D., Sheikh, I., Dhiman, A., Yadav, N., Yadav, A. N., Rastegari, A. A., Singh, K., & Saxena, A. K. (2019). Endophytic fungi: biodiversity, ecological significance, and potential industrial applications. In Fungal biology (pp. 1–62). https://doi.org/10.1007/978-3-030-10480-1_1

Rawat, A., Srivastava, A., Bhatnagar, A., & Gupta, A. K. (2022). Technological advancements for the treatment of steel industry wastewater: Effluent management and sustainable treatment strategies. Journal of Cleaner Production, 383, 135382. https://doi.org/10.1016/j.jclepro.2022.135382

Razzak, S. A., Faruque, M. O., Alsheikh, Z., Alsheikhmohamad, L., Alkuroud, D., Alfayez, A., Hossain, S. M. Z., & Hossain, M. M. (2022). A comprehensive review on conventional and biological-driven heavy metals removal from industrial wastewater. Environmental Advances, 7, 100168. https://doi.org/10.1016/j.envadv.2022.100168

Saaristo, M., Brodin, T., Balshine, S., Bertram, M. G., Brooks, B. W., Ehlman, S. M., McCallum, E. S., Sih, A., Sundin, J., Wong, B. B. M., & Arnold, K. E. (2018). Direct and indirect effects of chemical contaminants on the behaviour, ecology and evolution of wildlife. Proceedings of the Royal Society B Biological Sciences, 285(1885), 20181297. https://doi.org/10.1098/rspb.2018.1297

Saeed, M. U., Hussain, N., Javaid, M., & Zaman, H. (2023). Microbial remediation for environmental cleanup. In Elsevier eBooks (pp. 247–274). https://doi.org/10.1016/b978-0-323-95090-9.00010-8

Sahai, R. and Neelam. (1987). Effect of fertilizer factory and distillery effluents on the seed germination, seedling growth, pigment content and biomass of Phaseolus radiatus L. Indian J. Ecol, 14: 21–25. https://www.cabidigitallibrary.org/doi/full/10.5555/19890719679

Saidon, N. B., Szabó, R., Budai, P., & Lehel, J. (2023). Trophic transfer and biomagnification potential of environmental contaminants (heavy metals) in aquatic ecosystems. Environmental Pollution, 340, 122815. https://doi.org/10.1016/j.envpol.2023.122815

Saigl, Z., Tifouti, O., Alkhanbashi, B., Alharbi, G., & Algamdi, H. (2023). Chitosan as adsorbent for removal of some organic dyes: a review. Chemical Papers, 77(5), 2363–2405. https://doi.org/10.1007/s11696-022-02641-y

Salian, R., Wani, S., Reddy, R., & Patil, M. (2018). Effect of brewery wastewater obtained from different phases of treatment plant on seed germination of chickpea (Cicer arietinum), maize (Zea mays), and pigeon pea (Cajanus cajan). Environmental Science and Pollution Research, 25(9), 9145–9154. https://doi.org/10.1007/s11356-018-1218-9

Saravanakumar, K., De Silva, S., Santosh, S. S., Sathiyaseelan, A., Ganeshalingam, A., Jamla, M., Sankaranarayanan, A., Veeraraghavan, V. P., MubarakAli, D., Lee, J., Thiripuranathar, G., & Wang, M. (2022). Impact of industrial effluents on the environment and human health and their remediation using MOFs-based hybrid membrane filtration techniques. Chemosphere, 307, 135593. https://doi.org/10.1016/j.chemosphere.2022.135593

Saravanan, A., Kumar, P. S., Duc, P. A., & Rangasamy, G. (2022). Strategies for microbial bioremediation of environmental pollutants from industrial wastewater: A sustainable approach. Chemosphere, 313, 137323. https://doi.org/10.1016/j.chemosphere.2022.137323

Saravanan, A., Kumar, P. S., Hemavathy, R., Jeevanantham, S., Harikumar, P., Priyanka, G., & Devakirubai, D. R. A. (2021). A comprehensive review on sources, analysis and toxicity of environmental pollutants and its removal methods from water environment. The Science of the Total Environment, 812, 152456. https://doi.org/10.1016/j.scitotenv.2021.152456

Sathya, K., Nagarajan, K., Malar, G. C. G., Rajalakshmi, S., & Lakshmi, P. R. (2022). A comprehensive review on comparison among effluent treatment methods and modern methods of treatment of industrial wastewater effluent from different sources. Applied Water Science, 12(4). https://doi.org/10.1007/s13201-022-01594-7

Sethy, S. K., & Ghosh, S. (2013). Effect of heavy metals on germination of seeds. Journal of Natural Science Biology and Medicine, 4(2), 272. https://doi.org/10.4103/0976-9668.116964

Shahjahan, M., Taslima, K., Rahman, M. S., Al-Emran, M., Alam, S. I., & Faggio, C. (2022). Effects of heavy metals on fish physiology—A review. Chemosphere, 300, 134519. https://doi.org/10.1016/j.chemosphere.2022.134519

Sharma, P., & Kumar, S. (2021). Bioremediation of heavy metals from industrial effluents by endophytes and their metabolic activity: Recent advances. Bioresource Technology, 339, 125589. https://doi.org/10.1016/j.biortech.2021.125589

Shivanna, K. R. (2020). The Sixth Mass Extinction Crisis and its Impact on Biodiversity and Human Welfare. Resonance, 25(1), 93–109. https://doi.org/10.1007/s12045-019-0924-z

Sigmund, G., Ågerstrand, M., Antonelli, A., Backhaus, T., Brodin, T., Diamond, M. L., Erdelen, W. R., Evers, D. C., Hofmann, T., Hueffer, T., Lai, A., Torres, J. P. M., Mueller, L., Perrigo, A. L., Rillig, M. C., Schaeffer, A., Scheringer, M., Schirmer, K., Tlili, A., ... Groh, K. J. (2023). Addressing chemical pollution in biodiversity research. Global Change Biology, 29(12), 3240–3255. https://doi.org/10.1111/gcb.16689

Singare, P., Lokhande, R., & Pathak, P. (2010). Study on physico-chemical properties and heavy metal content of the soil samples from Thane Creek of Maharashtra, India, Interdisciplinary Environmental Review, 11(1), 38. https://doi.org/10.1504/ier.2010.034605

Singh VK, Singh J. Toxicity of industrial wastewater to the aquatic plant Lemna minor L. J Environ Biol. 2006 May;27(2 Suppl):385-90.

Singh, B. J., Chakraborty, A., & Sehgal, R. (2023). A systematic review of industrial wastewater management: Evaluating challenges and enablers. Journal of Environmental Management, 348, 119230. https://doi.org/10.1016/j.jenvman.2023.119230

Singh, J., Chawla, G., Naqvi, S. H. N., & Viswanathan, P. N. (1994). Combined effects of cadmium and linear alkyl benzene sulfonate on Lemna minor L. Ecotoxicology, 3(1), 59–67. https://doi.org/10.1007/bf00121388

Singh, V., Ahmed, G., Vedika, S., Kumar, P., Chaturvedi, S. K., Rai, S. N., Vamanu, E., & Kumar, A. (2024). Toxic heavy metal ions contamination in water and their sustainable reduction by eco-friendly methods: isotherms, thermodynamics and kinetics study. Scientific Reports, 14(1). https://doi.org/10.1038/s41598-024-58061-3

Sinha, R. K., Herat, S., & Tandon, P. (2007). Phytoremediation: Role of plants in contaminated site management. In Springer eBooks (pp. 315–330). https://doi.org/10.1007/978-3-540-34793-4_14

Soares, T. F. S. N., Dias, D. C. F. D. S., Oliveira, A. M. S., Ribeiro, D. M., & Dias, L. a. D. S. (2020). Exogenous brassinosteroids increase lead stress tolerance in seed germination and seedling growth of Brassica juncea L. Ecotoxicology and Environmental Safety, 193, 110296. https://doi.org/10.1016/j.ecoenv.2020.110296

Song, Z. H., & Huang, G. L. (2001). Effect of Triphenyltin on Duckweed Lemna minor. Bulletin of Environmental Contamination and Toxicology, 67(3), 368–375. https://doi.org/10.1007/s001280134

Sonone, S., Jadhav, S., Sankhla, M., & Kumar, R. (2020). Water Contamination by Heavy Metals and their Toxic Effect on Aquaculture and Human Health through Food Chain. Letters in Applied NanoBioScience, 10(2), 2148–2166. https://doi.org/10.33263/lianbs102.21482166

Sonune, A., & Ghate, R. (2004). Developments in wastewater treatment methods. Desalination, 167, 55–63. https://doi.org/10.1016/j.desal.2004.06.113

Sou Dakouré, M. Y., Mermoud, A., Yacouba, H., & Boivin, P. (2013). Impacts of irrigation with industrial treated wastewater on soil properties. Geoderma, 200–201, 31–39. https://doi.org/10.1016/j.geoderma.2013.02.008

Srivastava, P., & Yadav, U. (2023). Mode of the mechanism of biogenic nanomaterials involved in the adsorption of pollutants. In Elsevier eBooks (pp. 273–296). https://doi.org/10.1016/b978-0-323-91767-4.00009-4

Stolt, P., Asp, H., & Hultin, S. (2006). Genetic variation in wheat cadmium accumulation on soils with different cadmium concentrations. Journal of Agronomy and Crop Science, 192(3), 201–208. https://doi.org/10.1111/j.1439-037x.2006.00202.x

Suedel, B. C., Boraczek, J. A., Peddicord, R. K., Clifford, P. A., & Dillon, T. M. (1994). Trophic transfer and biomagnification potential of contaminants in aquatic ecosystems. Reviews of Environmental Contamination and Toxicology, 21–89. https://doi.org/10.1007/978-1-4612-2656-7_2

Syeda, S. R., Khan, E. A., Padungwatanaroj, O., Kuprasertwong, N., & Tula, A. K. (2021). A perspective on hazardous chemical substitution in consumer products. Current Opinion in Chemical Engineering, 36, 100748. https://doi.org/10.1016/j.coche.2021.100748

Talukdar, A., Kundu, P., Bhattacharya, S., & Dutta, N. (2024). Microplastic contamination in wastewater: Sources, distribution, detection and remediation through physical and chemical-biological methods. The Science of the Total Environment, 916, 170254. https://doi.org/10.1016/j.scitotenv.2024.170254

Tamjidi, S., Kamyab Moghadas, B., Abdollahi, P., Parnian, N., & Farhadi, A. (2024). Industrial By-Products as Potentially Economic and Promising Adsorbents for Removing Dyes from Effluents: A Review Study. Physical Chemistry Research, 12(2), 361-382. doi: https://doi.org/10.22036/pcr.2023.396003.2332

Tangahu, B. V., Abdullah, S. R. S., Basri, H., Idris, M., Anuar, N., & Mukhlisin, M. (2011). A Review on Heavy Metals (As, Pb, and Hg) Uptake by Plants through Phytoremediation. International Journal of Chemical Engineering, 2011, 1–31. https://doi.org/10.1155/2011/939161

Tchounwou, P. B., Yedjou, C. G., Patlolla, A. K., & Sutton, D. J. (2012). Heavy metal toxicity and the environment. EXS, 133–164. https://doi.org/10.1007/978-3-7643-8340-4_6

Tremblay, L., & Van Der Kraak, G. (1999). Comparison between the effects of the phytosterol β-Sitosterol and pulp and paper mill effluents on sexually immature rainbow trout. Environmental Toxicology and Chemistry, 18(2), 329–336. https://doi.org/10.1002/etc.5620180233

Tuomisto, H. L., Scheelbeek, P. F., Chalabi, Z., Green, R., Smith, R. D., Haines, A., & Dangour, A. D. (2017). Effects of environmental change on agriculture, nutrition and health: A framework with a focus on fruits and vegetables. Wellcome Open Research, 2, 21. https://doi.org/10.12688/wellcomeopenres.11190.2

Uddin, M. G., Nash, S., & Olbert, A. I. (2020). A review of water quality index models and their use for assessing surface water quality. Ecological Indicators, 122, 107218. https://doi.org/10.1016/j.ecolind.2020.107218

Vidu, R., Matei, E., Predescu, A. M., Alhalaili, B., Pantilimon, C., Tarcea, C., & Predescu, C. (2020). Removal of Heavy Metals from Wastewaters: A Challenge from Current Treatment Methods to Nanotechnology Applications. Toxics, 8(4), 101. https://doi.org/10.3390/toxics8040101

Vijayaraghavan, K., & Yun, Y. (2008). Bacterial biosorbents and biosorption. Biotechnology Advances, 26(3), 266–291. https://doi.org/10.1016/j.biotechadv.2008.02.002

Vijver, M. G., & Peijnenburg, W. J. (2011). Metals and metalloids in terrestrial systems: Bioa ccumulation,Biomagnification and subsequent adverse effects. In BENTHAM SCIENCE PUBLISHERS eBooks (pp. 43–62). https://doi.org/10.2174/978160805121210043

Vongdala, N., Tran, H., Xuan, T., Teschke, R., & Khanh, T. (2018). Heavy metal accumulation in water, soil, and plants of municipal solid waste landfill in Vientiane, Laos. International Journal of Environmental Research and Public Health, 16(1), 22. https://doi.org/10.3390/ijerph16010022

Voumik, L. C., & Sultana, T. (2022). Impact of urbanization, industrialization, electrification and renewable energy on the environment in BRICS: fresh evidence from novel CS-ARDL model. Heliyon, 8(11), e11457. https://doi.org/10.1016/j.heliyon.2022.e11457

Wang, W. (1991). Literature review on higher plants for toxicity testing. Water Air & Soil Pollution, 59(3–4), 381–400. https://doi.org/10.1007/bf00211845

Webber, J. (1981). Trace metals in agriculture. In Springer eBooks (pp. 159–184). https://doi.org/10.1007/978-94-009-8099-0_5

Western, D. (2001). Human-modified ecosystems and future evolution. Proceedings of the National Academy of Sciences, 98(10), 5458–5465. https://doi.org/10.1073/pnas.101093598

Wickerham, E. L., Lozoff, B., Shao, J., Kaciroti, N., Xia, Y., & Meeker, J. D. (2012). Reduced birth weight in relation to pesticide mixtures detected in cord blood of full-term infants. Environment International, 47, 80–85. https://doi.org/10.1016/j.envint.2012.06.007

Wong, J.W.C., Lai, K.M., Su, D.S. et al. Availability of Heavy Metals for Brassica Chinensis Grown in an Acidic Loamy Soil Amended with a Domestic and an Industrial Sewage Sludge. Water, Air, & Soil Pollution 128, 339–353 (2001). https://doi.org/10.1023/A:1010353107240

Xie, S. (2024). Biosorption of heavy metal ions from contaminated wastewater: an eco-friendly approach. Green Chemistry Letters and Reviews, 17(1). https://doi.org/10.1080/17518253.2024.2357213

Yadav, I. C., Devi, N. L., Syed, J. H., Cheng, Z., Li, J., Zhang, G., & Jones, K. C. (2014). Current status of persistent organic pesticides residues in air, water, and soil, and their possible effect on neighboring countries: A comprehensive review of India. The Science of the Total Environment, 511, 123–137. https://doi.org/10.1016/j.scitotenv.2014.12.041

Yousif, R., Choudhary, M. I., Ahmed, S., & Ahmed, Q. (2021). Review: Bioaccumulation of heavy metals in fish and other aquatic organisms from Karachi Coast, Pakistan. Nusantara Bioscience, 13(1). https://doi.org/10.13057/nusbiosci/n130111

Zaidi, A., Ahmad, E., Khan, M. S., Saif, S., & Rizvi, A. (2015). Role of plant growth promoting rhizobacteria in sustainable production of vegetables: Current perspective. Scientia Horticulturae, 193, 231–239. https://doi.org/10.1016/j.scienta.2015.07.020

Zhang, P., Yang, M., Lan, J., Huang, Y., Zhang, J., Huang, S., Yang, Y., & Ru, J. (2023). Water quality degradation due to Heavy metal contamination: Health impacts and Eco-Friendly Approaches for Heavy Metal Remediation. Toxics, 11(10), 828. https://doi.org/10.3390/toxics11100828

Zhang, X., Yan, L., Liu, J., Zhang, Z., & Tan, C. (2019). Removal of different kinds of heavy metals by novel PPG-NZVI beads and their application in simulated stormwater infiltration facility. Applied Sciences, 9(20), 4213. https://doi.org/10.3390/app9204213

Zhang, Z., Hinger, J., Audretsch, D. B., & Song, G. (2014). Environmental technology transfer and emission standards for industry in China. The Journal of Technology Transfer, 40(5), 743–759. https://doi.org/10.1007/s10961-014-9371-x

Zheng, R., Liu, Y., & Zhang, Z. (2023). Trophic transfer of heavy metals through aquatic food web in the largest mangrove reserve of China. The Science of the Total Environment, 899, 165655. https://doi.org/10.1016/j.scitotenv.2023.165655

Ziani, K., Ioniță-Mîndrican, C., Mititelu, M., Neacșu, S. M., Negrei, C., Moroșan, E., Drăgănescu, D., & Preda, O. (2023). Microplastics: a real global threat for environment and food Safety: A State of the art review. Nutrients, 15(3), 617. https://doi.org/10.3390/nu15030617

Ziegler, P., Appenroth, K. J., & Sree, K. S. (2023). Survival strategies of duckweeds, the world's smallest angiosperms. Plants, 12(11), 2215. https://doi.org/10.3390/plants12112215

Ziolko, D., Martin, O. V., Scrimshaw, M. D., & Lester, J. N. (2011). An evaluation of metal removal during wastewater treatment: the potential to achieve more stringent final effluent standards. Critical Reviews in Environmental Science and Technology, 41(8), 733–769. https://doi.org/10.1080/10643380903140299

Mechanistic Understanding of Xenobiotic Detoxification by Plants

Sahib Kaur, Sushma Rani, Sonal Mishra, and Gargi

Abstract Xenobiotics could be defined as any foreign substances that neither provide any nutritional benefit to the organism nor could be metabolized in order to release energy. With the progression of modernization, humans, animals, and plants are continually subjected to numerous chemical substances that produce toxicity. The xenobiotics lead to the production of reactive oxygen species, which cause oxidative stress, leading to damage in cellular components. Unlike animals, plants do not have an elaborate excretory system. However, they can detoxify the xenobiotics by generating enzymatic and non-enzymatic antioxidant defense mechanisms. Secondary metabolites play a vital role in detoxification and the defense against oxidative stress. Therefore, plants act as an imperative sink for chemicals that are harmful to the environment as well as to living beings. In plants, xenobiotic metabolism occurs in three stages. First is modification, which involves the conversion of toxic xenobiotics into less toxic ones by chemical methods. Second is conjugation: combining xenobiotics with other substances to increase their solubility, facilitating their elimination from the plant. Third, compartmentalization involves storage or isolation of conjugated xenobiotics in specific parts of the plant, preventing damage to cellular functions. This chapter provides detailed insights into the processes involved in the detoxification of xenobiotics by plants.

Keywords Detoxification mechanisms · Enzymatic detoxification · Phytoremediation · Phytotoxicity · Xenobiotics · Xenobiotic metabolism

S. Kaur · S. Mishra · Gargi
Department of Bioscience and Biotechnology, Banasthali Vidyapith, Rajasthan, India

S. Rani (✉)
School of Life and Basic Sciences, Jaipur National University, Jaipur, Rajasthan, India

S. Roy, V. Mandal (eds.), *Plant-Microbe Interaction under Xenobiotic Exposure*,
https://doi.org/10.1007/978-981-96-8260-7_8

1 Introduction

The term "xenobiotic" is derived from the Greek words "xenos" (foreign) and "bios" (life). Therefore, any substance foreign to the biological system can be termed as a xenobiotic. The compounds produced in one species can be foreign to another species. Besides, any substances present in higher concentrations than optimal limits in living beings or environments are categorized as xenobiotics (Sharma et al. 2016). A wide range of these xenobiotics tend to persist in the environment and accumulate in the food chain over a period of time due to their high stability and non-polar nature. This bioaccumulation of xenobiotics can induce toxicity, mutagenicity, and teratogenicity in living beings. They induce oxidative stress via reactive oxygen species (ROS), thereby damaging the nucleic acid, lipid, and protein components of cells (Vicidomini et al. 2024). The xenobiotics may cause toxicity by interfering with the enzyme activities. For instance, organochlorinated pesticides inhibit the activity of sodium and potassium ion ATPases. The xenobiotics originate from natural sources (volcanic eruptions, forest fires, etc.) as well as anthropogenic sources (Marghade et al. 2021). The pesticides, heavy metals along with metalloids, persistent organic pollutants including polyaromatic hydrocarbons, dioxins, and furans, constitute synthetic xenobiotics (Capozzi and Spagnuolo 2024). The chemical residues from pharmaceutical and personal care products have been titled as the new generation pollutants found prevalent in the soil and freshwater sources (Wei et al. 2023). These compounds can be categorized on the basis of their acute or chronic toxic effects on the organisms. In order to survive the adverse growing conditions, plants have developed an intricate mechanism for the detoxification of these xenobiotics. Although the uptake rate depends upon the structural differences, the metabolism of xenobiotics shares a similar pathway (Muerdter et al. 2023). The process of detoxification occurs in three phases. The first phase includes the conversion of the xenobiotic via enzyme-catalyzed (peroxidases, P450 mono-oxygenases, laccases, reductases, esterases, and dehydrogenases) transformation, oxidation, reduction, and hydrolysis of the xenobiotic. In the second phase, the contaminant gets conjugated with peptides, glutathione, organic acids, polysaccharides, etc., for enhanced solubility. Finally, the third phase is characterized by the compartmentation of the conjugated or unmodified xenobiotic into vacuoles or its direction toward the cell wall (Jha et al. 2020). Depending upon the nature of the contaminant, its metabolism may or may not involve all three phases (Fig. 1). The xenobiotics are usually lipophilic organic substances that need to be modified before compartmentation (Singh et al. 2016). Some plants have the tendency to accumulate the toxic xenobiotic in their parent form and are termed bioaccumulators. However, the response toward xenobiotics is highly species-specific (Espinosa-Vellarino et al. 2024). Henceforth, the phytotoxicity and the mechanism of xenobiotic detoxification have been discussed in detail.

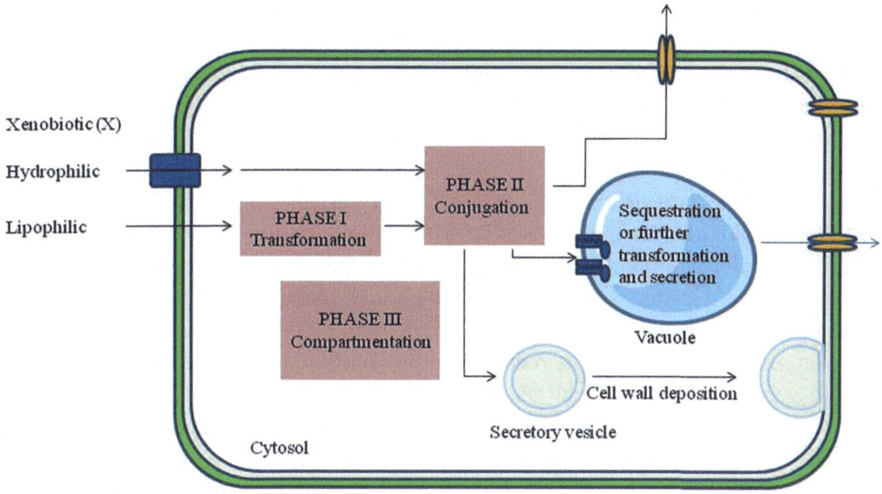

Fig. 1 Processes involved in xenobiotic detoxification in plants

2 Impacts of Xenobiotics on Plants

The xenobiotics are either transported from the soil (pyrene, bisphenol A, heavy metal ions, etc.) or taken up through stomata as air pollutants (carbendazim, polychlorinated biphenyls, phenanthrene, etc.). These not only have impacts on seed germination or plant growth but also affect the overall functioning of the plants. Many xenobiotics act as mutagens and induce genotoxicity (Miglani et al. 2022). The most common mechanism through which DNA damage occurs in plants is via the accumulation of reactive oxygen species (ROS). Although plants have an elaborate antioxidant system, the response toward abiotic as well as biotic stresses is highly genotype-specific. The oxidative stress induced by the contaminant leads to lipid peroxidation, thereby damaging the lipid, nucleic acid, and protein components of the cell (Rani et al. 2023). These xenobiotics accumulate in the leaves and directly impact the physiological processes by altering the photosynthetic efficiency of the plants (Rani and Kumari 2023). For instance, fluoranthene, which is a polyaromatic hydrocarbon, at higher concentrations (1 mg L^{-1}) leads to the degradation of photosynthetic pigments as well as obstructs electron transport by damaging the photosystem II (PSII) (Lonova et al. 2023). The typical symptoms of xenobiotic toxicity include the formation of necrotic lesions, chlorotic spots, and deformities in the leaves. They also damage the chloroplasts and thylakoid membranes and reduce the levels of CO_2 fixation in the cells. As a result of the impact on photosynthesis, there is a reduction in the overall accumulation of biomass, which thereby decreases the final productivity of plants (Ahammed et al. 2013). Bisphenol A is a known organic pollutant used for the production of epoxy resins and plastics. It not only causes a reduction in growth but also suppresses the reproductive development of plants (Jiao et al. 2017). Bisphenol A has been shown to induce inhibition of pollen germination in kiwifruit (Speranza et al. 2011). Some xenobiotics, like the metalloid antimony, can bind to the functional groups (thiol) of proteins that stabilize their tertiary or quaternary structures and thereby alter the activity of enzymes, including the antioxidants (glutathione-S-transferases, superoxide dismutase, ascorbate peroxidase, etc.). This further leads to toxicity in plants due to uncontrolled levels of ROS (Espinosa-Vellarino et al. 2024). Since the Green Revolution, pesticides have been used to protect plants against pests. However, they also induce toxicity in plants (Sharma et al. 2019). Besides the effects on photosynthesis, antioxidant system, and growth, insecticides like chlorpyrifos (organophosphorus) suppress the nitrogen metabolism in plants (Parween et al. 2012). The glyphosate-based herbicides increase the lipid peroxidation levels in plants and limit the growth of roots as well as shoots (Fernandez et al. 2022). Moreover, the toxic effects of xenobiotics include the downregulation of genes coding for proteins involved in the physiological processes, and therefore there is a reduction in the protein content in plant tissues. The reduction in ascorbate and glutathione content, stomatal conductance, and leaf area has also been observed due to xenobiotic toxicity (Sharma et al. 2019; Alengebawy et al. 2021).

3 Transformation and Conjugation of Xenobiotics

As mentioned above, the fate of the xenobiotics in plants is based on their structure, as some are subjected to transformation (phase I), while others may be conjugated (phase II) and stored in the vacuole or cell wall (phase III). Cytochrome P450 is one of the key enzyme families involved in the transformation of xenobiotics, including antibiotics, pesticides, and other drugs (Chakraborty et al. 2023). These are versatile in nature and can carry out hydroxylation (-N, C-H bonds), decarboxylation, isomerization, dealkylation, and epoxidation reactions. These enzymes are mostly located in the endoplasmic reticulum and utilize heme as a cofactor. Besides, peroxidase and reductase activities have been shown by some of these cytochromes (Esteves et al. 2021). The levels of extracellular peroxidases also increase upon exposure to xenobiotics in plants, which tend to prevent the entry of pollutants inside the cells. These enzymes use hydrogen peroxide for the oxidation of sulfates, indoles, aromatic amines, and phenolic contaminants (Gałązka et al. 2023; Pinto et al. 2021). The dehydrogenases also play a vital role in the oxidation of aldehydes and alcohols (Jiang et al. 2020). On the other hand, the aldo-keto reductases, peptidases, epoxide hydrolases, and carboxylesterases metabolize the aldehyde, ketone, amide, epoxide, and ester xenobiotics (Shabbir et al. 2022). The transformed products usually undergo conjugation by the addition of polar functional groups, thereby making the compound less reactive.

Glutathione is another detoxification agent in plants that reduces the toxicity of xenobiotics. The glutathione-S-transferase attaches the glutathione residue to the pollutant, which increases its solubility for easy transport and compartmentation.

The Tau class glutathione-S-transferases (GSTs) are generally involved in the detoxification of xenobiotics. They perform the conjugation of the reduced tripeptide glutathione with the electrophile (Sousa et al. 2021). The lipophilic xenobiotics have highly toxic effects, as these can easily move through the cells. Moreover, the electrophilic pollutants tend to react with the nucleophilic sites in proteins and nucleic acids, thereby inducing cellular toxicity. For instance, the polarized double bonds and aldehydes react with the sulfur in cysteinyl residues of proteins; the epoxides, alkyl sulfates, strained ring lactones, and alkyl halides react with the sulfur in methionyl residues and amino groups (histidine, lysine, arginine) in proteins; and arylcarbonium ions are reactive to the amino groups of purines, while alkylcarbonium ions react with the phosphate oxygen of nucleic acids. Here, it could be noted that electrophiles react with the nucleophiles of the same grade. This means that soft electrophiles like aldehydes react with soft nucleophiles, whereas strong electrophiles like alkylcarbonates react with strong nucleophilic sites (Schriever and Lamshoeft 2020). Glutathione is highly hydrophilic, attributed to amine, thiol, two carboxylic acid groups, and two peptide bonds. It is synthesized largely in the cytosol (and in small quantities in plastids) in an ATP-dependent two-step process from the amino acids glutamate, cysteine, and glycine. The former two forms glutamylcysteine in the plastids, which further gets converted into glutathione by the enzyme glutathione synthetase (Hossain et al. 2022). When conjugated with the xenobiotic, it reduces its lipophilicity, which changes the physio-chemical properties of the contaminants, thereby reducing the mobility across the biomembranes and facilitating the compartmentalization. The glutathione detoxification system has been known to be activated against chlorothalonil pesticide, pretilachlor herbicide, drugs (ethinylestradiol), antibiotics (trimethoprim, sulfamethoxazole, diclofenac), chlorinated herbicide atrazine, and many more (Christou et al. 2016; Jaiswal et al. 2021; Potęga 2022). The different classes of GSTs specific to plants include phi, tau, lambda, and dehydroascorbate reductases. Here, the former two are dimeric with serine in the active site, while the latter two are monomeric and have cysteine in the active site. The other two classes, theta and zeta, are found in both animals and plants. The lambda class GSTs have been known to protect plants from arsenic toxicity (Kumar and Trivedi 2018).

Besides glutathione, the xenobiotics and their transformed products are known to be conjugated with the sugar moieties. Here, the transferases could be categorized on the type of sugar that is used for conjugation and the site (-OH, -COOH, and –NH_2 groups) at which it is conjugated. The most common form used for conjugation is the activated nucleotide (uridine diphosphate, UDP) sugar carried out by the enzyme UDP-glycosyltransferase or UGT (Vaskova et al. 2023). Apart from UDP-glucose, which is the widely used activated form, the others include UDP-galactose, UDP-glucuronic acid, and UDP-xylose (Bartikova et al. 2015). The mechanism of glycosylation has been known to be deployed by plants in response to a wide array of xenobiotics, including pesticides and therapeutic drugs (Schroder and Collins 2002; Zhang and Yang 2021; Kantharaj et al. 2022). The photosynthesis-inhibiting broadleaf weeds and grass herbicide metribuzin get acted upon by the N-glucosyltransferase enzyme (Javaid et al. 2022). However, the toxicity of this contaminant is also reduced by glutathione conjugation in plant cells. Systemic pesticides like 2,4-dichlorophenoxyacetic acid and pentachlorophenol undergo phase II conjugation by O-malonyltransferases (Fernandez et al. 2022). The conjugation reaction in which the malonyl-coenzyme A acts as a donor of electrons is termed "malonylation." The malonyl conjugation is used to detoxify the systemic herbicide isoproturon in plants (Lu et al. 2015). Another class of proteins that are involved in the transfer of the sulfate group at the hydroxyl group of the xenobiotics by using phosphoadenosine-5′-phosphosulfate are sulfotransferases (Bartikova et al. 2015). The sulfotransferases are involved in the detoxification of chemical naphthalic anhydride that is used as a seed protectant and herbicide safener. It may be noted that herbicide safeners are chemicals that are used to alleviate the highly toxic effects of herbicides by inducing elevated detoxification responses in plants. Similarly, detoxification of xenobiotics is also carried out by methyltransferases (methyl group transfer) and acetyltransferases (acetyl-CoA transfer) (Bartikova et al. 2015; Chen et al. 2024). The phytochelatin synthases act upon the glutathione conjugates to release cysteine-rich peptides known as phytochelatins, which are known for chelating heavy metals and maintaining homeostasis. Therefore, these enzymes also form a part of xenobiotic metabolism and carry out the processing of the xenobiotic conjugates in the cytosol (Chia 2021). The xenobiotic conjugates can also be sequestered in the vacuole, where they may be treated by vacuolar lyases, peptidases, and other enzymes for further transformation and degradation.

The effluents from the petrochemical and pesticide industries, petroleum refineries, etc., contain up to 1000 mgL-1 of phenol, which is an organic pollutant. When taken up by the plants, the toxicity of phenol is reduced by hydroxylation with the enzyme peroxidase, and in a series of reactions, the compound fumaric acid is generated, which is used for ATP production in the cells via the citric acid cycle (Jha et al. 2013) (Fig. 2).

The use of animal manure in the agricultural fields predisposes the plants to various drugs and antibiotics. A wide range of antibiotics (metformin, penicillin, tetracycline, carbamazepine, ciprofloxacin, etc.) have been reported to be taken up by the crops. In plants, paracetamol is converted by the P450 mono-oxygenase into an intermediate, which in turn gets conjugated with glutathione via the glutathione-S-transferase enzyme. On the other hand, paracetamol may be directly conjugated with the glucoside residue from uridine-5-diphosphoglucose by the enzyme glucosyltransferase. The conjugated antibiotics get transported in the vacuole with the help of glutathione-conjugate or ATP-dependent glucoside-conjugate transporter. The conjugates can also get transported in the secretory vesicle toward the cell wall, where they bind to its lignin component (Jha et al. 2020) (Fig. 3).

It can be noted that a single xenobiotic could be detoxified by more than one pathway in plants. Trichloroethylene is a widely used chemical to produce paints, printing ink, refrigerants, etc. The high expression of the cytochrome P450 enzyme reduces the toxicity of trichloroethylene in plants and enhances their phytoremediation efficiency. These enzymes are also required to cope with benzene, chloroform, mercury, phenylurea herbicides, carbon tetrachloride, 4-chloro-N-methyl-aniline used for the production of azo dyes, chlordimeform pesticide, etc. (Chakraborty

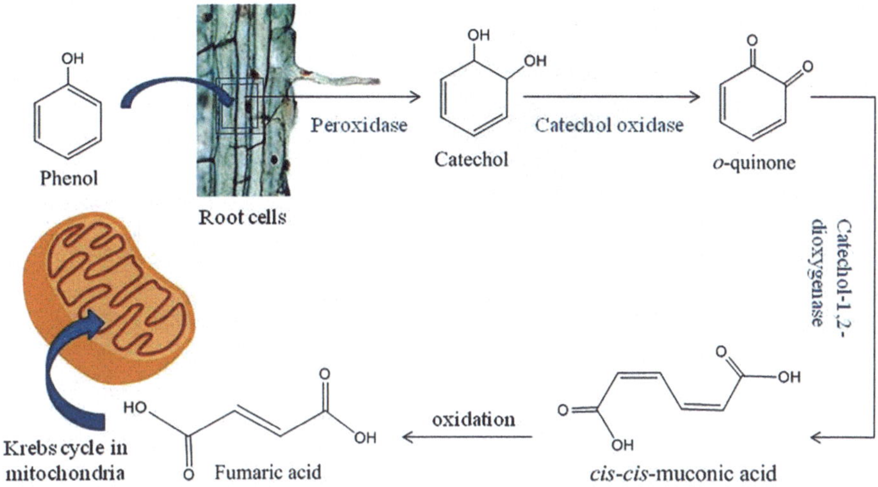

Fig. 2 The metabolism of toxic phenol into fumaric acid that is directed toward the Krebs cycle for ATP synthesis. (Adapted from Jha et al. 2020)

et al. 2023). The detoxification of the herbicide fluroxypyr-meptyl in plants is carried out by cytochrome P450, glycosyltransferases, and acetyltransferases (Chen et al. 2024).

4 Sequestration and Further Processing of Modified Xenobiotics in Plant Cell

The modified xenobiotic compounds have the potential to hinder the normal metabolic processes, and the conjugates might inhibit the activity of various enzymes by either negative feedback or by binding at the active sites (Chen 2020). These intermediates are not able to cross the plasma membrane and therefore require the aid of transporters characterized to be ATP-binding cassette (ABC) and multidrug and toxins extrusion (MATE) transporters (Nuruzzaman et al. 2014). Besides their role in plant growth and development, ABC transporters also participate in the metabolism of xenobiotics. These glycoproteins extend through the plasma membrane, having a couple of transmembrane and nucleotide-binding domains, and are termed full-size transporters. Since the superfamily of ABC transporters is classified into several subfamilies on the basis of their divergent functions, some of them are half-size (one domain each), while some are soluble, i.e., without the transmembrane domain (Hwang et al. 2016). The proteins embedded in the tonoplast carry out the active transport of conjugated xenobiotics. These are sub-categorized as ABCC transporters, formerly known as multidrug resistance-associated proteins, or MRPs. The ABC transporters provide resistance against a wide range of herbicides

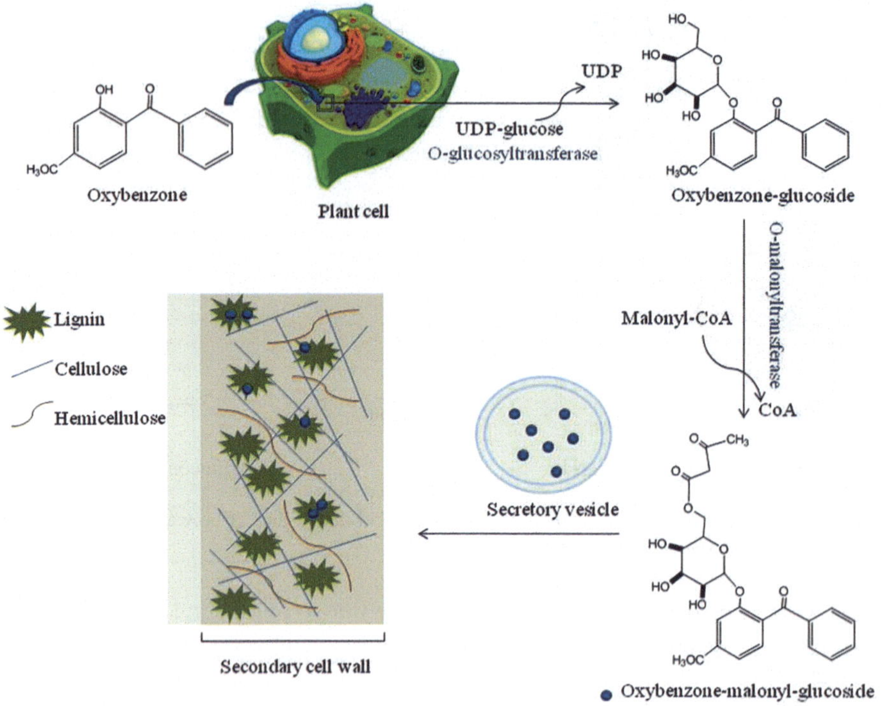

Fig. 3 The conjugation and deposition of the organic pollutant oxybenzone in the lignin component of plant cell wall. (Adapted from Jha et al. 2020)

(Goldberg-Cavalleri et al. 2023). For instance, the conjugated sulfonylurea herbicide chlorsulfuron gets accumulated in vacuolar vesicles in an ATP-driven manner by the ABC transporters. The herbicide may be conjugated with glucose or glutathione (Bartholomew et al. 2002; Hua et al. 2022). Therefore, there are different subtypes of transporters that recognize the different conjugation tags. This was also observed in the case of 1-chloro-2,4-dinitrobenzene in *Arabidopsis,* where AtMRP2 and AtMRP3 transporters carry out the transport of glutathione and glucuronide conjugates into the vacuoles, respectively (Devi et al. 2024; Hua et al. 2022). The ABC transporters have also been reported to be involved in the detoxification mechanism of heavy metals, such as imparting resistance toward arsenic toxicity by sequestering arsenic-phytochelating conjugates (Zhang et al. 2017; Jalmi 2022). The overexpression of the ABC gene in rice enhanced the accumulation of bentazone pesticide in cells (Qiao et al. 2024). Similarly, pesticide ametryn catabolism requires the ABC transporter proteins in the plasma membrane (Qiao et al. 2023).

Another family of transporters that conduct the detoxification of xenobiotics are MATE transporters. These are large transmembrane (12 domains) proteins that utilize the H+/Na + gradient to transfer the conjugates into the vacuole (Zhang et al. 2023; Radhakrishnan et al. 2024). Since these transporters maintain the balance of

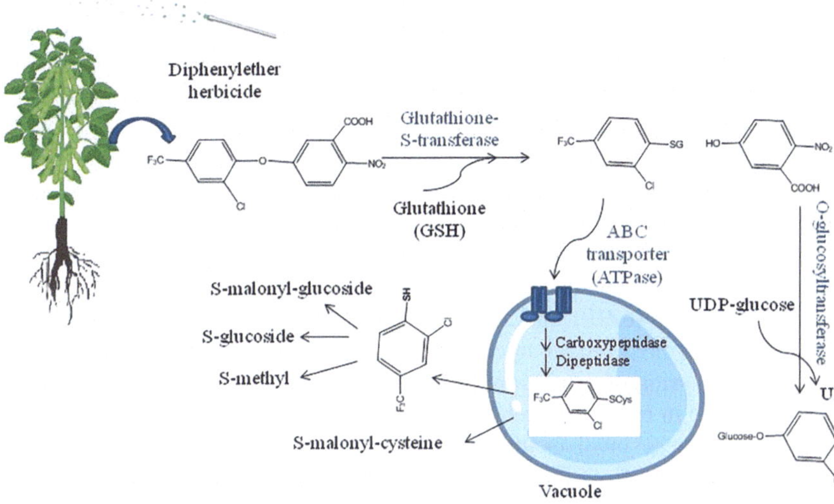

Fig. 4 The pathways involved in the detoxification of diphenyl ether herbicides in plants. (Adapted from Schroder and Collins 2002)

secondary metabolites during abiotic as well as biotic stresses, therefore they tend to regulate the oxidative stress in plant cells (Lu et al. 2019). The expression of these proteins can reduce aluminum toxicity in plants by chelating the aluminum ions using citrate efflux (Kar et al. 2022). Their role in the detoxification of antibiotic norfloxacin and herbicide atrazine has been elucidated in Arabidopsis (Alberto et al. 2022; Miyauchi et al. 2017). As mentioned earlier, after the accumulation in the vacuole, some of the xenobiotic conjugates may get further metabolized. Depending upon the xenobiotic, the glutathione conjugates, like those of herbicide atrazine and other diphenyl ether herbicides, might be acted upon by the carboxypeptidase that cleaves the glycine residue or by the dipeptidase that cleaves at the glutamic acid amino-terminal. The cysteinyl conjugates thus produced are effluxed again into the cytosol (Fig. 4), where these get further digested by lyases and are directed toward the apoplast to be tagged to the lignin and cellulose components via secretory vesicles. Apart from the cysteinyl products, many other soluble and insoluble derivatives, including methylthio, thiol, and thiolactyl, are obtained after the vacuolar processing (Malchi et al. 2022; Javaid et al. 2022).

5 Xenobiotic-Induced Signaling in Plants

It has been documented that different xenobiotics elicit drastically different responses, causing changes in the gene expression patterns in plants (Ekman et al. 2005; Chen 2020). On the contrary, it is also suggested by transcriptomics analysis that xenobiotics elicit similar classes of genes involved in antioxidant defense, cell repair, transformation, conjugation, compartmentation, transport, etc. (Pinto et al. 2021). It could be said that class-specific responses are induced in plants. For instance, the response toward polyaromatic hydrocarbons is different from that of herbicides (Zokm et al. 2022). However, the response of plants toward a particular xenobiotic depends upon the physiological conditions that might induce toxicity or could possibly lead to tolerance toward the pollutant. This could be observed in the case of the herbicide atrazine in *Arabidopsis*. The expression of genes like sigma factor binding protein and glutamyl-tRNA reductase induced tolerance, whereas although the genes for cytochromes P450 and peroxidases were unregulated under stress, they didn't participate in inducing tolerance against atrazine. The basic region/leucine zipper (bZIP) and zinc finger (ZAT12) transcription factors have been known to be involved in tolerance against atrazine (Alberto et al. 2022). Besides, the expression of cytochrome P450 has been shown to be dose-dependent and tissue-specific in response to chromium toxicity in plants (Yu et al. 2020). The detoxification process involves the mitogen-activated protein kinase (MAPK)-driven pathways. Apart from the basic leucine and zinc finger motifs, differentially expressed transcription factors include MYB, RING-H2, and WRKY in response to heavy metals (Jan et al. 2020).

The xenobiotic signaling pathways tend to overlap the pathways elicited under other biotic as well as abiotic stresses. There has been an increased involvement of salicylic acid and brassinosteroid phytohormones that tend to accelerate the metabolism of xenobiotics. Salicylic acid has been known to act as the detoxification agent for heavy metals (cadmium), herbicide safeners, pesticides clethodim, napropamide, etc. (Bocova et al. 2013; Lu et al. 2015). An exogenous application of salicylic acid alleviates the toxic effects of xenobiotics by enhancing the activities of peroxidases and glutathione-S-transferases, thereby improving the overall antioxidant system (Kumar et al. 2023). Triclosan is an antimicrobial agent used as a common constituent of soaps and body washes. The untreated wastewater contaminated with high levels of triclosan causes damage to the overall growth of plants and thereby reduces productivity. It has been reported that salicylic acid induces tolerance against triclosan in plants (Peng et al. 2021). The external application of brassinosteroids leads to the reduction in pesticide (fungicide chlorothalonil, insecticide imidacloprid) residues in food crops (Wang et al. 2017; Sharma et al. 2018). This phytohormone is also known to conduct the detoxification of cadmium and phenanthrene (polyaromatic hydrocarbon) by reducing oxidative stress, increasing antioxidant activity, and thereby enhancing the photosynthetic efficiency of plants (Ahammed et al. 2013). These brassinosteroids also regulate the ascorbate-glutathione cycle, which eventually helps in the alleviation of xenobiotic toxicity in plants (Li et al. 2022). A pesticide degradation factor has been characterized in rice on exposure to acetochlor and atrazine. The overexpression of the factor OsBR6ox that is involved in the brassinosteroid biosynthesis pathway leads to tolerance against the pesticides by up-regulating the expression of glycosyltransferase, laccase, and O-methyltransferase (Qiao et al. 2022). Another phytohormone that participates in the catabolism of xenobiotics is jasmonic acid and its derivatives (methyl jasmonate, oxylipin). The glutathione-assisted detoxification of the organochlorine fungicide chlorothalonil is induced by jasmonic acid (Yu et al. 2022). Besides, the foliar application of methyl jasmonate has been shown to elevate the chlorophyll content and growth and reduce pesticide residues, thereby preventing cellular damage in plants (Zhang and Yang 2021). The interaction between salicylic acid and jasmonic acid affects the heavy metal uptake capacity of bioaccumulator plants (Shi et al. 2024). The brassinosteroid and jasmonic acid application reduce lead toxicity in plants (Emamverdian et al. 2024). Therefore, these phytohormones provide an alternative sustainable approach for the alleviation of xenobiotic toxicity in plants.

6 Conclusion

Plants, being sessile in nature, are constantly exposed to a wide range of xenobiotics. These pollutants tend to have a detrimental impact on the physiological, biochemical, and molecular functioning of plants. Therefore, the plants have evolved the xenobiotic detoxification mechanisms characterized by three phases, namely, transformation, conjugation, and compartmentation. The sequestration of transformed or conjugated xenobiotics usually occurs in a vacuole, where it might get degraded further, or it could get deposited in the cell wall. It is imperative to increase the polarity of the lipophilic xenobiotics in order to sequester them inside the cells. Some pollutants may be completely metabolized into an endogenous product that could be utilized during natural processes. The ability of plants to metabolize and accumulate hazardous xenobiotics could be exploited for the remediation of polluted soil and water resources. Although there have been many studies regarding the enzymes and transporters, there is a need to characterize the regulatory proteins and transcription factors involved in the detoxification processes. Moreover, the identification of xenobiotic (especially pesticide) tolerance genes could be used to engineer tolerant crops producing higher yields.

References

Ahammed, G.J., Choudhary, S.P., Chen, S., Xia, X., Shi, K., Zhou, Y. and Yu, J., 2013. Role of brassinosteroids in alleviation of phenanthrene–cadmium co-contamination-induced photosynthetic inhibition and oxidative stress in tomato. *Journal of Experimental Botany*, *64*(1), pp. 99–213.

Alberto, D., Ramel, F., Sulmon, C., Gouesbet, G. and Couee, I., 2022. Differential effects of root-level exposure to triazine xenobiotics on root development plasticity in Arabidopsis thaliana. *Acta Physiologiae Plantarum*, *44*(11), pp. 111.

Alengebawy, A., Abdelkhalek, S.T., Qureshi, S.R. and Wang, M.Q., 2021. Heavy metals and pesticides toxicity in agricultural soil and plants: *Ecological risks and human health implications*. *Toxics*, *9*(3), pp. 42.

Bartholomew, D.M., Van Dyk, D.E., Lau, S.M.C., O'Keefe, D.P., Rea, P.A. and Viitanen, P.V., 2002. Alternate energy-dependent pathways for the vacuolar uptake of glucose and glutathione conjugates. *Plant Physiology*, *130*(3), pp. 1562–1572.

Bartikova, H., Skalova, L., Stuchlíkova, L., Vokral, I., Vanek, T. and Podlipna, R., 2015. Xenobiotic-metabolizing enzymes in plants and their role in uptake and biotransformation of veterinary drugs in the environment. *Drug Metabolism Reviews*, *47*(3), pp. 374–387.

Bocova, B., Huttova, J., Mistrik, I. and Tamas, L., 2013. Auxin signalling is involved in cadmium-induced glutathione-S-transferase activity in barley root. *Acta Physiologiae Plantarum*, *35*, pp. 2685–2690.

Capozzi, F. and Spagnuolo, V., 2024. The potential of plants to absorb xenobiotics. *Plants*, *13*(7), pp. 922.

Chakraborty, P., Biswas, A., Dey, S., Bhattacharjee, T. and Chakrabarty, S., 2023. Cytochrome P450 gene families: role in plant secondary metabolites production and plant defense. *Journal of Xenobiotics*, *13*(3), pp. 402–423.

Chen, C.H., 2020. Xenobiotic metabolic enzymes: bioactivation and antioxidant defense (pp. 109–118). Cham, Switzerland: Springer.

Chen, Z.J., Li, S.Y., Qu, Y.N., Ai, G., Wang, Y.H., Pan, D.J., Wang, H.W., Lu, D. and Liu, X.L., 2024. Comprehensive analyses show the enhancement effect of exogenous melatonin on fluroxypyr-meptyl multiple phase metabolisms in *Oryza sativa* for reducing environmental risks. *Pesticide Biochemistry and Physiology*, *203*, pp. 106021.

Chia, J.C., 2021. Phytochelatin synthase in heavy metal detoxification and xenobiotic metabolism, Mendes, K.F., Sousa, R., Mielke, M.C. (ed), *Biodegradation technology of organic and inorganic pollutants*, 10.

Christou, A., Antoniou, C., Christodoulou, C., Hapeshi, E., Stavrou, I., Michael, C., Fatta-Kassinos, D. and Fotopoulos, V., 2016. Stress-related phenomena and detoxification mechanisms induced by common pharmaceuticals in alfalfa (*Medicago sativa* L.) plants. *Science of the Total Environment*, *557*, pp. 652–664.

Devi, R., Goyal, P., Verma, B., Hussain, S., Chowdhary, F., Arora, P. and Gupta, S., 2024. A transcriptome-wide identification of ATP-binding cassette (ABC) transporters revealed participation of ABCB subfamily in abiotic stress management of *Glycyrrhiza glabra* L. *BMC genomics*, *25*(1), p. 315.

Ekman, D.R., Wolfe, N.L. and Dean, J.F., 2005. Gene expression changes in *Arabidopsis thaliana* seedling roots exposed to the munition hexahydro-1, 3, 5-trinitro-1, 3, 5-triazine. *Environmental Science and Technology*, *39*(16), pp. 6313–6320.

Emamverdian, A., Khalofah, A., Pehlivan, N., Zia-ur-Rehman, M., Li, Y. and Zargar, M., 2024. Exogenous application of jasmonates and brassinosteroids alleviates lead toxicity in bamboo by altering biochemical and physiological attributes. *Environmental Science and Pollution Research*, *31*(5), pp. 7008–7026.

Espinosa-Vellarino, F.L., Garrido, I., Casimiro, I., Silva, A.C., Espinosa, F. and Ortega, A., 2024. Enzymes involved in antioxidant and detoxification processes present changes in the expression levels of their coding genes under the stress caused by the presence of Antimony in tomato. *Plants*, *13*(5), pp. 609.

Esteves, F., Rueff, J. and Kranendonk, M., 2021. The central role of cytochrome P450 in xenobiotic metabolism—a brief review on a fascinating enzyme family. *Journal of xenobiotics*, *11*(3), pp. 94–114.

Fernandez, M.E., del Rosario Morel, M., Clebot, A.C., Zalazar, C.S. and de los Milagros Ballari, M., 2022. Effectiveness of a simple biomixture for the adsorption and elimination of 2, 4-dichlorophenoxyacetic acid (2, 4-D) herbicide and its metabolite, 2, 4-dichlorophenol (2, 4-DCP), for a biobed system. *Journal of Environmental Chemical Engineering*, *10*(1), pp. 106877.

Gałązka, A., Jankiewicz, U. and Szczepkowski, A., 2023. Biochemical characteristics of laccases and their practical application in the removal of xenobiotics from water. *Applied Sciences*, *13*(7), pp. 4394.

Goldberg-Cavalleri, A., Onkokesung, N., Franco-Ortega, S. and Edwards, R., 2023. ABC transporters linked to multiple herbicide resistance in blackgrass (*Alopecurus myosuroides*). *Frontiers in Plant Science*, 14, pp. 1082761.

Hossain, A., Bhutia, K.L., Pramanick, B., Maitra, S., Ibrahimova, U., Kumari, V.V., Ahmad, Z., Uzair, M. and Aftab, T., 2022. Glutathione in higher plants: biosynthesis and physiological mechanisms during heat and drought-induced oxidative stress. In Antioxidant defense in plants: molecular basis of regulation (pp. 181–214). Singapore: Springer Nature Singapore.

Hua, Y., Yuan, Y., Qin, Y., Zhang, C., Wang, X., Feng, S. and Lu, Y., 2022. Advances in the agro-environment migration of organic chemical pollutants and their biotransformation in crops. *Agronomy*, *12*(12), pp. 3009.

Hwang, J.U., Song, W.Y., Hong, D., Ko, D., Yamaoka, Y., Jang, S., Yim, S., Lee, E., Khare, D., Kim, K. and Palmgren, M., 2016. Plant ABC transporters enable many unique aspects of a terrestrial plant's lifestyle. *Molecular Plant*, *9*(3), pp. 338–355.

Jaiswal, D., Pandey, A. and Agrawal, S.B., 2021. Role of glutathione in enhancing plant tolerance to organic pollutants. In *Handbook of Bioremediation*, Academic Press, pp. 523–533

Jalmi, S.K., 2022. The role of ABC transporters in metal transport in plants. In Plant Metal and Metalloid Transporters (pp. 55–71). Singapore: Springer Nature Singapore.

Jan, S., Singh, R., Bhardwaj, R., Ahmad, P. and Kapoor, D., 2020. Plant growth regulators: a sustainable approach to combat pesticide toxicity. *3 Biotech*, 10(11), pp. 466.

Javaid, M.M., Mahmood, A., Bhatti, M.I.N., Waheed, H., Attia, K., Aziz, A., Nadeem, M.A., Khan, N., Al-Doss, A.A., Fiaz, S. and Wang, X., 2022. Efficacy of metribuzin doses on physiological, growth, and yield characteristics of wheat and its associated weeds. *Frontiers in Plant Science*, *13*, pp. 866793.

Jha, P., Jobby, R., Kudale, S., Modi, N., Dhaneshwar, A. and Desai, N., 2013. Biodegradation of phenol using hairy roots of *Helianthus annuus* L. *International Biodeterioration and Biodegradation*, 77, pp. 106–113.

Jha, P., Sen, R., Jobby, R., Sachar, S., Bhatkalkar, S. and Desai, N., 2020. Biotransformation of xenobiotics by hairy roots. *Phytochemistry*, *176*, pp. 112421.

Jiang, Y., Zhang, T., Kusumanchi, P., Han, S., Yang, Z. and Liangpunsakul, S., 2020. Alcohol metabolizing enzymes, microsomal ethanol oxidizing system, cytochrome P450 2E1, catalase, and aldehyde dehydrogenase in alcohol-associated liver disease. *Biomedicines*, *8*(3), pp. 50.

Jiao, L., Ding, H., Wang, L., Zhou, Q. and Huang, X., 2017. Bisphenol A effects on the chlorophyll contents in soybean at different growth stages. *Environmental Pollution*, 223, pp. 426–434.

Kantharaj, V., Ramasamy, N.K., Yoon, Y.E., Cheong, M.S., Kim, Y.N., Lee, K.A., Kumar, V., Choe, H., Kim, S.Y., Chohra, H. and Lee, Y.B., 2022. Auxin-glucose conjugation protects the rice (*Oryza sativa* L.) seedlings against hydroxyurea-induced phytotoxicity by activating UDP-glucosyltransferase enzyme. *Frontiers in Plant Science*, 12, pp. 767044.

Kar, D., Pradhan, A.A., Dutta, A., Bhagavatula, L. and Datta, S., 2022. The Multidrug and Toxic Compound Extrusion (MATE) family in plants and their significance in metal transport. In *Plant Metal and Metalloid Transporters* (pp. 151–177). Singapore: Springer Nature Singapore.

Kumar, A., Yadav, P.K., Singh, S. and Singh, A., 2023. An overview on the modulation of pesticide detoxification mechanism via salicylic acid in the plants. *Environmental Pollutants and Bioavailability*, 35(1), pp. 2242701.

Kumar, S. and Trivedi, P.K., 2018. Glutathione S-transferases: role in combating abiotic stresses including arsenic detoxification in plants. *Frontiers in Plant Science*, 9, pp. 751.

Li, B.B., Fu, Y.S., Li, X.X., Yin, H.N. and Xi, Z.M., 2022. Brassinosteroids alleviate cadmium phytotoxicity by minimizing oxidative stress in grape seedlings: Toward regulating the ascorbate-glutathione cycle. *Scientia Horticulturae*, 299, pp. 111002.

Lonova, K., Kalousek, P., Klems, M. and Fiserova, H., 2023. Changes in photosynthetic dispositions in pea plants caused by fluoranthene and flurochloridone: from the subcellular level to the anatomical changes. *Acta Physiologiae Plantarum*, *45*(1), pp. 11.

Lu, P., Magwanga, R.O., Kirungu, J.N., Hu, Y., Dong, Q., Cai, X., Zhou, Z., Wang, X., Zhang, Z., Hou, Y. and Wang, K., 2019. Overexpression of cotton a DTX/MATE gene enhances drought, salt, and cold stress tolerance in transgenic *Arabidopsis*. *Frontiers in Plant Science*, 10, pp. 299.

Lu, Y.C., Zhang, S. and Yang, H., 2015. Acceleration of the herbicide isoproturon degradation in wheat by glycosyltransferases and salicylic acid. *Journal of Hazardous Materials, 283*, pp. 806–814.

Malchi, T., Eyal, S., Czosnek, H., Shenker, M. and Chefetz, B., 2022. Plant pharmacology: Insights into in-planta kinetic and dynamic processes of xenobiotics. *Critical Reviews in Environmental Science and Technology*, *52*(19), pp. 3525–3546.

Marghade, D.T., Chahande, A.D., Tiwari, M.S. and Patil, P.D., 2021. Microbial degradation of xenobiotic compounds. *Recent advances in microbial degradation*, pp. 173–217.

Miglani, R., Parveen, N., Kumar, A., Ansari, M.A., Khanna, S., Rawat, G., Panda, A.K., Bisht, S.S., Upadhyay, J. and Ansari, M.N., 2022. Degradation of xenobiotic pollutants: an environmentally sustainable approach. *Metabolites*, *12*(9), pp. 818.

Miyauchi, H., Moriyama, S., Kusakizako, T., Kumazaki, K., Nakane, T., Yamashita, K., Hirata, K., Dohmae, N., Nishizawa, T., Ito, K. and Miyaji, T., 2017. Structural basis for xenobiotic extrusion by eukaryotic MATE transporter. *Nature Communications*, *8*(1), pp. 1633.

Muerdter, C.P., Powers, M.M., Webb, D.T., Chowdhury, S., Roach, K.E. and LeFevre, G.H., 2023. Functional group properties and position drive differences in xenobiotic plant uptake rates, but metabolism shares a similar pathway. *Environmental Science and Technology Letters*, *10*(7), pp. 596–603.

Nuruzzaman, M., Zhang, R., Cao, H.Z. and Luo, Z.Y., 2014. Plant pleiotropic drug resistance transporters: transport mechanism, gene expression, and function. *Journal of Integrative Plant Biology*, *56*(8), pp. 729–740.

Parween, T., Jan, S. and Fatma, T., 2012. Evaluation of oxidative stress in *Vigna radiata* L. in response to chlorpyrifos. *International Journal of Environmental Science and Technology*, *9*, pp. 605–612.

Peng, D., Wang, W., Liu, A., Zhang, Y., Li, X., Wang, G., Jin, C., Guan, C. and Ji, J., 2021. Comparative transcriptome combined with transgenic analysis reveal the involvement of salicylic acid pathway in the response of *Nicotiana tabacum* to triclosan stress. *Chemosphere*, *270*, pp. 129456.

Pinto, A.P., Lopes, M.E., Dordio, A. and Castanheiro, J.E.F., 2021. Microbe and Plant-Assisted Remediation of Organic Xenobiotics. *Handbook of Assisted and Amendment: Enhanced Sustainable Remediation Technology*, 437–475.

Potęga, A., 2022. Glutathione-mediated conjugation of anticancer drugs: an overview of reaction mechanisms and biological significance for drug detoxification and bioactivation. *Molecules*, *27*(16), pp. 5252.

Qiao, Y., Ma, L.Y., Chen, Z.J., Wang, Y., Gu, Y. and Yang, H., 2022. OsBR6ox, a member in the brassinosteroid synthetic pathway facilitates degradation of pesticides in rice through a specific DNA demethylation mechanism. *Science of The Total Environment*, 838, pp. 156503.

Qiao, Y., Wang, Y., Gu, Y., Zhang, N., Yang, H. and Liu, J., 2024. An ABC transporter-mediated transport and metabolism of the pesticide bentazone in rice (*Oryza sativa* L.). *Journal of Advanced Research*. https://doi.org/10.1016/j.jare.2024.08.020

Qiao, Y., Zhang, A.P., Ma, L.Y., Zhang, N., Liu, J. and Yang, H., 2023. An ABCG-type transporter intensifies ametryn catabolism by Phase III reaction mechanism in rice. *Journal of Hazardous Materials*, *457*, pp. 131804.

Radhakrishnan, A., Shakshi, Nandini, R., Kumar, A., Kapardar, R.K. and Srivastav, R., 2024. Role of MATE transporters in xenobiotics tolerance. *Microbes Based Approaches for the Management of Hazardous Contaminants*, 411–419.

Rani, S. and Kumari, N., 2023. Alteration in physio-biochemical attributes and chlorophyll fluorescence of mungbean caused by arsenic stress. *Vegetos*, *36*(1), pp. 79–86.

Rani, S., Kumari, N., and Sharma, V., 2023. Biosynthesized zinc oxide nanoparticles as a novel nanofertilizer influence growth, yield, antioxidant enzymes, and arsenic accumulation in mungbean (*Vigna radiata* L.) under arsenic stress. *Journal of Soil Science and Plant Nutrition*, *23*(2), 2360–2380.

Schriever, C. and Lamshoeft, M., 2020. Lipophilicity matters–a new look at experimental plant uptake data from literature. *Science of the Total Environment*, *713*, pp. 136667.

Schroder, P. and Collins, C., 2002. Conjugating enzymes involved in xenobiotic metabolism of organic xenobiotics in plants. International Journal of Phytoremediation, *4*(4), pp. 247–265.

Shabbir, A., Haider, K., Rehman, K., Akash, M.S.H. and Chen, S., 2022. Biochemical activation and functions of drug-metabolizing enzymes. In *Biochemistry of Drug Metabolizing Enzymes* (pp. 1–27). Academic Press.

Sharma, A., Kumar, V., Kumar, R., Shahzad, B., Thukral, A.K. and Bhardwaj, R., 2018. Brassinosteroid-mediated pesticide detoxification in plants: A mini-review. *Cogent Food and Agriculture*, *4*(1), pp. 1436212.

Sharma, A., Kumar, V., Thukral, A.K. and Bhardwaj, R., 2019. Responses of plants to pesticide toxicity: an overview. *Planta Daninha*, *37*, pp. e019184291.

Sharma, B., Singh, S., Jaiswal, S.K. and Siddiqi, N.J., 2016. Xenobiotics-mediated modulation of ATPases and biomedical implications. In: Regulation of Membrane Na+-K+ ATPase, (pp. 173–183) Springer Champ.

Shi, A., Liu, J., Zou, S., Rensing, C., Zhao, Y., Zhang, L., Xing, S. and Yang, W., 2024. Enhancement of cadmium uptake in *Sedum alfredii* through interactions between salicylic acid/jasmonic acid and rhizosphere microbial communities. *Science of The Total Environment*, *947*, pp. 174585.

Singh, S., Bashri, G., Singh, A. and Prasad, S.M., 2016. Regulation of xenobiotics in higher plants: signalling and detoxification. *Plant Responses to Xenobiotics*, 39–56.

Sousa, B., Lopes, J., Leal, A., Martins, M., Soares, C., Azenha, M., Fidalgo, F. and Teixeira, J., 2021. Specific glutathione-S-transferases ensure an efficient detoxification of diclofenac in *Solanum lycopersicum* L. plants. *Plant Physiology and Biochemistry*, *168*, 263–271.

Speranza, A., Crosti, P., Malerba, M., Stocchi, O. and Scoccianti, V., 2011. The environmental endocrine disruptor, bisphenol A, affects germination, elicits stress response and alters steroid hormone production in kiwifruit pollen. *Plant Biology*, *13*(1), pp. 209–217.

Vaskova, J., Kocan, L., Vasko, L. and Perjesi, P., 2023. Glutathione-related enzymes and proteins: a review. *Molecules*, *28*(3), pp. 1447.

Vicidomini, C., Palumbo, R., Moccia, M. and Roviello, G.N., 2024. Oxidative Processes and Xenobiotic Metabolism in Plants: Mechanisms of Defense and Potential Therapeutic Implications. Journal of Xenobiotics, 14(4), pp. 1541-1569.

Wang, Z., Jiang, Y., Peng, X., Xu, S., Zhang, H., Gao, J. and Xi, Z., 2017. Exogenous 24-epibrassinolide regulates antioxidant and pesticide detoxification systems in grapevine after chlorothalonil treatment. *Plant Growth Regulation*, *81*, pp. 455–466.

Wei, H., Tang, M. and Xu, X., 2023. Mechanism of uptake, accumulation, transport, metabolism and phytotoxic effects of pharmaceuticals and personal care products within plants: A review. *Science of The Total Environment*, *892*, pp. 164413.

Yu, G.B., Chen, R.N., Chen, Q.S., Chen, F.Q., Liu, H.L., Ren, C.Y., Zhang, Y.X., Yang, F.J. and Wei, J.P., 2022. Jasmonic acid promotes glutathione assisted degradation of chlorothalonil during tomato growth. *Ecotoxicology and Environmental Safety*, *233*, pp. 113296.

Yu, X.Z., Lu, C.J., Tang, S. and Zhang, Q., 2020. Transcriptomic analysis of cytochrome P450 genes and pathways involved in chromium toxicity in *Oryza sativa. Ecotoxicology*, *29*, pp. 503–513.

Zhang, J., Hwang, J.U., Song, W.Y., Martinoia, E. and Lee, Y., 2017. Identification of amino acid residues important for the arsenic resistance function of *Arabidopsis* ABCC 1. *FEBS Letters*, *591*(4), pp. 656–666.

Zhang, J., Li, Q., Li, C., Wang, Q., Hou, X., Zhao, C., Li, S. and Guo, Q., 2023. Research Progress on MATE Transporters in Plants. *Chinese Bulletin of Botany*, 58(3), pp. 461.

Zhang, J.J. and Yang, H., 2021. Metabolism and detoxification of pesticides in plants. *Science of the Total Environment*, *790*, pp. 148034.

Zokm, G.M.E., Ismail, M.M. and Okbah, M.A.E., 2022. Seaweed as bioindicators of organic micropollutants polycyclic aromatic hydrocarbons (PAHs) and organochlorine pesticides (OCPs). *Environmental Science and Pollution Research*, *29*(23), 34738–34748.

Plant Defense System Against Xenobiotics: Molecular Mechanisms and Metabolic Responses

Abhishek Singh, Pooran Chand, and Tapas Ranjan Das

Abstract Plants' complex defense systems, especially those against xenobiotics—defending themselves against foreign substances that can disrupt physiological processes—are closely related to their capacity to flourish in a variety of settings. Due to their continuous exposure to a variety of xenobiotics that may be detrimental to their internal organisms, plants have evolved extraordinarily sophisticated systems that comprise vast arrays of enzymes that degrade xenobiotics. Environmental xenobiotics, or different chemical contaminants, have gotten worse due to an increase in anthropogenic activities. Comprehending the interaction between genes and proteins in plant defense has important consequences for agriculture and environmental management as well as providing insight into the basic biological mechanisms supporting plant resilience. Researchers can improve crop yields, boost plant tolerance, and advance sustainable agricultural practices by developing strategies to clarify the pathways and processes that plants use to detect and detoxify xenobiotics. This chapter seeks to explain the complex metabolic reactions and molecular mechanism that plants use to protect themselves against dangerous xenobiotics, illuminating advanced strategies that ensure their survival and resilience.

Keywords Biochemical strategies · Detoxification pathways · Gene expression · Gene-protein interaction · Genetic basis · Xenobiotics

A. Singh · P. Chand
Department of Genetics and Plant Breeding, Sardar Vallabhbhai Patel University of Agriculture and Technology, Meerut, Uttar Pradesh, India

T. R. Das (✉)
Division of Genetics, ICAR-Indian Agricultural Research Institute, New Delhi, India

S. Roy, V. Mandal (eds.), *Plant-Microbe Interaction under Xenobiotic Exposure*,
https://doi.org/10.1007/978-981-96-8260-7_9

1 Introduction

The term xenobiotics, which comes from the Greek word meaning "foreign chemicals," refers to any chemical or other material that is not typically present in ecosystems and that cannot be used by plants to produce energy. In modern agriculture, xenobiotics—chemicals that are foreign to plants—are commonly used pesticides, herbicides, etc., which have tainted nearly every area of our surroundings. In the natural world, xenobiotics are secondary metabolites that are created by other plants and leach from aerial portions or enter the rhizosphere to enter the environment. Toxicology, ecotoxicology, and stress physiology are being used to characterize plant responses to various chemical challenges and stimuli. Through volatilization, spray drift, runoff, leaching, and absorption, phytochemical crop protection releases xenobiotic chemicals into the environment. These xenobiotics are absorbed and accumulate despite the fact that they cannot be eaten or used as fuel (Dévier et al. 2011). Comprehending the exposure of organisms to xenobiotic chemicals is crucial for comprehending environmental and toxicological chemistry. It is challenging to forecast the intricate biological effects of chemical pollution combinations based solely on the effects of individual contaminants (Serra et al. 2013).

Xenobiotics can affect an organism's growth and physiology, either individually or in combination, and this differs from species to species. Furthermore, in higher plants, the control and expression of certain genes may be altered via coordinated signaling networks. As sessile creatures, plants have evolved a sophisticated defense system that incorporates biochemical and molecular techniques to mitigate the impact of these potentially dangerous compounds. The initial level of operation of this defense system is its genetic base, which encodes a specific protein responsible for recognizing and responding to xenobiotic stress. It is evident that xenobiotics are chemicals that plants do not naturally produce, and that their exposure to these compounds is growing due to an increase in industrial pollution. Plants have developed various tolerance mechanisms and signal perceptions to deal with these xenobiotics. Plants seem to possess isoenzymes that are specific to secondary plant substrates or xenobiotics. This chapter aims to investigate the intricate relationship between expression of gene and protein function in the context of plant defense systems against xenobiotics. It also highlights recent advancements and possible directions for further study in this constantly changing field.

2 Environmental and Physiochemical Fate of Xenobiotic Compounds

Water and carbon dioxide are produced when a parent chemical oxidizes and degrades biologically. A particular enzyme produced by a degrading cell catalyzes each step of the degradation pathway. But the exact structure of the chemical which contains the necessary enzymes like oxygenase determines how the xenobiotics

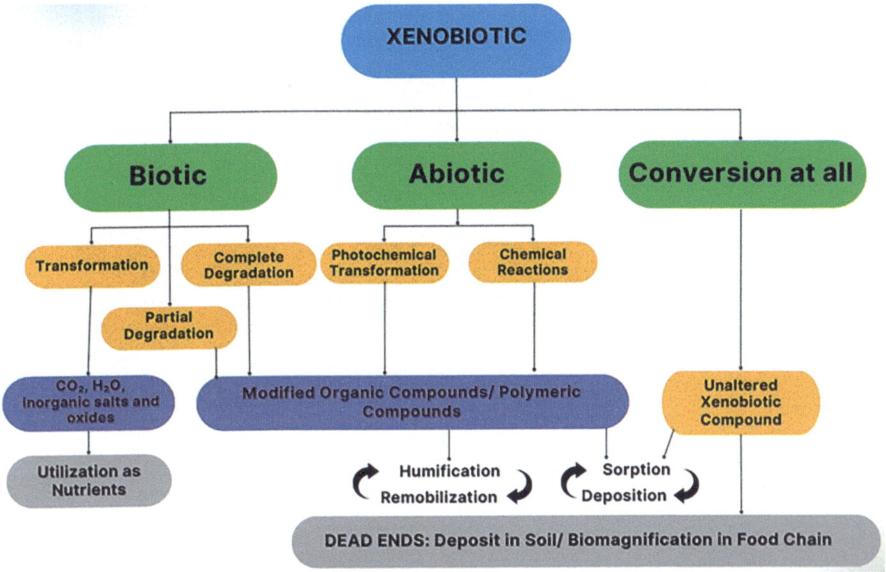

Fig. 1 Environment fate of xenobiotic compounds

degrade. Energy and reducing equivalents for the breakdown process are produced by the metabolism of these enzymes (Bonjoko 2014). This illustration is shown in Fig. 1. The environmental fate of xenobiotic compounds is a complex interplay of biotic and abiotic degradation processes, as well as humification and remobilization mechanisms, which determine the environmental destiny of xenobiotic chemicals. Comprehending these mechanisms is essential for evaluating the durability and possible ecological consequences of these substances in diverse settings. Additionally, humification processes can lead to the incorporation of these compounds into the soil organic matter, affecting their mobility and bioavailability. Remobilization further complicates the fate of xenobiotics, as previously bound compounds can be released back into the environment under changing conditions. Understanding these chemicals' mobility, persistence, and possible ecological effects depends on how they interact with environmental matrices through sorption and desorption processes. Through the clarification of these mechanisms, scientists can more accurately evaluate the hazards linked to xenobiotics and create plans to lessen their detrimental impacts on ecosystems.

Some xenobiotic compounds have complex chemical structures, with nonphysiological connections connecting some of their molecules. These chemicals and the natural substances they are generated from have some structural similarities (see Table 1). Xenobiotics are compounds with a high molecular weight that are insoluble or just weakly soluble in water and that condense into aromatic and polycyclic rings. Synthetic polymers, azo compounds, halogen-modified aliphatic chemicals, aromatic hydrocarbons, aromatic amines, and polycyclic hydrocarbons (PAHs) are examples of these xenobiotics.

Table 1 Some xenobiotic compounds and their chemical formula

Xenobiotic compounds	Chemical formula
Pesticides	
1. Deltamethrin.	$C_{22}H_{19}Br_2NO_3$
2. Paraquat.	$C_{12}H_{14}C_{12}N2$
3. Soman (nerve gas).	$C_7H_{16}F$
4. Methyl isocyanate.	C_2H_3NO
Food additive	
1. Vinegar	CH_3COOH
2. Citric acid	$C_6H_8O_7$
3. Methyl cyclopropane	C_4H_{10}
4. Lecithin	$C_{35}H_{66}NO_7P$
Pharmaceutical industry	
1. Paracetamol	$C_8H_9NO_2$
2. Ibuprofen	$C_{13}H_{10}O_2$
3. Caffeine	$C_8H_{10}N_4O_2$
Paint industry	
1. Xylene	C_8H_{10}
2. Toluene	C_7H_8
Wood preservatives	
1. Creosate	$(CH_3)C_6H_4(OH)$
2. Cupric chromate	$CrCuO_4$

3 Xenobiotics Signaling and Sensing

It is clear that xenobiotics are chemical substances that plants do not naturally make and as a result of increased industrial pollutants, plants are increasingly exposed to these compounds. Mutant analysis is essential to comprehending how plants respond to xenobiotic stress. Numerous tolerance conditions have been connected for promoter mutations that cause the overexpression of detoxification genes or to mutations in genes that encode target proteins or are linked to target organelles. Response phenotypes are altered by mutations that impact the development of the organelles that xenobiotics target, by initiating the stress adaption process. Mutations in "Happy on norflurazon" (hon) impact plastid homeostasis and confer resistance to bleaching herbicides. Overexpression or knockout mutants have been used to evaluate several stages of xenobiotic detoxification. Mutant analysis is the main technique for characterizing new genes and studying the signaling and regulatory networks that can govern xenobiotic reactions. TGA TFs, which are the members of D branch of the basic region leucine zipper (bZIP) family, possess the unique capacity to identify and attach to regulatory regions with TGACG as the primary recognition sequences. This makes it possible for TFs to control the expression of target genes and participate in a number of biological regulatory processes. They function as either negative or positive regulators of the stress response by binding to *as-1-type* elements of target gene promoter. A number of gene families linked to oxidative and abiotic stress responses were not produced by the TGA quadruple mutant, commonly referred to as 4tga. The TGA factors are consequently essential elements of xenobiotic signaling cascades (Beringer et al. 2011).

The allelochemical benzoxazolin-2(3H)-one and reactive electrophilic species oxylipins have been found to be favorably associated with herbicide safener-induced responses in *Arabidopsis thaliana*. Oxidative stress is associated with the transcription factors ZAT12 and bZIP60, which are connected to the atrazine tolerance response in *Arabidopsis thaliana*. Because transcriptome responses are similar, it is challenging to distinguish between various activities. Studies suggest the involvement of transcription factor interactions or signaling pathways (Riechers et al. 2010). A considerable structural and functional genome reorganization is associated with polyploidy, and these changes may have an impact on the adaptive dynamics of recently formed species (Alix et al. 2017a, b). Both genome reduction (from genome duplication per se) and genome merging (aka "genome shock") in allopolyploids provide unique genetics and epigenetics interactions that influence the evolution of organisms (Doyle et al., 2008a, b). The decline in photosynthetic activity is also consistent with reactive oxygen species' (ROS) known negative impacts on outcomes (Reddy et al. 2008). The Spartina xenome can be investigated using the current transcriptome and genomic techniques, which are based on 52 genes that were found in *A. thaliana* to be possibly involved in the detoxification of phenanthrene. The glutathione S-transferases (GSTs) are a group of dispersed enzymes that help detoxify organic xenobiotics, by examining their expression profile of specific genes and the various patterns of parental expression alterations using RT-qPCR. The neo-allopolyploid *Spartina anglica* appears to be a viable phytoremediation candidate in PAH-contaminated wetlands. The proteins that might be implicated seem to be poorly understood (Riechers et al. 2010). TSPO-related (AtTSPO) is associated with the mitochondrial translocator protein, peripheral benzodiazepine receptors in mammals, and bacterial tryptophan-rich sensory protein. The complete characterization of these large gene families will take a long time. However, it was not demonstrated that any plant gene or protein family has the capacity to function as a xenobiotic sensor. The perception of cytokinin and ethylene is mediated by the plant component systems that have been described thus far. Nonetheless, chemotaxis, monoaromatic chemical sensing, and abiotic stress signaling are all mediated by two-component kinases in prokaryotes.

Moreover, xenobiotics' effects on plants depend on their chemical makeup, the physiological state of the plant, and the genotype of the plant. This clearly implies that gene expression regulation and xenobiotic detection systems need to be combined. Similar to Yap1 in yeast, the AtGPX3 gene appears to monitor and transduce hydrogen peroxide signals to downstream components in plants. Furthermore, it is also challenging to differentiate between the perception of changes mediated by ROS and direct ROS sensing. The nonenzymatic production of oxylipins by ROS action is a well-known example of this kind of indirect sensing of damage brought on by xenobiotics. These phytoprostanes and hydroxy fatty acids have a major signaling role in plant stress responses by enhancing defense against oxidative stress and promoting the expression of genes related to stress responses.

4 Transcription Factors (TFs) Relate Xenobiotic Responses

According to transcriptomics research, transcription factor encoding gene expression regulation has a role in plants' responses to xenobiotics (Wang et al. 2012). Genes encoding transcription factors were shown to be activated in correlation with atrazine sensitivity and changes in the ROS balance. Accordingly, ROS distribution and gradients are regulated by UPBEAT1 as shown in Table 1, a basic helix-loop-helix transcription factor. Ethylene, a key phytohormone involved in plant defense and stress signaling, often acts in concert with reactive oxygen species (ROS) like H_2O_2 to modulate gene expression and defense pathways. The BTB/TAZ domain proteins, known for their role in transcriptional regulation and protein degradation, may serve as molecular hubs integrating environmental signals into adaptive transcriptional responses. Their heightened sensitivity to H_2O_2 further implies a role in fine-tuning plant responses to oxidative damage induced by xenobiotics, thereby contributing to detoxification and resilience mechanisms (Ramel et al. 2007). Alternate expression patterns, however, indicate that the transcription may connected to the formation of defenses against xenobiotic stress. In conditions that can withstand atrazine, CEJ1, which codes for the transcription factor ERF/AP2, is strongly expressed. Additionally, this gene is triggered by toxins such as cantharidin (Bajsa et al. 2011a, b) and ochratoxin A (Wang et al. 2012). Moreover, ochratoxin A activates the genes WRKY75 and ZAT12, which are linked to atrazine defensive responses. Several ROS-generating effectors, including methyl viologen, the *Alternaria alternata* toxin, 3-aminotriazole, and ozone, activate the gene that codes for WRKY75 (Gadjev et al. 2006). The coordinated gene networks found in transcriptomics investigations are probably regulated by transcription factors associated with xenobiotics (Ekman et al. 2005; Wang et al. 2012). Herbicide safeners also use class II TGA transcription to trigger xenobiotic defense responses. The majority of these xenobiotic-associated transcription factors, which are also involved in the response to various abiotic or biotic stimuli, highlight the interaction mechanisms between xenobiotic, abiotic, and biotic challenges (Gadjev et al. 2006).

5 Transcriptomic Evidence for Xenobiotic Stress Responses

The range of treatments (trinitrotoluene, royal demolition explosive, glyphosate, and phenanthrene) consistently provides information about the coordinated gene expression responses that follow xenobiotic exposure, even though there are very few studies of this kind. Previous studies investigated the effects of xenobiotics on the expression of the candidate gene; later studies included tissue-specific patterns and expression kinetics for further characterization. The polycyclic aryl hydrocarbons and herbicide explosions stimulate the GST AtGSTU24 (Atlg17170) gene (Weisman et al. 2010). Using genome-wide techniques, it was demonstrated that xenobiotic response gene networks, which involve the activation and repression of hundreds of genes in plants, are large-scale and coordinated, similar to other abiotic stress responses (Mittler et al. 2004). This was also true for the xenobiotic reaction in vertebrates (Wang 2010) and invertebrates (Chapman et al. 2011). Gene expression varies significantly among the different xenobiotic treatment types. Each of the aforementioned gene classes takes part in these various patterns. Response genes specific to PAHs were identified by a broad-scale comparison of herbicides, heavy metals, and PAHs (Weisman et al. 2010). Ramel et al. (2007) evaluated physiological scenarios where atrazine medication causes tolerance or a fatal outcome and is dependent on carbon status. Both conditions are associated with large-scale transcriptome alterations. Different expression patterns are produced by the same atrazine treatment depending on whether damage or tolerance is sustained. Sigma factors binding protein (SIB1; At3g56710) and glutamyl-tRNA reductase (At1g09940) are two genes that are induced in the setting of tolerance but are produced differently in the context of harm. Several transcriptomics studies reveal that safeners' protective qualities were linked to their signaling-level activities, which resulted in significant changes in gene expression. Therefore, the specific type of xenobiotic to which a plant is exposed plays a critical role in shaping its molecular response. These responses, which include alterations in gene expression, activation of defense pathways, and modulation of signaling networks, are essential in determining whether the plant exhibits sensitivity or develops tolerance to the xenobiotic. Different xenobiotics can trigger distinct transcriptomic and proteomic changes depending on their chemical structure, mode of action, and the physiological state of the plant at the time of exposure. As a result, plants may activate different sets of stress-related genes and detoxification mechanisms, such as glutathione-S-transferases, cytochrome P450s, or ABC transporters, to mitigate the damage. Understanding this dynamic interaction between xenobiotic type and plant response not only provides insight into stress adaptation but also offers potential strategies for improving crop resilience through genetic or biotechnological interventions. The intricate function of transcription factors requires further investigation (Ramel et al. 2012). Transcriptomics, candidate gene expression, and mutant analysis should all be a part of this study.

6 Plant Defense System

In order to defend themselves against xenobiotics—dangerous compounds that can interfere with a plant's physiological functions—plants have developed an advanced defensive mechanism. Detoxification enzymes are essential to this defense because they aid in the breakdown and removal of these xenobiotics from plant tissues. Enzymes function through intricate biochemical pathways, enabling an array of chemical reactions that argue the plant's capacity to detoxify deleterious substances. The network of signaling channels that make up the plant immune system also enables plants to recognize and react to xenobiotics and pathogens, initiating the proper defense mechanisms in the process. It highlights the intricate defensive

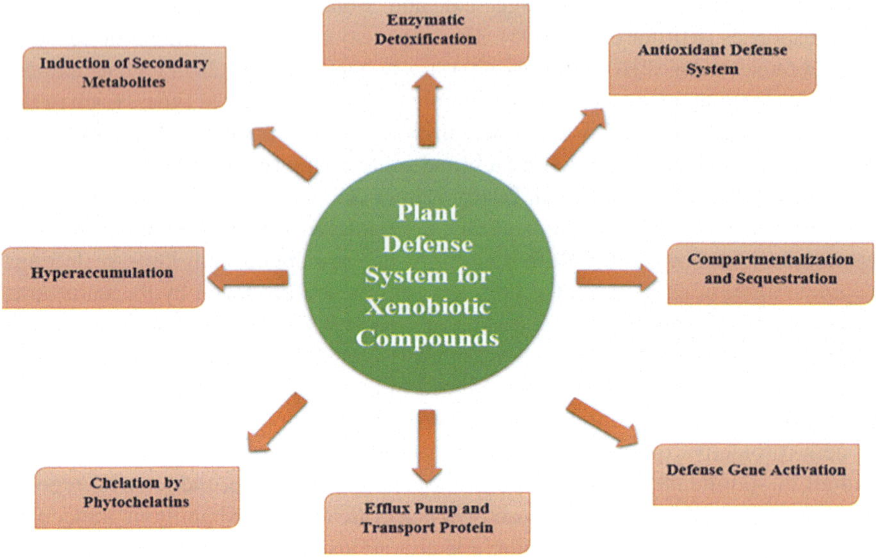

Fig. 2 Plant defense system for xenobiotics compounds

mechanisms of plants against foreign substances and their ability to adapt to harsh environments. The basic illustration of various types of plant defense systems for xenobiotic compounds is shown in Fig. 2.

6.1 Enzymatic Detoxification

To combat the phytotoxicity of the numerous manmade and natural substances (xenobiotics) found in the environment, plants have adaptable detoxifying systems. Frequent application of pesticides to protect crops against unwelcome invaders results in the environmental buildup of xenobiotic compounds. Plant communities are directly targeted by residual agricultural pollution, which can be absorbed through soils, water, or air. This is because plants are sessile organisms. Herbicides, including triazine, have the ability to make organophosphate insecticide more hazardous. The toxicity of a pesticide mixture is also influenced by the breakdown products that are produced. Aminomethylphosphonic acid (AMPA), the main breakdown product of the broad-spectrum nonselective herbicide glyphosate, is commonly found in freshwater habitats. Plants have therefore evolved a defense mechanism to handle the potential damage that xenobiotics may cause to their growth and development. Through complex, multistep processes that exhibit significant species variability, plants can detoxify herbicides. Their ability to metabolically detoxify xenobiotics is particularly noteworthy. Consequently, plants have developed a number of detoxifying techniques to help them live under these conditions.

6.1.1 Glutathione-Mediated Detoxification

The chemical alteration of the xenobiotic by covalent bonding to the endogenous glutathione is a significant detoxifying mechanism. There are numerous similarities between this detoxification process and the pathway that plants utilize to synthesize and deposit secondary metabolites like anthocyanin in their vacuoles. The tripeptide glutathione (GSH) has the structure Glu-Cys-Gly, which stands for the molecule of glutamate, cysteine, and glycine, respectively. Plants have high levels of GSH, usually more than 1 mM in the cytoplasm, and it is essential for the vital function of redox buffering and detoxification. Additionally, plants have a number of GSH-dependent detoxifying enzymes, such as glutathione S-transferases (GSTs), which catalyze the conjugation of electrophilic herbicides with GSH in order to detoxify them as shown in Fig. 3. The ABC transporters, which transfer GSH conjugates alone, flag the glutathionylated metabolites for vacuolar import (Singh et al. 2016).

6.1.2 Hormone-Mediated Detoxification

In reaction to xenobiotics such as pesticides, some genes involved in plant detoxification mechanisms express themselves differently (Ramel et al. 2012). When applied externally, BRs reduced the residues of major organophosphorus,

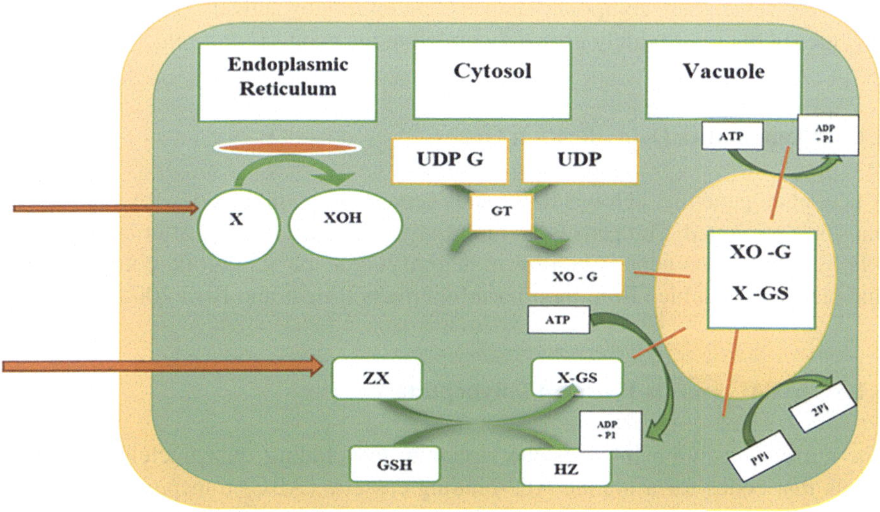

Fig. 3 Glutathione-mediated detoxification

organochlorine, and carbamate pesticides by 30–70% in cucumber, broccoli, tomato, and other plants (Zhou et al. 2015). There was a direct correlation between the degree of BRs and the amount of the "NIH shift" method, which involves the displacement of the Cl group, is how the hydroxylation process occurs. When a hydrogen atom on an aromatic ring shifts intramolecularly, typically as a result of a hydroxylation process, this is referred to as an NIH shift (Singh et al. 2016). For instance, when 2,4-dichlorophenoxy acetic acid (2,4-D) permeates plant cells, fourth-position hydroxylation takes place. The hydroxylation process is carried out using the "NIH shift" approach, which entails the displacement of the Cl group. Following glycosylation and malonylation, 4-OH-2,5-D is stored in the vacuole as O-(malonyl)glucosyl 4-OH-2,4-D.

6.1.3 Safener-Mediated Detoxification

One interesting feature of their xenobiotic detoxification mechanism is the infusibility of some synthetic substances that some monocotyledonous crop species call to as "safeners." The class of chemically diverse substances known as safeners has the special ability to shield grasses from herbicide damage without influencing the herbicide's effectiveness against the target weed species. To increase herbicide tolerance in rice (*Oryza sativa*), grain sorghum (*Sorghum bicolor*), and maize (*Zea mays*), a number of safener classes have been developed. The target plant is unaffected by safeners, which shield it from herbicide harm and encourage the expression of genes related to plant defense and detoxification, such as GSTs and P450s. Safeners have a signaling mechanism that facilitates the detoxification of endogenous toxins or xenobiotics (Riechers et al. 2005). Only cereals and a few non-domesticated grasses have shown herbicide-scavenging activity (Skipsey et al. 2011). However, these compounds can also preferentially activate xenomic response enzymes in the most widely grown dicotyledonous plant, *A. thaliana*.

6.1.4 Xenobiotics Detoxification Genes

A new plant gene that belongs to the mdr gene family which codes for ABC transporters is presented. The gene is preferentially expressed in roots after being inducible by NPPB. The gene is helpful in detoxifying some xenobiotic compounds so that plants are shielded from their harmful effects (Edgar and Bosl 2000).

6.1.5 p1PAC and Its Encoded Polypeptide

The current invention involves the isolation and cloning of a nucleic acid from plants that codes for a unique ATP-binding cassette (ABC) transporter. Here, the nucleic acid is called p1PAC. The nucleotide sequence of the exemplary p1PAC of the invention and a cDNA clone of the p1PAC from *Arabidopsis thaliana* is presented in example 1 as SEQ ID No:1. We call this nucleic acid molecule "ATPAC." Throughout the entire sequence, it is 51% similar to human mdr1 and 36% identical. According to the Genbank database, it is 50% identical to atpgp2, a near homolog of atpgp1, and 51% identical to the atpgpq1 gene. The ATPAC protein and the atpgp1 and atpgp2 proteins are 65% similar. This nucleic acid molecule's inducibility by certain substances, such as NPPB and herbicide, and its preferential expression in roots are among its distinctive characteristics when compared to other mdr-like genes from plants. According to Edgar (2001), residues make up the ATPAC promoter regulatory region.

6.2 Antioxidant Defense System

The reactive oxygen species (ROS) antioxidant interaction is being proposed as a metabolic interface for environmental and metabolic inputs in a redox homeostasis model. This interface regulates when to initiate acclimatization processes or, on the other hand, when to initiate cell death programs.

6.2.1 ROS Interactions in Redox Signaling

The knowledge of antioxidant status in plants has changed dramatically in the last several years, emphasizing the vital role it plays in both defensive mechanisms and developmental processes. In addition to being vital for scavenging ROS, antioxidants are also crucial for controlling how plants react to a range of xenobiotic stresses. Ascorbate oxidase and the absence of important reducing agents like NAD(P)H and glutathione are two of the apoplast's distinctive features. These elements combine to create a special oxidative environment that affects cellular signaling and defense reactions. The interplay between ROS and antioxidants, specifically ascorbate, not only supports metabolic functions but also signals transduction pathways that control how plants respond to environmental stressors. As research in this area continues, it is becoming more and more evident that improving crop performance in the face of environmental stresses and developing agricultural methods depend on an understanding of the complexity of antioxidant dynamics and their implications for plant resilience. Building on the complex interactions between ROS and antioxidants, it is crucial to take into account how these dynamics affect plant resilience as well as the evolution of species-specific adaptations. As a means of survival, plants exposed to long-term environmental stressors, for example, frequently have strengthened antioxidant systems, leading to phenotypic plasticity that facilitates increased flexibility (Horemans et al. 2000). This adaptive reaction may result in changes to growth patterns, methods of reproduction, and general fitness which may have an effect on community structure and ecological interaction.

6.2.2 Mechanism of Redox and Antioxidant Sensing

Thiol oxidation undoubtedly works a significant role in the activation of the kinase pathway in plants caused by ROS, even though the exact biochemical and structural basis of this process is yet unknown. The most appropriate redox signal transduction mechanism in plants is the stromal ferredoxin-thioredoxin system, which regulates photosynthetic carbon metabolism.

6.3 Compartmentalization and Sequestration

As an integrated part of a complex cellular network, vacuolar transporters enable the plant to store nutrients and energy for periods when they are plentiful, maintain optimal metabolic conditions in the cytosol, and react appropriately to shifting environmental conditions. The presence of at least two distinct vacuole types in plants—lytic and storage—has been proven beyond a reasonable doubt. This suggests that the plant's vacuole system is far more intricate. The soluble protein that vacuoles contain and the kind of aquaporins (Luu and Maurel 2005) that are found in their membrane can be used to classify vacuoles.

6.3.1 Secondary Compounds and Xenobiotics

Secondary plant chemicals and modified xenobiotics are also traditionally stored in vacuoles. Furthermore, transport experiments showed that an ABC-type transporter in isolated Arabidopsis vacuoles absorbs the flavonoid glucosyl saponarin, which is found in barley but not from the plant. These findings supported the hypothesis that plants have two different kinds of glycosylated chemical transporters: ABC transporters for foreign glycosylated molecules and antiporters for local compounds. This notion, however, is not true until the situation is determined. This implies that a secondary active transporter is responsible for the vacuolar absorption of glycosylated xenobiotics in this case. The subject of whether ABC transporters play a role in anthocyanin transport has been thoroughly studied.

6.4 Defense Gene Action

Plants identify xenobiotics by activating particular stress signaling pathways, which result in the activation of defense and detoxification genes.

6.4.1 Overview of Mitogen-Activated Protein Kinase (MAPK) Pathway

In MAPK scenario, the activated receptor triggers a cascade of signaling events, resulting in activation of MAPK that phosphorylates the transcriptional regulators. Although phosphorylation is the most prevalent change utilized to transfer information across signal transduction pathways, additional modifications such as proteolysis and dephosphorylation are also employed. The downstream MAPK pathway is connected to the Ras pathways. Xenobiotic factors attach to their receptors to bring the chains closer; the chains are phosphorylated to one another. Grb2, an adaptor protein with an SH2 domain that can identify a phosphorylated tyrosine residue in an active receptor, is drawn to the phosphorylation. Ras GEF, the guanine nucleotide exchange factor, is bounded by the opposite end of Grb2. When Ras comes into contact with SOS, it exchanges its bound GDP for a GTP and undergoes a conformational change. Ras is a small GTPase protein that adopts one shape when bound to GTP and another when bound to GDP. Ras triggers a kinase at the head of the so-called MAPK cascade in this altered configuration. When activated by Ras, the first kinase in this pathway—MAPK kinase kinase (MAPKKK), such as Raf—phosphorylates serine and threonine residues on the next kinase in the cascade, known as MAPK kinase (MAPKK), for example Mek. Once phosphorylated, Mek becomes active and subsequently phosphorylates MAP kinase (MAPK), thereby activating Erk. Activated Erk then phosphorylates a variety of substrates, including transcriptional activators such as Jun, which regulate the expression of specific genes, including interferon-β (Watson 2013).

6.5 *Efflux Pump and Transport Protein*

Efflux pumps are a feature of plants that actively move foreign substances or their metabolites out of cells, lowering the concentrations of these substances inside cells. Important carriers consist of ABC transporters which move conjugated foreign substances out of the cell or into vacuoles. The majority of ABC proteins, but not all, are symmetrically arranged membrane proteins known as ABC transporters. Amphipathic molecules of intermediate complexity provide the majority of transport substrates for known or highly implicated plant ABC transporters. Even though it is frequently possible to predict an ABC transporter's likely function based on its subfamily membership, many plant ABC transporters behave differently than would be expected based on the characteristics of even closely related species.

6.6 Hyperaccumulation

The distribution of hyperaccumulation species across a large number of distantly related families suggests that the hyperaccumulation characteristic underwent multiple distinct evolutionary events due to selected ecological variables. The analysis of transcriptomes of hyperaccumulation and non-hyperaccumulator or species showing constitutive overexpression of genes encoding a protein involved in heavy metal transfer across the plasma membrane and involved in excluding from the cytoplasm is also responsible at least partially for the sequestration trait. Cation diffusion facilitator (CDE) family, sometimes referred to as MTP, or the metal transporter protein, which promotes the cytosolic outflow of bivalent cations, are significant candidates. The leaves of Zn/Ni hyperaccumulators exhibit a substantial overexpression of MTP1, a gene codes for a protein located at the tonoplast (Hammond et al. 2006). MTP1 has been proposed to improve zinc absorption in addition to its function in zinc tolerance. In fact, the movement of zinc into vacuole in hyperaccumulator plants may trigger a reaction to zinc deficiency, which involves increasing the expression of ZIP transporter to enhance heavy metal absorption and translocation (Rascio and Navari-Izzo 2011).

6.7 Induction of Secondary Metabolites

With the help of large-scale genomics programs and recent advancements in molecular biology, a more comprehensive understanding of the enzyme machinery underlying the frequently intricate routes of plant natural product biosynthesis is being achieved. The majority are generated from the phenylpropanoid isoprenoid, fatty acid/polyketide, or alkaloid routes.

6.7.1 Genomic Methods for Deciphering and Changing Intricate Natural Product Chains for Plant Defense

Some ways to do this include chemical elicitation, which does not require the infecting organism, T-DNA activation tagging, which increases the expression of regulatory genes or pathway enzymes that limit rate, and taking advantage of differences in metabolite levels or composition between natural ecotypes and cultivar germplasm. Since this kind of analysis will make it easier to link gene sets or gene products to specific metabolic pathways, it will necessitate the development of better computational techniques for the comparative analysis of databases that contain information on relative changes in proteins, metabolites, and gene transcripts. Examples of such advanced analytical techniques include FTIC-MS (Fourier Transform Ion Cyclotron Resonance Mass Spectrometry); GC/MS (Gas Chromatography–Mass Spectrometry); LC/MS (Liquid Chromatography–Mass

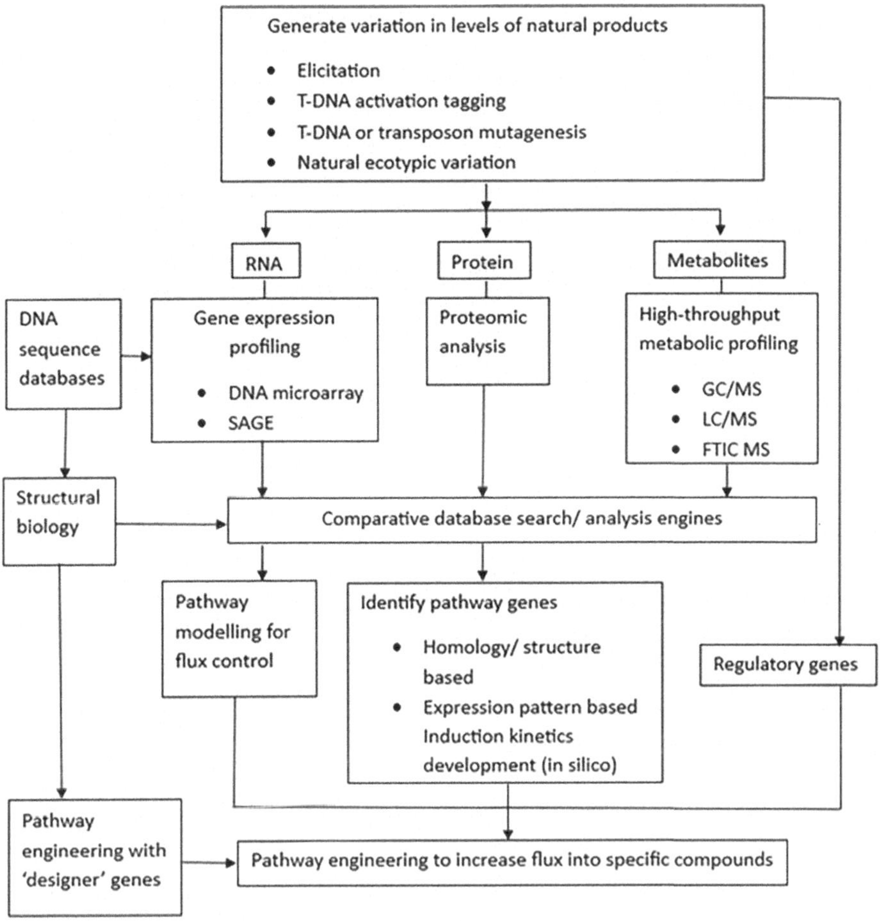

Fig. 4 Genomic approach for complex natural pathways

Spectrometry); and SAGE (Serial Analysis of Gene Expression), all of which are used to identify and quantify changes in metabolites, proteins, and gene transcripts (as illustrated in Fig. 4); SAGE, or serial analysis of gene expression.

7 Green Liver Concept

In terms of metabolic patterns, protein and DNA sequences, and enzyme classes, the metabolism of xenobiotics in plants is similar to that of mammalian liver. It seems that plants have isoenzymes that are unique to xenobiotics or secondary plant substrates. After plant cell suspension culture was used as a novel tool to study

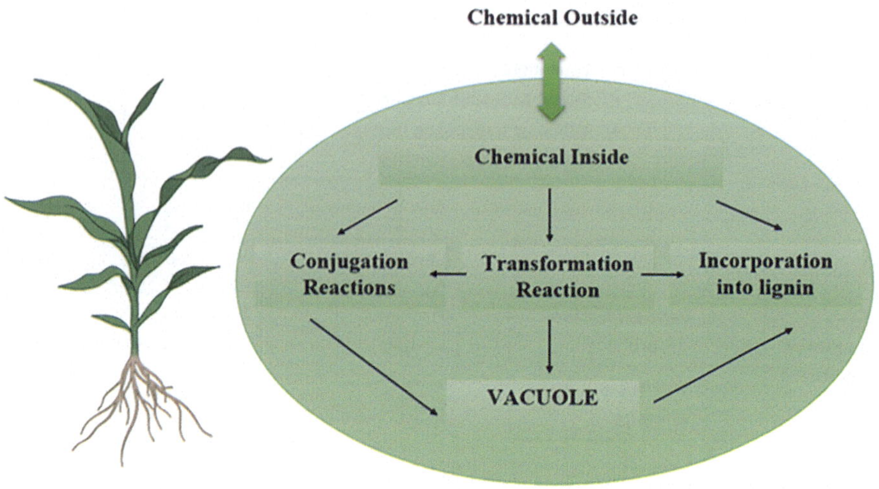

Fig. 5 Green liver concept

xenobiotic metabolism, the term "green liver" was coined (Sandermann et al. 1977). An overview of green liver concept is shown in Fig. 5.

7.1 Xenobiotics Metabolism

The first study of xenobiotic metabolism was performed by Knoop 90 years ago using whole animals. With regards to xenobiotic plant metabolism, a landmark discovery was made in the 1940s by L.P. Miller even before the radiochemical tracer techniques became available. Different xenobiotics, including 2-chlorophenol chlorohydrate and ethylene chlorohydrin, were either absorbed from the gas phase or infused into intact plant section, where they subsequently underwent metabolism to produce beta-D-gentiobioside and beta-D-glucosides. These metabolic products, which amounted to several grams per liter of plant extract, were separated and recognized as crystalline compounds. In plants, B-oxidation shortens fatty acid replaced with herbicidal moiety 2,4-D. Additionally, 2,4-D partially conjugates with amino acids, offering a general similarity to the finding of Knoop. A schematic summary of cellular plant metabolism is shown in Fig. 5.

7.1.1 Phase I: Transformation

During phase I reactions, xenobiotics become more polar and reactive by exposing or introducing functional groups. Cytochrome P450 enzymes catalyze oxidative reactions, which are the main means by which this is accomplished. For instance, research has demonstrated that plants may oxidize pesticides and produce a variety of metabolites (Fujisawa et al. 2009). Xenobiotic compounds usually go through oxidative or hydrophilic changes in phase I. Additionally, the enzymes of phase I participate in reductive processes, including generally under an oxygen-deficient state. These reactions typically result in the molecule's introduction of new functional groups or the exposure of preexisting functional groups which produce more water-soluble polar and typically derivatives with greater biological and chemical activity. Consequently, phase I of metabolism is frequently referred to as the bioactivation phase.

7.1.2 Phase II: Conjugation

Phase II is crucial for detoxifying compounds because it converts reactive intermediates into less hazardous forms. Phase II is where the real detoxification takes place, with the endogenous sugar or peptide molecule binding the result of the phase I transformation. Conjugation processes, also known as phase II reactions, yield the xenobiotic compound's glycosyl, amino acid, or peptide conjugates. The metabolites from phase I undergo conjugation with hydrophilic molecules like glutathione or glucose, enhancing their solubility for easier elimination. Except for aquatic plants, they generally do not have an excretory system like that of mammals. Pollutants and hazardous byproducts are stored in storage compartments found in plant cells.

7.1.3 Phase III: Compartmentation

Phase III involves the transport of these conjugated metabolites out of the plant cells, ensuring that they don't accumulate dangerous levels. The clearance of xenobiotics from the plant system is facilitated by both passive and active transport pathways. The ATP-driven pumps that are confined to membranes catalyze this active phase III process. It is characterized by the final removal of these conjugates into the cell vacuole or their incorporation into biopolymer, including lignin.

7.2 *Chemistry Behind the Green Liver Concept*

The polycyclic hydrocarbon DDT, benzo(α)pyrene, and the plasticiser chemical DEHP underwent metabolic transformations in plants that were comparable to those observed in mammalian metabolic pathways, with the notable exception that D-glucose, rather than D-glucuronic acid, was used for conjugation. Phase I involved conjugation with D-glucose. After being transformed into its isomeric quinones benzo alpyrene was then conjugated with either glutathione or lignin. The esterase cleavage and potential hydroxylation changed DEHP before conjugation with

D-glucose. There were five traits that plant microsomal cinnamic acid 4-hydroxylase seemed to share with liver microsomal cytochrome P450 system: (1) A spectral cytochrome P450 component is present; (2) inducibility via ageing, light, injury, and phytohormone exposure; (3) need for phospholipid cofactors as a requirement; (4) sensitivity to common inhibitors (*I-vinyl imidazole, SKF-525A*); and (5) the occurrence of an epoxide intermediate-induced hydrogen NIH shift. Using a glutathione trapping approach, the latter was initially identified in a plant microsomal hydroxylation process as shown in Fig. 5. Researchers have found that parsley and soybean cell cultures, as well as pea seedlings, contain soluble glutathione S-transferase for the double bond of cinnamic acid. Additionally, high activity of this enzyme reaction was seen in plant microsomal fraction. Chromatographical separation was used to distinguish the soluble pea glutathione S-transferase for cinnamic acid from the previously reported soluble pea transferase for fluorodifen.

7.3 The Molecular Aspects of "Green Liver" Concept

An extensive amount of fresh molecular information on metabolic isoenzymes and their genes served as the foundation for the second iteration of the "green liver" theory (Sandermann 1992). Plant metabolic isoenzymes were now clearly seen to be selective for either natural or xenobiotic substrates. Even while many of the relevant proteins had commonalities, such as molecular weight values within an enzyme class, many of them could be isolated. The fundamental finding is demonstrated by the fact that, out of the several wheat carboxylesterases, only one minor species was active in the presence of both the lipophilic standard substrate, naphthyl octanoate, and the persistent plasticizer chemical, DEHP.

7.4 The Evolution of Green Liver Concept

The green liver theory was first supported by the metabolite patterns and types of enzyme reactions that were similar to the xenobiotic metabolism of plants and the liver. With new parallels to enzymes of typical plant secondary metabolism, the similarity might then be expanded to the levels of enzyme protein structure and cDNA sequences. It is believed that gene duplication and mutation mechanisms led to the evolution of plant isoenzymes from ancestor genes; however, the selection pressure responsible for producing enzymes that exhibit apparent sensitivity for modern industrial xenobiotics remains unclear.

7.5 Purified Plant Isoenzymes for Xenobiotics

The study of isozymes or isoforms of enzymes that catalyze the same biological reaction but differ in structure and kinetic properties has received a lot of attention in the field of xenobiotic metabolism. Table 2 shows examples of purified plant isozymes for xenobiotics and certain natural substrates. A detailed study of highly pure plant enzymes involved in secondary metabolisms. From soybean, two highly specific O-glucosyltransferases have been isolated. The previously isolated O-glucosyltransferase isoenzyme for pentachlorophenol and quercetin, as well as the N-glucosyltransferase for chlorinated anilines and metribuzin, were not the same as these enzymes for 6-hydrogenation and DDA. Pea seeds were used to purify a homogenous GSH-dependent formaldehyde dehydrogenase. The

Table 2 Examples of newly purified plant isozymes for xenobiotics and some natural substrate

Enzyme	Plant species	Substrate	Structure, subunit molecular weight, isoelectric point	References
O-glucosyltransferase	Soybean	DDA	Monomer, 50 kDa; pI 4.9	Wetzel and Sandermann (1994)
O-glucosyltransferase	Soybean	6-hydroxybentazon	Monomer, 44 kDa; pI 5.8	Gallandt and Balke (1994)
O-glucosyltransferase	*Brassica*	Thiohydroximates	Monomer, 46 kDa; pI 4.3–4.6	Reed et al. (1993)
Putrescine *N*-methyl-transferase	Datura	1,4-diamines	Monomer, 40 kDa; pI 5.3	Walton et al. (1994)
GSH-transferase	Maize	Metolachlor	Dimer, 27 kDa; pI 5.75	Irzyk and Fuerst (1993)
Allene oxide synthase (cytochrome P450)	Flax seed	Fatty acid peroxides	Monomer, 48.5 kDa	Song and Brash (1991)
N-malonyltransferase	Mungbean	ACC, D-amino acid	Monomer, 55 kDa	Guo et al. (1992, 1993)
P450 oxidoreductase	*Catharanthus helianthus*	Cytochrome P450	Monomer, 80 kDa; pI 4.6	Shet et al. (1993), Benveniste et al. (1991) and Meijer et al. (1993)
Epioxide hydrolase	Soybean	Fatty acid epoxide	Dimer, 32 kDa; pI 5.4	Blée and Schuber (1992)

GSH-dependent plant enzyme had many biochemical characteristics with *E. coli*. In many plant species, allene oxide synthase seems to be the major cytochrome P450. Additionally, oxygenase and effective co-oxidation processes can be catalyzed by plant peroxidases. They are capable of catalyzing both one- and two-electron transfer processes, just like cytochrome P450 species. It is challenging to choose the appropriate enzyme for a particular oxidation process due to the wide range of catalytic potential exhibited by the plant oxygenases discussed.

7.6 Plant cDNA Sequences for Enzymes with Xenobiotic Substrate

According to Jakoby (1991), the plant sulfotransferase appeared to be highly homologous to the same enzyme class from animal liver. However, no xenobiotic substrate has so far been identified in plants. The NADPH cytochrome P450 reductase was also highly homologous to the animal enzymes. The plant enzyme could even substitute for the rat liver reductase in a reconstituted system. Generally speaking, these sequences support the idea that the so-called plant 'GST-like' formaldehyde dehydrogenases actually constitute a distinct subgroup within the medium-chain zinc-dependent alcohol dehydrogenase (ADH) enzyme superfamily. Specifically, type III ADHs—commonly referred to as GSH-dependent formaldehyde dehydrogenases (FDHs) or GSNORs—are thought to include homologous enzymes found across species, including E. coli, mammals, and plants. Within the alcohol dehydrogenase enzyme superfamily, type III is thought to include the GSH-dependent formaldehyde dehydrogenases found in *E. coli* and mammalian sources. The N-terminal sequences of type III formaldehyde dehydrogenases were discovered to be identical to those encoded by the Adhi and Adh2 genes of maize as well as other plant Adh genes. Nevertheless, it is still unknown what the actual maize gene for formaldehyde dehydrogenase is. The maize AdhI and Adh2 genes do not appear to encode formaldehyde dehydrogenase.

7.7 Cytochrome P450's Function in the Defense System

Protein superfamilies include cytochrome P450 (CYPs). Chemically speaking, they are heme-thiolate-harboring proteins that bind to carbon monoxide and have a maximal absorption band at 450 nm when reduced. Plant genomes contain a significant number of CYP genes. They constitute about 1% of all the genes encoding proteins. Numerous distinct plant species have been identified, such as the 245 genes found in Arabidopsis (Schuler & Werck-Reichhart 2006), the 316 genes found in grapevine (Jiu et al. 2020), the 332 genes found in soybean, the 334 genes found in rice (Nelson 2004), the 372 genes found in sorghum (Nelson et al. 2008), and the 270 genes found in tomatoes (Schuler 2015). Most plant CYPs reside in the endoplasmic reticulum (ER) and play a crucial role in the metabolism of xenobiotics, which causes resistance to antibiotics, drugs, insecticides, and herbicides; the biosynthesis of phytohormones and homeostasis; the metabolism of fats; the synthesis of cell wall components; and the production of a wide range of primary and secondary metabolites.

8 Conclusions and Future Aspects

Transcriptome profiling analysis is the main method used to analyze plant xenobiotics nowadays. This analysis provides more information on signaling pathways than sensors, such as TFs, protein kinase, metabolite crosstalk, and metabolite crosstalk. However, as was already noted, plants still have a lot of potential for detecting receptors and sensors, and the identification of mutants is primarily how novel environmental cue sensors are constantly discovered. Plants and other are known for ability to adapt various xenobiotics which can change an organism's development and growth. In order to defend themselves, plants have therefore developed a system for metabolism and detoxification of these xenobiotics. In case of hydrophilic chemicals, a distinct compartment exists where a large number of these xenobiotics are kept. The Disruptome method that uses mutant collection as those used in yeast research can identify regulatory genes, even constitutively expressed ones that are undetectable using gene expression methods. The xenobiotics are protected via a variety of methods in addition to compartmentalization, secondary metabolites, and detoxification; the precise mechanism behind these events is still not very much known. Thus, a great deal more research is needed to have a better knowledge of how xenobiotics act on plants.

References

Alix K, Gérard PR, Schwarzacher T, & Heslop-Harrison JSP. (2017a). Polyploidy and interspecific hybridization: partners for adaptation, speciation and evolution in plants. Annals of Botany, 120(2), 183–194. https://doi.org/10.1093/aob/mcx079.

Armand Cavé-Radet, Armel Salmon, Oscar Lima, Malika L. Ainouche, Abdelhak El Amrani. (2019). Increased tolerance to organic xenobiotics following recent allopolyploid in Spartina (Poaceae), Plant Science, Volume 280, Pages 143–154, ISSN 0168–9452, https://doi.org/10.1016/j.plantsci.2018.11.005.

Bajsa J, Pan Z, Duke SO. (2011a). Transcriptional responses to cantharidin, a protein phosphatase inhibitor, in Arabidopsis thaliana reveal the involvement of multiple signal transduction pathways. Physiologia Plantarum 143, 188–205.

Bajsa J, Pan Z, Duke SO, & Dayan FE. (2011b). Transcriptional responses to a natural protoporphyrinogen oxidase inhibitor in Arabidopsis. Frontiers in Plant Science, 2, 1–12. https://doi.org/10.3389/fpls.2011.00073

Benveniste I, Lesot A, Hasenfratz M-P, Kochs G, & Durst F. (1991). Multiple forms of NADPH-cytochrome P-450 reductase in higher plants. Biochemical and Biophysical Research Communications, 177, 105–112.

Beringer J, Rotenberg E, & Anderson MC. (2011). Deriving an improved estimate of the decadal trend in the Australian terrestrial carbon sink. Tellus B: Chemical and Physical Meteorology, 63(1), 31–40. https://doi.org/10.1111/j.1600-0889.2010.00506.x.

Blée B, & Schuber F. (1992). Occurrence of fatty acid epoxide hydrolases in soybean (Glycine max); purification and characterization of the soluble form. Biochemical Journal, 282, 711–714.

Bonjoko B (2014) Environmental pharmacology: an overview. In: Pharmacology and therapeutics (Monograph on the internet). Intech, pp 133–178. Available from: http://www.intechopen.com/books/pharmacology-andtherapeutics/environmental-pharmacology-an-overview.

Chapman RW, Mancia A, Beal M, Veloso A, Rathburn C, Blair A, Holland AF, Warr GW, Didinato G, Sokolova IM, Wirth EF, Duffy E, Sanger D. (2011). The transcriptomic responses of the eastern oyster, Crassostrea virginica, to environmental conditions. Molecular Ecology 20, 1431–1449.

Dévier MH, Mazellier P, Ait-Aissa S et al (2011). New challenges in environmental analytical chemistry: identification of toxic compounds in complex mixtures. Comptes Rendus Chimie 14:766–779.

Doyle JJ, Flagel LE, Paterson AH, Rapp RA, Soltis DE, Soltis PS, & Wendel JF. (2008a). Evolutionary genetics of genome merger and doubling in plants. Annual Review of Genetics, 42, 443–461. https://doi.org/10.1146/annurev.genet.42.110807.091524.

Doyle JJ, Flagel LE, Paterson AH, Rapp RA, Soltis DE, Soltis PS, Wendel JF. (2008b). Evolutionary genetics of genome merger and doubling in plants, Annu. Rev. Genet. 42 443–461

Edgar R. (2001). Gene Expression Omnibus: NCBI gene expression and hybridization array data repository. Nucleic Acids Research, 29(1), 207–210.

Edgar P. Spalding., Bosl, Noh. (2000). Xenobiotic detoxification gene from plants.

Ekman DR, Wolfe NL, Dean JF. (2005). Gene expression changes in Arabidopsis thaliana seedling roots exposed to the munition hexahydro-1,3,5-trinitro-1,3,5-triazine. Environmental Science and Technology 39, 6313–6320

Fujisawa M, Shima Y, & Ito Y. (2009). Coordination of fruit ripening and ethylene biosynthesis by the tomato transcription factors RIN and TOC. The Plant Cell, 21(4), 1233–1248.

Gadjev I, Vanderauwera S, Gechev TS, Laloi C, Minkov IN, Shulaev V, Apel K, Inze D, Mittler R, Van Breusegem F. 2006. Transcriptomic footprints disclose specificity of reactive oxygen species signaling in Arabidopsis. Plant Physiology 141, 436–445.

Gallandt ER, & Balke NE. (1994). Properties of a purified soybean UDP-glucose: 6-hydroxybentazon glucosyltransferase. Abstract Volume, IUPAC Pesticide Congress, Washington, DC. (In press).

Guo L, Arteca RN, Phillips AT, & Liu Y. (1992). Purification and characterization of 1-aminocy clopropane-1-carboxylate N-malonyltransferase from etiolated mung bean hypocotyls. Plant Physiology, 100, 2041–2045.

Guo L, Phillips AT, & Arteca RN. (1993). Amino acid N-malonyltransferases from mung beans. Journal of Biological Chemistry, 268, 25389–25394.

Hammond JP, et al. (2006). A comparison of the *Thlaspi caerulescens* and *Thlaspi arvense* shoot transcriptomes. *New Phytologist, 170*(2), 239–260.

Horemans N, Foyer CH, & Asard, H. (2000). Transport and action of ascorbate at the plant plasma membrane. Trends Plant Sci. 5, 263–267.

Irzyk GP, & Fuerst EP. (1993). Purification and characterization of a glutathione S-transferase from benoxacor-treated maize *(Zea mays)*. Plant Physiology, 102, 803–810.

Jakoby M, Weisshaar B, Droge-Laser W, Vicente-Carbajosa J, Tiedemann J, Kroj T, & Parcy F. (1991). bZIP transcription factors in Arabidopsis. Trends in Plant Science, 7(3), 106–111.

Jiu S, Xu Y, Wang J, & Wang L. (2020). The role of WRKY transcription factors in plant abiotic stresses. Frontiers in Plant Science, 11, 615760. https://doi.org/10.3389/fpls.2020.615760.

Karine Alix, Pierre R. Gérard, Trude Schwarzacher, J S Pat Heslop-Harrison, (2017b) Polyploidy and interspecific hybridization: partners for adaptation, speciation and evolution in plants, *Annals of Botany*, Volume 120, Issue 2, Pages 183–194, https://doi.org/10.1093/aob/mcx079

Luu DT, Maurel C. 2005. Aquaporins in a challenging environment: molecular gears for adjusting plant water status. Plant, Cell and Environment 28, 85–96.

Meijer AH, Lopes Cardoso MI, Voskuilen JT, de Waal A, Verpoorte R, & Hoge JHC. (1993). Isolation and characterization of a cDNA clone from Catharanthus roseus encoding NADPH: cytochrome P-450 reductase, an enzyme essential for reactions catalyzed by cytochrome P-450 mono-oxygenases in plants. The Plant Journal, 4, 47–60.

Mittler R, Vanderauwera S, Gollery M, Van Breusegem F. (2004). Reactive oxygen gene network of plants. Trends in Plant Science 9, 490–498.

Nelson DR. (2004). The cytochrome P450 homepage. Human Genomics, 1(4), 59–63.

Nelson DR, & Werck-Reichhart D. (2008). A P450-centric view of plant evolution. The Plant Journal, 66(1), 194–211.

Ramel F, Birtic S, Cuine S et al (2012) Chemical quenching of singlet oxygen by carotenoids in plants. Plant Physiol 158:1267–1278.

Ramel F, Sulmon C, Cabello-Hurtado F, Taconnat L, Martin-Magniette ML, Renou JP, El Amrani A, Couée I, Gouesbet G. (2007). Genome-wide interacting effects of sucrose and herbicide-mediated stress in Arabidopsis thaliana: novel insights into atrazine toxicity and sucrose-induced tolerance. BMC Genomics 8, 450

Rascio N, & Navari-Izzo F. (2011). Heavy metal hyperaccumulating plants: How and why do they do it? And what makes them so interesting? Plant Science, 180(2), 169–181.

Reddy ASN, Ali GS, & Celesnik H. (2008). Coping with stresses: Roles of calcium- and calcium/calmodulinregulated gene expression. The Plant Cell, 20(8), 2010–2037.

Reed DW, Davin L, Jain JC, Deluca V, Nelson L, & Underhill EW. (1993). Purification and properties of UDP-glucose: Thiohydroximate glucosyltransferase from Brassica napus L. seedlings. Archives of Biochemistry and Biophysics, 305, 526–532.

Riechers DE, Kreuz K, & Zhang Q. (2005). Detoxification without intoxication: Herbicide safeners activate stress-response genes. Plant Physiology, 138(3), 939–948.

Riechers DE, Kreuz K, Zhang Q. (2010). Detoxification without Intoxication: herbicide safeners activate plant defense gene expression. Plant Physiology 153, 3–13.

Sandermann H. (1992). Plant metabolism of xenobiotics. Trends in Biochemical Sciences, 17(2), 82–84.

Sandermann H et al. (1977) In: Plant Tissue Culture and its Biotechnological Application, (Barz W et al., eds.) pp. 178–196, Springer, Berlin, Germany

Schuler MA. (2015). The role of cytochrome P450 monooxygenases in plant-insect interactions. Plant Physiology and Biochemistry, 47(5), 320–329.

Schuler MA, & Werck-Reichhart D. (2006). Functional genomics of P450s. Annual Review of Plant Biology, 54, 629–667.

Serra AA, Nuttens A, Larvor V et al (2013) Low environmentally relevant levels of bioactive xenobiotics and associated degradation products cause cryptic perturbations of metabolism and molecular stress responses in Arabidopsis thaliana. J Exp Bot 64:2753–2766.

Shet MS, Sathasivan K, Arlotto MA, Mehdy MC, & Estabrook RW. (1993). Purification, characterization, and DNA cloning of an NADPH-cytochrome P-450 reductase from mung bean. Proceedings of the National Academy of Sciences of the United States of America, 90, 2890–2894.

Singh S, Bashri G, Singh A, Prasad SM. (2016). Regulation of Xenobiotics in Higher Plants: Signalling and Detoxification. In: Singh, A., Prasad, S., Singh, R. (eds) Plant Responses to Xenobiotics. Springer, Singapore. https://doi.org/10.1007/978-981-10-2860-1_3.

Skipsey M, Knight KM, Brazier-Hicks M, Dixon DP, Steel PG, Edwards R. (2011). Xenobiotic responsiveness of Arabidopsis thaliana to a chemical series derived from a herbicide safener. Journal of Biological Chemistry 286, 32268–32276.

Song W-C, & Brash AR. (1991). Purification of an allene oxide synthase and identification of the enzyme as a cytochrome P-450. Science, 253, 781–784.

Walton NJ, Peerless ACJ, Robins RJ, Thodes MJC, Boswel HD, & Robins DJ. (1994). Purification and properties of putrescine N-methyltransferase from transformed roots of Datura stramonium L. Planta, 193, 9–15.

Wang W. (2010). Comparative transcriptome analyses of drought stress responses in maize and Arabidopsis. Plant Molecular Biology, 72(4–5), 407–421.

Wang Y, Peng X, Xu W, Luo YB, Zhao W, Hao J, Liang Z, Zhang Y, Huang KL. (2012). Transcript and protein profiling analysis of OTA-induced cell death reveals the regulation of the toxicity response process in Arabidopsis thaliana. Journal of Experimental Botany 63, 2171–2187.

Watson JD. (2013). Molecular biology of the gene. W. A. Benjamin 7th Edition.

Weisman D, Alkio M, Colon-Carmona A (2010) Transcriptional responses to polycyclic aromatic hydrocarbon-induced stress in Arabidopsis thaliana reveal the involvement of hormone and defense signaling pathways. BMC Plant Biol https://doi.org/10.1186/1471-2229-10-59.

Wetzel A, & Sandermann H. (1994). Plant biochemistry of xenobiotics: Isolation and characterization of a soybean O-glucosyltransferase of DDT metabolism. Archives of Biochemistry and Biophysics, 314. (In press).

Zhou Y, Xia X, Yu G et al (2015) Brassinosteroids play a critical role in the regulation of pesticide metabolism in crop plants. Sci Rep 5:9018. https://doi.org/10.1038/srep09018

Microbe-Mediated Xenobiotic Degradation: Recent Status and Prospects

Supratim Mandal and Soumita Kumar

Abstract Xenobiotic pollution is an emerging global concern; these pollutants are foreign to biological systems and accumulate in the environment, causing notorious environmental effects and affecting humankind. Due to anthropogenic activity such as industry build-ups, excessive usage of pesticides, and harmful chemical introduction, these pollutants get bioaccumulated, leading to mutagenic, carcinogenic, and teratogenic effects on living entities. The existing conventional methods for xenobiotic degradation are limited because of the high cost, production of more toxic by-products and intermediates than the xenobiotic pollutant, and waste-disposal issues. On the other hand, the degradation of xenobiotics using microorganisms is sustainable, eco-friendly, and economically feasible. Various microorganisms, for example, bacteria, fungi, algae, and protozoa, are potential degraders of xenobiotic pollutants. To characterize these microbial communities for a better approach toward bioremediation, several advanced molecular technologies like genomics, metagenomics, transcriptomics, proteomics, metabolomics, and bioinformatics should be adapted to have a deep insight into the microbial communities exposed to the contaminated environments and thus give a satisfactory approach to resolving the issues related to bioremediation. This review article aims to provide a detailed overview of microbial bioremediation of xenobiotics with special emphasis on the "omics" approach to improve the bioremediation processes. Some prospects of xenobiotic bioremediation, like genetic engineering of microorganisms, bacteriophage-based bioremediation, and nanobioremediation of xenobiotics, have also been discussed to focus on the technologies to make bioremediation faster and more economical in the near future.

Keywords Microbial bioremediation · Genomics · Metabolomics · Metagenomics · Proteomics · Transcriptomics

S. Mandal (✉) · S. Kumar
Department of Microbiology, University of Kalyani, Nadia, West Bengal, India
e-mail: supratimmicro19@klyuniv.ac.in; soumitamicro24@klyuniv.ac.in

S. Roy, V. Mandal (eds.), *Plant-Microbe Interaction under Xenobiotic Exposure*,
https://doi.org/10.1007/978-981-96-8260-7_10

1 Introduction

The word xenobiotic is derived from two Greek words: "xeno," which means foreign, and "bios," which means life. As such, xenobiotics are substances foreign to a biological system and are neither naturally produced nor expected to be present in an organism; these could include pollutants, drugs, food additives, and pesticides (Singh et al. 2023). Pollution, the accumulation of harmful substances in the body, and ecosystem disruption due to xenobiotics threaten the environment and human health (Kumar Maurya and Malik 2016; Piwowarska and Kiedrzyńska 2022). Organ damage (Österreicher and Trauner 2012), nervous system disorders like Parkinson's disease (Bjørklund et al. 2020) or Alzheimer's disease (Yegambaram et al. 2015), as well as cancers (Nogacka et al. 2019), are but a few examples of what they could do to our bodies.

The types of xenobiotic pollution include pollutants like pesticides (Kumar Maurya and Malik 2016), pharmaceutical chemicals (Singh et al. 2023), personal care products (Eriksson et al. 2003), industrial chemicals (Kirici and Sokmen 2024), and nuclear waste. These chemicals can stay in the environment for an elongated period, thus becoming an issue of bioaccumulation and biomagnification, resulting in adverse impacts such as toxic, mutagenic, carcinogenic, and teratogenic effects on living organisms (Connell 2018). Possible ways to combat these xenobiotic pollutions are adsorption, photodegradation, advanced oxidation methods, biodegradation, bioremediation, and photodegradation (Karthigadevi et al. 2021).

The difficulties with traditional techniques to manage and break down these xenobiotic pollutants include diversity of compounds, slow degradation rates, generation of bad-smelling waste, waste-disposal complications, limited effectiveness, and enormous expenses (Bouwer et al. 1994). These activities are accompanied mainly by generating a large amount of waste, thus contributing to secondary environmental pollution. Apart from these, some of these xenobiotics are highly resistant to degradation. So, it is not always easy to clean up the environment when these substances are in a mixture of other chemicals. The above issues need to be resolved by developing more eco-friendly and effective technologies to handle xenobiotic species better.

This review paper focuses on a detailed picture of the role of xenobiotics as environmental hazards and the recent status of their remediation techniques, with special emphasis on microbe-mediated bioremediation. It highlights the limitations of present remediation techniques. It tries to provide further research and application insights for eco-friendly technologies like microbe-mediated biodegradation, careful management of xenobiotics, and some advanced molecular techniques that give a better approach toward bioremediation.

2 Sources and Types of Xenobiotics

2.1 Halocarbons

Halocarbons are xenobiotic compounds with halogen atoms like chlorine, bromine, fluorine, or iodine, which are not naturally occurring and are introduced into the environment through human activities. These compounds are frequently used in refrigeration industries as solvents and propellants. These are also present in pesticides, paints, and other household appliances (Patel et al. 2021).

2.2 Polychlorinated Biphenyls (PCBs)

Polychlorinated biphenyls (PCBs) are synthetic organic compounds that contain hydrogen, carbon, and chlorine. Since these compounds have chemical constancy, high boiling points, and protective properties, different industries are using them (Reddy et al. 2019). PCB contamination occurs in humans via foods like seafood, dairy products, and fish. Environmental exposure is also one of the reasons behind PCB contamination, and inhaling such pollutants may cause serious health hazards. Accumulation of PCBs can reduce fertility and affect the reproductive system, which can be transferred to the next generation (Montano et al. 2022).

2.3 Polycyclic Aromatic Hydrocarbons (PAHs)

Polycyclic aromatic hydrocarbons (PAHs) are produced from partial combustion of organic materials. Both natural and anthropogenic sources of PAH are there. Anthropogenic sources are coal gasification, liquefying plants, coal-tar pitch, and motor vehicle exhaust, and natural sources are fires and volcanic eruptions. These pollutants are lipid-soluble and can be deposited in tissues that contain fat. PAHs have mutagenic, carcinogenic, teratogenic, and genotoxic effects, and long-time exposure to PAHs may cause health problems like liver and kidney damage, breathing problems, decreased immune function, and skin inflammation (Abdel-Shafy and Mansour 2016).

2.4 Synthetic Polymers

Synthetic polymers are a major group of xenobiotics. Plastics and other artificial materials, such as polyethylene and polystyrene, are insoluble in water and have high molecular weights. Synthetic polymers are used in plastic industries for various purposes like packaging, transporting goods, etc. Bioaccumulation of these synthetic polymers has adverse effects on various ecosystems. (Hashmi et al. 2017).

2.5 Polybrominated Biphenyls (PBBs)

PBBs are toxic to humans. PBB enters the human body through skin absorption, ingestion, and inhalation; it can also be transferred from mother to fetus by breastfeeding or placenta. High exposure to PBB causes several diseases like hepatic and reproductive dysfunction and thyroid (Chang et al. 2020).

2.6 Pharmaceuticals

Pharmaceuticals are an indirect source of xenobiotics. Drugs and medications that are not naturally found in the environment but can enter through human and animal waste. The primary sources of pharmaceuticals, such as antibiotics, hormones, and anesthetics, are hospital effluent and wastewater. These undesirable compounds are incorporated in food chains via bioaccumulation, which harms living beings (Hashmi et al. 2017).

2.7 Pesticides

Pesticides are chemically produced substances used in agricultural fields to prevent pest attacks, increasing crop yield. However, unrestricted usage of pesticides has hazardous consequences on the environment and human health. DDTs (dichlorodiphenyltrichloroethane), DDD (dichlorodiphenyldichloroethane), HCHs (hexacyclochlorohexanes), and DDE (dichlorodiphenyldichloroethylene) are some examples of pesticides (Wang et al. 2022).

2.8 Heavy Metals

Heavy metals like Hg, Pb, Cd, and As have toxic effects on human health. Heavy metals are potential generators of nitrogen and reactive oxygen species, which interfere with the redox potential of the cell. This reactive oxygen species causes lipid peroxidation, because of which several diseases, like neurodegenerative and cardiovascular diseases and cancer, can occur.

2.8.1 Mercury (Hg)

Mercury is one of the toxic heavy metals that has many hazardous effects. Methyl mercury is the most toxic form; microorganisms present in water and soil methylate the inorganic mercuric ions and convert them into methyl mercury. Monomethyl mercury, dimethyl mercury, and phenyl mercury are the other forms of mercury. Methyl mercury is a neurotoxic compound; accumulation of mercury in the central nervous system and motor regions of the brain gives rise to a neurotoxic effect.

2.8.2 Lead (Pb)

Pb is another toxic metal found in the environment. Lead toxicity may occur from both natural sources and anthropogenic activities. Lead exposure from industrial sources like battery industries and lead smelters is a matter of concern nowadays. Long-term exposure to Pb leads to various dysfunctions like dyslexia, psychosis, autism, brain damage, kidney damage, and paralysis.

2.8.3 Arsenic (As)

Arsenic is a metalloid (intermediate between metal and non-metal) and a potent carcinogen for humans. It causes skin, lung, liver, and kidney cancers. Arsenocholine, arsenobetaine, and arsenosugars are organic forms of arsenic that are present in some foods.

2.8.4 Cadmium (Cd)

Humans get exposed to Cd by different sources, such as contaminated soil and water, which are contaminated by metal in phosphate fertilizers. Cd is also known for its nephrotoxicity, which can cause several kidney problems and may lead to renal failure (Dinka 2018).

3 Xenobiotics and Their Hazardous Effects

3.1 Xenobiotics as an Environmental Hazard

Xenobiotics are chemical compounds that do not originate from the body and can include drugs, pesticides, industrial substances, and cosmetics, which can find their way into the environment because of industrial effluent, agricultural runoff, and seasonal monsoonal rains. These substances contribute to the pollution of water, soil, and atmosphere. Bodies of water can be poisoned, especially by water flora, overshadowing other forms of aquatic life by endocrine disruptors. Soils can be polluted with pesticides and various industrial chemicals. Many of these xenobiotics also lead to air contamination, thus becoming dangerous to humans and the environment. These xenobiotic compounds become persistent in the environment because of their resistance to natural degradation processes. Therefore, these remain as persistent contaminants within water and soil ecosystems. Xenobiotics such as PAHs, 1,1,1-trichloro-2,2-bis(4-chlorophenyl)ethane (DDT), PCBs, and pesticides, which are used for crop protection, play a key role in soil contamination (Miglani et al. 2022). Pesticides, heavy metals, and pharmaceuticals (human drugs, veterinary drugs) contaminated the water bodies due to improper wastewater treatment and incorrect waste management policy (Piwowarska and Kiedrzyńska 2022). Xenobiotic accumulation in water bodies leads to eutrophication and increases oxidative stress in aquatic creatures (Miglani et al. 2022).

3.2 *Hazardous Effect of Xenobiotics on Humans*

Xenobiotics, including halocarbons, PCBs, PAHs, heavy metals, pesticides, synthetic polymers, and polybrominated biphenyls (PBBs), have notorious effects on human health (Table 1).

Table 1 Xenobiotics and their hazardous effect on humans

Name of the xenobiotic compound	Hazardous effects on human	References
Halocarbons	Carcinogenic, mutagenic, and teratogenic effect.	Yang et al. (2024b)
Polychlorinated Biphenyls (PCB)	Cancer, neuropsychological and neurobehavioral disorders, immune system dysfunctions, and dementia.	Montano et al. (2022)
Polycyclic aromatic hydrocarbons (PAH)	Mutagenic and carcinogenic effect.	Idowu et al. (2019)
Heavy metals	Carcinogenic, genotoxic, and immunotoxic effects.	Ankit et al. (2022)
Pesticides	Diabetes, cancer, and asthma.	Kim et al. (2017)
Synthetic polymers (Polystyrene, polyethylene)	Cancer.	Patel et al. (2021)
Polybrominated biphenyls (PBBs)	Carcinogenicity, neurodevelopmental toxicities, epigenetic toxicity, and immunotoxicity.	Yang et al. (2024a)

4 Remediation Techniques

Treatment methods like adsorption, coagulation, filtration, electrolysis, ozonation, and chemical precipitation have been used to degrade xenobiotic compounds. However, they have limitations due to their waste-disposal issue, high cost, and generation of more poisonous by-products than the compound itself. Compared with that, bioremediation is a sustainable process that detoxifies the xenobiotic compounds without producing toxic by-products and is economically feasible (Mishra et al. 2021).

4.1 Microbes as Bioremediation Agents

4.1.1 Bacterial Bioremediation

Bacteria constitute a significant part of bioremediation. They depend on oxidation, reduction, and hydrolysis methods to break up complex compounds. Bacteria can break down these pollutants both aerobically and anaerobically. Bacteria generally produce enzymes capable of disintegrating xenobiotics through mechanisms such as bioventing, cometabolism, and horizontal gene transfer. Enzymes like oxygenase produced by bacteria can degrade the xenobiotic compounds by oxidation reaction (PATIL and DESAI 2024). Aerobic bacteria such as *Rhodococcus, Serratia, Pseudomonas, Escherichia, Moraxella, Sphingobium, Bacillus, and Micrococcus* and anaerobic bacteria such as *Methanospirillum, Desulfovibrio, Desulfotomaculum, Pelatomaculum, Syntrophobacter, and Syntrophus* could degrade xenobiotic compounds (Patel et al. 2021) (Table 2).

Bioremediation is more effective when more than one type of bacteria works together in a consortium. Bacteria in the same consortia enhance each other's metabolic activity and growth, increasing bioremediation's efficacy. Cross-feeding, in which compounds produced by one microbial community serve as substrates for another community, benefits bioremediation. Here, synergistic interaction between two communities enhances the bioremediation process (PATIL and DESAI 2024).

Xenobiotic bioremediation by bacteria includes several metabolic pathways that transform these biotechnological compounds into less dangerous chemicals. The pathways include the following:

4.1.1.1 Aerobic Pathways

The aerobic mode of degradation of xenobiotics in bacteria is done either by oxidative or hydrolytic pathways. In oxidative pathways, bacteria use enzymes monooxygenase, dioxygenase, cytochrome P450 enzymes, and flavin-containing monooxygenases (FMOs) to introduce the oxygen atoms in xenobiotic molecules,

Table 2 Detailed list of the microorganisms involved in different xenobiotic bioremediation

Types of compounds	Name of compound	Microorganisms involved in bioremediation	References
Hydrocarbons	Crude oil	*Microbacterium deserti*	Mordecai et al. (2024)
		Priestia megaterium	Liu et al. (2024)
		Bacillus licheniformis	Srimathi et al. (2024)
		Acinetobacter junii	Gyasi et al. (2024)
		Alcanivorax xenomutans	Gyasi et al. (2024)
	TPHs	*Bacillus thuringiensis*	Raju et al. (2017)
		Pseudomonas sp.	Li et al. (2024)
	Aliphatic and aromatic hydrocarbons	*Candida tropicalis*	Hegazy et al. (2024)
	Toluene	*Brevibacillus parabrevis*	Heydarnezhad et al. (2024)
Polychlorinated biphenyl	Recalcitrant coplanar PCBs like 3,4,3′,4′-CB	*Paenibacillus* sp.	Sakai et al. (2005)
Heavy metals	Mercury (Hg)	*Alteromonas macleodii*	Zhang et al. (2024)
	Lead (Pb)	*Enterobacter kobei*	Abdollahi et al. (2020)
	Arsenic (As)	*Lysinibacillus* sp.	Mandal et al. (2022)
		Bacillus safensis	
	Cadmium (Cd)	*Ochrobactrum* sp.	Khadivinia et al. (2014)
Pesticides	4-Chloro-2-nitrophenol	*Bacillus* sp.	Arora (2020)
	Pyrethroid	*Bacillus cereus*	Huang et al. (2024)
Synthetic polymers	Polystyrene	*Lysinibacillus sp. and Pseudomonas* sp.	Wang et al. (2024)
	Polyethylene	*Rhodococcus* sp.	Rong et al. (2024)
PBBs	2,3′,4,4′,5-pentachlorobiphenyl (PCB 118)	*Methylorubrum* sp.	Wu et al. (2024)
	2,2′,4,4′-tetrabrominated diphenyl ether	*Burkholderia xenovorans*	Gu et al. (2023)

which makes them more reactive and more straightforward to break down aerobically. For example, the formation of catechols is involved during the breakdown of polyaromatic hydrocarbons that are further broken down via the ortho or meta-cleavage pathways. Meanwhile, in hydrolytic pathways, enzymes like esterases and amidases hydrolyze xenobiotics' ester and amide bonds into simpler compounds that can be further metabolized (Mishra et al. 2021).

4.1.1.2 Anaerobic Pathways

Anaerobic degradation of xenobiotics involves pathways like reductive dehalogenation and fermentation. Without oxygen, bacteria remove the halogen atoms from manufactured chemicals through reductive dehalogenation. This process is essential in the degradation of chlorinated chemicals such as PCBs and trichloroethylene (TCE). Some bacteria can ferment xenobiotics and get their products from organic acids, alcohols, and gases like hydrogen and methane.

4.1.1.3 Co-metabolic Pathways

Some bacteria degrade xenobiotics when they metabolize other primary substrates. This process, which is called cometabolism, is usually performed by enzymes that do not specifically evolve for xenobiotic degradation but are transformed as a side activity.

In addition, some bacteria use efflux pumps to expel the xenobiotic pollutants outside the cell and decrease the intracellular xenobiotic concentration. Some bacteria harbor auxiliary genes and plasmids that make them capable of xenobiotic degradation (Li et al. 2021).

4.1.2 Fungal Bioremediation

Fungi play a pivotal role in nature as decomposers because of their hyphal networks. Fungal bioremediation is a promising approach for eliminating xenobiotics. Fungi, mainly white rot fungi (WRF), are quite successful in this capacity because of the capacity to construct a variety of extracellular enzymes capable of breaking down complex organic pollutants. Laccases and peroxidases are ligninolytic enzymes produced by fungi; they degrade xenobiotic compounds like PAHs (PATIL and DESAI 2024). Peroxidases like manganese peroxidases (MnP) and lignin peroxidases (LiP) are also found in fungi that break down complex organic pollutants by generating free radicals. These enzymes are particularly effective against lignin-like structures and various aromatic compounds (Khan et al. 2023).

4.1.3 Algal Bioremediation

The process by which micro- and macro-algae remove contaminants from the environment is known as phycoremediation. Pollutants like xenobiotics and pharmaceuticals, which are found in wastewater, sometimes are not removed properly by conventional wastewater treatment. Microalgae-based treatments are effective, partially removing pharmaceuticals or xenobiotics from wastewater (8–70%) (Goswami et al. 2022). Algae remove the pollutants and release oxygen via photosynthesis, instigating aerobic bacteria to grow and degrade organic pollutants (PATIL and DESAI 2024). Microalgae has the potential to remove heavy metals by two mechanisms: by consuming heavy metals within cells at minimal concentrations, and another is biosorption by the adsorption process (Ankit et al. 2022).

4.1.4 Protozoal Bioremediation

The method of using protozoa to clean up environmental pollution is known as protozoan bioremediation. These single-celled organisms can ingest and break down complex organic materials, including pollutants (Holmes et al. 2013). Protozoa often form symbiotic relationships with bacteria which can degrade xenobiotics, which is supported by protozoa grazing on the bacteria, stimulating their growth and functioning (Kamika and Momba 2013). Such interactions result in enhanced degradation since protozoa will disperse bacterial cells, increasing surface area for degradation while reducing biofilm formation. Protozoal bioremediation can be used in various habitats such as soils, water bodies, and wastewater treatment plants where bacterial degradation alone is not effective enough. However, balancing between protozoans and bacteria remains an uphill task that requires a thorough understanding of complicated relationships among protozoan, bacterial, and xenobiotic groups to optimize this bioremediation technique.

5 Advanced Technologies to Identify and Characterize Microorganisms for Xenobiotic Degradation (the Omics Approach)

Genomics, metagenomics, transcriptomics, proteomics, metabolomics, and bioinformatics are advanced molecular techniques that are used to determine the genes, mRNA expression levels, proteins, and metabolic pathways in switching environments. By using these techniques, we can understand the potency of "viable but non-culturable (VBNC)" bacteria in the biodegradation of xenobiotics (Mishra et al. 2021).

Many culture-independent molecular techniques, for example, fluorescent in situ hybridization (FISH), terminal restriction fragment length polymorphism (T-RFLP), fatty acid methyl ester (FAME), denaturing and temperature gradient gel electrophoresis (DGGE/TGGE), etc., play a significant role in the identification and characterization of xenobiotic-degrading microorganisms (Bodor et al. 2009; Zhao et al. 2017). Advanced techniques like gene sequencing and analysis methods have been crucial for both the microorganisms used in the xenobiotic degradation process and the genes' character. Genomic and metagenomic techniques are beneficial for identifying and characterizing microorganisms used for xenobiotic degradation. Genomics is a technology that involves the sequencing and analysis of the genetic makeup of microorganisms in which the metabolic functions of the cells containing the degradation capacities, among others, are encoded. Thus, necessary information can also produce the metabolic routes and enzymes. The genomic technique is sequencing and the analysis of microorganisms' genomes to identify genes with degradation capabilities and the production of the corresponding maps of the metabolic and enzymatic shifts. In addition, metagenomics is a method of microbial community analysis conducted through the direct extraction of genomic DNA from an environmental sample. Thus, the information obtained reflects the collective genetic potential of the community, notwithstanding the microbial species. Transcriptomics involves RNA expression studies to know the expression level of genes. Proteomic studies, in this case, aim to determine how the relative abundance of the critical genes involved in the degradation of the xenobiotic is altered. In this method, enzymatic activity can also be investigated. On the other hand, metabolomics is the most rapidly developing and innovative method to better understand xenobiotic degradation, and bioinformatics is used to analyze the data generated from the above techniques. All these related methods, such as genomics, transcriptomics, proteomics, metabolomics, and bioinformatics, must be applied in the process of the metabolic monitoring system for xenobiotic degradation to obtain the global picture for the biotransformation of xenobiotics by microorganisms and new techniques for bioremediation that can be utilized (Fig. 1).

5.1 Genomics as a Tool to Solve the Xenobiotic Degradation Problem

The understanding and improvement of xenobiotic degradation may hinge on genomics.

5.1.1 Degradation Pathways Identification

The breakdown of xenobiotics can be studied by annotating metagenomic sequences followed by identifying specific genes and metabolic pathways that may have been involved in their degradation (Kneubehl and Iyer 2023). For instance, complete pathways for degrading aromatic compounds and halobenzoates exist in bacteria such as *Pseudomonas putida* and *Achromobacter xylosoxidans*. Functional genes and metabolic pathways responsible for xenobiotic bioremediation have also been revealed through metagenomics data (Vera et al. 2022). A recent study has evidenced the xenobiotic degradation potential of 166 metagenome-assembled

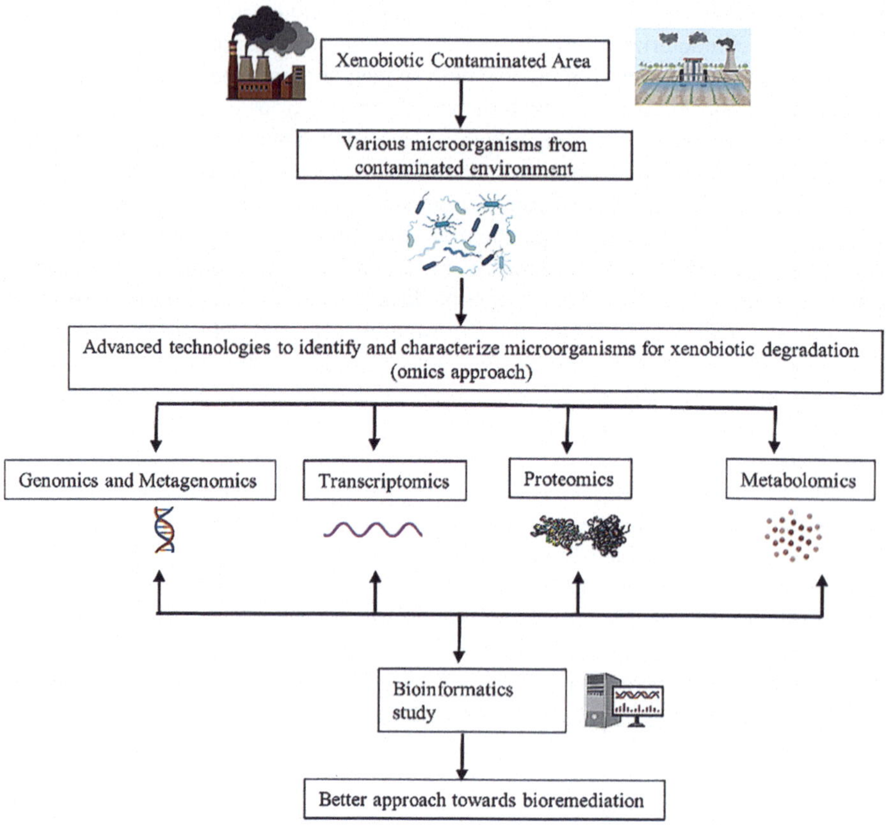

Fig. 1 Illustration of microbe-mediated xenobiotic degradation using omics approaches

genomes (MAGs) obtained from the metagenome sequences of xenobiotic-degrading consortium enriched from soils and lake sediments (Jia et al. 2024).

5.1.2 Exploration of the Diversity of Xenobiotic-Degrading Microbes

Some microbes may have an excellent potential for xenobiotic degradation, which is identified through diverse microbial communities in contaminated environments using metagenomics approaches (Mishra et al. 2021).

5.1.3 Gene Recruitment and Mobile Genetic Elements

Recent genome analyses indicate that some bacteria capable of degrading xenobiotics acquired these abilities by acquiring corresponding genes from other organisms. Plasmids and transposons are examples of mobile genetic elements involved in this process.

5.1.4 Bioremediation Potentiality

It is also possible to foster bioremediation by studying the genetic structure of microorganisms capable of breaking down xenobiotics. More specifically, it is possible to manipulate genes to enhance their degradation efficiency or the range of compounds they can break down by knowing the whole genome sequences of the degraders.

5.2 Transcriptomics to Better Understand Microbe-Mediated Bioremediation of Xenobiotics

One of the immediate advantages of transcriptomics in environmental biotechnology is that it serves as a helpful instrument for studying xenobiotics' degradation. It helps researchers analyze gene expression changes among microorganisms exposed to xenobiotic chemicals, thus identifying genes whose expressions were upregulated or downregulated accordingly. Knowledge gained from this provides insight into the metabolic pathways behind such degradation. By carrying out transcriptome analysis, researchers can locate genes and enzymes that are important during the breakdown process of xenobiotics. This is significant in comprehending biodegradation processes and engineering microorganisms with superior degradation properties. Transcriptomics can also be employed to examine several ecosystems, to assess the impact of xenobiotics on microbial communities, and to optimize bioremediation procedures. Knowing how effectively degraders express their genes can improve their activity and efficacy (Sanchez-Quant et al. 2023). Studies have shown how bacteria gene expression changes in *Pseudomonas* sp. TNT3 upon exposure to 100 mg/L of 2,4,6-trinitrotoluene TNT (TNT) (Cabrera et al. 2024). Resistance mechanisms against xenobiotic compounds like heavy metals have also been explored via transcriptomics (Kerdsomboon et al. 2024).

5.3 A Methodology Based on Proteomics for Solving the Issue with Bioremediation of Xenobiotics

Xenobiotics, which are foreign chemical entities occurring in biological systems, form proteomics' subject matter of great importance. It facilitates both the identification and characterization of proteins responsible for biodegradation by microbial communities, enabling insight into how these organisms break down poisonous compounds. Proteomics also involves studying the xenobiotics' effects on aquatic life by identifying stress or toxicity proteins that may aid in monitoring and mitigating pollutants' harmfulness to surface water environments. Furthermore, proteomics helps identify biomarkers for detecting xenobiotic exposure and assessing their impacts on living entities, which is particularly useful in environmental monitoring and public health. Besides that, by observing all alterations within protein expression levels together with modifications, proteomics provides information on toxicity mechanisms as well as how biological pathways get affected by xenobiotics (Sharma et al. 2022; Gohain et al. 2023).

5.4 The Use of Metabolomics in Addressing the Xenobiotics Degradation Issue

Metabolomics is an important tool used to study the breakdown of xenobiotic compounds in living organisms that have been exposed to them. This helps to identify the pathways implicated in the metabolic timetable and also helps understand how they are degraded. Metabolomics can be used to observe all alterations occurring within protein expression levels alongside modifications that provide insight into mechanisms underlying toxicity and biological pathways affected by these substances. (Sharma et al. 2022; Gohain et al. 2023).

5.5 The Use of Bioinformatics in Xenobiotic Degradation

Bioinformatics is an essential tool that combines both biological data and computational technology. Biological data that are generated from DNA, RNA, and protein sequences are analyzed through bioinformatics, so bioinformatics is important for a better understanding of the biodegradation of xenobiotic compounds. Some well-known software used to study biodegradation, such as the University of Minnesota Biocatalysts/Biodegradation Database (UM-BBD), has information about biodegradation pathways of xenobiotics and microbial catabolism. Introducing bioinformatics to the study of bioremediation gives it a better direction for eliminating pollutants. (Rodríguez et al. 2020; Mishra et al. 2021).

6 Future Prospects

6.1 Genetic Engineering of Microorganisms

A changing climate has made scientists turn their attention to using new technologies that alter the genetic makeup of microbes, intending to enhance their ability to cleanse polluted environments. This approach can involve the modification of these organisms in such ways that they would have additional pathways for degradation, sometimes leading up to superbug development or even introducing new ones via plasmids (Vishwakarma et al. 2020). To boost in-situ bacterial communities responsible for degrading different types of pollutants, the genetically engineered microbes must be inoculated into areas affected by contamination since they are designed specifically for that purpose. Moreover, plasmids allow sharing between communities, increasing global remediation.

The application of microorganisms has proven successful in using genetically modified bacteria for cleaning sites like those contaminated through oil spillage, removal of heavy metals, or treatment of industrial effluents. On the other hand, releasing genetically modified bacteria into ecosystems could spur concerns about possible effects on ecosystem dynamics and resistance phenomena. Therefore, regulatory frameworks and safety assessments are essential to adopt appropriate bioremediation strategies without compromising their security or efficacy.

Future studies will aim at maximizing the stability and effectiveness of engineered microorganisms existing in different environments, whereby future achievements in the fields of synthetic biology and metagenomics will be critical in the production of next-generation bioremediation (Priya et al. 2024).

6.2 Bacteriophage-Based Bioremediation

The use of bacteriophages in removing toxic compounds is a recent development in bioremediation and can be an important future aspect of bioremediation. It includes focused degradation, improved biodegradation, and diminished harmful by-products. Specific bacteriophages have been designed to degrade xenobiotics by targeting bacteria with similar metabolic pathways, leading to their effective remediation (Shivaram et al. 2023). Besides these, they secrete enzymes and other cellular components that help break down xenobiotics. This approach can be employed in diverse environmental settings, such as soils, aquatic systems, and wastewater treatment stations, making it an attractive alternative for solving contamination problems. However, one major challenge is that some bacteria may develop resistance against these viruses; thus, alternative solutions will be needed, including cocktail mixtures of phages or genetically modified ones. Achieving an optimal level of bioremediation requires a deep understanding of how different bacteria interact at the molecular level during the process.

6.3 Nano Bioremediation of Xenobiotics

The combination of nanotechnology and bioremediation is called nanobioremediation, which enhances the biodegradation of xenobiotics. With it, zero-valent iron nanoparticles, titanium dioxide nanoparticles, and silver nanoparticles can be used to accelerate the breakdown of xenobiotic compounds, which precipitate in this process via microbial enzymes (Patra Shahi et al. 2020; Bhatt et al. 2022). Since the particles' surface area to volume ratio is very high, they have more contact with pollutants and thus promote their faster degradation. Additionally, specific enzymes or microorganisms may be directly delivered to contaminated sites using engineered nanoparticles, improving the overall performance of such processes. This also increases the availability of xenobiotics by breaking them into smaller fragments that are easier for microorganisms to consume and utilize in a degradation pathway. For these reasons, nanobioremediation has a wide range of uses, including but not limited to soil remediation and water treatment technologies (Patra Shahi et al. 2020). Nevertheless, there remain worries about the potential toxicity posed by nanoparticles; thus, research is still underway to find safe materials for bioremediation purposes.

7 Conclusion

Microbial bioremediation is becoming important, especially with the impacts of xenobiotic pollutants on environmental well-being and human health. These conventional methods can be highly costly in addition to generating toxic residuals, whereas using microorganisms such as bacteria, fungi, and algae is much more cost-effective. Molecular techniques, including genomics and proteomics, are crucial for identifying and improving the microbial degradation of toxic wastes. New methods such as genetic engineering and bacteriophage-based bioremediation promise higher effectiveness as the concept develops. We can identify the genetic components that break down xenobiotics by integrating an omics approach. This opens up the possibility of designing microorganisms that can safely manage pollution while meeting regulatory standards.

Acknowledgement S.M. would like to acknowledge the financial grant received from the Department of Science & Technology and Biotechnology, Government of West Bengal (WBDST) (Grant No. STBT-11012(25)/2/2024-ST SEC). S.K. sincerely acknowledges the fellowship support received from the Department of Science & Technology and Biotechnology, Government of West Bengal (WBDST). S.M. is thankful to the Department of Microbiology, University of Kalyani, for providing the infrastructural facilities.

References

Abdel-Shafy HI, Mansour MSM (2016) A review on polycyclic aromatic hydrocarbons: Source, environmental impact, effect on human health and remediation. Egyptian Journal of Petroleum 25:107–123. https://doi.org/10.1016/J.EJPE.2015.03.011

Abdollahi S, Golchin A, Shahryari F (2020) Lead and cadmium-resistant bacterial species isolated from heavy metal-contaminated soils show plant growth-promoting traits. Int Microbiol 23:625–640. https://doi.org/10.1007/S10123-020-00133-1

Ankit, Bauddh K, Korstad J (2022) Phycoremediation: Use of Algae to Sequester Heavy Metals. Hydrobiology 2022, Vol 1, Pages 288–303 1:288–303. https://doi.org/10.3390/HYDROBIOLOGY1030021

Arora PK (2020) Bacilli-Mediated Degradation of Xenobiotic Compounds and Heavy Metals. Front Bioeng Biotechnol 8:570307. https://doi.org/10.3389/FBIOE.2020.570307/PDF

Bhatt P, Pandey SC, Joshi S, et al (2022) Nanobioremediation: A sustainable approach for the removal of toxic pollutants from the environment. J Hazard Mater 427. https://doi.org/10.1016/J.JHAZMAT.2021.128033

Bjørklund G, Dadar M, Chirumbolo S, Aaseth J (2020) The Role of Xenobiotics and Trace Metals in Parkinson's Disease. Mol Neurobiol 57:1405–1417. https://doi.org/10.1007/S12035-019-01832-1

Bodor A, Bounedjoum N, Vincze GE, et al (2009) Challenges of unculturable bacteria: environmental perspectives. Rev Environ Sci Biotechnol 19. https://doi.org/10.1007/S11157-020-09522-4

Bouwer E, Durant N, Wilson L, et al (1994) Degradation of xenobiotic compounds in situ: capabilities and limits. FEMS Microbiol Rev 15:307–317. https://doi.org/10.1111/J.1574-6976.1994.TB00142.X

Cabrera MÁ, Márquez SL, Pérez-Donoso JM (2024) New insights into xenobiotic tolerance of Antarctic bacteria: transcriptomic analysis of Pseudomonas sp. TNT3 during 2,4,6-trinitrotoluene biotransformation. Environ Sci Pollut Res Int 31:17256–17274. https://doi.org/10.1007/S11356-024-32298-X

Chang CJ, Terrell ML, Marcus M, et al (2020) Serum concentrations of polybrominated biphenyls (PBBs), polychlorinated biphenyls (PCBs) and polybrominated diphenyl ethers (PBDEs) in the Michigan PBB Registry 40 years after the PBB contamination incident. Environ Int 137:105526. https://doi.org/10.1016/J.ENVINT.2020.105526

Connell DW (2018) Achievements and challenges in bioaccumulation research. Bioaccumulation of Xenobiotic Compounds 209–213. https://doi.org/10.1201/9781351070126/BIOACCUMULATION-XENOBIOTIC-COMPOUNDS-DES-CONNELL/ACCESSIBILITY-INFORMATION

Dinka DD (2018) Environmental Xenobiotics and Their Adverse Health Impacts-A General Review. Journal of Environment Pollution and Human Health, Vol 6, 2018, Pages 77–88 6:77–88. https://doi.org/10.12691/JEPHH-6-3-1

Eriksson E, Auffarth K, Eilersen AM, et al (2003) Household chemicals and personal care products as sources for xenobiotic organic compounds in grey wastewater. Water SA 29:135–146. https://doi.org/10.4314/WSA.V29I2.4848

Gohain A, Sarma R, Singh P (2023) Recent Advancements in Bioremediation of Xenobiotics Using Microbes. Xenobiotics in Urban Ecosystems: Sources, Distribution and Health Impacts 355–374. https://doi.org/10.1007/978-3-031-35775-6_17

Goswami RK, Agrawal K, Verma P (2022) An exploration of natural synergy using microalgae for the remediation of pharmaceuticals and xenobiotics in wastewater. Algal Res 64. https://doi.org/10.1016/J.ALGAL.2022.102703

Gu C, Jin Z, Fan X, et al (2023) Comparative evaluation and prioritization of key influences on biodegradation of 2,2',4,4'-tetrabrominated diphenyl ether by bacterial isolate B. xenovorans LB400. J Environ Manage 331. https://doi.org/10.1016/J.JENVMAN.2023.117320

Gyasi SF, Sarfo MK, Kabo-Bah AT, et al (2024) In vitro assessment of crude oil degradation by Acinetobacter junii and Alcanivorax xenomutans isolated from the coast of Ghana. Heliyon 10:e24994. https://doi.org/10.1016/J.HELIYON.2024.E24994

Hashmi MZ, Kumar V, Varma A (eds) (2017) Xenobiotics in the Soil Environment. 49. https://doi.org/10.1007/978-3-319-47744-2

Hegazy GE, Soliman NA, Farag S, et al (2024) Isolation and characterization of Candida tropicalis B: a promising yeast strain for biodegradation of petroleum oil in marine environments. Microb Cell Fact 23. https://doi.org/10.1186/S12934-023-02292-Y

Heydarnezhad F, Hoodaji M, Shahriarinour M, Tahmourespour A (2024) Statistical modelling and optimization of medium for toluene degradation by Brevibacillus parabrevis ATHH40 immobilized cells. International Journal of Environmental Studies 81:724–733. https://doi.org/10.1080/00207233.2024.2331950

Holmes DE, Giloteaux L, Williams KH, et al (2013) Enrichment of specific protozoan populations during in situ bioremediation of uranium-contaminated groundwater. ISME J 7:1286–1298. https://doi.org/10.1038/ISMEJ.2013.20

Huang Y, Yang L, Pan K, et al (2024) Heavy metal-tolerant bacteria Bacillus cereus BCS1 degrades pyrethroid in a soil-plant system. J Hazard Mater 461. https://doi.org/10.1016/J.JHAZMAT.2023.132594

Idowu O, Semple KT, Ramadass K, et al (2019) Beyond the obvious: Environmental health implications of polar polycyclic aromatic hydrocarbons. Environ Int 123:543–557. https://doi.org/10.1016/J.ENVINT.2018.12.051

Jia C, Wu C, Li Y, et al (2024) Metagenome-assembled genomes from enrichment cultures grown on xenobiotic solvents. Microbiol Resour Announc 13. https://doi.org/10.1128/MRA.00107-24

Kamika I, Momba MNB (2013) Assessing the resistance and bioremediation ability of selected bacterial and protozoan species to heavy metals in metal-rich industrial wastewater. BMC Microbiol 13. https://doi.org/10.1186/1471-2180-13-28

Karthigadevi G, Manikandan S, Karmegam N, et al (2021) Chemico-nanotreatment methods for the removal of persistent organic pollutants and xenobiotics in water – A review. Bioresour Technol 324. https://doi.org/10.1016/J.BIORTECH.2021.124678

Kerdsomboon K, Techo T, Mhuantong W, et al (2024) Genomic and transcriptomic analyses reveal insights into cadmium resistance mechanisms of Cupriavidus nantongensis strain E324. Sci Total Environ 952. https://doi.org/10.1016/J.SCITOTENV.2024.175915

Khadivinia E, Sharafi H, Hadi F, et al (2014) Cadmium biosorption by a glyphosate-degrading bacterium, a novel biosorbent isolated from pesticide-contaminated agricultural soils. Journal of Industrial and Engineering Chemistry 20:4304–4310. https://doi.org/10.1016/J.JIEC.2014.01.037

Khan MF, Hof C, Niemcová P, Murphy CD (2023) Recent advances in fungal xenobiotic metabolism: enzymes and applications. World J Microbiol Biotechnol 39:296. https://doi.org/10.1007/S11274-023-03737-7

Kim KH, Kabir E, Jahan SA (2017) Exposure to pesticides and the associated human health effects. Sci Total Environ 575:525–535. https://doi.org/10.1016/J.SCITOTENV.2016.09.009

Kirici M, Sokmen TO (2024) Xenobiotics: Effects of Unnatural Chemicals on the Environment. Recent Applications and Biological Activities in Aquaculture and Agriculture 19–52. https://doi.org/10.69860/NOBEL.9786053359319.2

Kneubehl AR, Iyer R (2023) Comparative Genomics Analysis of the Aromatic and Xenobiotic Degradation Capacities and Heavy Metal Resistance in Seven Environmentally Derived Bacterial Isolates. Water, Air, & Soil Pollution 234. https://doi.org/10.1007/S11270-023-06495-2

Kumar Maurya P, Malik DS (2016) Bioaccumulation of xenobiotics compound of pesticides in riverine system and its control technique: a critical review

Li J, Jia C, Lu Q, et al (2021) Mechanistic insights into the success of xenobiotic degraders resolved from metagenomes of microbial enrichment cultures. J Hazard Mater 418. https://doi.org/10.1016/J.JHAZMAT.2021.126384

Li X, Zhang R, Liu Z, et al (2024) Enhanced indigenous bacteria-mediated bioremediation of aged petroleum-contaminated soil by Fenton pretreatment: Synergistic effect of soil environmental improvement and microbial community response. Chemical Engineering Journal 498. https://doi.org/10.1016/j.cej.2024.155771

Liu X, Ji J, Zhang X, et al (2024) Microbial Remediation of Crude Oil in Saline Conditions by Oil-Degrading Bacterium Priestia megaterium FDU301. Appl Biochem Biotechnol 196:2694–2712. https://doi.org/10.1007/S12010-022-04245-4

Mandal D, Sonar R, Saha I, et al (2022) Isolation and identification of arsenic resistant bacteria: a tool for bioremediation of arsenic toxicity. International Journal of Environmental Science and Technology 19:9883–9900. https://doi.org/10.1007/S13762-021-03673-9/FIGURES/8

Miglani R, Parveen N, Kumar A, et al (2022) Degradation of Xenobiotic Pollutants: An Environmentally Sustainable Approach. Metabolites 12. https://doi.org/10.3390/METABO12090818

Mishra S, Lin Z, Pang S, et al (2021) Recent Advanced Technologies for the Characterization of Xenobiotic-Degrading Microorganisms and Microbial Communities. Front Bioeng Biotechnol 9:632059. https://doi.org/10.3389/FBIOE.2021.632059/PDF

Montano L, Pironti C, Pinto G, et al (2022) Polychlorinated Biphenyls (PCBs) in the Environment: Occupational and Exposure Events, Effects on Human Health and Fertility. Toxics 10. https://doi.org/10.3390/TOXICS10070365

Mordecai J, Al-Thukair A, Musa MM, et al (2024) Bacterial Degradation of Petroleum Hydrocarbons in Saudi Arabia. Toxics 12:800. https://doi.org/10.3390/TOXICS12110800/S1

Nogacka A, Gómez-Martín M, Suárez A, et al (2019) Xenobiotics Formed during Food Processing: Their Relation with the Intestinal Microbiota and Colorectal Cancer. Int J Mol Sci 20:2051. https://doi.org/10.3390/IJMS20082051

Österreicher CH, Trauner M (2012) Xenobiotic-induced liver injury and fibrosis. Expert Opin Drug Metab Toxicol 8:571–580. https://doi.org/10.1517/17425255.2012.674511

Patel H, Trivedi S, Bhatt D, et al (2021) Microbe-Mediated Removal of Xenobiotics for Sustainable Environment. Microbial Technology for Sustainable Environment 483–497. https://doi.org/10.1007/978-981-16-3840-4_25

Patil R, Desai S (2024) Advances in bioremediation agents and processes for removal of persistent contaminants from environment. Int J Pharm Pharm Sci 42–47. https://doi.org/10.22159/IJPPS.2024V16I5.50724

Patra Shahi M, Kumari P, Mahobiya D, Kumar Shahi S (2020) Nano-bioremediation of environmental contaminants: applications, challenges, and future prospects. Bioremediation for Environmental Sustainability: Approaches to Tackle Pollution for Cleaner and Greener Society 83–98. https://doi.org/10.1016/B978-0-12-820318-7.00004-6

Piwowarska D, Kiedrzyńska E (2022) Xenobiotics as a contemporary threat to surface waters. Ecohydrology and Hydrobiology 22:337–354. https://doi.org/10.1016/J.ECOHYD.2021.09.003

Priya AK, Muruganandam M, Kumar A, et al (2024) Recent advances in microbial-assisted degradation and remediation of xenobiotic contaminants; challenges and future prospects. Journal of Water Process Engineering 60. https://doi.org/10.1016/J.JWPE.2024.105106

Raju MN, Scalvenzi L, Raju MN, Scalvenzi L (2017) Petroleum Degradation: Promising Biotechnological Tools for Bioremediation. Recent Insights in Petroleum Science and Engineering. https://doi.org/10.5772/INTECHOPEN.70109

Reddy AVB, Moniruzzaman M, Aminabhavi TM (2019) Polychlorinated biphenyls (PCBs) in the environment: Recent updates on sampling, pretreatment, cleanup technologies and their analysis

Rodríguez A, Castrejón-Godínez ML, Salazar-Bustamante E, et al (2020) Omics Approaches to Pesticide Biodegradation. Curr Microbiol 77:545–563. https://doi.org/10.1007/S00284-020-01916-5

Rong Z, Ding ZH, Wu YH, Xu XW (2024) Degradation of low-density polyethylene by the bacterium Rhodococcus sp. C-2 isolated from seawater. Sci Total Environ 907. https://doi.org/10.1016/J.SCITOTENV.2023.167993

Sakai M, Ezaki S, Suzuki N, Kurane R (2005) Isolation and characterization of a novel polychlorinated biphenyl-degrading bacterium, Paenibacillus sp. KBC101. Appl Microbiol Biotechnol 68:111–116. https://doi.org/10.1007/S00253-004-1848-3

Sanchez-Quant E, Richter ML, Colomé-Tatché M, Martinez-Jimenez CP (2023) Single-cell metabolic profiling reveals subgroups of primary human hepatocytes with heterogeneous responses to drug challenge. Genome Biol 24:1–29. https://doi.org/10.1186/S13059-023-03075-9/FIGURES/4

Sharma N, Shukla S, Shukla K, et al (2022) Recent advancements in microbial degradation of xenobiotics by using proteomics approaches. Omics for Environmental Engineering and Microbiology Systems 181–201. https://doi.org/10.1201/9781003247883-9/RECENT-ADVANCEMENTS-MICROBIAL-DEGRADATION-XENOBIOTICS-USING-PROTEOMICS-APPROACHES-NEHA-SHARMA-SMRITI-SHUKLA-KARTIKEYA-SHUKLA-AJIT-VARMA-VINEET-KUMAR-MENAKA-DEVI-SALAM-ARTI-MISHRA

Shivaram KB, Bhatt P, Applegate B, Simsek H (2023) Bacteriophage-based biocontrol technology to enhance the efficiency of wastewater treatment and reduce targeted bacterial biofilms. Sci Total Environ 862. https://doi.org/10.1016/J.SCITOTENV.2022.160723

Singh M, Mishra RC, Shah I, et al (2023) Xenobiotics: Sources, Pathways, Degradation, and Rish Associated with Major Emphasis on Pharmaceutical Compounds. Xenobiotics in Urban Ecosystems: Sources, Distribution and Health Impacts 87–106. https://doi.org/10.1007/978-3-031-35775-6_5

Srimathi M, Suganthi M, Sugitha S, et al (2024) Characterization of Crude Oil Degrading Marine Bacterium Bacillus licheniformis. Indian J Microbiol 64. https://doi.org/10.1007/S12088-024-01222-9

Vera A, Wilson FP, Cupples AM (2022) Predicted functional genes for the biodegradation of xenobiotics in groundwater and sediment at two contaminated naval sites. Appl Microbiol Biotechnol 106:835–853. https://doi.org/10.1007/S00253-021-11756-3

Vishwakarma GS, Bhattacharjee G, Gohil N, Singh V (2020) Current status, challenges and future of bioremediation. Bioremediation of Pollutants: From Genetic Engineering to Genome Engineering 403–415. https://doi.org/10.1016/B978-0-12-819025-8.00020-X

Wang W, Yao S, Zhao Z, et al (2024) Degradation and potential metabolism pathway of polystyrene by bacteria from landfill site. Environ Pollut 343. https://doi.org/10.1016/J.ENVPOL.2023.123202

Wang X, Sial MU, Bashir MA, et al (2022) Pesticides Xenobiotics in Soil Ecosystem and Their Remediation Approaches. Sustainability 2022, Vol 14, Page 3353 14:3353. https://doi.org/10.3390/SU14063353

Wu Y, Zhu M, Ouyang X, et al (2024) Integrated transcriptomics and metabolomics analyses reveal the aerobic biodegradation and molecular mechanisms of 2,3',4,4',5-pentachlorodiphenyl (PCB 118) in Methylorubrum sp. ZY-1. Chemosphere 356. https://doi.org/10.1016/J.CHEMOSPHERE.2024.141921

Yang L, Sun P, Tao L, Zhao X (2024a) An in silico study on human carcinogenicity mechanism of polybrominated biphenyls exposure. Chem Biol Interact 397. https://doi.org/10.1016/J.CBI.2024.111075

Yang X, Du J, Jia C, et al (2024b) Groundwater pollution risk, health effects and sustainable management of halocarbons in typical industrial parks. Environ Res 250. https://doi.org/10.1016/J.ENVRES.2024.118422

Yegambaram M, Manivannan B, Beach T, Halden R (2015) Role of environmental contaminants in the etiology of Alzheimer's disease: a review. Curr Alzheimer Res 12:116–146. https://doi.org/10.2174/1567205012666150204121719

Zhang D, Chu B, Yang Q, et al (2024) Degradation of organic mercury in high salt environments by a marine aerobic bacterium Alteromonas macleodii KD01. Bioresour Technol 402. https://doi.org/10.1016/J.BIORTECH.2024.130831

Zhao X, Ma F, Feng C, et al (2017) Complete genome sequence of Arthrobacter sp. ZXY-2 associated with effective atrazine degradation and salt adaptation. J Biotechnol 248:43–47. https://doi.org/10.1016/J.JBIOTEC.2017.03.010

Plant-Microbe Interaction Under Environmental Extremities with Special Reference to Rhizospheric Engineering

Malay Kumar Adak, Nilakshi Chakraborty, and Soumita Roy

Abstract The microbial consortium is the wealth of soil ecosystem where plants mostly get predominantly benefited in nutrients, mechanical support, and stress tolerance like essential life supports. Microbial population offers a significant biostimulation to plants for promising growth, and reactivity to environments, including moisture stress, salinity, temperature variation, and heavy metal/xenobiotic contamination. The biostimulants from rhizobacteria, nitrogen-fixing bacteria, and mycorrhizal association could enhance plant resistance with improved growth regulators, conversions of the nutrients into a simpler form, osmotic adjustment, and reactivity to toxic endogenous chemicals like reactive oxygen/ nitrogen/sulfur species, phenols, etc. Understanding the molecular signaling from microbes transduced into plants is based on recent advancements in genomics, transcriptomics, proteomics, metabolomics, and phenomics. These all together collected the data and expanded the insights of molecular communication between plant-microbe (specially rhizobacteria) for the development and improvement of stress resilience. Still, a significant knowledge gap exists in molecular signaling and the development of biostimulants inside the plants and their adoption for stress tolerance. The conversion of chemical signals into other molecular form at cellular and nuclear levels is improvised into gene regulation. Specific genes under environmental stresses are modulated in plants both for expression and regulation with rhizobacterial signaling. Precise mechanisms are still to be deciphered with gene regulation levels through microRNA and epigenetic control under induction of rhizospheric microbial consortium. This short, however, compact review is expected to underscore the current visions and future prospective in the field of potential for the application of biostimulants in sustainable agriculture.

Keywords Bacteria · PGPR · ROS · Secondary metabolites · Stress

M. K. Adak (✉) · N. Chakraborty · S. Roy
Department of Botany, University of Kalyani, Kalyani, Nadia, India

S. Roy, V. Mandal (eds.), *Plant-Microbe Interaction under Xenobiotic Exposure*,
https://doi.org/10.1007/978-981-96-8260-7_11

1 Introduction

The rhizosphere happens to be influenced by plant root exudates, action of soil borne microorganisms, and the physicochemical properties of soil subjected to edaphic variations. The rhizosphere is the most dynamic as well as a complex environment that is highly controlled by root activities, microbial metabolism, and those collectively regulate mineral nutrient availability and assimilation (Wang et al. 2024). The rhizospheric microbial consortium (RMC) is more influenced by the activities of soil-borne biota including algae, fungi, archaea, and eubacteria with their parasitic or mutualistic relationship. It is mostly the activities of RMC that govern modulations of mineral dynamics in soil-plant systems, accusation of minerals over the membrane, metabolism of minerals into organic residues, and sequestering the excess ion into vacuoles and other apoplastic spaces (Santoyo et al. 2021). These all may directly or indirectly be related to the signaling by the endophytic rhizobacteria (*Serratia liquefaciens*, *Bacillus cereus*, *Pseudomonas* sp.) where solubilization of minerals can be integrated into biomolecules and excess is transformed into reactive oxygen species (ROS) (Shaffique et al. 2022). Thus, microbial biostimulants specifically from plant growth promoting rhizobacteria (PGPR) have been increasing in beneficial roles in the sustainable development of crops. In comparison with mineral fertilizers, and caustic pesticides, fungicides, herbicides, the exercise of PGPR-based biostimulants would be a better substitute for mineral absorption, assimilation and overall growth and productivity of crops (Hamid et al. 2021). The RMC is more influenced by a mutualistic approach of roots exudates and bacterial exopolysaccharides-based secretion where nutrient uptake over the membrane is better facilitated as well as activated in plants' antioxidation mechanism against oxidative redox. This would be the key to tolerance against drought, salinity, anoxia/hypoxia, and metal/xenobiotic toxicity where the redox imbalance induces the degenerative effects of stress (Adoko et al. 2022). The PGPR also influences nitrogen fixation and mineral solubilization as compared to mycorrhizal improvement of nutrient uptake particularly in acid soil. Therefore, agricultural farming would be seeking any sustainable methods to support crop growth, and yield coupled with improved abiotic stress tolerance by biostimulant application as an effective alternative to chemical pesticides (Jiang et al. 2024). Abiotic stresses are the major determining factors to downregulate plant growth and productivity where the life processes including photosynthesis, energy-yielding metabolism, and osmotic adjustment are compromised. Osmotic stress, for example,. initially limits water availability ensuring physiological wilting, and disrupting the ionic balance. Reduced metabolic reactions, turnover of redox potential and thereby overall of plant growth and productivity are the results. The activity of PGPR-based biostimulants has been effective in every case in reducing the intensities of severity, developing alternative support for normal cellular activity and thereby imposing stress tolerance (Rai et al. 2021). In this context, RMC has proven a defined role and gaining focus for their commercial and economic exercises in crop practices. It is mostly due to global acceptance, eco-friendly, sustainable, and cost-effective without significant lethal impositions to the cropping environment. The PGPR-based accessibility for organic substances supporting biofertilizers must have synergistic improvement for phytohormone supplementation (Sani and Yong 2021). Under environmental extremities, plant-microbe interaction can induce the biosynthetic pathways of abscisic acid (ABA) for water stress, salicylic acid (SA) for biotic stress, ethylene (ET) for heat and submergence stress, and jasmonic acids (JA) for oxidative stress tolerance in different crops (Muhammad et al. 2024).

Now, the microbial biostimulants must have some molecular communication in the plant-rhizosphere consortium and that is the dearth of understanding of stress tolerance fully. In most of the cases, the reduced root exudates whether influence the soil generated signals due to fluctuation of minerals, moisture content, or redox can also induce the metabolic reactions are not properly elucidated. Reports on the exchange of signals in the rhizosphere in the form of sugars, nitrogen derivatives, and secondary metabolites residues supported the changes in gene expression in roots (Saud et al. 2023). Therefore, a synergistic action could be expected that communicates the rhizosphere-root consortium for developing specific chemicals like phytohormones, quorum-sensing residues, and volatile compounds which also activate plant defense (Li et al. 2024). Even a suitable redox existing in the rhizosphere could be the principle of PGPR colonization which undergoes perturbed abiotic stresses and therefore transmits any signal to root for adaptation (Yang et al. 2024). This not only enriches the insights for plant-microbe interaction at molecular levels for the signaling residues but also needs for optimization of biostimulants for better responses of plants. The development of the biostimulants from microbial populations could be induced also by some exogenous elicitors which are required to optimize the growth of bacteria, induction of osmoprotectants, and over-expression of genes for the development of organic residues (Olowe et al. 2020). Thus, beyond the categories of biostimulant application, this short, however, comprehensive review may offer a basic understanding of targeted biostimulant and their molecular roles tailored to crop species under specific stress conditions. Understanding this mechanism/pathway could suffice a profound background and advanced insights in improvising or simulating compatible residues with optimized plant responses substituting indigenous biostimulants against environmental extremities. Conclusively, the present topic would be focusing any biological, physical, or physicochemical modification in rhizospheric flora which otherwise supports crops against environmental extremities through plant-microbe interaction. Additionally, the genome assisted modification of microbial activities with special reference to epigenomic control to the mechanism in development of climate resilient crops.

2 Genome Constitution Is the Key to the Environmental Interaction of Rhizobacteria

Undoubtedly environmental pollutants significantly threaten the ecological balance due to industrialization and related anthropogenic activities. Microbes as usual with their inveterate enzymatic pathway degrade the xenobiotics and related pollutants in which rhizobacteria play a significant role. In most cases, genetically engineered rhizobacteria have been in good support for the bioremediation of toxic chemicals as well as under various abiotic stresses including water, heavy metals, heat, and oxidative redox in soil. These compatible features are supported by a significant genome plasticity that may allow alteration of cellular and biochemical pathways to adapt to the toxicity. Microbial genomes with their wider diversity of gene expression as well as stringent regulation are the basis for the flexibility of remediating the pollutants. The whole genome sequencing of microbial flora isolated from organic wastewater observed the specific genes to metabolize xenobiotics like toluene, naphthalene, and chlorinated hydrocarbon-like residues (Mishra et al. 2021). Rhizospheric contamination with heavy metals like cadmium, chromium, antimony, nickel, and lead is also downregulated by bio-sequestration as well as rhizo-degradation or rhizo-filtration. The genome analysis of a few strains of *Bacillus* (e.g., LZ-C) indicated a specific gene for polycyclic aromatic residues. Toxic chemicals like dimethylformamide (DMF) are reported with other genes in *Paracoccus* sp. (Arya et al. 2023). The flexibility of the genome with the identification of next-generation sequencing revealed many pathways for phenanthrene degradation along with important aromatic compounds like *p*-hydroxybenzoic acid, amines, toluene, and other mostly chlorinated hydrocarbons. Thus, the microbial genome with its wider flexibility of gene regulation is involved through altered metabolism to transform different xenobiotics into smaller residues for C and end source. Genome modification has been another tool for rhizobacteria-mediated bioremediation. This technique has been faithful in exercise to introduce novel genes increasing copy numbers or any tailoring of sequences that facilitate better degradation for xenobiotics. Microbes with their simplicity in the genome are quite amenable to modifying the genome editing, unlike higher organisms. Through horizontal gene transfer (HGT), a number of genes concerning bio-degradation have been reported for petroleum hydrocarbon (Salim and Busi 2024). Important catabolic genes like *nah*Ac were also found to degrade naphthalene in the petrochemical-contaminated zone. For PGPR, for growth promotion of sensitive species is also accompanied by the degradation of herbicide detoxification bacteria as referred to with 2,4-dichlorophenoxy acetic acid (2,4-D) residue. From phylogenetically distant strains of *Pseudomonas, a* significant gene transfer for coding sequences for ACR1/R3/R3 has been possible to involve arsenic tolerance with significant plant growth (Biswas and Ganesan 2024). In some investigations, the identification of transposon for xenobiotic tolerance has been reported as exampled by Tn21 for reductase and specific proteins (e.g., MerT, MerC, MerP) for other herbicides. Therefore, the rhizospheric bacteria are more amenable not only for plant growth but also a better support for soil decontamination following regulated pollutant uptake by plants. In *Ralstonia sp*, the metal accumulation within the membrane is governed by the duplication of metal-resistant genes in the genome. This genome duplication supports the bacteria for adaptation to high copper salt in soil. Therefore, the de novo existence of genes and contribution to metabolic pathways would be due to genome flexibility contributing to environmental remediation for metal uptake.

3 Biosynthesis of Special Metabolites Under Xenobiotic Stress in PGPR

Besides the salinity and other heavy/toxic metals, a number of pollutants mostly from chlorinated hydrocarbons, benzene derivatives benzopyrene, anthracene, and naphthalene are included to induce special metabolites development in bacteria. The common PGPR stimulates the resistance against crop species mostly by supporting the growth regulators and related moieties. Under xenobiotic toxicity, the major constraint that plants face is the redox potential inclined to oxidative stress. Therefore, xenobiotic stress that leads to the ROS at different cellular organelles can eventually block or downregulate the key metabolic cycles like carbon metabolism, oxidative phosphorylation, mineral assimilation, and protein synthesis (Sahu et al. 2022). Rhizobacteria can also cause acidification of the soil by secretion of a special volatile compound commonly called volatile organic compound (VOC) into the rhizosphere which otherwise enhances the iron (Fe) homeostasis. At the molecular level, Fe-deficit soil induces the crop for special types of metabolite generation including siderophore. This chemical belongs to hydroxymate [-C(=O)-NH-OH], phenoxybenzene that joins Fe binding sites and forms ligands with Fe oxides (Van Stempvoort et al. 2020). Specific genes are induced by TFs like Fe-induced TFs1 (FIT1) may coregulate adjoining ion channels high potassium transporter 1 (HKT1) to share resistance against common metals like potassium presented in Fig. 1. A few important enzymes like phosphoethanolamine, N-methyl transferase that regulate the synthesis of VOCs through the transfer of methyl groups are important in specific toxic residue homeostasis. Therefore, PGPR-treated rhizosphere can not only balance the redox of the root but also induce the VOCs which otherwise support both enzymatic and non-enzymatic systems (Grover et al. 2021). The VOC-induced tolerance to xenobiotics may be attributed to metabolism or chemical transformation by the rhizobacteria however, yet to be fully understood (Oyerinde et al. 2023). In a predicted model, three consecutive principal reactions are accounted for: i. signaling by VOC, ii. transportation over the membrane and sequestering, and iii. Metabolism or chemical transformation. The roles of VOC or any other special metabolites possibly induced a set of enzymatic systems which include alpha, beta hydrolases, ROS lysing class III peroxidases, esterases, and cytochrome oxidase (Cyt P_{450}) are important. The latter is responsible for hydroxylation by donation of electrons and thereby upregulates the solubility of xenobiotics over the membrane (Torres et al. 2003). A few PGPRs are also reported with conjugations of toxic residues, where VOCs are important to induce malonyl transferase (MTs), glycosyl transferase (GTs). Enzymes like MTs are most important for sequestering specifically for lipid peroxides, hydroperoxides, carbonyl derivatives initiating more peroxidation reactions of the membrane as a common consequence of xenobiotic stress. Rhizobacteria are also reported to sequester the xenobiotics over the membrane in the cell wall, apoplast-like non-cellular spaces to reduce phytotoxicity. Genome-wide scanning studies in *Arabidopsis* revealed a transcriptional change upon xenobiotic contamination when invaded with PGPR. The important enzyme

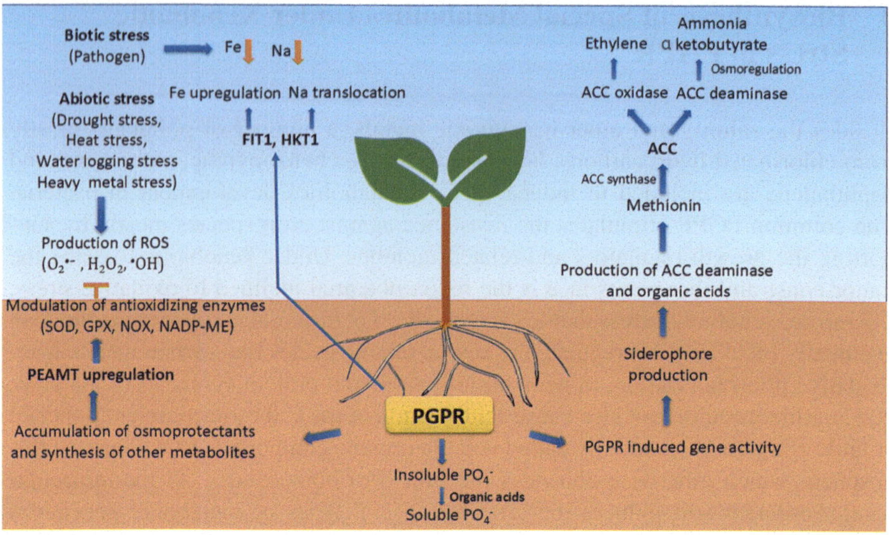

Fig. 1 Solubilization of minerals can induce rhizosphere to activate PGPR gene regulation for metabolites. Ethylene production through 1-aminocyclopropane 1-carboxylic acid (ACC) is aggravated by ACC oxidase and ACC deaminase. Activity of siderophores is based on the PGPR-plant reaction through upregulation of genes (*FIT1, HKT1*) are responsible for uptake of iron (Fe). Similarly, upregulations of stress responsive proteins (SOD, GPX, NOX, NADP-ME) can minimize oxidative damages of plants under abiotic stress. Dehydration is minimized by PGPR induced plant growth substances which otherwise act as osmoprotectant in roots by expression of *phosphoethanolamine N-methyltransferase 1* (*PEAMT*) gene. *PEAMT* can protect root apical growth by adjusting auxin modulation and ROS destabilization in dry soil of rhizosphere

over-expressed to transport the toxic residues into vacuole is done by ATP-binding cassette (ABC) transporter (Dahuja et al. 2021). Other xenobiotics like polycyclic aromatic hydrocarbons (PAH) are also important to be detoxified by specific groups of rhizobacteria *Pseudomonas aeruginosa*, *Burkholderia cepacian*, *Mycobacterium vanbaalenii* which can produce specific dehydrogenase to oxidize the residue. In this reaction, rhizospheric organisms also induce an oxidative burst which can destabilize the cellular redox of root cortical cells which otherwise induces antioxidative cascades. The expression of specific genes is also under control of soil characteristics like pH, aeration, temperature, and other edaphic characteristic features. In wheat plant application of rhizobacteria tolerant to salt can also upregulate specific ABA signaling pathways which amplifies salt overly sensitive (SOS) mechanism to detoxify the PAHs (Munir et al. 2022).

4 Role of Phytohormones in Detoxification for Pollutant

The growth substances including auxin (AUX), gibberellin (GA), cytokinin (CK), ethylene (ET), and abscisic acid (ABA) are commonly moderated when a plant is affected with rhizospheric bacteria, particularly under stressful conditions.

Xenobiotic substances being a strong source of salt stress can also induce the specific ion effect and osmotic deficits where ABA-like residues are common in overproduction (Kesawat et al. 2023). Rhizospheric bacteria can improve the overall vigor of the plants which is facilitated through hormonal stability. The biosynthesis of ET is also common due to overexpression of ACC deaminase from bacteria that reduces ET biosynthesis and regulates the earlier senescence of plants under toxicity (Singh et al. 2022). On the other hand, this is a definite mechanism for rhizospheric microbial consortium where the synergistic functioning of ACC deaminase can support the salt induced trehalose accumulation (del Carmen Orozco-Mosqueda et al. 2020) . This happens to be the criteria for PGPR as a bioinoculant for increasing productivity through other bacterial and fungal colonization in the rhizosphere to support the drought stress in crops. The growth regulation like ABA is mostly focused on endogenous over-expression of plant genotypes sensitized to PGPR. In a few cases, accumulations of ABA can also regulate the other osmolytes like trehalose, mannitol, and VOC to modulate plant responses. In other studies, PGPR-releasing phytohormones also alter the root metabolite flux where growth regulators are important. In orchid rhizosphere, *Mycobacterium, Cellulomonous, Azotobactor* are more common to increase the IAA production to support plant growth under water stress (Karnwal et al. 2024). Other classes of PGPRs commonly applied to reduce the redox imbalance include *Proteus, Klebsiella, Bacillus* which are common to support ABA generation under water stress. At the metabolic level growth hormones are well connected with rhizodeposits for some hormone-based compounds that are required for selective pressure of PGPR association with host (Giannelli et al. 2023). Rhizodeposits basically are carbon residues, low to high in molecular weights, and also are the by-products of primary metabolites. This compound is quite selective or specific for the construction of microbial communities for a particular rhizosphere and often helps the bacterial biofilm formation under wider coverage of PGPR. Rhizospheric exudates can interact with rhizodeposits in different signaling molecules like inositol pyrophosphate, diacyl glycerol, etc., to transmit the signals from drying soil. The important rhizobacterial communities covering *Pseudomonas, Rhizobium, Arthrobacter* are quite frequent to induce MAPK kinase pathways for overloading the signal into the nucleus following gene expression. Therefore, the basic function of rhizospheric metabolome supplemented with PGPR is the priming of cellular responses of sensitized plants to encounter xenobiotic contamination. The importance of monitoring and adjustment of adequate redox is now understood in the rhizospheric environment and its attachment to the bacterial colony. Optimization of redox potential thus may induce specific compounds generation in roots or by PGPR, and adjustment of soil pH to modulate the redox potential. Still, the crop system developed in hydroponic solution can benefit specific rhizospheric bacteria where pH and redox potential are regulated (Chiaranunt and White 2023). Crops like maize, wheat in cereals, and other vegetables under depleted soil moisture, alkaline pH, and hypoxic/anoxic conditions may be promoted for mineral acquisition as well as dry matter partitioning into biological and economic yield (Kumar et al. 2021). A developing body of investigation suggests the plant-associated microbial growth is typically based on signaling as

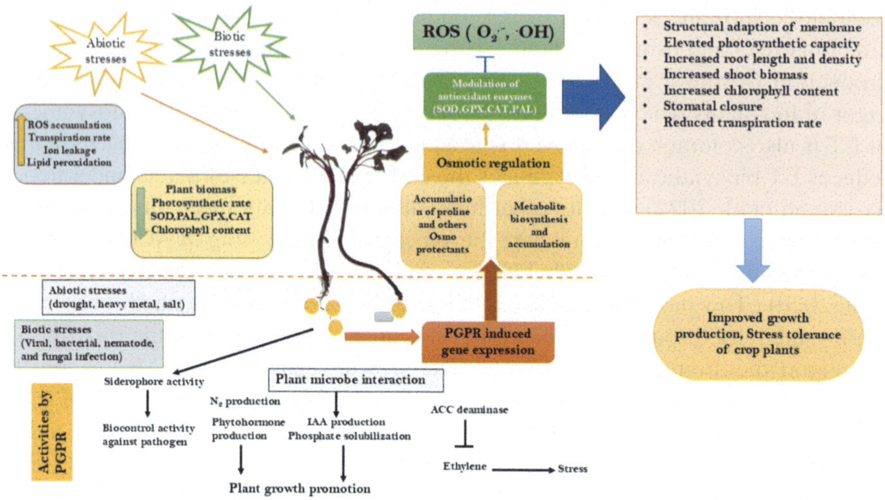

Fig. 2 Adjustment of plant growth regulators is more facilitated by rhizobacteria where IAA production is antagonized with ethylene. Biocontrol activity against the fungal pathogen may be induced by siderophore production which otherwise can control the viral, nematode and bacterial infection. PGPR-induced hormonal adjustment can also stabilize the metabolite flux from root through shoots which otherwise support the adjoining physiological attributes to improve stress tolerance in host plants

hormones, elicitors, ROS/RNS, and related inductive residues from plants which directly regulate the metabolism of microbial growth (Vishwakarma et al. 2020). This may support the host-specific application of PGPR for its practical application to support plant growth under environmental extremities or changing environment as outlined in Fig. 2.

5 Molecular Signaling Within Plant-Rhizobacteria Consortium to Stress

Plants may interact with diverse microbial communities and are hypothesized to be co-evolved within a microenvironment. The latter may depend on a mutualistic approach from plants and rhizobacteria where the stress resilience is most expected as beneficiary for the crops (Bhat et al. 2023). Microbes are more adapted to the environmental estimates, and thereby, they can influence the plants with support for basic nutrients like N, P, etc., and induce specific metabolites into plants covering growth substances, and secondary metabolites (Pattnaik et al. 2021). For example, the rhizospheric exudates as an amalgamation of chemical residues from plants and rhizobacteria can benefit other beneficial fungi to interact directly or indirectly.

Likewise, arbuscular mycorrhizal fungi (AMF) are referred to with specific phosphate solubilization when interacting with other rhizospheric bacteria in a common soil of plant growth. Plants can allow this interaction between microbes-AMF through a specific mechanism where post-transcriptional regulation of target gene expression is modified by microRNAs (miRNAs). Importantly, environmental stresses (moisture, temperature, anoxia/hypoxia) and interspecific interaction between related crop species would be more important for plant-microbe interaction for better crop growth. In most recently understood insights, the regulations of plant-PGPR interaction are based on the contribution of miRNAs. Classically, miRNAs are well established as a noncoding, stretch of $\approx$ 25 nucleotides which target the genes for post-transcriptional regulation however negatively. Though the role of miRNAs is well established for growth and development, still, in abiotic stress adaptation, a few have been well documented in different plants (Mondal et al. 2023). For the plant-microbe interaction, the regulation by miRNAs was observed to be critical in the promotion of two-way communications between the partners (Middleton et al. 2021). This was more illustrated with *Arabidopsis* where interference by miRNA773 can upregulate the specific methyl transferase gene (METII) where the resistance from *Plectosphaerrella* is important. This is also observed with miRNA396 where AMF infection is negatively regulated by targeting the gene like *GRF4* overexpression which substantially increased root growth. In apples, the PGPR (*Botryosphaeria*) can promote the expression of miRNA168 which may target the inhibition of *MhAGO1* gene (Kar and Raichaudhuri 2021). In this, a number of miRNAs are identified which can induce root exudation and morphological development to host PGPR, more illustrated with AMF. The direct infection of AMF can also induce the colonization of certain PGPR on the basis of specific changes in root morphology as well as some signaling residues. The host plants can control their regulation for microbial infestation where promoter activity for particular gene supporting protease-like inhibitors is downregulated by specific miRNA. For environmental extremities through drought, the heat response of plants through PGPR colonization is also modulated by other miRNAs. Likewise, the infection by *Botryosphaeria* sp. was reported to induce systematic resistance through salicylic acid with delay of bacterial infection syndromes by inhibiting senescence, loss of leaves, etc. (Sun et al. 2020). Though miRNAs play regulatory role against plant-microbial interaction still, activities of microbial genes are also targeted from the host plant genome. Likewise, fungal miRNA can restrict the mycelial growth inside the host plants by the formation of vesicles where lysis of the cell wall is restricted by the downregulation of chitinase-like genes (Regente et al. 2017). Still, interkingdom miRNA activity and regulation for plant-PGPR interaction is yet to be understood from the core background of molecular signaling as well as the epigenetic level. The possible interactive sessions with PGPRs against different abiotic stresses influence the major metabolic flux which directly or indirectly governs the physiological alterations in the host plant as deduced in Table. 1.

Table 1 Plants inhabiting rhizospheric bacteria with their attribution in stress tolerance those involved regulations of cellular and metabolic biosynthesis and turn over

Stress	PGPR	Plants	Tolerance strategy
Salt	*Bacillus subtilis*	*Arabidopsis thaliana*	Regulation of sodium transporter in tissue-specific manner
Salt	*Pseudomonas koreensis*	*Zea maize*	Induced phytohormonal regulation
Drought	*Rhizobium tropici*	*Phaseolus vulgaris* (common bean)	Regulate the drought tolerant genes
Zinc toxicity	*Pseudomonas aeruginosa*	*Triticum aestivum* (common wheat)	Nitrogen uptake and solubilization of protein
Salt	*Glomus fasciculatum*	*Glycine max* (soybean)	Carbohydrate accumulation
Drought	*Pseudomonas chlororaphis*	*Arabidopsis thaliana*	Butanediol production
Salinity	*Glomus clarum*	*Capsicum annuum*	Regulate stomatal permeability
Salt	*Escherichia coli*	*Oryza sativa*	Trehalose formation
Salt	*Achromobacter piechaudii*	*Solanum lycopersicum*	Increase sodium and potassium uptake
Salt	*Dietzia* sp.	*Triticum aestivum*	Involved in ABA signaling and SOS pathway

6 Impact of Changing Climate on Plant-PGPR-Mediated Reactions to Crop

Climate change particularly through frequent exposure to soil moisture deficits must have a significant effect on rhizobacterial colonization and infection through roots. In some cases, an elevated temperature to reduce the soil moisture tension can cause some anaerobic or hypo-aerobic stress for rhizobacteria (Paul 2013). This is more aggravated with increasing radiation coupled with temperature to anoxia to hypoxia for the sustainability of microbes under in rhizosphere. In frequency and harshness of drought stress, the changing temperature of soil as well as polluted ground water, the plant-microbiome is changed negatively for its growth and reflected in weak association with the host plant (Abdul Rahman et al. 2021). Still, climate change alteration of plant microorganisms mostly affects the host for water relation as well as metal acquisition through roots. The adaptation or resilience of plants to changing climate is thus manifested into long-term effects where adaptation may be driven by ecological evolutions from both microbes and plants. Plant microbiota is a complex ecological phenomenon where root exudation through different chemicals has a definite role in the association of rhizospheric bacteria, changing the immune reaction and therefore, establishing a synergistic relationship (Rojo et al. 2003). Microbial association regardless of seasons, geographical regions, and agroclimatic or edaphic characteristics may be stable or variable for a particular species (John and Ray 2023). Climate change is more likely to be less effective on communities of microbes at the phyllosphere. Contrarily, the internal microbial consortium, particularly, with endophytes is more sensitive to the changing environment and its reflection of plant's metabolic cycle to adjust rhizobacteria.

The bulk soil is a reservoir of heterogeneous microbes that colonizes directly with roots in comparison with rhizospheric microbiomes whose growth is determined by both the plant's response and external soil factors. As for example, under drought stress, host species selectively colonize the Gram-positive bacteria with adequate tolerance to dehydration stress. The rhizobacteria can also influence modification of the rhizosphere and roots with thickening of cell wall, cutin deposition, and other secondary metabolites biosynthesis (El Zemrany et al. 2007). Plant-microbe exchange is mostly based on chemical communication from microbes under changing environment where stressful condition may demand the development of different newly synthesized chemical residues. In a few cases, VOCs are most important to perceive the chemical changes of the rhizosphere and turn into a signal to amplify the plant response either by up/down-regulation of specific genes (Lin et al. 2007). Drought stress coupled with ozone exposure might be a direct involvement of a few microbiotas including *Sphignomonas, Pseudomonas, Cupriavidus* for the development of some biogenic volatile organic compounds (BVOCs) (Sharma et al. 2022). Root exudates complexed with bacterial polysaccharide-based complex residues, organic acids, sugar derivatives, alkaloids would be included in this category, however, not exclusively be volatile. In the roots of tomato, maize colonized by *Actinobacteria* would be an endosphere and rhizosphere-based organism to release this BVOC. Even *Rhizobium* can also be a PGPR with a better supplement of nitrogenous residues in Fabaceae with phytoremediation potential for heavy metals (Sánchez-Navarro et al. 2020). Therefore, the effects of microbial diversity on the resilience of abiotic stress out of climate change are important. Thus, for the major issues on plant-microbe complex influenced by climatic changes, a deep insight would be the key to understand the host-specific interaction through moder state-of-arts. Research with PGPRs as bio-controlling agents is still under investigation, particularly for any specific genes either horizontally or vertically transferred. The shortcoming of such bio-control technology would be mitigated by the identification of specific rhizo-microbiome and its modalities of interaction. As for example, PGPR produces hydrocyanic acid (HCN) in the promotion of plant growth against cross protections of bacteria. The inhibition is based on mitochondrial-based oxidative burst as well as to reduce the energy demands of the invading root tissues which otherwise may ensure the disintegration of fungal invasion. Rhizodeposition is another path where enrichment of minerals is facilitated by a combination of microbe and plant interaction through complex signaling pathways and other cellular secretions like strigolactone as depicted in Fig. 3.

7 Simulation of Plant Resilience Against Climate Change Through Microbial Assistance

As the climatic irregularities are in a gradual progression, it would be a cause for pathogen invasion to the crops if those are favorable for microbial growth. The climatic aberrations definitely hit the hydrological along with other bio-geological

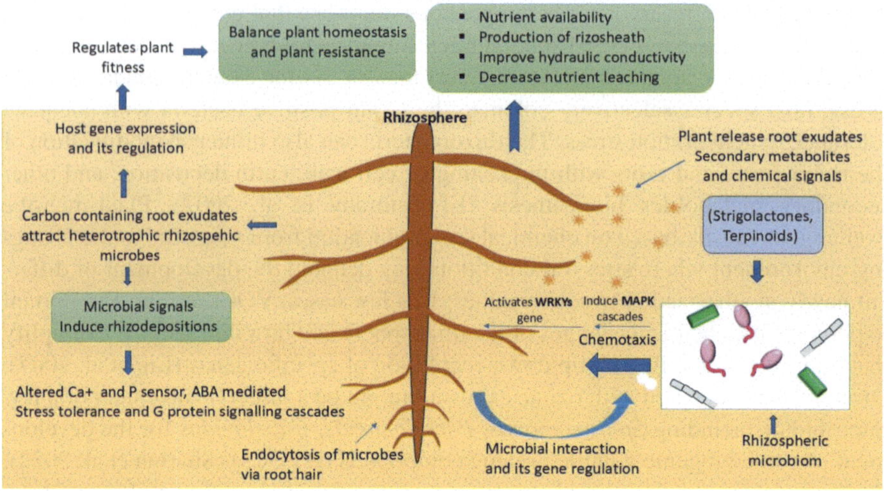

Fig. 3 Rhizospheric bacteria are more affiliated through chemotaxis where WARKY and MAP kinase cascade are involved. The association is invited particularly in drought stress. Plants releasing secondary metabolites like strigolactone can also induce roots for better rhizospheric solubilization of minerals, nutrients availability, and improved hydraulic conductivity. On the other hand, endocytosis of the microbes can also alter Ca^{2+} and P-signaling with G-protein cascade. Overall, mineral deposition by microbial signaling induces a rhizospheric deposition for different residues supposedly favoring plant growth

cycles; the water availability specifically for drought induction would be detrimental to crop growth (Zlatev and Lidon 2012). Rhizospheric engineering is mostly conceptualized for stress tolerance in crop where modification of microbial environment in rhizosphere can enhance the in-built tolerance or modulations of metabolic pathways improvising the crops resilience. At physiological level, the modification of rhizosphere includes enhancement of nutrient uptake, generation of growth regulators to sustain plant growth even under stress condition, allelopathic or antagonistic activities of invading pathogens into roots, biologically degradation or xenobiotics or pollutant. The root-microbe interaction is often targeted with development of osmo-protectant, heat tolerant proteins, growth supplements, bio-stimulants like residues. In some cases (Atriwal et al. 2021), at molecular level, specific metabolites (exopolysaccharides, stress responsive proteins), triggering of gene expression and even supporting the post-translational modification of transcription factors, are forwarded from plant-microbe interaction. Root development on microbial activity by specific growth hormones (e.g., indole-3-acetic acid) and thereby increasing hydraulic conductivity of membrane is important. The VOCs (e.g., ABA, hydrocyanic acids, ethylene, jasmonic acids) are important for induced systemic resistance against direct pathogens infection or/and secondary oxidative stress management. The molecular key players in systematic resistance induced by PGPR may encompass non-expression of pathogenesis related genes, which encodes chitinases, glucanases like antimicrobial protein (Lazarus and Easwaran 2024). Rhizospheric alteration by PGPR activity also invites to target modulation of few osmotic related genes and their regulatory elements. The most common example is dehydration response element binding/cold binding factors (DREB/CBF), abscisic acid binding factor (ABF), late embryogenesis abundant protein (LEA) are important (Ma et al. 2014). Dry soil and its repeated rewatering make a cycle that happens to be a factor for bacterial and other microfloral growth. Under this condition, the sequestering of carbon is a key for microbial decontamination where specific organic residues are concern. Those come under category of VOC to support plants resilience under drying soil (Ng et al. 2015). A shift for improved tolerance in bacterial growth would be induced by sufficient hydration of the soil which otherwise induces a quick colonization. Simulation may also be based on organic residue amendments in drying soil and invite more rhizobacterial communities. As for example, frequent growth of rhizobacteria commonly meets with drought tolerance where soil enzyme is reduced in activities (Ngumbi and Kloepper 2016). In drying soil, litter decomposition would be the rate limiting factor to sustain the bacteria growth as an alternative carbon source, particularly when the host plant restricts the allocation of carbon towards roots and leaching into the rhizosphere (Richardson et al. 2009). Temperature and high radiation may downregulate the carbon assimilation, thereby increasing the transformation of available carbon organic biomass. Elevated carbon dioxide is already reported with variable effects through C_3 and C_4 crops. The excess rainfall would be another cause for microbial aggression under waterlogged as well as submerged conditions of soil where anaerobic bacteria play an important role in microbiota (Xu et al. 2021). Irrespective of those conditions, the biotechnological utilization has been in progress for the mitigation of climate change through decarbonization, carbon budgeting, and carbon sequestration microflora including bacteria, algal biomass, fungal association, and lichen consortium which are important. The rhizospheric region offers a vast network of microbial activities; any changes in soil or edaphic factors must be reflected through nutrient recycling, dry matter partitioning, sequestration of carbon, and other plant-based benefits to the threatening environment. Rhizospheric bacteria inhabiting host-specific activity also offer some growth-promoting substances through rhizodeposition in favor of other beneficiary microbes (Kumar et al. 2013). Thus, the rhizosphere supporting the PGPR activities would be a two-way beneficiary where ample chances are there for engineering the microbiome. The rhizosphere engineering initially started with PGPR and was targeted for crop sustainability coupled with phytohormone production in the root (Hakim et al. 2021). The modifications of the microbiome are also dedicated to horizontal gene transfer where multiple pathways would be operative simultaneously for bacteria in a reciprocal manner that can suffice crop productivity along with climate resilience. Thus, it would be a synthetic biological approach for the improvement of soil characters, adaptation of roots to drought/salinity stress as well as overall increase in plant biomass with productivity. Therefore, combination of PGPR inoculants and different abiotic/biotic stressors is well competent in signaling pathways from environment and thereby modulation of plant response through gene expression. Principally, manipulation of those molecular functions by direct application of rhizospheric bacteria directly or any modifications by chemical, physical, or physicochemical treatment is expected for crop resilience with promising results in agricultural sustainability.

8 Conclusion and Future Prospective

Despite a long discussion about the rhizosphere modification, the precise nature of plant-microbe interaction against abiotic stress resilience to crops is still an enigma. Many evidences enriched the possible changes of plant metabolism supporting to stress tolerance, whether directly from microbial consortium or de novo synthesis exclusively in plant bodies is yet to be deciphered. This is also questionable whether environmental extremities and fluctuations are typified for sustainable growth of the rhizobacteria that otherwise may support the crop growth. The better understanding of plant-microbe assembly in rhizosphere or phyllosphere could be influenced by different factors is still obscure. The signaling mechanism adjoining microbial community to develop specific pathways for tolerance to soil drying, salinity, heavy metals or pollutants, ozone and harmful radiation exposure would be a factor with broader scope in future. At the cellular level, even endogenous bacteria would be sensitized for the development of antioxidants, compatible solutes, secondary messengers for growth and vis-a-vis to support plant resistance against stresses. The adverse effects of a changing environment are targeted to decipher the extent of changes in structural and functional ecological hierarchy or niches. Still, what are the key factors to substitute from the environment changing the plant-microbe interaction and its reflectance on the plant's response is most important for sustainable yield? Still, to what extent any specific pathways being altered in specific rhizospheric bacteria is yet to be understood and its manipulation through simulation of environmental cues. Recent identification of specific genes through both horizontal and vertical resistance for bacteria under drying soil may lead to the development of traits to induce any host plant for climate change resilience. Climate changes may influence on rhizosphiric microbiota that also governs microbial consortium for an wider ranges of ecological niche of vegetation. Likewise, temperature and irradiance altering the carbon assimilation and transpiration for drought-sensitive crops could be identified. The climatic conditions have real effects on the plant community for a significant duration; the rhizospheric bacterial identification may also suffice the environmental supports for plant sustainable growth. The adverse impact on crops from climate change is now focused on microbial structure-functional relationships in specific ecological niches. Still, the level or extent to which any plant rhizosphere consortium could be modified without hampering crop sustainability as well as maintaining eco-friendliness is still obscure. The host genetic features must be involved where horizontal gene transfer would be the better option for the development of new traits for microbial metabolism in rhizospheric habitation and to transduce chemical signals into roots for stress tolerance. Therefore, to explore broader and deeper insights into host-microbe interaction, a multi-omics technology under a specific rhizospheric niche is the tool to interlink the microbial community and environmental resilience that otherwise improvises plant tolerance to abiotic stresses.

Acknowledgements MKA is acknowledging the support from DST PURSE II applicable to the University of Kalyani. NC and SR are acknowledged for financial support from URS fellowship, UGC-NET JRF fellowship respectively.

References

Wang, Q., Wang, J., Zhang, Z., Li, M., Wang, D., Zhang, P., Li, N., & Yin, H. (2024). Microbial metabolic traits drive the differential contribution of microbial necromass to soil organic carbon between the rhizosphere of absorptive roots and transport roots. *Soil Biology and Biochemistry*, 197, 109529.

Santoyo, G., Guzmán-Guzmán, P., Parra-Cota, F. I., Santos-Villalobos, S. D. L., Orozco-Mosqueda, M. D. C., & Glick, B. R. (2021). Plant growth stimulation by microbial consortia. *Agronomy*, 11(2), 219.

Shaffique, S., Khan, M. A., Wani, S. H., Pande, A., Imran, M., Kang, S. M., Rahim, W., Khan, S. A., Bhatta, D., & Kwon, E. H. (2022). A review on the role of endophytes and plant growth promoting rhizobacteria in mitigating heat stress in plants. *Microorganisms*, 10, 1286.

Hamid, B., Zaman, M., Farooq, S., Fatima, S., Sayyed, R. Z., Baba, Z. A., Sheikh, T. A., Reddy, M. S., El Enshasy, H., Gafur, A., & Suriani, N. L. (2021). Bacterial plant biostimulants: A sustainable way towards improving growth, productivity, and health of crops. *Sustainability*, 13(5), 2856.

Adoko, M. Y., Noumavo, A. D. P., Agbodjato, N. A., Amogou, O., Salami, H. A., Aguégué, R. M., Adjovi Ahoyo, N., Adjanohoun, A., & Baba-Moussa, L. (2022). Effect of the application or coating of PGPR-based biostimulant on the growth, yield, and nutritional status of maize in Benin. *Frontiers in Plant Science*, 13, 1064710.

Jiang, Y., Yue, Y., Wang, Z., Lu, C., Yin, Z., Li, Y., & Ding, X. (2024). Plant biostimulant as an environmentally friendly alternative to modern agriculture. *Journal of Agricultural and Food Chemistry*, 72(10), 5107–5121.

Rai, N., Rai, S. P., & Sarma, B. K. (2021). Prospects for abiotic stress tolerance in crops utilizing phyto- and bio-stimulants. *Frontiers in Sustainable Food Systems*, 5, 754853.

Sani, M. N. H., & Yong, J. W. (2021). Harnessing synergistic biostimulatory processes: A plausible approach for enhanced crop growth and resilience in organic farming. *Biology*, 11(1), 41.

Muhammad, A., Kong, X., Zheng, S., Bai, N., Li, L., Khan, M. H. U., Fiaz, S., & Zhang, Z. (2024). Exploring plant-microbe interactions in adapting to abiotic stress under climate change: A review. *Frontiers in Plant Science*, 15, 1482739.

Saud, S., Jiang, Z., & Fahad, S. (2023). Significance and exploitation of rhizosphere chemical signaling metabolites for enhancing soil nutrient transformation. *Journal of Soil Science and Plant Nutrition*, 23(4), 4827–4842.

Li, Y., Chen, Y., Fu, Y., Shao, J., Liu, Y., Xuan, W., Xu, G., & Zhang, R. (2024). Signal communication during microbial modulation of root system architecture. *Journal of Experimental Botany*, 75(2), 526–537.

Yang, L., Qian, X., Zhao, Z., Wang, Y., Ding, G., & Xing, X. (2024). Mechanisms of rhizosphere plant-microbe interactions: Molecular insights into microbial colonization. *Frontiers in Plant Science*, 15, 1491495.

Olowe, O. M., Akanmu, A. O., & Asemoloye, M. D. (2020). Exploration of microbial stimulants for induction of systemic resistance in plant disease management. *Annals of Applied Biology*, 177(3), 282–293.

Mishra, S., Lin, Z., Pang, S., Zhang, W., Bhatt, P., & Chen, S. (2021). Recent advanced technologies for the characterization of xenobiotic-degrading microorganisms and microbial communities. *Frontiers in Bioengineering and Biotechnology*, 9, 632059.

Arya, C. K., Maurya, S., & Ramanathan, G. (2023). Insight into the metabolic pathways of *Paracoccus* sp. strain DMF: A non-marine halotolerant methylotroph capable of degrading aliphatic amines/amides. *Environmental Science and Pollution Research*, 30(60), 125947–125964.

Salim, S. A., & Busi, S. (2024). Aquatic biofilms and their dynamics. In *Aquatic Ecosystems and Microbial Biofilms: Significance, Dynamics, Prevention and Control* (p. 172).

Biswas, S., & Ganesan, M. (2024). Current perspectives of ACR3 (arsenite efflux system) toward the reduction of arsenic accumulation in plants. *Journal of Crop Science and Biotechnology*, 1–17.

Sahu, P. K., Jayalakshmi, K., Tilgam, J., Gupta, A., Nagaraju, Y., Kumar, A., Hamid, S., Singh, H. V., Minkina, T., Rajput, V. D., & Rajawat, M. V. S. (2022). ROS generated from biotic stress: Effects on plants and alleviation by endophytic microbes. *Frontiers in Plant Science*, 13, 1042936.

Van Stempvoort, D. R., Brown, S. J., Smyth, S. A., & Division, E. R. (2020). Detections of alkylphenoxy-benzenesulfonates in municipal wastewater. *Chemosphere*, 251, 126386.

Grover, M., Bodhankar, S., Sharma, A., Sharma, P., Singh, J., & Nain, L. (2021). PGPR mediated alterations in root traits: Way toward sustainable crop production. *Frontiers in Sustainable Food Systems*, 4, 618230.

Oyerinde, A. S., Selvaraju, V., Babu, J. R., & Geetha, T. (2023). Potential role of oxidative stress in the production of volatile organic compounds in obesity. *Antioxidants*, 12(1), 129.

Torres, E., Bustos-Jaimes, I., & Le Borgne, S. (2003). Potential use of oxidative enzymes for the detoxification of organic pollutants. *Applied Catalysis B: Environmental*, 46(1), 1–15.

Dahuja, A., Kumar, R. R., Sakhare, A., Watts, A., Singh, B., Goswami, S., Sachdev, A., & Praveen, S. (2021). Role of ATP-binding cassette transporters in maintaining plant homeostasis under abiotic and biotic stresses. *Physiologia Plantarum*, 171(4), 785–801.

Munir, N., Hasnain, M., Roessner, U., & Abideen, Z. (2022). Strategies in improving plant salinity resistance and use of salinity resistant plants for economic sustainability. *Critical Reviews in Environmental Science and Technology*, 52(12), 2150–2196.

Kesawat, M. S., Satheesh, N., Kherawat, B. S., Kumar, A., Kim, H. U., Chung, S. M., & Kumar, M. (2023). Regulation of reactive oxygen species during salt stress in plants and their crosstalk with other signaling molecules—current perspectives and future directions. *Plants*, 12(4), 864.

Singh, R. P., Ma, Y., & Shadan, A. (2022). Perspective of ACC-deaminase producing bacteria in stress agriculture. *Journal of Biotechnology*, 352, 36–46.

del Carmen Orozco-Mosqueda, M., Glick, B. R., & Santoyo, G. (2020). ACC deaminase in plant growth-promoting bacteria (PGPB): An efficient mechanism to counter salt stress in crops. *Microbiological Research*, 235, 126439.

Karnwal, A., Shrivastava, S., Al-Tawaha, A. R. M. S., Kumar, G., Kumar, A., & Kumar, A. (2024). PGPR-mediated breakthroughs in plant stress tolerance for sustainable farming. *Journal of Plant Growth Regulation*, 43(9), 2955–2971.

Giannelli, G., Potestio, S., & Visioli, G. (2023). The contribution of PGPR in salt stress tolerance in crops: Unravelling the molecular mechanisms of cross-talk between plant and bacteria. *Plants*, 12(11), 2197.

Chiaranunt, P., & White, J. F. (2023). Plant beneficial bacteria and their potential applications in vertical farming systems. *Plants*, 12(2), 400.

Kumar, A., Nayak, A. K., Hanjagi, P. S., Kumari, K., Vijayakumar, S., Mohanty, S., Tripathi, R., & Panneerselvam, P. (2021). Submergence stress in rice: Adaptive mechanisms, coping strategies and future research needs. *Environmental and Experimental Botany*, 186, 104448.

Vishwakarma, K., Kumar, N., Shandilya, C., Mohapatra, S., Bhayana, S., & Varma, A. (2020). Revisiting plant–microbe interactions and microbial consortia application for enhancing sustainable agriculture: A review. *Frontiers in Microbiology*, 11, 560406.

Bhat, M. A., Mishra, A. K., Jan, S., Bhat, M. A., Kamal, M. A., Rahman, S., Shah, A. A., & Jan, A. T. (2023). Plant growth promoting rhizobacteria in plant health: A perspective study of the underground interaction. *Plants*, 12(3), 629.

Pattnaik, S., Mohapatra, B., & Gupta, A. (2021). Plant growth-promoting microbe mediated uptake of essential nutrients (Fe, P, K) for crop stress management: Microbe–soil–plant continuum. *Frontiers in Agronomy*, 3, 689972.

Mondal, R., Das, A., & Bandyopadhyay, A. (2023). Implications of small RNAs in plant development, abiotic stress response and crop improvement in changing climate. *Nucleus*, 66(3), 321–339.

Middleton, H., Yergeau, É., Monard, C., Combier, J. P., & El Amrani, A. (2021). Rhizospheric plant–microbe interactions: miRNAs as a key mediator. *Trends in Plant Science*, 26(2), 132–141.

Kar, M. M., & Raichaudhuri, A. (2021). Role of microRNAs in mediating biotic and abiotic stress in plants. *Plant Gene*, 26, 100277.

Sun, X., Pan, B., Wang, Y., Xu, W., & Zhang, S. (2020). Exogenous calcium improved resistance to *Botryosphaeria dothidea* by increasing autophagy activity and salicylic acid level in pear. *Molecular Plant-Microbe Interactions*, 33(9), 1150–1160.

Regente, M., Pinedo, M., San Clemente, H., Balliau, T., Jamet, E., & De La Canal, L. (2017). Plant extracellular vesicles are incorporated by a fungal pathogen and inhibit its growth. *Journal of Experimental Botany*, 68(20), 5485–5495.

Paul, D. (2013). Osmotic stress adaptations in rhizobacteria. *Journal of Basic Microbiology*, 53(2), 101–110.

Abdul Rahman, N. S. N., Abdul Hamid, N. W., & Nadarajah, K. (2021). Effects of abiotic stress on soil microbiome. *International Journal of Molecular Sciences*, 22(16), 9036.

Rojo, E., Solano, R., & Sánchez-Serrano, J. J. (2003). Interactions between signaling compounds involved in plant defense. *Journal of Plant Growth Regulation*, 22, 82–98.

John, S. A., & Ray, J. G. (2023). Optimization of environmental and other variables in the application of arbuscular mycorrhizal fungi as an ecotechnological tool for sustainable paddy cultivation: A critical review. *Journal of Applied Microbiology*, 134(6), 111.

El Zemrany, H., Czarnes, S., Hallett, P. D., Alamercery, S., Bally, R., & Jocteur Monrozier, L. (2007). Early changes in root characteristics of maize (*Zea mays*) following seed inoculation with the PGPR *Azospirillum lipoferum* CRT1. *Plant and Soil*, 291, 109–118.

Lin, C., Owen, S. M., & Peñuelas, J. (2007). Volatile organic compounds in the roots and rhizosphere of *Pinus* spp. *Soil Biology and Biochemistry*, 39(4), 951–960.

Sharma, B., Singh, B. N., Dwivedi, P., & Rajawat, M. V. S. (2022). Interference of climate change on plant-microbe interaction: Present and future prospects. *Frontiers in Agronomy*, 3, 725804.

Sánchez-Navarro, V., Zornoza, R., Faz, Á., Egea-Gilabert, C., Ros, M., Pascual, J. A., & Fernández, J. A. (2020). Inoculation with different nitrogen-fixing bacteria and arbuscular mycorrhiza affects grain protein content and nodule bacterial communities of a fava bean crop. *Agronomy*, 10(6), 768.

Zlatev, Z., & Lidon, F. C. (2012). An overview on drought-induced changes in plant growth, water relations, and photosynthesis. *Emirates Journal of Food and Agriculture*, 24(1), 57–72.

Atriwal, T., Azeem, K., Husain, F. M., Hussain, A., Khan, M. N., Alajmi, M. F., & Abid, M. (2021). Mechanistic understanding of Candida albicans biofilm formation and approaches for its inhibition. *Frontiers in Microbiology, 12*, 638609.

Lazarus, H. P. S., & Easwaran, N. (2024). Molecular insights into PGPR Fluorescent Pseudomonads complex mediated intercellular and interkingdom signal transduction mechanisms in promoting plant's immunity. *Research in Microbiology*, 104218.

Ma, L. F., Zhang, J. M., Huang, G. Q., Li, Y., Li, X. B., & Zheng, Y. (2014). Molecular characterization of cotton C-repeat/dehydration-responsive element binding factor genes that are involved in response to cold stress. *Molecular biology reports*, *41*, 4369–4379.

Ng, E. L., Patti, A. F., Rose, M. T., Schefe, C. R., Smernik, R. J., & Cavagnaro, T. R. (2015). Do organic inputs alter resistance and resilience of soil microbial community to drying? *Soil Biology and Biochemistry*, 81, 58–66.

Ngumbi, E., & Kloepper, J. (2016). Bacterial-mediated drought tolerance: Current and future prospects. *Applied Soil Ecology*, 105, 109–125.

Richardson, A. E., Barea, J. M., McNeill, A. M., & Prigent-Combaret, C. (2009). Acquisition of phosphorus and nitrogen in the rhizosphere and plant growth promotion by microorganisms. *Advances in Agronomy*, 102, 41–81.

Xu, M., Wang, F., Sheng, H., Stedtfeld, R. D., Li, Z., Hashsham, S. A., Jiang, X., & Tiedje, J. M. (2021). Does anaerobic condition play a more positive role in dissipation of antibiotic resistance genes in soil? *Science of the Total Environment*, 757, 143737.

Kumar, D., Shivay, Y. S., Dhar, S., Kumar, C., & Prasad, R. (2013). Rhizospheric flora and the influence of agronomic practices on them: A review. *Proceedings of the National Academy of Sciences, India Section B: Biological Sciences*, 83, 1–14.

Hakim, S., Naqqash, T., Nawaz, M. S., Laraib, I., Siddique, M. J., Zia, R., Mirza, M. S., & Imran, A. (2021). Rhizosphere engineering with plant growth-promoting microorganisms for agriculture and ecological sustainability. *Frontiers in Sustainable Food Systems*, 5, 617157.

Plant Defence System Against Heavy Metal Stress: From Genes to Proteins

Kakan Ball and Sanjoy Sadhukhan

Abstract Biomagnification, followed by bioaccumulation of heavy metals (HMs) in the ecosystem, has seriously threatened all organisms, including plants. Heavy metals are hazardous to plants at excessive levels as they engage with various essential macromolecules of cells, including DNA and protein, resulting in an abnormal increase of reactive oxygen species (ROS). Due to the toxicity of heavy metals, plants would suffer from severe morphological, metabolic, and physiological abnormalities, including protein breakdown, lipid peroxidation, and chlorosis of the shoot. Plants use a range of defence mechanisms at cellular as well as physiological levels to withstand the stress caused by heavy metals. However, these natural defensive systems cannot combat the extreme stress experienced by HMs. Plants produce a variety of secondary messengers to initiate cell signaling, triggering several transcriptional reactions linked to plant defence. A range of transcription factor (TF) gene families, such as basic leucine zipper domain (bZIP), WRKY, basic helix-loop-helix (bHLH), Cys2-His2 (C2H2), mitogen-activated protein kinase (MAPK), myeloblastosis (MYB), and APETALA2/ethylene-responsive element-binding protein (AP2/EREBP), are essential for the hormone-mediated pathways that regulate stress under HM stress. Heavy metal stress tolerance of plants depends on two proteins, phytochelatins and metallothionein, which chelate the metal ions in the cytoplasm, followed by the removal and destruction of proteins that cannot function in their natural conformations, the restoration of proteins degraded by stress, and the segregation of metals in the vacuoles. Various strategies have been investigated to comprehend the plant defence mechanism when exposed to heavy metals. Nevertheless, analyzing these defence systems will require much more sophisticated molecular techniques. This book chapter discusses the genes and proteins linked to enhancing the plants' defence mechanism against HM stress.

Keywords Heavy metals · Plant defence · Genes · Proteins · Transcription factors · Transporters

K. Ball · S. Sadhukhan (✉)
Department of Botany, Raiganj University, Raiganj, West Bengal, India

S. Roy, V. Mandal (eds.), *Plant-Microbe Interaction under Xenobiotic Exposure*,
https://doi.org/10.1007/978-981-96-8260-7_12

1 Introduction

The characteristics of heavy metals include higher density, high atomic number, and/or weight. Due to their tendency to dissolve in water and their bioaccumulative characteristics, heavy metals (HMs) can have detrimental impacts on a broad range of ecosystems via food chains and water (Oladoye et al. 2022). Arsenic (As), cadmium (Cd), lead (Pb), chromium (Cr), and mercury (Hg) are regarded as the most hazardous heavy metals, as they are hazardous to living beings (Keyster et al. 2020; Mitra et al. 2022). Increased concerns have been raised by the increased levels of heavy metal (HM) poisoning in the environment brought on by industrial processes such as mineral extraction, vehicle exhaust, waste disposal, chemical fertilizer, and urban wastewater treatment. Agricultural productivity is gravely threatened by this contamination, highlighting the urgent need for effective mitigation measures. (Tiwari and Lata 2018; Riyazuddin et al. 2021). One of the most harmful abiotic stresses on crops is HM stress, which interacts with essential chemicals and cellular processes to produce toxicity. (Li et al. 2022). A protein's function is fundamentally determined by its structure. Its conformation is primarily influenced by its environment's physical and chemical conditions, which can impact HM ions, reactive chemicals, high temperatures, and other stressors. These factors can cause newly synthesized proteins to misfold and cause existing proteins to fold incorrectly. By binding with protein and changing its conformation, HMs are believed to impede its biological functions (Hossain and Komatsu 2013). Plant productivity is negatively impacted when HMs continuously stress the growth and development of plants, and that causes irreversible damage to the plant's internal systems (Husen 2021).

As a result, plants employ a multitude of intricate signaling pathways to modify their physiology and biochemistry in response to a range of environmental stressors (Haak et al. 2017; Kumar et al. 2020). Cell membrane receptors sense the external stress signals, amplified by a cascade of pathways. Specific genes linked to stress are controlled in the nucleus, and stress-generated signals activate effector proteins like kinase or certain TFs (Mondal et al. 2022; Hao et al. 2021a). The key to HM detoxification is also believed to produce metal-binding cysteine-rich peptides, which immobilize, sequester, and detoxify the metal ions. The chelation of metal ions in the cytoplasm with phytochelatins and metallothioneins, subsequent metal compartmentalization in the vacuoles, reconstruction of stress-induced proteins, and elimination and degradation of proteins incapable of regaining their native conformations are all necessary for plants to be able to withstand HM stress. (Hasan et al. 2017). Many protective genes in the HM stress are activated by TFs, including the AP2/ERF family, MYB family, and bZIP. However, there has been conflicting and unclear knowledge about the function of TFs in regulating gene expression brought on by HMs. One possible explanation for the intricacy of TF's function in plants' defensive and adaptable responses to HM stress is their involvement in the signaling pathways triggered by the activity of other abiotic factors. (Zaikina et al. 2019). This book chapter aims to shed light on the genes and protein regulation system in plants that govern their ability to withstand heavy metal exposure.

2 Plant-Heavy Metal Interactions and Detoxification

The agroecosystem's toxicity to HMs has become a significant global concern—some heavy metals like As, Ni, Cd, Hg, platinum group, Co, Cr, etc., are present in trace amounts naturally in the environment. However, if they are in excess, they could significantly change crop output and quality, endangering the stability of the ecosystem. (Ball et al. 2022). All plant species most likely have a minimum tolerance to HMs. Consequently, they manage a sophisticated system that includes chelation, transport/sequestration, and uptake/efflux. The homeostasis of vital metal micronutrients is closely correlated with these essential components. Based on the overall traits of these components, the kingdom of plants is divided into two categories: (i) non-accumulating plants and (ii) hyperaccumulating plants (Viehweger 2014). The majority of plants are classified as non-accumulators.

Plants must withstand heavy metals to grow in metalliferous soils and for nutritional reasons (Hall 2002; Clemens 2006). Oxidative stress results from harmful metal ions producing ROS at the cellular level (Li et al. 2016). They affect protein function and activity, nutritional balance, and the integrity of membrane function. HMs impede repair systems or encourage DNA damage (Tamás et al. 2014). Plants employ defensive mechanisms as their primary means of defence to limit or stop the uptake of metal ions. Among these methods are blocking long-distance transport and attaching metal ions to cellular exudates or the cell wall. Microbes and root exudates impact the rhizosphere's metal availability and mobility. The compounds discharged into the medium when healthy, intact plant roots are nearby are called root exudates.

Among the most crucial mechanisms for obtaining nutritional components is the discharge of exudates with chelator properties from the roots into the rhizosphere. Because of this complex formation and increased metal solubility, the plant can absorb the metals more effectively. Recent findings suggest flavonoids may help *A. thaliana* resist heavy metals. In addition to creating intricate structures, these substances can interact with redox-active metals to change their redox condition reductively, which leads to speciation. There are two sides to this intricate process of non-essential HM reduction and formation: On one hand, it can be employed as a defensive tactic to create less soluble metal complexes inappropriate for plant entry. However, on the other hand, it can also compete with the processes necessary to obtain key metals and stabilize unstable redox states. The proper cellular component will instantly bind HMs if they get into the cytoplasm. According to literature analysis, a variety of compounds, including enzymes, peptides, nucleotides, organic acids, vitamins, fungal stimulators, attractants, nucleotides, inhibitors, and other unknown molecules, are released by healthy roots (Bais et al. 2006; Dakora and Phillips 2002). In order to absorb metal ions and inorganic nutrients from the soil, higher plants have sophisticated systems. There are just a few transport mechanisms on which these systems depend. It is commonly recognized that heavy metals can cotransport across the roots' plasma membrane (Krämer et al. 2007).

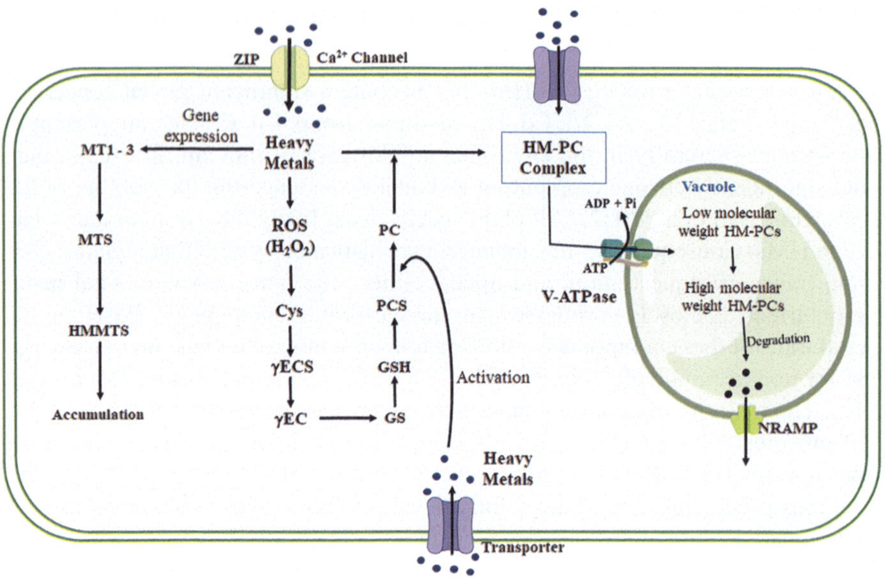

Fig. 1 HM triggers Phytochelatin Synthase (PCS) and MTs expression. HM-MTs and HM-PC complexes are formed in the cytoplasm. ATP-binding cassette, PTB-ATPase, and V-ATPase transporters transport these complexes to the vacuole through the tonoplast. PC acts as a ligand and chelates the HMs in the cytosol. Enhanced activity of GSH metabolic enzymes, γ-ECS and GS, contributes to the induction of PC synthesis by HM and a significant GSH flux. These enzymes are not directly active by HM but rather by ROS, which is produced by the response of heavy metals. HM is transported by ZIP across the cell membrane and by NRAMP via the tonoplast

Conversely, when elevated amounts of HMs are present, cells start an intricate network of detoxification and storage processes, which include vacuolar transporters that move and sequester substances into the vacuole (Fig. 1), as well as the cytoplasmic ligation of metal ions with metallothioneins and phytochelatins (Yang and Chu 2011). Metals can enter cells using cation carriers with a wide range of substrate selectivity. There are no unique transporters for hazardous HMs like Pb and Cd because the exact mechanisms that carry them have no known biological purpose (Silver et al. 2001).

As mentioned earlier, plants have evolved various strategies at the molecular level to counteract HM stress. Understanding HM detoxification at the molecular mechanism based on protein and gene levels is essential.

3 Heavy Metal Protein Transporters in Metal Detoxification

Plants have many metal transporter proteins essential for metal uptake and overall homeostasis. As a result, they may be necessary for tolerance, which includes heavy metal ATPase, cation diffusion facilitator (CDF), natural resistance-associated

Table 1 List of plants, along with their genes involved in metal resistance

Gene	Name of the plant	Involvement in metal resistance	Reference
HMA3	*A. thaliana*	Cd deposition within the vacuole and restriction of its movement toward the xylem.	Morel et al. (2009)
ZmHMA3	*Zea mays*	Increases maize's ability to withstand Zn and Cd stressors via absorption and transferring Cd and Zn ions.	Liao et al. (2023)
SpHMA3	*Sedum plumbizincicola*	Young leaves and stems are detoxified of cadmium by being sequestered into vacuoles.	Liu et al. (2017)
GmHMA3	*G. max*	High seed Cd levels in vacuoles due to a single nucleotide mutation in GmHMA3 from A to G at nucleotide position 1823.	Wang et al. (2012)
HMA3	*Brassica nigra* and *B. juncea*	Transport Cu and compartmentalize the metal in chloroplast and vacuole.	Meraklı and Memon (2023a)
HMA2	*B. juncea*	Contributes significantly to the sequestration and subcellular detoxification of zinc.	Meraklı and Memon (2023b)
HMA4	*A. thaliana*	In addition to acting as a chelator (sensor) of Zn^{2+} and Cd^{2+}, the C-terminal domain of AtHMA4 regulates their efficient export.	Bækgaard et al. (2010)
HMA4	*Noccaea caerulescens*	A significant contribution to CD tolerance.	Craciun et al. (2012)
HMA4	*B. juncea*	Sequestration and subcellular detoxification of Zn.	Meraklı and Memon (2023b)
HMA1	*Brassica nigra* and *B. juncea*	As Cu levels increased, HMA1 rose gradually in both plant species' foliage.	Meraklı and Memon (2023a)

macrophage protein (NRAMP), and zinc-iron permease (ZIP) family. Plants include metal transport proteins and chelators that regulate the homeostasis of metal. The *HMA3, HMA4*, and *HMA* genes encode the PIB-type of proteins (Table 1).

3.1 Heavy Metal ATPase in Heavy Metal Resilience

3.1.1 NRAMP Family

NRAMPs are another crucial metal transporter for various purposes in various species. The uptake, dispersion, and retention of metallic elements in different plant tissues depend on transporter proteins. A detailed analysis of the NRAMP family genes in various plants is still necessary despite these genes being fully discovered in many species. (Hussain et al. 2024). The highly conserved class of essential membrane proteins known as NRAMPs is crucial for accessing divalent metals across cell membranes. *NRAMP* genes are present in numerous species of microbes, fungi, plants, and animals (Segond et al. 2009). These proteins have a consensus transport signature and typically have 10–12 transmembrane domains (TMD) (Tan et al. 2023). The majority of NRAMP proteins transport a wide range of metal ions. In *Populus trichocarpa,* the *PtNRAMP* genes responded differently to metal stressors, as well as iron and manganese deficiency. This was revealed by gene expression analysis (Ma et al. 2023). Forty-eight *StNRAMP* genes from many species were discovered, including Arabidopsis, potato, tomato, tobacco, rice, and pepper, further categorized into 4 subfamilies. Tissue specificity is the characteristic feature of NRAMP gene expression in *Solanum tuberosum.* These gene family members respond significantly when exposed to heavy metals (Tian et al. 2021). *SpNRAMP1, SpNRAMP2*, and *SpNRAMP3* are discovered as transporter proteins and found in the plasma membrane. In yeast, it is observed that the main functions of these proteins are the absorption and transportation of Cd. Cd, Fe, Mn, and Cd are absorbed by the over-expression of *SpNRAMP1* in *Spirodela polyrhiza* (Chen et al. 2021a).

3.1.2 CDF Family

The family of CDF proteins, which participate in the transfer of zinc, copper, and cobalt, has also been found in select plants. Some CDF family members may accelerate efflux, whereas others are believed to be involved in HM absorption. Some of the CDF family members are localized in intracellular membranes, while others are found to be associated with plasma membranes. Cells include a ubiquitous family of transmembrane transporters called CDF proteins that maintain divalent transition metal cation homeostasis. The apo-open V-shaped dimer condenses into a more compact structure when metal cations attach to the transporter CDF protein's cytoplasmic C-terminal domain. Then, a conformational change in the transmembrane domain enables the movement of the metal cation (Lopez-Redondo et al. 2018). Typically, CDF proteins use the proton motive force during conformational changes to export divalent cations from the cytoplasm, such as Mn^{2+}, Zn^{2+}, Fe^{2+}, Co^{2+}, and Cd^{2+}. All known CDF proteins have the transmembrane domain, which has six helices per monomer and a well-defined metal-binding site that facilitates the transport of metal cations. A typical characteristic of CDF proteins is the CTD domain, which takes on a metallochaperone-like structure and undergoes conformational changes linked to metal interaction. In a recent study, 12 ZmMTPs were found and described in the complete maize genome. During evolution, these *ZmMTP* genes were separated into three subfamilies: Zn-CDF, Zn/Fe-CDF, and Mn-CDF. These subfamilies displayed varying expression patterns in various maize tissues. *ZmMTP11* is a Pb accumulation-related MTP component in maize found in the cell membrane and capable of transporting Pb ions. It was found using gene-based association studies. *ZmMTP11* displayed a generally lower expression with the Pb treatment than the usual settings (Liu et al. 2024).

3.1.3 ZIP Family

Another vital metal transporter is the *ZIP* gene family, first discovered in plants. This family transports iron, cadmium, zinc, manganese, and other cations. The name of the protein comes from the first member discovered, "ZRT, IRT-like protein." Arabidopsis cation transporter IRT1 or iron-deficient plants express an iron-regulated transporter. (Eide et al. 1996). The mechanism, character, and structure of plant ZIP family transporters' Zn transport are mostly unclear. Only few information is available about the structure of the family. The plant ZIP family transporter is assumed to have 6–9 transmembrane (TM) domains (α-helices), with 8 being the most common form. The protein sequence of the zinc transporters varied from 322 to 478 amino acids, while their molecular weight ranged from 33.1 to 51.4 kDa (Vatansever et al. 2016). There are currently about 25 members of the ZIP family known to exist. Because of their comparable amino acid composition, these genes can be categorized into two subfamilies. Fifteen genes from plants (11 from Arabidopsis, 2 from tomato, 1 from pea, and 1 from rice), two yeast genes (*ZRT1 and ZRT2*), and 1 gene from the protozoan *Trypanosoma brucei* are included in subfamily I. Subfamily II comprises two human genes, one in *Drosophila* and eight in the nematode *Caenorhabditis elegans* (Guerinot 2000). 13 ZIP proteins, which were predicted to be primarily prevalent in the plasma membrane, were discovered in Tartary buckwheat (*Fagopyrum tataricum*). The qRT-PCR gene expression investigation revealed that some plant organs' *FtZIP* genes differed considerably. However, in comparison to customary conditions, excessive Zn^{2+}, Fe^{2+}, Mn^{2+}, and Cd^{2+} enhanced the transcript of *FtZIP5* or *FtZIP13*, although the relative expression of *FtZIP10* was noticeably raised in seedlings under the $CdCl_2$ treatment (Zhang et al. 2024). In the event of a deficiency in Cu, Fe, or Mn, Zn, as well as in Cd exposure, *NtZIP5B* is engaged in maintaining multi-metal homeostasis. The promoter activity of *NtZIP5B* is thought to be an indication of a defensive mechanism against cadmium's detrimental effects (Palusińska et al. 2024).

3.1.4 ATP-Binding Cassette

ATP-binding cassettes (ABCs), transporters, belong to a broad protein family, primarily found in many different organisms. ATP hydrolysis activates ABC transporters, which can function as importers and exporters. These proteins are found in the membranes of vacuoles, peroxisomes, mitochondria, and chloroplasts. Many biological processes in plants, such as nutrient intake, development, growth, resilience to metal toxicity, stomatal closure, pollen particle size and shape, and phytohormone transport, are regulated by ABC proteins. ABC transporters located in tonoplasts play a part in the intracellular detoxification of metal ions. Preventing damage to DNA and sustaining cellular proliferation (Dahuja et al. 2021). In plants, ABC transporters are grouped into eight subfamilies: A, B, C, D, E, F, G, and I (Gong and Wang 2022). When Arabidopsis is exposed to metal stress, it has been discovered that both ABC transporters, *AtABCC1* and *AtABCC2*, take part in the vacuolar transfer of phytochelatin (PC). Several reports exhibit the crucial importance of ABC transporters in imparting resistance against Cd and Hg (Stein et al. 2006). According to certain other studies on cadmium toxicity, ABCC-type transporters significantly translocate Cd within the central vacuole for detoxification (Yang et al. 2021). Recent research shows how soybean (*Glycine max* L.) biochemical and physiological characteristics are altered by cadmium and mercury exposure and how certain ABC transporter genes react to these stressors. Under Cd and Hg stress, the subfamily of ABC transporter genes *MDR/ABAB, MRP/ABCC*, and *PDR/ABCG* was significantly increased (Naaz et al. 2023).

Apart from metal transporters, several proteins like metal tolerance proteins (MTPs), metallochaperones, chelating proteins like metallothioneins, and phytochelatins participate in metal transport and plant homeostasis. In order to control the intake, distribution, and storage of vital nutritional components, plants have developed a homeostatic mechanism. Plants employ MTPs to preserve this equilibrium. Chelating MTPs with heavy metal ions may achieve plant detoxification. A recent study on rice found that OsMTPs were significant for mineral transport and gallium (Ga) tolerance (Zhan et al. 2024). Previous research revealed that when exposed to excessive levels of Cd, Cu, Ag, and Zn ions, the internal synthetic metal chelators—such as organic acids, amino acids, and short peptides—might bind free metal ions (Cobbett and Goldsbrough 2002). A class of distinct protein families known as metallochaperones was first discovered to interact with cellular metal ions by delivering metal to particular target proteins, like metal enzymes. Copper is transported to internal copper transporters by Atx1-like chaperones and zinc superoxide dismutase by CCS chaperones (Rosenzweig 2001). However, the detailed mechanisms of phytochelatin and metallothioneins are described below.

4 Phytochelatin and Its Response to Heavy Metal Stress

The primary cause of HM accumulation in organisms, including plants, is anthropogenic activity. Because of their molecular resemblance to the metals required for cellular functioning, these metals have the potential to have extremely harmful effects on cells. At hazardous concentrations, metals can participate with several essential cellular macromolecules, which include DNA and nuclear proteins, which can result in an overabundance of ROS (Emamverdian et al. 2015). Various defensive mechanisms against metal toxicity exist in higher plants, algae, certain yeasts, and animals. These consist primarily of similar cysteine-rich polypeptides and phytochelatins (PCs), which chelate metals (Rea 2012). PC synthases, also called γ-glutamylcysteine (γ-Glu-Cys) dipeptidyl transpeptidases, catalyze the production of PCs from glutamine, cysteine, and glycine (Rea 2012; Mendoza-Cózatl et al. 2008). They were thought to be the byproducts of a metabolic route because of their structural similarity to glutathione (GSH). Several genetics, physiology, and biochemistry investigations have identified that GSH is the substrate for PC production (Rauser 1999). PCs have the general structure (γ-Glu-Cys)n-Gly, where n is often limited to five repetitions of the dipeptide Glu-Cys but can vary from two to eleven (Pivato et al. 2014). Other amino acid residues, except glycine, are on the C-terminal end of (γ-Glu-Cys) $_n$ peptides. Plant genotypes tolerant to HMs have a significantly greater PC accumulation than non-tolerant lineages. PC, which chelates the metals, is synthesized using glutathione (GSH) as the substrate. The metal-PC chelatins' inactive toxic metal ions are then transferred from the cytoplasm to the vacuole before they have a chance to cause damage to vital metabolic pathway enzymes. In order to accomplish heavy metal detoxification and reduce the amount of HMs in the cytosol, they are thereafter temporarily stored there. (Srivastava 2016).

5 Metallothioneins and Heavy Metal Stress

The term "metallothionein" was given for the first time when the renal cortex protein of a horse facilitated the identification of Cd2þ-binding protein. Although plant metallothioneins (MT) have been identified over the past four decades, their specific physiological role remains unknown (Liu et al. 2015). Metallothioneins are a diverse family of transition metal ion-associated proteins in all cells. Their molecular weights range from 4 to 8 kDa. In addition, they have a high cysteine residue (approximately 30% of their total amino acid composition), which is organized in distinctive patterns. The term "early cysteine-labeled," or Ec, protein refers to the observation that during an early stage of germination phases of *Triticum aestivum*, it was found that nascent proteins combined into a single protein accounted for about 20–25% of the overall cysteine content (Freisinger 2011). Numerous studies have shown that plant metallothioneins are crucial for preserving metal equilibrium during vegetative development and senescence, especially for Cu.

According to all the findings, MTs are necessary to maintain critical ion homeostasis and facilitate shipment. Still, they are not essential for the plant to successfully complete its life cycle. The primary characteristics of the metal-MT complex are low kinetic stability and high thermodynamics. It permits a portion of the metal ion to be transferred to other proteins and provides a strong binding (Wolfgang 2000). Better metal tolerance is achieved when MT genes are elevated in transgenic and genetically engineered plants with distinct metal dispersion and accumulation mechanisms (Gu et al. 2015; Liu et al. 2015; Tomas et al. 2015). It is assumed that the final detoxification of HMs occurs in plants. There is evidence of metal chelation by MTs, but the transportation mechanism of metal from cytosol to vacuole remains unclear (Yang et al. 2011). From the beginning of their discovery, the physiological roles of MTs have been obscure. According to several studies, MTs are essential for preserving metal ion homeostasis or chelation as well as cellular redox equilibrium under various stressors (Kang 2006). Cellular macromolecules, basically the proteins, get damaged by ROS, which is produced by heavy metal stress and causes physiological disorders or senescence.

Remarkably, MTs have been thought to deliver an extra line of defence for plants against stress-induced oxidative damage. Various studies demonstrate that MTs act as ROS scavengers responding to abiotic stress. However, the exact mechanism of ROS scavenging by MTs remains unclear (Hassinen et al. 2011). Metals are thought to be released from MTs at the time of ROS scavenging, and the Cys residues of MTs restrict the ROS particles. Numerous researchers have also suggested that the liberated metals may initiate the signaling cascade needed to scavenge ROS.

6 Heat Shock Protein

Both eukaryotic and prokaryotic species contain significant heat shock proteins (HSPs). Although the name of this protein is Heat Shock Protein, its function is not limited to protection from heat exposure (Shafikova et al. 2024). HSPs, also referred to as “stress proteins,” can be triggered by HMs, oxidants, proteins, viruses, free radicals, and other stressors (Adhikari 2024). They can be grouped according to the size of their molecules and are found in all classes of living beings, as was previously mentioned.

Furthermore, it is recognized that they can react to a range of stressors, including HM stresses, by expressing themselves. HSPs serve as molecular chaperones in the normal folding and assembly of proteins, but they may also protect and repair proteins under stressful circumstances. Various stressors cause the production of chloroplast small heat shock proteins (smHSPs). To find out whether HMs cause chloroplast smHSPs, whether smHSPs protect the electron transport chain, and whether there are any effects on chloroplast smHsp and photosynthesis, maize plants were exposed to varying soil concentrations of Cu, Ni, Pb, and Zn. The findings are among the first to pinpoint certain pathways that HSPs protect under metal stress and show that HSPs shield photosynthesis and photosynthetic machinery from heavy metals. The members of the HSP70 family among the HSPs are well-studied. Analysis of HSP70 showed that different plant species accumulate this HPS70 as a response to environmental stress. The particular members of this family are constitutively expressed and regulated to preserve cellular homeostasis. Research on proteomics aids in discovering the gene functions or proteins engaged in response to stress from heavy metals. Transcript analysis of numerous plants shows that the HSP70 is strongly expressed under diverse metal stress. The molecular mechanisms behind HSP70 activity under stressful conditions are still poorly understood despite multiple studies showing a positive association between the over-expression of HSP70 genes and the emergence of resistance to various stressors, including HMs (Wang et al. 2004).

HM stress signals produce distinct intermediate molecules, which trigger various TFs. MAPKs are the significant and extensively conserved signaling molecules in higher organisms. They catalyze protein phosphorylation, a crucial signal transduction pathway in plants (Sharma et al. 2020). Every cell has different MAPK pathways. Every pathway is separate and interconnected and is critical for transmitting cell signals. The MAPK cascade is linked to many signal transduction pathways, resulting in plant molecular-level interactions. (Sharma et al. 2021)—several cellular processes in plants include growth and developmental responses to biotic and abiotic stresses. Under stress conditions, the MAPK pathway regulates different TFs like WRKY, bZIP, MYB, and MYC (Andreasson and Ellis 2010).

7 Regulation of Heavy Metal Tolerance by Transcriptional Factors (TFs)

Numerous phosphorylation sites are present in TFs, which can influence the expression of downstream genes to regulate HM stress. Additionally, they are crucial to the regulatory systems that control tolerance and detoxification from heavy metals (Fig. 2). Plants include many TFs associated with HM resistance and detoxification.

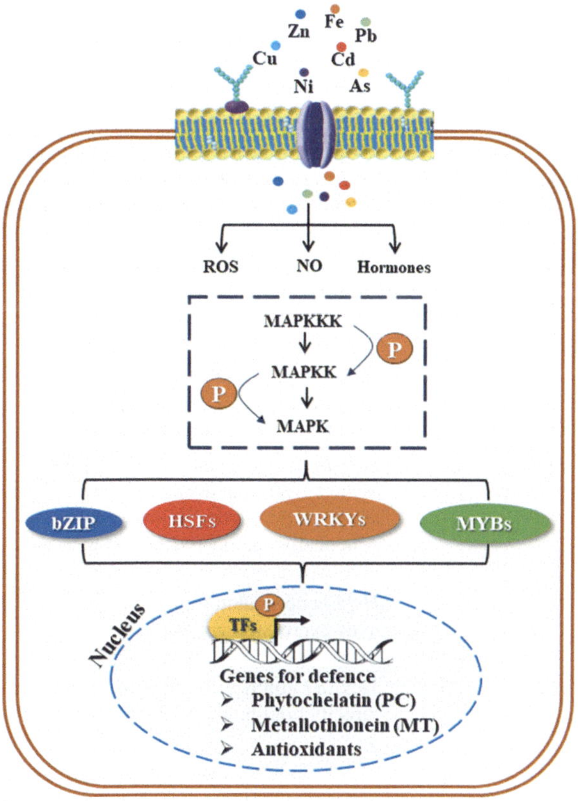

Fig. 2 Heavy metals activate several signals, such as phytohormones, NO, and ROS. MAPK cascade gets activated by such a signaling pathway. MAPK cascade subsequently phosphorylated and triggers associated TFs like bZIP, WRKY, MYB, and HSF. Eventually, these TFs provide plants with a high level of heavy metal accumulation or tolerance by expressing the genes for metal transporters, PC, MTS, and antioxidants [adapted from (Li et al. 2022)]

Table 2 List of genes associated with heavy metals and their functions

Family	Gene	Heavy metal	Function	References
WRKY	*ZmWRKY64*	Cd	A cadmium-sensitive phenotype is brought on by a notable deposition of cadmium in leaf and root cells as a result of *ZmWRKY64* gene activation.	Gu et al. (2024)
	TaWRKY70	Cd	*TaWRKY70* controls plant response to cadmium stress by binding to the TaCAT5 promoter.	Jia et al. (2021)
	GmWRKY172	Cd	*GmWRKY172*over-expression lines showed decreased cadmium concentration in shoots and increased cadmium tolerance.	He et al. (2023)
	AtWRKY6	As	WRKY6 transcription factor limits transposon activation and absorption of arsenate in *Arabidopsis*.	Castrillo et al. (2013)
Basic helix-loop-helix (bHLH)	*bHLH39*, *bHLH104*	Cd	Elevating Fe transporter gene expression improves the tolerance of Cd.	Meng et al. (2022)
	bHLH104	Cd	*BHLH104* positively regulates cd-tolerant genes *MTP3*, *HMA3*, *IREG2*, and *NAS4* play a significant role in Cd tolerance.	Yao et al. (2018)
MYB	*AtMYB4*	Cd	The expression of *PCS1* and *MT1C* in Arabidopsis regulates the plant's resistance to Cd stress.	Agarwal et al. (2020)
	SbMYB15	Cd, Ni	In transgenic tobacco, the tolerance of Cd and Ni increased due to the over-expression of *SbMYB15*.	Sapara et al. (2019)
bZIP	*SpbZIP*	Cd	SpbZIP TFs in *Sedum plumbizincicola* control plant responses to Cd stress.	Lu et al. (2022)
	bZIP43	Cd	Under Cd stress, *bZIP43* might have an advantageous regulatory function.	Yuan et al. (2018)
Hsf	*SaHsfA4c*	Cd	*SaHsfA4c* can enhance ROS and HSP-related genes and decrease ROS accumulation under Cd toxicity in genetically engineered Arabidopsis and non-hyperaccumulator *Sedum alfredii*.	Chen et al. (2020)
	HsfA4a	Cd	*HsfA4a* upregulates the metallothionein gene expression and performs better Cd tolerance.	Shim et al. (2009)
Others	*VuNAR1*	Al	*VuNAR1* controls the Al resistance by directly binding to the WAK1 promoter.	Lou et al. (2020)
	ZAT6	Cd	*ZAT6* stimulates the expression of the PC gene, which regulates Cd tolerance and accumulation in Arabidopsis.	Chen et al. (2016)

Among these, the known TFs that control resistance and detoxification of HM in plants include WRKY, basic leucine zipper (bZIP), myeloblastosis protein (MYB), ethylene-responsive transcription factor (ERF), and others (Table 2).

7.1 WRKY

Multiple plant genome databases contained WRKY genes. The WRKY protein's unique characteristics include a zinc finger motif and the WRKY domain (WRKYGQK), allowing it to identify downstream genes' cis-acting W-box sequences. There are 126 WRKY genes in the radish (*Raphanus sativus*) genome database. According to an RT-qPCR study, thirty-six *RsWRKY* exhibit significant changes under more than one HM stress. In particular, treatments with Cd and Pb, respectively, activate 20 and 24 *RsWRKY* transcripts (Karanja et al. 2017). The transcriptome study of soybeans under Cd stress reveals 29 Cd-responsive *WRKY* genes. In *A. thaliana* and *Glycine max*, the elevated expression of GmWRKY142 reduced Cd absorption. It efficiently controlled Cd tolerance. A subsequent investigation discovered that GmWRKY142 activated the transcription of AtCDT1, GmCDT1–1, and GmCDT1-2 by directly binding to the W-box element in their promoters. Cysteine (Cys)-rich CDT1 protein is a significant chelator of Cd (Cai et al. 2020). By modulating the downstream functional genes, WRKY can improve plant resistance to Cd or restore the equilibrium of metal ions. Using direct binding to the W-box of the GSH1 promoter and indirect repression of phytochelatin synthesis-related gene expression, *AtWRKY12* adversely controls Cd resistance in Arabidopsis. *AtWRKY47* controls the expression of genes that modify cell walls (XTH17, ELP, etc.), which contributes to maintaining the equilibrium of aluminum (Al) within ectoplasts as well as symplasts and enhances resistance to Al (Li et al. 2020).

7.2 bZIP

bZIPs, or basic leucine zipper domain plant TFs, are linked to several biological processes and environmental challenges. By analyzing the radish's (*Raphanus sativus*) whole genome and transcriptome, 135 bZIP-encoding genes were found. More specifically, under a range of HM pressures, including Cr, Cd, and lead (Pb) strains, *RsbZIP010* revealed decreased expression (Fan et al. 2019). Sixty-six members of the *GubZIP* gene family were found in the *Glycyrrhiza uralensis* genome using various bioinformatics techniques, depending on the Hidden Markov Model (HMM). The expression of 45 and 51 *GubZIP* genes varied in the roots as well as leaves under 0.02 $g{\cdot}kg-{}^{1}$ Cd stress, respectively. (Han et al. 2021). In response to chromium (Cr6+) stress in Arabidopsis, the bZIP transcription factor TGA3 increased the transcription of L-cysteine desulfhydrase (LCD) via a calcium (Ca^{2+})/calmodulin2 (CaM2)-mediated process, which in turn caused the production of hydrogen sulfide (H_2S). H_2S can set off various defence reactions and aid in reducing HM accumulation in plants (Fang et al. 2017). In the HM accumulator *Brassica juncea*, plants treated with Cd for 6 h have higher levels of the TGA3 homolog gene *BjCdR1*, suggesting that the transcription factor *BjCdR15* is crucial for controlling Cd absorption and long-distance transport. Western blot study revealed that in plant shoots treated with Cd, the quantity of AtPCS1 protein increased considerably.

Furthermore, because it controls the expression of several metal transporters and the synthesis of phytochelatin synthase, its over-expression in tobacco and *A. thaliana* confers resistance and Cd accumulation. In order to control Cd uptake, the bZIP transcription factor mediates downstream transcriptional factors (TFs) or interacts with other TFs (Zhang et al. 2019b). By binding zinc (Zn) ions to the zinc sensor motif, *bZIP19* and *bZIP23*, the two members of group F of the bZIP TFs, may centrally regulate the zinc deficiency reaction of *A. thaliana* (Lilay et al. 2021).

7.3 MYB

Among the largest groups of multifunctional TFs found in plants is MYB, which is crucial for regulating several transcription-level functions, plant growth and development, the transmission of phytohormone signals, and reactions to biotic and abiotic stimuli. A unique MYB domain is present in the N-terminal of MYB. This domain is further divided into subgroups depending on how many R repetitions are present. However, because the C-terminal domains differ significantly, they exhibit a variety of functions. Numerous studies have demonstrated that MYB alone could bind to MYBCORE and AC-box or in the conjugation of protein, which is activated by environment-induced signals. Thus, MYB can contribute to the regulation of downstream target gene expression, which in turn regulates the stress tolerance of plants.

Additionally, MYB-regulated signaling pathways in response to biotic and abiotic stresses include those for ROS, jasmonic acid (JA), abscisic acid (ABA), and brassinolide (BR) (HU et al. 2024). Cd stress is positively regulated by *OsMYB45* in rice *(Oryza sativa)*. The mutant leaves had higher H_2O_2 concentrations and decreased catalase (CAT) activity compared to the wild-type. *SbMYB15* is an HM-responsive gene that is derived from *Salicornia brachiata* Roxb. In transgenic tobacco, the accumulation of Cd and Ni gets reduced due to the over-expression of this *SbMYB15* gene (Sapara et al. 2019). AtMYB4 controls tolerance to Cd by improving defence against oxidative damage and increasing the expression of PCS1 and MT1C (Agarwal et al. 2020). It is thought that *JrMYB2* from *Juglans regia* enhances plants' resistance to $CdCl_2$ by acting as an upstream regulator of JrVHAG1. In rice, OsARM1 (ARSENITE-RESPONSIVE MYB1), an R2R3 MYB TF, controls As (III) absorption and its movement from root to aerial part via modulating the transporters related to As (OsLsi1, OsLsi2, and OsLsi6). As an essential regulator that imparts As stress resistance and lower arsenic accumulation, *AtMYB40* of Arabidopsis has negatively regulated PHT1 expression, and ABCC2, ABCC1, and PCS1 expression were positively regulated. (Chen et al. 2021b). MYB TFs absorb or maintain essential elements like iron (Fe), zinc (Zn), As, and Cd. MYB72 helps Arabidopsis maintain its metal equilibrium, and compared to the wild-type, its knockout variety was more susceptible to severe zinc or iron deprivation. (Van De Mortel et al. 2008).

7.4 *Other Transcription Factors*

In addition to WRKY, bZIP, and MYB TFs, the plants also have a number of additional TF families that control the HM stress response. It is commonly known that the heat shock TF can react to surrounding heat stress. It is reported that the member of Class A participates in the HM stress-related responses. The Cd/Zn/Pb hyperaccumulator *Sedum alfredii* contained 22 Hsf members, divided into three groups based on phylogenetic analysis: SaHsfC, SaHsfB, and SaHsfA. Specifically, 18 SaHsfs responded to the Cd stress (Chen et al. 2018). In *Arabidopsis thaliana,* another vital TF is the basic Helix-Loop-Helix (bHLH), the most significant TF gene family. This family has a highly conserved bHLH motif throughout eukaryotic organisms. Two conserved motifs are present in members of this family: a helix-loop-helix (HLH) region and a basic DNA binding region (Hao et al. 2021b). In tobacco plants, the increased resistance to aluminum toxicity occurred due to the over-expression of *GmbHLH30* (Song et al. 2014). The TFs present show a complex network in between. For instance, the ABI5-MYB49-HIPP regulatory network mediated by ABA inhibited the absorption of Cd heavy metal in Arabidopsis (Zhang et al. 2019a). Another crucial TF family, APETALA2/ETHYLENE-RESPONSIVE FACTOR (AP2/ERF), is vital for plants' growth and development as well as their resistance to biotic and abiotic stressors. Research on *Solanum tuberosum* L., which is exposed to Cd, Pb, Zn, Ni, and Cu heavy metals. After identifying 181 putative StAP2/ERF genes, it was shown that the genes *StAP2075*, *StAP2077*, and *StAP2126* improved yeast growth and Cd accumulation (Tian et al. 2022). A transcriptional pathway for Cd stress is formed in *Phaseolus vulgaris* by combining metal response element-binding transcription factor (MTF) PvMTF-1 and ethylene-responsive factors (PvERF15).

8 Future Perspective and Conclusion

Among the many abiotic stresses that plants encounter, heavy metal toxicity is a worldwide problem. HM stress is a significant issue for commercially substantial crops. To combat such problems, plants trigger several gene expressions of genes, producing proteins functional for HM stress responses. The primary workhorses of living cells are the proteins, which are basically agile macromolecules. Transcription factors involved in HM stress control gene expression—current findings regarding the function of proteins and TFs in HM stress resistance and increased plant productivity. Understanding the precise mechanism of HM detoxification in the plant cell is essential. However, almost every plant has a minimal metal tolerance characteristic. Unlike non-accumulating plants, some plant species can hyperaccumulate and have unique tolerance responses. Plants respond to HM stress in a variety of ways, including by repairing proteins damaged by stress, chelating detrimental ions in the cytosol with cysteine-rich peptides like PCs and MTs, and compartmentalizing metals in the vacuole using transporters present in tonoplasts.

Nonetheless, sequestration, chelation, and exclusion processes provide the foundation of general tolerance mechanisms. Research has demonstrated that recent technical developments in transcriptomics, metabolomics, and genomics have made it feasible to identify and molecularly characterize the HMs stress response. The abundance of differently expressed proteins, TFs, genes, and metabolites, such as phytohormones, lipids, and osmolytes, can be used in conjunction with OMICS approaches to build economically significant crops that are resistant to the toxicity of HMs and promote sustainable development. The exact mechanism underlying the succeeding signaling cascades that control metal buildup is unknown. Developing strategies to affect plant cellular protein quality control systems to reduce metal uptake and translocation seems like a potential approach that could ensure increased safety and production. It is essential to give careful attention to the altered expression of genes that cause HMs to spread from the rhizosphere to the aerial part in economically important crops. Monogenic features impact both abiotic and biotic stress, while several genes regulate diverse metabolic and signaling pathways in abiotic stress. Thus, managing and designing HM-tolerant plants becomes a challenging issue (Vinocur and Altman 2005). Therefore, OMICS techniques can be used to get the molecular characterization of metabolites, proteins, and genes that lead to resistance to HM stress. Consequently, it is crucial to develop plants that can respond to environmental stimuli like HMs and function better by utilizing cutting-edge functional genomics, transcriptomics, and phenomics techniques.

References

Adhikari D (2024) Plant Heat Shock Proteins-Origin, Evolution and Mechanism of Action for Biotechnological Implications in Crop Improvement. International Journal of Innovative Research in Science, Engineering and Technology 13: 2758–2762

Agarwal P, Mitra M, Banerjee S, Roy S (2020) MYB4 transcription factor, a member of R2R3-subfamily of MYB domain protein, regulates cadmium tolerance via enhanced protection against oxidative damage and increases expression of *PCS1* and *MT1C* in *Arabidopsis*. Plant Science 297: 110501

Andreasson E, Ellis B (2010) Convergence and specificity in the *Arabidopsis* MAPK nexus. Trends in Plant Science 15: 106–113

Bækgaard L, Mikkelsen MD, Sørensen DM, Hegelund JN, Persson DP, Mills RF, Yang Z, Husted S, Andersen JP, Buch-Pedersen MJ (2010) A Combined Zinc/Cadmium Sensor and Zinc/Cadmium Export Regulator in a Heavy Metal Pump. Journal of Biological Chemistry 285: 31243–31252

Bais HP, Weir TL, Perry LG, Gilroy S, Vivanco JM (2006) The Role of Root Exudates in Rhizosphere Interactions with Plants and Other Organisms. Annual Review of Plant Biology 57: 233–266

Ball K, Tiru Z, Chakraborty AP, Mandal P, Sadhukhan S (2022) Heavy Metal Toxicity and Phytoremediation by the Plants of Brassicaceae Family: A Sustainable Management. *In* T Aftab, ed, Sustainable Management of Environmental Contaminants: Eco-friendly Remediation Approaches. Springer, Cham, pp. 237–264

Cai Z, Xian P, Wang H, Lin R, Lian T, Cheng Y, Ma Q, Nian H (2020) Transcription Factor GmWRKY142 Confers Cadmium Resistance by Up-Regulating the Cadmium Tolerance 1-Like Genes. Frontiers in Plant Science 11: 724

Castrillo G, Sánchez-Bermejo E, de Lorenzo L, Crevillén P, Fraile-Escanciano A, Tc M, Mouriz A, Catarecha P, Sobrino-Plata J, Olsson S (2013) WRKY6 Transcription Factor Restricts Arsenate Uptake and Transposon Activation in *Arabidopsis*. The Plant Cell 25: 2944–2957

Chen J, Yang L, Yan X, Liu Y, Wang R, Fan T, Ren Y, Tang X, Xiao F, Liu Y (2016) Zinc-finger transcription factor ZAT6 positively regulates cadmium tolerance through the glutathione-dependent pathway in Arabidopsis. Plant physiology 171: 707–719

Chen S-S, Jiang J, Han X-J, Zhang Y-X, Zhuo R-Y (2018) Identification, Expression Analysis of the Hsf Family, and Characterization of Class A4 in *Sedum alfredii* Hance under Cadmium Stress. International Journal of Molecular Sciences 19: 1216

Chen S, Yu M, Li H, Wang Y, Lu Z, Zhang Y, Liu M, Qiao G, Wu L, Han X (2020) SaHsfA4c From *Sedum alfredii* Hance Enhances Cadmium Tolerance by Regulating ROS-Scavenger Activities and Heat Shock Proteins Expression. Frontiers in Plant Science 11: 142

Chen Y, Li G, Yang J, Zhao X, Sun Z, Hou H (2021a) Role of *Nramp* transporter genes of *Spirodela polyrhiza* in cadmium accumulation. Ecotoxicology and Environmental Safety 227: 112907

Chen Y, Wang H-Y, Chen Y-F (2021b) The transcription factor MYB40 is a central regulator in arsenic resistance in *Arabidopsis*. Plant Communications 2: 100234

Clemens S (2006) Toxic metal accumulation, responses to exposure and mechanisms of tolerance in plants. Biochimie 88: 1707–1719

Cobbett C, Goldsbrough P (2002) Phytochelatins and Metallothioneins: Roles in Heavy Metal Detoxification and Homeostasis. Annual Review of Plant Biology 53: 159–182

Craciun AR, Meyer C-L, Chen J, Roosens N, De Groodt R, Hilson P, Verbruggen N (2012) Variation in HMA4 gene copy number and expression among *Noccaea caerulescens* populations presenting different levels of Cd tolerance and accumulation. Journal of Experimental Botany 63: 4179–4189

Dahuja A, Kumar RR, Sakhare A, Watts A, Singh B, Goswami S, Sachdev A, Praveen S (2021) Role of ATP-binding cassette transporters in maintaining plant homeostasis under abiotic and biotic stresses. Physiologia Plantarum 171: 785–801

Dakora FD, Phillips DA (2002) Root exudates as mediators of mineral acquisition in low-nutrient environments. *In* JJ Adu-Gyamfi, ed, Food security in nutrient-stressed environments: exploiting plants' genetic capabilities, Vol 95. Springer, Dordrecht, pp. 201–213

Eide D, Broderius M, Fett J, Guerinot ML (1996) A novel iron-regulated metal transporter from plants identified by functional expression in yeast. Proc Natl Acad Sci U S A 93: 5624–5628

Emamverdian A, Ding Y, Mokhberdoran F, Xie Y (2015) Heavy metal stress and some mechanisms of plant defense response. Scientific World Journal 2015: 756120

Fan L, Xu L, Wang Y, Tang M, Liu L (2019) Genome- and Transcriptome-Wide Characterization of bZIP Gene Family Identifies Potential Members Involved in Abiotic Stress Response and Anthocyanin Biosynthesis in Radish (*Raphanus sativus* L.). International Journal of Molecular Sciences 20: 6334

Fang H, Liu Z, Long Y, Liang Y, Jin Z, Zhang L, Liu D, Li H, Zhai J, Pei Y (2017) The Ca^{2+}/calmodulin2-binding transcription factor TGA 3 elevates *LCD* expression and H_2S production to bolster Cr^{6+} tolerance in Arabidopsis. The Plant Journal 91: 1038–1050

Freisinger E (2011) Structural features specific to plant metallothioneins. JBIC Journal of Biological Inorganic Chemistry 16: 1035–1045

Gong X, Wang S (2022) New Insights into Evolution of the ABC Transporter Family in *Mesostigma viride*, a Unicellular Charophyte Algae. Current Issues in Molecular Biology 44: 1646–1660

Gu C-S, Liu L-Q, Deng Y-M, Zhu X-D, Huang S-Z, Lu X-Q (2015) The Heterologous Expression of the *Iris lactea* var. *chinensis* Type 2 Metallothionein IlMT2b Gene Enhances Copper Tolerance in *Arabidopsis thaliana*. Bulletin of Environmental Contamination and Toxicology 94: 247–253

Gu L, Hou Y, Sun Y, Chen X, Wang G, Wang H, Zhu B, Du X (2024) The maize WRKY transcription factor ZmWRKY64 confers cadmium tolerance in Arabidopsis and maize (Zea mays L.). Plant Cell Reports 43: 44

Guerinot ML (2000) The ZIP family of metal transporters. Biochimica et Biophysica Acta (BBA)-Biomembranes 1465: 190–198

Haak DC, Fukao T, Grene R, Hua Z, Ivanov R, Perrella G, Li S (2017) Multilevel Regulation of Abiotic Stress Responses in Plants. Frontiers in Plant Science 8: 1564

Hall JL (2002) Cellular mechanisms for heavy metal detoxification and tolerance. Journal of Experimental Botany 53: 1–11

Han Y, Hou Z, He Q, Zhang X, Yan K, Han R, Liang Z (2021) Genome-Wide Characterization and Expression Analysis of bZIP Gene Family Under Abiotic Stress in *Glycyrrhiza uralensis*. Frontiers in Genetics 12: 754237

Hao S, Wang Y, Yan Y, Liu Y, Wang J, Chen S (2021a) A Review on Plant Responses to Salt Stress and Their Mechanisms of Salt Resistance. Horticulturae 7: 132

Hao Y, Zong X, Ren P, Qian Y, Fu A (2021b) Basic Helix-Loop-Helix (bHLH) Transcription Factors Regulate a Wide Range of Functions in Arabidopsis. International Journal of Molecular Sciences 22: 7152

Hasan MK, Cheng Y, Kanwar MK, Chu X-Y, Ahammed GJ, Qi Z-Y (2017) Responses of Plant Proteins to Heavy Metal Stress—A Review. Frontiers in Plant Science 8: 1492

Hassinen V, Tervahauta A, Schat H, Kärenlampi S (2011) Plant metallothioneins–metal chelators with ROS scavenging activity? Plant Biology 13: 225–232

He G, Saleem M, Deng T, Zhong Z, He T, Wu J (2023) Unraveling the Mechanism of StWRKY6 in Potato (*Solanum tuberosum*)'s Cadmium Tolerance for Ensuring Food Safety. Foods 12: 2303

Hossain Z, Komatsu S (2013) Contribution of proteomic studies towards understanding plant heavy metal stress response. Frontiers in Plant Science 3: 310

Hu Y-D, Wu G-Q, Liu C, Wei M (2024) Roles of MYB Transcription Factor in Regulating the Responses of Plants to Stress. Biotechnology Bulletin 40: 5–22

Husen A (2021) The Harsh environment and resilient plants: an overview. *In* A Husen, ed, Harsh environment and plant resilience: Molecular and functional aspects. Springer, Cham, pp. 1–23

Hussain Q, Ye T, Shang C, Li S, Khan A, Nkoh JN, Mustafa AE-ZM, Elshikh MS (2024) NRAMP gene family in *Kandelia obovata*: genome-wide identification, expression analysis, and response to five different copper stress conditions. Frontiers in Plant Science 14: 1318383

Jia Z, Li M, Wang H, Zhu B, Gu L, Du X, Ren M (2021) *TaWRKY70* positively regulates *TaCAT5* enhanced Cd tolerance in transgenic *Arabidopsis*. Environmental and Experimental Botany 190: 104591

Kang YJ (2006) Metallothionein Redox Cycle and Function. Experimental Biology and Medicine 231: 1459–1467

Karanja BK, Fan L, Xu L, Wang Y, Zhu X, Tang M, Wang R, Zhang F, Everlyne M'mbone Muleke, Liu L (2017) Genome-wide characterization of the WRKY gene family in radish (*Raphanus sativus* L.) reveals its critical functions under different abiotic stresses. Plant Cell Reports 36: 1757–1773

Keyster M, Niekerk L-A, Basson G, Carelse M, Bakare O, Ludidi N, Klein A, Mekuto L, Gokul A (2020) Decoding Heavy Metal Stress Signalling in Plants: Towards Improved Food Security and Safety. Plants 9: 1781

Krämer U, Talke IN, Hanikenne M (2007) Transition metal transport. FEBS Letters 581: 2263–2272

Kumar K, Raina SK, Sultan SM (2020) *Arabidopsis* MAPK signaling pathways and their cross talks in abiotic stress response. Journal of Plant Biochemistry and Biotechnology 29: 700–714

Li CX, Yan JY, Ren JY, Sun L, Xu C, Li GX, Ding ZJ, Zheng SJ (2020) A WRKY transcription factor confers aluminum tolerance via regulation of cell wall modifying genes. Journal of Integrative Plant Biology 62: 1176–1192

Li M, Ahammed GJ, Li C, Bao X, Yu J, Huang C, Yin H, Zhou J (2016) Brassinosteroid Ameliorates Zinc Oxide Nanoparticles-Induced Oxidative Stress by Improving Antioxidant Potential and Redox Homeostasis in Tomato Seedling. Frontiers in Plant Science 7: 615

Li S, Han X, Lu Z, Qiu W, Yu M, Li H, He Z, Zhuo R (2022) MAPK Cascades and Transcriptional Factors: Regulation of Heavy Metal Tolerance in Plants. International Journal of Molecular Sciences 23: 4463

Liao C, Li Y, Wu X, Wu W, Zhang Y, Zhan P, Meng X, Hu G, Yang S, Lin H (2023) *ZmHMA3*, a Member of the Heavy-Metal-Transporting ATPase Family, Regulates Cd and Zn Tolerance in Maize. International Journal of Molecular Sciences 24: 13496

Lilay GH, Persson DP, Castro PH, Liao F, Alexander RD, Aarts MG, Assunção AG (2021) *Arabidopsis* bZIP19 and bZIP23 act as zinc sensors to control plant zinc status. Nature Plants 7: 137–143

Liu H, Zhao H, Wu L, Liu A, Zhao F-J, Xu W (2017) Heavy metal ATPase 3 (HMA3) confers cadmium hypertolerance on the cadmium/zinc hyperaccumulator *Sedum plumbizincicola*. New Phytologist 215: 687–698

Liu J, Shi X, Qian M, Zheng L, Lian C, Xia Y, Shen Z (2015) Copper-induced hydrogen peroxide upregulation of a metallothionein gene, *OsMT2c*, from *Oryza sativa* L. confers copper tolerance in *Arabidopsis thaliana*. Journal of Hazardous Materials 294: 99–108

Liu P, An L, Ma L, Zou L, Du S, Shen Y (2024) MTP family analysis and association study reveal the role of *ZmMTP11* in lead (Pb) accumulation. Plant Physiology and Biochemistry 212: 108740

Lopez-Redondo ML, Coudray N, Zhang Z, Alexopoulos J, Stokes DL (2018) Structural basis for the alternating access mechanism of the cation diffusion facilitator YiiP. Proceedings of the National Academy of Sciences 115: 3042–3047

Lou HQ, Fan W, Jin JF, Xu JM, Chen WW, Yang JL, Zheng SJ (2020) A NAC-type transcription factor confers aluminium resistance by regulating cell wall-associated receptor kinase 1 and cell wall pectin. Plant, Cell & Environment 43: 463–478

Lu Z, Qiu W, Jin K, Yu M, Han X, He X, Wu L, Wu C, Zhuo R (2022) Identification and Analysis of bZIP Family Genes in *Sedum plumbizincicola* and Their Potential Roles in Response to Cadmium Stress. Frontiers in Plant Science 13: 859386

Ma X, Yang H, Bu Y, Zhang Y, Sun N, Wu X, Jing Y (2023) Genome-wide identification of the *NRAMP* gene family in *Populus trichocarpa* and their function as heavy metal transporters. Ecotoxicology and Environmental Safety 261: 115110

Mendoza-Cózatl DG, Butko E, Springer F, Torpey JW, Komives EA, Kehr J, Schroeder JI (2008) Identification of high levels of phytochelatins, glutathione and cadmium in the phloem sap of *Brassica napus*. A role for thiol-peptides in the long-distance transport of cadmium and the effect of cadmium on iron translocation. The Plant Journal 54: 249–259

Meng X, Li W, Shen R, Lan P (2022) Ectopic expression of IMA small peptide genes confers tolerance to cadmium stress in *Arabidopsis* through activating the iron deficiency response. Journal of Hazardous Materials 422: 126913

Meraklı N, Memon A (2023a) Expression of heavy metal ATPases (HMA1 and HMA3) in *Brassica nigra* and *B. juncea* grown at different Cu levels. Botanica Serbica 47: 309–316

Meraklı N, Memon A (2023b) Gene Expression of Metal Transporting P-Type ATPases HMA2 and HMA4 in *Brassica juncea* L. Grown at Different Zn Levels. Russian Journal of Plant Physiology 70: 34

Mitra S, Chakraborty AJ, Tareq AM, Emran TB, Nainu F, Khusro A, Idris AM, Khandaker MU, Osman H, Alhumaydhi FA (2022) Impact of heavy metals on the environment and human health: Novel therapeutic insights to counter the toxicity. Journal of King Saud University-Science 34: 101865

Mondal S, Pramanik K, Ghosh SK, Pal P, Ghosh PK, Ghosh A, Maiti TK (2022) Molecular insight into arsenic uptake, transport, phytotoxicity, and defense responses in plants: a critical review. Planta 255: 87

Morel M, Crouzet J, Gravot A, Auroy P, Leonhardt N, Vavasseur A, Richaud P (2009) AtHMA3, a P_{1B}-ATPase Allowing Cd/Zn/Co/Pb Vacuolar Storage in Arabidopsis. Plant Physiology 149: 894–904

Naaz S, Ahmad N, Jameel MR, Al-Huqail AA, Khan F, Qureshi MI (2023) Impact of Some Toxic Metals on Important ABC Transporters in Soybean (*Glycine max* L.). ACS omega 8: 27597–27611

Oladoye PO, Olowe OM, Asemoloye MD (2022) Phytoremediation technology and food security impacts of heavy metal contaminated soils: A review of literature. Chemosphere 288: 132555

Palusińska M, Barabasz A, Antosiewicz DM (2024) *NtZIP5A/B* is involved in the regulation of Zn/Cu/Fe/Mn/Cd homeostasis in tobacco. Metallomics: mfae035

Pivato M, Fabrega-Prats M, Masi A (2014) Low-molecular-weight thiols in plants: Functional and analytical implications. Archives of Biochemistry and Biophysics 560: 83–99

Rauser WE (1999) Structure and function of metal chelators produced by plants: the case for organic acids, amino acids, phytin, and metallothioneins. Cell Biochemistry and Biophysics 31: 19–48

Rea PA (2012) Phytochelatin synthase: of a protease a peptide polymerase made. Physiologia Plantarum 145: 154–164

Riyazuddin R, Nisha N, Ejaz B, Khan MIR, Kumar M, Ramteke PW, Gupta R (2021) A Comprehensive Review on the Heavy Metal Toxicity and Sequestration in Plants. Biomolecules 12: 43

Rosenzweig AC (2001) Copper Delivery by Metallochaperone Proteins. Accounts of Chemical Research 34: 119–128

Sapara KK, Khedia J, Agarwal P, Gangapur DR, Agarwal PK (2019) *SbMYB15* transcription factor mitigates cadmium and nickel stress in transgenic tobacco by limiting uptake and modulating antioxidative defence system. Functional Plant Biology 46: 702–714

Segond D, Dellagi A, Lanquar V, Rigault M, Patrit O, Thomine S, Expert D (2009) NRAMP genes function in *Arabidopsis thaliana* resistance to *Erwinia chrysanthemi* infection. The Plant Journal 58: 195–207

Shafikova T, Maksimova L, Omelichkina YV (2024) Heat Shock Proteins in Plant Immunity. Russian Journal of Plant Physiology 71: 100

Sharma D, Verma N, Pandey C, Verma D, Bhagat PK, Noryang S, Singh K, Tayyeba S, Banerjee G, Sinha AK (2020) MAP kinase as regulators for stress responses in plants: An overview. *In* GK Pandey, ed, Protein Kinases and Stress Signaling in Plants: Functional Genomic Perspective. Wiley-Blackwell, Hoboken, NJ, pp. 369–392

Sharma SS, Kumar V, Dietz K-J (2021) Emerging Trends in Metalloid-Dependent Signaling in Plants. Trends in Plant Science 26: 452–471

Shim D, Hwang J-U, Lee J, Lee S, Choi Y, An G, Martinoia E, Lee Y (2009) Orthologs of the Class A4 Heat Shock Transcription Factor HsfA4a Confer Cadmium Tolerance in Wheat and Rice. The Plant Cell 21: 4031–4043

Silver S, Phung LT, Rosen BP (2001) Arsenic metabolism: resistance, reduction, and oxidation. *In* WTJ Frankenberger, ed, Environmental Chemistry of Arsenic. Taylor & Francis, New York, pp. 247–272

Song Q, Qian S-F, Chen X-Q, Chen L-M, Li K-Z (2014) Study on the Function of Transcription Factor GmbHLH30 on Aluminum Tolerance Preliminary in Tampa Black Soybean. Life Science Research 18: 332–337

Srivastava N (2016) Role of phytochelatins in phytoremediation of heavy metals contaminated soils. *In* A Ansari, S Gill, R Gill, G Lanza, L Newman, eds, Phytoremediation: Management of Environmental Contaminants, Volume 3. Springer, Cham, pp. 393–419

Stein M, Dittgen J, Sánchez-Rodríguez C, Hou B-H, Molina A, Schulze-Lefert P, Lipka V, Somerville S (2006) *Arabidopsis* PEN3/PDR8, an ATP Binding Cassette Transporter, Contributes to Nonhost Resistance to Inappropriate Pathogens That Enter by Direct Penetration. The Plant Cell 18: 731–746

Tamás MJ, Sharma SK, Ibstedt S, Jacobson T, Christen P (2014) Heavy Metals and Metalloids As a Cause for Protein Misfolding and Aggregation. Biomolecules 4: 252–267

Tan Z, Li J, Guan J, Wang C, Zhang Z, Shi G (2023) Genome-Wide Identification and Expression Analysis Reveals Roles of the *NRAMP* Gene Family in Iron/Cadmium Interactions in Peanut. International Journal of Molecular Sciences 24: 1713

Tian W, He G, Qin L, Li D, Meng L, Huang Y, He T (2021) Genome-wide analysis of the NRAMP gene family in potato (*Solanum tuberosum*): Identification, expression analysis and response to five heavy metals stress. Ecotoxicology and Environmental Safety 208: 111661

Tian W, Huang Y, Li D, Meng L, He T, He G (2022) Identification of *StAP2/ERF* genes of potato (*Solanum tuberosum*) and their multiple functions in detoxification and accumulation of cadmium in yest: Implication for Genetic-based phytoremediation. Science of The Total Environment 810: 152322

Tiwari S, Lata C (2018) Heavy Metal Stress, Signaling, and Tolerance Due to Plant-Associated Microbes: An Overview. Frontiers in Plant Science 9: 452

Tomas M, Pagani M, Andreo C, Capdevila M, Atrian S, Bofill R (2015) Sunflower metallothionein family characterisation. Study of the Zn (II)-and Cd (II)-binding abilities of the HaMT1 and HaMT2 isoforms. Journal of Inorganic Biochemistry 148: 35–48

Van De Mortel JE, Schat H, Moerland PD, Van Themaat EVL, Van Der Ent S, Blankestijn H, Ghandilyan A, Tsiatsiani S, Aarts MG (2008) Expression differences for genes involved in lignin, glutathione and sulphate metabolism in response to cadmium in *Arabidopsis thaliana* and the related Zn/Cd-hyperaccumulator *Thlaspi caerulescens*. Plant, Cell & Environment 31: 301–324

Vatansever R, Özyiğit İİ, Filiz E (2016) Comparative and phylogenetic analysis of zinc transporter genes/proteins in plants. Turkish Journal of Biology 40: 600–611

Viehweger K (2014) How plants cope with heavy metals. Botanical Studies 55: 1–12

Vinocur B, Altman A (2005) Recent advances in engineering plant tolerance to abiotic stress: achievements and limitations. Current Opinion in Biotechnology 16: 123–132

Wang W, Vinocur B, Shoseyov O, Altman A (2004) Role of plant heat-shock proteins and molecular chaperones in the abiotic stress response. Trends in Plant Science 9: 244–252

Wang Y, Yu K-F, Poysa V, Shi C, Zhou Y-H (2012) A Single Point Mutation in GmHMA3 Affects Cadimum (Cd) Translocation and Accumulation in Soybean Seeds. Molecular Plant 5: 1154–1156

Wolfgang M (2000) The Function of Zinc Metallothionein: A Link between Cellular Zinc and Redox State. The Journal of Nutrition 130: 1455S–1458S

Yang J, Wang Y, Liu G, Yang C, Li C (2011) *Tamarix hispida* metallothionein-like *ThMT3*, a reactive oxygen species scavenger, increases tolerance against Cd^{2+}, Zn^{2+}, Cu^{2+}, and NaCl in transgenic yeast. Molecular Biology Reports 38: 1567–1574

Yang YH, Wang CJ, Li RF, Yi YJ, Zeng L, Yang H, Zhang CF, Song KY, Guo SJ (2021) Transcriptome-based identification and expression characterization of RgABCC transporters in *Rehmannia glutinosa*. Plos one 16: e0253188

Yang Z, Chu C (2011) Towards understanding plant response to heavy metal stress. *In* A Shanker, B Venkateswarlu, eds, Abiotic Stress in Plants - Mechanisms and Adaptations, Vol 10. InTech, Rijeka, pp. 59–78

Yao X, Cai Y, Yu D, Liang G (2018) bHLH104 confers tolerance to cadmium stress in *Arabidopsis thaliana*. Journal of Integrative Plant Biology 60: 691–702

Yuan J, Bai Y, Chao Y, Sun X, He C, Liang X, Xie L, Han L (2018) Genome-wide analysis reveals four key transcription factors associated with cadmium stress in creeping bentgrass (*Agrostis stolonifera* L.). PeerJ 6: e5191

Zaikina EA, Rumyantsev SD, Sarvarova ER, Kuluev BR (2019) Transcription factor genes involved in plant response to abiotic stress factors. Ecological genetics 17: 47–58

Zhan H, Li C-Z, Kang Y, Yu X-Z (2024) The Role of Metal Tolerance Proteins (MTPs) Associated with the Homeostasis of Divalent Mineral Elements in Ga-Treated Rice Plants. Toxics 12: 831

Zhang H, Yang J, Li W, Chen Y, Lu H, Zhao S, Li D, Wei M, Li C (2019b) PuHSFA4a Enhances Tolerance to Excess Zinc by Regulating Reactive Oxygen Species Production and Root Development in *Populus*. Plant Physiology 180: 2254–2271

Zhang P, Wang R, Ju Q, Li W, Tran L-SP, Xu J (2019a) The R2R3-MYB Transcription Factor MYB49 Regulates Cadmium Accumulation. Plant Physiology 180: 529–542

Zhang X, Kong J, Yu L, Wang A, Yang Y, Li X, Wang J (2024) Functional characterization of *Fagopyrum tataricum ZIP* gene family as a metal ion transporter. Frontiers in Plant Science 15: 1373066

A Critical Analysis of Spatio-Temporal Expression Regulation of Genes and Proteins Upon Xenobiotic Exposure in Plants

Rudrakshi Chowdhury, Amit Ghosh, and Subhendu Bandyopadhyay

Abstract Increased anthropogenic activities in the last century have aggravated diverse chemical pollutants in the environment. Among these pollutants, xenobiotics are chemicals or other substances that plants cannot utilize, leading to toxic effects or altering the normal function of metabolic processes. Xenobiotics originate from various sources, such as environmental pollutants, pesticides, herbicides, household cleaning products, and many more. Different xenobiotics affect plants in other ways, and it is now an important area of research to understand coordinated signaling pathways that alter the gene expression and regulation of higher plants. In this respect, gene expressions and the roles of different enzymes for the mobilization and detoxification of xenobiotics must be explored further. Therefore, in this chapter, the various roles of genes and enzymes in the signaling and detoxification processes of xenobiotics in plant systems are discussed in detail.

Keywords Anthropogenic activities · Detoxification · Environmental pollutants · Gene expression · Signaling pathways · Xenobiotics

1 Introduction

Over the years, extensive research on plant sciences, emphasizing especially their adaptations to the diverse environmental conditions in their surroundings, has revealed the prevalence of innumerable mechanisms, enzyme systems, and pathways that are utilized by the plant to either adapt to or combat one of the significant challenges faced on account of uncontrollable expansion due to industrialization, overproduction, and illegitimate release of xenobiotics. The term xenobiotics

R. Chowdhury · A. Ghosh (✉) · S. Bandyopadhyay (✉)
Department of Biological Sciences, School of Life Science and Biotechnology, Adamas University, Kolkata, West Bengal, India
e-mail: rudrakshi.chowdhury@stu.adamasuniversity.ac.in; amit.ghosh1@adamasuniversity.ac.in; subhendu.bandyopadhyay@adamasuniversity.ac.in

S. Roy, V. Mandal (eds.), *Plant-Microbe Interaction under Xenobiotic Exposure*,
https://doi.org/10.1007/978-981-96-8260-7_13

originates from two Greek words, “xenos,” meaning foreign, and “bios,” meaning life. Population growth over the decades, leading to rapid urbanization and demand for technological advancement, has resulted in the profuse exploitation of natural resources (Štefanac et al. 2021a). In addition to this, the industrial revolution of the mid-eighteenth century, ushering in the modern era, introduced, among several other positive impacts, the negative impact of environmental pollution through the illegitimate release of chemical by-products, toxins, etc., thus familiarizing humankind with the term xenobiotics, which Dr. Howard Mason coined in the 1960s (Goel 2021). Scientifically, xenobiotics generally refers to a foreign chemical substance found but not naturally produced in an organism or environment; this may include antibiotics, heavy metals, pesticides, benzene, dyes, pigments, etc., the prevalence of which increases with unregulated industrial activities. However, even naturally occurring chemical compounds (endobiotics) present in abnormal concentrations than usual can be treated as xenobiotics. Generally, the number of xenobiotics found in the environment ranges from ng/L to g/L, and prolonged exposure to even minute doses of xenobiotics can have toxic, mutagenic, carcinogenic, or teratogenic effects on humans and animals over time. The harmful consequences of xenobiotic exposure manifest either through blockage of enzyme active sites or by affecting the enzyme in an allosteric manner, thereby inhibiting the subsequent metabolic reaction. Chlordecone, dichlorodiphenyltrichloroethane (DDT), and dichlorodiphenyldichloroethylene (DDE), like xenobiotics, often bioaccumulate, negatively impacting intracellular signaling pathways, even though low-level chronic exposures (Singh et al. 2023). The fundamental properties contributing to the harmful impacts of the xenobiotics are their physicochemical structures, such as small molecular size, ionizability, water solubility, lipophilicity, polarity, and volatility, making them difficult to identify, quantify, and remove. The complex chemical structures of xenobiotics make it difficult even to degrade them such that their accumulation or partial degradation can lead to threatening consequences.

With every passing decade, as society progresses to uncover greater technological advancement, a parallel expansion in the environmental xenobiotic levels is witnessed, which is high enough to allow the entry of these harmful elements into the food chain, making way for drastic and non-negotiable outcomes in animals and plants alike, contributing to the phenomenon of xenobiotic toxicity. In plants, phytotoxic effects of N-heterocyclic derivatives of polycyclic aromatic hydrocarbons (NPAHs) are more pronounced than their corresponding polycyclic aromatic hydrocarbons like phenanthrene, anthracene, and fluorene. Chemicals like acridine, phenanthridine, and benzo[h]quinoline exhibit strong inhibitory signals on germination and growth of plants. These xenobiotic compounds also induce significant activities of detoxification and antioxidant enzymes like glutathione reductase and glutathione peroxidase at nanomolar to low micromolar concentrations and lipid peroxidation, therefore hinting at a possible occurrence of oxidative stress and consequent Reactive Oxygen Species (ROS) generation in plant metabolism (Pašková et al. 2006).

In recent times, excessive deposits of veterinary antibiotics (VAs) like tetracycline, norfloxacin, chloramphenicol, etc., in agricultural fields due to their profuse

use through wastewater irrigation and manure application are associated with high levels of phytotoxicity adversely affecting seed germination and root elongation in plants like tomatoes, lettuce, carrots, and cucumbers. The physicochemical property that governs the negative impacts of these compounds on seeds is their hydrophobicity, which is evident from their participation in polar necrosis mechanisms in plants (Pan and Chu 2016).

In light of these notable adversities due to xenobiotic toxicity, both plants and animals have been shown to evolve various responses to different measures. For example, in an experiment where bulbs of onion were exposed to increasing concentrations of trichloroethylene for 48 h, the activities of several selected cytochrome P450 isozymes like ethoxyresoruffin-o-deethylase (EROD), pentoxyresorufin-o-deethylase (PROD), and N-nitrosodimethylamine demethylase (NDMA-d) were measured. The observations revealed a 22-fold increase in enzymatic activity of PROD and an 11-fold and ninefold increase of the same for NDMA-d and EROD compared to the control, which indicates that these selected cytochrome-P450 isozymes can act as potent biomarkers in plant systems for assessing TCE pollution (Tabrez and Ahmad 2013) (Fig. 1).

This chapter navigates through the various avenues of novel plant adaptations to anthropogenic stress factors, with special attention to the role of xenobiotics. The discussion will further elaborate on plant responses to xenobiotic toxicity and molecular adaptations, enabling some plant systems to modify and metabolize these xenobiotics chemically.

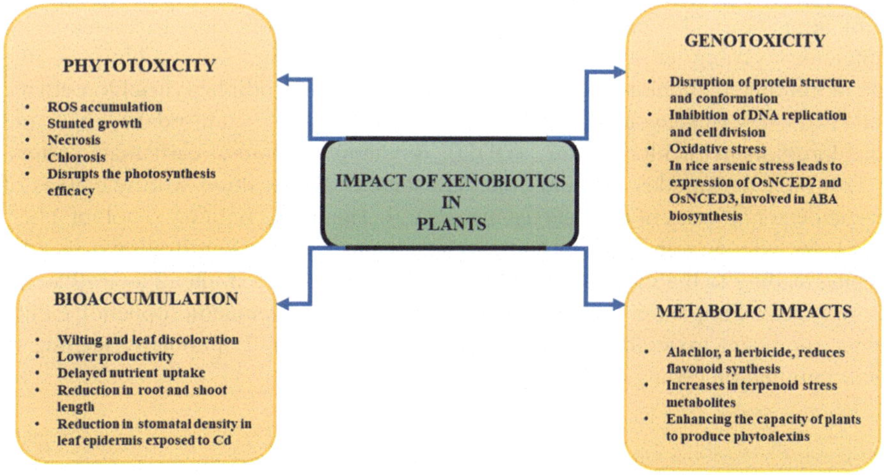

Fig. 1 Impact of xenobiotics on plants: Xenobiotics can really throw plants off balance. They lead to things like poisoning, accumulation of harmful substances, genetic damage, and a general mess in their metabolism. These foreign substances stress plants out, stunt their growth, tweak their internal processes, and even damage their DNA, ultimately impacting their health and how well they produce

2 Sources of Xenobiotics in Plant System

Countless instances of human negligence and lack of awareness concerning rising incidences of pollution, improper disposal of industrial chemical wastes, and indiscriminate use of drugs introduce new and harmful xenobiotics to the environment regularly without being checked. Pollution occurs as artificially manufactured compounds cannot be degraded very easily and rapidly by natural processes in the environment. To date, no absolute artificial intervention has been proven to be effective in degrading these compounds (Singh et al. 2016). Prolonged persistence of these compounds in nature, air, soil, and especially in water through ineffective wastewater treatment has resulted in increased exposure and uptake of these xenobiotics in the plant system through roots and other routes, therefore serving as potential sources of contamination. Further investigations into the plausible sources of xenobiotic contaminations in the plant system have led to the classification of these sources into the below-mentioned categories:

2.1 *Agricultural Sources*

In an attempt to increase agricultural yield and crop production, farmers extensively use chemical fertilizers and pesticides to reduce insect pest attacks. The uncontrolled use of pesticides and their delayed natural elimination has resulted in the bioaccumulation and biomagnification of these chemical xenobiotics in several levels of the major food chains, particularly in plants (Wang et al. 2022b). These contaminating pesticides can be further classified into insecticides, herbicides, and fungicides (Wang et al. 2022b). Insecticides, including organochlorines like Dichlorodiphenyldichloroethylene (DDE), Dichlorodiphenyltrichloroethane (DDTs), Hexacyclochlorohexanes (HCHs), chlordane, and Dichlorodiphenyldichloroethane (DDD), organophosphates, carbamate esters, pyrethroids, acetamides, triazoles, etc., are some of the mention-worthy classes of pesticide xenobiotics of high persistence in soil. The most common xenobiotic herbicide includes atrazine, excessive use of which may lead to its infiltration in healthy plants, leading to the inhibition of photosynthetic processes in them, as well as the target weeds. Fungicides include hexachlorobenzene and pentachlorophenol, exhibiting significant toxicological properties like profound hydrophobicity leading to bioaccumulation (Wang et al. 2022b).

2.2 *Industrial Sources*

In order to foster continuous improvement in the quality of life, several industries, like the pharmaceutical and cosmetic industries, fuel manufacturing industries, and other essential chemical industries, are emphasizing increased production to meet the growing needs of the rising population (Miglani et al. 2022). As a result, the generation rate of xenobiotic compounds and their uncontrolled release in industrial waste has also increased, leading to ecosystem contamination and subsequent contamination of plants and animals along the food chain. Possible types of xenobiotics arising from industrial activities may be classified as follows:

Pharmaceutical residues like antibiotics, amoxicillin, ciprofloxacin, endocrine disruptors, NSAIDS, psychotic drug components, beta-blockers, analgesics, etc., are all classified under pharmaceutically active compounds (PhACs); illicit drugs like hallucinogenic drugs, inhibitors, and stimulants; chemicals used in cosmetics like parabens and azodyes like indigold and azo-dibenzothiophene; food additives like methyl glyoxal and food contaminants like aflatoxin B1; heavy metal contamination like that of arsenic and cadmium in soil and water from mining activities; synthetic polymers like PVCs, etc.; solvents and chemical effluents from fuel and chemical industries like petroleum, hydrocarbons, polychlorinated biphenyls, trinitrotoluene, etc.

2.3 Atmospheric Sources

The release of pollutants from chimneys and industrial vents into the atmosphere contributes to air pollution and the spread of harmful xenobiotics. Some of these pollutants may disperse over long distances as suspended particulate matter. At the same time, the heavier particles may descend and settle on leaf surfaces, where accumulation may reduce the photosynthetic surface area. The xenobiotic pollutants then may enter the plant vascular system through the stomatal pores and travel to different parts and tissues, manifesting varied physiological and molecular consequences. A few examples of these airborne xenobiotic pollutants contaminating plants include polycyclic aromatic hydrocarbons (PAHs), fluorine, benzo[α]pyrene, polychlorinated biphenyls (PCBs), polychlorinated dibenzo-p-dioxins (PCDD), and polychlorinated dibenzofurans (PCDF) (Donner et al. 2010).

2.4 Waste Disposal

Inadequate treatment of sewage and indiscriminate release of waste through canals and other water bodies, as well as runoff from agricultural fields, may contribute to the remainder of xenobiotic pollution and its entry into the root system of the plants. Waste materials like municipal solid waste, household organic waste, inorganic waste, wastes from thermal and electrical power plants, e-waste, and nuclear waste can all be sources of harmful toxicants in plants. Examples may include plasticizers, like dibutyl phthalates (DBPs), phthalates, etc. Fragmentation of these waste products into minute particles through several levels of degradation, as well as transmission in the form of aerosols and chemical solvents through air and water, respectively, may get these waste-generated xenobiotic pollutants absorbed into the plant tissue (Wołejko et al. 2021).

3 Impact of Xenobiotics on Plants

Rising industrialization due to population growth leading to overproduction and indiscriminate release of xenobiotics in soil, water, and air has been overburdening enough on the natural flora, so much so that the impacts on the physiological and morphological aspects of the plant system are becoming non-negotiable. The long-term persistence of chemical and plastic xenobiotics and their corresponding uptake and circulation by the plant root and vascular system have been shown to impose threats on regulatory networks of plant hormonal signaling. It is possible that some of the xenobiotic elements may act as structural analogs to the plant hormones and interfere with their metabolism, transport, and signaling (Serra et al. 2017). Moreover, the emergence of tiny plastic particles called microplastics (less than 5 mm in size) and nanoplastics (between 1 nm and 1 micron in size) and their subsequent spread to deeper layers of the soil and then finally within the plant system by way of roots have resulted in uncountable consequences ranging from pore blockage in the plant cell wall and hindrances in water and nutrient uptake due to microplastic accumulation on plant roots and germinating seed surfaces to reduction in rates of germination, compromised photosynthetic activity, and slow growth. A recent study observed that polystyrene nanoplastics (PS-NPs) affect the physiology of rice seedlings, impeding normal RNA biosynthesis and hindering the plants' responses to external stresses. Polyvinyl chloride-based MPs (PVC-MPs) have been shown to portray an enormous intensity of toxicity compared to PS-MPs, largely impacting metabolism, ionic homeostasis, and growth of crop plants (Roy et al. 2023). Hence, the notable threats to plant systems arising from exposure to the limitless array of chemical toxicants and xenobiotics have been classified here into four sublevels.

3.1 Phytotoxicity

Uptake and circulation followed by the accumulation of various xenobiotic compounds like commercial nanoparticles, microplastics, unusual amounts of heavy metals, fungicides, insecticides, etc., have imposed multitudes of adversities on plant metabolism and growth, endangering its survival and propagation, thereby giving rise to the phenomenon of phytotoxicity. The widespread presence of nanoparticles in sewage sludges with loads up to 1.01–2380 micrograms per kg per year of sludge treated, waste treatment effluent (0.001–15 ng/L/yr), atmospheric emissions from industry and laboratory exhausts, and nano-pesticides and nano-fertilizers has led to frequent exposures of both terrestrial and aquatic plants and plankton. It has been confirmed through a study utilizing 15–21 nm Pbs nanoparticles on maize plants under hydroponic treatments for 15 days that Pbs NPs, after penetrating through the root epidermis, travel through the apoplastic and symplastic pathways to accumulate in maize tissues, including edible kernels, showcasing their phytotoxicity on seed germination, root elongation, biomass of roots and shoots, and root morphology, thereby significantly impacting the quality of crop plants (Ullah et al. 2020). Extensive application of fungicides on plants has also been considered a significant factor in inducing phytotoxicity. Fungicides have been commercially classified into two broad categories, namely contact fungicides like Bordeaux mixture with copper carbonate and systemic fungicides including dicarboximide, vinclozolin, metomeclan, etc., both of which have been shown to drastically impact plant physiology through reduction in plant growth, perturbation of reproductive organ development, and alteration of nitrogen and/or carbon metabolism. Modifications of photosynthetic activity and chlorophyll *a* fluorescence have been observed through physiological studies on plants following fungicide application, demonstrating effects on the light reaction of photosynthesis. For the very first time, it was shown by Bader and Abdel-Basset (Ullah et al. 2020) that triphorin-like fungicides inhibited electron transport reactions of chloroplasts, while triazole and dithiocarbamate contact fungicide application affected the quantum yield of PSII and maximum quantum efficiency of PSII through decreased photochemical quenching. However, the most crucial impact reported was the decreased assimilation of net CO2 and intracellular CO2 concentrations in plants like *Vitis vinifera* after treatment with a nonsystemic fungicide, fludioxonil. Stomatal closures, alterations in stomatal conductance, disruption of RUBISCO carboxylation, decrease of RuBisCO content, and/or reduction of the ribulose 1.5 bisphosphate regeneration are some notable changes ensuring reduced CO2 assimilation (Dias 2012). On the other hand, heavy metal xenobiotics like lead, arsenic, mercury, and copper, which are also predominant environmental pollutants, cause an overwhelming intensity of DNA damage by either reacting directly with the DNA molecule or indirectly opposing DNA repair enzymes by production of ROS and consequent induction of damage like strand breaks (SSBs and DSBs), DNA-protein cross-links, base and sugar lesions, abasic site production, and DNA modifications upon interacting and accumulating in plant systems. All these significant consequences ultimately deteriorate the quality of food crops, projecting a threat on human health (Faisal et al. 2020).

3.2 Genotoxicity

The broad term "genotoxicity" refers to mutagenicity, that is, the induction of mutation in the genetic material, along with damages to cellular components involved in cell division like spindle apparatus, DNA polymerase, etc., and DNA damages, which may be reversed by DNA repair processes or other cellular processes or result in cell death without altering the structure or information content of the surviving cell or its progeny. It has been thoroughly analyzed and reported that some xenobiotics also function as genotoxins in plants, inducing genetic instability through extensive DNA damage. This category includes heavy metals, like zinc, that express their toxic potential through chromosomal damage and mitotic irregularities. It also induces clastogenic and aneugenic effects in root meristems of barley seedlings (Truta et al. 2013). A unique genotoxic feature that has been observed and reported through experimentations is the DNA adduct formations due to overuse of pesticide formulations like hexanal and 4-hydroxy-2-nonenal, pointing toward a probable occurrence of oxidative stress and lipid peroxidation in crop plants. Upon 32P post-labeling studies of isolates of DNA extracted from various crop plants like grapes, bush beans, soybeans, pumpkins, cucumbers, etc., radio isogram analysis revealed elevated adduct levels of different types of adduct products of DNA bases with pesticide molecules or their metabolites in pesticide formulation-treated plants than in the untreated ones (Boerth et al. 2008). In light of scientific progress and technological advancement, scientists are continually synthesizing engineered nanoparticles, which constitute essential components of day-to-day products. Due to extensive demands on the grounds of rising utility, the production of these engineered nanoparticles is also increasing, leading to their consequent indiscriminate release in soil and water, from where these are easily passed to the plant tissues by different uptake pathways through the plant root system. Accumulating these nanoparticles has given rise to extensive damages in different parts and aspects of the plants, which can be confirmed through transcriptomic and proteomic analysis using various experimental plant systems like *Allium cepa* and *Zea mays* (Thienpont et al. 2024). Optical and fluorescence microscopic analysis of the effects of titanium oxide nanoparticles in *Allium cepa* roots, in a dose range of 0–100 mg/L, has shown chromosomal aberrations including chromosomal breaks, chromosome stickiness during metaphase, multiple micronucleus formation, as well as the occurrence of binucleate cells in a dose-dependent manner, and an additional confocal microscopic imaging confirmed the existence of chromosomal bridges alongside a distorted and notched nucleus (Pakrashi et al. 2014). Besides utilizing optical and confocal laser scanning microscopic analysis for determination of chromosomal aberrations, other alternative strategies that are used to confirm genotoxicity are electrophoresis-based methods like comet assay to detect GDNA damages, rapid amplified polymorphic DNA (RAPD) to highlight mutational events due to nanoparticle exposure, and conduction of real-time PCR to observe copy number variation in plastid or mitochondrial DNA (Marmiroli et al. 2022). Experimental utilization of the comet assay for measuring the genotoxic effects of Ni NP (0–2 g/L) on *Solanum lycopersicum* L. seedlings led to the conclusive increase in genomic DNA damage and a notable increase in the number of apoptotic (21.8%) and necrotic (24.0%) cells (Faisal et al. 2013). Another study reported that in *A. thaliana* transgenic lines, exposure to ZnO NPs and their corresponding dissolution to Zn2+ ions led to an increase in homologous recombination (particularly in the *Rad54-GFP-*GUS gene expression) and a reduction in transcriptional gene silencing in leaves at the root level (Faisal et al. 2013).

3.3 Bioaccumulation

The term bioaccumulation refers to the progressive increase in the concentration of an exogenous substance like xenobiotics, including pesticides, drugs, heavy metals, etc., in an organism, in this case, plants or parts of that organism like a specific plant tissue, due to an increased rate of natural uptake that is able to overcome the rates of elimination of that compound from the organism. Bioaccumulation of chemical pollutants and intoxicants in different strata of the various ecosystem types has often shown signs of deterioration in that stratum and the further strata ahead. Hence, a natural consequence of bioaccumulation at a particular trophic level of a food chain may often lead to biomagnification down the food chain, also known as food web biomagnification. It is the bioaccumulation of xenobiotic elements, after all, due to their rapid uptake through roots, that is held responsible for the innumerable biochemical, metabolic as well as genotoxic effects in plants, ultimately contributing to immense phytotoxicity (Maurya and Malik 2016). Exquisite detrimental ecological consequences have been reported on account of bioaccumulation of heavy metals and metalloids like arsenic, chromium, and lead in plants, which seems to impose its impacts by disrupting supportive soil composition and altering the behavior of the surrounding rhizosphere soil microbial community. High concentrations of heavy metals have been projected to reduce leaf growth, leading to its wilting and change in coloration to bluish-purple. Due to initial exposure to roots in the soil, heavy metals are frequently associated with impediments in root extension and proliferation. On the other hand, high phosphate levels in plant roots lead to increased heavy metal tolerance and hyperaccumulation. Some other notable impacts of bioaccumulation of heavy metals include alteration of biochemical pathways, delayed nutrient uptake, and eventual stunted plant growth and low productivity in terms of biomass. Together with heavy metals, microplastics, which are considered ubiquitous contaminants owing to their continuous discharge and inadequate management, make way for longstanding adverse effects in plant physiology and metabolism, like reduction in photosynthetic activity, chlorophyll content, root and shoot length, and induction of ROS (reactive oxygen species). It has even been observed that polyethylene facilitates the bioaccumulation of copper and lead in rapeseed plants; hence the impacts of exposure and bioaccumulation of heavy metals and microplastics in plants are quite intertwined with one another (Kumar et al. 2022). Successful experimentation with *Brassica napas* (oilseed rape) grown in reconstituted soil contaminated with cadmium resulted in the observation of chlorosis in plants. It was suggested through this experiment that interference of Cd with chloroplast replication and cell division, leading to a reduction in the number of chloroplasts per cell and a vivid change in its size, attributed to a consequent decrease in chloroplast density, leads to leaf chlorosis, and not the direct interaction of Cd with the chlorophyll biosynthetic pathway. Further studies, including depth profiling of photosynthetic pigments by phase-resolved photoacoustic spectroscopy, showed a decrease in pigment content that was more profound at the leaf surface (stomatal guard cells) compared to the leaf interior (at the mesophyll cells) in response to Cd. This data was found in accordance with the light transmission and fluorescence microscopic analyses that reported a reduction in stomatal density in the epidermis of Cd-exposed leaves. As a result, estimations from gas exchange measurements also revealed a strong reduction in stomatal conductance in the presence of Cd (Baryla et al. 2001). Through these above-mentioned ways, and probably many other uncountable aspects, bioaccumulation of the most prevalent xenobiotics in nature impacts and damages the plant systems, thereby also indirectly impacting members of the higher trophic levels.

4 Plant Defense Mechanisms Against Xenobiotics

Xenobiotic medications and other environmental pollutants can infiltrate plant organisms, posing a risk to them. Because of this, plants have developed detoxification mechanisms to protect themselves from potentially harmful substances. These mechanisms work by activating the compounds after they are absorbed, allowing certain functional groups to couple with hydrophilic molecules. The tonoplast transporters identify the resultant conjugates, which are then either moved to the apoplast or sequestered into vacuoles (Puglisi et al. 2013; Koschmieder et al. 2021). The functions of xenobiotic-metabolizing enzymes are essential for the impact of xenobiotics on exposed organisms as well as on the environment at large. They aid in the detoxification of possibly hazardous xenobiotics.

In most cases, the physical-chemical characteristics of the parent material and its metabolite or conjugate are distinct, as are their biological activity, which includes toxicity and actions toward living things. The presence and activity of enzymes that break down xenobiotics dictate how many xenobiotics are digested. Therefore, the biological effects of both positive and negative xenobiotics are influenced mainly by the enzymes that break down xenobiotics. An organism may express more xenobiotic-metabolizing enzymes to protect itself from chemical stress if it has more significant contact with xenobiotics. Both positive and negative effects may result from this action. In plants, higher activity of the enzymes that break down xenobiotics causes the xenobiotics to deactivate more quickly. This can lead to better detoxification of environmental pollutants as well as herbicide resistance (Powles and Yu 2010; Dai et al. 2024).

The biotransformation and transportation of xenobiotics are often included in their metabolism. There are two stages in the biotransformation of xenobiotics. Phase I of biotransformation is characterized by xenobiotic oxidation, reduction, or hydrolysis. Reactive and hydrophilic groups are added to or revealed in the xenobiotics' structures in this stage. Xenobiotics or their phase I metabolites may conjugate with endogenous substances in phase II. In plants, glutathione and saccharides are the primary conjugation agents. The current consensus is that an essential component of xenobiotic metabolism is the active transport of substrates, metabolites, or conjugates across membranes via the action of certain protein transporters. Table 1 provides an overview of the primary enzymes in plants that metabolize xenobiotics.

Table 1 Xenobiotic-metabolizing enzymes

Xenobiotic metabolism	Enzymes	Reference
Biotransformation (Phase I)	Cytochromes P450	Deng et al. (2022)
	Peroxidases	Dubrovskaya et al. (2014)
	Ascorbate peroxidases (peroxidases I-class)	Ramel et al. (2012) and Bártíková et al. (2015)
	Peroxidases III-class	Kidwai et al. (2020)
	Alcohol dehydrogenases	Hansen and Shane (2019)
	Aldehyde dehydrogenases	Sauvêtre et al. (2020)
	Short-chain dehydrogenases/ reductases	Testa (2008) and Moummou et al. (2012)
	Aldo-keto reductase	Och (2019)
	Carboxylesterases	Gershater and Edwards (2007) and Lin et al. (2017)
	Epoxide hydrolases	Och (2019)
Biotransformation (Phase II)	Glutathione S-transferases	Sun et al. (2019)
	Glycosyltransferases	Dimunová et al. (2022)
	Malonyltransferases	Sauvêtre et al. (2020)
	Sulfotransferases	Hirschmann et al. (2014)
	Methyltransferases	Sun et al. (2019)
Transporters	ABC transporters	Zhang and Yang (2021a)
	MATE transporters	Tiwari et al. (2014)

4.1 Phase I Enzymes

The most significant enzymes involved in phase I xenobiotic metabolism are cytochromes P450 (CYPs), heme proteins found in all living things. Through a hydrogen abstraction–oxygen rebound process, CYPs catalyze the scission of the O–O bond of dioxygen molecules, allowing a single oxygen atom to be inserted into the inactivated substrate. This process is employed in the production of several endogenous chemicals as well as the detoxification of xenobiotics (Esteves et al. 2021). One of the biggest families of plant proteins is made up of CYPs, which are categorized into several families and subfamilies according to the sequence homology of their genes. Thousands of distinct CYPs have been found (Hansen et al. 2021; Chakraborty et al. 2023a). The intricate and adaptable chemistry that higher plants have developed to synthesize a wide range of natural chemicals that allow plants to interact with one another, attract pollinators, and repel diseases and herbivores is undoubtedly connected to the evolution of this enormous superfamily of enzymes.

In addition to their usual function as monooxygenases, CYPs can also function as reductases or peroxidases, and they can catalyze hydroxylations, epoxidations, dealkylations, decarboxylations, and isomerizations of substrates. All of these reactions are essential for the synthesis of the enormous variety of natural compounds found in plants and are also crucial for the metabolism of xenobiotics (Štefanac et al. 2021b). Plant CYP enzyme identification may aid in the development of herbicide-resistant plants or plants for phytoremediation, although comprehensive information about the metabolism of xenobiotics by plant CYPs has not yet been sufficiently provided (Pandian et al. 2020).

Other significant enzymes engaged in the first step of xenobiotic metabolism in plants are peroxidases. These heme-containing proteins catalyze the redox reaction between hydrogen peroxide and different substrates, which exhibit a broad structural variety. Numerous cellular activities in plants, including development and stress responses, are mediated by peroxidases. Three classes—I, II, and III—of the peroxidase superfamily may be distinguished based on variations in primary structure. Plants, bacteria, and yeast all have internal enzymes called class I peroxidases, sometimes called ascorbate peroxidases. Class III peroxidases are the conventional peroxidases of higher plants, whereas class II peroxidases are often extracellular fungus peroxidases, which defines them as secretory peroxidases (Pandey et al. 2017).

Alcohol dehydrogenases (ADHs) and aldehyde dehydrogenases (ALDHs) oxidize alcohols and aldehydes, both xenobiotic and eobiotic. We know a great deal about the structure and catalytic activity of animal ADHs, which helps us comprehend the structure and function of plant ADHs. There are just a few tiny ADH families identified in plants, and these family members contribute to the creation of distinct aromas that draw in animals, act as pollinators or seed dispersers, and provide defense against herbivores. ALDHs are far more significant enzymes in the metabolism of xenobiotics than ADHs (Tola et al. 2020).

In plants, a variety of xenobiotics, including amides, epoxides, and esters, go through hydrolytic biotransformation. These xenobiotics can be hydrolyzed by a number of hydrolases, including peptidases, carboxyl esterases, and epoxide hydrolases, whose main role is in endogenous metabolism (Morisseau 2022). Carboxylesterases accelerate the hydrolysis of natural and xenobiotic compounds, significantly changing their biological activity. There are a lot of proteins that have carboxylesterase activity in plants (Ghodke and Punekar 2022). In addition to the several additional hydrolases that may break amide and thioester linkages, classical plant carboxylesterases are classified into three different superfamilies (Wang et al. 2022a).

4.2 Phase II Enzymes

Phase II biotransformation is used to combine some endogenous molecules with the majority of metabolites generated in phase I and xenobiotics possessing suitable functional groups. A xenobiotic molecule becomes more polar and less active by conjugation, which adds hydrophilic ionizable functional groups.

Enzymes from the superfamily of glutathione S-transferases (GSTs) facilitate the conjugation of xenobiotics with glutathione. Eight classes comprise the GST superfamily in plants, seven of which are soluble, and one is microsomal. These enzymes have been well established as phase II detoxification enzymes that carry out several other crucial tasks in plant growth and development since their discovery in plants in 1970. Most GSTs function as dimers and have a variety of catalytic and noncatalytic functions in endogenous metabolism, oxidative damage prevention, and xenobiotic detoxification (Estévez and Hernández 2020). The roles that plant GSTs play in herbicide tolerance have garnered significant interest (Li et al. 2017). The primary function of plant-specific Tau-class GSTs is to aid in the metabolism of herbicides. Numerous investigations revealed that Tau-class GSTs are involved in the metabolism of many xenobiotics, including herbicides, in various plant species (Martínez-Márquez et al. 2017).

Glycosyltransferases (UGTs) are important plant enzymes that help detoxify toxic xenobiotics such as organochlorines from wastewater. Furthermore, a notable upregulation of UGT in the leaves of *Phragmites australis* plants subjected to sublethal quantities of organochlorines was noted, suggesting a crucial function for UGTs in the plant's efficient response to chemical stress (Gharabli et al. 2023). Elevated expression of these UGTs may enhance plant protection, as some can deactivate specific mycotoxins (De Zutter et al. 2016).

Despite their identification, a limited number of plant sulfotransferases (STs) have been thoroughly described (Faraji et al. 2021). According to Hirschmann et al. (2014), 21 genes in the completely sequenced genome of Arabidopsis thaliana are probably responsible for producing ST proteins. One of them was described as a protein that catalyzes the sulfation of 12-hydroxyjasmonate, which inactivates plants' excess jasmonic acid (Faraji et al. 2021).

In both xenobiotics' metabolism and natural compounds' biosynthesis, adding a methyl moiety to a minor molecule is a frequent change. S-adenosyl-L-methionine (SAM)-dependent methyltransferases (MTs) are most frequently responsible for catalyzing these methylation processes. The electron-rich, methyl-accepting atom—typically O, N, C, or S—is used to classify MTs. Many structurally different substrates, such as cofactors, pigments, prosthetic groups, signaling and host defense chemicals, and cell membrane and cell wall components, are modified by MTs. Notably, methylation affects acceptor molecule bioavailability, bioactivity, and reactivity, and as a result, it plays a crucial part in the functional outcome of several endogenous metabolic processes. Furthermore, plants' ability to detoxify different xenobiotics depends critically on their methylation (Hansen and Shane 2019; Zhang and Yang 2021b). There is currently no concrete evidence to support the theory that other enzymes, such as N- or O-acetyltransferases, which catalyze the production of secondary plant metabolites, may also play a role in the detoxification of xenobiotics in plants.

4.3 Transporters of Xenobiotics

Numerous metabolites, conjugates, and xenobiotics are impervious to lipid membranes. Because of this, they need to be carried by specific transport proteins known as drug transporters. Due to their significant therapeutic importance, these proteins have been well investigated in bacteria, yeast, and mammalian cells; however, although plant genomes include a substantial number of genes encoding potential drug transporters, these proteins have not been thoroughly researched in plant systems. Drug transporters may carry a variety of structurally different substances and usually have broad substrate specificities. Plant drug transporters have been linked to the transfer of several endogenous substances, such as auxins, flavonoids, glutathione conjugates, metal chelators, and xenobiotics (Bártíková et al. 2015). Particular attention has been paid to two protein families involved in transporting a broad variety of chemical substances: ATP-binding cassette (ABC) transporters and multidrug and toxins extrusion (MATE).

ABC transporters are a large protein superfamily involved in detoxification processes. These glycoproteins consist of two transmembrane and two nucleotide-binding domains and are divided into various families and subfamilies with diverse functions (Wu et al. 2019). In *Arabidopsis thaliana*, over 100 ABC transporters participate in xenobiotic metabolism, chlorophyll biosynthesis, Fe/S cluster formation, stomatal movement, and plant growth. Some transporters, like those in the ABCC subfamily, originally known as multidrug resistance-related proteins (MRPs), were initially identified as vacuolar pumps but are now understood to play roles in responses to pathogen stress. The ABCG subfamily is unique to fungi and plants and is involved in detoxification, water loss prevention, and the transport of phytohormones and secondary metabolites. Due to their role in regulating the transport and storage of xenobiotics, ABC transporters have potential applications in phytoremediation strategies (Gräfe and Schmitt 2021).

MATE transporters are highly conserved secondary active transporters that use electrochemical gradients to transport substrates. Though their functions in plants are not fully understood, they play key roles in metal detoxification and phenol metabolism by storing or extruding these compounds at the plasma membrane. MATE transporters help maintain homeostasis by excreting metabolic waste and xenobiotics. They are expressed in developing and reproductive organs and are localized to the plasma membrane. MATE transporters influence plant growth, development, and disease resistance, with two characterized members, MATE1 and MATE2, showing differences in substrate selectivity. MATE2, in particular, is associated with malonyltransferases, and its loss leads to reduced flavonoid levels, affecting leaf and flower pigmentation. Despite the importance of MATE transporters in drug metabolism in mammals, their role in plant drug transport remains unexplored (Upadhyay et al. 2019; Radhakrishnan et al. 2024).

5 Impact of Xenobiotic on Plant Signaling

5.1 Stress Responses

Xenobiotic metabolism can trigger stress signaling pathways, producing ROS and activating stress-related genes. Various xenobiotics are known to induce ROS production, with important biochemical consequences. However, ROS can function as cellular second messengers that are likely to modulate many different genes and proteins, thus leading to various responses (Martin et al. 2022). ROS produced by membrane-bound NAD(P)H oxidases in plant cells activates plasma membrane calcium channels, further increasing cytosolic calcium (Mori and Schroeder 2004; Ravi et al. 2023). This increase activates downstream signaling involving the MAPK pathway, with ROS seemingly exerting a direct and central role in the induction and stabilization of the MAPK cascade (Jagodzik et al. 2018). Thus, the vast network of MAPKs is involved in relaying hydrogen peroxide signals to their targets, which consist of various transcription factors. Therefore, the involvement of ROS-mediated signaling in plant responses to abiotic stress is well established. Researchers have characterized the importance of heat-shock transcription factors in the transduction of ROS signaling and their potential role as direct hydrogen peroxide sensors (Verma et al. 2023). Several genes seem to sense and transduce hydrogen peroxide signals to plants' downstream components and potentially work as a primary ROS sensor.

Disruption of photosynthesis by some xenobiotics leads to singlet oxygen production (Sachdev et al. 2021), which has a very short half-life, powerfully oxidizes various targets, including carotenoids, and triggers significant cell death processes. Cell death associated with 1O2-induced oxidative stress in Arabidopsis has been ascribed to activating a genetic program requiring a chloroplast protein named EXECUTER1, rather than physicochemical damage. This protein interacts with the EXECUTER2 modulator to transfer 1O2-derived plastid signals to the nucleus (Lee et al. 2007). However, the comparison of different transcriptomic studies performed in Arabidopsis plants (Xu et al. 2015; Baek et al. 2019) and the npq1lut2 mutant (Baek et al. 2019) reflects the complexity of CO_2 signaling pathways. Even if all of these situations lead to the production and accumulation of CO2, cellular responses present major differences with the activation of different gene networks, thus highlighting the specificities of xenobiotics and stresses.

5.2 *Hormonal Balance*

Xenobiotics can mimic or interfere with plant hormones like auxins, cytokinins, and gibberellins, affecting growth and development. Plant CYP enzymes tightly regulate the biosynthesis and catabolism of plant hormones like auxins, gibberellins, cytokinins, and abscisic acid. These enzymes' hydroxylation, epoxidation, and oxidative cleavage of hormones can alter their bioactivity or functions. Recent developments in molecular genetic analysis led to identifying and characterizing genes involved in producing plant hormones. In Arabidopsis, CYP79B2/B3 and CYP83 control the production of auxin. Gibberellin metabolism involves several CYPs in different plants, including CYP88A and CYP714A in Arabidopsis and CYP701A8 and CYP714B in rice (Jiu et al. 2020). Moreover, the Arabidopsis CYP735A1 and CYP735A2 genes code for cytokinin hydroxylases that catalyze trans-zeatins, and the CYP707A gene codes for an enzyme called ABA 80-hydroxylase that regulates the amount of abscisic acid in Arabidopsis and barley. The polyhydroxysteroid class, known as brassinosteroids (BRs), has been recognized as the sixth class of plant hormones. Numerous CYPs control BR biosynthesis; for instance, AtCYP72B1 and AtCYP72C1 control photomorphogenesis and plant steroid signal transduction by deactivating BR, and AtCYP85A controls the production of BR C-6 oxidase. AtCYP90A1 and AtCYP90B1 enhance vegetative growth and seed yield by regulating the biosynthesis of BRs. On the other hand, CYP90B3 and CYP724B2 catalyze the initial C-22 hydroxylation reactions in the BR biosynthesis pathway of tomatoes (Bak et al. 2011). Additionally, experiments have found that in rice, ABA biosynthesis is triggered due to the expression of OsNCED2 and OsNCED3 in response to arsenic stress. (Dutta et al. 2018).

5.3 *Defense Mechanisms*

Plants may activate defense pathways, such as the production of secondary metabolites, to counteract the effects of xenobiotics. Eukaryotes contain a variety of cytochrome P450 monooxygenases comprising a vast and diversified class of enzymes with more than 35,000 members (Nelson et al. 2004). CYPs participate in various biosynthetic and xenobiotic processes, including incorporating carbon sources, xenobiotic detoxification, and creating secondary metabolites (Nelson 2009). The CYP superfamily is one of the most extensive gene families in plants; it has 245 genes in Arabidopsis, 332 full-length genes, and 378 pseudogenes in soybean (Guttikonda et al. 2010). CYPs play a pivotal role in a wide range of cellular functions that profoundly impact plant growth and development. CYPs synthesize secondary metabolites, including terpenoids, flavonoids, steroids, alkaloids, phenylpropanoids, glucosinolate, and cyanogenic glycosides. Plant CYP is critical in numerous metabolic processes and can bind to various biological compounds. As a result of these reactions, different fatty acid conjugates are produced, plant hormones, secondary metabolites, lignin, and many protective chemicals (Tang et al. 2016).

5.4 *Gene Expression*

The presence of xenobiotics can lead to changes in the expression of genes involved in metabolism, stress responses, and signaling pathways. Long and low herbicide exposure also results in rapid evolution of herbicide resistance for exposed populations, as demonstrated by for Lolium rigidum in the presence of diclofop-methyl. At the molecular level, Serra et al. (2015) demonstrated by genome-wide expression profiling that five commercial herbicide formulations at concentrations producing a 50% reduction in shoot dry weight (EC50, sublethal levels) specifically affected the expression of genes related to ribosome biogenesis and translation, secondary metabolism, cell wall modification, and growth. Another study showed that complex mechanisms of low-intensity herbicide stress responses may exist (Aliferis and Chrysayi-Tokousbalides 2011). Complex mixtures of xenobiotics at realistic environmental levels produced different effects than those caused by compounds alone or by simply adding effects (Lydy et al. 2004). It has also been found that the proportions of unigenes and reads related to "response to salicylic acid," "MAPK cascade," "intracellular protein kinase cascade," and "regulation of hydrogen peroxide metabolic process" were negatively affected by all the chemical stressors. The Differential Expression Seq analysis in another study revealed that many xenobiotic-responsive genes are involved in signaling and correspond to signal transduction through protein kinases and protein phosphatases. This was the case for the CIPK9-annotated gene, which encodes a Calcineurin B-like Protein (CBL)-interacting protein kinase highly conserved between monocot and dicot species. This serine/threonine protein kinase annotated unigene, also known as sucrose non-fermenting (SNF)-related kinase family of serine/threonine kinase 3.12 (SNRK3.12), as well as a calcium-sensing receptor-annotated transcript, were found to be differentially expressed. Transcriptomic and signaling effects of xenobiotics in plants encompass coordinated gene networks and defense responses but also incomplete responses and even deleterious and cell death responses (Lehotai et al. 2012). Table 2 provides a detailed description of the plethora of genes involved in xenobiotic metabolism and description.

Table 2 List of some of the genes involved in the metabolism and detoxification of xenobiotics in plants

Plant species	Name of genes	Functions	Useful traits	Reference
Arabidopsis thaliana	CYP74A1	Converts 13-hydroperoxylinolenic acid (13-HPOT) to plant defence hormone jasmonic acid by allene oxide synthase (AOS)	Xenobiotic tolerance and metabolism	Chakraborty et al. (2023b)
Zea mays	CYP81A9	Metabolism of the acetolactate synthase (ALS) inhibitor herbicide nicosulfuron	Herbicide resistance	Chakraborty et al. (2023b)
Zea mays	CYP88A	Synthesis of Gibberellins that integrate multiple hormone signaling pathways in response to heavy metal toxicity stress	Heavy metal resistance	Chakraborty et al. (2023b)
Arabidopsis thaliana	GST-U24 and GST-U25	Overexpression leads to enhance ability to withstand and detoxify TNT	Xenobiotic resistance	Hernández Estévez and Rodríguez Hernández (2020)

(continued)

Table 2 (continued)

Plant species	Name of genes	Functions	Useful traits	Reference
Oryza sativa	OsGSTL2	It is an L-type GST. Expression shows high tolerance to glyphosate and chlorsulfuron, which often contaminate agricultural fields, and reduced superoxide levels with simultaneous GST and GPX expression.	Herbicide tolerance	Hernández Estévez and Rodríguez Hernández (2020)
Arabidopsis thaliana *Zea mays* *Carex rigescens*	UGT79B2 and UGT79B3 UGT78D2 UGT87A1	Involved in secondary metabolism via glycosylation of flavonoids, a kind of natural polyphenolic secondary metabolite with antioxidant and free radical scavenging functions in different tissues	Xenobiotic metabolism and ROS control	([CSL STYLE ERROR: reference with no printed form.])
Echinochlo colona	C-class EcABCC8 (ABC) transporter	Active efflux of glyophosphate out of plant cells resulting in herbicide resistance	Herbicide resistance	Goldberg-Cavalleri et al. (2023)
Cajanus cajan	CYP75B165	Biosynthesis of flavonoids		Chakraborty et al. (2023b)
Panax ginseng	CYP71	Biosynthesis of secondary metabolites, flavonoids and alkaloids	Heavy metal stress resistance	Chakraborty et al. (2023b)
Pisum sativum	CYP88A	Biosynthesis of gibberellin	Heavy metal stress tolerance	Chakraborty et al. (2023b)
Arabidopsis thaliana	AtPRX71 (Class III Peroxidases)	Protects plants from oxidative stress by scavenging ROS	Heavy metal resistance	Raggi et al. (2015)
Arabidopsis thaliana	AtPRX33 (Class III Peroxidases)	Involved in cell wall metabolism, particularly lipid biosynthesis.		Shigeto and Tsutsumi (2016)
Arabidopsis thaliana	At2g37770 (aldo-keto reductase) At1g54870 and At3g04000 (aldehyde reductase)	Reduction of an aldehyde group of saturated aldehydes and methylglyoxal as well as α,β-unsaturated aldehydes in chloroplasts	Detoxification of reactive carbonyls	Liu et al. (2024)

(continued)

Table 2 (continued)

Plant species	Name of genes	Functions	Useful traits	Reference
P. halotolerans ANSP101	AKR4 (NADPH dependent aldo/keto reductase)	Involved in the conversion of 3-keto-DON into 3-*epi*-DON (deoxynivalenol)	Xenobiotic biotransformation into less toxic forms	Liu et al. (2024)
Nicotiana tabacum	TrxR1, TrxR2 (thioredoxin reductase)	Catalyzes the reduction of thioredoxins, which are small proteins involved in various redox reactions	ROS detoxification in xenobiotic metabolism	([CSL STYLE ERROR: reference with no printed form.])
Oryza sativa	Cinnamate 4-hydroxylase (C4H), 4-coumarate-CoA ligase (4CL), caffeic acid O-methyltransferase (COMT)	Involved in immobilizing and detoxifying certain xenobiotics by incorporating them into the lignin structure	Xenobiotic detoxification	Alber and Ehlting (2012)

6 Future Directions

The scientific community has focused on discovering methods to detoxify contaminated plants and mitigate the harmful effects of xenobiotic pollutants. Research is centered on enabling plants to either convert toxic substances into less harmful forms or eliminate them altogether, potentially through genetic engineering. Successful research is being developed into sustainable practices for environmental protection.

6.1 *Phytoremediation*

Phytoremediation is a more specific and emphasized sub-branch of bioremediation, which refers to the use of any biological system, plant, or microorganism to treat soil, air, and water to reduce or eliminate the noxious impacts of pollutants on the environment. Therefore, phytoremediation only specifically concentrates on using plants as the ultimate means of detoxification. Conventional techniques of remediation are associated with excessive amounts of waste generation and side effects, and therefore, phytoremediation emerges as an essential alternative remediation strategy, an important application of green technology, which has been found to be eco-friendly, generating limited secondary waste and cost-effective (Singh et al. 2022). Some plant species have been shown to possess high absorption capacity and are able to accumulate large amounts of chemical compounds taken up from the soil, together with capacity for high biomass production and tolerance to toxic effects of metals and other xenobiotics without showing symptoms of phytotoxicity. These plants serve as the ideal candidates exhibiting potential for phytoremediation and are called hyperaccumulator plants. Such plants are also easily cultivable, extremely resistant to adverse environmental conditions, and, above all, should not be attractive for herbivory. Sometimes scientific interventions also aid in increasing the overall phytoremediation potential through soil amendments using biochar, ethylene diamine tetraacetic acid (EDTA), application of endophytic bacteria, arbuscular mycorrhiza, and even through usage of specialized transgenic plants with genes that support effective phytoremediation (Kafle et al. 2022). The most important criteria that should be emphasized while defining hyperaccumulator plants are its shoot-to-root ratio for heavy metals, which should be greater than 1, indicating great transport ability; the shoot-to-soil ratio being greater than 1, pointing to a good uptake capacity; and finally, the metal concentration optimum in the shoots that is different for different metals. The strategies that are being heavily practiced, especially for remediation of heavy metal-contaminated soil, aiming at the eradication of heavy metal toxicity from soil include the following.

6.2 Phytostabilization

It refers to the underground immobilization of heavy metals to reduce their bioavailability and spread to the higher trophic levels of the food chains in the ecosystem by using metal-resistant plant varieties. This process can be carried out through the precipitation of heavy metals or reduction in metal valence in the rhizosphere, absorption and sequestration within root tissues, or adsorption onto root cell walls. The plant species that are most appropriate for this action have been found to exhibit a dense rooting system (owing to the critical role that is played by the roots in immobilization of the heavy metals in the soil and prevention of soil erosion), excessive biomass production, and rapid growth of dense vegetative cover that ensures stabilization of underground heavy metals and minimizes their chances of leaching to underground water or dispersion by wind.

6.2.1 Phytoextraction

It is a more permanent strategy than phytostabilization, whereby plants uptake, translocate, and accumulate the heavy metals as biomass in the plant parts above the ground. The several key factors determining the success of phytoextraction involve plant selection, plant performance, heavy metal bioavailability, and soil and rhizosphere properties.

6.2.2 Phytovolatilization

It is the uptake of pollutants from the soil through plant roots, their subsequent conversions to less toxic volatile forms, and their release into the atmosphere via leaf transpiration. Although it is a partial elimination strategy that does not entirely eradicate the pollutant but only transfers it from soil to air, the advantage of this strategy over others is that it doesn't require plant harvesting and disposal. For example, detoxification in the case of inorganic selenium is achieved through its assimilation in plants in the form of organic selenoamino acids selenocysteine (SeCys) and selenomethionine (SeMet), followed by its biomethylation to dimethyl selenide, which is volatile and less toxic, and finally its release into the air. Members of the Brassicaceae family, such as *Brassica juncea*, are known for their good volatilizing capacity for selenium. Phytovolatilization can be applied to the eradication of organic pollutants and heavy metals like Se, Hg, and As.

6.2.3 Phytofiltration

It is a phytoremediation strategy for the removal of pollutants from contaminated surface waters or wastewater through plant roots (rhizofiltration), shoots (caulofiltration), or plant seedlings (blastofiltration). Rhizofiltration-performing plants are known to be initially grown hydroponically in clean water to develop an extensive root system, followed by transfer to polluted water for acclimatization and then a final transfer to the contamination site after acclimatization for remediation. After the roots are saturated with heavy metals, the plants are harvested and disposed of. The notable plant species that are fit to participate in phytofiltration are aquatic species such as hyacinth, *Azolla*, duckweed, cattail, and poplar for remediation in wetland waters and terrestrial plants such as Indian mustard (*B. juncea*) and sunflower (*H. annuus*), which exhibit an extensive system of hairy roots with high bioaccumulation and tolerance capacity for this function.

6.3 Agricultural Practices

Awareness regarding the uncontrolled spread and interactions of chemical xenobiotics with the plants, both terrestrial and aquatic, and the multitudes of adverse impacts it imposes on the healthy functioning plant system, resulting ultimately in the reduction of the overall quality of the crop plants, making them unsuitable for consumption, has led many scientists to come up with novel approaches and strategies that are aimed at the restoration of the yield and the quality of these plants. Due to the rising concerns about chemical pollution, scientists have already started working on developing transgenic plant varieties that show resistance to harmful pesticides, weedicides, radionuclides, heavy metals, etc. Research has revealed the identity of several genes whose products exhibit the capacity to degrade herbicides; the two most important gene products in this regard are the CYP450s and GSTs. Genome sequence analysis has put forward the existence of over 1000 plant P450 sequences, some of which are capable of secondary metabolism and herbicide tolerance. The CYP450 monooxygenases are involved in the monooxygenation, hydroxylations, oxidative dealkylations, and epoxidations of lipophilic herbicides into water-soluble nontoxic or less toxic metabolites so that they can be easily tolerated and eliminated from the plant system. Although many plants, like Arabidopsis, Japonica, and Indian rice varieties, inherently express several P450-mediated oxidations in plant microsomes, like the oxidation of chlorotoluron in maize and wheat, linuron in wheat and maize, atrazine in tulip, and isoproterenol in yam bean, constructing transgenic varieties of crop plants with human CYP450s namely CYP1, CYP2, CYP3, and CYP4 families capable of metabolizing large scores of xenobiotic chemicals, has promised improved herbicide resistance by displaying a broad and overlapping substrate specificity toward lipophilic foreign chemicals, ensuring cleanup of residual agrochemicals. With this goal in mind, scientists have successfully created a transgenic variety of rice with a human gene for CYP1A1, a cytochrome P450 monooxygenase that metabolizes xenobiotic chemicals. The expression of human CYP1A1 in rice under the control of modified CaMV 35S has shown profound herbicide resistance to 10 out of 13 tested herbicides of different chemical families in germination tests, and these herbicides are aryloxyphenoxypropionate (quizalofopethyl), benzamide (isoxaben), carbamate (chlorpropham), dinitroaniline (pendimethalin), oxyacetamide (mefenacet), phosphoamidate (amiprophos-methyl), pyridazinone (norflurazon), thiocarbamate (pyributicarb), and urea (chlorotoluron and diuron). As compared to the non-transgenic Nipponbare plants in the greenhouse, these transgenic varieties showed absolutely normal morphology, like normal height, leaf color, flowering time, fertility, and seed size, with the exceptional feature of resistance, however, against a myriad of herbicides of several categories, like the acetyl-CoA carboxylase-inhibiting herbicide quizalofop-ethyl, very long-chain fatty acid (VLCFA) synthesis-inhibiting herbicide mefenacet, microtubule assembly/organization-inhibiting herbicides amiprophos-methyl, chlorpropham, and pendimethalin, etc., all of which adversely affected the growth of the Nipponbare plants but not the CYP1A1 rice plants, as these plants were able to metabolize during and after germination, thereby maintaining the concentration of herbicides in plant tissues under the lethal threshold, and showcased cross-resistance to several herbicides with different chemical structures and different modes of action. Above all, these transgenic varieties of rice showed greater metabolism of the photosynthesis-inhibiting herbicide chlorotoluron and the carotenoid biosynthesis inhibitor norflurazon via CYP1A1 action, while the growth of the non-transgenic control varieties (although not inhibited during germination upon exposure to these photosynthesis inhibitors, as they might have derived their nutrients from the seeds or possibly the amount of ROS generated wasn't enough to kill the plants) was seriously impacted under non-germination stages (Kawahigashi et al. 2007).

7 Conclusion

Apart from developing transgenic crop varieties, scientists have also made efforts to search for naturally resistant crop varieties that can be used in fields and accessed easily at a much lower cost than transgenic ones. Besides these advancements, considerable labor has been put into revising the conventional agriculture management strategies through the introduction and use of less toxic varieties of chemical fertilizers, shifting to the use of more sustainable biofertilizers, preventing discharge of contaminated agricultural runoff into nearby water bodies by ensuring proper drainage systems, and practicing effective land management strategies, etc. Hence, it implies that the progress that has already been made in this direction and shall continue to be made in the future promises a more pollutant-free, less toxic environment by putting forward more alternatives to the use of harmful chemicals in fields.

References

Alber A, Ehlting J (2012) Chapter 4 – Cytochrome P450s in Lignin Biosynthesis. In: Jouanin L, Lapierre C (eds) Advances in Botanical Research. Academic Press, pp 113–143

Aliferis KA, Chrysayi-Tokousbalides M (2011) Metabolomics in pesticide research and development: review and future perspectives. Metabolomics 7(1):35–53. https://doi.org/10.1007/s11306-010-0231-x

Baek YS, Goodrich LV, Brown PJ, James BT, Moose SP, Lambert KN, Riechers DE (2019) Transcriptome profiling and genome-wide association studies reveal GSTs and other defense genes involved in multiple signaling pathways induced by herbicide safener in grain sorghum. Frontiers in plant science 10:192

Bak S, Beisson F, Bishop G, Hamberger B, Höfer R, Paquette S, Werck-Reichhart D (2011) Cytochromes P450. The Arabidopsis Book/American Society of Plant Biologists 9

Bártíková H, Skálová L, Stuchlíková L, Vokřál I, Vaněk T, Podlipná R (2015) Xenobiotic-metabolizing enzymes in plants and their role in uptake and biotransformation of veterinary drugs in the environment. Drug metabolism reviews 47(3):374–387

Baryla A, Carrier P, Franck F, Coulomb C, Sahut C, Havaux M (2001) Leaf chlorosis in oilseed rape plants (Brassica napus) grown on cadmium-polluted soil: causes and consequences for photosynthesis and growth. Planta 212(5–6):696–709. https://doi.org/10.1007/s004250000439

Boerth DW, Eder E, Stanks JR, Wanek P, Wacker M, Gaulitz S, Skypeck D, Pandolfo D, Yashin M (2008) DNA Adducts as Biomarkers for Oxidative and Genotoxic Stress from Pesticides in Crop Plants. J Agric Food Chem 56(15):6751–6760. https://doi.org/10.1021/jf072816q

Chakraborty P, Biswas A, Dey S, Bhattacharjee T, Chakrabarty S (2023a) Cytochrome P450 gene families: role in plant secondary metabolites production and plant defense. Journal of Xenobiotics 13(3):402–423

Chakraborty P, Biswas A, Dey S, Bhattacharjee T, Chakrabarty S (2023b) Cytochrome P450 Gene Families: Role in Plant Secondary Metabolites Production and Plant Defense. Journal of Xenobiotics 13(3):402–423. https://doi.org/10.3390/jox13030026

Dai Z, Wu Y, Xiong Y, Wu J, Wang M, Sun X, Ding X, Yang L, Sun X, Ge G (2024) CYP1A inhibitors: Recent progress, current challenges, and future perspectives. Med Res Rev 44(1):169–234. https://doi.org/10.1002/med.21982

De Zutter N, Audenaert K, Arroyo-Manzanares N, De Boevre M, Van Poucke C, De Saeger S, Haesaert G, Smagghe G (2016) Aphids transform and detoxify the mycotoxin deoxynivalenol via a type II biotransformation mechanism yet unknown in animals. Scientific reports 6(1):38640

Deng Z, Zhang Y, Gao C, Shen W, Wang S, Ni X, Liu S, Li X (2022) A transposon-introduced G-quadruplex motif is selectively retained and constrained to downregulate *CYP321A1*. Insect Science 29(6):1629–1642. https://doi.org/10.1111/1744-7917.13021

Dias MC (2012) Phytotoxicity: An Overview of the Physiological Responses of Plants Exposed to Fungicides. Journal of Botany 2012:1–4. https://doi.org/10.1155/2012/135479

Dimunová D, Matoušková P, Podlipná R, Boušová I, Skálová L (2022) The role of UDP-glycosyltransferases in xenobioticresistance. Drug Metabolism Reviews 54(3):282–298. https://doi.org/10.1080/03602532.2022.2083632

Donner E, Eriksson E, Holten-Lützhøft H-C, Scholes L, Revitt M, Ledin A (2010) Identifying and Classifying the ssSources and Uses of Xenobiotics in Urban Environments. In: Fatta-Kassinos D, Bester K, Kümmerer K (eds) Xenobiotics in the Urban Water Cycle. Springer Netherlands, Dordrecht, pp 27–50

Dubrovskaya EV, Polikarpova IO, Muratova AYu, Pozdnyakova NN, Chernyshova MP, Turkovskaya OV (2014) Changes in physiological, biochemical, and growth parameters of sorghum in the presence of phenanthrene. Russ J Plant Physiol 61(4):529–536. https://doi.org/10.1134/S1021443714040074

Dutta S, Mitra M, Agarwal P, Mahapatra K, De S, Sett U, Roy S (2018) Oxidative and genotoxic damages in plants in response to heavy metal stress and maintenance of genome stability. Plant Signal Behav 13(8):e1460048. https://doi.org/10.1080/15592324.2018.1460048

Esteves F, Rueff J, Kranendonk M (2021) The central role of cytochrome P450 in xenobiotic metabolism—a brief review on a fascinating enzyme family. Journal of xenobiotics 11(3):94–114

Estévez IH, Hernández MR (2020) Plant glutathione S-transferases: an overview. Plant gene 23:100233

Faisal M, Saquib Q, Alatar AA, Al-Khedhairy AA (eds) (2020) Cellular and Molecular Phytotoxicity of Heavy Metals. Springer International Publishing, Cham

Faisal M, Saquib Q, Alatar AA, Al-Khedhairy AA, Hegazy AK, Musarrat J (2013) Phytotoxic hazards of NiO-nanoparticles in tomato: a study on mechanism of cell death. Journal of hazardous materials 250:318–332

Faraji S, Heidari P, Amouei H, Filiz E, Abdullah, Poczai P (2021) Investigation and computational analysis of the sulfotransferase (SOT) gene family in potato (Solanum tuberosum): insights into sulfur adjustment for proper development and stimuli responses. Plants 10(12):2597

Gershater MC, Edwards R (2007) Regulating biological activity in plants with carboxylesterases. Plant Science 173(6):579–588

Gharabli H, Della Gala V, Welner DH (2023) The function of UDP-glycosyltransferases in plants and their possible use in crop protection. Biotechnology advances 67:108182

Ghodke VM, Punekar NS (2022) Environmental role of aromatic carboxylesterases. Environmental Microbiology 24(6):2657–2668. https://doi.org/10.1111/1462-2920.15774

Goel A (2021) Short review on xenobiotics and human health. Int J Nutr Lifestyle 1(3):87–92

Goldberg-Cavalleri A, Onkokesung N, Franco-Ortega S, Edwards R (2023) ABC transporters linked to multiple herbicide resistance in blackgrass (Alopecurus myosuroides). Front Plant Sci 14. https://doi.org/10.3389/fpls.2023.1082761

Gräfe K, Schmitt L (2021) The ABC transporter G subfamily in Arabidopsis thaliana. Journal of Experimental Botany 72(1):92–106

Guttikonda SK, Trupti J, Bisht NC, Chen H, An Y-QC, Pandey S, Xu D, Yu O (2010) Whole genome co-expression analysis of soybean cytochrome P450 genes identifies nodulation-specific P450 monooxygenases. BMC Plant Biol 10(1):243. https://doi.org/10.1186/1471-2229-10-243

Hansen CC, Nelson DR, Møller BL, Werck-Reichhart D (2021) Plant cytochrome P450 plasticity and evolution. Molecular Plant 14(8):1244–1265

Hansen LG, Shane BS (2019) Xenobiotic metabolism. In: Basic environmental toxicology. CRC Press, pp 49–105

Hernández Estévez I, Rodríguez Hernández M (2020) "Plant Glutathione S-transferases: An overview." Plant Gene 23:100233. https://doi.org/10.1016/j.plgene.2020.100233

Hirschmann F, Krause F, Papenbrock J (2014) The multi-protein family of sulfotransferases in plants: composition, occurrence, substrate specificity, and functions. Frontiers in plant science 5:556

Jagodzik P, Tajdel-Zielinska M, Ciesla A, Marczak M, Ludwikow A (2018) Mitogen-Activated Protein Kinase Cascades in Plant Hormone Signaling. Front Plant Sci 9. https://doi.org/10.3389/fpls.2018.01387

Jiu S, Xu Y, Wang J, Wang L, Liu X, Sun W, Sabir I ali, Ma C, Xu W, Wang S (2020) The cytochrome P450 monooxygenase inventory of grapevine (Vitis vinifera L.): genome-wide identification, evolutionary characterization and expression analysis. Frontiers in genetics 11:44

Kafle A, Timilsina A, Gautam A, Adhikari K, Bhattarai A, Aryal N (2022) Phytoremediation: Mechanisms, plant selection and enhancement by natural and synthetic agents. Environmental Advances 8:100203

Kawahigashi H, Hirose S, Ohkawa H, Ohkawa Y (2007) Herbicide resistance of transgenic rice plants expressing human CYP1A1. Biotechnology advances 25(1):75–84

Kidwai M, Ahmad IZ, Chakrabarty D (2020) Class III peroxidase: an indispensable enzyme for biotic/abiotic stress tolerance and a potent candidate for crop improvement. Plant Cell Rep 39(11):1381–1393. https://doi.org/10.1007/s00299-020-02588-y

Koschmieder J, Wüst F, Schaub P, Álvarez D, Trautmann D, Krischke M, Rustenholz C, Mano J, Mueller MJ, Bartels D, Hugueney P, Beyer P, Welsch R (2021) Plant apocarotenoid metabolism utilizes defense mechanisms against reactive carbonyl species and xenobiotics. Plant Physiol 185(2):331–351. https://doi.org/10.1093/plphys/kiaa033

Kumar R, Ivy N, Bhattacharya S, Dey A, Sharma P (2022) Coupled effects of microplastics and heavy metals on plants: Uptake, bioaccumulation, and environmental health perspectives. Science of the Total Environment 836:155619

Lee KP, Kim C, Landgraf F, Apel K (2007) EXECUTER1- and EXECUTER2-dependent transfer of stress-related signals from the plastid to the nucleus of Arabidopsis thaliana Proceedings of the National Academy of Sciences 104(24):10270–10275. https://doi.org/10.1073/pnas.0702061104

Lehotai N, Kolbert Z, Pet\Ho A, Feigl G, Ördög A, Kumar D, Tari I, Erdei L (2012) Selenite-induced hormonal and signalling mechanisms during root growth of Arabidopsis thaliana L. Journal of experimental botany 63(15):5677–5687

Li D, Gao Q, Xu L, Pang S, Liu Z, Wang C, Tan W (2017) Characterization of glutathione S-transferases in the detoxification of metolachlor in two maize cultivars of differing herbicide tolerance. Pesticide biochemistry and physiology 143:265–271

Lin Q, Chen S, Chao Y, Huang X, Wang S, Qiu R (2017) Carboxylesterase-involved metabolism of di-n-butyl phthalate in pumpkin (Cucurbita moschata) seedlings. Environmental Pollution 220:421–430

Liu Y, Ma M, Tang Y, Huang Z, Guo Y, Ma Q, Zhao L (2024) A NADPH-Dependent Aldo/Keto Reductase Is Responsible for Detoxifying 3-Keto-Deoxynivalenol to 3-epi-Deoxynivalenol in Pelagibacterium halotolerans ANSP101. Foods 13(7):1064. https://doi.org/10.3390/foods13071064

Lydy M, Belden J, Wheelock C, Hammock B, Denton D (2004) Challenges in regulating pesticide mixtures. Ecology and society 9(6)

Marmiroli M, Marmiroli N, Pagano L (2022) Nanomaterials induced genotoxicity in plant: Methods and strategies. Nanomaterials 12(10):1658

Martin RE, Postiglione AE, Muday GK (2022) Reactive oxygen species function as signaling molecules in controlling plant development and hormonal responses. Current opinion in plant biology 69:102293

Martínez-Márquez A, Martínez-Esteso MJ, Vilella-Antón MT, Sellés-Marchart S, Morante-Carriel JA, Hurtado E, Palazon J, Bru-Martínez R (2017) A tau class glutathione-S-transferase is involved in trans-resveratrol transport out of grapevine cells. Frontiers in Plant Science 8:1457

Maurya PK, Malik DS (2016) Bioaccumulation of xenobiotics compound of pesticides in riverine system and its control technique: a critical review. J Ind Pollut Control 32(2):580–594

Miglani R, Parveen N, Kumar A, Ansari MA, Khanna S, Rawat G, Panda AK, Bisht SS, Upadhyay J, Ansari MN (2022) Degradation of xenobiotic pollutants: an environmentally sustainable approach. Metabolites 12(9):818

Mori IC, Schroeder JI (2004) Reactive oxygen species activation of plant Ca2+ channels. A signaling mechanism in polar growth, hormone transduction, stress signaling, and hypothetically mechanotransduction. Plant physiology 135(2):702–708

Morisseau C (2022) The role of hydrolases in biology and xenobiotics metabolism. International journal of molecular sciences 23:4870

Moummou H, Kallberg Y, Tonfack LB, Persson B, Van Der Rest B (2012) The Plant Short-Chain Dehydrogenase (SDR) superfamily: genome-wide inventory and diversification patterns. BMC Plant Biol 12(1):219. https://doi.org/10.1186/1471-2229-12-219

Nelson DR (2009) The Cytochrome P450 Homepage. Hum Genomics 4(1):59. https://doi.org/10.1186/1479-7364-4-1-59

Nelson DR, Schuler MA, Paquette SM, Werck-Reichhart D, Bak S (2004) Comparative Genomics of Rice and Arabidopsis. Analysis of 727 Cytochrome P450 Genes and Pseudogenes from a Monocot and a Dicot Abstract. Plant Physiol 135(2):756–772. https://doi.org/10.1104/pp.104.039826

Och O (2019) Enzymes involved in xenobiotic metabolism. Evaluation of environmental contaminants and natural products: A human health perspective :165

Pakrashi S, Jain N, Dalai S, Jayakumar J, Chandrasekaran PT, Raichur AM, Chandrasekaran N, Mukherjee A (2014) In vivo genotoxicity assessment of titanium dioxide nanoparticles by Allium cepa root tip assay at high exposure concentrations. PloS one 9(2):e87789

Pan M, Chu LM (2016) Phytotoxicity of veterinary antibiotics to seed germination and root elongation of crops. Ecotoxicology and Environmental Safety 126:228–237

Pandey VP, Awasthi M, Singh S, Tiwari S, Dwivedi UN (2017) A comprehensive review on function and application of plant peroxidases. Biochem Anal Biochem 6(1):308

Pandian BA, Sathishraj R, Djanaguiraman M, Prasad PV, Jugulam M (2020) Role of cytochrome P450 enzymes in plant stress response. Antioxidants 9(5):454

Pašková V, Hilscherová K, Feldmannová M, Bláha L (2006) Toxic effects and oxidative stress in higher plants exposed to polycyclic aromatic hydrocarbons and their *N* -heterocyclic derivatives. Enviro Toxic and Chemistry 25(12):3238–3245. https://doi.org/10.1897/06-162R.1

Powles SB, Yu Q (2010) Evolution in action: plants resistant to herbicides. Annu Rev Plant Biol 61:317–347. https://doi.org/10.1146/annurev-arplant-042809-112119

Puglisi I, Lo Cicero L, Lo Piero AR (2013) The glutathione S-transferase gene superfamily: an in silico approach to study the post translational regulation. Biodegradation 24(4):471–485. https://doi.org/10.1007/s10532-012-9604-3

Radhakrishnan A, Shakshi, Nandini R, Kumar A, Kapardar RK, Srivastav R (2024) Role of MATE Transporters in Xenobiotics Tolerance. In: Kumar A, Shukla L, Singh J, Ferreira LFR (eds) Microbes Based Approaches for the Management of Hazardous Contaminants, 1st edn. Wiley, pp 411–419

Raggi S, Ferrarini A, Delledonne M, Dunand C, Ranocha P, De Lorenzo G, Cervone F, Ferrari S (2015) The Arabidopsis Class III Peroxidase AtPRX71 Negatively Regulates Growth under Physiological Conditions and in Response to Cell Wall Damage. Plant Physiology 169(4):2513–2525. https://doi.org/10.1104/pp.15.01464

Ramel F, Sulmon C, Serra A-A, Gouesbet G, Couée I (2012) Xenobiotic sensing and signalling in higher plants. Journal of Experimental Botany 63(11):3999–4014

Ravi B, Foyer CH, Pandey GK (2023) The integration of reactive oxygen species (ROS) and calcium signalling in abiotic stress responses. Plant Cell & Environment 46(7):1985–2006. https://doi.org/10.1111/pce.14596

Roy T, Dey TK, Jamal M (2023) Microplastic/nanoplastic toxicity in plants: an imminent concern. Environ Monit Assess 195(1):27. https://doi.org/10.1007/s10661-022-10654-z

Sachdev S, Ansari SA, Ansari MI, Fujita M, Hasanuzzaman M (2021) Abiotic stress and reactive oxygen species: Generation, signaling, and defense mechanisms. Antioxidants 10(2):277
Sauvêtre A, Eichhorn P, Pérez S (2020) Metabolism of Pharmaceuticals in Plants and Their Associated Microbiota. In: Pérez Solsona S, Montemurro N, Chiron S, Barceló D (eds) Interaction and Fate of Pharmaceuticals in Soil-Crop Systems. Springer International Publishing, Cham, pp 221–264
Serra A, Alberto D, Ramel F, Gouesbet G, Sulmon C, Couée I (2017) Perturbation and Disruption of Plant Hormone Signaling by Organic Xenobiotic Pollution. In: Pandey GK (ed) Mechanism of Plant Hormone Signaling under Stress, 1st edn. Wiley, pp 199–221
Serra A-A, Couée I, Heijnen D, Michon-Coudouel S, Sulmon C, Gouesbet G (2015) Genome-wide transcriptional profiling and metabolic analysis uncover multiple molecular responses of the grass species Lolium perenne under low-intensity xenobiotic stress. Frontiers in Plant Science 6:1124
Shigeto J, Tsutsumi Y (2016) Diverse functions and reactions of class III peroxidases. New Phytologist 209(4):1395–1402. https://doi.org/10.1111/nph.13738
Singh A, Prasad SM, Singh RP (eds) (2016) Plant Responses to Xenobiotics. Springer Singapore, Singapore
Singh BSM, Dhal NK, Kumar M, Mohapatra D, Seshadri H, Rout NC, Nayak M (2022) Phytoremediation of 137Cs: factors and consequences in the environment. Radiat Environ Biophys 61(3):341–359. https://doi.org/10.1007/s00411-022-00985-3
Singh M, Mishra RC, Shah I, Wadhwa V, Mor V (2023) Xenobiotics: Sources, Pathways, Degradation, and Risk Associated with Major Emphasis on Pharmaceutical Compounds. In: Singh R, Singh P, Tripathi S, Chandra KK, Bhadouria R (eds) Xenobiotics in Urban Ecosystems. Springer International Publishing, Cham, pp 87–106
Štefanac T, Grgas D, Landeka Dragičević T (2021a) Xenobiotics—division and methods of detection: a review. Journal of xenobiotics 11(4):130–141
Štefanac T, Grgas D, Landeka Dragičević T (2021b) Xenobiotics—division and methods of detection: a review. Journal of xenobiotics 11(4):130–141
Sun C, Dudley S, McGinnis M, Trumble J, Gan J (2019) Acetaminophen detoxification in cucumber plants via induction of glutathione S-transferases. Science of the Total Environment 649:431–439
Tabrez S, Ahmad M (2013) Cytochrome P450 system as potential biomarkers of certain toxicants: comparison between plant and animal models. Environ Monit Assess 185(4):2977–2987. https://doi.org/10.1007/s10661-012-2765-z
Tang C, Yang M, Fang Y, Luo Y, Gao S, Xiao X, An Z, Zhou B, Zhang B, Tan X (2016) The rubber tree genome reveals new insights into rubber production and species adaptation. Nature plants 2(6):1–10
Testa B (2008) Biotransformation reactions and their enzymes. In: The practice of medicinal chemistry. Elsevier, pp 655–673
Thienpont A, Cho E, Williams A, Meier MJ, Yauk CL, Rogiers V, Vanhaecke T, Mertens B (2024) Unlocking the Power of Transcriptomic Biomarkers in Qualitative and Quantitative Genotoxicity Assessment of Chemicals. Chem Res Toxicol 37(3):465–475. https://doi.org/10.1021/acs.chemrestox.3c00318
Tiwari M, Sharma D, Singh M, Tripathi RD, Trivedi PK (2014) Expression of OsMATE1 and OsMATE2 alters development, stress responses and pathogen susceptibility in Arabidopsis. Scientific reports 4(1):3964
Tola AJ, Jaballi A, Germain H, Missihoun TD (2020) Recent development on plant aldehyde dehydrogenase enzymes and their functions in plant development and stress signaling. Genes 12(1):51
Truta E, Gherghel D, Bara ICI, Vochita GV (2013) Zinc-induced genotoxic effects in root meristems of barley seedlings. Notulae Botanicae Horti Agrobotanici Cluj-Napoca 41(1):150–156
Ullah H, Li X, Peng L, Cai Y, Mielke HW (2020) In vivo phytotoxicity, uptake, and translocation of PbS nanoparticles in maize (Zea mays L.) plants. Science of the total environment 737:139558

Upadhyay N, Kar D, Deepak Mahajan B, Nanda S, Rahiman R, Panchakshari N, Bhagavatula L, Datta S (2019) The multitasking abilities of MATE transporters in plants. Journal of experimental botany 70(18):4643–4656

Verma S, Verma PK, Chakrabarty D (2023) Plant transcription factors: important factors controlling oxidative stress in plants. In: Plant transcription factors. Elsevier, pp 383–417

Wang L, Xie X, Xu Y, Li Z, Xu G, Cheng L, Yang J, Li L, Pu W, Cao P (2022a) Comprehensive analysis of the carboxylesterase gene reveals that NtCXE22 regulates axillary bud growth through strigolactone metabolism in tobacco. Frontiers in Plant Science 13:1019538

Wang X, Sial MU, Bashir MA, Bilal M, Raza Q-U-A, Ali Raza HM, Rehim A, Geng Y (2022b) Pesticides xenobiotics in soil ecosystem and their remediation approaches. Sustainability 14(6):3353

Wołejko E, Jabłońska-Trypuć A, Butarewicz A, Wydro U (2021) Heavy metals, hydrocarbons, radioactive materials, xenobiotics, pesticides, hazardous chemicals, explosives, pharmaceutical waste and dyes bioremediation. Rhizomicrobiome dynamics in bioremediation :269–298

Wu C, Chakrabarty S, Jin M, Liu K, Xiao Y (2019) Insect ATP-binding cassette (ABC) transporters: roles in xenobiotic detoxification and Bt insecticidal activity. International journal of molecular sciences 20(11):2829

Xu W, Di C, Zhou S, Liu J, Li L, Liu F, Yang X, Ling Y, Su Z (2015) Rice transcriptome analysis to identify possible herbicide quinclorac detoxification genes. Frontiers in genetics 6:306

Zhang JJ, Yang H (2021a) Metabolism and detoxification of pesticides in plants. Science of the Total Environment 790:148034

Zhang JJ, Yang H (2021b) Metabolism and detoxification of pesticides in plants. Science of the Total Environment 790:148034

Understanding the Mechanisms of Heavy Metal Detoxification in Land Plants

Priya Borah, Uddhav Sarania, and Hemen Deka

Abstract Heavy metal (HM) contamination is an escalating environmental concern emerging mainly due to anthropogenic processes such as agricultural activities, urbanization and industrialization. The HMs pose a serious threat to both ecosystems and life form due to its persistent and bioaccumulative nature. Unlike the biodegradable pollutants, HMs do not degrade over time and can only be detoxified into their least toxic forms. This makes detoxification of HMs a crucial process to mitigate metal-associated biological and environmental impacts. In such cases, phytoremediation is an emerging eco-friendly and low-cost strategy that can be effective for detoxification of HMs. Hence, this chapter explores the various physiological, biochemical and evolutionary metal transport mechanisms and detoxification processes in bryophytes to the higher plants. Bryophytes serve as excellent bioindicators due to their metal adsorption abilities, while the higher plants deploy complex mechanisms such as metal compartmentalization, chelation and antioxidant defences to detoxify HMs. Besides, the chapter also seeks to provide a comprehensive overview of HM detoxification in different plants emphasizing their potential role in sustainable environmental management.

Keywords Phytoremediation · Contamination · Sequestration · Phytochelatin · Metallothionein

---

P. Borah · U. Sarania · H. Deka (✉)
Ecology and Environmental Remediation Laboratory, Department of Botany, Gauhati University, Guwahati, Assam, India
e-mail: priyaborah7@gauhati.ac.in; uddhav@gauhati.ac.in; hemendeka@gauhati.ac.in

S. Roy, V. Mandal (eds.), *Plant-Microbe Interaction under Xenobiotic Exposure*,
https://doi.org/10.1007/978-981-96-8260-7_14

1 Introduction

The metal contamination in an environment, more particularly in soil and water bodies, is an emerging issue worldwide (Ali et al. 2021). Human-induced activities, particularly urbanization and industrial processes, have significantly partaken in increased instances of heavy metal (HM) contamination, thereby resulting in widespread ecosystem degradation and threat to every life form (Chen and Ding 2023). The HMs such as Ni, Cr, Hg, Cd, As and Pb do not degrade over time and can only be transformed from toxic forms to least toxic and/or non-toxic forms. Moreover, occurrence of the toxic forms of HMs is very much predominant in the soil systems (Li et al. 2022b). The toxic forms of HMs are responsible for hampering the microbial activities and ecological processes (Rahman and Singh 2020) in soil. Hence, leveraging different techniques for detoxification of HMs in polluted soil is very crucial in the present scenario (Rahman 2022). One such promising, economically viable and environment-friendly technique is phytoremediation, which utilizes plants to remove different HM pollutants from various soil systems and water bodies (Yadav et al. 2022). Plants adopt various physiological, biochemical and molecular strategies for HM removal and detoxifications that help them to adapt in environments adulterated with HM pollutants (Pirzadah et al. 2015). Different strategies that plants can choose during phytoremediation include metal exclusion, compartmentalization and sequestration within specific plant tissues (Yang et al. 2022). Unlike the traditional methods for HM removal including excavation, soil washing or chemical treatments, phytoremediation is minimally invasive for improving soil fertility and biodiversity (Guidi Nissim et al. 2023).

The HM detoxification ability of plants has evolved over millions of years and can differ across various plant species ranging from bryophytes to higher plants (Fasani et al. 2022). Bryophytes are among the early colonizers of terrestrial environments having unique adaptations that help the plants to sustain within harsh environments including HM contaminations (De Agostini 2022). Due to the lack of vascular networks and cuticles, bryophytes are particularly sensitive to HM pollutant as they can readily adsorb metals through the plant's entire surface area. This particular physiological attribute makes bryophytes an excellent bioindicators of HMs (Hussain et al. 2022). Despite suffering visible damage, bryophytes can accumulate significant amount of HMs suggesting the presence of efficient metal detoxification mechanism (Joshi 2021). In case of higher plants including pteridophyte, gymnosperms and angiosperms, a more sophisticated set of mechanism is observed. Unlike bryophytes, higher plants have defined vascular systems that facilitate the transportation of HMs to aerial parts via roots (Yi et al. 2023). Many higher plants possess well-developed specialized mechanisms in order to detoxify and compartmentalize HMs. For example, HMs can be chelated by organic acids, peptides or enzymes known as metallothioneins (MTs) and phytochelatins (PCs) by binding and rendering the HMs into their non-toxic counterparts (Tarigholizadeh et al. 2024). These metal chelates are further transported and stored in the vacuoles, thereby reducing the toxic effects (Sharma et al. 2016). Certain plants can also exhibit tolerance to HMs through a variety of modifications including alteration of cell wall (CW) composition, production of proteins that can bind with HMs and elevation of antioxidant activities to counteract the metal exposure-induced oxidative stress (Emamverdian et al. 2015). Additionally, certain plants have also evolved extraordinarily to tolerate and accumulate extreme levels of certain HMs in their above-ground tissues and are called as metal hyperaccumulators. For instance, *Brassica juncea* (Indian mustard) is recognized for its ability to effectively accumulate Pb, Se and Hg (Ahmad 2015). However, this can be also problematic as accumulation of such HMs in excess amounts can be toxic to most plants (DalCorso 2012). Here, an effort has been made in order to consolidate the current understanding concerned with HM accumulation and detoxification strategies employed by plants and their potential use in metal remediation techniques.

2 Metal Toxicity in Plants: An Overview

Many studies have tried to investigate the toxicity and tolerance against HMs in plants (Table 1). Accordingly, the differences in HM toxicity and tolerance levels have been noted depending on the type of plant species, metal concentrations, types and sources of contaminations (Yadav 2010). Severe HM stress in plants causes deterioration such as oxidative injury through generation of reactive oxygen species (ROS) and/or interrupting enzymatic functions by replacing essential elements as co-factors (Oves et al. 2016). Although alterations in cell metabolic activities permit plants to encounter HM stress, reduced plant growth is one of the key factors that limits their success to overcome HM toxicity. Moreover, plant response to excess level of HMs also depends on the plant's growth stage (Noor et al. 2022) and other associated parameters related to plant developments. This corroborates with the study conducted by Peralta-Videa et al. (2004) where alfalfa was found to be more resistant to HMs at an initial stage rather than at a later stage of growth. Conversely, plant adaptation to HM toxicity at a later stage is not so efficient and flexible (Chandrakar et al. 2020).

It has been suggested that responses and sensitivity of plants towards metal toxicity vary across different taxa which include bryophytes, pteridophytes and angiosperms (Fasani et al. 2022). Bryophytes comprising of liverworts, horned liverworts (Shaw et al. 2011) lack a well-developed vascular network making the bryophytes very susceptible to environmental fluctuations including HM contamination (Joshi 2021). For example, notable metal toxicity in terms of reduced growth, chlorophyll content and lower photosynthetic efficiency was seen in *Marchantia polymorpha* (Ares et al. 2018). In some mosses such as *Funaria hygrometrica*, exposure to higher levels of Zn and Cu leads to reduced growth (Basile et al. 2008). Zn and Cu in extreme amounts that can inhibit notable enzymes involved in respiration and photosynthesis ultimately lead to reduced plant vigour and energy production (Riyazuddin et al. 2021). Furthermore, *Hypnum cupressiforme* is a useful bioindicator of air quality due to its ability to accumulate HMs from the atmosphere;

Table 1 Some plant species of different plant groups having toxic effects on HM exposure

Plant group	Plant species	Heavy metal(s)	Phytotoxicity effect	Reference(s)
Bryophytes	*Pleurochaete squarrosa*	Ni, Pb and Cu	Reduced chlorophyll content	Aydoğan et al. (2017)
	Timmiella barbuloides	Ni	Reduced chlorophyll content	Aydoğan et al. (2017)
	Sphagnum palustre	Pb	Reduced chlorophyll content, growth inhibition	Carroll et al. (2009)
	Marchantia polymorpha	Cd	Decreased photosynthesis efficiency, oxidative stress	Ares et al. (2018)
Pteridophytes	*Pteris vittata*	As	Slight reduced growth rate at higher concentration	Drava et al. (2012)
	Equisetum hyemale	Zn	Stunted growth and chlorosis	Silva et al. (2022)
	Ceratopteris thalictroides	Cd	Root damage, chlorosis, reduced gametophytic growth	Lu et al. (2022)
Gymnosperms	*Pinus sylvestris*	Cu	Needle chlorosis, root damage, reduced growth	Możdżeń et al. (2017)
	Picea abies	Ni	Leaf necrosis, impaired root growth, reduced photosynthesis, chlorosis	Radotić et al. (2003)
Angiosperms	*Helianthus annuus* L.	Cr	Reduced seed germination and root elongation, stunted growth, chlorosis	Zhao et al. (2021)
	Lactuca sativa	Ni, Cu	Inhibition of root elongation and seed germination	Di Salvatore et al. (2008)
	Tritichum aestivum L	Pb	Reduced shoot length and root elongation	Mohamed et al. (2017)
	Trifolium repens	Cd	Reduced shoot and root growth	Stravinskienė and Račaitė (2014)
	Lolium perenne	Pb	Affects seed germination	Jelea et al. (2015)

however, this makes the plant species very sensitive to metal toxicity (Mahapatra et al. 2019). In case of pteridophytes, certain ferns like *Pteris vittata* have gained significant attention for the ability to hyperaccumulate arsenic (Niazi et al. 2016). Despite the plant's known tolerance when grown in highly contaminated soils, *P. vittata* greatly suffers through reduced growth rates, chlorosis and oxidative injury (Drava et al.2012). In a study conducted by Silva et al. (2022), it has been reported that *Equisetum* sp. when exposed to high levels of Pb and Cd shows clear signs of reduced growth, impaired photosynthesis and chlorophyll degradation (Silva et al. 2022). However, plants may counteract this by production of an increased number of antioxidants such as glutathione and PCs in order to mitigate HM-induced oxidative injury (Silva et al. 2022).

Despite the evolutionary advancements, higher plants are not immune to the phytotoxic effects of HMs. Cu toxicity in particular has been reported in several gymnosperms including *Pinus sylvestris* (Możdżeń et al. 2017). In conifers like *Picea abies*, Pb poisoning leads to disruption of nutrient uptake especially by interfering with the absorption of calcium and magnesium elements essential for proper growth and development (Saladin 2015). Pb poisoning also hampers with the root elongation leading to diminished biomass production and stunted growth (Zulfiqar et al. 2019). Similar reports have indicated that Cd is highly toxic to various angiosperms including crop plants, *Oryza sativa* and *Zea mays*. Cadmium disrupts the cell homeostasis by replacing essential elements like calcium and zinc (Zn) leading to impaired metabolic pathways (Riaz et al. 2021).

3 HM Detoxification in Plant: A Mechanistic Insight

3.1 Subcellular Compartmentalization

Subcellular compartmentalization is a key strategy in plants for mitigating the toxic effects of metals. This process involves transporting and sequestering metal ions into some specific cellular compartments like vacuole and CW, which effectively separates metals from other sensitive cellular components (Sharma et al. 2016). By compartmentalizing, plants can effectively reduce the bioavailability of metals, thereby preventing the interferences with essential cellular processes (Viehweger 2014). The strategy like subcellular compartmentalization enables the plants to accumulate and tolerate HM toxicity (De Caroli et al. 2020) without causing substantial damage to its different parts.

3.1.1 Vacuolar Sequestration

The vacuoles are storage site in plant cell where HMs possess less reactive form. The mechanism of HM sequestration in vacuoles involves their transportation from the cytoplasm into the vacuole, isolating them from sensitive cellular components and reducing the toxicity. This sequestration process typically involves the binding of metal ions to various chelators or ligands before its entry into the vacuole (Clemens 2019). PCs and organic acids are the key players in this process as they can form stable metal complexes with HMs like Zn, Cd and Pb (Yadav 2010). Such metal-chelate complexes are subsequently transported across the vacuolar membrane via active transport mechanisms. Specific transport proteins such as members of the ATP-binding cassette (ABC) transporter family and cation/ H+ antiporters like the cation exchanger (CAX) family facilitate this process (He et al. 2022). One of the primary advantages of vacuolar sequestration is that it reduces the free concentration of HMs in the cytosol, thereby minimizing their potential to disrupt cellular processes or generate ROS (Riyazuddin et al. 2021). By sequestering HMs in the vacuole, plants can effectively neutralize their toxic effects and prevent such HMs from interfering with vital cellular activities.

Considering the mechanism of sequestration, once HMs entered inside the vacuole, it can be stored in various forms. HMs can be chelated by organic molecules such as PCs, MTs and the organic acids within the vacuole (Anjum et al. 2015) in order to prevent the damages in cell. Chelation reduces the bioavailability of HMs and prevents its interference with cellular components. In some cases, HMs can precipitate within the vacuole by formation of less toxic insoluble compounds (Abd Elnabi et al. 2023). Metal transport proteins (MTPs) are integral membrane proteins that play a crucial role in the transport of essential metals, particularly Zn and Ni (Dhir 2021). Such proteins are localized to both the plasma membrane and the tonoplast of the plants (Muro et al. 2024). Again, the ABC transporters constitute a family of proteins involved in the active efflux of HMs from the cytoplasm to the vacuole. Moreover, heavy metal-transporting ATPases (HMAs) are also an ATP-dependent pumps that can passage HM ions inside the vacuole (Sharma et al. 2016). Additionally, CAX family can exchange calcium ions that facilitated the sequestration of HMs into the vacuole. Another transporter involved in the vacuolar compartmentalization process is the natural resistance-associated macrophage proteins (NRAMPs) that show divalent cation transporters that can transport HMs into the vacuole (Ishida and Corcino 2022). For example, cadmium (Cd) can be sequestered in the vacuole as a Cd-PC complex where the Cd/H^+ antiporter and an ATP-dependent PC transporter are involved (Shahid et al. 2016). Therefore, the vacuolar sequestration can be considered as a one of the highly effective mechanisms for the detoxification of HMs by the plants which can minimize metal's harmful effects and shows critical aspects in maintaining cellular homeostasis. Understanding the molecular mechanisms involved in vacuolar sequestration can deliver valuable insights to develop tolerance strategies of the plant against HM pollution.

3.1.2 CW Binding

The plant CW serves as a crucial barrier by playing a pivotal role in a plant's response to HM stress (Shi et al. 2019). The ability to bind and sequester metals by plant CW can significantly influence its tolerance to these toxic substances. However, the exact mechanisms involved and the extent of their effectiveness are still being explored. The CW acts as a metal barrier by direct metal binding and metal sequestration process against such toxic metals (Parrotta et al. 2015). The CW rich in various functional groups like hydroxyl groups (–OH), carboxyl groups (–COOH) and thiol group (–SH) can directly bind to the divalent and trivalent metal cations (Kopittke et al. 2014). Pectin, a major component of the CW, is particularly effective in binding metals due to its abundance of carboxyl groups (Meychik et al. 2014). By binding to metals, the CW can impede their entry into the cytoplasm, effectively sequestering them and mitigating their toxic effects on cellular processes

(Kosakivska et al. 2021). Further, the CW can act as a physical barrier, restricting the movement of metals from the external environment to the intracellular space (Schwab et al. 2016), and help to prevent excessive metal uptake and accumulation. Generally, the mechanism of metal binding and sequestration with CW shows three types of interactions: (a) *pectin-metal interactions*, where the carboxyl groups of the pectin molecules are the key players in the metal binding interaction (Baghdadi et al. 2023); (b) *protein-metal interactions*, where CW proteins participate in metal binding process, particularly those with high affinities for specific metals (Wu et al. 2021); and (c) *silicate-metal interactions*, where silicates within the CW can bind to metals and provide an additional mechanism for metal sequestration (Sarkar et al. 2021).

While the CW is often seen as a stable repository for metals, previous studies recorded that some metal-bound CW components, such as low methyl-esterified pectins, can be internalized through endocytosis (Krzesłowska 2011). This suggests that the CW's role in metal sequestration is not entirely static. Again, plants can form callose layers between the CW and plasma membrane, which can potentially restrict metal efflux from cellular compartments to the cytoplasm (Kulich et al. 2015). For example, hyperaccumulator plants like *Phragmites australis* that can tolerate and accumulate extreme levels of metals often exhibit enhanced CW-mediated metal sequestration (Jiang and Wang 2008). The ability to sequester metals in the CW is a key adaptation of such plants to thrive in metal-contaminated environments.

Besides CW, the plasma membrane also serves as a vital defence line against toxic metals. By regulating the metal influx and efflux, plasma membrane serves as a selective barrier in maintaining intracellular metal homeostasis (Zangi and Filella 2012). It is involved in signal transduction processes that alert the cell against various metal stress condition, such as the presence of HMs (Tiwari and Lata 2018). When metal ions enter the cell through the plasma membrane, they can disrupt cellular functions leading to oxidative stress and cellular damage. Therefore, the integrity of the plasma membrane is crucial for plant health as it must effectively manage ion imbalances and initiate repair mechanisms when damaged, for that plasma membrane contains various transport proteins to facilitate the uptake of essential and non-essential metal ions (Chen et al. 2023). These include co-transporters, proton pumps and carrier proteins that create electrochemical gradients necessary for metal transport. Moreover, the plasma membrane also holds receptors that can detect metal stress and triggers signalling pathways which lead to the expression of detoxification genes (Jain et al. 2018).

3.2 Chelation

Chelation is a process where a substance known as a chelating agent or chelator binds to metal ions forming a stable, ring-like complex (Song et al. 2010, 2019). It is a crucial process that enables plants to detoxify HMs and mitigate their harmful effects particularly in environments contaminated with such pollutants. To survive in such extreme conditions, plants have developed intricate mechanisms like chelation to detoxify and tolerate the harmful elements. Two key players that take part in this type of defence strategy are PCs and MTs.

3.2.1 Role of PCs and MTs in Metal Detoxification

3.2.1.1 Phytochelatins

PCs are cysteine-rich peptides synthesized by plants in response to HM stress (Sethi 2023). These peptides act as chelating agents binding to HM ions through their thiol groups (Faizan et al. 2024). This binding process effectively neutralizes the toxic effects of the metals by reducing their bioavailability. The primary role of PCs is to detoxify HMs by chelating them, forming metal-PC complexes (Pal and Rai 2010). Such complexes can be safely sequestered in cellular compartments like vacuole by effectively neutralizing the free metal ions within the cytoplasm and protecting essential cellular processes. The thiol (-SH) groups of cysteine residues in PCs are mainly responsible for binding with the metal ions (Satofuka et al. 2001). The degree of polymerization of PCs increases with higher intracellular metal concentrations, enhancing the stability of metal-PC complexes (Vatamaniuk et al. 2000). For example, Cd can form complexes with PCs that involve the binding of sulphur atoms from the cysteine residues to the metal ion, resulting in amorphous complexes with a stoichiometry of approximately four sulphur atoms to one Cd ion (Pal and Rai 2010). In contrast, Zn forms a simpler 1:1 complex with PCs, while arsenic (As) preferentially forms a more stable trihedral coordination with three thiol groups, rather than forming complexes with fewer thiol groups (Kobayashi and Yoshimura 2006). By formation of complexes with such metals, PCs can prevent the metals from disrupting vital cellular processes that can occur when metal ions bind inappropriately to thiol groups or other functional groups in proteins.

When the HMs such as Cd and Cu entered the plant cells, PCs bind to these metal ions forming stable metal-thiolate complexes. This chelation process effectively reduces the bioavailability of the metals preventing them from interacting with essential cellular components and causing oxidative damage (Gupta et al. 2013). Following the formation of a PC-Cd complex, it is transported into the vacuole where the metal is deposited and often bound to the organic acids (Ahmad et al. 2019). The PC is then released back into the cytoplasm to continue its detoxification role. The acidic environment of the vacuole with a pH typically between 3.5 and 5.0 facilitates the separation of the metal from the PC preventing the formation of the PC-Cd complex at such low pH levels (Ahmad et al. 2019). PCs are not only effective against Cd but also exhibit an important role in detoxifying diverse array of HMs including Pb, As, Hg Ag, Zn, Ni and Cu. The availability of metals in the environment can influence the production of PCs in plants (Figueroa et al. 2008). The mutants of *Arabidopsis thaliana* (cad1–3) that are deficient in PCs were found to be more sensitive to Cd and As than the wild-type plants (Liu et al. 2015).

However, aquatic plants like *Vallisneria spiralis* and *Hydrilla verticillata* can synthesize various types of PCs to bind and detoxify mercury (Shrivastava and Srivastava 2021). Studies have demonstrated that Cu can induce PC biosynthesis both in vivo and in vitro; however, PC-deficient mutants exhibit relatively lesser sensitivity to copper (Srivastava 2016). For example, the copper-tolerant *Mimulus guttatus* supports the role of PCs in copper tolerance (Macnair et al. 2020). However, in other plants like *Silene vulgaris*, both copper-tolerant and copper-sensitive ecotypes produced similar levels of PCs when exposed to copper, suggesting that other mechanisms may contribute to copper tolerance (Hall 2002). Further research has revealed the formation of PC-metal complexes in the roots of plants like *Rubia tinctorum* when exposed to various HMs including Ag, As, Cd, Cu, Pb and Hg (Shaw et al. 2005). Interestingly, the PC complexes identified in vivo were primarily associated with Cd, Ag and Cu ions. In some cases, PC complexes induced by lead and arsenate were found to contain copper ions rather than the metals that triggered their synthesis (Gupta et al. 2013). However, the humic substances in soil can reduce the bioavailability of metals leading to decreased PC production in *Ricinus communis* (Figueroa et al. 2008).

In addition to the role in detoxification, PCs also contribute to metal homeostasis, antioxidant properties and sulphur metabolism within the cell (Seregin and Kozhevnikova 2023). Interestingly, high levels of PCs in transgenic plants have been associated with increased metal accumulation without necessarily enhancing metal tolerance. In fact, these plants often exhibit hypersensitivity to HMs (Manara 2012), indicating that while PCs play a protective role, their overproduction can lead to detrimental effects. Finally, PCs can be considered as a vital component in the HM detoxification in plants having ability to bind and sequester metals like Cd, Cu and others into vacuoles. However, the complexity of their roles including the potential for metal exchange within PC complexes and the varying responses to different metals highlights the intricate nature of metal detoxification in plants.

3.2.1.2 Metallothioneins

MTs are another class of cysteine-rich proteins essential for metal homeostasis in plants (Sunitha et al. 2013). While PCs are enzymatically produced peptides, the MTs are gene-encoded polypeptides synthesized through mRNA translation (Anjum et al. 2015). MTs bind to HMs through the –SH (thiol) groups of their cysteine residues forming stable complexes that sequester the metals and prevent them from causing cellular damage (Shi et al. 2019). While plant PCs are primarily responsible to detoxify Cd, the MTs exhibit broader metal binding affinity including Cd, As, Zn and Cu (Emamverdian et al. 2015). Such proteins are involved in various physiological functions including maintaining the metal homeostasis, detoxifying toxic HMs like Pb, Cd and Hg and shielding cells from oxidative injury caused by ROS (Mansoor et al. 2023). MT gene expression in plants is transcriptionally regulated and can be induced by various factors including HMs, hormones, oxidative stress and environmental stresses like extreme temperatures, drought and nutrient

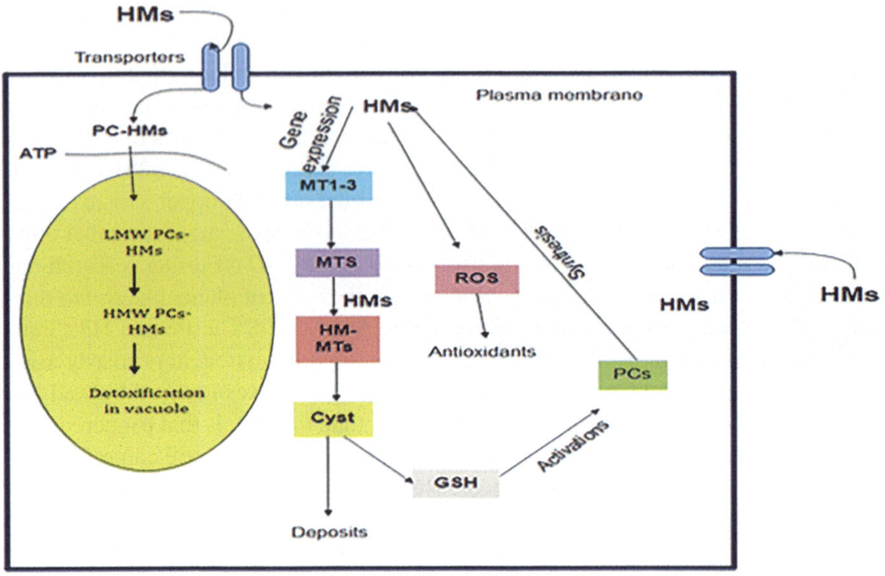

Fig. 1 Schematic diagram of cellular mechanism of MTs and PCs in HM detoxification

deficiency (Kim and Kang 2018). For example, different MT genes are differentially regulated against the HM exposure in *Brassica rapa* (Ahn et al. 2012). MT1 and MT3 genes are upregulated in response to Mn stress, while MT2 is downregulated under Zn stress (Joshi et al. 2016). Moreover, MTs possesses a vital role in metal detoxification process by sequestering metal ions and preventing them from participating in harmful redox reactions that could generate ROS and cause oxidative stress. In addition to the HM detoxification capacity, MTs also involve in various cellular processes such as DNA repair, maintenance of the redox balance and cell proliferation (Zalewska et al.2014).

Both PCs and MTs represent two distinct but complementary systems that plants use to cope with HM stress. PCs are more specialized in chelating and sequestering metals like Cd, while MTs have a broader role in maintaining metal homeostasis, detoxifying a wider range of metals and protecting against oxidative damage. Both systems are crucial for plant survival in HM-contaminated environment, thus contributing to the overall metal tolerance of plants (Fig. 1).

3.2.2 Other Metal Binding Compounds

Plants utilize various strategies to tolerate and detoxify HMs that can be broadly categorized into external exclusion and internal tolerance strategies (Hossain et al. 2012). In external exclusion, the plant's roots release organic acids that bind with HM ions, which creates stable metal-ligand complexes reducing the portability and bioavailability of the metals by preventing the entry of HMs into the plant cell and accumulation in sensitive root tissues (Jan and Parray 2016). Internally, organic acids can chelate HMs within the cytosol by transforming them into less harmful forms. Plants produce various ligands such as amino acids, carboxylic acids and phosphate derivatives which play critical roles in HM tolerance and detoxification process (Dubey et al. 2018).

Besides PCs and MTs, plants utilize other metal binding compounds to maintain metal balance and protect against toxicity. Such compounds exhibit important roles in metal transport, storage and detoxification ensuring the plant's health and growth. One of the most prominent groups of metal binding compounds in plants is *organic acids* such as malic acid, oxalic acid, citric acid and tartaric acid that help in metal ion chelation (Kocaman 2023). For example, citric acid can form stable complexes with aluminium (Al) and iron (Fe) reducing their toxicity (Vance et al. 2020). In hyperaccumulator plants, organic acids can sequester metal ions in vacuoles by isolating them from the cytosol preventing them from disrupting cellular metabolism (Seregin and Kozhevnikova 2024). This sequestration is particularly important for the plants found to be growing in the soils with high metal concentrations. Another significant metal binding compound in plants is the *histidine* that strongly binds to metal ions like Zn and Ni (Sharma and Dietz 2006). Histidine's imidazole group binds to metal ions to form stable complexes which transport such metals to specific storage sites within the plant (Clemens 2019). This is particularly important in Ni hyperaccumulator plants where histidine plays a significant role in transporting nickel to vacuoles and prevents it from interfering with cellular functions. By binding and compartmentalizing Ni, histidine not only assists in metal tolerance but also contributes to the plant's ability to accumulate elevated level of Ni without suffering from toxicity (Seregin et al. 2022). *Ferritins* are another critical component of the plant's metal management system particularly concerning to Fe homeostasis (Briat et al. 2010) which can storage Fe ions into non-toxic, soluble form. This capability is essential for maintaining iron balance within the plant as excess free iron can catalyse the ROS production which further creates oxidative stress and cellular damage (Sultana and Ghosh 2024). By storing Fe in a controlled manner, ferritins help to prevent the negative effects of Fe overload and ensure that it is available when needed for physiological processes of the plant (Rai et al. 2021). Moreover, compounds like *polyamines* that include putrescine, spermidine and spermine also display major contributions in metal detoxification process (Raychaudhuri et al. 2021). Such small positively charged molecules can bind to metal ions like Cd, Cr, Cu and Zn and form complexes to reduce the metals' availability for participation in harmful reactions (Gupta et al. 2024). The capacity of polyamines to protect cellular components from metal-induced damage and mitigate oxidative stress particularly in the presence of high concentration of toxic metals is vital for the plant's defence against metal toxicity (Spormann et al. 2021).

3.3 Antioxidant-Mediated Defence

Plants when exposed to toxic levels of HMs either directly or indirectly lead to heightened generation of ROS in their cells (Berni et al. 2019). For example, HMs can induce redox reactions, which ultimately can lead to the production hydroxyl ion from either superoxide anions or hydrogen peroxide (H_2O_2) by the process of Fenton reaction or Haber-Weiss (Miller et al. 2017), whereas redox-inactive HMs, namely, Pb and Cd, may indirectly trigger the accumulation of ROS by inhibiting the antioxidant production impairing the respiratory chain or by facilitating production of enzymes linked with ROS production (Al Mahmud et al. 2019). In addition, HMs can disrupt the electron transport chain (ETC) situation in the mitochondria resulting in the release of electrons and formation of superoxide anions (Sun et al. 2022). In plant cells, chloroplast, mitochondria and peroxisome are the main reservoirs of ROS and any disruption in their functions will ultimately result in overproduction of ROS (Corpas et al. 2015). The oxidative damage induced due to HM exposure leads to a variety of cellular injury. In addition to that, lipid peroxidation triggered by ROS attack can further propagate the oxidative injury. ROS generation causes protein oxidation, mutations, DNA strand breakage and base modification contributing to cancer and other such related diseases (Sytar et al. 2013).

In order to mitigate the radical stress injury caused due to exposure of HMs, the plants can employ a range of non-enzymatic and enzymatic antioxidant defence strategies (Ansari et al. 2024). The antioxidants play a crucial part in regulation of redox equilibrium by detoxifying HM-induced ROS and reversing oxidative injury (Hossain et al. 2012). One of the notable enzymatic antioxidants is superoxide dismutase (SOD) which is accountable for dismutation of superoxide ions into H_2O_2 and oxygen (Ighodaro and Akinloye 2018). The various isoforms of SOD include mitochondrial Mn-SOD, cytosolic Zn/Cu SOD and extracellular SOD, each of which provides protection in the respective cellular compartments (Giménez et al. 2021). When exposed to HMs, SOD is found to be upregulated reflecting its importance in managing oxidative stress (Bhaduri and Fulekar 2012). An investigation conducted by Dubey et al. (2014) found high concentrations of SOD in response to Cr (VI) stress in rice crops. Another such enzyme is catalase (CAT), which is pivotal in detoxification of H_2O_2 into water and oxygen. This prevents excessive accumulation of hydroxyl radicals (·OH) produced via Fenton reaction (Ighodaro and Akinloye 2018). This enzymatic activity is crucial in protecting plant cells from oxidative injury under metal stress conditions (Gebicka and Krych-Madej 2019). Likewise, glutathione peroxidase (GPX) and ascorbate peroxidase (APX) can also reduce H_2O_2 into H_2O, thereby protecting plant's plasma membrane from lipid peroxidation by reducing lipid hydroperoxides (Pandey et al. 2017). Another enzymatic antioxidant called glutathione reductase (GR) induces the continuous activity of GPX and ascorbate-glutathione cycle enhancing the ability of plants to detoxify HMs and the generated ROS (Pang and Wang 2010). Plants also rely on non-enzymatic antioxidants to mitigate and detoxify HM-induced oxidative stress (Mansoor et al. 2023). For instance, glutathione (GSH) is a compound composed of three peptides, namely, glycine, cysteine and glutamine. GSH plays a multifaceted role by directly scavenging ROS, serving as a substrate for APX and GPX in reducing lipid hydroperoxides and hydrogen peroxide (Rai et al. 2023). Moreover, GSH can also synthesize PCs by forming complexes with HMs via the antioxidant's thiol groups. These complexes are then compartmentalized inside the vacuoles limiting the availability of free HM ions in the cytoplasm and their potential toxic effects (Yadav 2010). Likewise, polyphenols and flavonoids may either directly scavenge ROS or chelate metal ions preventing their active participation in redox reactions that could further generate ROS (Dangles 2012). Hence, when forming complexes with HMs, flavonoids can reduce the metal bioavailability and toxicity (Flora et al. 2013). Besides, these compounds could also modulate with the activity of enzymatic antioxidants maximizing plant's overall health under metal-generated oxidative stress (Imam et al. 2017). In ascorbate-glutathione cycle, for instance, ascorbate and glutathione function together with antioxidant enzymes such as APX and GR to detoxify hydrogen peroxidase (Roychoudhury and Basu 2012).

3.4 Metal Transport Mechanisms

3.4.1 ATP-Dependent Processes

Plants have adopted an intricate array of strategies to cope with the toxic effects of HMs. One crucial strategy involves the active transport of metal ions across cellular membrane. ABC transporters play a vital role during such active transportation of metals (Dahuja et al. 2021). It can be characterized by their ability to use ATP hydrolysis to power the transfer of various molecules across membranes (Thomas and Tampé 2020). In plants, ABC transporters inculpate in different physiological process like nutrient uptake, stress response and hormonal transport (Li et al. 2021). ABC transporters can actively pump HM ions out of cells or organelles, preventing their accumulation to toxic levels (Vishwakarma et al. 2019). This metal efflux mechanism is particularly important in protecting sensitive cellular components. Besides, some ABC transporters can sequester HM ions bound to chelators like PCs into intracellular compartments by transforming them into less harmful forms (Sharma et al. 2016). This metal-chelate transportation process is essential for the elimination of toxic metal-chelate complexes.

Some of the specific ABC transporters involved in metal detoxification are as follows:

1. *ABCC Subfamily*: Members of the ABCC subfamily such as AtABCC2 and AtABCC1 in *Arabidopsis thaliana* involve in vacuolar sequestration of HM-PC complexes (Kim et al. 2007).
2. *ABCG Subfamily*: The ABCG transporters also known as the pleiotropic drug resistance (PDR) subfamily include several transporters involved in metal efflux. For example, AtPDR8 and AtPDR12 in *A. thaliana* and OsABCG36 in rice plant

involved in metal efflux mechanism (Zhang et al. 2023; Lee et al. 2005; Fu et al. 2019).

3. *ABCB Subfamily*: ABCB transporters also contribute to metal detoxification process particularly by transporting metals across the plasma membrane of plants (Jalmi 2022).

ATP-dependent metal transporters particularly ABC transporters can carry a diverse array of HMs including Zn, Cu, Cd, As and Fe (Yang et al. 2022). The diversity and versatility of ABC transporters let plants to effectively cope with a wide range of HM contaminants. Again, ABC transporters are expressed in different plant tissues enabling them to function in specific locations where metal detoxification is most critical. Moreover, the activity of ABC transporters can be regulated in response to HM stress, ensuring that detoxification mechanisms are activated when needed (Kosakivska et al. 2021).

Another type of ATP-dependent metal transporter is the HMAs which is a group of CPx-ATPases characterized by a conserved sequence of amino acids (Kaur et al. 2019). Such ATPases are essential for transporting HM ions including Pb, Cd, Zn and Cu across plasma membrane against their concentration gradient (Kaur et al. 2022). Depending on metal substrate specificity, HMAs are primarily categorized into two groups such as Cu^{2+}/Ag^{2+} and the $Pb^{2+}/Zn^{2+}/Co^{2+}/Cd^{2+}$ group (Li et al. 2024). *A. thaliana* possesses eight HMAs of which AtHMA5, AtHMA6, AtHMA7 and AtHMA8 belong to the first group of HMAs, while AtHMA1, AtHMA2, AtHMA3 and AtHMA4 belong to the second group (Takahashi et al. 2012). The HMA transporters particularly AtHMA4 and AtHMA2 in *A. thaliana* involve in Zn^{2+} and Cd^{2+} transport through plasma membrane (Sterckeman and Thomine 2020; Escudero et al. 2022). Particularly, the AtHMA4 serves dual function: dispersing Zn^{2+} under normal conditions and exporting excess Cd^{2+} and Zn^{2+} when levels become toxic (Pasricha et al. 2021). A study on maize plant by Liao et al. (2023) revealed the involvement of ZmHMA3 which can regulate the Cd and Zn stress by transporting such metal ions. Similarly, TcHMA3 expression in *Thlaspi caerulescens* facilitated the Cd sequestration mechanisms in extreme contamination by Cd (Ueno et al. 2011). Again, OsHMA3 in rice plant was identified as a key player in sequestering Cd within root vacuoles preventing translocation of Cd to shoots (Zhang et al. 2022).

3.4.2 Non-ATP-Dependent Processes

Plants possess a diverse array of adaptations to transport metal ions across membranes, some of which do not directly rely on ATP hydrolysis. Such mechanisms include secondary active transporters and ion channels which utilize ion gradients or membrane potentials instead of ATP to drive the metal ion transportation.

3.4.2.1 NRAMP Transporters

NRAMPs are the metal transporter proteins found both in the plant and in the animal cell (Manatschal and Dutzler 2022). NRAMPs primarily transfer divalent metal ions such as Mn^{2+}, Fe^{2+}, Cd^{2+}, Cu^{2+}, Co^{2+}, Ni^{2+} and Zn^{2+} across cell membranes by utilizing the proton gradients (Courville et al. 2006). This secondary active transport mechanism allows for the efficient uptake of metals without direct ATP utilization. NRAMP transporters are located primarily in the shoots and roots in case of plants. Specific NRAMPs like OsNRAMP1 and OsNRAMP5 in rice plant and AtNRAMP1, AtNRAMP3, AtNRAMP4 and AtNRAMP6 in *A. thaliana* are involved in the movement and acquisition of metals like Fe, Mn and Cd (Kanwal et al. 2024; Ishimaru et al. 2012). Moreover, TcNRAMP3 and TcNRAMP4 that were found to be localized in vacuoles of hyperaccumulator plants such as *T. caerulescens* are crucial for Fe^{2+} transport and metal tolerance (Oomen et al. 2009).

3.4.2.2 Zinc-Iron Permeases Transporters

The zinc-iron permease (ZIP) transporter primarily facilitates the uptake and transport of Zn systematically in the entire plant (Ajeesh Krishna et al. 2020). However, it also involves in the uptake of Fe, Zn, Mn and Cd especially under metal deficiency (Palusińska et al. 2020). ZIP transporters typically possess eight transmembrane domains with a histidine-rich region important for metal binding and transport (Hu 2021). ZIP transporters like AtZIP1, AtZIP2 in *A. thaliana* and OsZIP11 in rice plant are crucial for keeping the metal homeostasis inside the plant body (Wang et al. 2023; Rekha et al. 2021).

3.4.2.3 Cation Diffusion Facilitators

Cation diffusion facilitators (CDFs) also known as metal tolerance proteins are involved in the detoxification and homeostasis of metal ions by facilitating their efflux from the cytoplasm into cellular compartments like vacuoles (Jogawat et al. 2021). CDFs classified into three subgroups depending on their substrate specificity such as Fe/Zn-CDFs mainly transport Fe^{2+} or Zn^{2+}, Zn-CDF family consists of Zn^{2+} and Co^{2+} transporters, and Mn-CDF family involves in Mn^{2+} transportation (Zou et al. 2021a). Notable examples include AtMTP1 and AtMTP3 transporters in *Arabidopsis* that are implicated in Zn and Mn transport (Gu et al. 2021).

3.4.2.4 Cation/H^+ Antiporters

Cation/H^+ antiporters function by exchanging cations such as Mn^{2+}, Cd^{2+} and Zn^{2+} with protons across plasma membranes and often into the vacuoles (Khoudi 2021). This process helps sequester potentially toxic metal ions away from the cytoplasm,

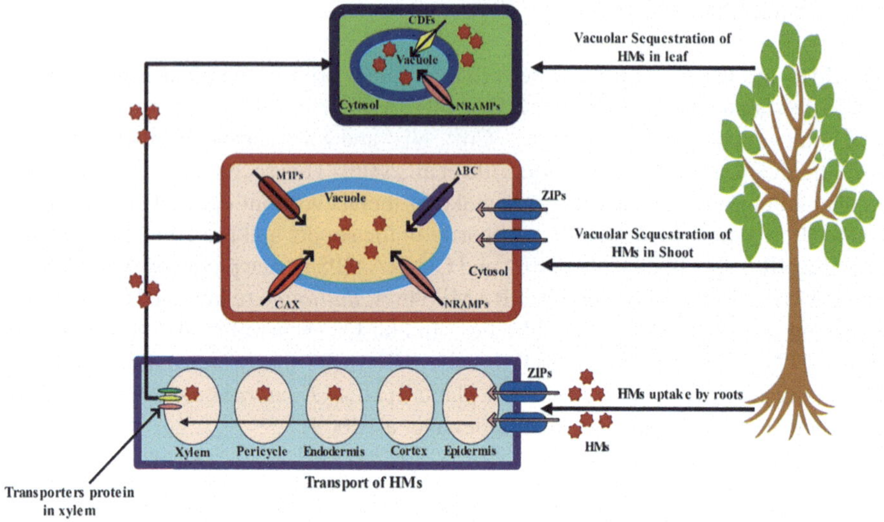

Fig. 2 Schematic diagram representing different HM transporters in metal detoxification process in higher plant

thereby contributing to metal tolerance. For example, the CAX family of transporters is known to confer tolerance to metal ions like Cd^{2+}, Zn^{2+} and Mn^{2+} by transporting such metal ions into vacuoles (He et al. 2022). Similarly, AtCAX2 and AtCAX4 transporters found in *A. thaliana* regulate the transportation of Mn, Cd and Zn within the plant (Zou et al. 2021b) (Fig. 2).

3.5 *Coping Mechanism of Land Plants Against Metal Toxicity: Comparative Analysis*

Different plant groups have thrived diverse tactics to counteract the various metal toxicity. These strategies involve detoxification processes targeting to reduce harmful impact of such metal ions.

3.5.1 Bryophytes

Bryophytes, the simplest land plants, possess a simple life cycle with a dominant haploid phase and dispersing via production of spores or vegetative propagules (Gibby 2023). Bryophytes thrive in varied habitats like soil, tree trunks and rock surfaces absorbing nutrients from the moist substrates (Bahuguna et al. 2013). Bryophytes, lacking cuticles and possessing a high surface-to-volume ratio, are highly sensitive to environmental changes making them ideal biomonitors for detecting toxic pollutants (De Agostini 2022). Bryophytes have evolved remarkable strategies to thrive in diverse habitats including those polluted with different metal pollutants. Mosses and liverworts have been widely studied for their remarkable capability to absorb significant amounts of HMs from their surroundings leaving out visible signs of damage (Khalid and Singh 2023). This makes them excellent bioindicators for monitoring environmental pollution. For example, mosses such as *Warnstorfia fluitans* against As and *Funaria hygrometrica* against Pb can accumulate a significant amount of metal ions to detoxify metal contamination (Basile et al. 2001; Sandhi et al. 2018). Besides, the liverworts *Marchantia polymorpha* can accumulate copper, Zn and Cd from metal-contaminated environment (Ares et al. 2018). Bryophytes have developed several strategies to tolerate HM toxicity. It can block HMs from entering the cells by creating physical barriers in the CWs (Glime 2024). Again, they have the ability to exchange ions and chelate metals using molecules like glutathione (GSH) and PCs reducing their toxicity. For example, GSH aids in Cd detoxification in *Leptodictyum riparium* (Maresca et al. 2022), while PCs are essential in *Marchantia polymorpha* (Li et al. 2022a). Moreover, to counteract oxidative stress caused by HMs bryophytes can produce antioxidant enzymes such as CAT and POX (Dutta et al. 2024).

3.5.2 Pteridophytes

Pteridophytes comprising of ferns and their relatives have a long evolutionary history and have developed diverse strategies to tolerate HMs. Many species from the group have been reported to thrive in contaminated soils and accumulate significant amounts of pollutants (Praveen and Pandey 2020). Ferns have developed detoxification mechanisms such as compartmentalization and chelation to tolerate HM stress (Sytar et al. 2021). Notably, aquatic ferns like *Salvinia* and *Azolla* can accumulate significant amounts of metals like Au, Cd, Pb and Cr making them effective for wastewater depuration (Fasani et al. 2022). Additionally, several land ferns especially from the *Pityrogramma* and *Pteris* genera have been identified in metal-contaminated soils and mining areas indicating their potential for ecological indicators and mine rehabilitation (Niazi et al. 2011). Among leptosporangiate ferns, Polypodiales has been most studied for metal accumulation with genera like *Adiantum, Asplenium* and *Pteris* showing significant variation in metal uptake (Sridokchan et al. 2005). *Pteris vittata, Pityrogramma calomelanos* and *Athyrium yokoscense* are notable for their exceptional metal accumulation capacities (Claveria et al. 2019). Particularly, *A. yokoscense* exhibits capacity for elevated accumulation of Pb, Cd, Zn, Cu and As (Van et al. 2006), while *P. calomelanos* and *P. vittata* are evidenced to accumulate high concentration of As (Campos et al. 2018). Hence, *P. vittata* with its high bioconcentration and translocation factors has become a model species for studying As hyperaccumulation at the molecular level with key transporters and enzymes involved in arsenate and arsenite uptake, reduction and storage (Gumaelius et al. 2004).

3.5.3 Seed-Bearing Plants: Gymnosperms and Angiosperms

3.5.3.1 Gymnosperms

Gymnosperms are valuable for assessing environmental contamination by tracking HM deposition on their leaves and bark over time. Gymnosperms show significant variability in sensitivity to metalloid ions depending on the species, element and environmental conditions (Fasani et al. 2022). Gymnosperms have been studied less extensively and also show varying degrees of tolerance to HMs, while angiosperms represent the majority of plant diversity and exhibit a wide range of responses to HM pollution. Gymnosperms have developed strategies such as chelation, antioxidant defence and compartmentalization to withstand the stress caused by HMs (Sethi 2023). The species-specific responses in gymnosperms include antioxidant activity including POX and SOD which counteract metal-induced oxidative stress (Kumar and Aery 2016). PCs can regulate metal detoxification processes with their production being induced by metals like Cd in various conifers (Saladin 2015). Understanding such tolerance mechanisms is crucial for selecting appropriate species for reforestation and phytoremediation particularly in polluted sites. For example, *Picea abies* shows tolerance to substantial concentrations of Zn, Pb, Cd and Cu making it suitable for reforestation in various contaminated areas (Konieczyński et al. 2018). Similarly, *Pinus sylvestris* accumulates HMs like Cd, Pb, Zn and Fe in its leaves and exhibits strong antioxidant activity to detoxify such metals (Kandziora-Ciupa et al. 2016).

3.5.3.2 Angiosperms

Angiosperms have developed several sophisticated mechanisms to detoxify metals enabling them to survive in contaminated environments. One key mechanism is the production of PCs and MTs which bind to toxic metal ions such as Pb and Cd by converting them into less harmful states (Yadav 2010). Additionally, angiosperms enhance the antioxidant enzyme activity including GPX, SOD and CAT to combat oxidative stress due to metal-induced ROS (Das and Roychoudhury 2014). *A. thaliana* possesses a complex network of metal transporters such as ABC, HMAs, NRAMPs and ZIPs to mediate the intracellular detoxification of multiple HMs (Komal et al. 2015). Similarly, MTs and PCs involve in metal detoxification process in rice plant to counteract the toxic effects of HMs including Cr, As and Cd (Kaur et al. 2021). Hyperaccumulator species like *Thlaspi caerulescens* are particularly specialized at accumulating and detoxifying elevated concentrations of metals including Zn and Ni by compartmentalizing them in leaf tissues which allows the plant to thrive in metal-rich soils (Cosio 2004). Another hyperaccumulator species such as *Brassica juncea* can tolerate and accumulate significant amounts of As and Cr making it a valuable species for phytoremediation (Mourato et al. 2015).

3.6 Similarities and Differences in Metal Detoxification Strategies Between Bryophytes and Vascular Plants

While bryophytes and vascular plants share some common metal detoxification strategies, bryophytes have evolved unique cellular mechanisms adapted to their simple structure and specific habitat requirements (Wang et al. 2022). The biomonitoring potential of bryophytes is also greater due to their distinctive morphological and physiological characteristics adopted over the years (Sinha et al. 2021).

Bryophytes and vascular plants employ the following several similar strategies to cope with metal toxicity.

3.6.1 Chelation

Both bryophytes and vascular plants use chelation mechanisms to bind HMs for reduction of their toxicity. The process involves the production of chelating compounds like PCs and MTs.

3.6.2 Antioxidant Responses

To counteract the oxidative stress produced by HMs, both bryophytes and vascular plants reported to be activated antioxidant enzymatic systems such as SOD, APX, CAT and GSH.

3.6.3 Compartmentalization

Sequestration of HMs in vacuoles and CWs is an important detoxification mechanism in both bryophytes and vascular plants. This prevents the metals from interfering with cellular processes.

However, there are some differences in metal detoxification strategies between bryophytes and vascular plants.

3.6.4 Anatomical Adaptations

Vascular plants have developed specialized anatomical features like a thick cuticle, stomatal regulation and endodermis in roots to restrict HM entry into the cell (Li et al. 2023). However, bryophytes lack such adaptations against HM exposure due to their simple structure.

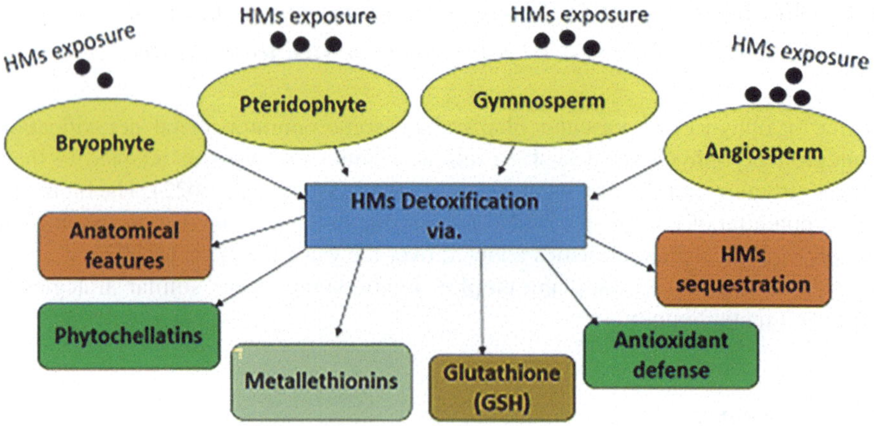

Fig. 3 Schematic diagram representing different ways of metal detoxifications in land plants

3.6.5 PC Synthase Function

In vascular plants like *A. thaliana*, PC synthases and PCs are involved in detoxifying various HMs (Pal and Rai 2010). However, in bryophytes PC synthases that have a narrower function primarily regulate the Fe/Zn homeostasis and Cd detoxification (Petraglia et al. 2014).

3.6.6 Glutathione-Based Detoxification

Some studies suggested that bryophytes rely more on glutathione-based complexation of HMs like Cd (Varela et al. 2018) followed by vacuolar sequestration compared to PC-based mechanisms.

3.6.7 Biomonitoring Potential

Due to their poikilohydric nature, high surface-to-volume ratio and lack of a root system, the bryophytes are more frequently used as biomonitors of HM pollution compared to the vascular plants (Chaudhuri and Roy 2024) (Fig. 3).

4 Evolutionary Adaptations of Land Plants Against Metal Toxicity

In an attempt to survive, plants must adapt themselves to the HM-stressed conditions in which the plants are subjected (Ghori et al. 2019). There are certain plants that bear the capacity to survive and reproduce normally in metal-rich soil and are hypertolerant against a concentration that is otherwise toxic for optimum growth and development (Macnair et al. 2020). Plants can achieve hypertolerance either by excluding toxic HMs from getting adsorbed or by restricting the entry of metals ions for root-shoot translocation (Bohra et al. 2015), while there are other plants that have evolved a peculiar capacity to hyperaccumulate HMs in their aerial tissues in comparison to non-accumulating or metal-sensitive plants growing in the same environment. These types of plants are known as hyperaccumulators (Sytar et al. 2021). The threshold concentration for categorizing plants as hyperaccumulator has been established for various metals, semimetals and metalloids (Pollard et al. 2014). According to recent suggested criteria, plants are considered as hyperaccumulators when such plants can bioconcentrate >10 mg/g of Mn, >3 mg/g of Zn, >1 mg/g of As, Pb and Cr, >0.3 mg/g of Cu and Co and >0.1 mg/g of Ti, Cd and Se in the above-ground parts of the plant (Manara et al. 2020). Such plants concentrate metals in the aerial parts and hence can remove HMs from the root tissue (Sytar et al. 2021). This is because of some physiological process which includes metal uptake from roots, xylem loading and unloading, symplast mobility (Sytar et al. 2021). These physiological responses are linked with the plant's capacity to combat the toxic effects of metals stored in above-ground parts via vacuolar sequestration and chelation (Riyazuddin et al. 2021). Since the discovery of these two traits in phytoremediation, a huge momentum has been gained regarding the study of rapid evolution and natural selection of hyperaccumulating and hypertolerating traits (Srivastava and Goyal 2010). An extensive study of a variety of behaviours of different plant species and populations has concluded hypertolerance and hyperaccumulation to be of independent genetic traits (Manara et al. 2020). As per the studies of Baker et al. (2020) followed by Krämer (2010), hyperaccumulation is an evolutionary character that has evolved several times. Hyperaccumulation is shown by a wide range of plant species belonging to remotely related families of bryophytes, pteridophytes, gymnosperms and angiosperms (Manara et al. 2020). Such hyperaccumulating plants are generally briefly lived perennial or biennial herbs, shrubs or small trees that are predominantly found in metalliferous soil in both temperate and tropical zones worldwide (Baker et al. 2020). Moreover, the distribution of such plants seems to be constricted to metal-contaminated or metalliferous soils where these plants become dominant over time (Borymski et al. 2018). Besides, such plants appear as pure population in metal-contaminated soil as these plants become dominant over other species due to their selective advantages often acting as strict metallophytes (Manara et al. 2020). Some facultative metallophytes also known as pseudo-metallophytes can thrive on non-metalliferous soil; however, such plants rarely compete with the adjacent non-tolerant plant species (Giorni 2017). This might be due to the fact that huge amount of energy and resource allocation is required to combat the toxic effects of HMs and carry out metal bio-accumulation (Maestri et al. 2010). In certain plant species, hyperaccumulation is reported as a constitutive character found in almost all the different populations as seen in *Noccaea caerulescens* which is a notable Zn hyperaccumulator (Escarré et al. 2013). A similar trend of hyperaccumulation where all the populations of species hyperaccumulate Zn irrespective of the substrate involved is recorded in *Arabidopsis halleri* (Stein et al. 2017).

Substantial evidence indicates that the HM tolerance trait in facultative metallophytes might have evolved rapidly due to prolonged exposure of high concentration of metals (Boquete et al. 2021). Besides, conspecific plant populations which are geographically distant could have evolved via independent evolution (Thakur et al. 2016; Manara et al. 2020). Recent investigations have pinpointed that elevated levels of Zn under controlled experimental conditions can be considered as a selective agent as it can negatively impact reproduction and survival capacity of plants (Nowak et al. 2018). Nowak et al. (2018) reported that just after few growth generations, experimental individuals of *N. caerulescens* grown in Zn-spiked soil have adapted physiological characters essential for tolerance against high levels of Zn metal. However, the underlying mechanism of metal-plant interactions still requires further investigations (Manara et al. 2020). An alternative hypothesis suggests metal-triggered gene methylation occurs in plants that is vertically inherited as a trans-generation adaptation to stress (Manara et al. 2020). However, concrete evidence regarding the inheritance of methylated gene induced as a result of different environmental stressors including HM toxicity is still very limited (Weinhold 2018).

5 Conclusion

The ability of plants to detoxify HMs suggests a promising and sustainable solution for mitigating metal contamination. Diverse mechanisms like HM uptake, sequestration transport and detoxification have evolved from bryophytes to angiosperms enabling plants to thrive in metal-contaminated environments. Harnessing such in-built capacities of plants via phytoremediation offers an alternative to conventional remediation methods eliminating further soil deterioration. Looking ahead, advancing the existing knowledge about plant responses to HM-induced stress especially through genetic engineering and molecular biology holds a significant promise for enhancement of phytoremediation efficiency of land plants. Moreover, extensive research can be focused on identifying new hypertolerants and hyperaccumulators for optimum plant-metal interactions and engineering of genetically modified plants with superior metal accumulating and tolerance traits. Additionally, integrating phytoremediation with other remediation technologies and adapting strategies will be crucial in development of sustainable technologies for metal remediation. Addressing ecological, ethical and regulatory concerns surrounding genetically modified organisms is also vital to ensure the safe and effective application of phytoremediation. As we confront the increasing challenges of metal pollution, the future of phytoremediation lies in a multidisciplinary approach that leverages plant diversity and innovation to protect and restore the environment sustainably (Table 2).

Table 2 Different detoxification strategies in plants to mitigate HM toxicity

Plant species	Metal detoxification strategies	Phylum	HMs	References
Funaria hygrometrica	Cellular accumulation	Bryophytes	Pb and As	Itouga et al. (2017)
Warnstorfia fluitans	Cellular accumulation	Bryophytes	Pb and As	Sandhi et al. (2018)
Hypnum plumaeforme	Antioxidant enzymes (CATs and POD)	Bryophytes		Sun et al. (2009)
Leptodictyum riparium	GSH	Bryophytes	Cd	Bellini et al. (2020)
Phragmites australis	Vacuolar sequestration	Angiosperm	Zn	Jiang and Wang (2008)
Alyssum serpyllifolium	Vacuolar sequestration	Angiosperm	Cd	de Knecht et al. (1994)
Silene vulgaris	Metals accumulated in the epidermal CWs	Angiosperm	Fe, Ni, Cu, Zn	Bringezu et al. (1999)
Sedum alfredii	PC synthesis	Angiosperm	Cd	Zhang et al. (2010)
Brassica rapa	MTs	Angiosperm	Cu, Zn, Mn	Ahn et al. (2012)
Sphagnum squarrosum	Glutathione-mediated detoxification	Bryophytes	Cu, Cd	Saxena and Saxena (2012)
Salvinia auriculata	Cellular accumulation	Pteridophytes	Fe, Cu, Cr, Cd, Cu and Ni	Dhir et al. (2011)
Azolla caroliniana	Cellular accumulation	Pteridophytes	Hg, Cr	Bennicelli et al. (2004)
Pinus sylvestris	SOD induction	Gymnosperm	Zn	Ivanov et al. (2012)
Ginkgo biloba	Production of PCs	Gymnosperm	Cd	Gekeler et al. (1989)
Cycas revoluta	Production of PCs	Gymnosperm	Cd	Gekeler et al. (1989)
Arabidopsis thaliana	PC accumulation	Angiosperm	Cd	Yamaguchi et al. (2020)
Lunularia cruciata	Vacuolar sequestration	Bryophytes	Cd	Degola et al. (2014)

References

Abd Elnabi, MK Elkaliny, NE Elyazied et al (2023) Toxicity of heavy metals and recent advances in their removal: a review. Toxics, 11(7):580. https://doi.org/10.3390/toxics11070580

Ajeesh Krishna TP, Maharajan T, Victor Roch G et al (2020) Structure, function, regulation and phylogenetic relationship of ZIP family transporters of plants. Frontiers in Plant Science, 11:662. https://doi.org/10.3389/fpls.2020.00662

Ahmad J, Ali AA, Baig MA et al (2019) Role of phytochelatins in cadmium stress tolerance in plants. In: Hasanuzzaman M, Narasimha M, Fujita VPM (eds) Cadmium toxicity and tolerance in plants, Academic press, 185–212. https://doi.org/10.1016/B978-0-12-814864-8.00008-5

Ahmad ZU (2015) Phytoremediation of heavy metal contaminated soil using Indian mustard and marigold plant. Dissertation, Bangladesh University of Engineering and Technology

Ahn YO, Kim SH, Lee J et al (2012) Three Brassica rapa metallothionein genes are differentially regulated under various stress conditions. Molecular biology reports, 39:2059–2067. https://doi.org/10.1007/s11033-011-0953-5

Al Mahmud J, Bhuyan MB, Anee TI et al (2019) Reactive Oxygen Species Metabolism and Antioxidant Defense in Plants Under Metal/Metalloid Stress. In: Hasanuzzaman M, Hakeem K, Nahar K, Alharby H (eds) Plant Abiotic Stress Tolerance. Springer, Cham. https://doi.org/10.1007/978-3-030-06118-0_10

Ali MM, Hossain D, Al-Imran A et al (2021) Environmental pollution with heavy metals: A public health concern. In: Nazal MK, Zhao H (eds) Heavy metals-their environmental impacts and mitigation, 771–783, IntechOpen, London, United Kingdom

Anjum NA, Hasanuzzaman M, Hossain MA et al (2015) Jacks of metal/metalloid chelation trade in plants—an overview. Frontiers in Plant Science, 6:192. https://doi.org/10.3389/fpls.2015.00192

Ansari MKA, Iqbal M, Ahmad M et al (2024) Heavy metal stress and cellular antioxidant systems of plants: A review. Agricultural Reviews, 45(3):400–409. https://doi.org/10.18805/ag.RF-321

Ares Á, Itouga M, Kato Y et al (2018) Differential metal tolerance and accumulation patterns of Cd, Cu, Pb and Zn in the liverwort Marchantia polymorpha L. Bulletin of environmental contamination and toxicology, 100(3):444–450. https://doi.org/10.1007/s00128-017-2241-0

Aydoğan S, Erdağ B, Aktaş L (2017) Bioaccumulation and oxidative stress impact of Pb, Ni, Cu, and Cr heavy metals in two bryophyte species, Pleurochaete squarrosa and Timmiella barbuloides. Turkish Journal of Botany, 41(5):464–475. https://doi.org/10.3906/bot-1608-33

Baghdadi, F, Nayebzadeh K, Aminifar M et al (2023) Pectin purification from plant materials. Macromolecular Research, 31(8), pp.753–770. https://doi.org/10.1007/s13233-023-00167-0

Baker AJ, McGrath SP, Reeves RD et al (2020) Metal hyperaccumulator plants: a review of the ecology and physiology of a biological resource for phytoremediation of metal-polluted soils. In: Terry N, Banuelos GS (eds) Phytoremediation of contaminated soil and water, pp.85–107. CRC Press

Basile A, Cogoni AE, Bassi P et al (2001) Accumulation of Pb and Zn in gametophytes and sporophytes of the moss Funaria hygrometrica (Funariales). Annals of Botany, 87(4):537–543. https://doi.org/10.1006/anbo.2001.1368

Bennicelli R, Stępniewska Z, Banach A et al (2004) The ability of Azolla caroliniana to remove heavy metals (Hg (II), Cr (III), Cr (VI)) from municipal waste water. Chemosphere, 55(1):141–146. https://doi.org/10.1016/j.chemosphere.2003.11.015

Berni R, Luyckx M, Xu X et al (2019) Reactive oxygen species and heavy metal stress in plants: Impact on the cell wall and secondary metabolism. Environmental and Experimental Botany, 161:98–106. https://doi.org/10.1016/j.envexpbot.2018.10.017

Bhaduri AM, Fulekar MH (2012) Antioxidant enzyme responses of plants to heavy metal stress. Reviews in Environmental Science and Bio/Technology, 11:55–69. https://doi.org/10.1007/s11157-011-9251-x

Bohra A, Sanadhya D, Chauhan R (2015) Heavy Metal Toxicity and Tolerance in Plants with Special Reference to Cadmium: A Review. Journal of Plant Science Research, 31(1)

Boquete MT, Lang I, Weidinger M et al (2021) Patterns and mechanisms of heavy metal accumulation and tolerance in two terrestrial moss species with contrasting habitat specialization. Environmental and Experimental Botany, 182:104336. https://doi.org/10.1016/j.envexpbot.2020.104336

Borymski S, Cycoń M, Beckmann M et al (2018) Plant species and heavy metals affect biodiversity of microbial communities associated with metal-tolerant plants in metalliferous soils. Frontiers in Microbiology, 9:1425. https://doi.org/10.3389/fmicb.2018.01425

Briat JF, Ravet K, Arnaud N et al (2010) New insights into ferritin synthesis and function highlight a link between iron homeostasis and oxidative stress in plants. Annals of botany, 105(5):811–822. https://doi.org/10.1093/aob/mcp128

Bringezu K, Lichtenberger O, Leopold I et al (1999) Heavy metal tolerance of Silene vulgaris. Journal of Plant Physiology, 154(4):536–546. https://doi.org/10.1016/S0176-1617(99)80295-8

Bahuguna YM, Gairola S, Semwal DP et al (2013) Bryophytes and ecosystem. In: Gupta RK, Kumar M (eds) Biodiversity of lower plants, pp 279–296. IK International Publishing House Pvt. Lt, New Delhi, India

Basile A, Sorbo S, Bassi P et al (2008) Effects of heavy metals on protonemal development and ultrastructure in populations of the moss Funaria hygrometrica Hedw.(Bryophyta) from a mine and an unpolluted site. Fresenius Environmental Bulletin, 17:1956–1963

Bellini E, Maresca V, Betti C et al (2020) The moss Leptodictyum riparium counteracts severe cadmium stress by activation of glutathione transferase and phytochelatin synthase, but slightly by phytochelatins. *International journal of molecular sciences, 21*(5):1583. https://doi.org/10.3390/ijms21051583

Campos NV, Arcanjo-Silva S, Freitas-Silva L et al (2018) Arsenic hyperaccumulation in Pityrogramma calomelanos L.(Link): adaptive traits to deal with high metalloid concentrations. Environmental Science and Pollution Research, 25:10720–10729. https://doi.org/10.1007/s11356-017-1085-9

Carroll J, Caporn s, Eades P (2009) Sphagnum in the Peak District: Current status and potential for restoration. moors for the future partnership

Chandrakar V, Yadu B, Xalxo R et al (2020) Mechanisms of Plant Adaptation and Tolerance to Metal/Metalloid Toxicity. In: Hasanuzzaman M (ed) Plant Ecophysiology and Adaptation under Climate Change: Mechanisms and Perspectives II. Springer, Singapore. https://doi.org/10.1007/978-981-15-2172-0_6

Chaudhuri S, Roy M (2024) Global ambient air quality monitoring: Can mosses help? A systematic meta-analysis of literature about passive moss biomonitoring. Environment, Development and Sustainability, 26(3):5735–5773. https://doi.org/10.1007/s10668-023-03043-0

Chen S, Ding Y (2023) Tackling heavy metal pollution: evaluating governance models and frameworks. Sustainability, 15(22):15863. https://doi.org/10.3390/su152215863

Chen X, Zhao Y, Zhong Y et al (2023) Deciphering the functional roles of transporter proteins in subcellular metal transportation of plants. Planta, 258(1):17. https://doi.org/10.1007/s00425-023-04170-8

Claveria RJR, Perez TR, Apuan MJB et al (2019) Pteris melanocaulon Fée is an As hyperaccumulator. Chemosphere, 236:124380. https://doi.org/10.1016/j.chemosphere.2019.124380

Clemens S (2019) Metal ligands in micronutrient acquisition and homeostasis. Plant, cell & environment, 42(10):2902–2912. https://doi.org/10.1111/pce.13627

Corpas FJ, Gupta DK, Palma JM (2015) Production Sites of Reactive Oxygen Species (ROS) in Organelles from Plant Cells. In: Gupta DK, Palma JM, Corpasn FJ (eds) Reactive Oxygen Species and Oxidative Damage in Plants Under Stress. Springer, Cham. https://doi.org/10.1007/978-3-319-20421-5_1

Cosio C (2004) Phytoextraction of heavy metals by hyperaccumulating and non hyperaccumulating plants: comparison of cadmium uptake and storage mechanisms in the plants (No. 2937). Dissertation, EPFL, Louis Pasteur University, France

Courville P, Chaloupka R, Cellier MFM (2006) Recent progress in structure–function analyses of Nramp proton-dependent metal-ion transporters. Biochemistry and Cell Biology, 84(6):960–978. https://doi.org/10.1139/o06-193

Dahuja A, Kumar RR, Sakhare A et al (2021) Role of ATP-binding cassette transporters in maintaining plant homeostasis under abiotic and biotic stresses. Physiologia Plantarum, 171(4):785–801. https://doi.org/10.1111/ppl.13302

DalCorso G (2012) Heavy Metal Toxicity in Plants. In: Furini A (ed) Plants and Heavy Metals. SpringerBriefs in Molecular Science, Springer, Dordrecht. https://doi.org/10.1007/978-94-007-4441-7_1

Dangles O (2012) Antioxidant activity of plant phenols: chemical mechanisms and biological significance. Current Organic Chemistry, 16(6):692–714. https://doi.org/10.2174/138527212799957995

Das K, Roychoudhury A (2014) Reactive oxygen species (ROS) and response of antioxidants as ROS-scavengers during environmental stress in plants. Frontiers in environmental science, 2:53. https://doi.org/10.3389/fenvs.2014.00053

De Caroli M, Furini A, DalCorso G et al (2020) Endomembrane reorganization induced by heavy metals. Plants, 9(4):482. https://doi.org/10.3390/plants9040482

De Agostini A (2022) Ecophysiological responses of bryophytes and orchids to environmental pollution: an integrated approach. Dissertation, Università degli Studi di Cagliari, Italy

de Knecht JA, van Dillen M, Koevoets PL et al (1994) Phytochelatins in cadmium-sensitive and cadmium-tolerant Silene vulgaris (chain length distribution and sulfide incorporation). Plant Physiology, 104(1):255–261. https://doi.org/10.1104/pp.104.1.255

Degola F, De Benedictis M, Petraglia A et al (2014) A Cd/Fe/Zn-responsive phytochelatin synthase is constitutively present in the ancient liverwort Lunularia cruciata (L.) Dumort. Plant and Cell Physiology, 55(11):1884–1891. https://doi.org/10.1093/pcp/pcu117

Dhir B (2021) Role of transporters of copper, manganese, zinc, and nickel in plants exposed to heavy metal stress. In: Roychoudhury A, Tripathi DK, Deshmukh R (eds) Metal and Nutrient Transporters in Abiotic Stress, pp. 145–168, Academic Press. https://doi.org/10.1016/B978-0-12-817955-0.00008-9

Dhir B, Sharmila P, Saradhi PP et al (2011) Heavy metal induced physiological alterations in Salvinia natans. Ecotoxicology and environmental safety, 74(6):1678–1684. https://doi.org/10.1016/j.ecoenv.2011.05.009

Di Salvatore M, Carafa AM, Carratu G (2008) Assessment of heavy metals phytotoxicity using seed germination and root elongation tests: A comparison of two growth substrates. Chemosphere, 73(9):1461–1464. https://doi.org/10.1016/j.chemosphere.2008.07.061

Drava G, Roccotiello E, Minganti V et al (2012) Effects of cadmium and arsenic on Pteris vittata under hydroponic conditions. Environmental Toxicology and Chemistry, 31(6):1375–1380. https://doi.org/10.1002/etc.1835

Dubey S, Shri M, Gupta A et al (2018) Toxicity and detoxification of heavy metals during plant growth and metabolism. Environmental Chemistry Letters, 16:1169–1192. https://doi.org/10.1007/s10311-018-0741-8

Dubey S, Shri M, Misra P et al (2014) Heavy metals induce oxidative stress and genome-wide modulation in transcriptome of rice root. Functional & integrative genomics, 14:401–417. https://doi.org/10.1007/s10142-014-0361-8

Dutta P, Prasad P, Indoilya Y et al (2024) Unveiling the molecular mechanisms of arsenic tolerance and resilience in the primitive bryophyte Marchantia polymorpha L. Environmental Pollution, 346:123506. https://doi.org/10.1016/j.envpol.2024.123506

Emamverdian A, Ding Y, Mokhberdoran F et al (2015) Heavy metal stress and some mechanisms of plant defense response. The scientific world journal, 2015(1):756120. https://doi.org/10.1155/2015/756120

Escarré J, Lefèbvre C, Frérot H et al (2013) Metal concentration and metal mass of metallicolous, non metallicolous and serpentine Noccaea caerulescens populations, cultivated in different growth media. Plant and soil, 370:197–221. https://doi.org/10.1007/s11104-013-1618-z

Escudero V, Ferreira Sánchez D, Abreu I et al (2022) Arabidopsis thaliana Zn2+-efflux ATPases HMA2 and HMA4 are required for resistance to the necrotrophic fungus Plectosphaerella cucumerina BMM. Journal of Experimental Botany, 73(1):339–350. https://doi.org/10.1093/jxb/erab400

Faizan M, Alam P, Hussain A et al (2024) Phytochelatins: key regulator against heavy metal toxicity in plants. Plant Stress, 11:100355. https://doi.org/10.1016/j.stress.2024.100355

Fasani E, Li M, Varotto C et al (2022) Metal detoxification in land plants: From bryophytes to vascular plants. STATE of the art and opportunities. Plants, 11(3):237. https://doi.org/10.3390/plants11030237

Figueroa JAL, Wrobel K, Afton S et al (2008) Effect of some heavy metals and soil humic substances on the phytochelatin production in wild plants from silver mine areas of Guanajuato, Mexico. Chemosphere, 70(11):2084–2091. https://doi.org/10.1016/j.chemosphere.2007.08.066

Flora SJS, Shrivastava R, Mittal M (2013) Chemistry and pharmacological properties of some natural and synthetic antioxidants for heavy metal toxicity. Current medicinal chemistry, 20(36):4540–4574. https://doi.org/10.2174/09298673113209990146

Fu S, Lu Y, Zhang X et al (2019) The ABC transporter ABCG36 is required for cadmium tolerance in rice. Journal of experimental botany, 70(20):5909–5918. https://doi.org/10.1093/jxb/erz335

Gebicka L, Krych-Madej J (2019) The role of catalases in the prevention/promotion of oxidative stress. Journal of inorganic biochemistry, 197:110699. https://doi.org/10.1016/j.jinorgbio.2019.110699

Gekeler W, Grill E, Winnacker EL et al (1989) Survey of the plant kingdom for the ability to bind heavy metals through phytochelatins. Zeitschrift für Naturforschung C, 44(5–6):361–369. https://doi.org/10.1515/znc-1989-5-605

Ghori NH, Ghori T, Hayat MQ et al (2019) Heavy metal stress and responses in plants. International journal of environmental science and technology, 16:1807–1828. https://doi.org/10.1007/s13762-019-02215-8

Giménez VMM, Bergam I, Reiter RJ et al (2021) Metal ion homeostasis with emphasis on zinc and copper: Potential crucial link to explain the non-classical antioxidative properties of vitamin D and melatonin. Life sciences, 281:119770. https://doi.org/10.1016/j.lfs.2021.119770

Giorni E (2017) Responses to environmental stresses in metallicolous and non metallicolous Silene paradoxa L. populations. Dissertation, Università degli Studi di Firenze, Italy

Gibby M (2023) Bryophytes and Pteridophytes: spore-bearing land plants. In: Maclean N (ed) The Living Planet: The State of the World's Wildlife, 37-64. Cambridge university press

Glime JM (2024) Roles of bryophytes in Forest sustainability—positive or negative?. Sustainability, 16(6):2359. https://doi.org/10.3390/su16062359

Gu D, Zhou X, Ma Y et al (2021) Expression of a Brassica napus metal transport protein (BnMTP3) in Arabidopsis thaliana confers tolerance to Zn and Mn. Plant Science, 304:110754. https://doi.org/10.1016/j.plantsci.2020.110754

Guidi Nissim W, Castiglione S, Guarino F et al (2023) Beyond cleansing: ecosystem services related to phytoremediation. Plants, 12(5):1031. https://doi.org/10.3390/plants12051031

Gumaelius L, Lahner B, Salt DE et al (2004) Arsenic hyperaccumulation in gametophytes of Pteris vittata. A new model system for analysis of arsenic hyperaccumulation. Plant physiology, 136(2):3198–3208. https://doi.org/10.1104/pp.104.044073

Gupta DK, Vandenhove H, Inouhe M (2013) Role of Phytochelatins in Heavy Metal Stress and Detoxification Mechanisms in Plants. In: Gupta D, Corpas F, Palma J (eds) Heavy Metal Stress in Plants. Springer, Berlin, Heidelberg. https://doi.org/10.1007/978-3-642-38469-1_4

Gupta A, Khan F, Pandey P et al (2024) A comprehensive review on the role of biosurfactants in remediation of heavy metals from contaminated environment. Bioremediation Journal, 1–27. https://doi.org/10.1080/10889868.2024.2427076

Hall JÁ (2002) Cellular mechanisms for heavy metal detoxification and tolerance. Journal of experimental botany, 53(366):1–11. https://doi.org/10.1093/jexbot/53.366.1

He F, Shi YJ, Li JL et al (2022) Genome-wide analysis and expression profiling of Cation/H+ exchanger (CAX) family genes reveal likely functions in cadmium stress responses in poplar. International journal of biological macromolecules, 204:76–88. https://doi.org/10.1016/j.ijbiomac.2022.01.202

Hossain MA, Piyatida P, da Silva JAT et al (2012) Molecular mechanism of heavy metal toxicity and tolerance in plants: central role of glutathione in detoxification of reactive oxygen species and methylglyoxal and in heavy metal chelation. Journal of botany, 2012(1):872875. https://doi.org/10.1155/2012/872875

Hu J (2021) Toward unzipping the ZIP metal transporters: structure, evolution, and implications on drug discovery against cancer. The FEBS journal, 288(20):5805–5825. https://doi.org/10.1111/febs.15658

Hussain S, Bharali P, Koushik B et al (2022) Bioaccumulation of metals in lichens and mosses understanding atmospheric deposition, metal-induced modifications and their suitability as biomonitors and bioremediators. In: Malik A, Kidwai MK, Garg VK (eds) Bioremediation of Toxic Metal(loid)s. pp. 57–80, CRC Press

Ighodaro OM, Akinloye OA (2018) First line defence antioxidants-superoxide dismutase (SOD), catalase (CAT) and glutathione peroxidase (GPX): Their fundamental role in the entire antioxidant defence grid. Alexandria journal of medicine, 54(4):287–293. https://doi.org/10.1016/j.ajme.2017.09.001

Imam MU, Zhang S, Ma J et al (2017) Antioxidants mediate both iron homeostasis and oxidative stress. Nutrients, 9(7):671. https://doi.org/10.3390/nu9070671

Ishimaru Y, Takahashi R, Bashir K et al (2012) Characterizing the role of rice NRAMP5 in manganese, iron and cadmium transport. Scientific reports, 2(1):286. https://doi.org/10.1038/srep00286

Ishida JK, Corcino DSM (2022) Role of natural resistance-associated macrophage proteins in metal ion transport in plants. In: Upadhyay SK (ed) Cation Transporters in Plants, pp. 337–356, Academic Press. https://doi.org/10.1016/B978-0-323-85790-1.00003-8

Itouga M, Hayatsu M, Sato M et al (2017) Protonema of the moss Funaria hygrometrica can function as a lead (Pb) adsorbent. PLoS One, 12(12). https://doi.org/10.1371/journal.pone.0189726

Ivanov YV, Savochkin YV, Kuznetsov VV (2012) Scots pine as a model plant for studying the mechanisms of conifers adaptation to heavy metal action: 2. Functioning of antioxidant enzymes in pine seedlings under chronic zinc action. Russian Journal of Plant Physiology, 59:50–58. https://doi.org/10.1134/S1021443712010098

Jan S, Parray JA (2016) Approaches to heavy metal tolerance in plants, pp. 1-18. Springer, Singapore. https://doi.org/10.1007/978-981-10-1693-6

Jain S, Muneer S, Guerriero G et al (2018) Tracing the role of plant proteins in the response to metal toxicity: a comprehensive review. Plant signaling & behavior, 13(9), e1507401. https://doi.org/10.1080/15592324.2018.1507401

Jalmi SK (2022) The Role of ABC Transporters in Metal Transport in Plants. In: Kumar K, Srivastava S (eds) Plant Metal and Metalloid Transporters. Springer, Singapore. https://doi.org/10.1007/978-981-19-6103-8_3

Jelea SG, Jelea M, Jelea OC (2015) Phytotoxicity of Lead on Lolium perenne L and Lactuca sativa var. capitata L: Effects of Germination and Growth. Bulletin of University of Agricultural Sciences and Veterinary Medicine Cluj-Napoca. Agriculture, 72(1):129–135. https://doi.org/10.15835/buasvmcn-agr:11187

Jiang X, Wang C (2008) Zinc distribution and zinc-binding forms in Phragmites australis under zinc pollution. Journal of Plant Physiology, 165(7):697–704. https://doi.org/10.1016/j.jplph.2007.05.011

Jogawat A, Yadav B, Chhaya et al (2021) Metal transporters in organelles and their roles in heavy metal transportation and sequestration mechanisms in plants. Physiologia plantarum, 173(1):259–275. https://doi.org/10.1111/ppl.13370

Joshi R, Pareek A, Singla-Pareek SL (2016) Plant metallothioneins: classification, distribution, function, and regulation. In: Ahmed P (ed) Plant metal interaction, pp. 239–261. Elsevier. https://doi.org/10.1016/B978-0-12-803158-2.00009-6

Joshi S (2021) BIOMONITORING OF ATMOSPHERIC HEAVY METALS DEPOSITION USING BRYOPHYTES: A REVIEW. Asian Journal of Plant and Soil Sciences, 6(1):46–58. https://doi.org/10.56557/ajopss/2021/v6i146

Kandziora-Ciupa M, Ciepał R, Nadgórska-Socha A et al (2016) Accumulation of heavy metals and antioxidant responses in Pinus sylvestris L. needles in polluted and non-polluted sites. Ecotoxicology 25:970–981. https://doi.org/10.1007/s10646-016-1654-6

Kanwal F, Riaz A, Ali S et al (2024) NRAMPs and manganese: Magic keys to reduce cadmium toxicity and accumulation in plants. Science of The Total Environment, 921:171005. https://doi.org/10.1016/j.scitotenv.2024.171005

Kaur H, Kohli SK, Khanna K et al (2022) Deciphering the role of metal binding proteins and metal transporters for remediation of toxic metals in plants. In: Malik A, Kidwai MK, Garg VK (eds) Bioremediation of Toxic Metal (loid)s, pp. 257–272, CRC Press

Kaur R, Das S, Bansal S et al (2021) Heavy metal stress in rice: Uptake, transport, signaling, and tolerance mechanisms. Physiologia plantarum, 173(1):430–448. https://doi.org/10.1111/ppl.13491

Kaur R, Yadav P, Kohli SK et al (2019) Emerging trends and tools in transgenic plant technology for phytoremediation of toxic metals and metalloids. In: Prasad MNV (ed) Transgenic plant technology for remediation of toxic metals and metalloids, 63–88. https://doi.org/10.1016/B978-0-12-814389-6.00004-3

Khalid Z, Singh B (2023) Looking at moss through the bioeconomy lens: biomonitoring, bioaccumulation, and bioenergy potential. Environmental Science and Pollution Research, 30(54):114722–114738. https://doi.org/10.1007/s11356-023-30633-2

Khoudi H (2021) Significance of vacuolar proton pumps and metal/H+ antiporters in plant heavy metal tolerance. Physiologia Plantarum, 173(1):384–393. https://doi.org/10.1111/ppl.13447

Kim YO, Kang H (2018) Comparative expression analysis of genes encoding metallothioneins in response to heavy metals and abiotic stresses in rice (Oryza sativa) and Arabidopsis thaliana. Bioscience, Biotechnology, and Biochemistry, 82(9):1656–1665. https://doi.org/10.1080/09168451.2018.1486177

Kim JJ, Kim YS, Kumar V (2019) Heavy metal toxicity: An update of chelating therapeutic strategies. Journal of Trace elements in Medicine and Biology, 54:226–231. https://doi.org/10.1016/j.jtemb.2019.05.003

Kobayashi R, Yoshimura E (2006) Differences in the binding modes of phytochelatin to Cadmium (II) and Zinc (II) ions. Biological trace element research, 114:313–318. https://doi.org/10.1385/BTER:114:1:313

Kocaman A (2023) Combined interactions of amino acids and organic acids in heavy metal binding in plants. Plant signaling & behavior, 18(1):2064072. https://doi.org/10.1080/15592324.2022.2064072

Komal T, Mustafa M, Ali Z et al (2015) Heavy Metal Uptake and Transport in Plants. In: Sherameti I, Varma A (eds) Heavy Metal Contamination of Soils. Soil Biology. Springer, Cham. https://doi.org/10.1007/978-3-319-14526-6_10

Konieczyński P, Moszczyński J, Wesołowski M (2018) Comparison of Fe, Zn, Mn, Cu, Cd and Pb concentration in spruce needles collected in the area of Gdansk and Gdynia in northern Poland. Environmental Protection and Natural Resources, 29(4):1–9. https://doi.org/10.2478/oszn-2018-0015

Kopittke PM, Menzies NW, Wang P et al (2014) The rhizotoxicity of metal cations is related to their strength of binding to hard ligands. Environmental Toxicology and Chemistry, 33(2):268–277. https://doi.org/10.1002/etc.2435

Kosakivska IV, Babenko LM, Romanenko KO et al (2021) Molecular mechanisms of plant adaptive responses to heavy metals stress. Cell Biology International, 45(2):258–272. https://doi.org/10.1002/cbin.11503

Krämer U (2010) Metal hyperaccumulation in plants. Annual review of plant biology, 61(1):517–534. https://doi.org/10.1146/annurev-arplant-042809-112156

Krzesłowska M (2011) The cell wall in plant cell response to trace metals: polysaccharide remodeling and its role in defense strategy. Acta physiologiae plantarum, 33:35–51. https://doi.org/10.1007/s11738-010-0581-z

Kulich I, Vojtíková Z, Glanc M et al (2015) Cell wall maturation of Arabidopsis trichomes is dependent on exocyst subunit EXO70H4 and involves callose deposition. Plant Physiology, 168(1):120–131. https://doi.org/10.1104/pp.15.00112

Kumar A, Aery NC (2016) Impact, Metabolism, and Toxicity of Heavy Metals in Plants. In: Singh A, Prasad S, Singh R (eds) Plant Responses to Xenobiotics. Springer, Singapore. https://doi.org/10.1007/978-981-10-2860-1_7

Lee M, Lee K, Lee J et al (2005) AtPDR12 contributes to lead resistance in Arabidopsis. Plant physiology, 138(2):827–836. https://doi.org/10.1104/pp.104.058107

Li J, Zhang Z, Shi G (2024) Genome-Wide Identification and Expression Profiling of Heavy Metal ATPase (HMA) Genes in Peanut: Potential Roles in Heavy Metal Transport. International Journal of Molecular Sciences, 25(1):613. https://doi.org/10.3390/ijms25010613

Li M, Leso M, Buti M et al (2022a) Phytochelatin synthase de-regulation in Marchantia polymorpha indicates cadmium detoxification as its primary ancestral function in land plants and provides a novel visual bioindicator for detection of this metal. Journal of Hazardous Materials, 440, 129844. https://doi.org/10.1016/j.jhazmat.2022.129844

Li P, Luo T, Pu X et al (2021) Plant transporters: roles in stress responses and effects on growth and development. Plant Growth Regulation, 93:253–266. https://doi.org/10.1007/s10725-020-00684-3

Li Q, Wang Y, Li Y et al (2022b) Speciation of heavy metals in soils and their immobilization at micro-scale interfaces among diverse soil components. Science of the Total Environment, 825, 153862. https://doi.org/10.1016/j.scitotenv.2022.153862

Li S, Lu S, Wang J et al (2023) Responses of physiological, morphological and anatomical traits to abiotic stress in woody plants. Forests, 14(9):1784. https://doi.org/10.3390/f14091784

Liao C, Li Y, Wu X et al (2023) ZmHMA3, a Member of the Heavy-Metal-Transporting ATPase Family, Regulates Cd and Zn Tolerance in Maize. International Journal of Molecular Sciences, 24(17), p.13496. https://doi.org/10.3390/ijms241713496

Liu W, Zhang X, Liang L et al (2015) Phytochelatin and Oxidative Stress Under Heavy Metal Stress Tolerance in Plants. In: Gupta D, Palma J, Corpas F (eds) Reactive Oxygen Species and Oxidative Damage in Plants Under Stress. Springer, Cham. https://doi.org/10.1007/978-3-319-20421-5_8

Lu Z, Yin L, Li W et al (2022) Low concentrations of silver nanoparticles inhibit spore germination and disturb gender differentiation of Ceratopteris thalictroides (L.) Brongn. Nanomaterials, 12(10), 1730. https://doi.org/10.3390/nano12101730

Macnair MR, Tilstone GH, Smith SE (2020) The genetics of metal tolerance and accumulation in higher plants. In: Terry N, Banuelos GS (eds) Phytoremediation of contaminated soil and water, pp. 235–250, CRC Press

Maestri E, Marmiroli M, Visioli G et al (2010) Metal tolerance and hyperaccumulation: costs and trade-offs between traits and environment. Environmental and Experimental Botany, 68(1):1–13. https://doi.org/10.1016/j.envexpbot.2009.10.011

Mahapatra B, Dhal NK, Dash AK et al (2019) Perspective of mitigating atmospheric heavy metal pollution: using mosses as biomonitoring and indicator organism. Environmental Science and Pollution Research, 26(29):29620–29638. https://doi.org/10.1007/s11356-019-06270-z

Manara A, Fasani E, Furini A et al (2020) Evolution of the metal hyperaccumulation and hypertolerance traits. Plant, Cell & Environment, 43(12):2969–2986. https://doi.org/10.1111/pce.13821

Manara A (2012) Plant Responses to Heavy Metal Toxicity. In: Furini A (eds) Plants and Heavy Metals, Springer, Dordrecht. https://doi.org/10.1007/978-94-007-4441-7_2

Manatschal C, Dutzler R (2022) The Structural Basis for Metal Ion Transport in the SLC11/NRAMP Family. Chimia, 76(12):1005–1010. https://doi.org/10.2533/chimia.2022.1005

Mansoor S, Ali A, Kour N et al (2023) Heavy metal induced oxidative stress mitigation and ROS scavenging in plants. Plants, 12(16):3003. https://doi.org/10.3390/plants12163003

Maresca V, Bellini E, Landi S et al (2022) Biological responses to heavy metal stress in the moss Leptodictyum riparium (Hedw.) Warnst. Ecotoxicology and Environmental Safety, 229:113078. https://doi.org/10.1016/j.ecoenv.2021.113078

Meychik N, Nikolaeva Y, Kushunina M et al (2014) Are the carboxyl groups of pectin polymers the only metal-binding sites in plant cell walls?. Plant and soil, 381:25–34. https://doi.org/10.1007/s11104-014-2111-z

Miller CJ, Wadley S, Waite TD (2017) Fenton, photo-Fenton and Fenton-like processes In: Stefan MI (ed) Advanced Oxidation Processes for Water Treatment: Fundamentals and Applications, 297. IWA publishing, London, UK

Mohamed BA, Ellis N, Kim CS et al (2017) The role of tailored biochar in increasing plant growth, and reducing bioavailability, phytotoxicity, and uptake of heavy metals in contaminated soil. Environmental Pollution, 230, 329–338. https://doi.org/10.1016/j.envpol.2017.06.075

Mourato MP, Moreira IN, Leitão I et al (2015) Effect of heavy metals in plants of the genus Brassica. International journal of molecular sciences, 16(8):17975–17998. https://doi.org/10.3390/ijms160817975

Możdżeń K, Wanic T, Rut G et al (2017) Toxic effects of high copper content on physiological processes in Pinus sylvestris L. Photosynthetica, 55:193–200. https://doi.org/10.1007/s11099-016-0229-3

Muro K, Segami S, Kawachi M et al (2024) Localization of the MTP4 transporter to trans-Golgi network in pollen tubes of Arabidopsis thaliana. Journal of Plant Research, 137(5):939–950. https://doi.org/10.1007/s10265-024-01559-8

Niazi NK, Bashir S, Bibi I et al (2016) Phytoremediation of Arsenic-Contaminated Soils Using Arsenic Hyperaccumulating Ferns. In: Ansari A, Gill S, Gill R, Lanza G, Newman L (eds) Phytoremediation. Springer, Cham. https://doi.org/10.1007/978-3-319-40148-5_19

Niazi NK, Singh B, Van Zwieten L et al (2011) Phytoremediation potential of Pityrogramma calomelanos var. austroamericana and Pteris vittata L. grown at a highly variable arsenic contaminated site. International Journal of Phytoremediation, 13(9):912–932. https://doi.org/10.1080/15226514.2011.568023

Noor I, Sohail H, Sun J et al (2022) Heavy metal and metalloid toxicity in horticultural plants: Tolerance mechanism and remediation strategies. Chemosphere, 303:135196. https://doi.org/10.1016/j.chemosphere.2022.135196

Nowak J, Frérot H, Faure N et al (2018) Can zinc pollution promote adaptive evolution in plants? Insights from a one-generation selection experiment. Journal of experimental botany, 69(22):5561–5572. https://doi.org/10.1093/jxb/ery327

Oomen RJ, Wu J, Lelièvre F et al (2009) Functional characterization of NRAMP3 and NRAMP4 from the metal hyperaccumulator Thlaspi caerulescens. New Phytologist, 181(3):637–650. https://doi.org/10.1111/j.1469-8137.2008.02694.x

Oves M, Saghir Khan M, Huda Qari A et al (2016) Heavy metals: biological importance and detoxification strategies. Journal of Bioremediation and Biodegradation, 7(2):1–15. https://doi.org/10.4172/2155-6199.1000312

Pal R, Rai JPN (2010) Phytochelatins: peptides involved in heavy metal detoxification. Applied biochemistry and biotechnology, 160:945–963. https://doi.org/10.1007/s12010-009-8565-4

Palusińska M, Barabasz A, Kozak K et al (2020) Zn/Cd status-dependent accumulation of Zn and Cd in root parts in tobacco is accompanied by specific expression of ZIP genes. BMC plant biology, 20:1–19. https://doi.org/10.1186/s12870-020-2255-3

Pandey S, Fartyal D, Agarwal A et al (2017) Abiotic stress tolerance in plants: myriad roles of ascorbate peroxidase. Frontiers in plant science, 8:581. https://doi.org/10.3389/fpls.2017.00581

Pang CH, Wang BS (2010) Role of Ascorbate Peroxidase and Glutathione Reductase in Ascorbate–Glutathione Cycle and Stress Tolerance in Plants. In: Anjum N, Chan MT, Umar S (eds) Ascorbate-Glutathione Pathway and Stress Tolerance in Plants. Springer, Dordrecht. https://doi.org/10.1007/978-90-481-9404-9_3

Parrotta L, Guerriero G, Sergeant K et al (2015) Target or barrier? The cell wall of early-and later-diverging plants vs cadmium toxicity: differences in the response mechanisms. Frontiers in plant science, 6:133. https://doi.org/10.3389/fpls.2015.00133

Pasricha S, Mathur V, Garg A et al (2021) Molecular mechanisms underlying heavy metal uptake, translocation and tolerance in hyperaccumulators-an analysis: Heavy metal tolerance in hyperaccumulators. Environmental Challenges, 4:100197. https://doi.org/10.1016/j.envc.2021.100197

Peralta-Videa JR, De la Rosa G, Gonzalez JH et al (2004) Effects of the growth stage on the heavy metal tolerance of alfalfa plants. Advances in Environmental Research, 8(3–4):679–685. https://doi.org/10.1016/S1093-0191(03)00040-6

Petraglia A, De Benedictis M, Degola F et al (2014) The capability to synthesize phytochelatins and the presence of constitutive and functional phytochelatin synthases are ancestral (plesiomorphic) characters for basal land plants. Journal of Experimental Botany, 65(4):1153–1163. https://doi.org/10.1093/jxb/ert472

Pirzadah TB, Malik B, Tahir I et al (2015) Phytoremediation: an eco-friendly green technology for pollution prevention, control and remediation. In: Hakeem KR, Sabir M, Ozturk M, Mermut AR (eds) Soil Remediation and Plants: Prospects and Challenges, 5:107–126. Academic press

Pollard AJ, Reeves RD, Baker AJ (2014) Facultative hyperaccumulation of heavy metals and metalloids. Plant Science, 217:8–17. https://doi.org/10.1016/j.plantsci.2013.11.011

Praveen A, Pandey VC (2020) Pteridophytes in phytoremediation. Environmental geochemistry and health, 42(8):2399–2411. https://doi.org/10.1007/s10653-019-00425-0

Rahman A (2022) How to remediate heavy metal contamination in soil?. Science Insights, 41(4):669–674. https://doi.org/10.15354/si.22.re082

Rahman Z, Singh VP (2020) Bioremediation of toxic heavy metals (THMs) contaminated sites: concepts, applications and challenges. Environmental Science and Pollution Research, 27(22):27563–27581. https://doi.org/10.1007/s11356-020-08903-0

Rai GK, Kumar P, Choudhary SM et al (2023) Antioxidant potential of glutathione and crosstalk with phytohormones in enhancing abiotic stress tolerance in crop plants. Plants, 12(5):1133. https://doi.org/10.3390/plants12051133

Rai S, Singh PK, Mankotia S et al (2021) Iron homeostasis in plants and its crosstalk with copper, zinc, and manganese. Plant Stress, 1:100008. https://doi.org/10.1016/j.stress.2021.100008

Radotić K, Dučić T, Prodanović R et al (2003) Toxicity of nickel and cadmium in spruce seedlings: Effect of separated and combined treatments on peroxidase and superoxide-dismutase activity. Jugoslovenska medicinska biohemija, 22(1):41–52

Raychaudhuri SS, Pramanick P, Talukder P et al (2021) Polyamines, metallothioneins, and phytochelatins—Natural defense of plants to mitigate heavy metals. In: Rahman A (ed) Studies in natural products chemistry, 69:227–261, Elsevier. https://doi.org/10.1016/B978-0-12-819487-4.00006-9

Rekha K, Usha B, Keeran NS (2021) Role of ABC transporters and other vacuolar transporters during heavy metal stress in plants. In: Roychoudhury A, Tripathi DK, Deshmukh R (eds) Metal and Nutrient Transporters in Abiotic Stress, Academic Press. https://doi.org/10.1016/B978-0-12-817955-0.00003-X

Riaz M, Kamran M, Rizwan M et al (2021) Cadmium uptake and translocation: selenium and silicon roles in Cd detoxification for the production of low Cd crops: a critical review. Chemosphere, 273:129690. https://doi.org/10.1016/j.chemosphere.2021.129690

Riyazuddin R, Nisha N, Ejaz B et al (2021) A comprehensive review on the heavy metal toxicity and sequestration in plants. Biomolecules, 12(1):43. https://doi.org/10.3390/biom12010043

Roychoudhury A, Basu S (2012) Ascorbate-glutathione and plant tolerance to various abiotic stresses. In: Anjum NA, Umar S, Ahmad A (eds) Oxidative stress in plants: causes, consequences and tolerance, pp.177–258. I.K. International Publishing House Pvt. Ltd., New Delhi

Saladin G (2015) Phytoextraction of Heavy Metals: The Potential Efficiency of Conifers. In: Sherameti I, Varma A (eds) Heavy Metal Contamination of Soils. Soil Biology, vol 44. Springer, Cham. https://doi.org/10.1007/978-3-319-14526-6_18

Sandhi A, Landberg T, Greger M (2018) Phytofiltration of arsenic by aquatic moss (Warnstorfia fluitans). Environmental Pollution, 237:1098–1105. https://doi.org/10.1016/j.envpol.2017.11.038

Sarkar B, Mukhopadhyay R, Ramanayaka S et al (2021) The role of soils in the disposition, sequestration and decontamination of environmental contaminants. Philosophical Transactions of the Royal Society B, 376(1834):20200177. https://doi.org/10.1098/rstb.2020.0177

Satofuka H, Fukui T, Takagi M et al (2001) Metal-binding properties of phytochelatin-related peptides. Journal of Inorganic biochemistry, 86(2–3):595–602. https://doi.org/10.1016/S0162-0134(01)00223-9

Saxena A, Saxena A (2012) Bioaccumulation and glutathione-mediated detoxification of copper and cadmium in Sphagnum squarrosum Crome Samml. Environmental monitoring and assessment, 184:4097–4103. https://doi.org/10.1007/s10661-011-2246-9

Schwab F, Zhai G, Kern M et al (2016) Barriers, pathways and processes for uptake, translocation and accumulation of nanomaterials in plants–Critical review. Nanotoxicology, 10(3):257–278. https://doi.org/10.3109/17435390.2015.1048326

Seregin IV, Kozhevnikova AD (2023) Phytochelatins: Sulfur-containing metal (loid)-chelating ligands in plants. International Journal of Molecular Sciences, 24(3):2430. https://doi.org/10.3390/ijms24032430

Seregin IV, Kozhevnikova AD (2024) The role of low-molecular-weight organic acids in metal homeostasis in plants. International Journal of Molecular Sciences, 25(17):9542. https://doi.org/10.3390/ijms25179542

Seregin IV, Kozhevnikova AD, Schat H (2022) Nickel tolerance and accumulation capacities in different populations of the hyperaccumulator Noccaea caerulescens. Russian Journal of Plant Physiology, 69(4):70. https://doi.org/10.1134/S1021443722040148

Sethi S (2023) Phytochelatins: Heavy Metal Detoxifiers in Plants. In: Shah MP (ed) Advanced and Innovative Approaches of Environmental Biotechnology in Industrial Wastewater Treatment. Springer, Singapore. https://doi.org/10.1007/978-981-99-2598-8_16

Shahid M, Dumat C, Khalid S et al (2016) Cadmium Bioavailability, Uptake, Toxicity and Detoxification in Soil-Plant System. In: de Voogt P (ed) Reviews of Environmental Contamination and Toxicology, Volume 241. Springer, Cham. https://doi.org/10.1007/398_2016_8

Sharma SS, Dietz KJ (2006) The significance of amino acids and amino acid-derived molecules in plant responses and adaptation to heavy metal stress. Journal of experimental botany, 57(4):711–726. https://doi.org/10.1093/jxb/erj073

Sharma SS, Dietz KJ, Mimura T (2016) Vacuolar compartmentalization as indispensable component of heavy metal detoxification in plants. Plant, Cell & Environment, 39(5):1112–1126. https://doi.org/10.1111/pce.12706

Shaw AJ, Szövényi P, Shaw B (2011) Bryophyte diversity and evolution: windows into the early evolution of land plants. American journal of botany, 98(3):352–369. https://doi.org/10.3732/ajb.1000316

Shaw BP, Prasad MNV, Jha VK et al (2005) Detoxification/Defense Mechanisms in Metal-Exposed Plants. In: Prasad MNV, Sajwan KS, Naidu R (eds) Trace Elements in the Environment: Biogeochemistry, Biotechnology, and Bioremediation. CRC Press. https://doi.org/10.1201/9781420032048

Shi W, Zhang Y, Chen S et al (2019) Physiological and molecular mechanisms of heavy metal accumulation in nonmycorrhizal versus mycorrhizal plants. Plant, Cell & Environment, 42(4):1087–1103. https://doi.org/10.1111/pce.13471

Shrivastava M, Srivastava S (2021) Application and research progress of Hydrilla verticillata in ecological restoration of water contaminated with metals and metalloids. Environmental Challenges, 4:100177. https://doi.org/10.1016/j.envc.2021.100177

Silva RMGD, Marques de Oliveira Moraes V, Marinho Dos Santos VH et al (2022) Heavy metal accumulation efficiency and subsequent of cytogenotoxicity evaluation in the medicinal plant Equisetum hyemale. Journal of Toxicology and Environmental Health, Part A, 85(24):989–1001. https://doi.org/10.1080/15287394.2022.2139313

Sinha S, Singh A, Sinha D et al (2021) A review on bryophytes as key bio-indicators to monitor heavy metals in the atmosphere. International Journal of Plant and Environment, 7(01):49–62. https://doi.org/10.18811/ijpen.v7i01.5

Spormann S, Soares C, Teixeira J et al (2021) Polyamines as key regulatory players in plants under metal stress—A way for an enhanced tolerance. Annals of Applied Biology, 178(2):209–226. https://doi.org/10.1111/aab.12660

Song WY, Park J, Mendoza-Cózatl DG et al (2010) Arsenic tolerance in Arabidopsis is mediated by two ABCC-type phytochelatin transporters. Proceedings of the National Academy of Sciences, 107(49), 21187–21192. https://doi.org/10.1073/pnas.1013964107

Sridokchan W, Markich S, Visoottiviseth P (2005) Arsenic tolerance, accumulation and elemental distribution in twelve ferns: a screening study. Australasian Journal of Ecotoxicology, 11(2):101–110.

Srivastava N (2016) Role of Phytochelatins in Phytoremediation of Heavy Metals Contaminated Soils. In: Ansari A, Gill S, Gill R, Lanza G, Newman L (eds) Phytoremediation. Springer, Cham. https://doi.org/10.1007/978-3-319-40148-5_13

Srivastava S, Goyal P (eds) (2010) Novel biomaterials: decontamination of toxic metals from wastewater. Springer Science & Business Media.

Stravinskienė V, Račaitė M (2014) Impact of cadmium and zinc on the growth of white clover (Trifolium repens L.) shoots and roots. Pol J Environ Stud, 23(4):1355–1359.

Stein RJ, Höreth S, de Melo JRF et al (2017) Relationships between soil and leaf mineral composition are element-specific, environment-dependent and geographically structured in the emerging model Arabidopsis halleri. New phytologist, 213(3):1274–1286. https://doi.org/10.1111/nph.14219

Sterckeman T, Thomine S (2020) Mechanisms of cadmium accumulation in plants. Critical Reviews in Plant Sciences, 39(4):322–359. https://doi.org/10.1080/07352689.2020.1792179

Sultana F, Ghosh A (2024) Exploring the evolutionary landscape and structural resonances of ferritin with insights into functional significance in plant. Biochimie, 227:217–230. https://doi.org/10.1016/j.biochi.2024.07.014

Sun Q, Li Y, Shi L et al (2022) Heavy metals induced mitochondrial dysfunction in animals: Molecular mechanism of toxicity. Toxicology 469: 153136. https://doi.org/10.1016/j.tox.2022.153136

Sun SQ, He M, Cao T et al (2009) Response mechanisms of antioxidants in bryophyte (Hypnum plumaeforme) under the stress of single or combined Pb and/or Ni. Environ Monit Assess 149(1):291–302. https://doi.org/10.1007/s10661-008-0203-z

Sunitha MS, Prashant S, Kumar SA et al (2013) Cellular and molecular mechanisms of heavy metal tolerance in plants: a brief overview of transgenic plants overexpressing phytochelatin synthase and metallothionein genes. Plant Cell Biotechnol Mol Biol 14(2): 33–48

Sytar O, Ghosh S, Malinska H et al (2021) Physiological and molecular mechanisms of metal accumulation in hyperaccumulator plants. Physiol Plant 173(1):148–166. https://doi.org/10.1111/ppl.13285

Sytar O, Kumar A, Latowski D et al (2013) Heavy metal-induced oxidative damage, defense reactions, and detoxification mechanisms in plants. Acta Physiol Plant 35: 985–999.

Takahashi R, Bashir K, Ishimaru Y et al (2012) The role of heavy-metal ATPases, HMAs, in zinc and cadmium transport in rice. Plant Signal Behav 7(12): 1605–1607. https://doi.org/10.4161/psb.22454

Tarigholizadeh S, Heydari R, Sushkova S et al (2024) Metallotolerant Microbes for Improving the Health of Heavily Polluted Soil. In: Ranjan A, Rajput VD, Chauhan A, Prazdnova EV, Minkina T, Zargar SM (eds) Extremophiles for Sustainable Agriculture and Soil Health Improvement. Springer, Cham. https://doi.org/10.1007/978-3-031-70203-7_8

Thakur S, Singh L, Wahid ZA et al (2016) Plant-driven removal of heavy metals from soil: uptake, translocation, tolerance mechanism, challenges, and future perspectives. Environ Monit Assess 188:1–11. https://doi.org/10.1007/s10661-016-5211-9

Thomas C, Tampé R (2020) Structural and mechanistic principles of ABC transporters. Annu Rev Biochem 89(1):605–636. https://doi.org/10.1146/annurev-biochem-011520-105201

Tiwari S, Lata C (2018) Heavy metal stress, signaling, and tolerance due to plant-associated microbes: an overview. Front Plant Sci 9:452. https://doi.org/10.3389/fpls.2018.00452

Ueno D, Milner MJ, Yamaji N et al (2011) Elevated expression of TcHMA3 plays a key role in the extreme Cd tolerance in a Cd-hyperaccumulating ecotype of Thlaspi caerulescens. Plant J 66(5):852–862. https://doi.org/10.1111/j.1365-313X.2011.04548.x

Van TK, Kang Y, Fukui T et al (2006) Arsenic and heavy metal accumulation by Athyrium yokoscense from contaminated soils. Soil Sci Plant Nutri 52(6): 701–710. https://doi.org/10.1111/j.1747-0765.2006.00090.x

Vance GF, Stevenson FJ, Sikora FJ (2020) Environmental chemistry of aluminum–organic complexes. In: Sposito G (ed) The environmental chemistry of aluminium, 2nd edn. CRC Press, London

Varela Z, Debén S, Saxena DK et al (2018) Levels of antioxidant compound glutathione in moss from industrial areas. Atmosphere 9(7): 284. https://doi.org/10.3390/atmos9070284

Vatamaniuk OK, Mari S, Lu YP et al (2000) Mechanism of heavy metal ion activation of phytochelatin (PC) synthase: blocked thiols are sufficient for PC synthase-catalyzed transpeptidation of glutathione and related thiol peptides. J of Biol Chem 275(40): 31451–31459. https://doi.org/10.1074/jbc.M002997200

Viehweger K (2014) How plants cope with heavy metals. Bot Stud 55:1–12. https://doi.org/10.1186/1999-3110-55-35

Vishwakarma K, Mishra M, Patil G et al (2019) Avenues of the membrane transport system in adaptation of plants to abiotic stresses. Crit Rev Biotechnol 39(7): 861–883. https://doi.org/10.1080/07388551.2019.1616669

Wang QH, Zhang J, Liu Y et al (2022) Diversity, phylogeny, and adaptation of bryophytes: insights from genomic and transcriptomic data. J Exp Bot 73(13): 4306–4322. https://doi.org/10.1093/jxb/erac127

Wang X, Ai S, Liao H (2023) Deciphering interactions between phosphorus status and toxic metal exposure in plants and rhizospheres to improve crops reared on acid soil. Cells 12(3)9: 441. https://doi.org/10.3390/cells12030441

Weinhold A (2018) Transgenerational stress-adaption: an opportunity for ecological epigenetics. Plant Cel Rep 37(1):3–9. https://doi.org/10.1007/s00299-017-2216-y

Wu D, Saleem M, He T et al (2021) The mechanism of metal homeostasis in plants: A new view on the synergistic regulation pathway of membrane proteins, lipids and metal ions. Membranes 11(12):984. https://doi.org/10.3390/membranes11120984

Yadav R, Singh S, Kumar A et al 2022) Phytoremediation: A wonderful cost-effective tool. In: Kathi S, Devipriya S, Thamaraiselvi K (eds) Cost effective technologies for solid waste and wastewater treatment, Elsevier, Netherlands. https://doi.org/10.1016/B978-0-12-822933-0.00008-5

Yadav SK (2010) Heavy metals toxicity in plants: an overview on the role of glutathione and phytochelatins in heavy metal stress tolerance of plants. S. Afr J, Botany 76(2):167–179. https://doi.org/10.1016/j.sajb.2009.10.007

Yamaguchi C, Khamsalath S, Takimoto Y et al (2020) SLIM1 transcription factor promotes sulfate uptake and distribution to shoot, along with phytochelatin accumulation, under cadmium stress in Arabidopsis thaliana. Plts 9(2):163. https://doi.org/10.3390/plants9020163

Yang Z, Yang F, Liu JL et al (2022) Heavy metal transporters: Functional mechanisms, regulation, and application in phytoremediation. Sci Total Environ 809:151099. https://doi.org/10.1016/j.scitotenv.2021.151099

Yi Y, Liu H, Chen G et al (2023) Cadmium Accumulation in Plants: Insights from Phylogenetic Variation into the Evolution and Functions of Membrane Transporters. Sustainability 15(16): 12158. https://doi.org/10.3390/su151612158

Zangi R, Filella M (2012) Transport routes of metalloids into and out of the cell: a review of the current knowledge. Chem-Biol Interac 197(1): 47–57. https://doi.org/10.1016/j.cbi.2012.02.001

Zalewska M, Trefon J, Milnerowicz H (2014) The role of metallothionein interactions with other proteins. Proteomics, 14(11):1343–1356. https://doi.org/10.1002/pmic.201300496

Zhang J, Zhu Y, Yu L et al (2022) Research advances in cadmium uptake, transport and resistance in rice (Oryza sativa L.). Cells 11(3): 569. https://doi.org/10.3390/cells11030569

Zhang X., Ma Y, Lai D et al (2023) RsPDR8, a member of ABCG subfamily, plays a positive role in regulating cadmium efflux and tolerance in radish (Raphanus sativus L.). Plant Physiol Biochem 205: 108149. https://doi.org/10.1016/j.plaphy.2023.108149

Zhang ZC, Chen BX, Qiu BS (2010) Phytochelatin synthesis plays a similar role in shoots of the cadmium hyperaccumulator Sedum alfredii as in non-resistant plants. Plant Cell Environ 33(8): 1248–1255. https://doi.org/10.1111/j.1365-3040.2010.02144.x

Zhao X, Joo JC, Kim JY (2021) Evaluation of heavy metal phytotoxicity to Helianthus annuus L. using seedling vigor index-soil model. Chemosphere, 275, 130026. https://doi.org/10.1016/j.chemosphere.2021.130026

Zou T, Lin R, Pu L et al (2021a) Genome-wide identification, structure characterization, expression pattern profiling, and substrate specificity of the metal tolerance protein family in Canavalia rosea (Sw.) DC. Plts 10(7): 1340. https://doi.org/10.3390/plants10071340
Zou W, Chen J, Meng L et al (2021b) The rice cation/H+ exchanger family involved in Cd tolerance and transport. International Journal of Molecular Sciences, 22(15): 8186. https://doi.org/10.3390/ijms22158186
Zulfiqar U, Farooq M, Hussain S et al (2019) Lead toxicity in plants: Impacts and remediation. Journal of Environmental Management, 250: 109557. https://doi.org/10.1016/j.jenvman.2019.109557

Plant Nutrition and Soil Fertility Management in Heavy Metals Contaminated Soils

Seyed Majid Mousavi, Azadeh Sedaghat, Vahid Alah Jahandideh Mahjen Abadi, and Seyed Ali Ghaffari Nejad

Abstract Over the last decades, with the rapid development of industrialization and overpopulation, a large volume of different contaminants has been produced and entered the soil. Heavy metals (HMs) and metalloids are known as important environmental contaminants. When HMs accumulate to toxic levels in agricultural soils, these non-biodegradable elements adversely affect crop health and productivity. There are different management strategies for mitigating the adverse effects of HMs in the soil and environment. By using the procedures related to soil fertility and plant nutrition management, including optimal application of fertilizers and beneficial and essential nutrients to plants, using hyperaccumulator plants, and using the potentials of soil bacteria known as plant growth-promoting rhizobacteria (PGPRs), this issue may be alleviated. This chapter discusses the position of HMs in the research of the recent decade, the usual procedures and the importance of plant nutrition and soil fertility management, and the functions and mechanisms related to the soil bacteria in the remediation of HM-contaminated soils.

Keywords Plant nutrition · Soil bacteria · Soil contamination · Soil fertility

1 Introduction

Because soil fertility and plant nutrition have a direct correlation with food security and public health, the status of soil fertility and plant nutrition under heavy metal stress is a critical global concern, particularly in regions experiencing rapid industrialization and agricultural intensification. The persistence of HMs poses significant threats to soil fertility and plant nutrition, which in turn threatens food security and human health, making it a topic of notable importance in environmental science and policy discussions worldwide. As it was mentioned in literature reviews, some

S. M. Mousavi (✉) · A. Sedaghat · V. A. J. M. Abadi · S. A. G. Nejad
Soil and Water Research Institute (SWR), Agricultural Research, Education and Extension Organization (AREEO), Karaj, Iran

S. Roy, V. Mandal (eds.), *Plant-Microbe Interaction under Xenobiotic Exposure*,
https://doi.org/10.1007/978-981-96-8260-7_15

HMs (Fe, Cu, and Zn) are essential for plants and animals (Wintz et al. 2002); their availability in the environment varies, and metals such as Cu, Zn, Fe, Mn, Mo, Ni, and Co are essential micronutrients whose uptake in excess of the plant requirements results in toxic effects (Monni et al. 2000). Some other HMs (Hg, Pb, Cd, and As) are extremely harmful to both plants and people (Mansoor et al. 2021). They may enter plant systems and pollute the food chain, which is extremely dangerous for the safety of human health and the quality of food. The industrial revolution and economic globalization, anthropogenic activities such as mining, smelting, and agricultural fertilization significantly increased the soil burden of HMs and are considered the main sources of soil toxic metal pollution in the environment, soil, and food chains (Mousavi 2022; Mousavi et al. 2023a). Research indicates that contaminated soils not only inhibit plant growth and nutrient uptake but also pose risks of bioaccumulation in the food chain, thereby impacting public health and safety (Rai et al. 2019). Risky concentrations of HMs in the soil can affect biodiversity by affecting microbial activity and soil fertility (Alengebawy et al. 2021). These non-biodegradable substances hurt crop health and productivity when they build up to hazardous levels in agricultural soils. They can damage cell membranes, affect organelle structure and function, and disrupt various physiological and biochemical processes (Zhu et al. 2021).

The concentration range of HMs in natural water, sediments, and agricultural soils; the methods by which they enter the bodies of living creatures; contamination of some food crops with HMs; and related health risk assessment were discussed in detail in the previous works (Mousavi 2022; Mousavi et al. 2023a). By using different management strategies related to agriculture, the entrance of HMs to the environment may be decreased. For example, optimal fertilization based on the results of soil testing and plant analysis, the use of high-quality fertilizers, the better management of using different biosolids and organic fertilizers, and monitoring the quality of wastewater used for irrigation are some management practices for mitigating the entrance of HMs in agricultural systems. Furthermore, there are different methods and technologies for mitigating the adverse effects of HMs on public health and remediating the HM-contaminated soils, including the use of different biologic, organic, and chemical amendments. This chapter tries to discuss the status, importance, and effects of HMs on public health and food security; management and remediation strategies for HM-contaminated soils; the importance and efficacy of plant nutrition and soil fertility management practices to mitigate HM stress; and the importance of using soil bacteria as an ingredient of integrated management of soil fertility and plant nutrition to control the HMs stress in agricultural soils.

2 Historical Background of Heavy-Metal-Contaminated Soils

The concentrations of HMs in the soil have been the subject of numerous studies. By using the Scopus database, a total of 13,053 papers on the topic of "heavy-metal-contaminated soils," which have been published in the last 10 years (2015–2025),

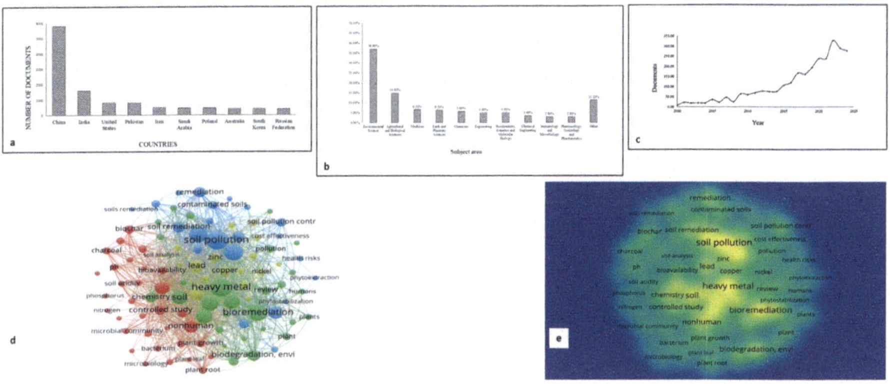

Fig. 1 Results obtained from data extracted from the Scopus site, (**a**) Number of articles published by different countries in the field of heavy-metal-contaminated soils; (**b**) Articles published by the subject area; (**c**) Articles published by year; (**d**) Co-occurrence of the top 100 keywords from existing research on HMs topics; (**e**) Item density visualization for the studies reviewed

were detected (Fig. 1a). HM contamination is mainly determined by the high level of pollution existing in this area.

Of the total number of articles obtained, roughly 36.9% have been studied in the fields of agriculture and the environment, together accounting for about 51.7% of all studies conducted in these two areas, with the remaining 48.3% being related to other fields (Fig. 1b).

Using the keywords "heavy metals," "soil contamination," and "bioremediation" in the period from 2000 to 2024, more information that is thorough was taken out of the Scopus database to better visualize the problem. In the past 24 years, about 2758 articles have been published in this field. As shown in Fig. 1c, research and publication of articles in this field have increased since 2000 because of the subject's importance in the agriculture and environment field and its impact on food safety and human health. An information map was created (Fig. 1d). Figure 1d displays the co-occurrence of the top 100 keywords from existing research on HM topics. The size of a node represents the number of times a keyword appears, while the length of the connection between two nodes represents their relationship.

Among the studies found, soil contamination with lead, copper, nickel, and zinc is among the most frequently studied topics. The existing literature focuses on the risk of HM toxicity in plants, plant growth, plant leaves, and plant roots, as well as human and non-human health. The importance of soil remediation using methods such as bioremediation, bioavailability, biochar, charcoal, phosphorus, microbial community, bacterium, biodegradation, and microbiology is highlighted by the keywords in these studies (Fig. 1d). Figure 1e shows that each point in the visualization has a color indicating the density of items at that point, with default colors ranging from blue to yellow. The closer the number of items in a neighborhood and the higher the weights of neighboring items, the closer the point is to yellow. The other way around, the smaller the number of items in the neighborhood of a point and the lower the weights of the neighboring items, the closer the color of the point is to blue. According to Fig. 1e, paying attention to soil pollution with HMs, particularly lead, is critical, and bioremediation is the most important method of remediation for HM-contaminated soils.

3 Risks of HMs for Public Health and Food Security

Considering that non-biodegradable metals (HMs) are persistent in the environment, they present serious health risks (Shah et al. 2012). The type of soil, pH, particle size, organic carbon content, CEC, root exudation, and other physicochemical parameters are factors that affect the accumulation of HMs in plants (Mousavi et al. 2013). The kind of HMs, their molecular structure, and the way their valences interconvert define how harmful they are to humans and how they affect food chains (Çelebi et al. 2014). The main and important pathways for exposing humans to HMs were explained in recent works (Mousavi 2022; Mousavi et al. 2023a). HM contamination in soils can harm ecosystems and human health by disrupting plant physiological processes, increasing oxidative damage, and affecting essential nutrient uptake (Rai et al. 2019). Chaney's concept of the "soil-plant barrier" suggests that certain elements, like Cd and As, have high transport capacity in plants and low solubility in soil (Rl 1980). In heavily metal-contaminated areas, residents and consumers of locally produced food are at risk, as certain goods, like rice and vegetables, fail to meet food quality standards (Emam and Soliman 2022; Zheng et al. 2020).

Vegetables, a staple worldwide, can accumulate high amounts of HMs, posing health risks to animals and humans (Collado-López et al. 2022). Roggeman et al. (2014) found that cow's livers and kidneys exceeded European limits for Cd and recommended that individuals weighing 70 kg should not consume more than 150 g of cow meat per day due to chromium levels in the muscles. Leung et al. (2014) found that edible fish species in the Pearl River Delta, China, contain varying concentrations of HMs (HMs/metalloids) in their muscles. The researchers recommend regular determination of HM concentrations in various fish species to mitigate health risks associated with food intoxication, both acute and chronic.

A study in South China (Zheng et al. 2013) found elevated concentrations of HMs in food, house dust, and water, except for drinking water. The study highlighted non-carcinogenic risks for adults and carcinogenic risks for young children, primarily from house dust exposure. The human body cannot degrade HMs, leading to malnourishment, diseases, and cancers. Nutrient deficiency and excessive intake of HMs can lead to nutrient deficiency and various health issues (Shah et al. 2012; Mousavi et al. 2022). Humans are exposed to HMs daily, which can have adverse effects on public health due to the lack of suitable removal methods. Contaminated foods and smoking cigarettes are the main sources of Cd entry into the human body (Mousavi et al. 2013). Iran's wheat, with a Cd concentration of 0.011 mg/kg, has a lower level compared to other countries, indicating high-quality wheat production in terms of Cd content (Shahbazi et al. 2021). Cd compounds have also been associated with an increased risk of various cancers and Alzheimer's disease in humans (Itoh et al. 2014).

Head and neck cancers (HNC) are the most prevalent types of cancer worldwide (Khandekar et al. 2006), and their incidence can be influenced by HMs found in tobacco and alcohol (Choi et al. 2020). High dosages of As ingestion can be fatal, while lower levels can cause systemic effects (Çelebi et al. 2014). Long-term exposure to inorganic As can increase the risk of certain cancers, including bladder, lung, and skin cancer (Leonardi et al. 2012). Selenium, a crucial element for selenoprotein biosynthesis and function (Mickiewicz et al. 2014), is believed to be a cancer-protective trace element (Schrauzer and Rhead 1971). Its anticarcinogenic activity is linked to reactive oxygen species generation, protein-thiol modification, and methionine mimicry. However, some studies suggest selenium supplementation doesn't significantly reduce prostate cancer incidence (Vinceti et al. 2018).

4 Management Strategies to Control HMs' Entrance to the Environment and Remediation Strategies for the HMs' Contaminated Soils

Soil is vital for Earth's life and biodiversity, but contamination by organic and inorganic pollutants disrupts production and recovery processes, interferes with vital organisms, and can damage plants, posing a threat to biodiversity. HMs are non-degradable metals with significant physiological effects on living organisms, even at low concentrations (Mousavi 2022; Mousavi et al. 2023a). Therefore, it is crucial to understand these sources and the management strategies for lowering their entrance to the soil and apply effective strategies for remediation of the HM-contaminated soils (Fig. 2).

The importance and vital role of agricultural soils in food security and, therewith, public health is an inevitable fact. Therefore, understanding the parameters and factors that affect soil quality and soil health is important. As was mentioned, one of the known environmental challenges for agricultural soils is contamination with HMs. Although soil is a crucial element in mitigating the effects of HMs, its management through optimal application of chemical and organic fertilizers (in general, agrochemicals) in agriculture, monitoring the quality of irrigation water, and efficient management of different solid wastes and wastewater are some important strategies related to agriculture for reducing HMs' entry into the environment, soil, and food chains.

Soil contaminants can be remedied using chemical, physical, and biological methods, with two approaches: in situ and ex situ (Fig. 2). In situ methods mean suitable plants and microorganisms as hyper-accumulators of HMs, such as bioremediation and amendments like biochar, organic and inorganic compounds (Mousavi et al. 2017b, 2018, 2023b). Ex situ methods use physical and chemical

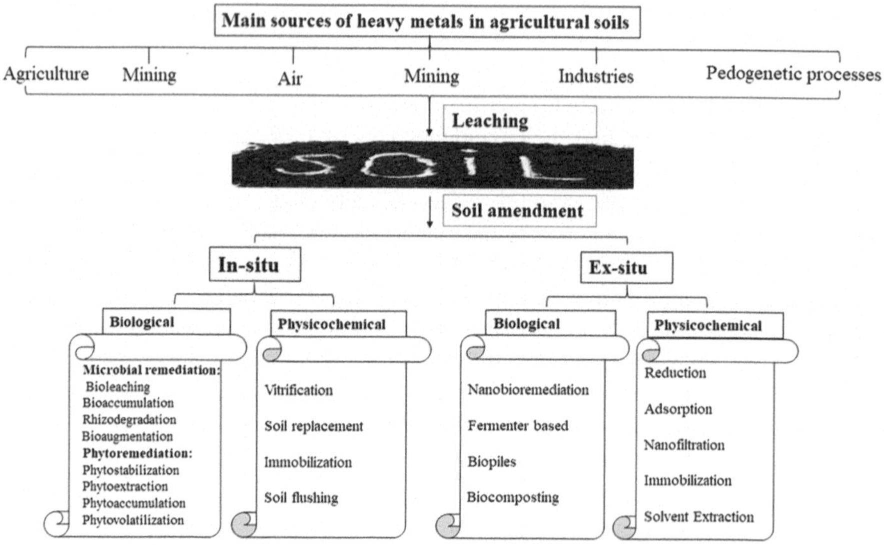

Fig. 2 Main sources of HMs in agricultural soils and the usual strategies for remediation of the HMs contaminated soils

remediation processes (Mousavi et al. 2022; Kuhad et al. 2004). Prioritizing preventive measures is crucial, as remedial methods are often more cost-effective and easier to implement.

In recent years, advanced and effective remediation strategies for HM-contaminated soils have developed rapidly. Effective strategies should be used to lessen human exposure to HMs (Table 1).

5 Using Plant Nutrition and Soil Fertility Management to Mitigate HM Stress

Numerous human activities, together with improper disposal or recycling methods, present a serious threat to human health and the environment. The majority of procedures are costly and logistically difficult, so the majority of research efforts are devoted to creating affordable techniques for removing contaminants. HM contamination significantly affects plant growth, causing physiological, morphological, and biochemical changes. There are different aspects related to soil fertility and plant nutrition management that significantly affect the status of HM in the soil, plant, and food chains (Table 2). This efficacy means a double effect: on the one hand, soil fertility and plant nutrition management practices may cause the entrance of HM to the environment, and on the other hand, by using these practices in the HM-contaminated soils, the HM stress may be ameliorated. For the first condition, the application of the right source of fertilizers at the right rate, at the right time, and

Table 1 Management strategies to remediate HMs contaminated soils

Strategies	Results	Reference
Low-metal cultivar selection/breeding	Nine out of 72 popular wheat genotypes were recommended for planting in slightly to moderately Cd-contaminated soil	Liu et al. (2020)
	A manganese transporter located at the cytoplasm membrane of rice can reduce Cd uptake and accumulation in rice	Sasaki et al. (2014)
	Se/Si addition increased the Cd sequestered in the vacuole	Wan et al. (2019)
	Foliar spraying with Si or Se during the tillering and heading phases considerably reduced Cd levels by 72% and 62%, respectively	Gao et al. (2018)
Water management	Optimized water management influences Cd and As bioavailability in paddy soils, and affects Cd and As absorption and translocation from roots to shoots	Bao et al. (2022)
	The adoption of sprinkler irrigation greatly minimizes the Cd, Pb, and As contents in 26 rice cultivars	Spanu et al. (2020)
	Rice grain contains 98% Cd from root uptake during filling, and 95% As from remobilization of As in previously accumulated parts during vegetative growth periods	Huang et al. (2022)
Soil amendment	Phosphorous materials can immobilize Cd and Pb through increased pH, electrostatic interaction, ion exchange, and surface complexation.	Garau et al. (2007)
	Sepiolite significantly increases soil pH, reduces acid-soluble Cd content, and reduces Cd content in spinach shoots	Bashir et al. (2020)
	Chicken or swine manure significantly reduces extractable Cd and Pb in soil and rice grain.	Wan et al. (2020)
	Biochar certainly increases crop yields while reducing the mobility and accumulation of HMs in soils and plants.	He et al. (2019)
Phytoremediation	*Pteris vittata L.* is highly resistant to As, accumulating 1442–752 mg kg^{-1} in contaminated soils and up to 27,000 mg kg^{-1} under hydroponic conditions	Ma et al. (2001)
	Sedum alfredii H. is an efficient Cd/Zn hyperaccumulator capable of accumulating thousands of mg kg − 1 Zn and Cd in its shoots	Yang et al. (2004)
	The Cd, Pb, and Zn contents of Brassica napus aboveground were significantly increased by the presence of Enterobacter sp. JYX7	Jing et al. (2014)

at the right place is important. Previous studies have shown that chemical fertilizers (especially phosphorus fertilizers) (Mousavi et al. 2024a), application of organic fertilizers like sewage sludge and municipal solid waste compost, and using wastewater for irrigation are the main sources of HMs in agriculture (Mousavi et al. 2013, 2017a, 2022, 2023a).

Table 2 Some strategies related to plant nutrition and soil fertility management for mitigating HMs stress in agriculture

Management strategy	Heavy metal	Type of plant	Type of soil	Result	Reference
Plant nutrition	Zn + B + Mo	French beans	Acidic soil	General improvement in crop health and nutrition.	Kumar et al. (2016)
	Mo + Co	Green gram	Acidic soil	The grain yield and total biomass increased.	Pattanayak et al. (2000)
Biochar	Cd + Pb + Ni	–	Silt loam	Biochar can potentially mitigate the harmful effects of HMs.	Mousavi et al. (2023b)
Vermicompost and biochar	Cd	–	Soil with PH 7.5	The biochemical indicators of soil improved, and the risk of HMs reduced.	Wang et al. (2022)
Boron and biochar	Cd	Wheat	Sandy loam	Increased the amount of chlorophyll, biomass produced, and root and shoot growth.	Hussain et al. (2022)
Vermicompost and biochar	Cd	Barberry	Sandy loam	Significant increase in plant biomass, height, chlorophyll content, relative water content (RWC), and antioxidant levels.	Khosropour et al. (2022)
Biochar and zeolite	Ni	Sunflower and maize	–	The Ni accumulation reduced, and biomass production and grain yield increased.	Shahbaz et al. (2018)
H_2S	Hg	Rice	–	Preventing oxidative damage and significantly promoting rice growth.	Chen et al. (2017)
Treatment with Jasmonic, Salicylic Acids, and Proline	Pb	Maize	Clay loamy	The ability of maize plants to overcome oxidative damage caused by Pb toxicity increased.	Sofy et al. (2020)

For the second condition, the studies have shown that by using optimal management of soil fertility and plant nutrition, the soil contamination by HMs and the adverse effects of HM stress on plants can be mitigated. Iron (Fe) is a crucial micronutrient, serving as a major component of cell redox systems like heme proteins, cytochromes, hemoglobin, peroxidase, catalase, and Fe-S clusters like ferredoxin and superoxide dismutase. It interacts with HMs and can improve crop growth and nutritional quality under HMs stress (Emamverdian et al. 2015). The vital roles of the application of the Si compounds in HM-contaminated soils were discussed in the previous work (Mousavi 2022). Huang et al. (2021) found that Si and Se could alleviate Cd toxicity in rice by regulating gene expression, sequestering Cd into root cell walls and vacuoles, and decreasing Cd translocation. Si and Zn also improve the ascorbate-glutathione cycle and antioxidant enzyme activities (Jiang et al. 2023). However, field studies found that the combined application of Si and Se was less effective, possibly due to the large colloidal particle size (Gao et al. 2018).

According to studies by Yang et al. (2011), iron phosphate can lessen the stress that rice seedlings experience from Cd, Pb, and Se. Zinc can also lessen the toxicity of Cd and Cu by lowering the uptake of these elements in winter wheat (Zhao et al. 2011). Combining ZnO nanoparticles (100 mg/L) with biochar (1.0% w/w) decreased the concentration of Cd in rice shoots and roots to 39% and 38%, respectively, while enhancing other growth parameters, according to Ali et al. (2019). According to the research of Wu et al. (2016), the exogenous supply of Se considerably decreased the levels of Pb and Cd in the roots and shoots of oilseed rape, preventing the oxidative damage that Pb and Cd cause. Se also enhanced the ability of oilseed rape plants to withstand HM stress by increasing superoxide free radicals, H_2O_2, and lipid peroxidation while decreasing glutathione peroxidase activities. Se application significantly mitigated Pb stress-induced damage in fava beans, enhancing chlorophyll and carotenoid content and preserving biochemical parameters. Pb stress damage caused by P-deficient conditions affected plant growth, highlighting the importance of selenium in plant health (Mroczek-Zdyrska et al. 2017).

Biochar, a byproduct of biomass pyrolysis from agriculture, sewage, and animal wastes, has been found to sequester pollutants in soils (Mousavi et al. 2023b). HMs, which are non-biodegradable and persistent, are difficult to remove from soils. Various techniques, like lime and compost, have been used to stabilize HMs (Komárek et al. 2013). Biochar has also been shown to effectively remediate polluted soils and reduce their bioavailability (Ippolito et al. 2012). Biochar produced from a variety of feedstocks, such as agricultural manure, animal dung, or woody plants, can immobilize HMs (Mousavi et al. 2023b). Biochar, when applied at higher rates, improves soil pH, particularly in acidic soil due to its alkaline nature (Yu et al. 2019). Higher pH influences HMs' mobility and complexation, reducing the desorption of HMs, particularly Pb, from soils, thus reducing HMs' toxicity and promoting soil health (Jiang et al. 2012). Biochar application increases soil CEC due to its high CEC (Zhang et al. 2017). However, studies suggest that biochar's cation exchange sites decrease solubility and leachability (Bashir et al. 2018). It also improves the fixation of Cu and Pb in soil (Ma et al. 2010). Biochar surfaces form mineral phases due to minerals like Ca, P, Mg, and K (Rees et al. 2014), which help increase the adsorption of HMs by soil (Bian et al. 2014). However, oxides of Si, Ca, and Mn are not completely reduced, leading to sorption sites with high energy for HMs (Rees et al. 2014). Organic carbon (OC) in soil helps reduce the mobility of HMs (Zhu et al. 2016), which originate from biochar. OC derived from biochar converts labile forms to less mobile forms (Abdelhafez et al. 2014), reducing the uptake of metal cations by plants.

6 Importance and Functions of Soil Bacteria in the Management and Reclamation of HM-Contaminated Soils in Agriculture

The conventional methods for remediating HM-contaminated soils, such as soil excavation, chemical treatment, and the use of energy-intensive physical processes, have significant limitations. These approaches can be costly, disruptive to the soil ecosystem, and often fail to achieve long-term, sustainable solutions (Zhou et al. 2023; Jin et al. 2018).

As a result, there is a growing need for innovative, eco-friendly, and cost-effective strategies to manage and reclaim HM-polluted agricultural lands. In this context, the potential of soil microorganisms to mitigate and reclaim HM-contaminated soils has emerged as a promising, eco-friendly, and cost-effective solution known as bioremediation. As an integral component of the soil microbiome, soil bacteria possess remarkable abilities to interact with, transform, and sequester HMs, making them a vital asset in the management of contaminated agricultural lands (Zhou et al. 2023; Jin et al. 2018).

By leveraging the intricate relationships between soil bacteria and HM dynamics, a holistic, microbiome-based approach can be employed to address the pressing challenge of HM pollution in agriculture. Continued research and the development of innovative, bacteria-based bioremediation strategies will be essential in paving the way for a more resilient and environmentally responsible agricultural future, where contaminated lands can be restored to productive and sustainable use.

Soil bacteria play a critical role in the remediation of HM-contaminated agricultural soils through a variety of specialized mechanisms (Table 3). These microbial-driven remediation strategies offer a sustainable and environmentally friendly approach to reclaiming polluted soils and restoring their productivity (Tarfeen et al. 2022). These include metal biosorption, bioaccumulation, mobilization, precipitation, and transformation (Ma et al. 2016a). The specific mechanisms employed by soil bacteria depend on factors such as the bacterial strains, the type and concentration of HMs, and the prevailing soil conditions. These microbial-mediated processes can effectively immobilize and remove HMs from the soil, thereby mitigating their bioavailability and potential for uptake by crops.

Beyond their direct role in HM remediation, soil bacteria can also enhance the phytoremediation potential of crops grown on contaminated soils. Plant-growth-promoting bacteria (PGPB) can improve the overall health and survival of plants by increasing nutrient availability, mitigating oxidative stress, and stimulating root development—all of which can lead to more effective HM extraction and accumulation by the plants (Ma et al. 2016b).

To overcome the limitations of metal toxicity and improve the effectiveness of phytoremediation, researchers have explored the use of PGPB as a complementary strategy. The symbiotic relationship between plants and PGPB creates a synergistic effect, where both organisms benefit and work together to overcome the challenges of HM remediation (Wang et al. 2022). Recent studies have shown that PGPB can

Table 3 The HMs removal from contaminated soils by bacterial strains

Mechanisms	Heavy metal	Strain	Result	Reference
Biosorption	Cd	*Bacillus subtilis*	Biosorption capacity of 56.497 mg/g	Huang et al. (2013)
	Pb	*Bacillus subtilis* DBM	Biosorption capacity of 77.1 mg/g	Bai et al. (2014)
	Cd	*Bacillus cereus* RC-1	Biosorption capacity of 31.95 mg/g	Huang et al. (2013)
	Pb	*Bacillus thioparans* U3	Biosorption capacity of 200.1 mg/g	Rodríguez-Tirado et al. (2012)
Bioaccumulation	Cr	*Klebsiella pneumoniae* MB361	83.51% uptake from 200 μg/ml	Aslam et al. (2020)
	Pb	*Stenotrophomonas* sp. MB339	85.30% uptake from 200 μg/ml	Aslam et al. (2020)
	Ni	*Stenotrophomonas* sp. MB339	48.78% uptake from 200 μg/ml	Aslam et al. (2020)
	Cr	*Microbacterium paraoxydans* VSVM IIT (BHU)	91.62% uptake from 50 μg/ml	Singh et al. (2023)
	Cd	*Microbacterium paraoxydans* VSVM IIT (BHU)	89.29% uptake from 50 μg/ml	Singh et al. (2023)
	Pb	*Microbacterium paraoxydans* VSVM IIT (BHU)	83.29% uptake from 50 μg/ml	Singh et al. (2023)
Biotransformation	Cr	*Bacillus* sp. SFC 500-1E	Transforming Cr(VI) to Cr(III)	Thatoi et al. (2014)
	Hg	*Pseudomonas* sp. B50A	Transforming Hg(II) to Hg(0)	Giovanella et al. (2016)
Biomineralization	Pb	*B. subtilis*	Formation of insoluble Pb-phosphate minerals	Li et al. (2017)
	Cd	*P. putida*	Formation of insoluble Cd-carbonate and phosphate minerals	Lin et al. (2016)

significantly improve plant growth, biomass, and metal uptake capacity in HM-contaminated soils (Bruno et al. 2021; Pan et al. 2019; Etesami 2018; Mousavi et al. 2024b). PGPB can also indirectly contribute to phytoremediation through the production of antibiotics, competition with indigenous microorganisms, induction of systemic resistance, and the release of volatile compounds (Morales-Cedeño et al. 2021; Saeed et al. 2021). These diverse beneficial effects of PGPB can enhance the efficiency of phytoremediation in metal-contaminated soils by improving plant growth, metal uptake, and overall remediation performance.

By harnessing the capabilities of PGPB, researchers can enhance the efficiency and success of phytoremediation strategies, paving the way for the reclamation of heavily polluted agricultural soils.

6.1 *Biosorption*

Biosorption is the reversible adsorption of HM ions onto the surface of bacterial cells. Numerous bacterial species, such as *Bacillus*, *Pseudomonas*, and *Arthrobacter*, exhibit high biosorption capacities for a range of HMs, including Pb, Cd, Cu, and Cr, from contaminated soil environments (Fomina and Gadd 2014; Oves et al. 2012). Tarfeen et al. (2022) reported that *B. subtilis* absorbs Cd(II) at about 56.497 mg/g onto its cell surfaces, effectively sequestering the toxic HM ions from the environment. While the specific biosorption capacities depending on the bacterial strain, metal type, and experimental conditions varied among the studies, the efficiency of biosorption is largely determined by the specific functional groups present on the bacterial cell surface, such as carboxyl, hydroxyl, amine, phosphonate, and sulfhydryl, which can bind to and sequester HM ions. *B. subtilis* demonstrated an impressive biosorption capacity, achieving up to 96% removal of Cr at a concentration of 25 μg mL^{-1}. In contrast, *Pseudomonas aeruginosa* effectively biosorbed a maximum of 91.0% of Cd at a concentration of 400 μg mL^{-1} (Rizvi et al. 2020).

6.2 *Bioaccumulation*

This strategy involves the production of metal-binding proteins (metallothioneins and phytochelatins), the compartmentalization of metals in specialized organelles (e.g., vacuoles), and the upregulation of efflux pumps to maintain cellular homeostasis (Nnaji et al. 2023). Bioaccumulation allows bacteria to remove HMs from the soil solution, reducing their bioavailability and potential for uptake by plants. Studies have shown that *P. putida* is capable of intracellularly and periplasmically accumulating Cd ions, suggesting it possesses metal-binding and/or efflux mechanisms within the cellular components that mediate resistance against toxic metal concentrations (Rani et al. 2009). Such metal sequestration abilities of the indigenous bacterial strains indicate they can effectively remediate polluted environments through intracellular accumulation of toxic metals like Ni, Cr, and Pb, thereby reducing metal mobility and availability in soils.

6.3 *Biotransformation*

Soil bacteria can also transform HMs into less toxic or mobile forms through enzymatic processes, a strategy known as biotransformation. Numerous studies have characterized the enzymatic pathways and genetic determinants responsible for these bacterial-mediated transformations, which can significantly alter the speciation and mobility of HMs in contaminated soils (Giovanella et al. 2016; Thatoi et al.

2014). For example, Cr (VI)-tolerant *Bacillus* sp. SFC 500-1E has been shown to reduce Cr (VI) to Cr (III) through the action of NADH-dependent reductase, effectively decreasing mobility and toxicity (Thatoi et al. 2014). Similarly, *Pseudomonas* sp. B50A has been reported to transform 86% of mercuric ions (Hg(II)) to less toxic elemental mercury (Hg(0)) through the action of the mercury reductase enzyme (Giovanella et al. 2016). These bacterial-driven biotransformation processes can significantly alter the speciation, mobility, and toxicity of HMs in contaminated soils.

6.4 Biomineralization

The precipitation of HMs into insoluble mineral forms through the process of biomineralization is another important bioremediation strategy employed by soil bacteria. Researchers have identified bacterial species capable of producing specific metabolites, such as sulfides, phosphates, and carbonates, that can react with HM ions to form stable mineral complexes, effectively immobilizing the contaminants and reducing their environmental mobility and bioavailability (Tarfeen et al. 2022; Ma et al. 2016b). *B. subtilis* strain FZUL-33 has been shown to promote the precipitation of Pb through the formation of phosphate (PO_4^{3-}) ions. This study demonstrated that FZUL-33 is capable of degrading the organophosphate pesticide acephate. During the degradation process, PO_4^{3-} ions are released, which facilitate the precipitation of insoluble Pb-phosphate minerals, thereby immobilizing Pb in contaminated soils (Li et al. 2017). Similarly, *P. putida* has been reported to accelerate the remediation of Cd through the degradation of organophosphates. The degradation pathway results in the production of carbonate and phosphate compounds, which drive the formation of insoluble Cd-mineral phases, reducing the availability and mobility of Cd (Lin et al. 2016). These findings highlight the potential of certain bacteria to contaminate degradation with mineral precipitation, providing an effective mechanism for in situ immobilization of toxic HMs in polluted soils.

The integration of these mechanisms highlights the multifaceted and versatile nature of soil bacteria in addressing HM contamination in agricultural soils. The scientific literature provides strong evidence for the efficacy of these bacterial-mediated strategies, underscoring the importance of a microbiome-based approach to the management and reclamation of HM-impacted agricultural ecosystems.

7 Conclusion

HMs have always been considered the most important mineral pollutants of the soil, environment, and food chains. Knowing the sources of these contaminants helps to control their entry into the food chains. However, by using management strategies, it is possible to reduce their entry into the environment, especially agricultural soils. Some important and efficient strategies include managing the application of agrochemicals, especially organic and chemical fertilizers, according to the real needs of the plant and soil conditions, and managing the entry of wastewater and solid wastes into the soil and environment. At the same time, when their entry into the soil, as the point of entry of elements into the food chain, exceeds the permissible limit, there are effective management solutions that can significantly reduce the effects of HM stress on the soil and plants. In agricultural soils, with optimal management of soil fertility and plant nutrition, the use of organic and inorganic compounds as soil conditioners, the use of beneficial and essential nutrients that increase plant resistance to HM stress, and the use of soil's inherent potentials, such as PGPRs that are isolated from contaminated soils with HMs and have high resistance to HMs and, by producing various metabolites and also their inherent capabilities, stabilize these contaminants and control their entry into food chains, this problem is solved significantly. However, each of these methods has its own advantages and disadvantages, which can be chosen according to the conditions of the soil, climate, cultivated plant, type of HMs, and economic feasibility. In general, based on literature reviews, the use of integrated management of soil fertility and plant nutrition (optimal application of chemical, organic, and biological fertilizers based on the needs of the plant and the content of nutrients in the soil) has the highest effect in supplying the required nutrients to plants, increasing the plant growth and yield, and promoting the soil health and quality and can be a general recommendation for agricultural soils to control the entry of mineral contaminants into the soil and food chains.

References

Abdelhafez, A. A., Li, J. & Abbas, M. H. 2014. Feasibility of biochar manufactured from organic wastes on the stabilization of heavy metals in a metal smelter contaminated soil. *Chemosphere,* 117, 66–71.

Alengebawy, A., Abdelkhalek, S. T., Qureshi, S. R. & Wang, M.-Q. 2021. Heavy metals and pesticides toxicity in agricultural soil and plants: Ecological risks and human health implications. *Toxics,* 9, 42.

Ali, S., Rizwan, M., Noureen, S., Anwar, S., Ali, B., Naveed, M., Abd Allah, E. F., Alqarawi, A. A. & Ahmad, P. 2019. Combined use of biochar and zinc oxide nanoparticle foliar spray improved the plant growth and decreased the cadmium accumulation in rice (Oryza sativa L.) plant. *Environmental Science and Pollution Research,* 26, 11288–11299.

Aslam, F., Yasmin, A. & Sohail, S. 2020. Bioaccumulation of lead, chromium, and nickel by bacteria from three different genera isolated from industrial effluent. *International Microbiology,* 23, 253–261.

Bai, J., Yang, X., Du, R., Chen, Y., Wang, S. & Qiu, R. 2014. Biosorption mechanisms involved in immobilization of soil Pb by Bacillus subtilis DBM in a multi-metal-contaminated soil. *Journal of Environmental Sciences,* 26, 2056–2064.

Bao, Q., Bao, W., Ding, Y. & Huang, Y. 2022. Effects of optimized water management on the uptake and translocation of cadmium and arsenic in Oryza sativa L. in two contaminated soils. *Environmental Science and Pollution Research,* 29, 2853–2865.

Bashir, S., Ali, U., Shaaban, M., Gulshan, A. B., Iqbal, J., Khan, S., Husain, A., Ahmed, N., Mehmood, S. & Kamran, M. 2020. Role of sepiolite for cadmium (Cd) polluted soil restoration and spinach growth in wastewater irrigated agricultural soil. *Journal of environmental management,* 258, 110020.

Bashir, S., Shaaban, M., Mehmood, S., Zhu, J., Fu, Q. & Hu, H. 2018. Efficiency of C3 and C4 plant derived-biochar for Cd mobility, nutrient cycling and microbial biomass in contaminated soil. *Bulletin of environmental contamination and toxicology,* 100, 834–838.

Bian, R., Zhang, A., Li, L., Pan, G., Zheng, J., Zhang, X., Zheng, J., Joseph, S. & Chang, A. 2014. Effect of Municipal Biowaste Biochar on Greenhouse Gas Emissions and Metal Bioaccumulation in a Slightly Acidic Clay Rice Paddy. *BioResources,* 9.

Bruno, L. B., Anbuganesan, V., Karthik, C., Kumar, A., Banu, J. R., Freitas, H., & Rajkumar, M. (2021). Enhanced phytoextraction of multi-metal contaminated soils under increased atmospheric temperature by bioaugmentation with plant growth promoting Bacillus cereus. Journal of Environmental Management, 289, 112553. https://doi.org/10.1016/j.jenvman.2021.112553

Çelebi, A., Şengörür, B. & Kløve, B. 2014. Human health risk assessment of dissolved metals in groundwater and surface waters in the Melen watershed, Turkey. *Journal of Environmental Science and Health, Part A,* 49, 153–161.

Chen, Z., Chen, M. & Jiang, M. 2017. Hydrogen sulfide alleviates mercury toxicity by sequestering it in roots or regulating reactive oxygen species productions in rice seedlings. *Plant Physiology and Biochemistry,* 111, 179–192.

Choi, J. W., Lee, M. H. & Fujii, T. 2020. Effect of high alcohol intake on heavy metal levels in the blood, urine cotinine metabolism, and pulmonary function according to the severity of smoking. *Int J Clin Exp Med,* 13, 7700–7708.

Collado-López, S., Betanzos-Robledo, L., Téllez-Rojo, M. M., Lamadrid-Figueroa, H., Reyes, M., Ríos, C. & Cantoral, A. 2022. Heavy metals in unprocessed or minimally processed foods consumed by humans worldwide: a scoping review. *International journal of environmental research and public health,* 19, 8651.

Emam, W. W. & Soliman, K. M. 2022. Geospatial analysis, source identification, contamination status, ecological and health risk assessment of heavy metals in agricultural soils from Qallin city, Egypt. *Stochastic Environmental Research and Risk Assessment,* 36, 2437–2459.

Emamverdian, A., Ding, Y., Mokhberdoran, F. & Xie, Y. 2015. Heavy metal stress and some mechanisms of plant defense response. *The scientific world journal,* 2015, 756120.

Etesami H, Beattie GA. 2018. Mining halophytes for plant growth-promoting halotolerant bacteria to enhance the salinity tolerance of non-halophytic crops. Frontiers in microbiology. Feb 8;9:148.

Fomina, M. & Gadd, G. M. 2014. Biosorption: current perspectives on concept, definition and application. *Bioresource technology,* 160, 3–14.

Gao, M., Zhou, J., Liu, H., Zhang, W., Hu, Y., Liang, J. & Zhou, J. 2018. Foliar spraying with silicon and selenium reduces cadmium uptake and mitigates cadmium toxicity in rice. *Science of the Total Environment,* 631, 1100–1108.

Garau, G., Castaldi, P., Santona, L., Deiana, P. & Melis, P. 2007. Influence of red mud, zeolite and lime on heavy metal immobilization, culturable heterotrophic microbial populations and enzyme activities in a contaminated soil. *Geoderma,* 142, 47–57.

Giovanella, P., Cabral, L., Bento, F. M., Gianello, C. & Camargo, F. A. O. 2016. Mercury (II) removal by resistant bacterial isolates and mercuric (II) reductase activity in a new strain of Pseudomonas sp. B50A. *New biotechnology,* 33, 216–223.

He, L., Zhong, H., Liu, G., Dai, Z., Brookes, P. C. & Xu, J. 2019. Remediation of heavy metal contaminated soils by biochar: Mechanisms, potential risks and applications in China. *Environmental pollution,* 252, 846–855.

Huang, B.-Y., Zhao, F.-J. & Wang, P. 2022. The relative contributions of root uptake and remobilization to the loading of Cd and As into rice grains: Implications in simultaneously controlling grain Cd and As accumulation using a segmented water management strategy. *Environmental Pollution,* 293, 118497.

Huang, F., Dang, Z., Guo, C.-L., Lu, G.-N., Gu, R. R., Liu, H.-J. & Zhang, H. 2013. Biosorption of Cd (II) by live and dead cells of Bacillus cereus Rc-1 isolated from cadmium-contaminated soil. *Colloids and surfaces b: biointerfaces,* 107, 11–18.

Huang, H., Li, M., Rizwan, M., Dai, Z., Yuan, Y., Hossain, M. M., Cao, M., Xiong, S. & Tu, S. 2021. Synergistic effect of silicon and selenium on the alleviation of cadmium toxicity in rice plants. *Journal of Hazardous Materials,* 401, 123393.

Hussain, S., Irfan, M., Sattar, A., Hussain, S., Ullah, S., Abbas, T., Ur-Rehman, H., Nawaz, F., Al-Hashimi, A. & Elshikh, M. S. 2022. Alleviation of cadmium stress in wheat through the combined application of boron and biochar via regulating morpho-physiological and antioxidant defense mechanisms. *Agronomy,* 12, 434.

Ippolito, J. A., Laird, D. A. & Busscher, W. J. 2012. Environmental benefits of biochar. *Journal of environmental quality,* 41, 967–972.

Itoh, H., Iwasaki, M., Sawada, N., Takachi, R., Kasuga, Y., Yokoyama, S., Onuma, H., Nishimura, H., Kusama, R. & Yokoyama, K. 2014. Dietary cadmium intake and breast cancer risk in Japanese women: a case–control study. *International journal of hygiene and environmental health,* 217, 70–77.

Jiang, T.-Y., Jiang, J., Xu, R.-K. & Li, Z. 2012. Adsorption of Pb (II) on variable charge soils amended with rice-straw derived biochar. *Chemosphere,* 89, 249–256.

Jiang, Y., Wei, C., Jiao, Q., Li, G., Alyemeni, M. N., Ahmad, P., Shah, T., Fahad, S., Zhang, J. & Zhao, Y. 2023. Interactive effect of silicon and zinc on cadmium toxicity alleviation in wheat plants. *Journal of Hazardous Materials,* 458, 131933.

Jin, Y., Luan, Y., Ning, Y. & Wang, L. 2018. Effects and mechanisms of microbial remediation of heavy metals in soil: a critical review. *Applied Sciences,* 8, 1336.

Jing, Y. X., Yan, J. L., He, H. D., Yang, D. J., Xiao, L., Zhong, T., Yuan, M., Cai, X. D. & Li, S. B. 2014. Characterization of bacteria in the rhizosphere soils of Polygonum pubescens and their potential in promoting growth and Cd, Pb, Zn uptake by Brassica napus. *International journal of phytoremediation,* 16, 321–333.

Khandekar, S., Bagdey, P. & Tiwari, R. 2006. Oral cancer and some epidemiological factors: A hospital based study. *Indian Journal of Community Medicine,* 31, 157.

Khosropour, E., Weisany, W., Tahir, N. A.-R. & Hakimi, L. 2022. Vermicompost and biochar can alleviate cadmium stress through minimizing its uptake and optimizing biochemical properties in Berberis integerrima bunge. *Environmental Science and Pollution Research,* 29, 17476–17486.

Komárek, M., Vaněk, A. & Ettler, V. 2013. Chemical stabilization of metals and arsenic in contaminated soils using oxides–a review. *Environmental Pollution,* 172, 9–22.

Kuhad, R. C., Johri, A. K., Singh, A. & Ward, O. P. 2004. Bioremediation of pesticide-contaminated soils. *Applied bioremediation and phytoremediation.* Springer.

Kumar, M., Jha, A., Hazarika, S., Verma, B., Choudhury, B., Ramesh, T., Moirangthem, P., Kumar, R., Brajendra & Rajkhowa, D. 2016. Micronutrients (B, Zn, Mo) for improving crop production on acidic soils of northeast India. *National Academy Science Letters,* 39, 85–89.

Leonardi, G., Vahter, M., Clemens, F., Goessler, W., Gurzau, E., Hemminki, K., Hough, R., Koppova, K., Kumar, R. & Rudnai, P. 2012. Inorganic arsenic and basal cell carcinoma in areas of Hungary, Romania, and Slovakia: a case–control study. *Environmental health perspectives,* 120, 721–726.

Leung, H., Leung, A., Wang, H., Ma, K., Liang, Y., Ho, K., Cheung, K., Tohidi, F. & Yung, K. 2014. Assessment of heavy metals/metalloid (As, Pb, Cd, Ni, Zn, Cr, Cu, Mn) concentrations in edible fish species tissue in the Pearl river delta (PRD), China. *Marine pollution bulletin,* 78, 235–245.

Li, X., Yin, X. & Lian, B. 2017. The degradation of dimethoate and the mineral immobilizing function for Cd2+ by Pseudomonas putida. *Geomicrobiology Journal,* 34, 346–354.

Lin, W., Huang, Z., Li, X., Liu, M. & Cheng, Y. 2016. Bio-remediation of acephate–Pb (II) compound contaminants by Bacillus subtilis Fzul-33. *Journal of Environmental Sciences,* 45, 94–99.

Liu, N., Huang, X., Sun, L., Li, S., Chen, Y., Cao, X., Wang, W., Dai, J. & Rinnan, R. 2020. Screening stably low cadmium and moderately high micronutrients wheat cultivars under three different agricultural environments of China. *Chemosphere,* 241, 125065.

Ma, L., Xu, R. & Jiang, J. 2010. Adsorption and desorption of Cu (II) and Pb (II) in paddy soils cultivated for various years in the subtropical China. *Journal of Environmental Sciences,* 22, 689–695.

Ma, L. Q., Komar, K. M., Tu, C., Zhang, W., Cai, Y. & Kennelley, E. D. 2001. A fern that hyperaccumulates arsenic. *Nature,* 409, 579–579.

Ma, Y., Oliveira, R. S., Freitas, H. & Zhang, C. 2016a. Biochemical and molecular mechanisms of plant-microbe-metal interactions: relevance for phytoremediation. *Frontiers in plant science,* 7, 918.

Ma, Y., Rajkumar, M., Zhang, C. & Freitas, H. 2016b. Beneficial role of bacterial endophytes in heavy metal phytoremediation. *Journal of environmental management,* 174, 14–25.

Mansoor, S., Kour, N., Manhas, S., Zahid, S., Wani, O.A., Sharma, V., Wijaya, L., Alyemeni, M.N., Alshali, A.A., El-Serehy, H.A. and Paray, B.A., 2021. Biochar as a tool for effective management of drought and heavy metal toxicity. *Chemosphere, 271*, p. 129458.

Mickiewicz, B., Villemaire, M. L., Sandercock, L. E., Jirik, F. R. & Vogel, H. J. 2014. Metabolic changes associated with selenium deficiency in mice. *Biometals,* 27, 1137–1147.

Monni, S., Salemaa, M. & Millar, N. 2000. The tolerance of Empetrum nigrum to copper and nickel. *Environmental pollution,* 109, 221–229.

Morales-Cedeño, L. R., Del Carmen Orozco-Mosqueda, M., Loeza-Lara, P. D., Parra-Cota, F. I., de Los Santos-Villalobos, S. & Santoyo, G. 2021. Plant growth-promoting bacterial endophytes as biocontrol agents of pre-and post-harvest diseases: Fundamentals, methods of application and future perspectives. *Microbiological Research,* 242, 126612.

Mousavi, S. M. 2022. Silicon and nano-silicon mediated heavy metal stress tolerance in plants. *Silicon and nano-silicon in environmental stress management and crop quality improvement.* Elsevier.

Mousavi, S. M., Bahmanyar, M. A. & Pirdashti, H. 2013. Phytoavailability of some micronutrients (Zn and Cu), heavy metals (Pb, Cd), and yield of rice affected by sewage sludge perennial application. *Communications in soil science and plant analysis,* 44, 3246–3258.

Mousavi, S. M., Bahmanyar, M. A., Pirdashti, H. & Moradi, S. 2017a. Nutritional (Fe, Mn, Ni, and Cr) and growth responses of rice plant affected by perennial application of two bio-solids. *Environmental Monitoring and Assessment,* 189, 1–10.

Mousavi, S. M., Brodie, G., Payghamzadeh, K., Raiesi, T. & Strivastava, A. 2022. Lead Bioavailability in the Environment: Its Exposure and and Effects. *Journal of Advances in Environmental Health Research,* 10, 1–14.

Mousavi, S. M., Motesharezadeh, B., Hosseini, H. M., Alikhani, H. & Zolfaghari, A. A. 2017b. Geochemical fractions and phytoavailability of zinc in a contaminated calcareous soil affected by biotic and abiotic amendments. *Environmental geochemistry and health,* 40, 1221–1235.

Mousavi, S. M., Motesharezadeh, B., Hosseini, H. M., Alikhani, H. & Zolfaghari, A. A. 2018. Root-induced changes of Zn and Pb dynamics in the rhizosphere of sunflower with different plant growth promoting treatments in a heavily contaminated soil. *Ecotoxicology and environmental safety,* 147, 206–216.

Mousavi, S. M., Raiesi, T., Sedaghat, A. & Srivastava, A. K. 2023a. Potentially Toxic Metals: Their Effects on the Soil-Human Health Continuum. *Journal of Advances in Environmental Health Research,* 12, 86–101.

Mousavi, S. M., Srivastava, A. K. & Cheraghi, M. 2023b. Soil health and crop response of biochar: an updated analysis. *Archives of Agronomy and Soil Science,* 69, 1085–1110.

Mousavi, S. M., Shahbazi, K., Basirat, M. & Esmaeili, M. 2024a. Spraying or fertigation? looking for the better method in terms of the lower concentration of Pb and Ni in greenhouse cucumber. *Environmental Geochemistry and Health,* 46, 405.

Mousavi, S. M., Tohidtalab, P., Abolfathi, P. 2024b. The Importance and Necessity to Consider Cadmium (Cd) as a Problematic Element in Soil Fertility and Plant Nutrition. In: T. Aftab (ed.),

Plant Responses to Cadmium Toxicity, Environmental Science and Engineering; Pp 43–88; https://doi.org/10.1007/978-3-031-73266-9_2

Mroczek-Zdyrska, M., Strubińska, J. & Hanaka, A. 2017. Selenium improves physiological parameters and alleviates oxidative stress in shoots of lead-exposed Vicia faba L. minor plants grown under phosphorus-deficient conditions. *Journal of Plant Growth Regulation,* 36, 186–199.

Nnaji, N., Onyeaka, H., Miri, T. & Ugwa, C. 2023. Bioaccumulation for heavy metal removal: a review. SN Applied Sciences, 5 (5), 125.

Oves, M., Khan, M. S., Zaidi, A. & Ahmad, E. 2012. *Soil contamination, nutritive value, and human health risk assessment of heavy metals: an overview*, Springer.

Pan, Q., Shikano, I., Hoover, K., Liu, T. X., & Felton, G. W. (2019). Enterobacter udwigii, isolated from the gut microbiota of Helicoverpa zea, promotes tomato plant growth and yield without compromising anti-herbivore defenses. Arthropod-Plant Interactions, 13, 271–278.

Pattanayak, S., Dash, D., Jena, M. & Nayak, R. 2000. Seed treatment of green gram with molybdenum and cobalt: effect on nodulation, biomass production and N uptake in an acid soil. *Journal of the Indian Society of Soil Science,* 48, 769–773.

Rai, P. K., Lee, S. S., Zhang, M., Tsang, Y. F. & Kim, K.-H. 2019. Heavy metals in food crops: Health risks, fate, mechanisms, and management. *Environment international,* 125, 365–385.

Rani, A., Souche, Y. S. & Goel, R. 2009. Comparative assessment of in situ bioremediation potential of cadmium resistant acidophilic Pseudomonas putida 62BN and alkalophilic Pseudomonas monteilli 97AN strains on soybean. *International Biodeterioration & Biodegradation,* 63, 62–66.

Rees, F., Simonnot, M.-O. & Morel, J.-L. 2014. Short-term effects of biochar on soil heavy metal mobility are controlled by intra-particle diffusion and soil pH increase. *European Journal of Soil Science,* 65, 149–161.

Rizvi, A., Ahmed, B., Zaidi, A. & Khan, M. S. 2020. Biosorption of heavy metals by dry biomass of metal tolerant bacterial biosorbents: an efficient metal clean-up strategy. *Environmental monitoring and assessment,* 192, 1–21.

Rl, C. 1980. Health risks associated with toxic metal in municipal sludge. *Sludge-Health Risks of Land Application*, 59–83.

Rodríguez-Tirado, V., Green-Ruiz, C. & Gómez-Gil, B. 2012. Cu and Pb biosorption on Bacillus thioparans strain U3 in aqueous solution: Kinetic and equilibrium studies. *Chemical Engineering Journal,* 181, 352–359.

Roggeman, S., de Boeck, G., de Cock, H., Blust, R. & Bervoets, L. 2014. Accumulation and detoxification of metals and arsenic in tissues of cattle (Bos taurus), and the risks for human consumption. *Science of the Total Environment,* 466, 175–184.

Saeed, Q., Xiukang, W., Haider, F. U., Kučerik, J., Mumtaz, M. Z., Holatko, J., Naseem, M., Kintl, A., Ejaz, M. & Naveed, M. 2021. Rhizosphere bacteria in plant growth promotion, biocontrol, and bioremediation of contaminated sites: a comprehensive review of effects and mechanisms. *International journal of molecular sciences,* 22, 10529.

Sasaki, A., Yamaji, N. & Ma, J. F. 2014. Overexpression of OsHMA3 enhances Cd tolerance and expression of Zn transporter genes in rice. *Journal of experimental botany,* 65, 6013–6021.

Schrauzer, G. & Rhead, W. 1971. Interpretation of the methylene blue reduction test of human plasma and the possible cancer protecting effect of selenium. *Experientia,* 27, 1069–1071.

Shah, M., Ara, J., Muhammad, S., Khan, S. & Tariq, S. 2012. Health risk assessment via surface water and sub-surface water consumption in the mafic and ultramafic terrain, Mohmand agency, northern Pakistan. *Journal of geochemical exploration,* 118, 60–67.

Shahbaz, A. K., Lewińska, K., Iqbal, J., Ali, Q., Iqbal, M., Abbas, F., Tauqeer, H. M. & Ramzani, P. M. A. 2018. Improvement in productivity, nutritional quality, and antioxidative defense mechanisms of sunflower (Helianthus annuus L.) and maize (Zea mays L.) in nickel contaminated soil amended with different biochar and zeolite ratios. *Journal of environmental management,* 218, 256–270.

Shahbazi K, Hashemi Nasab Zavareh K, Yeganeh M, Zare AA, Bazargan K, Esfandiyaripour E & Al, E. 2021. Investigation the Concentration of Heavy Metals (Cadmium and Lead) and Some

Nutrients in Wheat, Bride and Flour Produced in Iran. Karaj, Iran: Soil and Water Research Institute.

Singh, V., Singh, J., Singh, N., Rai, S. N., Verma, M. K., Verma, M., Singh, V., Chivate, M. S., Bilal, M. & Mishra, V. 2023. Simultaneous removal of ternary heavy metal ions by a newly isolated Microbacterium paraoxydans strain VSVM IIT (BHU) from coal washery effluent. *BioMetals,* 36, 829–845.

Sofy, M. R., Seleiman, M. F., Alhammad, B. A., Alharbi, B. M. & Mohamed, H. I. 2020. Minimizing adverse effects of pb on maize plants by combined treatment with jasmonic, salicylic acids and proline. *Agronomy,* 10, 699.

Spanu, A., Valente, M., Langasco, I., Leardi, R., Orlandoni, A. M., Ciulu, M., Deroma, M. A., Spano, N., Barracu, F. & Pilo, M. I. 2020. Effect of the irrigation method and genotype on the bioaccumulation of toxic and trace elements in rice. *Science of The Total Environment,* 748, 142484.

Tarfeen, N., Nisa, K. U., Hamid, B., Bashir, Z., Yatoo, A. M., Dar, M. A., Mohiddin, F. A., Amin, Z., Ahmad, R. A. A. & Sayyed, R. 2022. Microbial remediation: a promising tool for reclamation of contaminated sites with special emphasis on heavy metal and pesticide pollution: a review. *Processes,* 10, 1358.

Thatoi, H., Das, S., Mishra, J., Rath, B. P. & Das, N. 2014. Bacterial chromate reductase, a potential enzyme for bioremediation of hexavalent chromium: a review. *Journal of Environmental Management,* 146, 383–399.

Vinceti, M., Filippini, T., del Giovane, C., Dennert, G., Zwahlen, M., Brinkman, M., Zeegers, M. P, Horneber, M., D'amico, R. & Crespi, C. M. 2018. Selenium for preventing cancer. *Cochrane database of systematic reviews.*

Wan, Y., Huang, Q., Wang, Q., Yu, Y., Su, D., Qiao, Y. & Li, H. 2020. Accumulation and bioavailability of heavy metals in an acid soil and their uptake by paddy rice under continuous application of chicken and swine manure. *Journal of hazardous materials,* 384, 121293.

Wan, Y., Wang, K., Liu, Z., Yu, Y., Wang, Q. & Li, H. 2019. Effect of selenium on the subcellular distribution of cadmium and oxidative stress induced by cadmium in rice (Oryza sativa L.). *Environmental Science and Pollution Research,* 26, 16220–16228.

Wang, Y., Narayanan, M., Shi, X., Chen, X., Li, Z., Natarajan, D. & Ma, Y. 2022. Plant growth-promoting bacteria in metal-contaminated soil: Current perspectives on remediation mechanisms. *Frontiers in Microbiology,* 13, 966226.

Wintz, H., Fox, T. & Vulpe, C. 2002. Responses of plants to iron, zinc and copper deficiencies. *Biochemical Society Transactions,* 30, 766–768.

Wu, Z., Yin, X., Bañuelos, G. S., Lin, Z.-Q., Liu, Y., Li, M. & Yuan, L. 2016. Indications of selenium protection against cadmium and lead toxicity in oilseed rape (Brassica napus L.). *Frontiers in Plant Science,* 7, 1875.

Yang, L., Li, Y., Yang, X., Xiao, H., Peng, H. & Deng, S. 2011. Effects of iron plaque on phosphorus uptake by Pilea cadierei cultured in constructed wetland. *Procedia Environmental Sciences,* 11, 1508–1512.

Yang, X., Long, X., Ye, H., He, Z., Calvert, D. & Stoffella, P. 2004. Cadmium tolerance and hyperaccumulation in a new Zn-hyperaccumulating plant species (Sedum alfredii Hance). *Plant and soil,* 259, 181–189.

Yu, M., Meng, J., Yu, L., Su, W., Afzal, M., Li, Y., Brookes, P. C., Redmile-Gordon, M., Luo, Y. & Xu, J. 2019. Changes in nitrogen related functional genes along soil ph, C and nutrient gradients in the charosphere. *Science of the Total Environment,* 650, 626–632.

Zhang, R.-H., Li, Z.-G., Liu, X.-D., Wang, B.-C., Zhou, G.-L., Huang, X.-X., Lin, C.-F., Wang, A.-H. & Brooks, M. 2017. Immobilization and bioavailability of heavy metals in greenhouse soils amended with rice straw-derived biochar. *Ecological Engineering,* 98, 183–188.

Zhao, A., Tian, X., Lu, W., Gale, W., Lu, X. & Cao, Y. 2011. Effect of zinc on cadmium toxicity in winter wheat. *Journal of plant nutrition,* 34, 1372–1385.

Zheng, J., Chen, K.-H., Yan, X., Chen, S.-J., Hu, G.-C., Peng, X.-W., Yuan, J.-G., Mai, B.-X. & Yang, Z.-Y. 2013. Heavy metals in food, house dust, and water from an e-waste recycling area in South China and the potential risk to human health. *Ecotoxicology and environmental safety,* 96, 205–212.

Zheng, S., Wang, Q., Yuan, Y. & Sun, W. 2020. Human health risk assessment of heavy metals in soil and food crops in the Pearl River Delta urban agglomeration of China. *Food chemistry,* 316, 126213.

Zhou, B., Zhang, T. & Wang, F. 2023. Microbial-based heavy metal bioremediation: Toxicity and eco-friendly approaches to heavy metal decontamination. *Applied Sciences,* 13, 8439.

Zhu, H., Zhong, H. & Wu, J. 2016. Incorporating rice residues into paddy soils affects methylmercury accumulation in rice. *Chemosphere,* 152, 259–264.

Zhu, T., Li, L., Duan, Q., Liu, X. & Chen, M. 2021. Progress in our understanding of plant responses to the stress of heavy metal cadmium. *Plant Signaling & Behavior,* 16, 1836884.

Accumulation of Arsenic in the Food Chain Through Groundwater Soil Environment System: Its Alleviation Strategies

Md Masidur Alam and Santanu Gupta

Abstract Arsenic (As) is a common element found initially in the earth's crust. In nature, arsenic is usually found in other metal ores. It is a heavy metal that exists at low concentrations in soil and groundwater from where it enters the plant body. The groundwater in some states of India has been reported to contain arsenic in much higher levels, ranging from 50 to 3700 μg As/L. The arsenic contamination in groundwater in these regions is thought to be geogenic. The incessant use of unhygienic groundwater for irrigation is one of the chief reasons for the increase in the soil. Various organized studies have been accomplished so far to examine the effect of arsenic in groundwater, used as an irrigation resource, on the soil-plant-human range, but these have been limited. This article has tried to collect arsenic mobilization mechanisms within groundwater and how they affect the food chain through the consumption of food crops and groundwater soil environment systems. Furthermore, different approaches deployed to alleviate arsenic toxicity from the soil are also included in this article.

Keywords Alleviation · Arsenic contamination · Groundwater · Toxicity

1 Introduction

Arsenic (As) is a polluted trace metalloid that poses a serious threat to the environment due to its occurrence in water, soil, plants, animal, and human continuum. Xenobiotics are compounds that may have toxic or lethal effects on living organisms. Arsenic is one of the heavy metals referred to in the class of xenobiotics due to its precarious effect on the environment. Arsenic is found in dissolved oxyanions, such as arsenites and arsenates, or both in soil and groundwater. In addition to its

M. M. Alam (✉) · S. Gupta
Department of Botany, Malda College, Malda, West Bengal, 732101, India

S. Roy, V. Mandal (eds.), *Plant-Microbe Interaction under Xenobiotic Exposure*,
https://doi.org/10.1007/978-981-96-8260-7_16

organic forms, the chemical form of arsenic and whether it exists in an inorganic or organic combination determine its solubility, bioavailability, mobility, and, ultimately, soil-crop system toxicity. The presence and elevated toxicity level of this substance in the biosphere have also caused significant concern among the public and political sectors. Bangladesh has the highest degree of arsenic contamination in its groundwater among 20 countries worldwide, followed by India's West Bengal (Sanyal et al. 2015). The severity and unusual nature of the problem expose the population in the Bengal Delta region to significant threats. The significant levels of arsenic found in the groundwater in some areas of West Bengal, particularly in five districts surrounding the Bhagirathi River and the adjacent districts of Bangladesh, are a huge concern (Sarkar et al. 2022). In addition to the Bengal Delta Basin, there has been a detection of significant contamination of some regions of the nation's groundwater with arsenic, such as Bihar, Uttar Pradesh, Jharkhand, and Punjab, where the levels are above the permitted limit (WHO 2001) of 50 μg As/L (Sanyal 2017; Rahman et al. 2019; Kumar et al. 2021). The WHO established the acceptable limit of arsenic in drinking water in momentary guidelines in 1993 (WHO 1993), allowing a level of 10μgAs/L. This was preferred because accurately measuring lower concentrations for widespread use proved challenging. In contrast, some countries, like India and Bangladesh, have established their own national limits (MAC) for arsenic in drinking water, following an older WHO recommendation.

Contamination of groundwater by arsenic has far-reaching effects. Not only does arsenic enter the food chain, but it also leads to social problems, health risks, and economic hardship. Furthermore, its unpredictable movement and the continued use of contaminated groundwater exacerbate the issue. Arsenic is present in several inorganic androgenic forms in the natural environment. The absorption of arsenic by plants from polluted soil is a notable issue, leading to a considerable rise in human exposure to this hazardous substance. Arsenite is the prevalent form of arsenic found in agricultural fields, making up approximately 63% of the total arsenic content in soil. Arsenate accounts for about 36% of the total arsenic, although methylated arsenic species are also present (Abedin et al. 2002a). Most of these arsenic species are absorbed into the plant's body by specialized transporter proteins and help mitigate arsenic from soil (Zaidi et al. 2024; Beniwal et al. 2023). Arsenate disrupts the process of phosphate metabolism in plants; arsenite forms chemical bonds with sulfhydryl groups found in proteins, which leads to changes in their structures and disrupts their catalytic functions (Zhao et al. 2010). Arsenic's phytotoxic effect in plants is verified through various physiological changes, such as chlorosis in leaves, decreased root growth, shrinkage, and necrosis in above-ground plant structures (Carbonell-Barrachina et al. 1998). Arsenic, when transported throughout the entire plant, from the underground roots to the above-ground shoots, can significantly hinder their growth by impeding biomass buildup and lowering their reproductive potential through decreased fertility, crop yield, and fruit output (Garg and Singla 2011). Common signs of toxicity in different monocot plants cultivated in soil consist of developing brown spots, hindered growth, and leaf scorching (Bakhat et al. 2017). In their study, Duxbury and Panaullah (2007) found that raising the arsenic content in soil from 12.0 to 60.0 mg/kg in traditional rice fields in Bangladesh led to a decline in total rice yields from 7.5 to 2.5 t/ha. Moreover, when present in higher concentrations, arsenic has a detrimental impact on multiple metabolic characteristics, negatively affecting plant metabolism and ultimately leading to the death of the plant (Fig. 1). The addition of arsenic resulted in a decrease in the rate of plant transpiration (Stoeva and Bineva 2003). Arsenic, as a redox-active metalloid, has been found to induce the generation of ROS (reactive oxygen species) in plants, resulting in stress caused by oxidation (Tripathi et al. 2012). The prevalence of arsenite over arsenate is ascribed to the deteriorating circumstances in wet soils, which harm vegetation growth (Bakhat et al. 2017). Plants have developed enzymatic and non-enzymatic antioxidants to combat the oxidative stress caused by arsenic and ensure their survival (Gupta et al. 2013). The roots of a plant are primarily responsible for absorbing and accumulating arsenic, which can affect the plant's growth and reproduction. Typically, the concentration of arsenic decreases as it moves from the roots to the above-ground portion of plants, primarily by translocation. The wetland of Taiwan exhibited a decrease in arsenic concentration across different portions of the *Kandelia obovata* plant, with the roots containing the highest concentration (19.74 mg/kg), followed by the stems (1.76 mg/kg), leaves (1.71 mg/kg), and seedlings (0.48 mg/kg) (Liu et al. 2014). In the *Ratna* variety of Aman rice, arsenic bioaccumulation was seen in the following decreasing order: root > basal portion of stem > median part of stem > apical part of stem > leaves > grains (Bhattacharya et al. 2014; Liu et al. 2004). Approximately 40% of the arsenic that is transferred to different species is detected as AsV (Zhao et al. 2009; Rascio and NavariIzzo 2011). In plants such as rice, oxygen from the leaves travels downwards to the roots, aerating the rhizosphere. This aeration and the

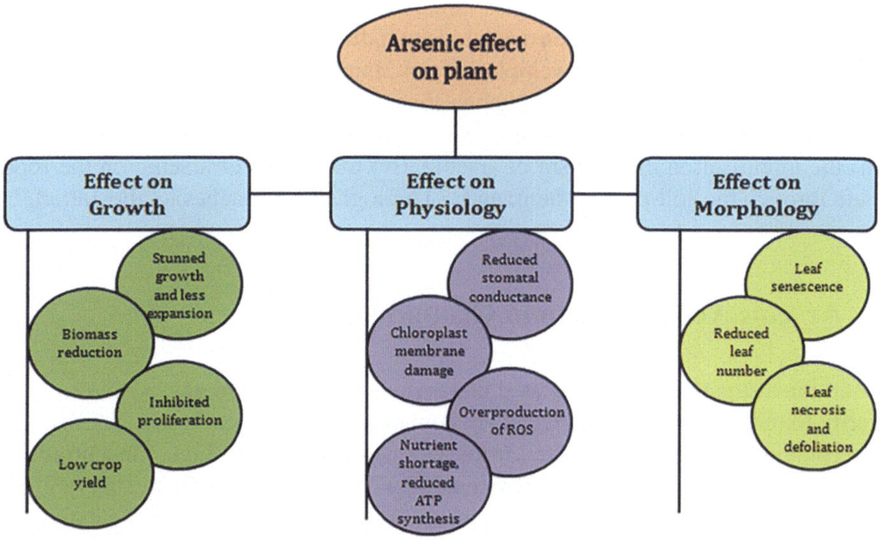

Fig. 1 Effect of arsenic in plant

presence of a bacterial community facilitate oxidation at the root level (Rahman and Hassler 2014; Chatterjee et al. 2017).

Food crops cultivated in water contaminated with arsenic are being transported to uncontaminated locations, where people may be exposed to arsenic through direct or indirect consumption of the contaminated food. This might potentially pose a fresh danger to the community. Several measures, remedial actions, and research approaches have been implemented and executed in West Bengal to combat the arsenic threat; however, they are limited to other states. The primary aim of the procedures and measures has been to supply all locals with drinking water free of arsenic affected by arsenic pollution. This is achieved by the establishment of alternative freshwater sources and the treatment of contaminated groundwater. Over the past few decades, a significant amount of research and development (R&D) has been conducted to improve understanding of the following:

- Origin and underlying cause of arsenic pollution in the groundwater.
- The extent and intensity of the pollution in the underground water.
- The process by which arsenic is released from soil and enters into groundwater.
- Suitable technology for arsenic removal from groundwater for community development.
- Methods utilized for identifying arsenic in groundwater by analytical techniques.

The objective is to ensure the provision of uncontaminated groundwater suitable for residential and agricultural purposes in regions now impacted by arsenic or susceptible to contamination. A critical evaluation is needed to assess the effectiveness of government-initiated remedial measures in restoring groundwater quality, resolving the issue, and achieving long-term sustainability. This evaluation should include understanding the underlying physical processes involved. Effective implementation of these solutions in other arsenic-contaminated areas depends on such a critical assessment. A comprehensive understanding of all aspects of arsenic contamination is crucial for developing urgent, interim, and long-term strategies to address this problem. The present chapter thus deals mainly with (i) the occurrence of arsenic in groundwater, (ii) arsenic toxicity in the groundwater soil environment, (iii) the mobilization mechanism of arsenic, (iv) the effects of arsenic on the food chain through the soil-plant system, and (v) biological approaches to alleviation.

2 Arsenic Accumulation in Groundwater

Several investigations have indicated that arsenic-contaminated groundwater is primarily found in the Ganges delta, which contains floodplain reservoirs. These aquifers contain sediments transported from regions in Bihar with high concentrations of sulfide minerals and other nearby regions within the basin (Bhattacharya et al. 1997). Nevertheless, in recent investigations, it has been shown that the extensive Indo-Gangetic alluvium, which stretches toward the west, as well as the Brahmaputra alluvium, situated in the late Quaternary as well as Holocene aquifers, has elevated amounts of arsenic. Arsenic is generated due to the breakdown of sulfide minerals that usually attach to iron oxyhydroxides' surface. These iron oxyhydroxides are formed under the prevalent oxidizing conditions that occur during the accumulation of sediments from the Holocene period (Amaya 2002). However, in the sediments, redox reactions cause the reduction of iron oxides, which leads to the movement of a significant amount of arsenic into the water via numerous biogeochemical interactions and exchanges (Smedley and Kinniburgh 2002). The sediments from the late Quaternary period are composed of sand, silt, and clay. The deposition of sediments is the cause of the increased quantities of arsenic present in the groundwater of the Ganga-Brahmaputra River basin beside the rivers in the Holocene or late Quaternary era, which occurred within the past 12,000 years. A substantial deposit of recent alluvium consisting of sand, silt, and clay has been extensively transported by various rivers originating from the northern and northeastern Himalayas (Bora and Borah 2024). The sediments have a mineralogical composition, including feldspars, quartz, illite, and kaolinite. Moreover, the fine-grained overbank facies contain a significant amount of diverse organic matter and particles (Sikdar and Banerjee 2003). The majority of environmental issues related to arsenic pollution have been identified; thus, the distance covered is a consequence of its movement in natural environments. Arsenic has been found in the Gangetic plains and BDP groundwater due to geological factors resulting from its mobilization under natural hydrogeological circumstances.

3 Arsenic Toxicity in the Groundwater Soil Environment

Groundwater and soil both contain dissolved oxyanions, a type of arsenic, specifically arsenate (pentavalent arsenic) or arsenites (trivalent arsenic), or both together with organic forms. The movement, capacity to dissolve, availability to living organisms, and arsenic's detrimental effects on the soil along with the growth of crops are determined mainly by their chemical qualities, namely, the oxidation state or valency of arsenic and whether the arsenic is present in an organic or inorganic compound state. The concentration of arsenic in various types of rocks and other materials (Table 1) and non-groundwater water sources (Table 2) can exhibit significant variation. The following is the order in which the toxicity levels of arsenic compounds are found in soil and/or groundwater environments: Arsine ≥ organoarsine material ≥ native arsenic metal ≥ arsenite oxides (trivalent arsenic form) ≥ arsenates (pentavalent arsenic form) ≥ arsonium metals (monovalent arsenic form) (Bhattacharyya and Alam 2020). Arsenites have more toxicity, solubility, and mobility than arsenates in water and soil environments. Dimethylarsinic acid (DMA) or cacodylic acid, which can be reduced to produce volatile di–/trimethylarsines, can also be found in the soil. Arsenic, in its organic forms, is often found to be less or non-toxic. Mono-methyl arsinic acid (MMA) is another type of organic compound that can be found in groundwater and soil (Dilpazeer et al. 2023). While the less hazardous pentavalent forms of arsenic are prevalent in aerobic

Table 1 Arsenic quantity in different types of rocks and some other materials (Source: Sanyal 2017)

Materials	Types	Subtypes	Quantity of arsenic (mg As./kg)
Rocks	(i). Igneous rocks	Ultrabasic (*Dunite, Peridotite, Serpentine*)	0.3–15.8
		Basic (*Gabbro, Basalt*)	0.18–113
		Intermediate (*Andesite, Latite, Trachyte*)	0.5–5.8
		Acidic (*Granite,Rhyolite*)	3.2–5.4
	(ii). Metamorphic rocks	Quartzite	2.2–7.6
		Slate/Phyllite	0.5–143
		Schist/Gneiss	0.0–185
	(iii). Sedimentary rocks	Shale/Claystone (near-shore)	4.0–25
		Shale/Claystone (offshore)	3.0–490
		Carbonates	0.1–20.1
		Phosphorites	0.4–188
		Sandstone	0.6–9.0
		Shales (Nonmarine)	3.0–12
		Claystone (Nonmarine)	3.0–10
Coal			Up to 2000
Crustal average			2.0

Table 2 Arsenic quantity in water other than groundwater. (Source: Sanyal 2017)

Source	Amount of arsenic (μg As. /L)
Snow and rainwater	< 0.002–0.59
Rivers	0.20–264
Lakes	0.38–1.00
Seawater	0.15–6.00
Canals (West Bengal, India)	40–150
Ponds (West Bengal, India)	4–70

cultivation, the submerged growth of rice is primarily characterized by the prevalence of both the more soluble and toxic trivalent versions.

4 Mobilization Mechanism of Arsenic

Studies in the majority of the analyzed regions indicated that groundwater with elevated levels of arsenic was not correlated with places where the source rock had high concentrations of arsenic. Two factors have been identified: firstly, specific biogeochemical triggers facilitate the movement of arsenic from solid matter to groundwater, and secondly, the released arsenic requires an adequate amount of duration for accumulation, meaning it needs to be retained in the aquifer and not flushed away (Smedley and Kinniburgh 2002). Several investigators robustly thought that the oxidation of sulfide minerals was the main factor responsible for groundwater pollution with arsenic in West Bengal (Das et al. 1995). According to the theory, arsenic in the shallow aquifer is released by the oxidation process from sulfide minerals, i.e., arsenopyrite (Mandal et al. 1998). The discharge of arsenic is caused by the depletion of the water table resulting from groundwater management for irrigation purposes. A research investigation elucidated that the dissolution or desorption of arsenic from iron oxides is thought to cause arsenic dispersion in water (Sanyal 2017). A discharge of arsenic from the source should occur rapidly with the time it takes for groundwater to flush through. The phenomena by which arsenic is released into groundwater are referred to as mobilization. The process of mobilization is as follows: (i) mineral dissolution, (ii) desorption and dissolution of arsenic under reducing conditions, (iii) arsenic desorption in an alkaline and oxidizing environment, (iv) reduction in bond strength between arsenic and holt mineral surface, and (v) reduction of oxide mineral surface area (Smedley and Kinniburgh 2002). Overall, it can be said that major reactions occur under reducing conditions, leading to the release of arsenic. These reactions involve the conversion of arsenic from its oxidized form-As (V) to its diminished form-As (III). As (III) exhibits less adsorption to iron oxides than As (V) in numerous circumstances. Investigators have shown that excessive irrigation water use and using fertilizers can cause phosphate movement from the fertilizers to the shallow aquifers. This shifting of phosphate has led to the release of arsenic because of anion exchange with reactive mineral surfaces (Michaela and Voss 2008). The occurrence of arsenic-contaminated groundwater in the region results from an intricate interplay of multiple variables. The factors contributing to groundwater replenishment are rainfall, natural water flow patterns, groundwater-soil interaction, seepage of arsenic-enriched irrigation water, organic matter's breakdown, and the iron as well as manganese oxides dissolution (Sikdar and Chakraborty 2008). One possible explanation for the emergence of high-arsenic groundwater is the dislocation of arsenic by dissolved bicarbonate, as described in a different hypothesis (Smedley and Kinniburgh 2002). Another potentially significant process is the impulsive breakdown of iron oxides in reducing circumstances. Several hypotheses exist, each with its inconsistencies and limitations, to explain the physical processes (Majumdar et al. 2002). Multiple research studies have been conducted on the modeling of groundwater arsenic in order to obtain a thorough comprehension of the chemical reactions of arsenic, including its occurrence, discharge, dissolution, mechanisms, sorption, and so on, in the soil-water phases within a local area (Paul and Sikdar 2008; NIH and CGWB 2009). Nevertheless, the results obtained from numerical modeling have been limited and uncertain due to a lack of comprehensive knowledge regarding the chemical behavior of arsenic in geochemical and geo-environmental processes. The modeling studies largely emphasize the following criteria: (i) The presence of arsenic in groundwater is derived from naturally formed deposits and is seen in specific regions. The magnitude of arsenic release and its subsequent mobility is significantly controlled by the volume of water extracted from the groundwater system. (ii) By applying careful aquifer management, it is possible to get arsenic-free groundwater for an extended period without the risk of disturbing areas contaminated with arsenic. (iii) By tapping into less polluted aquifers or freshwater sources, in addition to surface water sources, it is possible to obtain a dependable water supply that is free from arsenic (Agarwal et al. 2024). This water can be used for both drinking and irrigation purposes. While several hypotheses describe the chemical mechanisms behind groundwater arsenic poisoning, the most widely accepted processes include dissolution. Among the various hypotheses presented to explain the potential process of arsenic groundwater pollution, a consensus has been reached among numerous scientists on two hypotheses: (i) reductive dissolution of arsenic from soils and (ii) the process of oxidizing arsenopyrite or pyrite with a high concentration of arsenic in soil layers. Based on arsenic geochemistry, three ideas have been proposed regarding the likely mechanisms of arsenic mobilization in groundwater, namely in Holocene aquifers such as those found in West Bengal and Bangladesh (Bose and Sharma 2002). These are as follows: (i) the oxidation of pyrite minerals containing arsenic leads to the mobilization of arsenic. Arsenic-containing minerals that cannot dissolve in a liquid, such as arsenopyrite (FeAsS), undergo fast oxidation upon exposure to the atmosphere, releasing soluble As (III), sulfate, and ferrous ions. The decomposition of these rock elements containing arsenic is significantly reliant on the presence of oxygen and the speed at which sulfide is oxidized. The As (III) that has been liberated undergoes partial oxidation to As (V) through microbial-mediated processes. (ii) The dissolution of iron oxy-hydroxides (FeOOH) containing high levels of arsenic occurs when decreasing conditions originate in the subsoil. When iron (Fe) is present under oxidizing circumstances, inorganic forms of arsenic are mostly kept in the solid phase of soil by binding with FeOOH coatings. Under certain circumstances, reducing conditions might result in the deterioration of FeOOH coatings. (iii) When fertilizers are applied to the soil subsurface, phosphate ions permeate aquifers and are actively exchanged with adsorbed materials, releasing them into the minerals. The most likely reason for the excessive arsenic buildup in groundwater is the subsequent process of FeOOH dissolution under decreasing conditions.

5 Effects of Arsenic on Food Chain Through the Soil-Plant System

There are limited works available that demonstrate arsenic consumption from water and a nutritious diet on a regular basis among individuals residing in locations with contaminated groundwater and how this relates to arsenic biomarkers. The investigation revealed that a higher concentration of trivalent arsenic was accumulated in the edible grains of the transplanted autumn paddy, while pentavalent arsenic was found to be predominant in rice plant straws (Sinha and Bhattacharyya 2014). Furthermore, adding organic manure to the soil reduced the amount of arsenic that accumulated in rice grains and straw, as indicated by the decrease in the amount of inorganic arsenic being loaded. Moreover, it was discovered that the process of parboiling and milling of rice led to an increase in the amount of arsenic present in rice, regardless of whether it was high-yielding or a traditional variety (Sanyal et al. 2012). Additionally, the researcher studied the toxicity profile of arsenic in rice grown in the contaminated rural area of Bengal and the possible health consequences after its consumption (Ghosh et al. 2012; Mridha et al. 2022). The hazard posed by contaminated drinking water in rural Bengal is the possible risk of ingesting arsenic by consuming rice and other contaminated food. Application of organically modified and enhanced phosphate, along with targeted micronutrient fertilization such as zinc or iron salts, has demonstrated significant efficacy in reducing the amount of inorganic and total arsenic that accumulates in rice and other crops, hence reducing the danger that comes with eating them (Mukhopadhyay et al. 2002). Physicians and agronomists surveyed West Bengal to determine demographic variables and the overall daily consumption of arsenic from water and food. The findings unequivocally demonstrated that, in some regions of West Bengal, India, there is a strong correlation between the use of contaminated groundwater for agricultural irrigation and excessive dietary arsenic consumption. However, this correlation was significantly lower in the adjoining areas, where the groundwater was pure and uncontaminated (Das et al. 2016). Arsenic concentrations were detected at higher-than-normal levels in the urine of individuals residing in an arsenic-endemic region, demonstrating noteworthy consumption of food and water contaminated with arsenic (Sanyal et al. 2015). Attempts were made to lower the allowed limit of arsenic in drinking water to 50μgAs/L, yet this exposure still happened. Individuals with skin lesions had higher hair and urine arsenic levels than those without skin lesions (Guha Mazumder ct al. 2014). A separate investigation demonstrated a significant and positive correlation between the quantities of arsenic consumed daily through water and diet and the levels of arsenic found in urine in a region of West Bengal known for its high-arsenic content. This correlation persisted even when individuals in the area were using water containing a high level of arsenic, i.e., 50μgAs/L, which is considered safe for cooking and drinking. Even though the WHO considers the arsenic levels in drinking water safe, they were still reduced to 10μgAs/L. There was still a significant association between the quantity of arsenic obtained from the diet and the secretion of arsenic in urine. Nevertheless, no notable correlation was identified between the dosage of arsenic from consuming water within this specific group (Halder et al. 2012; Guha Mazumder et al. 2013). Moreover, when individuals were just subjected to arsenic ingestion via their food, there was a direct correlation between the amount of arsenic in their urine and the amount of arsenic they consumed through their food. Therefore, when considering all of this information collectively, it is clear that providing drinking water with arsenic levels below 10μgAs/L to the residents of the countryside of Bengal areas can be seen as a viable solution. Therefore, to deal with chronic arsenic toxicity, it is necessary to implement a comprehensive and coordinated intervention. Key elements should include efforts to prevent the introduction of arsenic into the food chain and reduce the amount of arsenic present in drinking water to amounts considered below the accepted safety standards (Sanyal et al. 2015; Golui et al. 2017).

6 Biological Approaches to Alleviation

Arsenic exhibits properties that are midway between those of metals and nonmetals. Nevertheless, nonmetals are more linked to ionization energy and electronegativity properties. Plants assimilate arsenates via phosphate transport proteins; arsenite and its methylated arsenic derivatives are taken in through nodulin-26-like intrinsic aquaporin channels (Zhao et al. 2010). Arsenate undergoes a gradual conversion into arsenite inside the plant's structures and is eliminated by forming complexes with peptides rich in thiol, such as vacuolar sequestration and phytochelatins. Diverse measures to reduce or alleviate the impact, including agricultural practices, plant breeding, and modification of genes, can be employed to diminish the absorption of arsenic in food and cultivated crops (Fig. 2). An experiment indicated that the *Oryza sativa* cultivar of rice contains 2PCS genes, namely, OsPCS1 and OsPCS2. These transcripts of genes are present in the endosperms of grains and help decrease the amount of arsenic in them (Das 2015). *Pantoea* sp. (EA106) is a strain of microorganisms that can improve food quality by decreasing the hazardous buildup of arsenic in the above-ground sections of the species. In a study conducted by Lakshman (2015), proteobacterial strains isolated from the rhizosphere of the rice crop were discovered to possess the capacity to stimulate growth and increase the oxidizing state in the area surrounding the rice root. In another investigation on the expression of OsMATE2, a rice MATE family transporter lowers the amount of arsenic found in rice grains (Das et al. 2014). Their research revealed a novel

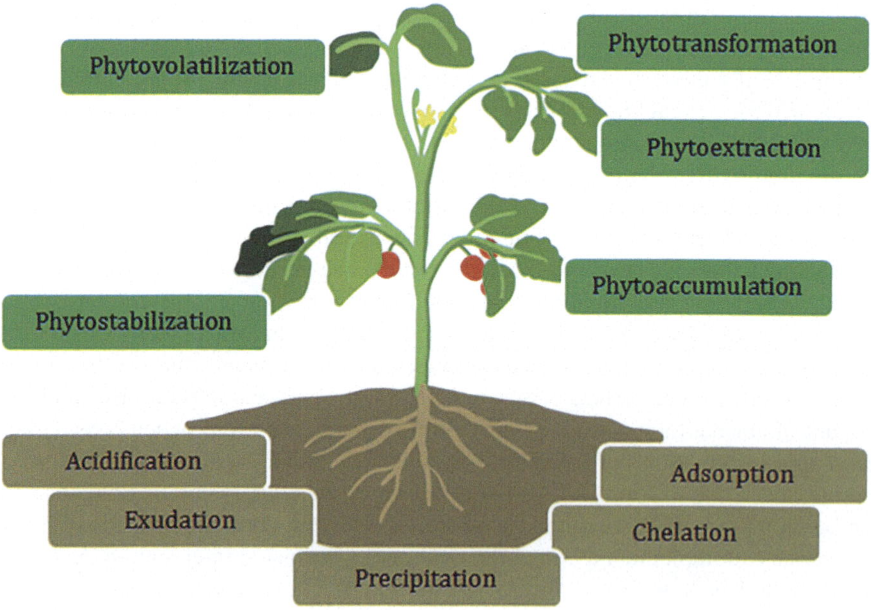

Fig. 2 Phytoremediation processes of arsenic

approach of mitigating arsenic levels by utilizing a hydrological control system and implementing a silicon fertilization process. Zhao et al. (2010) documented *Oryza sativa*'s efficiency in dispersing arsenite in a flooded rice field. The lowest amounts of arsenic were found to be deposited in rice grains under aerobic conditions, and straw was shown to be present during the whole vegetative and developmental stages of reproduction. Additional biological methods involve using specific plants with monocot or dicot roots that function as hyper-accumulators. *Ipomoea alpina, Thlaspi caerulescens, Haumaniastrum robertii, Cistus ladanifer, Thlaspi rotundifolium, Macadamia neurophylla, Psychotria douarret, Salix* sp., etc., have the potential to remove the existence of hazardous heavy metals from soil. These plants exhibit tolerance to high levels of metal contaminants in the soil. Compared to other crops, certain plants can accumulate heavy metals that are 10–500 times greater (Abedin et al. 2002b). An EDTA application was observed to enhance the process of metal extraction by hyper-accumulators (Dotaniya 2014). Arsenic subtraction was enhanced by phosphate rock through the performance of hyper-accumulator *Pteris vittata* (Fayiga and Ma 2006). Arsenic is a prevalent metallic element in the environment and occurs in varying amounts throughout different regions. The plants involved in phytoremediation can collect high levels of metals, but they might generate a smaller amount of biomass, which reduces their overall efficiency in extracting metals from the environment. Numerous high-biomass hyper-accumulating plants that have the potential for the establishment of regeneration practices comprise Indian sunflower (*Helianthus annuus*), tomato (*Lycopersicon esculentum*), mustard (*Brassica juncea*), and yellow poplar. Arsenic exposure in *Artemisia annua* was found to modulate physiological attributes and artemisinin biosynthesis (Naeem et al. 2019). Some monocot plants have been discovered to extract certain metallic elements from the soil by converting them through phyto-volatilization. Several plants have been seen to utilize the phytostabilization process to prevent metal pollution of groundwater and lower soil layers by deactivating metals that can dissolve in the rhizosphere (Pigna et al. 2010). Researchers have demonstrated that utilizing the "Bio-metal slurry reactor" method and the bio-surfactants that are made by microorganisms can extract metals from soil (Wang and Mulligan 2004). At the field level, though, the likelihood of these methods hasn't been proven yet. Arsenic is classified as a hazardous metalloid that is unnecessary and frequently found in the natural environment. Arsenic concentrations in soil that are higher than permitted harm plant growth and development as well as agricultural production (Nabi et al. 2020). Because of both natural and human activity, arsenic is present in the environment and poses a serious risk to global ecosystems and human health due to its high toxicity (Nriagu et al. 2007).

Arsenic exhibits intricate chemistry in natural environments, similar to phosphorus, and predominantly exists as anions (Fitz and Wenzel 2002). Arsenates are more frequently encountered than arsenic in soil with pH-Eh values within the usual range (Cao et al. 2009). Also controlling the quantity of these complexes is crucial for managing soil toxicity and the presence of water and/or sediment in the soil (Farré and Barceló 2003). The average arsenic concentration in natural soils is below 10 mg/kg, whereas in the earth's crust, it is approximately 1.8 mg/kg (Adriano

2001; Deliyanni et al. 2003). Arsenic usually accumulates in soils due to its limited movement within the environment (Basu et al. 2015). Because organic compounds exhibit lower solubility in soil, they often originate as an inorganic complex (Nriagu et al. 2007; Bhattacharya et al. 2007). The soil's arsenic content fluctuates due to its dependence on factors such as the primary concentration of soil type, parent material, and natural biogeochemical cycles (Fendorf et al. 2004). As a result, arsenic levels in the subsurface vary globally, with mean values ranging from 0.1 to 80 mg/kg. Plants can absorb arsenic and other trace metals through their roots. However, there is still a lack of sufficient explanation for arsenic's precise biological role in living things. Notably, plants have been shown to regularly contain amounts of arsenic ranging from 1.0 to 1.7 mg/kg (Yamaguchi et al. 2011). In terrestrial ecosystems, soil conditions largely control the bioavailability, transport, and potential toxicity or insufficiency of most elements in plants and microbes (Beesley and Marmiroli 2011; Majumdar et al. 2013). Grasses *(Agrostis stolonifera* and *A. tenuis),* the dominant species, had arsenic levels in their leaves above 1.0 g As/kg (dry weight). A total of 72 species of plants originated from 15 Rhodesian arsenical mine dumps, with the total arsenic level ranging from 0.200 to 30.000 gAs/kg. The reduction in arsenic content is believed to have led to an increase in both the quantity of plant species and their concentration. The following criteria influence the availability of arsenic in soil: The sample comprises the capacity to exchange cations, iron oxides, clay concentration, and calcium carbonate along with pH as well as organic matter concentration (Rooney et al. 2006; Chatterjee et al. 2013). Using modern methods, researchers have successfully generated certain plants with high-biomass production capabilities, known as phyto-accumulators. This was achieved by introducing specific genes from bacteria, animals, and hyper-accumulators into non-accumulator plants (Singh et al. 2003). The survey result indicates that the mine dumping Banshee (30.0 gAs/kg) was the area that experienced significant devastation and had minimal impact on the growth of plant life. The weed species were found to be the most prevalent species in plant abundance in the mine dumps. *Cynodon dactylon* is the most significant grass species in soils containing 0.2–30.00 mg As/kg (Alloway 1995). Arsenic and other heavy metals found in various mine wastes can hinder the process of stabilizing soil vegetation. However, certain plants, such as *Cynodon dactylon*, can withstand high levels of arsenic and can be beneficial in stabilizing soil in mine dumps. These methods may also provide a viable and cost-effective solution for large mine waste piles where other treatment methods are not feasible. Numerous atmospheric arsenic emissions are observed during the smelting of ores of Au, Cu, Pb, and Zn (Acharya 1997). The study of plants capable of absorbing significant quantities of arsenic from soils is necessary. Hyper-accumulator plants, such as *Pteris vittata* ferns and *Brassica* sp. mustard, are examples of plants that possess the capability to gather high levels of certain substances. The brake fern (*Pteris vittata*) and mustard are suitable options to address the issue in areas where it is impractical to supply irrigation with minimal amounts of arsenic. *Eichhornia crassipes*, commonly called water hyacinth, is an aquatic plant renowned for its capacity to hyper-accumulate specific chemicals. However, it is important to note that water hyacinth can only extend its roots in water, limiting its effectiveness in removing arsenic from irrigation water but not from the soil's arsenic content. The feasibility of utilizing hyper-accumulating plants to restore paddy soils must be evaluated. Additionally, methods for safely managing large volumes of plant residues with elevated concentrations of arsenic need to be developed to reduce potential hazards to the health of children, wildlife, and cattle that consume these plants or are exposed to dust from burned materials. Plants that absorb arsenic are typically removed from fields after the crop is harvested. An analysis was conducted on the records of several irrigation water sources located near the district, subdivisions, and municipalities. This research specifically examined the physical and chemical properties and the concentration of heavy metals.

7 Conclusion

As a result of water contamination with arsenic, a large number of individuals are experiencing severe arsenic toxicity. Its chemical form determines the toxicological properties of arsenic. Arsenic within the body alters cellular metabolic activity and disrupts proper cellular development. Therefore, there is a significant and immediate worry for all living organisms, including humans, over the deadly consequences of consuming contaminated water and the subsequent buildup of toxins in crop plants. There are numerous difficulties related to arsenic in crops and the factors that influence arsenic absorption, accumulation, and toxicity levels. Moreover, it is a subject of great concern that not only are the people living in the district polluted with arsenic at risk, but also individuals dwelling in nearby cities or districts are not safe due to their consumption of arsenic-infected food and food items produced in these areas. Multiple actions need to be implemented simultaneously to address this issue. These actions include integrating the water supply with a suitable water management system, studying the presence of arsenic in the food chain across diverse ecosystems, establishing effective protocols for responding to arsenic contamination in new regions, overcoming political and financial barriers related to alleviation efforts, implementing a strategic research plan to understand the interactions between arsenic exposure and health effects, conducting research on the geological and chemical aspects of water sources, and identifying or developing crop varieties that accumulate low levels of arsenic and require less water but still have high yields. Additional strategies such as employing arsenic-resistant bacteria for bioremediation, implementing rainwater collecting for crop irrigation, utilizing natural arsenic chelators, employing genetically engineered plants, and engaging in aerobic farming techniques have been beneficial in reducing arsenic pollution. The introduction of molecular biology technology has provided numerous opportunities to enhance the safety of grain diets by reducing arsenic accumulation. Nevertheless, alterations in cultivation practices, such as the implementation of spray irrigation techniques, can potentially reduce the accumulation of arsenic in crops. The transition from growing procedures to the subsequent application of plants in the field is crucial for prospective plant development. Thorough investigations are necessary to develop a sustainable method for treating soil polluted with arsenic. This method should be used carefully to reduce the accumulation of arsenic in economically significant crops to meet the need for human food.

References

Abedin MJ, Cotter-Howells J, Meharg AA (2002a) Arsenic uptake and accumulation in rice (*Oryza sativa* L.) irrigated with contaminated water. Pl Soil 240:311–319

Abedin MJ, Feldmann J, Meharg AA (2002b) Uptake kinetics of arsenic species in rice plants. Plant Physiol 128:1120–1128

Acharya SK (1997) Arsenic in groundwater-Geological overview. Consultation on Arsenic in Drinking water and Resulting Arsenic Toxicity in India and Bangladesh. World Health Organization, New Delhi

Adriano DC (2001) Trace elements in terrestrial environments. Biogeochemistry, bioavailability and risk of metals, (2nd edn). Springer, New York

Agarwal V, Kumar M, Pandey DP et al (2024) Unlocking the potential of remote sensing for arsenic contamination detection and management: Challenges and Perspectives. Current Op Env Sc & Hel 100578

Alloway BJ (1995) Heavy metals in soils. Blackie Academic & Professional, London

Amaya A (2002) Arsenic in groundwater of alluvial aquifers in Nawalparasi and Kathmandu districts of Nepal. Master's thesis, Dept. of Land and Water Resources Engineering, Kungl TekniskaHogskolan, Stockholm

Bakhat HF, Zia Z, Fahad S et al (2017) Arsenic uptake, accumulation and toxicity in rice plants: Possible remedies for its detoxification: A review. Environ Sci Pollut Res 24:9142–9158

Basu B, Kundu M, Hedayatullah M et al (2015) Mitigation of Arsenic in rice through deficit irrigation in field and use of filtered water in kitchen. Int J Env Sc Tech 12:2065–207

Beesley L, Marmiroli M (2011) The immobilization and retention of soluble arsenic, cadmium and zinc by biochar. Env Poll 159:474–480

Beniwal R, Yadav R, Ramakrishna W (2023) Multifarious effect of arsenic on plant and strategies for mitigation. Agriculture 13(2): 401

Bhattacharya P, Chatterjee D, Jacks G (1997) Occurrence of As contaminated groundwater in alluvial aquifers from the Delta Plains, eastern India: option for safe drinking water supply. Int J Water Res Dev 13:79–92

Bhattacharya P, Ewlch AH, Stollenwerk KG et al (2007) Arsenic in the environment: biology and chemistry. Sci Total Env 379:109–120

Bhattacharya P, Jovanovic D, Polya D (2014) Best practice guide on the control of arsenic in drinking water, IWA Publishing, London, UK

Bhattacharyya P, Alam MM (2020) Arsenic-contaminated soil toxicity and its mitigation through monocot crops, In: Contaminants in Agriculture Sources, Impacts and Management, M. Naeem, A.A. Ansari and S.S. Gill (eds), Springer Nature Switzerland AG, pp. 327–338

Bora G and Borah T (2024) Arsenic contamination of groundwater: Review on arsenic contamination of Assam. World J Adv Eng Tech Sc 11(02): 238–244

Bose P, Sharma A (2002) Role of iron in controlling speciation and mobilization of arsenic in subsurface environment. Water Research Vol. 36:4916–4926

Cao Q, Hu Q-H, Baisch C et al (2009) Arsenate toxicity for wheat and lettuce in six Chinese soils with different properties. Environ Toxicol Chem 28:1946–1950

Carbonell-Barrachina A, Aarabi MA, Delaune RD et al (1998) Bioavailability and uptake of arsenic by wetland vegetation: Effects on plant growth and nutrition. J Environ Sci Health 33:45–66

Chatterjee M, Sarkar S, Debnath S et al (2013) Genotypic difference in temporal variation of arsenic accumulation and expression of silicon efflux transporter (LSi2) gene in the field grown rice. Ind J Gene 73:94–97

Chatterjee S, Sharma S, Gupta DK (2017) Arsenic and its effect on major crop plants: Stationary awareness to paradigm with special reference to rice crop. In: Arsenic Contamination in the Environment; D.K. Gupta, and S. Chatterjee (eds) Springer International Publishing AG, Cham, Switzerland, pp. 123–143

Das D, Chatterjee A, Mandal BK et al (1995) Arsenic in groundwater in six districts of West Bengal, India: the biggest arsenic calamity in the world. Part II: Arsenic concentration in drinking water, hair, nail, urine, skin scale and liver tissues (biopsy) of the affected people. Analyst 120:917–924

Das I, Ghosh K, Das DK et al (2014) Transport of Arsenic in some affected soils of Indian subtropics. Soil Res 52:822–832

Das I, Sanyal SK, Koushik G et al (2016) Arsenic mitigation in soil-plant system through zinc application in West Bengal soils. Bioremed J 20:24–37

Das N (2015) Experiment with rice reporting two PCSgenes OsPCS1 OsPCS2 in *Oryza sativa* cultivar. Envn Int 242:1037–1050

Deliyanni EA, Bakoyannakis DN, Zouboulis AI et al (2003) Removal of arsenic and cadmium by alaganeite fixed beds. Sci Technol 38:3967–3981

Dilpazeer F, Munir M, Shafiq I et al. (2023) A Comprehensive Review of the Latest Advancements in Controlling Arsenic Contaminants in Groundwater. Water 15 (9): 1781

Dotaniya ML (2014) Impact of tannery effluent irrigation on heavy metal build up in soil and groundwater in Kanpur. Agrotech 2:77

Duxbury JM, Panaullah GM (2007) Remediation of Arsenic for agriculture sustainability, food security and health in Bangladesh; FAO: Rome, Italy, pp. 1–28

Farré M, Barceló D (2003) Toxicity testing of wastewater and sewage sludge by biosensors, bioassays and chemical analysis. Trends Anal Chem 22:299–310

Fayiga AO, Ma LQ (2006) Using phosphate rock to immobilize metals in soil and increase arsenic uptake by hyperaccumulator *Pteris vittata*. Sci Total Env 15:17–25

Fendorf S, La Force MJ, Li G (2004) Heavy metals in the environment. Temporal changes in soil partitioning and bioaccessibility of arsenic, chromium, and lead. J Environ Qual 33:2049–2055

Fitz WJ, Wenzel WW (2002) Arsenic transformations in the soil/rhizosphere/plant system: fundamentals and potential application to phytoremediation. J Biotechnol 99:259–278

Garg N, Singla P (2011) Arsenic toxicity in crop plants: Physiological effects and tolerance mechanisms. Env Chem Lett 9:303–321

Ghosh K, Das I, Das DK et al (2012) Evaluation of humic and fulvic acid extracts of compost, oil cake, and soils on complex formation with arsenic. Soil Res 50:239–248

Golui D, Guha Mazumder DN, Sanyal SK et al (2017) Safe limit of arsenic in soil in relation to dietary exposure of arsenicosis patients from Malda district, West Bengal-A case study. Ecotox Env Saf 144:227–235

Guha Mazumder DN, Deb D, Biswas A et al (2013) Evaluation of dietary arsenic exposure and its biomarkers: A case study of West Bengal, India. J Envi Sc Health Part A 48:896–904

Guha Mazumder DN, Deb D, Biswas A et al (2014) Dietary arsenic exposure with low level of arsenic in drinking water and biomarker: A study in West Bengal. J Envi Sc Health Part A 49:555–564

Gupta DK, Inouhe M, Rodríguez-Serrano M et al (2013) Oxidative stress and arsenic toxicity: Role of NADPH oxidases. Chemosphere 90:1987–1996

Halder D, Bhowmick S, Biswas A et al (2012) Consumption of brown rice: a potential pathway for arsenic exposure in rural Bengal. Env Sci Tech 46:4142–4148

Kumar A, Kumar R, Rahman MS et al (2021) Assessment of arsenic exposure in the population of Sabalpur village of Saran District of Bihar with mitigation approach. Env Sci Pollut Res https://doi.org/10.1007/s11356-021-13521-5

Lakshman V (2015) Proteobacterial strains promote rice growth and enhance the oxidizing environment surrounding rice root. Env Int 74:221–230
Liu CW, Chen YY, Kao YH et al (2014) Bioaccumulation and translocation of arsenic in the ecosystem of the Guandu Wetland, Taiwan. Wetlands 34:129–140
Liu ZJ, Boles E, Rosen BP (2004) Arsenic trioxide uptake by hexosepermeases in Saccharomyces cerevisiae. J Biol Chem 279:17312–17318
Majumdar A, Bhattacharyya K, Kole SC et al (2013) Efficiency of indigenous soil microbes in arsenic mitigation from contaminated alluvial soil of India. Env Sc Poll Res 20:5645–53
Majumdar PK, Ghosh NC, Chakravorty B (2002) Analysis of arsenic-contaminated groundwater domain in the Nadia district of West Bengal (India), Special issue: Towards Integrated Water Resources Management for Sustainable Development. Hydr Sci J 47(S): S55–S66
Mandal BK, Chowdhury TR, Samanta G et al (1998) Impact of Safe Water for Drinking and Cooking on Five ArsenicAffected Families for 2 Years in West Bengal, India. The Sc Tot Env 218:185–201
Michaela HA, Voss CI (2008) Evaluation of the sustainability of deep groundwater as an arsenic-safe resource in the Bengal Basin, Proceedings of National Academy of Science, USA, 105:8531–8536
Mridha D, Ray I, Sarkar J et al. (2022) Effect of sulfate application on inhibition of arsenic bioaccumulation in rice (*Oryza sativa* L.) with consequent health risk assessment of cooked rice arsenic on human: a pot to plate study. Environ Pollut 293:118561
Mukhopadhyay D, Mani PK, Sanyal SK (2002) Effect of phosphorus, Arsenic and farmyard manure on arsenic availability in some soils of West Bengal. J Ind Soc Soil Sci 50:56–61
Nabi A, Naeem M, Aftab T et al (2020) Intimidating effects of heavy metals on mentha species and their mitigation using scientific approaches, In: Contaminants in Agriculture Sources, Impacts and Management, M. Naeem A.A. Ansari and S.S. Gill (eds), Springer Nature Switzerland AG, pp. 305–326
Naeem M, Aftab T, Ansari AA et al (2019) Arsenic exposure modulates physiological attributes and artemisinin biosynthesis in *Artemisia annua* L. Int J Herb Med 7:19–26
National Institute of Hydrology (NIH) and Central Ground Water Board (CGWB) (2009) Vision document on "Mitigation and Remedy of Groundwater Arsenic Menace in India" prepared under the aegis of Ministry of Water Resources, GOI (under finalization)
Nriagu JO, Bhattacharya P, Mukherjee AB et al (2007) Arsenic in soil and groundwater: an overview. Trace Metals and other Contaminants in the Environment 9:3–60
Paul PK, Sikdar PK (2008) Assessment and Management of groundwater resources of English Bazar block, Malda district, West Bengal. Unpublished Report of Department of Science and Technology, New Delhi, pp. 279
Pigna M, Cozzolino V, Giandonato Caporale A et al (2010) Effects of phosphorus fertilization on arsenic uptake by wheat grown in polluted soils. J Soil Sc Plant Nutr 10: 428–422
Rahman MA, Hassler C (2014) Is arsenic biotransformation a detoxification mechanism for microorganisms? Aquat Toxicol 146: 212–219
Rahman MS, Kumar A, Kumar R et al (2019) Comparative quantification study of arsenic in the groundwater and biological samples of Simri village of Buxar District, Bihar, India. Indian J Occup Env Med 23:126–132
Rascio N, NavariIzzo F (2011) Heavy metal hyperaccumulating plants: How and why do they do it? And what makes them so interesting? Plant Sci 180:169–181
Rooney CP, Zhao FJ, Mc GRath SP (2006) Soil factors controlling the expression of copper toxicity to plants in a wide range of European soils. Environ Toxicol Chem 25:726–732
Sanyal SK (2017) A Textbook of Soil Chemistry. Daya Publishing House, A Division of Astral International Pvt. Limited, New Delhi
Sanyal SK, Gupta SK, Kukal SS (2015) Soil degradation, pollution and amelioration. In: State of Indian Agriculture-Soil H. Pathak, S. K. Sanyal, and P. N. Takkar (eds) National Academy of Agricultural Sciences, New Delhi, pp. 2342–66

Sanyal SK, Jeevan Rao K, Sadana Upkar S (2012) Toxic elements and other pollutants-A threat to nutritional quality. In: Soil Science in the Service of Nation. N.N. Goswami et al (eds) Indian Society of Soil Science, New Delhi, pp. 266–291

Sarkar A, Paul B, Darbha GK (2022) The ground water arsenic contamination in the Bengal Basin-A review in brief. Chemosphere 299: 134369

Sikdar PK, Banerjee S (2003) Genesis of Arsenic in groundwater of Ganga delta: An anthropogenic model. J Hum Sett April:10–24

Sikdar PK, Chakraborty S (2008) Genesis of Arsenic in groundwater of North Bengal Plain using PCA: A case study of English Bazar Block, Malda District, West Bengal, India. Hydrol Proc 22:1796–1809

Singh OV, Labana S, Pandey G et al (2003) Phytoremediation: an overview of metallic ion decontamination from soil. Appl Microbiol Biotech 61:405–412

Sinha B, Bhattacharyya K (2014) Arsenic toxicity in rice with special reference to speciation in Indian grain and its implication on human health. J Sc Food and Agr 95:1435–1444

Smedley PL, Kinniburgh DG (2002) A review of the source, behavior and distribution of arsenic in natural waters. Appl Geochem 17:517–568

Stoeva N, Bineva T (2003) Oxidative changes and photosynthesis in Oat plants grown in As- contaminated soil. Bulg J Plant Physiol 29:87–95

Tripathi P, Mishra A, Dwivedi S et al (2012) Differential response of oxidative stress and thiol metabolism in contrasting rice genotypes for arsenic tolerance. Ecotoxicol Env Saf 79:189–198

Wang S, Mulligan CN (2004) Rhamnolipid foam enhanced remediation of cadmium and nickel contaminated soil. Water Air Soil Pollut 157:315–330

WHO (1993) Guidelines for Drinking-Water Quality, Second edition, Volume 1

WHO (2001) http://www.who.int/inf-fs/en/fact210.html

Yamaguchi N, Nakamura T, Dong D et al (2011) Arsenic release from flooded paddy soils is influenced by speciation, Eh, pH, and iron dissolution. Chemosphere 83:925–932

Zaidi S, Hayat S, Pichtel J et al (2024) Arsenic-induced plant stress: Mitigation strategies and omics approaches to alleviate toxicity. Pl. Physiol. Biochem 213: 108811

Zhao FJ, Ma JF, McGrath SP et al (2009) Arsenic uptake and metabolism in plants. New Phytol 181:777–794

Zhao FJ, McGrath SP, Meharg AA (2010) Arsenic as a food chain contaminant: Mechanisms of plant uptake and metabolism and mitigation strategies. Annu Rev Plant Biol 61:535–559

Metal Detoxification in Land Plants: From Bryophytes to Vascular Plants

Vikas Kumar, Nitin Sharma, Somvir Singh, and Rajat Laller

Abstract Metal detoxification is a crucial evolutionary adaptation in land plants, enabling survival in environments with varying heavy metal concentrations. From early bryophytes to complex vascular plants, detoxification mechanisms have evolved significantly in complexity and efficiency. Bryophytes, such as mosses and liverworts, primarily rely on passive strategies such as adsorption, sequestration, and immobilization of metals within cell walls or vacuoles, preventing metabolic interference. They also utilize organic acids and phenolic compounds to chelate metals, reducing toxicity. In vascular plants, such as ferns, gymnosperms, and angiosperms, detoxification strategies are more advanced. They synthesize metal-binding proteins such as metallothioneins and phytochelatins, which form less toxic metal complexes that are sequestered into vacuoles. Enhanced vacuolar compartmentalization further protects cells from metal-induced damage. Moreover, vascular plants have robust antioxidant defence systems, including reactive oxygen species (ROS)-scavenging enzymes like superoxide dismutase (SOD) and catalase, which mitigate oxidative stress caused by metal toxicity. Understanding the evolutionary progression of these mechanisms from bryophytes to vascular plants offers valuable insights into plant resilience and adaptability. These detoxification processes not only allow plants to thrive in metal-contaminated environments but also play a critical role in phytoremediation applications. By harnessing and enhancing

V. Kumar (✉)
Department of Biotechnology, University Institute of Biotechnology, Chandigarh University, Mohali, Punjab, India

N. Sharma
Department of Biotechnology, Chandigarh Group of Colleges, Mohali, Punjab, India

S. Singh
Department of Biosciences, University Institute of Biotechnology, Chandigarh University Gharuan, Mohali, Punjab, India

R. Laller
School of Biological and Environmental Sciences, Shoolini University of Biotechnology and Management Sciences, Sultanpur, India

S. Roy, V. Mandal (eds.), *Plant-Microbe Interaction under Xenobiotic Exposure*,
https://doi.org/10.1007/978-981-96-8260-7_17

these natural processes, researchers can develop sustainable strategies to manage heavy metal contamination in agricultural and natural ecosystems.

Keywords Bryophytes · Detoxification · Ecosystem · Reactive oxygen species · Superoxide dismutase

1 Introduction

Metal detoxification is a vital process enabling land plants to thrive in heavy metal-contaminated environments. This mechanism protects cellular structures, sustains essential metabolic activities, and ensures plant survival. These detoxification pathways not only help plants to adapt to diverse habitats but also contribute to ecosystem stability. Moreover, they are integral to phytoremediation, where plants are used to clean contaminated soils, offering them invaluable tools for environmental management (Priya et al. 2023). In agriculture, effective metal detoxification is essential for ensuring the safety and productivity of crops by preventing the accumulation of toxic metals in the food chain and maintaining soil health—key factors for sustainable farming and food security (Vasilachi et al. 2023). Heavy metal stress poses significant challenges to plant health, disrupting physiological, biochemical, and molecular processes. Elevated levels of heavy metals such as cadmium (Cd), lead (Pb), mercury (Hg), and arsenic (As) can impair plant growth and development (Hasanuzzaman et al. 2022). Excessive nickel and cobalt exposure hinders the growth of leguminous crops, resulting in reduced yields. At the cellular level, heavy metals weaken cell walls by interacting with components like pectin and cellulose, thereby reducing membrane integrity, increasing permeability, and disrupting normal functions. These effects collectively impair plant growth and development (Yuan et al. 2024). Biochemically, heavy metal exposure leads to excessive production of reactive oxygen species (ROS) and methylglyoxal (MG), which cause peroxidation of lipids, oxidation of proteins, inactivation of enzymes, and DNA damage. To mitigate this oxidative stress, plants activate antioxidant defense systems involving enzymes such as superoxide dismutase (SOD), catalase (CAT), and peroxidase (POD). However, prolonged metal stress can overwhelm these defenses, resulting in cellular damage and impaired metabolic functions (Yang et al. 2022; Rahman et al. 2023). At the molecular level, metal stress alters gene expression, protein synthesis, and signaling pathways. Plants upregulate stress-responsive genes, including those encoding metal-binding proteins like metallothioneins and phytochelatins, to sequester and detoxify heavy metals. Additionally, heavy metals can also disrupt hormonal regulation and other critical pathways, further impairing plant functions (Ghori et al. 2019).

Heavy metal accumulation in soils, primarily due to industrial activities and improper waste disposal, threatens agricultural productivity and environmental health. Contaminated soils reduce crop yields and compromise food safety due to the potential metal transfer into the food chain. The persistence and resistance of heavy metals to degradation exacerbate these challenges, highlighting the need for effective remediation strategies (Hasanuzzaman et al. 2020). Understanding plant responses to heavy metal stress is crucial for developing strategies to enhance their tolerance and mitigate the associated risks to agriculture and ecosystems (Yuan et al. 2024). Plants display diverse responses to heavy metal exposure, primarily categorized into non-accumulators and hyperaccumulators. Non-accumulator plants employ strategies to avoid metal uptake or detoxify absorbed metals through chelation and vacuolar sequestration (Kosakivska et al. 2021). In contrast, hyperaccumulator plants actively absorb and concentrate metals in their tissues at toxic levels for most species. They utilize specialized mechanisms for metal uptake, such as overexpression of heavy metal ATPases (HMAs) and metal tolerance proteins (MTPs) to facilitate metal sequestration into vacuoles or transport into aerial parts (Ghori et al. 2019). Advances in genetic engineering also help in enhancing the growth rates and metal-accumulating capacities of these plants, offering sustainable alternatives to traditional mining practices (Selwal et al. 2023). These varied responses among plant groups highlight the complexity of plant adaptations to heavy metal stress and underscore the potential applications of hyperaccumulators in environmental remediation and sustainable metal extraction (Liang et al. 2024).

The evolutionary transition from bryophytes to vascular plants represents a significant milestone in plant evolution, marked by significant advancements in structure, function, and ecological adaptability. Bryophytes, such as liverworts, mosses, and hornworts, lack specialized vascular tissues and rely on simple diffusion for water and nutrient transport, restricting them to moist environments. The development of vascular tissues in ferns, gymnosperms, and angiosperms enabled efficient resource transport, structural support, and colonization of diverse habitats, including arid regions (Yadav et al. 2023). Lignified tissues in vascular plants further enhanced their ecological success (Konrad et al. 2019). Recent studies suggested that bryophytes and vascular plants diverged from a complex common ancestor during the Cambrian period, approximately 515–494 million years ago. While bryophytes experienced significant genome reduction, losing genes linked to vascular development and complex stomatal structures, vascular plants retained and expanded these features, leading to their extensive diversification (Harris et al. 2022). Some bryophytes, such as *Polytrichum commune*, exhibit advanced features such as vascular functions enabling them to transport water under tension and leaves that can regulate transpiration, permitting photosynthetic gas exchange without causing cavitation in the vascular system. These findings suggest their convergent evolution (Atwood and Buck 2020). The evolution of vascular tissues also facilitated the development of advanced adaptations to environmental stresses. Vascular plants have evolved sophisticated mechanisms for metal detoxification, allowing them to colonize a broader range of habitats, including those with challenging soil chemistries. This adaptability has been crucial for their widespread distribution and ecological success (Chen et al. 2024). In summary, the transition from bryophytes to vascular plants involved significant innovations in plant structure and function. The development of vascular tissues not only enabled efficient resource transport and structural support but also opened avenues for diversification into various habitats. While bryophytes have remained relatively constrained in their ecological roles, some have independently evolved complex traits, underscoring the dynamic evolutionary pathways that have shaped the plant kingdom (Brodribb et al. 2020). The present chapter explores the mechanism of metal detoxification in bryophytes and vascular plants, emphasizing their evolutionary adaptations and potential applications in environmental remediation and sustainable agriculture.

2 Metal Detoxification Mechanisms in Bryophytes

2.1 Metal Uptake and Sequestration

Bryophytes, such as mosses, liverworts, and hornworts, are well adapted to metal-contaminated environments due to their ability to absorb and tolerate metals like copper (Cu), lead (Pb), zinc (Zn), and cadmium (Cd), which are toxic at high concentrations (Fasani et al. 2022). Their simple structure, which lacks a protective cuticle, allows direct absorption of metals from their surroundings. Once absorbed, these metals are often sequestered in vacuoles, where they are stored in a non-toxic form bound to organic acids or phytochelatins (Fasani et al. 2022). Phytochelatins, which are small, cysteine-rich peptides, chelate metal ions, preventing them from disrupting cellular processes (Seregin and Kozhevnikova 2023).

2.2 Complexation with Organic Molecules

A key strategy for metal detoxification in bryophytes involves the complexation of metal ions with organic molecules. Bryophytes produce organic acids, such as citric and malic acid, which bind to metal ions, forming stable, non-toxic complexes. This reduces the bioavailability of metals within the cells and facilitates their transport and storage in less harmful forms (González and Pokrovsky 2014).

2.3 Antioxidant Defense System

Bryophytes mitigate oxidative stress caused by heavy metal exposure through an enhanced antioxidant defense system. Heavy metals can catalyze the production of ROS, which cause cellular damage (Berni et al. 2019). To counteract this, bryophytes increase the activity of antioxidant enzymes like SOD, CAT, and POD, which neutralize ROS, thereby protecting cells from oxidative damage and aiding the detoxification process (Dutta et al. 2024).

2.4 Gene Expression and Metal Transporters

Bryophytes respond to metal stress by expressing specific genes involved in metal transport and detoxification. Metal transporters, such as those in the ATP-binding cassette (ABC) transporter family, play a critical role in pumping excess metals out of cells or into vacuoles, reducing their cytoplasmic concentration. These transporters are regulated at the transcriptional level in response to metal exposure, ensuring efficient and adaptive detoxification (Martins et al. 2011).

2.5 Bioremediation Potential

Due to their high tolerance and ability to accumulate heavy metals, bryophytes are increasingly studied for their potential in bioremediation. Mosses, for example, are effective bioindicators of air and water pollution due to their metal accumulation capabilities (González and Pokrovsky 2014). Their use in phytoremediation strategies could provide an eco-friendly solution for decontaminating polluted sites, especially in areas where higher plants cannot survive due to extreme metal concentrations.

3 Evolutionary Milestones in Metal Detoxification

The transition from bryophytes to vascular plants represents a significant evolutionary leap in metal detoxification mechanisms (Table 1). Bryophytes, which are among the earliest land plants, exhibit relatively simple detoxification strategies. These include passive adsorption of metals onto cell walls and sequestration into vacuoles, which were sufficient for survival in simpler environments but limited their ability to thrive in metal-rich soils (Bellini et al. 2021). The emergence of vascular plants marked the development of more advanced detoxification mechanisms, driven by specialized tissues and increased metabolic complexity. One key evolutionary milestone was the development of metal-binding proteins such as phytochelatins and metallothioneins. These proteins bind heavy metals, forming fewer toxic complexes that are then sequestered into vacuoles, allowing vascular plants to tolerate higher concentrations of metals (Thakur et al. 2022). Another significant advancement was the enhanced compartmentalization within vacuoles, isolating toxic metals from the cytoplasm to prevent cellular damage (De Caroli et al. 2020). Additionally, the evolution of sophisticated antioxidant defense systems, including

Table 1 Key characteristics, adaptations, and innovations from different time periods in plant groups

Plant groups	Time period	Key characteristics	Adaptation and innovations	References
Bryophytes	Ordovician (~479 MYA)	Non-vascular, simple structure, lack of true roots, stem, and leaves	Dominated by gametophytes, spore-based reproduction	Kenrick and Crane (1997) and Goffinet (2008)
Early vascular plants (e.g., Rhyniophytes)	Silurian (~430 MYA)	Primitive vascular tissues, simple dichotomous branching	Development of xylem and phloem, cuticle, and stomata	Kenrick and Crane (1997)
Lycophytes	Devonian (~400 MYA)	Microphyll leaves, vascular tissues, simple roots	Formation of true roots, specialized vascular system	Gensel (2008) and Taylor et al. (2009)
Ferns and seed plants	Carboniferous (~360 MYA)	Complex vascular systems, megaphyll leaves, seeds (in seed plants)	Development of seeds, complex leaf structures	Stewart and Rothwell (1993) and Taylor et al. (2009)

enzymes that scavenge ROS, further protected vascular plants from metal-induced oxidative stress. These advancements significantly contributed to the success of vascular plants in colonizing diverse and metal-contaminated environments (Priya et al. 2023).

4 Structural and Functional Advancements

The evolution from bryophytes to vascular plants also involved significant structural and functional advancements that allowed plants to adapt to a wide range of challenging environments. Bryophytes, being non-vascular, rely on simple diffusion for water and nutrient transport, which limits their size and habitat range (Vitt et al. 2014). The development of vascular tissues, particularly xylem and phloem, was a critical structural advancement in vascular plants. These tissues enabled efficient long-distance transport of water, nutrients, and photosynthates, which allowed plants to grow taller and colonize a wider range of environments (Raven and Edwards 2001). Another key structural advancement was the evolution of lignified cell walls, which provided mechanical support and allowed vascular plants to maintain an upright structure. This was essential for growth and survival in terrestrial habitats (Meents et al. 2018). Additionally, the development of roots in vascular plants enhanced their ability to absorb water and nutrients from the soil, anchoring plants more securely and improving their access to essential resources (Beerling and Kelly 1997; Ma et al. 2024). Functionally, vascular plants exhibit greater complexity in their physiological processes, including advanced mechanisms for coping with environmental stresses such as drought, high light, and heavy metal toxicity. The evolution of more sophisticated stomatal regulation allowed vascular plants to better control water loss and gas exchange, optimizing photosynthesis and water use efficiency (Haworth et al. 2021). These structural and functional advancements played a crucial role in the success and diversification of vascular plants, allowing them to dominate terrestrial ecosystems (Beck 2010).

5 Metal Detoxification Mechanisms in Vascular Plants

Vascular plants often encounter environments where toxic metals like cadmium, lead, arsenic, mercury, and aluminum are present. These metals can be highly detrimental, disrupting plant cellular processes and leading to oxidative stress. To survive in such hostile conditions, vascular plants have evolved various detoxification mechanisms to avoid or tolerate metal toxicity (Fasani et al. 2022).

5.1 Exclusion Mechanism

Vascular plants reduce metal toxicity by preventing harmful metals from entering the plant system. They modify the rhizosphere by secreting organic acids such as citrate and malate, which chelate metal ions, making them less available for absorption by plant roots (Verbruggen et al. 2009). Additionally, changes in soil pH induced by root exudates can further reduce metal solubility and uptake (Ma et al. 2001; Haworth et al. 2021).

5.2 Metal Chelation and Sequestration

Once metals enter the plant, chelating agents like phytochelatins, metallothioneins, and organic acids neutralize them. Phytochelatins, derived from glutathione, are particularly crucial for binding heavy metals like cadmium and arsenic (Sethi 2023). Metallothioneins, cysteine-rich proteins, also play a vital role in binding and sequestering metals to prevent damage to cellular components (Raychaudhuri et al. 2021). These metal-chelate complexes are then transported into vacuoles for sequestration, effectively isolating them from the rest of the cell (Sethi 2023).

5.3 Detoxification through Antioxidant Defense

Toxic metals often lead to the production of ROS, causing oxidative stress. Plants mitigate this by enhancing their antioxidant defenses, including enzymes such as SOD, catalase, and glutathione peroxidase, along with non-enzymatic antioxidants like ascorbate and glutathione. These antioxidants work together to neutralize ROS, protecting plant cells from oxidative damage caused by metal toxicity (Zulfiqar and Ashraf 2022).

5.4 Efflux and Hyperaccumulation

Vascular plants may also use efflux mechanisms to pump toxic metals out of their cells. Transporter proteins like HMAs and ABC transporters help remove excess metals from the cytoplasm, reducing intracellular metal concentrations (Jalmi 2022). Conversely, some hyperaccumulator plants can tolerate and accumulate high levels of metals in their tissues, using mechanisms such as enhanced uptake, efficient chelation, and compartmentalization. These plants are valuable in phytoremediation efforts (Sharma et al. 2023).

5.5 Gene Regulation and Signaling Pathways

The regulation of these detoxification processes is controlled at the genetic level, with transcription factors like bZIP, MYB, and WRKY modulating the expression of genes involved in metal chelation, sequestration, and antioxidant defense. Additionally, signaling pathways involving plant hormones like abscisic acid (ABA), ethylene, and jasmonic acid coordinate the response to metal stress, activating appropriate detoxification mechanisms as needed (Mitra et al. 2023).

6 Comparative Analysis of Bryophytes vs. Vascular Plants

Bryophytes and vascular plants are two distinct classifications within the plant kingdom, each exhibiting unique characteristics in their structure, physiology, reproductive strategies, and ecological functions. A comparative analysis of these groups underscores the differences and similarities, offering valuable insights into their evolutionary development and adaptability to various environments (Patiño and Vanderpoorten 2018).

6.1 Structural Organization

Bryophytes, which include mosses, liverworts, and hornworts, are non-vascular plants, meaning they lack specialized tissues for the transport of water and nutrients. Their structure is relatively simple, typically consisting of a thallus (as seen in liverworts and hornworts) or leafy shoots (in the case of mosses), and they have rhizoids that help anchor them to their growing surface. In contrast, vascular plants, which encompass ferns, gymnosperms, and angiosperms, possess complex tissues like xylem and phloem. The xylem is responsible for transporting water and nutrients, while the phloem distributes organic compounds throughout the plant. This vascular system allows these plants to grow larger and thrive in a broader range of habitats compared to bryophytes (Budke et al. 2018).

6.2 Water Relations and Desiccation Tolerance

Bryophytes are heavily dependent on water for their survival, reproduction, and physiological processes. Because they lack a vascular system, water and nutrient transport occurs via diffusion, which limits their size and growth potential. However, bryophytes have evolved remarkable desiccation tolerance, enabling them to survive in dry conditions by entering a dormant state during periods of water scarcity. On the other hand, vascular plants have developed sophisticated water regulation systems, including roots, stems, and leaves optimized for water uptake and transpiration reduction. Stomata in vascular plants facilitate gas exchange and help control water loss, balancing the needs for photosynthesis and hydration (He et al. 2016).

6.3 Reproductive Strategies

Bryophytes exhibit a life cycle that is primarily gametophyte-dominant, where the haploid plant produces gametes. The diploid sporophyte is usually dependent on the gametophyte for nutrients and has a shorter lifespan, limiting the bryophytes' dispersal and colonization abilities. In contrast, vascular plants have a dominant sporophyte stage, which is independent and capable of producing numerous spores or seeds. The ability to produce seeds, in particular, has allowed vascular plants to colonize a wide range of environments, as seeds offer protection and nourishment to the developing embryo (Pandey et al. 2022).

6.4 Ecological Roles and Adaptations

Bryophytes are predominantly found in moist, shaded environments, where they play essential roles in soil formation, water retention, and nutrient cycling. Their simple structure allows them to inhabit harsh environments, such as rocky surfaces and tree trunks, where vascular plants may not thrive. However, their reliance on water constrains their distribution. Vascular plants, with their advanced structural and reproductive adaptations, dominate most terrestrial ecosystems. They contribute significantly to carbon fixation, soil stabilization, and habitat creation, supporting a diverse array of organisms (Hu et al. 2023).

6.5 Evolutionary Significance

Bryophytes are among the earliest land plants, acting as a transitional group between aquatic algae and more complex vascular plants. Their simple structure and life cycle offer valuable insights into the initial adaptations required for terrestrial life. Vascular plants represent a significant evolutionary advancement, with specialized tissues, organs, and reproductive strategies that have enabled them to become the dominant flora in most terrestrial ecosystems (Harris et al. 2022).

7 Ecological and Agricultural Implications

7.1 Metal Detoxification and Ecological Balance

The ability of land plants, ranging from bryophytes to vascular plants, to detoxify metals plays a crucial role in maintaining ecological balance and enhancing agricultural sustainability. The strategies these plants use to manage toxic metal accumulation contribute not only to environmental detoxification but also provide essential insights for improving crop resilience and food safety (Liao et al. 2024).

7.2 Ecological Implications

Bryophytes, as some of the most primitive land plants, are highly effective bioindicators of heavy metal pollution due to their simple structure and extensive surface area, which facilitates the absorption of metals like cadmium, lead, and zinc from their surroundings (Singh et al. 2020). This ability makes bryophytes invaluable for monitoring pollution levels in various ecosystems and highlights their potential use in bioremediation efforts, where they can help restore contaminated environments.

Moving up the evolutionary scale, ferns and gymnosperms demonstrate more complex metal detoxification mechanisms. For example, ferns such as *Pteris vittata* are known for their ability to hyperaccumulate arsenic, making them essential for phytoremediation projects aimed at cleaning up arsenic-contaminated soils. Gymnosperms, although less studied, also contribute to ecological detoxification, particularly in forest ecosystems, where they can withstand heavy metal stress and prevent the spread of pollutants through runoff and soil erosion (Tulay 2024). Angiosperms, or flowering plants, represent the most diverse and extensively studied group in the context of metal detoxification. Hyperaccumulators like *Arabidopsis halleri* and *Thlaspi caerulescens* have evolved advanced mechanisms to sequester metals in vacuoles and cell walls, thereby minimizing their toxic effects. These plants not only assist in cleaning up contaminated environments but also prevent the entry of heavy metals into the food chain, thus protecting both wildlife and human health (Kumar et al. 2019).

7.3 Agricultural Implications

Heavy metal contamination of agricultural soils is a significant concern, as it can lead to reduced crop yields and pose substantial health risks if metals accumulate in edible plant parts (Ali and Khan 2019). Understanding the mechanisms of metal detoxification in plants is crucial for developing crop varieties that can tolerate or exclude heavy metals, ensuring both productivity and safety. For instance, genetically engineered crops that overexpress metal-binding proteins or enhance metal transporter activity can thrive in contaminated soils, thereby reducing the risk of metal uptake in food crops. Using hyperaccumulator plants in crop rotation or as cover crops offers a practical solution for soil decontamination. These plants can absorb and store metals from the soil, lowering contamination levels before sensitive crops are cultivated. This approach not only enhances soil health but also supports sustainable agricultural practices in areas affected by industrial pollution. Moreover, insights from studies on bryophytes and ferns can be applied to crop improvement programs. By transferring genes responsible for metal tolerance from these plants into crop species, it is possible to develop varieties that are resilient to metal stress and capable of contributing to soil detoxification efforts (Rai et al. 2021).

8 Future Perspectives and Conclusion

The evolution of the metal detoxification mechanism in land plants, from primitive bryophytes to advanced vascular species, reflects a remarkable trajectory of adaptation and resilience. As environmental pollution, particularly with heavy metals like lead, cadmium, and arsenic, continues to escalate, it is crucial to understand the mechanism of these detoxification processes to develop effective remediation strategies and ensure environmental sustainability. Bryophytes, being the most primitive land plants, rely on fundamental but effective metal detoxification strategies, such as metal sequestration in vacuoles and chelation with organic acids. Despite their simplicity, these mechanisms demonstrate the evolutionary ingenuity of these plants in surviving metal-rich environments. Future research on bryophytes might reveal novel metal-binding proteins, stress-responsive genes, and metabolic pathways that may provide critical insights into the core principles of metal detoxification. These early land plants could serve as models for understanding the origins and evolution of metal tolerance mechanisms, with potential applications in bioremediation.

Vascular plants exhibit more complex and efficient metal detoxification mechanisms, leveraging advanced compartmentalization, enhanced antioxidant defenses, and the synthesis of metal-binding molecules such as phytochelatins and metallothioneins. These mechanisms enable vascular plants to thrive in highly contaminated environments, making them pivotal in the field of phytoremediation. Recent progress in genetic engineering and synthetic biology offers promising opportunities to enhance these natural capabilities. By manipulating genes associated with metal uptake, transport, and sequestration, scientists can design hyperaccumulator plants capable of extracting and storing larger quantities of metals from polluted soils, thereby improving the efficiency of phytoremediation efforts. The application of omics technologies, such as genomics, proteomics, and metabolomics, holds great promise for advancing our understanding of plant responses to heavy metal stress. These technologies offer a molecular-level view of the detoxification mechanism, facilitating the identification of regulatory genes, signaling pathways, and critical enzymes involved in metal tolerance. Such knowledge could pave the way for creating plants with tailored traits for specific environmental conditions, offering innovative solutions for addressing heavy metal stress.

The evolution of metal detoxification strategies across plant lineages not only underscores the increasing complexities and adaptability of these mechanisms but also highlights their ecological and agricultural importance. By safeguarding plants from metal toxicity, these processes enhance biodiversity, ecosystem stability, and agricultural productivity. In agriculture, metal detoxification is critical for preventing toxic metal accumulation in crops, ensuring food safety, and maintaining soil health. Sustainable practices that integrate phytoremediation with crop management can further boost food security and environmental conservation. In conclusion, the study of the metal detoxification mechanism in land plants provides valuable insights into their evolutionary adaptation and resilience. Continued research, coupled with advancements in biotechnology, offers the potential to develop innovative strategies for managing heavy metal contamination, improving soil health, and supporting both ecological and agricultural sustainability. Using the natural detoxification abilities of plants, we can take significant steps toward mitigating the environmental and agricultural challenges posed by heavy metal pollution.

References

Ali H, Khan E (2019) Trophic transfer, bioaccumulation, and biomagnification of non-essential hazardous heavy metals and metalloids in food chains/webs—concepts and implications for wildlife and human health. Hum Ecol Risk Assess 25(6):1353–1376

Atwood JJ, Buck WR (2020) Recent literature on bryophytes—123 (3). Bryologist 123(3):547–583

Beck CB (2010) An introduction to plant structure and development: plant anatomy for the twenty-first century. Cambridge University Press, Cambridge

Beerling DJ, Kelly CK (1997) Stomatal density responses of temperate woodland plants over the past seven decades of CO2 increase: a comparison of Salisbury (1927) with contemporary data. Am J Bot 84(11):1572–1583

Bellini E, Betti C, Sanità di Toppi L (2021) Responses to cadmium in early-diverging streptophytes (Charophytes and bryophytes): current views and potential applications. Plants 10(4):770

Berni R, Luyckx M, Xu X, Legay S, Sergeant K, Hausman JF, Lutts S, Cai G, Guerriero G (2019) Reactive oxygen species and heavy metal stress in plants: impact on the cell wall and secondary metabolism. Environ Exp Bot 161:98–106

Brodribb TJ, Carriquí M, Delzon S, McAdam SA, Holbrook NM (2020) Advanced vascular function discovered in a widespread moss. Nature Plants 6(3):273–9

Budke JM, Bernard EC, Gray DJ, Huttunen S, Piechulla B, Trigiano RN (2018) Introduction to the special issue on bryophytes. Crit Rev Plant Sci 37(2–3):102–112

Chen G, Qin Y, Wang J, Li S, Zeng F, Deng F, Chater C, Xu S, Chen ZH (2024) Stomatal evolution and plant adaptation to future climate. Plant Cell Environ 47(9):3299–3315

De Caroli M, Furini A, DalCorso G, Rojas M, Di Sansebastiano GP (2020) Endomembrane reorganization induced by heavy metals. Plants 9(4):482

Dutta P, Prasad P, Indoilya Y, Gautam N, Kumar A, Sahu V, Kumari M, Singh S, Asthana AK, Bag SK, Chakrabarty D (2024) Unveiling the molecular mechanisms of arsenic tolerance and resilience in the primitive bryophyte Marchantia polymorpha L. Environ Pollut 346:123506

Fasani E, Li M, Varotto C, Furini A, DalCorso G (2022) Metal detoxification in land plants: from bryophytes to vascular plants. State of the art and opportunities. Plants 11(3):237

Gensel PG (2008) The earliest land plants. Annu Rev Ecol Evol Syst 39(1):459–477

Ghori NH, Ghori T, Hayat MQ, Imadi SR, Gul A, Altay V, Ozturk M (2019) Heavy metal stress and responses in plants. Int J Environ Sci Technol 16:1807–1828

Goffinet B (2008) Bryophyte biology. Cambridge University Press, Cambridge

González AG, Pokrovsky OS (2014) Metal adsorption on mosses: toward a universal adsorption model. J Colloid Interface Sci 415:169–178

Harris BJ, Clark JW, Schrempf D, Szöllősi GJ, Donoghue PC, Hetherington AM, Williams TA (2022) Divergent evolutionary trajectories of bryophytes and tracheophytes from a complex common ancestor of land plants. Nat Ecol Evol 6(11):1634–1643

Hasanuzzaman M, Bhuyan MHMB, Zulfiqar F, Raza A, Mohsin SM, Mahmud JA, Fujita M, Fotopoulos V. (2020) Reactive Oxygen Species and Antioxidant Defense in Plants under Abiotic Stress: Revisiting the Crucial Role of a Universal Defense Regulator. Antioxidants (Basel) 9(8):681

Hasanuzzaman M, Nahar K, García-Caparrós P, Parvin K, Zulfiqar F, Ahmed N, Fujita M (2022) Selenium supplementation and crop plant tolerance to metal/metalloid toxicity. Front Plant Sci 12:792770

Haworth M, Marino G, Loreto F, Centritto M (2021) Integrating stomatal physiology and morphology: evolution of stomatal control and development of future crops. Oecologia 197(4):867–883

He X, He KS, Hyvönen J (2016) Will bryophytes survive in a warming world? Perspect Plant Ecol Evol Syst 19:49–60

Hu X, Gao Z, Li XY, Wang RZ, Wang YM (2023) Structural characteristics of the moss (bryophyte) layer and its underlying soil structure and water retention characteristics. Plant Soil 490(1):305–323

Jalmi SK (2022) The role of ABC transporters in metal transport in plants. In: Plant metal and metalloid transporters. Springer Nature Singapore, Singapore, pp. 55–71

Kenrick P, Crane PR (1997) The origin and early evolution of vascular plants. Nature 389(6646):127–136

Konrad W, Katul G, Roth-Nebelsick A, Jensen KH (2019) Xylem functioning, dysfunction and repair: a physical perspective and implications for phloem transport. Tree Physiol 39(2):243–261

Kosakivska IV, Babenko LM, Romanenko KO, Korotka IY, Potters G (2021) Molecular mechanisms of plant adaptive responses to heavy metals stress. Cell Biology International. 45(2):258–72

Kumar S, Prasad S, Yadav KK, Shrivastava M, Gupta N, Nagar S, Bach QV, Kamyab H, Khan SA, Yadav S, Malav LC (2019) Hazardous heavy metals contamination of vegetables and food chain: role of sustainable remediation approaches—a review. Environ Res 179:108792

Liang Z, Neményi A, Kovács GP, Gyuricza C (2024) Incorporating functional traits into heavy metals phytoremediation: the future of field-based phytoremediation. Ecol Indic 166:112262

Liao K, Chen C, Ye W, Zhu J, Li Y, She S, Wang P, Tao Y, Lv A, Wang X, Chen L (2024) The adaptability, distribution, ecological function and restoration application of biological soil crusts on metal tailings: a critical review. Sci Total Environ 172169

Ma Y, Zhang Y, Xu J, Zhao D, Guo L, Liu X, Zhang H (2024) Recent advances in response to environmental signals during Arabidopsis root development. Plant Physiol Biochem 109037

Ma JF et al. (2001) Rhizosphere acidification and its role in phytoremediation. Plant Soil 237(1):61–68

Martins M, Mourato MP, Azevedo RA (2011) Metal accumulation in different tissues of woody plants in contaminated environments. Environ Sci Pollut Res 18(7):1207–1218

Meents MJ, Watanabe Y, Samuels AL (2018) The cell biology of secondary cell wall biosynthesis. Ann Bot 121(6):1107–1125

Mitra M, Agarwal P, Roy S (2023) Plant response to heavy metal stress: an insight into the molecular mechanism of transcriptional regulation. In: Plant transcription factors. Academic Press, pp. 337–367

Pandey S, Moradi AB, Dovzhenko O, Touraev A, Palme K, Welsch R (2022) Molecular control of sporophyte-gametophyte ontogeny and transition in plants. Front Plant Sci 12:789789

Patiño J, Vanderpoorten A (2018) Bryophyte biogeography. Crit Rev Plant Sci 37(2–3):175–209

Priya AK, Muruganandam M, Ali SS, Kornaros M (2023) Clean-up of heavy metals from contaminated soil by phytoremediation: a multidisciplinary and eco-friendly approach. Toxics 11(5):422

Rahman SU, Li Y, Hussain S, Hussain B, Riaz L, Ashraf MN, Khaliq MA, Du Z, Cheng H (2023) Role of phytohormones in heavy metal tolerance in plants: a review. Ecol Indic 146:109844

Rai KK, Pandey N, Meena RP, Rai SP (2021) Biotechnological strategies for enhancing heavy metal tolerance in neglected and underutilized legume crops: a comprehensive review. Ecotoxicol Environ Saf 208:111750

Raven JA, Edwards D (2001) Roots: evolutionary origins and biogeochemical significance. Journal of experimental botany 52(suppl_1):381–401

Raychaudhuri SS, Pramanick P, Talukder P, Basak A (2021) Polyamines, metallothioneins, and phytochelatins—natural defense of plants to mitigate heavy metals. Stud Nat Prod Chem 69:227–261

Selwal N, Tabassum Z, Rahayu F, Yulia ND, Sugiono S, Endarto O, Riajaya PD, Djajadi D, Khamidah A, Wani AK (2023) Therapeutic potential and phytoremediation capabilities of the tobacco plant: advancements through genetic engineering and cultivation techniques. Biocatal Agric Biotechnol 52:102845

Seregin IV, Kozhevnikova AD (2023) Phytochelatins: sulfur-containing metal(loid)-chelating ligands in plants. Int J Mol Sci 24(3):2430

Sethi S (2023) Phytochelatins: heavy metal detoxifiers in plants. In: Advanced and innovative approaches of environmental biotechnology in industrial wastewater treatment. Springer Nature Singapore, Singapore, pp. 361–379

Sharma P, Rai S, Gautam K, Sharma S (2023) Phytoremediation strategies of plants: challenges and opportunities. In: Plants and their interaction to environmental pollution, pp. 211–229

Singh R, Joshi H, Singh A (2020) Bryophytes: natural biomonitors. In: Natural products chemistry. Apple Academic Press, pp. 139–153

Stewart WN, Rothwell GW (1993) Paleobotany and the evolution of plants. Cambridge University Press, Cambridge

Taylor EL, Taylor TN, Krings M (2009) Paleobotany: the biology and evolution of fossil plants. Academic Press, London

Thakur M, Praveen S, Divte PR, Mitra R, Kumar M, Gupta CK, Kalidindi U, Bansal R, Roy S, Anand A, Singh B (2022) Metal tolerance in plants: molecular and physicochemical interface determines the "not so heavy effect" of heavy metals. Chemosphere 287:131957

Tulay EZ (2024) Bryophytes in biotechnology. In: Biotechnology in action: unveiling nature's potential 2, pp. 347

Vasilachi IC, Stoleru V, Gavrilescu M (2023) Analysis of heavy metal impacts on cereal crop growth and development in contaminated soils. Agriculture 13(10):1983

Verbruggen N, Hermans C, Schat H (2009) Molecular mechanisms of metal hyperaccumulation and tolerance in plants. Plant Soil 285(1–2):29–47

Vitt DH, Crandall-Stotler B, Wood A (2014) Bryophytes: survival in a dry world through tolerance and avoidance. In: Plant ecology and evolution in harsh environments, pp. 267–295

Yadav S, Basu S, Srivastava A, Biswas S, Mondal R, Jha VK, Singh SK, Mishra Y (2023) Bryophytes as modern model plants: an overview of their development, contributions, and future prospects. J Plant Growth Regul 42(11):6933–6950

Yang L, Huang SQ, Liu Y, Rensing C, Fan ZL, Feng RW (2022) Selenate regulates the activity of cell wall enzymes to influence cell wall component concentration and thereby affects the uptake and translocation of Cd in the roots of Brassica rapa L. Sci Total Environ 821:153156

Yuan Z, Cai S, Yan C, Rao S, Cheng S, Xu F, Liu X (2024) Research progress on the physiological mechanism by which selenium alleviates heavy metal stress in plants: a review. Agronomy 14(8):1787

Zulfiqar F, Ashraf M (2022) Antioxidants as modulators of arsenic-induced oxidative stress tolerance in plants: an overview. J Hazard Mater 427:127891

Pteridophytes: Potential Phytoremediator for a Sustainable Environment

Jinnatunnesa Khatun, Nishika Jaishee, Ayan Roy, and Rohini Lama

Abstract Heavy metals or metalloids that contaminate soil by natural or anthropogenic activities have a deleterious impact on the ecosystem. Soil contamination with heavy metals and associated hazards is a grave concern and demands necessary actions. Accumulation of heavy metals such as chromium, nickel, zinc, lead, copper, mercury and cobalt in the ecosystem destroys the food chains immensely and elevates the risk of various fatal diseases such as kidney disease, bone and cardiovascular abnormalities, hypertension, low birth weight, cancer and neurodegenerative diseases. Besides the contamination of soil, these contaminants also heavily contaminate the groundwater which in turn has led to significant agricultural loss and compromised food quality. Numerous strategies such as soil replacement, electrokinetic remediation, immobilization, encapsulation and soil washing by using various reagents are currently employed to restore the heavy metal contaminated soil. However, the most affordable and eco-friendly method being extensively explored in recent times is phytoremediation. Amongst the various plants used as heavy metal remediators, pteridophytes such as *Pteris, Pityrogramma, Adiantum, Nephrolepis, Asplenium, Athyrium, Azolla, Blechnum* and *Pellaea* have been revealed to effectively restore contaminated soil. Among heavy metal hyperaccumulating ferns, *Pteris vittata* L. has been extensively studied. Removal of heavy metal defilements with pteridophytes can be achieved using phytoremediation techniques such as phytoextraction, rhizofiltration and phytostabilization. Moreover, the symbiotic association of microbes and ferns has been shown to significantly enhance the phytoremediation process. Several metal-binding protein (MBP) coding genes (MT2 and PCS1) and metal transfer genes (SmNhaD and ACR3) are upregulated

J. Khatun · N. Jaishee (✉)
Department of Botany, The University of Burdwan, Bardhaman, India
e-mail: njaishee@bot.buruniv.ac.in

A. Roy
Department of Biotechnology & Bioinformatics, Asian University for Women, Chattogram, Bangladesh

R. Lama
Department of Botany, Darjeeling Govt. College, Darjeeling, India

S. Roy, V. Mandal (eds.), *Plant-Microbe Interaction under Xenobiotic Exposure*,
https://doi.org/10.1007/978-981-96-8260-7_18

during the heavy metal phytoremediation process in pteridophytes. In this chapter, we have made an effort to comprehensively compile some heavy metal accumulating pteridophytes and portray their relevant phytoremediation mechanism.

Keywords Metal-binding protein · Phytoremediation · Pteridophytes · *Pteris vittata*

1 Introduction

Metals are generally categorized as essential and non-essential depending on the requirements of plants for their growth. Essential metals include iron (Fe), manganese (Mn), zinc (Zn), copper (Cu), magnesium (Mg), molybdenum (Mo) and nickel (Ni), while non-essential heavy metals (HMs) are arsenic (As), cadmium (Cd), antimony (Sb), chromium (Cr), lead (Pb), cobalt (Co), silver (Ag), selenium (Se) and mercury (Hg) (Singh et al. 2016). Any non-essential elements more than 5 g/cm^3 and highly toxic for living organisms are defined as heavy metals. However, in recent times 'potentially toxic elements' are preferred by researchers for these toxic elements (Pourret and Hursthouse 2019). Land erosion, chemical weathering of minerals, volcanic activity, mining, fuel and energy generation, industrialization, wastewater disposal and treatments accumulate enormous amounts of heavy metals in soil and water (Zhang et al. 2010). Further, the alarming rise in the population has resulted in excessive utilization of agrochemicals to harvest enhanced amounts of crops for meeting the growing population, augmented demands for industrialization and mining products subsequently causing contamination of soil by heavy metals (Foley et al. 2011; Amundson et al. 2015; Mueller et al. 2012; Nriagu 1996; Reichl et al. 2018). Pollution of soil and water by heavy metal (HM) is of grave concern worldwide because of its deleterious effect on the food chain endangering human health (Hu and Zhang 2012; Khalid et al. 2017).

Remediation techniques such as soil replacement and electrokinetic remediation (Yao et al. 2012), vitrification (Miner 2016), immobilization (Sun et al. 2016), encapsulation (Ucaroglu and Talini 2012) and soil washing using various reagents and extractants (Guo et al. 2016) are employed to decrease heavy metal contaminants in soil. Almost all these techniques are known to have an adverse effect on the texture and composition of the soil. For example, in soil washing technology, high chelating agent solution is used to wash the soil surface deeply which concurrently reduces soil nutrients completely as there is no washing effluents treatment policy (Dermont et al. 2008; Jelusic and Lestan 2015). Thus, phytoremediation using green plants is one of the most extensively practised techniques which has attracted much attention from engineers and scientists across the globe in recent years. Plants can extract or stabilize heavy metals without having a negative impact on the food chain and soil properties by utilizing various eco-friendly techniques (Mosa et al. 2016). The advantages of phytoremediation over traditional physical and chemical methods are non-invasive, simplicity, eco-friendly and cost-effectiveness (Lee et al. 2017; Liu et al. 2020a; Liu et al. 2020b; Mosa et al. 2016). Plants have an immense capability to exceptionally absorb and accumulate potential toxic elements (PTEs) in their parts for the removal of contaminants, yet without being adversely affected (Gill et al. 2021). Phytoremediation depends upon soil pH, aeration, plant capacity, microbial assistance and type of heavy metals (Wang et al. 2020b).

Besides angiospermic plants, pteridophytes have also been extensively explored as a potent phytoremediator in the last few decades. These are the most ancient vascular land plants which have incredibly adapted to the changing environment and have luxuriantly flourished till date. The adaptive capabilities of pteridophytes to anthropogenic actions specify their evolutionary discreteness and resilience to stresses caused by various abiotic factors like high concentrations of heavy metals (Sajeev et al. 2022). This has further promoted the application of pteridophytes for in situ or ex situ remediation of polluted soil or water. So far, approximately about 13,000 pteridophytes or fern species have been identified (Drăghiceanu et al. 2014; Praveen and Pandey 2020; Prabhu et al. 2016; Tiwari et al. 2013). However, the role of these groups of species in the animal and human food chain is limited. Additionally, the presence of spores facilitates rapid growth of the ferns (Ko et al. 2004). Thus, being abundantly available and with a scarce role in the food chain, it serves as a potential resource for phytoremediator species.

Pteridophytes have an implausible ability to phytoremediate potentially hazardous and toxic contaminants from the environment. Klopper (2011) has reported 450 varieties out of 45 families to be mostly nickel (Ni) accumulators. It can effectively and safely remove heavy metals from the ecosystem. *Azolla*, an aquatic fern can efficiently uptake heavy metals like Ni and Cu thereby being a potent phytoremediator of contaminated water (Banerjee and Roychoudhury 2022). Another rapidly growing fern, *Pteris vittata* L. is reported to accumulate a maximum amount of As in fronds (tolerance level up to 1500 ppm) (Ronzan et al. 2017; Cantamessa et al. 2016). Generally, heavy metal accumulating plants employ diverse phytoremediation technologies like phytostabilization, phytodegradation, phytoextraction, rhizodegradation, phytofiltration, phytosequestration, phytohydraulics and phytovolatilization. Amongst all, phytoextraction unalters the soil properties and is the most safe and efficient phytoremediation technology (Ashraf et al. 2019; Singh et al. 2019). This chapter attempts to collate information on the usage of various pteridophytic species as potential phytoremediators and an elucidated mechanism studied so far in some species. Highlights on different metal contaminant of soil and their hazardous health implication have also been presented.

2 Heavy Metal Contamination in Soil and Health Hazards in Living Organisms

Natural and human activities like volcanic activity, weathering, industrialization, extensive application of agrochemicals and mining are the primary sources of soil contamination by heavy metals or metalloids (As, Pb, Zn, Cu, Hg, Cd, etc.) (Smolders et al. 2003; Pandey and Rattan 2021; Onder and Dursun 2006; Gimeno-García et al. 1996) (Fig. 1). When the density of an element is more than 5 g cm^{-3}, it is considered a heavy metal and 53 are reported as heavy metals (Ashraf et al. 2019; Prieto et al. 2018). However, in 2005 about 30 metalloids and 62 heavy metals were reported (Mingho 2005). Heavy metal persists in the ecosystem for a considerably prolonged period of time, resulting in bioaccumulation in the food chain and causing extremely harmful effects on both wildlife and human health. Higher amounts of heavy metals like As, Cd, Cr, Cu, Hg, Ni, Selenium, Ag, Pb and Zn are associated with numerous health problems (Alengebawy et al. 2021; Salem et al. 2000). Similarly, Be, B, Li, Al, Ti, V, Cr, Mn, Co, Ni, Cu, As, Se, Sr, Mo, Pd, Ag, Cd, Sn, Sb, Te, Cs, Ba, W, Pt, Au, Hg, Pb and Bi at higher concentration affects environmental property causing complications in living organisms.

Though heavy metals in trace amounts are advantageous for plants' normal growth, in excess can alter enzymatic activity and increase oxidative stress, ultimately leading to damage of the plant cell and reduction in the plant growth (Djingova and Kuleff 2000; Jadia and Fulekar 1999). Further, it greatly affects the quality of crops and food safety (Tóth et al. 2016). The range of heavy metals in land

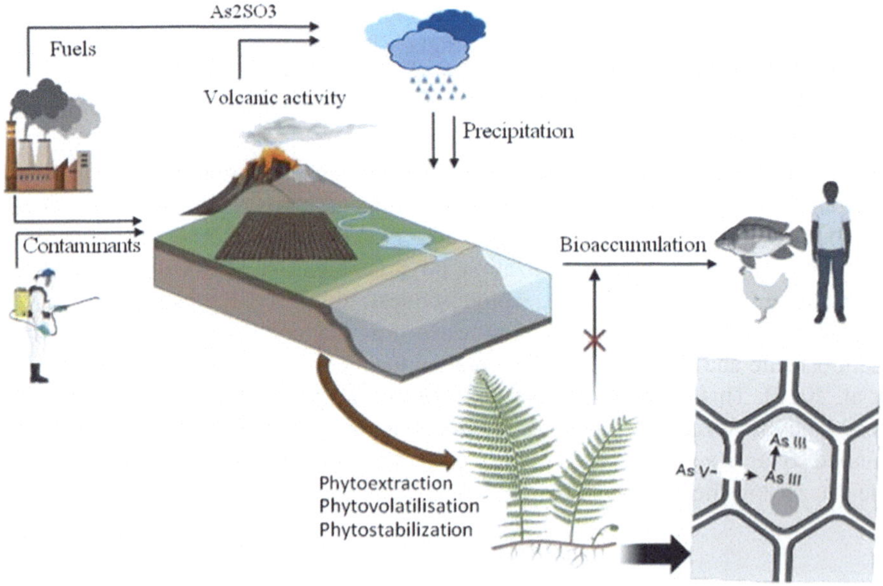

Fig. 1 Proposed arsenic contamination in soil and water, bioaccumulation and phytoremediation

plants varies as follows: arsenic 0.02–7; cadmium 0.1–2.4; mercury 0.005–0.02; lead 1–13; antimony 0.02–0.06; cobalt 0.05–0.5; chromium 0.2–1; Cu 4.15; iron 140; Mn 15–100; Mo 1–10; Ni 1; Sr 0.30 and Zn 8–100 in $\mu g\ g^{-1}$ dry wt (Misra and Mani 1991). As mentioned earlier, long-term exposure to heavy metals leads to several health issues like gastrointestinal, respiratory, cardiovascular, reproductive, renal, hemopoietic and neurological disorders in humans (Gerhardsson and Kazantzis 2015; Garza-Lombo et al. 2018). High concentrations of chromium VI can even change microbial metabolism (Huang et al. 2009). On the other hand, both the states of arsenic (arsenite III and arsenate V) are extremely toxic. However, arsenite III is comparatively more toxic than arsenate V. Inhalation of 10 ppm of arsine gas (AsH_3) can lead to death (Kuivenhoven and Mason 2021). Though the consequences of heavy metal pollution are multifarious, they are beyond the purview of this chapter.

3 Pteridophytes in Phytoremediation

Ferns and fern allies have been blessed with enormous adaptive and resilient capabilities making them widely available throughout the globe. The last few decades have seen massive growth in the application of pteridophytes as a potent phytoremediator of hazardous toxic soil contaminants like arsenic, cadmium, chromium and copper. Pteridophytes like *Pteris vittata, P. umbrosa, P. cretica, Pityrogramma calomelanos* and *Adiantum capillus-veneris* can accumulate higher concentration of arsenic either in the leaves or roots. Among ferns, *P. vittata* was the first species to be identified as a potent arsenic accumulator. It has arsenous acid transfer protein (on vacuole membrane) that codes gene *ACR3* which aids in decreasing arsenic in the soil. It has also been noted that due to the higher concentration of accumulated heavy metals, most of the ferns are avoided by the herbivores which concurrently increases the survivability of ferns (Indriolo et al. 2010). Pteridophytes have specific heavy metal phytostabilization abilities. *Athyrium wardii* (mining and non-mining ecotypes) has been reported to accumulate Cd more effectively in the underground parts of the mining ecotype (ME) in comparison to the non-mining ecotypes (NME), with the addition of humic substances. Comparatively, NME revealed significantly higher Cd accumulation in the above-ground parts while ME in the underground parts. Thus, the mining ecotype exhibited better remediation especially the underground parts for phytostabilization of cadmium. This study displays that ecotype plants undergo phytostabilization in different parts of the plants depending upon the physical and chemical properties of soil (Zhan et al. 2016). *Tectaria impressa, T. herpetocaulos, Lygodium* spp., *Adiantum caudatum* L., *A. philippense, Angiopteris erecta, Colysis pothifolia, Lindsaea ensifolia, Pteridium aquilinum* (Supaporn et al. 2008), *Asplenium adylterium* (Vogt 1942), *A. septentrionale* (Page 1988) and *Azolla filiculoides* (Sela et al. 1989) are heavy metal hyperaccumulating ferns. Similarly, *Nephrolepis cordifolia* and *Hypolepis muelleri* phytostabilize copper, lead, zinc and nickel and *Dennstaedtia vallioides* copper and zinc while *Polypodium cambricum* can phytostabilize only zinc. Phytoextraction is another mechanism used by ferns like *Adiantum philippense* and *Adiantum caudatum* to remediate lead (Pb) and nickel (Ni). *Blechnum orientale* is another potent fern which can significantly accumulate toxic metals and hazardous polycyclic aromatic hydrocarbons (PAHs). These ferns play a potential role in controlling heavy metal contamination in the ecosystem (Zhu et al. 2013). Some, Polypodiales *such as Adiantum, Asplenium, Athyrium, Azolla, Blechnum, Nephrolepis, Pellaea* and *Petris* have an exceptional potential to accumulate heavy metal or metalloid. *Athyrium yokoscense* can accumulate high Cd, Pb and Zn, while *Pteris vittata* (Dhir 2018) and *Pteridium aquilinum* are accumulators of arsenic (Onyia et al. 2021). Several species of *Azolla* (aquatic ferns) can hyperaccumulate heavy metals like Cr, Pb, Cu, Hg, Zn and Ni (Pandey 2012; Rai 2008). Likewise, *Salvinia minima, S. molesta* and *S. natans* of Salviniaceae accumulate Pb, As, Zn, Cu, Ni and Cr (Hoffmann et al. 2004; Da Silva et al. 2018; Dhir and Srivastava 2011). *Pityrogramma* and *Petris* are good indicators of heavy metal pollution growing naturally in mining area (Claveria et al. 2019; Ha et al. 2019; Prasetia et al. 2017).

Metal-binding protein (MBP) coding genes are upregulated in plants to chelate, transport or detoxify heavy metals (Feki et al. 2021; Sharma et al. 2021). Plants with rapid growth, high biomass and hyperaccumulation capacity perform a pivotal role in the reduction of soil contaminants without any alteration in soil properties (Kramer 2018). Upregulation of several genes such as *PvGRX5, PvECS, PvACR3, PvACR3;1, MT2* and *PCS1* cause an increase in metal uptake capacity, transporter protein activities, enzymes and metal chelating protein activities to provide resistance against heavy metal stress (Singh et al. 2019; Chaudhary et al. 2018; Mathur and Chauhan 2020). Metals around the roots are dissolved by root exudates which are further taken up by the plants thereby preventing soil and groundwater contamination (Ashraf et al. 2019). Plants secrete some chemicals and nutritional substances like sugars, alcohols and organic acids which are advantageous for root-assisted microbe interactions. This crosstalk is reported to enhance hyperaccumulation properties. Hyperaccumulator plants can effectively and proficiently uptake heavy metals and then transfer them to the above-ground part via xylem translocation (Rascio and Navari-Izzo 2011).

3.1 Phytoremediation Strategies in Pteridophytes

Phytoremediation strategies like phytostabilization, phytodegradation, phytoextraction, rhizodegradation, phytofiltration, phytosequestration, phytohydraulics and phytovolatilization have been reported to be frequently employed by different heavy metal accumulator plants. Phytoextraction, however, is considered to be the most efficient and safe technology with no effect on soil properties (Ashraf et al. 2019; Singh et al. 2019). Phytoremediation is cost-effective and eco-friendly on-site or in

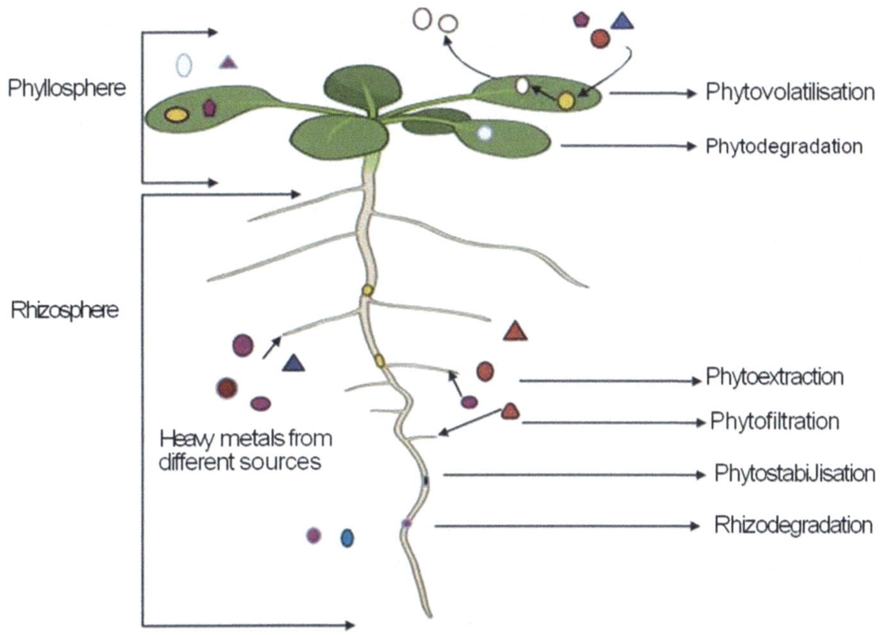

Fig. 2 Proposed general phytoremediation mechanisms in plants

situ process through which plants can uptake, transport, immobilize, volatilize or degrade soil or water contaminants (Pathak et al. 2024; Jan et al. 2024; Shah et al. 2024; Sharma et al. 2023). Although several types of phytoremediation are reported, depending upon the contaminant's bioavailability and plant species, phytoremediation is generally categorized into phytoextraction, phytostabilization and phytovolatilization (Chunduri et al. 2024; Sahu et al. 2024). Some general phytoremediation mechanisms of heavy metal and organic contaminants removal from contaminated soils have been depicted in Fig. 2.

Nevertheless, phytoremediation methods either singly or in combination exploited by the plants to remediate soil and water vary depending upon the contaminant types, forms and media. Soil, sediments or sludge contaminations are remediated through phytoextraction, phytodegradation, phytostabilization, rhizodegradation or phytovolatilization. Surface and wastewater contaminations can be treated by rhizofiltration, phytodegradation or rhizodegradation while contaminated ground through phytodegradation, phytovolatilization, rhizofiltration, rhizodegradation and phytodegradation (Jacob et al. 2018). Pteridophytes have been reported to employ mostly phytoextraction followed by phytostabilization, phytovolatilization (less reported in ferns) and rhizofiltration mostly found in aquatic ferns for phytoremediation (Fig. 3).

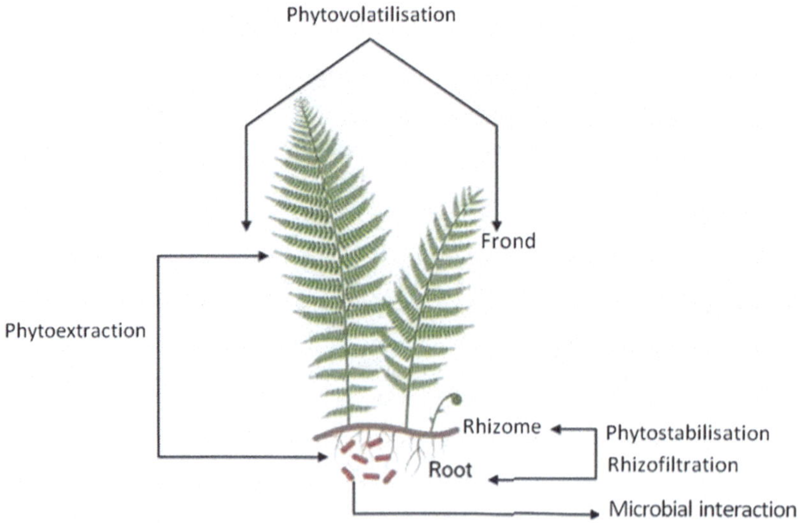

Fig. 3 Proposed phytoremediation process in pteridophytes

3.1.1 Phytoextraction

Phytoextraction technique uptakes heavy metals in roots based on their chemical states and loads them into the xylem for transportation to shoot or other harvestable plant parts. Heavy metals interact with the phytochelatins or ligands within the accumulating parts. Plants are harvested when saturated with heavy metals (Lajayer et al. 2019; Saini and Dhania 2020). In the harvestable parts, heavy metals are transferred either for storage or exclusion (Li et al. 2022; Sarwar et al. 2017; Yanitch et al. 2020). The shoot-to-root ratio of heavy metal (translocation factor = TF) and the root-to-soil ratio of heavy metal (bioconcentration factor = BCF) are important for the phytoextractibility ($1 < TF$ and $1 < BCF$) of a plant. Both factors should be less than 1 for better phytoextractibility ($1 < TF$ and $1 < BCF$).

Several studies report phytoextraction to be an efficient mechanism of heavy metals remediation from the soil. Interestingly, microorganisms present in the rhizospheric zone have been reported to substantially assist in arsenic uptake in *P. vittata.* Bacterial AsV-reductase coded by *arsC* gene facilitates the reduction of AsV to AsIII facilitating arsenic uptake. Additionally, bacterial siderophore also aids in arsenic uptake capability, while enhancing the overall phytoextraction ability of *P. vittata* (Patten and Glick 2020; Rajkumar et al. 2012; Kidd et al. 2009; Kohda et al. 2021). Bacteria-derived phytohormones like indole acetic acid, siderophores and AsV-reductase enhance plant growth and As phytoextraction (Patten and Glick 2020; Rajkumar et al. 2012).

Pteris vittata L. and other species like *P. cretica* L., *P. longifolia* L., *P. umbrosa* L., *P. argyraea* L., *P. quadriaurita* L., *P. ryiunkensis* L., *P. biaurita* and *Pityrogramma calomelanos* L. can hyperaccumulate As (Zhao et al. 2002; Srivastava et al. 2005).

Petris vittata has a high potentiality of As phytoextraction. Likewise, *Athyrium yokoscense, Pityrogramma calomelanos* and *Pteris vittata* are hyperaccumulators of other potentially toxic elements (Cd, Cr, Hg, Pb as well as Ni, Cu and Zn) besides arsenic. *A. yokoscense* has been reported to tolerate a number of PTEs and simultaneously can accumulate higher amounts of PTEs like Cd about 1000 mg kg^{-1}, Pb 10,000 mg kg^{-1}, Zn 9000 mg kg^{-1} and Cu 3300 mg kg^{-1}. However, *A. yokoscense* was reported to accumulate a lower amount of As in comparison to *P. calomelanos* and *Pteris vittata. Pityrogramma calomelanos* was reported to accumulate >8000 mg kg^{-1} dry mass whereas above-ground organs and fronds of *P. vittata* accumulated up to 7500 mg kg^{-1} and > 22,000 mg kg^{-1} arsenic accumulation, respectively (Claveria et al. 2019; Kamachi et al. 2005; Van et al. 2006; Remigio et al. 2021; Danh et al. 2014). *Asplenium adylterium* is an indicator of nickel (Vogt 1942). *A. adylterium* and *A. septentrionale* both can hyperaccumulate lead and copper (Vogt 1942). *Blechnum orientale* accumulates toxic metals and hazardous PAHs (Zhu et al. 2013). *Pteris umbrosa* and *P. cretica* accumulate As in leaves. *Adiantum philippense* and *Adiantum caudatum* can accumulate Pb and Ni through phytoextraction process (Pandey 2020; Zhu et al. 2013).

Aquatic ferns *Salvinia biloba* and *S. minima* of Salviniaceae family can reduce Pb by 90% through phytoextraction mechanism (Loría et al. 2019). Salviniaceae family member *Azolla filiculoides* can accumulate Cd, Pb and Cu (Mukhtar and Abdullahi 2017; Hazra et al. 2015) while *Azolla pinnata* remediates Cd, Zn Pb, Hg and Ni specifically from wastewater by using phytoextraction method (Ali et al. 2020; Rai 2008; Talebi et al. 2019). Additionally, *other* members of Salviniaceae like *S. minima* accumulate more than 34 mg/g dw of lead and 0.5 mg/g dw of arsenic, *S. molesta* accumulates about As 148.63 g/g dw in leaves and *S. natans* could phytoremediate more than 50% of Zn, copper, nickel and chromium (Loría et al. 2019; Hoffmann et al. 2004; Da Silva et al. 2018; Dhir and Srivastava 2011).

3.1.2 Phytostabilization

Neutralization, immobilization and detoxification of heavy metals using plants to reduce bioavailability, limit leaching and prevent biomagnifications in the food chain are known as phytostabilization, phytorestoration, phytoimmobilization or phytosequestration (Sylvain et al. 2016; Khalid et al. 2017). *Athyrium wardii* from Athyriaceae family can reduce bioavailability and detoxify heavy metals like lead and cadmium through phytostabilization technology (Zhao et al. 2016). *Nephrolepis cordifolia, Hypolepis muelleri, Dennstaedtia vallioides* and *Polypodium cambricum* are the phytostabilizer of Zn while Pb and Ni are also phytostabilized by *Nephrolepis cordifolia* and *Hypolepis muelleri* (Pandey 2020, Zhu et al. 2013). However, both *D. vallioides* and *N. Cordifolia* can also effectively phytoremediate Cu through phytostabilization process. *Polypodium cambricum* can make plants resistant against Zn probably through phytostabilization technology. On the other hand, higher concentrations of Zn adversely affect the growth or even cause death of *Pteris vittata* (Kachenko et al. 2007; Roccotiello et al. 2010).

3.1.3 Rhizofiltration

Rhizofiltration is another important technology where contaminants are adsorbed onto plant roots or into the plant roots from the solution surrounding the rhizospheric zone (Javed et al. 2019; Tiwari et al. 2019). It is usually used to decontaminate surface water as well as groundwater by preventing the mobilization of heavy metals (Abhilash et al. 2009; Benavides et al. 2018). Rhizofiltration by *Azolla pinnata* can remove Cu from contaminated water (Kumar et al. 2020; Padmavathiamma and Li 2007). The ever-increasing human population, coal or ore mining and deforestation due to rapid urbanization and industrialization are the prime causes of the enhancement of toxic heavy metal or metalloid repositioning into the soil and water. This polluted soil becomes hazardous for agriculture and increases pest and pathogen attacks which ultimately result in various fatal diseases and eventually collapse the food chain of an ecosystem. In India, heavy metal pollution greatly affects agriculture thereby negatively impacting economic status also (Kumar et al. 2019; Milicevic et al. 2021; Wang et al. 2020). Some phytochemicals released by the plant roots attract microbes and these microbes depending on the temperature, pH, soil conditions, air, etc., degrade soil pollutants (Saravanan et al. 2019; Oberoi and Khanna 2018; Sathishkumar et al. 2008). A few endophytic bacterial strains of *Pseudomonas* and *Ochrobacterium* are reported to produce Cr-III from Cr-VI in the plant roots and further facilitate the precipitation of these metals in the roots (Dimitroula et al. 2015). Hence, several studies reveal that microbial assistance can effectively aid in rhizofiltration (Rane et al. 2022). 16S rRNA gene sequence analysis report reveals that *Agrobacterium* and *Bacillus* are the most predominant arsenite-resistant bacteria among *Agrobacterium, Stenotrophomonas, Pseudomonas, Rhodococcus* and *Bacillus* of Proteobacteria, Actinobacteria and Firmicutes phyla in *P. vittata*. Indole-3-acetic acid (IAA) synthesis, as well as *arsB* and *ACR3(2)* genes, is highly expressed during As stress (Yang et al. 2019). Another genomic DNA analysis of endophyte *Pseudomonas* spp. PG-12 from *P. vittata* also reveals that metal efflux transporters like PbrA, CadA2 and CzcR enhance phosphate transfer while inhibiting Pb transport. Besides developing heavy metal resistance in *P. vittata*, several phytohormones like IAA and gibberellins, and siderophores are produced, which assist in increasing the growth of the plant. This implies that microbe-plant interactions have twofold advantages eradicating pollutants from the soil and increasing plant growth via enhancing the production of phytohormones, siderophores, metal transporters activity and gene-like *arsB* and *ACR3(2)* expression in response to heavy metal contamination (Manzoor et al. 2019).

Soil mineral composition and pollutants have a great impact on the microbial population which in turn determines the tolerance ability, growth, productivity, immigration and acclimatization of the plants (Vincent et al. 2024; Yin et al. 2023; Zulfiqar et al. 2023). The role of microorganisms in ecosystem and plant growth and development is indispensable. Moreover, microbes could be induced efficiently to degrade or remove soil contaminants for the improvement of its phytoremediation property. Some advanced approaches such as genomics, transcriptomics, proteomics and metabolomics are frequently employed to understand and target the molecular pathways of phytoremediation processes. Of late, CRISPR/Cas9 has emerged as one of the promising technologies to modify the target genome for enhancing the rate of phytoremediation (Oubohssaine and Dahmani 2023).

3.1.4 Phytovolatilization

Potentially toxic elements after uptake by plants are made less toxic before releasing it in a volatile form mainly through the leaf surfaces. Ferns like *Azolla caroliniana* (aquatic) and *Pteris vittata* have the ability to remediate contaminants by phytovolatilization (Pilon-Smits et al. 1999). Metals like mercury and arsenic taken up by the plants are transported and converted into less harmful forms through phytovolatilization process before being released through stomatal aperture (Pang et al. 2023; Sharma et al. 2023; Shehu et al. 2023; Sharma et al. 2024). *Pteris vittata* leaves are reported to volatilize As and facilitate arsenic decontamination from soil (Sakakibara et al. 2010). Similarly, sulphate transporter of plants transport selenate and release it after being volatilized through enzymatic process into less harmful volatile compounds like dimethyl diselenide [$(CH_3)_2Se$], dimethyl selenone [$(CH_3)_2SeO_2$], dimethyl selenylsulphide [$(CH_3)_2SeS$] and methaneselenol (CH_3Se) (Sharma et al. 2015; Chen et al. 2019).

3.2 *Mechanism of Phytoremediation in Pteridophytes*

Numerous ferns grow naturally on the waste or contaminated field and act as a good indicator of soil pollution (Chang et al. 2009). Significant roles of phytochelatins (PCs), metallothioneins (MTs) and transporter proteins in phytoremediation have been reported. In pteridophytic species, *Azolla pinnata* and *A. filiculoides*, upregulation of phytochelatin synthase and metallothionein coding gene (*MT2* and *PCS1*) expression is known to enhance the capability to chelate and detoxify Ni, Zn, Cu and Cd. On the other hand, an increase in the Na^+/H^+ antiporter (gene *SmNhaD*) activity augmented Pb transportation in *Salvinia minima* Baker (Chaudhary et al. 2018; Mathur and Chauhan 2020; Talebi et al. 2019). Reports on the possible mechanism utilized for remediation by pteridophytes are scarce. Thus, we focus on the most extensively explored species *P. vittata* to discuss the mechanism understood or unravelled so far.

Although the arsenic tolerance level of *P. vittata* is 1500 mg As kg^{-1} dry soil, it has been observed to accumulate 22,000 mg arsenic kg^{-1} dry weight in the fronds which is exceptionally highest than in any other arsenic tolerant plant (Watanabe et al. 2007). In the fronds, arsenic activates glutathione (GSH) activity which in turn induces the expression of γ-glutamylcysteine synthetase encoding gene (*PvECS*) (Sakai et al. 2010). Simultaneously, higher activity of superoxide dismutase (SOD) in fronds may contribute to hyperaccumulation of As (Srivastava et al. 2005). In the roots and fronds, arsenate (AsV) is converted to arsenite by arsenate reductase and

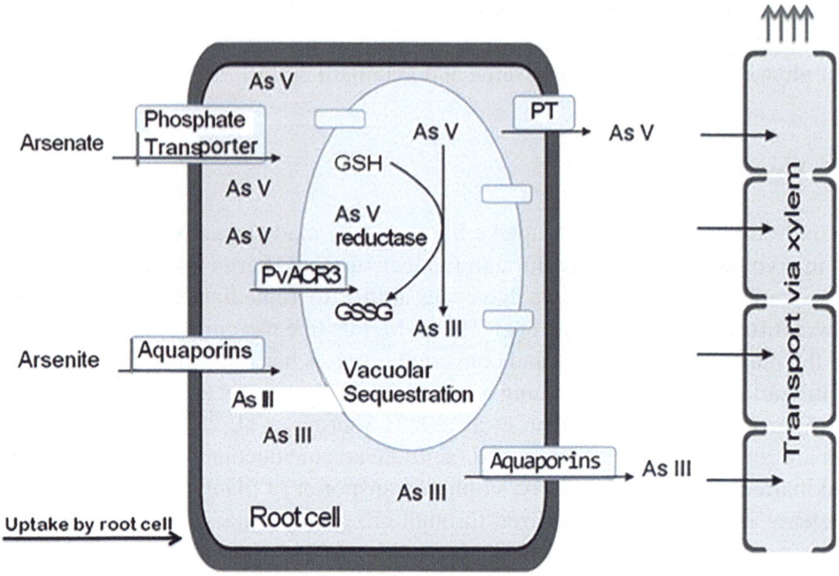

Fig. 4 Mechanism of arsenic uptake by plant root cells

then transported with the help of arsenite transporter into the vacuoles for sequestration (Indriolo et al. 2010). *Pteris vittata* accumulate comparatively higher As than any other plants. Heavy metal like arsenate (AsV) is transported through phosphate transporters while arsenite (AsIII) is transported through PsPvH (nodulin 26-like intrinsic proteins) transporter, aquaglyceroporins and hexose permeases (Li et al. 2016; Mukhopadhyay et al. 2014). Aquaporins can transport AsIII form of As in the root cell. As[V] to As[III] conversion [$(AsO)^{3-}$ + 2GSH → $(AsO_3)^{3-}$ + GSSG] occurs inside the vacuole of the root cell in the presence of arsenate reductase (AR) and glutathione (Abbas et al. 2018). Figure 4 illustrates the mechanism of arsenic uptake by plant root cells.

Studies report that the transfer of As through phosphate transporters is basically due to the structural resemblance of arsenate (AsV) with phosphate (Li et al. 2016, Mukhopadhyay et al. 2014). A higher concentration of arsenite has no significant effect on phosphate uptake but on the contrary, increased arsenate concentration can inhibit phosphate transfer in the root itself (Xie et al. 2009). Arsenic stress in *P. vittata* results in the expression of thiobarbituric acid-reacting substances at low levels which triggers the production of superoxide dismutase, catalase and ascorbate peroxidase to detoxify the generated reactive oxygen species (Srivastava et al. 2005). GSH-mediated pathway is observed to be activated in *P. vittata* during arsenic stress. In this pathway, the arsenic–glutathione complex $As(GSH)_3$ is transported into the tonoplast via ABC transporters and subsequently degraded even in the absence of Glutathione S-transferase (GST) in *P. vittata,* probably via an alternative pathways or enzymes (Yan et al. 2019; Leslie et al. 2004).

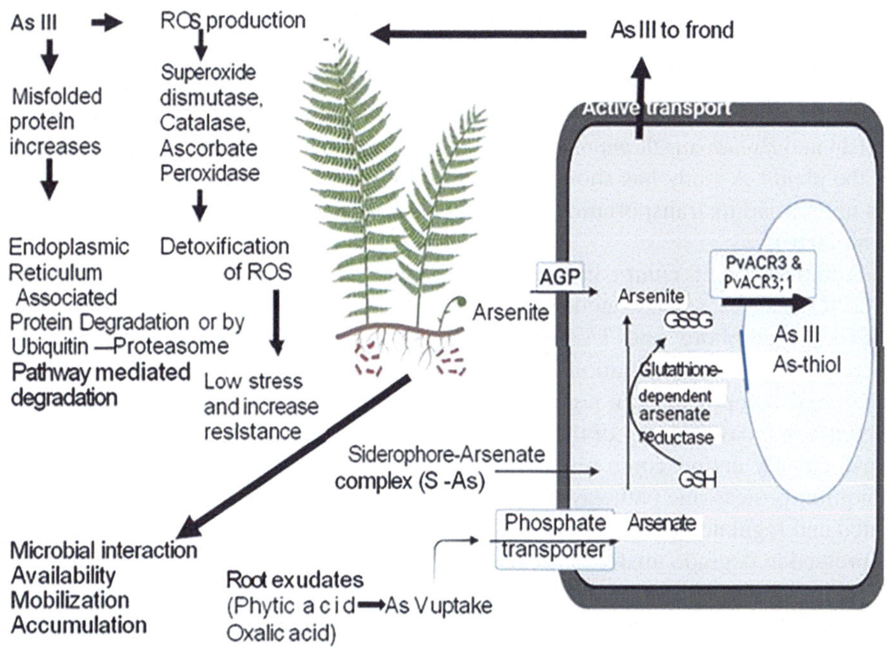

Fig. 5 An illustration of the mechanism of As-uptake and transport in *P. vittata*, for its potential use in phytoremediation

Glutathione-dependent arsenate reductase is present mostly in the roots of *P. vittata* (Duan et al. 2005). Roots uptake arsenates and then convert it into arsenite in the presence of thiols (Abbas et al. 2018). Arsenical resistance protein Acr3 (*ACR3*), the ABC superfamily, P-type ATPase, the major facilitator superfamily (MFS), MIP and nitrate transporter 3.1 (NRT3.1) are gene families of As transporter. Arsenic is transported via these active transporters from root to shoot in *P. vittata* (Leslie et al. 2004). Two vacuole transporters, PvACR3 and PvACR3;1, were identified from *P. vittata*. PvACR3 transports arsenic to vacuole and reduces cytotoxicity while PvACR3;1 has no such role reported so far. Glutaredoxin encoding gene *PvGRX5* was also identified in *P. vittata* which may contribute primarily to the role of arsenic sequestration (Indriolo et al. 2010; Sundaram et al. 2008). Several other genes like *PvGRX5, PvECS, PvACR3* and *PvACR3;1* are reported to be expressed which significantly assist in improving heavy metal tolerance and accumulation (Indriolo et al. 2010; Chen et al. 2013; Chen et al. 2017). Besides PvACR3 and PvACR3;1, another transporter, namely, PvTIP4;1 has been reported to be involved in As transportation (Chen et al. 2013; Chen et al. 2017; Yan et al. 2019). Figure 5 provides the detailed mechanism involved in arsenic uptake and transport in *P. vittata.*

Furthermore, root exudates of *P. vittata* composed of phytic acid and oxalic acid induce bacterial growth and also affect soil pH which has a significant role in increasing the solubility of arsenic (Tu et al. 2004; Yang et al. 2022). Root exudates can induce specific bacterial growth. *Bacillus, Pseudomonas, Lysinibacillus,*

Acinetobacter and *Arthrobacter* are found to be associated with plant roots which are grown in As contaminated soil or water (Yang et al. 2019; Singh et al. 2015). Quite interestingly, microbial associations have been reported to increase higher As uptake in *P. vittata* (Vacheron et al. 2013). *Bacillus* sp. MPV12, *Variovorax* sp. P4III4 and *Pseudoxanthomonas* sp. P4V6 increase the hyperaccumulating activities of the plants. A study has shown that *P. vittata* inoculated with *B. pumilus* induces As uptake and its transportation mostly in the above-ground plant parts (Srivastava et al. 2016).

Additionally, *P. vittata* inoculated with plant growth-promoting rhizobacteria (PGPR) reduces arsenic-induced oxidative stress and cell damage (Yizhu et al. 2020). Siderophore and IAA-producing As-reducing bacteria are reported to increase As phytoremediation by 35% in *P. vittata* via root (Yang et al. 2012). Microbial compounds can significantly cause alteration in soil pH leading to an increase in bioavailability of the heavy metals (Sullivan and Gadd 2019). Two pathways namely endoplasmic reticulum-associated protein degradation (ERAD) and ubiquitin-proteasome pathway (UPP) regulated genes that were reported to be activated and regulate the arsenic uptake in the roots of *P. vittata.* Numerous genes are expressed to degrade misfolded proteins in the ER of root cells in arsenic-mediated stress conditions (Yan et al. 2019).

Symbiotic association of mycorrhizal fungi and plants also induce metal resistance activities (Sarwar et al. 2017; Mthembu et al. 2020). Several studies have been done to understand the heavy metal uptake and detoxification mechanisms of phytoremediation by plant or plant-microbial assistance (Yang et al. 2016; Yang et al. 2021). Another report reveals that the symbiotic association with the AFM facilitates *P. Vittata* L. to hyperaccumulate As (Bettiol et al. 2012; Cantamessa et al. 2016). Plant growth-promoting endophytic (PGPE) bacteria and plant growth-promoting rhizobacteria (PGPR) secrete some organic anions and protons which increase the solubility of heavy metals in the rhizosphere and their further accumulation. Plants easily uptake, neutralize and sequester these heavy metals (Gill et al. 2021).

Needless to say, ferns have remarkable abilities to hyperaccumulate heavy metals and are the potential resources of phytoremediation. However, the most important concern is about the disposal of these hyperaccumulators safely. Phytoremediation involves the accumulation of contaminants in the crop biomass throughout plant growth (Attinti et al. 2017). Thus, extreme precautions should be taken while harvesting these plants as they create highly contaminated bio-waste, if not handled properly (Gong et al. 2018). These bio-wastes are tremendously hazardous waste and could easily become a secondary source of pollution (Gong et al. 2018). Sub- and supercritical water treatments have been utilized to treat the heavy metals accumulating in ferns for arsenic removal. Several by-products such as methylcyclopentenones, phenols and guaiacols may find their application in the production of fragrance, herbicide precursors and antiseptics (Carrier et al. 2011; Cao et al. 2010). Though several other disposal and utilization methods such as heat treatment, extraction treatment, microbial treatment, compression landfill and synthesis of nanomaterials to phytoremediation plants with heavy metals have been utilized yet, a lot needs to be refined for efficient disposal of these heavy metal accumulator plants. However, discussion on the detailed disposal methods is out of the horizon of this chapter.

4 Future Approaches of Pteridophytes in Phytoremediation

Several studies report augmentation in the phytoremediation capability of *P. vittata* with microbial association; however, the actual mechanism still remains unclear. Thus, metagenomic analysis could be beneficial to understand the metabolic and structural changes in the roots, microbes and the activities of plant rhizosphere which could eventually be advantageous to unravel the actual mechanism involved (Visioli et al. 2015). Pteridophytes blessed with diverse capabilities like phytochelation, phytoextraction, phytostabilization, etc., contribute largely to establish a favourable soil environment. Table 1 provides the list of ferns with its probable mechanism of action.

Metabolites-enriched healthy soil condition stimulates microbial-fern association which may generate conditions favourable for agriculture in mining areas (Rout et al. 2024). Further, microbial association is known to induce mineral uptake in plants, increase heavy metal accumulation from metal contaminated area by ferns thereby indicating heavy metal contamination in soil (Abd Elnabi et al. 2023). *Pteris vittata* is a hyperaccumulator of arsenic; however, other ferns like *Christella dentata, Adiantum caudatum, Pteris* sp., *Marsilea quadrifolia* and *Azolla* sp. also accumulate higher concentration of arsenic in rhizome from soil and groundwater (Samal et al. 2023). *Pteridium aquilinum*, an indicator of chromium [Cr(VI)] contamination, has been reported to hyperaccumulate up to 11,854 mg of $Cr{\cdot}kg^{-1}$ DW in the sporophyte and up to 915 mg of $Cr{\cdot}kg^{-1}$ DW in gametophyte (Eslava-Silva et al. 2023). Ferns like *Astrolepis integerrima, A. sinuata, Myriopteris notholaenoides, Notholaena affinis* and *N. sulphurea* accumulate Cd, Pb and As in higher concentration in fronds from polluted soil in mine areas (Guzmán-Cornejo et al. 2023). Different strategies are employed by plants to compartmentalize and detoxify accumulated mercury. *Blechnum serrulatum* Rich. is known to accumulate and tolerate high concentration of mercury (de Freitas et al. 2024; Bakshe and Jugade 2023).

Phytoremediation metagenomics has emerged as an efficient acceptable omics technology. Plant-microbes interaction can control soil contamination and omics data of such microbes can be a potent phytoremediation tool for agricultural land (Maddheshiyaa et al. 2024). Plant microbial interactions render other pathogenic microbes from symbiosis and control heavy metal transport through plant parts like root and shoot (Bhuvaneswari et al. 2023). Genome analysis of these microbial communities renders benefit in heavy metal remediation as microbial organisms can potentially convert and mineralize several complex compounds or soil contaminants from soil (Kumar et al. 2018; Hou et al. 2017). Taxonomic, metagenomic, functional, and meta-transcriptomic data can be analyzed using MEGAN (MetaGenome Analyzer) software, which provides valuable information for understanding the

Table 1 List of ferns with probable phytoremediation mechanism

Name	Heavy metal	Mechanism	Reference
Nephrolepis cordifolia	Cu, Pb, Zn and Ni	Phytostabilization	Zhu et al. (2013) and Pandey (2020)
Blechnum serrulatum	Hg	Compartmentalization	de Freitas et al. (2024)
Pteris vittata, P. umbrosa, P. cretica	As Cd, Cr, Hg, Pb, Ni, Cu and Zn.	Phytoextraction, Phytovolatilization, Rhizofiltration	Zhu et al. (2013), Claveria et al. (2019), Kamachi et al. (2005), Van et al. (2006), Remigio et al. (2021), Danh et al. (2014), Pandey (2020), Manzoor et al. (2019) and Sakakibara et al. (2010)
Athyrium wardii	Cd	Phytostabilization	Zhan et al. (2016)
Adiantum philippense, A. caudatum	Pb, Ni	Phytoextraction	Zhu et al. (2013)
Hypolepis muelleri	Cu, Pb, Zn and Ni	Phytostabilize	Pandey (2020) and Zhu et al. (2013)
Dennstaedtia vallioides	Cu and Zn	Phytostabilize	Pandey (2020) and Zhu et al. (2013)
Polypodium cambricum	Zn	Phytostabilize	Pandey (2020) and Zhu et al. (2013)
Athyrium yokoscense,	As Cd, Cr, Hg, Pb, Ni, Cu, and Zn	Phytoextraction	Claveria et al. (2019), Kamachi et al. (2005), Van et al. (2006), Remigio et al. (2021), and Danh et al. (2014)
Pityrogramma calomelanos	As Cd, Cr, Hg, Pb, Ni, Cu, and Zn	Phytoextraction	Claveria et al. (2019), Kamachi et al. (2005), Van et al. (2006), Remigio et al. (2021), and Danh et al. (2014)
Asplenium adylterium, A. adylterium and *A. septentrionale*	Ni, Pb and Cu	Phytoextraction	Vogt (1942)
Astrolepis integerrima, A. sinuata	Cd, Pb and As	Phytoextraction	Guzmán-Cornejo et al. (2023)
Myriopteris notholaenoides	Cd, Pb and As	Phytoextraction	Guzmán-Cornejo et al. (2023)
Notholaena affinis and *N. sulphurea*	Cd, Pb and As	Phytoextraction	Guzmán-Cornejo et al. (2023)
Marsilea quadrifolia	As	Phytoextraction	Samal et al. (2023)
Christella dentata	As	Phytoextraction	Samal et al. (2023)

characteristics of soil microbes andtheir interactions with plants. Additionally, the Integrated Microbial Genomes with Microbiome Samples (IMG/M) system is very helpful as it can be used to compare several metagenomic data (Hadidi et al. 2013; Harrington et al. 2010; Markowitz et al. 2012).

Of late, nanoparticles (1 to 100 nm) are gaining much attention for being highly mobile and eco-friendly thereby enabling plants to efficiently uptake mineral and interact through roots for better plant growth (Arora et al. 2024). Biocompatible nanoparticles can induce mineral uptake, water absorption, photosynthesis, symbiotic interactions with microbes and overcome stresses, thereby improving phytoremediation efficiency, especially in hyperaccumulator plants (Kumar et al. 2021a; Kumar et al. 2021b; Gulzar and Mazumder 2022; Ditta et al. 2021; Etesami et al. 2022). The catalytic activity of high surface charge-bearing nanoparticles can immobilize soil contaminants (Wuana and Okieimen 2021). Nanoparticles like zero-valent iron (ZVI), titanium dioxide (TiO_2), carbon nanotubes (CNTs) and iron oxide nanoparticles can potentially increase phytoremediation of soil contaminants because of their high surface charge activity (Shahid et al. 2021; Rizwan et al. 2018; Arora et al. 2023). It is reported that titanium dioxide can efficiently induce photosynthesis rate by positively regulating electron transport chain during water stress (Alabdallah and Hasan 2021; Silva et al. 2022).

Transcriptome and/or proteome study is another significant tool to understand the interaction between plants and related microorganisms present in the soil responsible for the increase in phytoremediation process and various progresses in the advancement of the techniques developed (Mhlongo et al. 2018). RNA Seq data of both plants and microbes could gather information about gene regulation in arsenic-mediated stressed plants. The combined study of metagenomics and computer-based technology has been recommended to investigate complicated plant-microbial relations and activities during heavy metal stress in *P. vittata* or other ferns. Scientific study on transcriptomics and proteomic level can foster to understanding proper mechanism of phytoremediation in plants (Lowe et al. 2017; Franzosa et al. 2015). Another frequently employed phytoremediation approach in recent days is the nano-bioremediation technology. Here, silica oxides, zero-valent iron nanoparticles and carbon nanotubes are involved to induce microbial activities and degrade heavy metals (Rizwan et al. 2014; Pandey 2018). Application of nanoparticles may facilitate an increase in phytoremediation efficiency in heavy metal accumulating ferns like *P. vittata*. Several studies have reported an increase in the phytoremediation ability of *P. vittata* by plant growth-promoting rhizobacteria (PGPR) (Sessitsch et al. 2013; Ojuederie and Babalola 2017). However, much study is needed to understand the complex mechanism involved in phytoremediation processes.

5 Conclusion

Pteridophytes are the most ancient incredibly dynamic plant kingdom which has adapted to several evolutionary changes making it one of the most resilient plants to numerous stresses on the earth. Thus, pteridophytes have been one of the most utilized resources for phytoremediation. The role of several genes during heavy metal stresses has been identified. Upregulation of genes such as *ACR3*, ABC superfamily, P-type ATPase, MFS, MIP and *NRT3.1* in *P. vittata* to uptake and neutralize arsenic either in the root or shoot is well established. Further, interactions of the rhizospheric microbial population and PGPR have been effectively involved in increasing heavy metal removal processes. However, detailed scientific investigations are required to understand the proper mechanism of pteridophyte-mediated phytoremediation. The involvement of some heavy metal transporters like phosphate transporters and aquaporins to transport different atomic forms of arsenic in the roots is well established. Likewise, accumulation of heavy metals in harvestable plant parts though may be safe but disposal of these plants without damaging soil properties is still the major concern globally. Thus, the application of metagenomics, transcriptomics, proteomics and nanotechnology may be utilized to effectively understand and increase the beneficial plant-microbe interactions, photosynthetic rate, overcome stresses and immobilization of soil pollutants, ultimately leading to an increase in biomass and efficiency of the phytoremediation process by the plants. The targeted omics modification using advanced genetic engineering technologies could be done to enhance the phytoremediation process.

References

Abbas G, Murtaza B, Bibi I, Shahid M, Niazi NK, Khan MI, Amjad M, Hussain M, Natasha (2018) Arsenic Uptake, Toxicity, Detoxification, and Speciation in Plants: Physiological, Biochemical, and Molecular Aspects. *Int. J. Environ. Res. Public Health* 2018, *15*, 59. [CrossRef] [PubMed]

Abd Elnabi MK, Elkaliny NE, Elyazied MM, Azab SH, Elkhalifa SA, Elmasry S, Mouhamed MS, Shalamesh EM, Alhorieny NA, Abd Elaty AE, Elgendy IM, Etman AE, Saad KE, Tsigkou K, Ali SS, Kornaros M, Mahmoud YAG (2023) Toxicity of heavy metals and recent advances in their removal: A review. Toxics. 2023 Jul;11(7):580.

Abhilash P, Jamil S, Singh N (2009) Transgenic plants for enhanced biodegradation and phytoremediation of organic xenobiotics. Biotechnol. Adv. 2009, 27, 474–488. [CrossRef] [PubMed]

Alabdallah NM and Hasan MM (2021) Plant-based green synthesis of silver nanoparticles and its effective role in abiotic stress tolerance in crop plants. Saudi Journal of Biological Sciences, 28(10), 5631–5639.

Alengebawy A, Abdelkhalek ST, Qureshi SR, Wang, M-Q (2021). Heavy metals and pesticides toxicity in agricultural soil and plants: Ecological risks and human health implications. Toxics 9, 42. https://doi.org/10.3390/toxics9030042.

Ali S, Abbas Z, Rizwan M, Zaheer I, Yavas I Ünay A, Abdel-Daim M, Bin-Jumah M, Hasanuzzaman M, Kalderis D (2020) Application of floating aquatic plants in phytoremediation of heavy metals polluted water: A review. *Sustainability* 2020, 12, 1927. [CrossRef]

Amundson, R. Berhe AA, Hopmans JW, Olson C, Sztein AE, Sparks DL (2015) Soil and human security in the 21st century. *Science* Vol 348, Issue 6235. https://doi.org/10.1126/science.1261071.

Arora D, Arora A, Bala R, Panghal V Kumar S (2023) Enhancement in phytoremediation efficiency of Tagetus erecta with the application of nano-scale zero valent iron (nZVI) for the restoration of lead contaminated soil: An approach toward sustainability. Water, Air, & Soil Pollution, 234(8), 535.

Arora D, Arora A, Panghal V, Singh A, Bala R, Kumari S, Kumar S (2024) Unleashing the Feasibility of Nanotechnology in Phytoremediation of Heavy Metal–Contaminated Soil: A Critical Review Towards Sustainable Approach. *Water Air Soil Pollut* (2024) 235:57 https://doi.org/10.1007/s11270-023-06874-9

Ashraf S, Ali Q, Zahir ZA, Ashraf S, Asghar HN (2019) Phytoremediation: environmentally sustainable way for reclamation of heavy metal polluted soils. Ecotoxicology and Environmental Safety, 174, 714–727.

Attinti R, Barrett KR, Datta R, Sarkar D (2017) Ethylenediaminedisuccinic acid (EDDS) enhances phytoextraction of lead by vetiver grass from contaminated residential soils in a panel study in the field. *Environ. Pollut.* 225, 524–533.

Bakshe P and Jugade R (2023) Phytostabilization and rhizofiltration of toxic heavy metals by heavy metal accumulator plants for sustainable management of contaminated industrial sites: a comprehensive review. J Hazard Mat Adv 10:100293. https://doi.org/10.1016/j.hazadv.2023.100293.

Banerjee A and Roychoudhury A (2022) Assessing the rhizofiltration potential of three aquatic plants exposed to fluoride and multiple heavy metal polluted water. Vegetos 35:1158–64. https://doi.org/10.1007/s42535-022-00405-3.

Benavides LCL, Pinilla LAC, Serrezuela RR, Serrezuela WFR (2018) Extraction in Laboratory of Heavy Metals Through Rhizofiltration using the Plant Zea Mays (maize). *Int. J. Appl. Environ. Sci.* 2018, 13, 9–26.

Bettiol C, Minello F, Gobbo L, Rigo C, Bedini S, Bona E, Berta G, Argese E (2012) Phytoremediation potential of the arsenic hyperaccumulator Pteris vittata: Preliminary results from a field study. *Sci. Ca' Foscari* 2012, 1, 25–31. [CrossRef]

Bhuvaneswari S, Illakiya KB, Rajkumari K (2023) Recent advances in bioremediation by metagenomics-based approach for pharmaceutical derived pollutants, in: Metagenomics to Bioremediation, Academic Press, 2023, pp. 55–80. https://doi.org/10.1016/B978-0-323-96113-4.00012-3.

Cantamessa S, D'agostino G, Berta G (2016) Hydathode structure and localization in Pteris vittata fronds and evidence for their involvement in arsenic leaching. *Plant Biosyst.* 2016, 150, 1208–1215. [CrossRef]

Cao X, Ma L, Shiralipour A, Harris W (2010) Biomass reduction and arsenic transformation during composting of arsenic-rich hyperaccumulator Pteris vittata L. *Environ Sci Pollut Res* 17:586–594.

Carrier M, Loppinet-Serani A, Absalon C, Marias F, Aymonier C, Mench M (2011) Conversion of fern (Pteris vittata L.) biomass from a phytoremediation trial in sub- and supercritical water conditions. *Biomass Bioenerg* 35:872–883.

Chang JS, Yoon IH, Kim KW (2009) Heavy metal and arsenic accumulating fern species as potential ecological indicators in As-contaminated abandoned mines. *Ecol. Indic.* 2009, 9, 1275–1279. [CrossRef]

Chaudhary K, Agarwal S, Khan S (2018) "Role of phytochelatins (PCs), metallothioneins (MTs), and heavy metal ATPase (HMA) genes in heavy metal tolerance," in Mycoremediation and environmental sustainability (Cham: Springer), 39 60. https://doi.org/10.1007/978-3-319-77386-5_2.

Chen QQ, Yin XB, Wang ZM, Yuan LX, Lu XQ, Li F, Liu ZK (2019) "Volatile organic selenium in atmosphere: A mini review," in Selenium research for environment and human health: Perspectives, technologies and advancements (eds.) G. Bañuelos, Z.-Q. Lin, D. Liang and X.-B. Yin (London: CRC Press), 25–26. https://doi.org/10.1201/9780429423482.

Chen Y, Hua C, Jia MR, Fu J, Liu X, Han YH, Liu Y, Rathinasabapathi B, Cao Y, Ma LQ (2017) Heterologous expression of Pteris vittata arsenite antiporter PvACR3;1 reduces arsenic accumulation in plant shoots, Environ. Sci. Technol. 51 (2017) 10387–10395.

Chen Y, Xu W, Shen H, Yan H, Xu W, He Z, Ma M (2013) Engineering arsenic tolerance and hyperaccumulation in plants for phytoremediation by a PvACR3 transgenic approach. *Environ Sci Technol* 47(16):9355–9362.

Chunduri V, Kapoor P, Kumari A, Kumar A, Sharma S, Sharma N, Kaur S, Garg M (2024) Biotechnological Advancements in Phytoremediation. Aquatic Contamination. John Wiley & Sons, Ltd, pp. 165–188. https://doi.org/10.1002/9781119989318.ch11.

Claveria RJR, Perez TR, Perez REC, Algo JLC, Robles PQ (2019) The identification of indigenous Cu and As metallophytes in the Lepanto Cu-Au Mine, Luzon, Philippines. Environ. Monit. Assess. 2019, 191, 185. [CrossRef] [PubMed]

Da Silva AA, De Oliveira JA, De Campos FV, Ribeiro C, Farnese FDS, Costa AC (2018) Phytoremediation potential of Salvinia molesta for arsenite contaminated water: Role of antioxidant enzymes. *Theor. Exp. Plant Physiol.* 2018, 30, 275–286. [CrossRef]

Danh LT, Truong P, Mammucari R, Foster N (2014) A Critical Review of the Arsenic Uptake Mechanisms and Phytoremediation Potential of Pteris Vittata. Int. J. Phytoremediation 2014, 16, 429–453. [CrossRef]

de Freitas F, Solera K, Lopes VJS, Córdova MO, Cavalheiro L, Moreno MIC, Battirola LD, de Andrade RLT (2024) Native accumulator plants with a differential mercury phytoremediation potential in a region in Southern Amazon. Environmental Science and Pollution Research (2024) 31:63120–63135 https://doi.org/10.1007/s11356-024-35407-y.

Dermont G, Bergeron M, Mercier G, Richer-Laflèche M (2008) Soil washing for metal removal: a review of physical/chemical technologies and field applications. *J. Hazard. Mater.* 152,1–31 (2008).

Dhir B (2018) Role of Ferns in Environmental Cleanup. In Current Advances in Fern Research; Fernández, H., Ed.; *Springer International Publishing: Cham, Switzerland,* 2018; pp. 517–531; ISBN 978-3-319-75103-0.

Dhir B and Srivastava S (2011) Heavy metal removal from a multi-metal solution and wastewater by Salvinia natans. *Ecol. Eng.* 2011, 37, 893–896. [CrossRef]

Dimitroula H, Syranidou E, Manousaki E, Nikolaidis NP, Karatzas GP, Kalogerakis N (2015) Mitigation measures for chromium-VI contaminated groundwater–the role of endophytic bacteria in rhizofiltration. *J. Hazard. Mater.* 281, 114–120.

Ditta A, Khalid N, Bibi S, Wang L, Sun Y, Sarwar M (2021) Applications of nanotechnology in plant nutrition: Current perspectives and future prospects. Journal of Nanobiotechnology, 19(1), 213.

Djingova R and Kuleff I (2000) "Instrumental techniques for trace analysis," in Trace Elements: Their Distribution and Effects in the Environment. J. P. Vernet, Ed., Elsevier, London, UK.

Drăghiceanu OA, Dobrescu CM, Soare LC (2014) Applications of pteridophytes in phytoremediation. *Curr. Trends Nat. Sci.* 2014, 3, 68–73.

Duan GL, Zhu YG, Tong YP, Cai C, Kneer R (2005) Characterization of arsenate reductase in the extract of roots and fronds of Chinese brake fern an arsenic hyperaccumulator. *Plant Physiol* 138(1):461–469.

Etesami H, Adban A, Farhoudi R, Khoshgoftarmanesh AH (2022) Effects of engineered nanoparticles on plant growth and nutrient uptake: A review. Environmental Science and Pollution Research, 29(2), 1125–1145.

Feki K, Tounsi S, Mrabet M, Mhadhbi H, Brini F (2021) Recent advances in physiological and molecular mechanisms of heavy metal accumulation in plants. *Environ. Sci. pollut. Res.* 28, 64967–64986. https://doi.org/10.1007/s11356-021-16805-y.

Eslava-Silva FdJ, Muñíz-Díaz de León ME, Jiménez-Estrada M (2023) *Pteridium aquilinum* (Dennstaedtiaceae), a Novel Hyperaccumulator Species of Hexavalent Chromium. *Applied Sciences*. 2023; 13(9):5621. https://doi.org/10.3390/app13095621.

Foley JA, Ramankutty N, Brauman KA, Cassidy ES, Gerber JS, Johnston M, Mueller ND, O'Connell C, Ray DK, West PC, Balzer C, Bennett EM, Carpenter SR, Hill J, Monfreda C, Polasky S, Rockström J, Sheehan J, Siebert S, Tilman D, Zaks DPM (2011) Solutions for a cultivated planet. *Nature* 478, 337–342 (2011). https://doi.org/10.1038/nature10452.

Franzosa EA, Hsu T, Sirota-Madi A, Shafquat A, Abu-Ali G, Morgan XC, Huttenhower C (2015) Sequencing and beyond: integrating molecular'omics' for microbial community profiling. *Nat Rev Microbiol* 13(6):360–372.

Garza-Lombo C, Posadas Y, Quintanar L, Gonsebatt ME, Franco R (2018) Neurotoxicity linked to dysfunctional metal ion homeostasis and xenobiotic metal exposure: Redox Signaling and Oxidative Stress. Antioxid. Redox Signal. 2018, 28, 1669–1703. [CrossRef] [PubMed]

Gerhardsson L and Kazantzis G (2015) Chapter 23: Diagnosis and treatment of metal poisoning: General aspects. In Handbook on the Toxicology of Metals, 4th ed.; Academic Press: Cambridge, MA, USA, 2015; pp. 487–505. [CrossRef]

Gill RA, Ahmar S, Ali B, Saleem MH, Khan MU, Zhou W, Liu S (2021) The role of membrane transporters in plant growth and development, and abiotic stress tolerance. *Int. J. Mol. Sci.* 22:12792. https://doi.org/10.3390/ijms222312792.

Gimeno-García E, Andreu V, Boluda R (1996) Heavy metals incidence in the application of inorganic fertilizers and pesticides to rice farming soils. *Environ. Pollut.* 1996, 92, 19–25. [CrossRef]

Gong XM, Huang DL, Liu YG, Zeng GM, Wang RZ, Wei JJ, Huang C, Xu P, Wan J, Zhang C (2018) Pyrolysis and reutilization of plant residues after phytoremediation of heavy metals contaminated sediments: for heavy metals stabilization and dye adsorption. *Bioresour. Technol.* 253, 64–71.

Gulzar ABM, Mazumder PB (2022) Helping plants to deal with heavy metal stress: The role of nanotechnology and plant growth promoting rhizobacteria in the process of phytoremediation. Environmental Science and Pollution Research, 29(27), 40319–40341.

Guo X, Wei Z, Wu Q, Li C, Qian T, Zheng WJ (2016) Chemosphere, 2016; 147:412-419.

Guzmán-Cornejo L, Pacheco L, Camargo-Ricalde SL, Ma. del Carmen González-Chávez (2023) Endorhizal fungal symbiosis in lycophytes and metal(loid)- accumulating ferns growing naturally in mine wastes in Mexico, International Journal of Phytoremediation, 25:4, 538-549, https://doi.org/10.1080/15226514.2022.2092060.

Ha NTH, Nga TTH, Minh NN, Anh BTK, Hang NTA, Duc NA, Nhuan MT, Kim KW (2019) Uptake of arsenic and heavy metals by native plants growing near Nui Phao multi-metal mine, northern Vietnam. *Appl. Geochem.* 2019, 108, 104368. [CrossRef]

Hadidi ME, Ruscheweyh HJ, Huson D (2013) Improved metagenome analysis using MEGAN5, Joint 21st Annual International Conference on Intelligent Systems for Molecular Biology (ISMB) and 12th European Conference on Computational Biology (ECCB), 2013. https://www.researchgate.net/publication/266157580_Improved_metagenome_analysis_using_MEGAN5.

Harrington ED, Arumugam M, Raes J, Bork P, Relman DA (2010) Smash-cell: a software framework for the analysis of single-cell amplified genome sequences, Bioinformatics 26 (23) (2010) 2979–2980.

Hazra M, Avishek K, Pathak G (2015) Phytoremedial Potential of Typha latifolia, Eichornia crassipes and Monochoriahastata found in Contaminated Water Bodies Across Ranchi City (India). Int. J. Phytoremediation 2015, 17, 835–840. [CrossRef]

Hoffmann TL, Kutter C, Santamaria JM (2004) Capacity of Salvinia minima Baker to Tolerate and Accumulate As and Pb. *Eng. Life Sci.* 2004, 4, 61–65. [CrossRef]

Hou J, Liu W, Wu L, Hu P, Ma T, Luo Y, Christie P (2017) Modulation of the efficiency of trace metal phytoremediation by Sedum plumbizincicola by microbial community structure and function, Plant Soil 421 (1–2) (2017) 285–299.

Hu J and Zhang M (2012) Research progress on remediation technology of heavy metal polluted water. *Light Ind. Sci. Technol.* 78–79.

Huang S-H, Bing P, Yang Z-H, Chai L-Y, Zhou L-C (2009) Chromium accumulation, microorganism population and enzyme activities in soils around chromium-containing slag heap of steel alloy factory. Transact. Nonferr. Metal. Soc. China 19, 241–248.

Indriolo E, Na G, Ellis D, Salt DE, Banks JA (2010) A vacuolar arsenite transporter necessary for arsenic tolerance in the arsenic hyperaccumulating fern Pteris vittata is missing in flowering plants. *The Plant Cell,* 22(6), 2045–2057.

Jacob JM, Karthik C, Saratale RG, Kumar SS, Prabakar D, Kadirvelu K, Pugazhendhi A (2018) Biological approaches to tackle heavy metal pollution: a survey of literature. J Environ Manag 217:56–70.

Jadia CD and Fulekar MH (1999) "Phytoremediation of heavy metals: recent techniques," *African Journal of Biotechnology*, vol. 8, no. 6, pp. 921–928. [14] L. Taiz and E. Zeiger (2002) Plant Physiology, Sinauer Associates, Sunderland, Mass, USA.

Jan M, Mir TA, Khare RK (2024) Phytoremediation. Aquatic Contamination. John Wiley & Sons, Ltd, pp. 145–164. https://doi.org/10.1002/9781119989318.ch10.

Javed MT, Tanwir K, Akram MS, Shahid M, Niazi NK, Lindberg S (2019) "Chapter 20—Phytoremediation of cadmium-polluted water/sediment by aquatic macrophytes: role of plant-induced pH changes," in Cadmium Toxicity and Tolerance in Plants, eds M. Hasanuzzaman, Prasad MNV, Fujita M () (London: Academic Press), 495–529. https://doi.org/10.1016/B978-0-12-814864-8.00020-6.

Jelusic M and Lestan D (2015) Remediation and reclamation of soils heavily contaminated with toxic metals as a substrate for greening with ornamental plants and grasses. *Chemosphere* 138, 1001e1007.

Kachenko AG, Singh B, Bhatia NP (2007) Heavy metal tolerance in common ferns species. *Australian Journal of Botany*, 300: 207-219.

Kamachi H, Komori I, Tamura H, Sawa Y, Karahara I, Honma Y, Wada N, Kawabata T, Matsuda K, Ikeno S, Noguchi M, Inoue H (2005) Lead tolerance and accumulation in the gametophytes of the fern Athyrium yokoscense. *J. Plant Res.* 2005, 118, 137–145. Digital Object Identifier https://doi.org/10.1007/s10265-005-0202-x [CrossRef]

Khalid S, Shahid M, Niazi NK, Murtaza B, Bibi I, Dumat C (2017) A comparison of technologies for remediation of heavy metal contaminated soils. *J. Geochem. Explor.* 182, 247–268.

Kidd P, Barceló J, Bernal MP, Navari-Izzo F, Poschenrieder C, Shilev S, Clemente R, Monterroso C (2009) Trace Element Behaviour at the Root–Soil Interface: Implications in Phytoremediation. Environ. Exp. Bot. 2009, 67, 243–259. [CrossRef]

Klopper RR (2011) The use of ferns in phytoremediation. Pteridoforum, 96, 1–5.

Ko WH, Ko SS, Chen M (2004) Origin and control of fern weeds in orchid production in greenhouses in Hawaii. Crop Prot 24(5):487-490.

Kohda YH-T, Qian Z, Chien M-F, Miyauchi K, Endo G, Suzui N, Yin Y-G, Kawachi N, Ikeda H, Watabe H, Kikunaga H, Kitajima N, Inoue C (2021) New Evidence of Arsenic Translocation and Accumulation in Pteris Vittata from Real-Time Imaging Using Positron-Emitting 74As Tracer. Sci. Rep. 2021, 11, 12149. [CrossRef] [PubMed]

Kramer U (2018) The plants that suck up metal. German Research, 40(3), 18–23.

Kuivenhoven M and Mason K (2021) Arsenic toxicity. In StatPearls; StatPearls Publishing: Treasure Island, FL, USA, 2021.

Kumar A, Deka B, Mahto TK (2021a) Nanomaterials for environmental remediation: A review. Journal of Environmental Chemical Engineering, 9(2), 105314.

Kumar S, Nehra M, Dilbaghi N, Marrazza G, Hassan AA (2021b) Nanotechnology for phytoremediation: Recent advances and future perspectives. Environmental Science and Pollution Research, 28(33), 44648–44670.

Kumar V, Almomin S, Al-Aqeel H, Al-Salameen F, Nair S, Shajan A (2018) Metagenomic analysis of rhizosphere microflora of oil-contaminated soil planted with barley and alfalfa, PLoS One 13 (8) (2018) e0202127. https://doi.org/10.1371/journal.pone.0202127.

Kumar V, Kumar P, Singh J, Kumar P (2020) Potential of Water Fern (Azolla pinnata R.Br.) in Phytoremediation of Integrated Industrial Effluent of SIIDCUL, Haridwar, India: Removal of Physicochemical and Heavy Metal Pollutants. *Int. J. Phytoremediation* 2020, 22, 392–403. [CrossRef]

Kumar V, Sharma A, Kaur P, Singh GPS, Bali AS, Bhardwaj R, Thukral AK, Cerda A (2019) Pollution assessment of heavy metals in soils of India and ecological risk assessment: a state-of-the-art. Chemosphere 216: 449–462. https://doi.org/10.1016/j.chemosphere.2018.10.066.

Lajayer BA, Moghadam NK, Maghsoodi MR, Ghorbanpour M, Kariman K (2019) Phytoextraction of heavy metals from contaminated soil, water and atmosphere using ornamental plants: mechanisms and efficiency improvement strategies. *Environ. Sci. Pollut. Res.* 26, 8468–8484.

Lee J, Kim JR, Jeong S, Cho J, Kim JY (2017) Long-term performance of anaerobic digestion for crop residues containing heavy metals and response of microbial communities. *Waste Manag.* 59, 498–507.

Leslie EM, Haimeur A, Waalkes MP (2004) Arsenic transport by the human multidrug resistance protein 1 (MRP1/ABCC1) Evidence that a tri-glutathione conjugate is required. *J Biol Chem* 279(31):32700–32708.

Li C, Yang G, Liu Z, Cai J (2022) "Overview of phytoremediation technology for heavy metal contaminated soil," in E3S Web of Conferences (EDP Sciences) ed R. Prasad (Cham: Springer), 01006. https://doi.org/10.1051/e3sconf/202235001006.

Li N, Wang J, Song WY (2016) Arsenic uptake and translocation in plants. *Plant Cell Physiol* 57(1):4–13.

Liu ZC, Chen BN, Wang LA, Urbanovich O, Nagorskaya L, Li X, Tang L (2020a) A review on phytoremediation of mercury contaminated soils. J. Hazard. Mater. 400, 123138.

Liu ZC, Wang LA, Xiao HY, Guo XW, Urbanovich O, Nagorskaya L, Li X (2020b) A review on control factors of pyrolysis technology for plants containing heavy metals. *Ecotoxicol. Environ. Saf.* 191, 110181.

Loría KC, Emiliani J, Bergara CD, Herrero MS, Salvatierra LM, Pérez LM (2019) Effect of daily exposure to Pb-contaminated water on Salvinia biloba physiology and phytoremediation performance. *Aquat. Toxicol.* 2019, 210, 158–166. [CrossRef]

Lowe R, Shirley N, Bleackley M, Dolan S, Shafee T (2017) Transcriptomics technologies. *PLoS Comput Biol* 13(5):1.

Maddheshiyaa R, Dhiman S, Sahu DM, Gupta H (2024) Role of phytoremediation and metagenomics in healing the wounded (polluted) planet: Ecofriendly and sustainable approach. https://doi.org/10.1016/B978-0-443-13659-7.00009-6.

Manzoor M, Abid R, Rathinasabapathi B, De Oliveira LM, da Silva E, Deng F, Rensing C, Arshad M, Gul I, Xiang P, Ma LQ (2019) Metal tolerance of arsenic-resistant bacteria and their ability to promote plant growth of Pteris vittata in Pb-contaminated soil. *Sci Total Environ.* 2019 Apr;660:18–24.

Markowitz VM, Chen IMA, Chu K, Szeto E, Palaniappan K, Grechkin Y, Kyrpides NC (2012) IMG/M: the integrated metagenome data management and comparative analysis system, Nucleic Acids Res. 40 (D1) (2012) 123–129.

Mathur J and Chauhan P (2020) "Mechanism of toxic metal uptake and transport in plants," in Sustainable solutions for elemental deficiency and excess in crop plants K. P. Mishra, K. Tandon and S. Srivastava (Singapore: *Springer*), 335–349.

Mhlongo MI, Piater LA, Madala NE, Labuschagne N, Dubery IA (2018) The chemistry of plant–microbe interactions in the rhizosphere and the potential for metabolomics to reveal signaling related to defense priming and induced systemic resistance. *Front Plant Sci* 9:112.

Milicevic T, Anicic Urosevic M, Relic D, Jovanovic G, Nikolic D, Vergel K, Popovic A (2021) Environmental pollution influence to soil-plant-air system in organic vineyard: bioavailability, environmental, and health risk assessment. *Environ Sci Pollut Res Int* 28(3):3361–3374. https://doi.org/10.1007/s11356-020-10649-8.

Miner DPJ (2016) Process, 2016; 9:239-245.

Mingho Y (2005) ISBN 1-56670-670-2, Boca Raton, USA.

Misra SG and Mani D (1991) Soil pollution. Ashish Publishing House, Punjabi Bagh.

Mosa KA, Saadoun I, Kumar K, Helmy M, Dhankher OP (2016) Potential biotechnological strategies for the cleanup of heavy metals and metalloids. *Front. Plant Sci.* 7:303. https://doi.org/10.3389/fpls.2016.00303 [PMC free article] [PubMed] [CrossRef] [Google Scholar]

Mthembu MS, Odinga CA, Bux F, Swalaha FM (2020) Constructed wetlands: an eco-sustainable phytotechnology for degradation and detoxification of industrial wastewaters. In G. Saxena,

& R. N. Bharagava (Eds.), Bioremediation of industrial waste for environmental safety (pp. 291–312). Singapore: *Springer* Nature Singapore Private Ltd.
Mueller ND, Gerber JS, Johnston M, Ray DK, Ramankutty N, Foley JA (2012) Closing yield gaps through nutrient and water management. *Nature* 490, 254–257 (2012). https://doi.org/10.1038/nature11420.
Mukhopadhyay R, Bhattacharjee H, Rosen BP (2014) Aquaglyceroporins: Generalized metalloid channels. Biochimica et Biophysica Acta (BBA) – General Subjects 2014;1840:1583–1591 [PubMed: 24291688].
Mukhtar AA, Abdullahi IL (2017) Heavy metals phytoremediation using Typha domingensis flourishing in an industrial effluent drainage in Kano, Nigeria. *Bayero J. Pure Appl. Sci.* 2017, 10, 277–280. [CrossRef]
Nriagu JO (1996) A history of global metal pollution. *Science* 272, 223–224 (1996).
Oberoi M and Khanna V (2018) Rhizoremediation—plant microbe interactions in the removal of pollutants. *Int J Curr Microbiol App Sci* 7(1):2280–2287. https://doi.org/10.20546/ijcmas.2018.701.276.
Ojuederie OB and Babalola OO (2017) Microbial and plant-assisted bioremediation of heavy metal polluted environments: a review. *Int J Environ Res Public Health* 14(12):1504.
Onder S and Dursun S (2006) Air borne heavy metal pollution of Cedrus libani (A. Rich.) in the city centre of Konya (Turkey). *Atmos. Environ.* 2006, 40, 1122–1133. [CrossRef]
Onyia PC, Ozoko DC, Ifediegwu SI (2021) Phytoremediation of arsenic-contaminated soils by arsenic hyperaccumulating plants in selected areas of Enugu State, Southeastern, Nigeria, *Geology, Ecology, and Landscapes*, 5:4, 308–319, https://doi.org/10.1080/24749508.2020.1809058.
Oubohssaine M and Dahmani I (2023) Phytoremediation: Harnessing plant power and innovative technologies for effective soil remediation. *Plant Stress journal.* https://doi.org/10.1016/j.stress.2024.100578.
Padmavathiamma PK and Li LY (2007) Phytoremediation Technology: Hyper-Accumulation Metals in Plants. *Water Air Soil Pollut.* 2007, 184, 105–126. [CrossRef]
Page CN (1988) Ferns: their habitats in British and Irish landscape. New Naturalist series. London: Collins. Pp. 1–45.
Pandey G (2018) Prospects of Nanobioremediation in Environmental Cleanup. Orient J Chem 34(6):2838–2850.
Pandey V and Rattan VK (2021) Assessment of monthly variation in heavy metal characteristics of Electroplating industrial untreated wastewater at selected sites of Chandigarh. *J. Appl. Nat. Sci.* 2021, 13, 250–257. [CrossRef]
Pandey VC (2012) Phytoremediation of heavy metals from fly ash pond by Azolla caroliniana. Ecotoxicol. Environ. Saf. 2012, 82, 8–12. [CrossRef] [PubMed]
Pandey VC (2020) Pteridophytes in phytoremediation Ashish Praveen. Environ Geochem Health (2020) 42:2399–2411 https://doi.org/10.1007/s10653-019-00425-0.
Pang YL, Quek YY, Lim S, Shuit SH (2023) Review on phytoremediation potential of floating aquatic plants for heavy metals: a promising approach. Sustainability. 15, 1290. https://doi.org/10.3390/su15021290.
Pathak HK, Chauhan PK, Seth CS, Dubey G, Upadhyay SK (2024) Mechanistic and future prospects in rhizospheric engineering for agricultural contaminants removal, soil health restoration, and management of climate change stress. Sci. Total Environ. 927, 172116 https://doi.org/10.1016/j.scitotenv.2024.172116.
Patten CL and Glick BR (2020) Role of Pseudomonas putida indoleacetic acid in development of the host plant root system. *Appl. Environ. Microbiol.* 68(8), 3795–3801 (2020).
Pilon-Smits EAH, de Souza MP, Hong G et al (1999) Selenium volatilization and accumulation by twenty aquatic plant species. J Environ Qual Madison 28:1011. https://doi.org/10.2134/jeq1999.00472425002800030035x.

Pourret O and Hursthouse A (2019) It's time to replace the term "heavy metals" with "potentially toxic elements" when reporting environmental research. *Int. J. Environ. Res. Public Health* 2019, 16, 4446. [CrossRef] [PubMed]

Prabhu SG, Srinikethan G, Hegde S (2016) Potential of Pteridophytes in Heavy Metal Phytoremediation. *Int. J. Res. Eng. Technol.* 2016, 5, 1–9.

Prasetia H, Sakakibara M, Takehara A, Sueoka Y (2017) Heavy metals accumulation by Athyrium yokoscence in a mine area, Southwestern Japan. IOP Conf. Ser. *Earth Environ. Sci.* 2017, 71, 12025. [CrossRef]

Praveen A and Pandey VC (2020) Pteridophytes in phytoremediation. *Environ. Geochem. Health* 2020, 42, 2399–2411. [CrossRef]

Prieto MJ, Acevedo SOA, Prieto GF (2018) Phytoremediation of soils contaminated with heavy metals. *International Journal of Biodiversity*, 2(4), 362–376.

Rai PK (2008) Technical Note: Phytoremediation of Hg and Cd from Industrial Effluents using an Aquatic Free Floating Macrophyte Azolla pinnata. *Int. J. Phytoremediat.* 2008, 10, 430–439. [CrossRef]

Rajkumar M, Sandhya S, Prasad MNV, Freitas H (2012) Perspectives of plant-associated microbes in heavy metal phytoremediation. *Biotechnol. Adv.* 30, 1562–1574. https://doi.org/10.1016/j.biotechadv.2012.04.011 (2012).

Rane NR, Tapase S, Kanojia A, Watharkar A, Salama E-S, Jang M, Yadav KK, Amin MA, Cabral-Pinto MM, Jadhav JP (2022) Molecular insights into plant–microbe interactions for sustainable remediation of contaminated environment. Bioresour. *Technol.* 344, 126246.

Rascio N and Navari-Izzo F (2011) Heavy metal hyperaccumulating plants: how and why do they do it? And what makes them so interesting? *Plant Science,* 180, 169–181.

Reichl C, Schatz M, Zsak G (2018) World Mining Data. http://www.wmc.org.pl/?q=node/49/(International Organizing Committee for the World Mining Congress, 2018).

Remigio AC, Edraki M, Baker AJM, van der Ent A (2021) Root responses to localised soil arsenic enrichment in the fern Pityrogramma calomelanos var. austroamericana grown in rhizoboxes. Plant Physiol. Biochem. 2021, 164, 147–159. [CrossRef]

Rizwan M, Ali S, Rehman MZ, Adrees M, Arshad M, Qayyum MF, Ok YS (2018) Mechanisms of silicon-mediated alleviation of heavy metal toxicity in plants: A review. Ecotoxicology and Environmental Safety, 147, 881–896.

Rizwan Md, Singh M, Mitra CK, Morve RK (2014) Ecofriendly application of nanomaterials: nanobioremediation. *J Nanomater* 2014:1–7.

Roccotiello E, Manfredi A, Drava G, Minganti V, Mariotti MG, Berta G, Cornara L (2010) Zinc tolerance and accumulation in the ferns Polypodium cambricum L. and Pteris vittata L. *Ecotoxicology and Environmental Safety* 73(6), 1264–1271.

Ronzan M, Zanella L, Fattorini L, Della Rovere F, Urgast D, Cantamessa S, Nigro A, Barbieri M, Sanità di Toppi L, Berta G, Feldmannb J, Altamuraa MM, Falasca G (2017) The morphogenic responses and phytochelatin complexes induced by arsenic in Pteris vittata change in the presence of cadmium. *Environ. Exp. Bot.* 2017, 133, 176–187. https://doi.org/10.1016/j.envexpbot.2016.10.011. [CrossRef]

Rout Y, Swain SS, Ghana M, Dash D, Nayak S (2024) Perspectives of pteridophytes microbiome for bioremediation in agricultural applications. Open Life Sciences 2024; 19: 20220870. https://doi.org/10.1515/biol-2022-0870.

Sahu SS, Kumar A, Prasad MNV, Maiti SK (2024) Chapter 4 – Aromatic, medicinal, and energy plantations on metalliferous/contaminated soil—Bioremediation and bioeconomy. In: Prasad, M.N.V. (Ed.), Bioremediation and Bioeconomy, Second Edition. Elsevier, pp. 81–98. https://doi.org/10.1016/B978-0-443-16120-9.00006-6.

Saini S and Dhania G (2020) Cadmium as an environmental pollutant: ecotoxicological effects, health hazards, and bioremediation approaches for its detoxification from contaminated sites. In R. N. Bharagava, & G. Saxena (Eds.), Bioremediation of industrial waste for environmental safety (pp. 357–387). *Singapore: Springer Nature Singapore Private Ltd.*

Sakai Y, Watanabe T, Wasaki J, Senoura T, Shinano T, Osaki M (2010) Influence of arsenic stress on synthesis and localization of low-molecular-weight thiols in Pteris vittata. *Environ Pollut* 158:3663–3669.

Sakakibara M, Watanabe A, Inouhe M, Sano S, Kaise T (2010) Phytoextraction And Phytovolatilization Of Arsenic From As-Contaminated Soils By Pteris vittata. Proceedings of the Annual International Conference on Soils, Sediments, Water and Energy.

Salem HM, Eweida EA, Farag A (2000) In ICEHM 2000; 1:542–556.

Samal AC, Bhattacharya P, Maity JP, Mallick A, Santra SC (2023) Phytoremediation of arsenic—contaminated soil and water through some hyperaccumulator pteridophytic plants Arsenic in the Environment: Bridging Science to Practice for Sustainable Development—van der Wal, Ahmad, Petrusevski, Weijma, Savic, van der Wens, Beerendonk, Bhattacharya, Bundschuh & Naidu (Eds) © 2023 the Author(s), ISBN 978-1-032-32928-4.https://doi.org/10.1201/9781003317395-159.

Saravanan A, Jeevanantham S, Narayanan VA, Kumar PS, Yaashikaa PR, Muthu CMM (2019) Rhizoremediation—A promising tool for the removal of soil contaminants: a review. *J Environ Chem Eng*. https://doi.org/10.1016/j.jece.2019.103543.

Sarwar N, Imran M, Shaheen MR, Ishaque W, Kamran MA, Matloob A, Rehim A, Hussain S (2017) Phytoremediation strategies for soils contaminated with heavy metals: modifications and future perspectives. *Chemosphere* 171, 710–721. https://doi.org/10.1016/j.chemosphere.2016.12.116.

Sathishkumar M, Binupriya AR, Baik SH, Yun SE (2008) Biodegradation of crude oil by individual bacterial strains and a mixed bacterial consortium isolated from hydrocarbon contaminated areas. Clean: Soil, Air, Water 36(1):92–96.

Sela M, Garty J, Telor E (1989) The accumulation and effect of heavy metals on the water fern. Azolla filliculoides. New Phytologist, 112: 7–12.

Sessitsch A, Kuffner M, Kidd P, Vangronsveld J, Wenzel WW, Fall mann K, Puschenreiter M (2013) The role of plant-associated bacteria in the mobilization and phytoextraction of trace elements in contaminated soils. Soil Biol Biochem 60:182–194.

Shah D, Kamili A, Sajjad N, Tyub S, Majeed G, Hafiz S, Noor W, Yaqoob S, Maqbool I (2024) Phytoremediation of pesticides and heavy metals in contaminated environs. Aquatic Contamination. John Wiley & Sons, Ltd, pp. 189–206. https://doi.org/10.1002/9781119989318.ch12.

Shahid M, Dumat C, Khalid S, Schreck E, Xiong T, Niazi NK, Sohail MI (2021) nanoparticles role in phytoremediation of metal-contaminated soils: A review. Environmental Science and Pollution Research, 28(10), 11483–11504.

Sharma JK, Kumar N, Singh NP, Santal AR (2023) Phytoremediation technologies and their mechanism for removal of heavy metal from contaminated soil: An approach for a sustainable environment. Front. Plant Sci. 14.

Sharma K, Devi P, Dey SR, Kumar P (2024) Chapter 31 – Mercury phytovolatilization: an overview of the mechanism and mitigation. In: Srivastav, A. L., Grewal, A.S., Markandeya Pham, T.D. (Eds.), Role of Green Chemistry in Ecosystem Restoration to Achieve Environmental Sustainability, Advances in Pollution Research. Elsevier, pp. 325–331. https://doi.org/10.1016/B978-0-443-15291-7.00031-6.

Sharma P, Pandey AK, Udayan A, Kumar S (2021) Role of microbial community and metal-binding proteins in phytoremediation of heavy metals from industrial wastewater. Bioresource Technol. 326, 124750. https://doi.org/10.1016/j.biortech.2021.124750.

Sharma S, Singh B, Manchanda VK (2015) Phytoremediation: role of terrestrial plants and aquatic macrophytes in the remediation of radionuclides and heavy metal contaminated soil and water. Environ. Sci. pollut. Res. 22, 946–962. https://doi.org/10.1007/s11356-014-3635-8.

Shehu AH, Gumi AM, Ismail N (2023) Phytoremediation of heavy metals: reaction mechanisms and selected efficient technologies of heavy metal contamination. In: Singh, R.P., Singh, P., Srivastava, A. (Eds.), Heavy Metal Toxicity: Environmental Concerns, Remediation and Opportunities. *Springer Nature*, Singapore, pp. 245–269. https://doi.org/10.1007/978-981-99-0397-9_12.

Silva S, Dias MC, Silva AM (2022) Titanium and zinc based nanomaterials in agriculture: A promising approach to deal with (a) biotic stresses? Toxics, 10(4), 172. https://doi.org/10.3390/toxics10040172.

Singh R, Jha AB, Misra AN, Sharma P (2019) Adaption mechanisms in plants under heavy metal stress conditions during phytoremediation. In Phytomanagement of polluted sites. The Netherlands: Elsevier Inc. https://doi.org/10.1016/B978-0-12-813912-7.00013-2.

Singh S, Parihar P, Singh R, Singh VP, Prasad SM (2016) Heavy metal tolerance in plants: role of transcriptomics proteomics metabolomics and ionomics. Front Plant Sci 6:1–36.

Singh S, Shrivastava A, Barla A, Bose S (2015) Isolation of arsenic-resistant bacteria from Bengal delta sediments and their efficacy in arsenic removal from soil in association with Pteris vittata. *Geomicrobiol. J.* 32(8), 712–723. https://doi.org/10.1080/01490451.2015.1004141 (2015).

Smolders AJP, Lock RAC, Velde GVD, Hoyos RIM, Roelofs JGM (2003) Effects of mining activities on heavy metal concentrations in water, sediment, and macroinvertebrates in different reaches of the Pilcomayo River, South America. *Arch. Environ. Contam. Toxicol.* 2003, 44, 314–323. [CrossRef] [PubMed]

Srivastava M, Ma LQ, Singh N, Singh S (2005) Antioxidant responses of hyper-accumulator and sensitive fern species to arsenic. *J Exp Bot* 56(415):1335–1342.

Srivastava S, Singh M, Paul AK (2016) Arsenic bioremediation and bioactive potential of endophytic bacterium Bacillus pumilus isolated from Pteris vittata L. *IJABAR* 7(1):77–92.

Sudha Sajeev, P. T. Roshni, Rachel Carmelita Mathias, Shaiesh Morajkar, Smruthi Prabhu, and Smitha Hegde (2022) Pteridophytes: Effective Agents of Phytoremediation. J. Marimuthu et al. (eds.), Ferns, https://doi.org/10.1007/978-981-16-6170-9_27.

Sullivan TS, Gadd GM (2019) Metal bioavailability and the soil micro biome. Adv Agron 155:79–120.

Sun L, Wu Q, Liao K, Yu P, Cui Q, Rui Q, Wang D *Chemosphere*, 2016; 144:2392–2400.

Sundaram S, Rathinasabapathi B, Ma LQ, Rosen BP (2008) An arsenate-activated glutaredoxin from the arsenic hyperaccumulator fern Pteris vittata L. Regulates Intracellular Arsenite. *J Biol Chem* 283:6095–6101.

Supaporn P, Savent P, Nimit S (2008) Heavy Metal Accumulation in Soil and Some Fern Species at Phu Soi Dao National Park, Phitsanulok Province, Thailand. *NU Science Journal,* 5 (2): 151—164.

Sylvain B, Mikael MH, Florie M, Emmanuel J, Marilyne S, Sylvain B, Domenico M (2016) Phytostabilization of As, Sb and Pb by two willow species (S. viminalis and S. purpurea) on former mine technosols. Catena, 136, 44–52.

Talebi M, Tabatabaei B ES and Akbarzadeh H (2019) Hyperaccumulation of Cu, Zn, Ni, and Cd in Azolla species inducing expression of methallothionein and phytochelatin synthase genes. *Chemosphere* 230, 488–497. https://doi.org/10.1016/j.chemosphere.2019.05.098.

Tiwari J, Ankit, Sweta, Kumar S, Korstad J, Bauddh K (2019) Chapter 5 – Ecorestoration of Polluted Aquatic Ecosystems Through Rhizofiltration. In Phytomanagement of Polluted Sites, Pandey, V.C., Bauddh, K., Eds. Elsevier: 2019, 179–201, https://doi.org/10.1016/B978-0-12-813912-7.00005-3.

Tiwari S, Sarangi BK, Anusha P, Pandey RA (2013) Metal Hyperaccumulating Ferns: Progress and Future Prospects. In Recent Advances towards Improved Phytoremediation of Heavy Metal Pollution; Leung, D.W.M., Ed.; Bentham Science Publishers: Sharjah, United Arab Emirates, 2013; ISBN 978-1-60805-787-0.

Tóth G, Hermann T, Da Silva MR, Montanarella L (2016) Heavy metals in agricultural soils of the European Union with implications for food safety. *Environ. Pollut.* 88, 299–309. https://doi.org/10.1016/j.envint.2015.12.017.

Tu S, Ma L, Luongo T (2004) Root exudates and arsenic accumulation in arsenic hyperaccumulating Pteris vittata and non-hyper accumulating Nephrolepis exaltata. *Plant Soil* 258, 9–19. https://doi.org/10.1023/B:PLSO.0000016499.95722.16 (2004).

Ucaroglu S and Talini I (2012) *J. Environ. Manag.,* 2012; 105:131–137.

Vacheron J, Desbrosses G, Bouffaud ML, Touraine B, Moënne-Loccoz Y, Muller D, Prigent-Combaret C (2013) Plant growth-promoting rhizobacteria and root system functioning. *Front Plant Sci* 4(356):1–19.

Van TK, Kang Y, Fukui T, Sakurai K, Iwasaki K, Aikawa Y, Phuong NM (2006) Arsenic and heavy metal accumulation by Athyrium yokoscense from contaminated soils. Soil Sci. *Plant Nutr.* 2006, 52, 701–710. [CrossRef]

Vincent WF, Kumagai M, Couture RM (2024) Chapter 27 – Sediments and Microbiomes. In: Jones, I.D., Smol, J.P. (Eds.), Wetzel's Limnology, Fourth Edition. Academic Press, San Diego, pp. 893–937. https://doi.org/10.1016/B978-0-12-822701-5.00027-6.

Visioli G, D'Egidio S, Sanangelantoni AM (2015) The bacterial rhizobiome of hyperaccumulators: future perspectives based on omics analysis and advanced microscopy. *Front Plant Sci* 5:752.

Vogt T (1942) Geochemical and geobotanical ore prospecting III. Selsk, 15: 21–24.

Wang L, Hou D, Shen Z, Zhu J, Jia X, Ok YS, Tack F, Rinklebe J (2020b) Field trials of phytomining and phytoremediation: A critical review of influencing factors and effects of additives. Crit. Rev. *Environ. Sci. Technol.* 50, 2724–2774. https://doi.org/10.1080/10643389.2019.1705724.

Wang Z, Wang H, Wang H, Li Q, Li Y (2020) Effect of soil washing on heavy metal removal and soil quality: a two-sided coin. Ecotoxicol Environ Saf 203:110981. https://doi.org/10.1016/j.ecoenv.2020.110981.

Watanabe T, Broadley MR, Jansen S, White PJ, Takada J, Satake K, Takamatsu T, Tuah SJ, Osaki M (2007) Evolutionary control of leaf element composition in plants. New Phytol 174:516 523.

Wuana RA and Okieimen FE (2021). Nanoparticles in remediation technologies for soil and groundwater pollution. Journal of Environmental Management, 285, 112083.

Xie QE, Yan XL, Liao XY, Li X (2009) The arsenic hyperaccumulator fern Pteris vittata L. Environ Sci Technol 43(22):8488–8495.

Yan H, Gao Y, Wu L, Wang L, Zhang T, Dai C, He Z (2019) Potential use of the Pteris vittata arsenic hyperaccumulation-regulation network for phytoremediation. *J Hazard Mater* 368:386–396.

Yang C, Han N, Inoue C, Yang YL, Nojiri H, Ho YN, Chien MF (2022). Rhizospheric plant-microbe synergistic interactions achieve efficient arsenic phytoextraction by Pteris vittata. *J. Hazard. Mater.* 434:128870. https://doi.org/10.1016/j.jhazmat.2022.128870.

Yang C, Ho YN, Makita R, Inoue C, Chien MF (2019) A multifunctional Rhizobacterial strain with wide application in different ferns facilitates arsenic phytoremediation. Sci. Total Environ. 712, 134504. https://doi.org/10.1016/j.scitotenv.2019.134504 (2019).

Yang Q, Tu S, Wang G, Liao X, Yan X (2012) Effectiveness of applying arsenate reducing bacteria to enhance arsenic removal from polluted soils by Pteris vittata L. *Int. J. Phytoremed.* 14, 89–99. https://doi.org/10.1080/15226510903567471 (2012).

Yang Q, Yang C, Yu H, Zhao Z, Bai Z (2021) The addition of degradable chelating agents enhances maize phytoremediation efficiency in cd-contaminated soils. Chemosphere 269, 129373. https://doi.org/10.1016/j.chemosphere.2020.129373.

Yang Y, Liang Y, Han X, Chiu T-Y, Ghosh A, Chen H, Tang M (2016) The roles of arbuscular mycorrhizal fungi (AMF) in phytoremediation and tree-herb interactions in Pb contaminated soil. *Sci. Rep.* 6, 1–14. https://doi.org/10.1038/srep20469.

Yanitch A, Kadri H, Frenette-Dussault C, Joly S, Pitre FE, Labrecque M (2020) A four-year phytoremediation trial to decontaminate soil polluted by wood preservatives: phytoextraction of arsenic, chromium, copper, dioxins and furans. *Int. J. Phytoremediation* 22, 1505–1514. https://doi.org/10.1080/15226514.2020.1785387.

Yao Z, Li J, Xie H, Yu (2012) C Procedia Environ. Sci., 2012; 16:722–729.

Yin Y, Wang X, Hu, Y, Li F, Cheng H (2023) Soil bacterial community structure in the habitats with different levels of heavy metal pollution at an abandoned polymetallic mine. J. Hazard. Mater. 442, 130063 https://doi.org/10.1016/j.jhazmat.2022.130063.

Yizhu L, Imtiaz M, Ditta A, Rizwan MS, Ashraf M, Mehmood S, Ijaz R (2020) Response of growth, antioxidant enzymes and root exudates production towards As stress in Pteris vittata and in Astragalus sinicus colonized by arbuscular mycorrhizal fungi. Environ Sci Pollut Res 27:2340–2352.

Zhan J, Li T, Yu H, Zhang X, and Zhao L (2016) The influence of humic substance on Cd accumulation of phytostabilizer Athyrium wardii (Hook.) grown in Cd-contaminated soils. Environ. Sci. pollut. Res. 23, 18524–18532. https://doi.org/10.1007/s11356-016-7037-y.

Zhang CX, Liu S, Huang BC, Su YI (2010) Geophys. *J. Int.*, 2010; 181:1381–1394. 10.

Zhao FJ, Dunham SJ, Mcgrath SP (2002) Arsenic hyperaccumulation by different fern species. New Phytologist, 156: 27–31.

Zhao L, Li T, Zhang X, Chen GD, Zheng ZC, Yu HY (2016) Pb uptake and phytostabilization potential of the mining ecotype of Athyrium wardii (Hook.) grown in Pb contaminated soil. CLEAN–Soil Air Water 2016, 44, 1184–1190. [CrossRef]

Zhu X, Kuang Y, Xi D, Li J, Wang F (2013) Absorption of hazardous pollutants by a medicinal fern Blechnum orientale L. *BioMed Research International*, 2013, 192986.

Zulfiqar U, Haider FU, Maqsood MF, Mohy-Ud-Din W, Shabaan M, Ahmad M, Kaleem M, Ishfaq M, Aslam Z, Shahzad B (2023) Recent advances in microbial assisted remediation of cadmium-contaminated Soil. Plants 12, 3147. https://doi.org/10.3390/plants12173147.

Microbes in Tackling Recalcitrant Xenobiotics

Madhumita G. Chaki, Samarpita Adhikary, and Ayon Pal

Abstract Contaminants of emerging concern and novel synthetic compounds have accumulated significantly in the environment. Their recalcitrance and outstanding stability have disrupted human health and environments. Hyperpigmentation reduction, exfoliation of dead cells, hair damage, and itchy scalp are reduced by chemical constituents such as kojic acid, niacinamide, glycolic acid, mandelic acid, sulfates, glycol, ammonium chloride, and sodium benzoate present in personal care products. Also, dependency on non-steroidal anti-inflammatory medicines and other lifesaving drugs has become imperative currently. This chapter primarily examines bacteria-mediated bioremediation and the metabolic degradation pathways of various xenobiotics, including phthalate-based chemical plasticizers, skin-whitening agents, surfactants, sunscreens, and non-steroid-based anti-inflammatory drugs. This chapter also addresses the prospect of omics-based techniques in identifying specific upregulated genes and gene clusters associated with xenobiotic degradation.

Keywords Bacteria-mediated xenobiotic degradation · Personal care products · Pharmaceuticals · NSAIDs degradation · Xenobiotics

1 Introduction

The modern era of rapid industrialization and swift urban growth has introduced several dangerous contaminants into the natural ecosystem. Unparalleled developments have flourished in the medical and cosmetics industries rapidly in the post-industrial age, resulting in the profuse use of synthetic medications and toiletries,

M. G. Chaki · S. Adhikary · A. Pal (✉)
Microbiology and Computational Biology Laboratory, Department of Botany, Raiganj University, Raiganj, West Bengal, India

S. Roy, V. Mandal (eds.), *Plant-Microbe Interaction under Xenobiotic Exposure*,
https://doi.org/10.1007/978-981-96-8260-7_19

thereby contaminating soil, sediment, as well as surface water (Nguyen et al. 2019). Xenobiotics are recalcitrant novel synthetic compounds with long-term persistence ability in ecosystems (Marghade et al. 2021). The practice of routinely utilizing synthetic compounds, like pesticides, nitroaromatics, personal care products (PCPs), pharmaceutically active compounds (PhACs), phenolics, and polycyclic aromatic hydrocarbons (PAHs) (Miglani et al. 2022; Mishra et al. 2021), has expanded due to many factors, including technological advancements, enormous population expansion, and improved access to medication. Apart from these synthetic compounds, the International Nomenclature of Cosmetics Ingredients (INCI) lists over a thousand natural components, primarily derived from herbs, of which 27% are deemed detrimental to people in addition to their surroundings (Bucheli et al. 2018; Klaschka 2015). Some of the naturally occurring substances, like terpenes, terpenoids, cyanogenic glucosides, quinones, strychnine, atropine, aconitine, phytoestrogens, sterols, isoflavones, and peroxides, are found in significant levels in cosmetics, such as soaps, shampoos, creams, and lotions, and could be perilous to both humankind and the aquatic environment (Bucheli et al. 2018). The pervasive nature of xenobiotics is brought about by the presence of chemically stable functional groups along with endurance to the actions of microbial enzymes, thereby providing durability and resilience (Marghade et al. 2021). Moreover, the high molecular weight and branching of straight chains, phenolic ring structure, enhanced stability, and insolubility in water are the additional causes of their non-degradability and endurance (Phale et al. 2019).

Personal care goods encompass all items required to maintain health and hygiene, predominantly chemical xenobiotics in significant proportions (Piwowarska and Kiedrzyńska 2022). It includes niacinamide, glycolic acid, mandelic acid, and kojic acid, a potent antioxidant widely used in cosmetics such as serums, creams, and body lotions to reduce hyperpigmentation. It lightens and brightens new skin cells that appear after the exfoliation of dead cells (Nowak-Lange et al. 2022). Hair cleansers, like shampoos and conditioners, include sodium dodecyl sulfate (SDS), citronellol, benzyl salicylate, cetrimonium chloride, keratin, hexyl cinnamal, and disodium EDTA. Hair conditioners contain positively charged quaternary ammonium species, such as behentrimonium chloride, as the chief ingredient (Onuche et al. 2016).

One well-known class of contaminants of emerging concern (CECs) critical to human health maintenance and sustainability is pharmaceuticals, which guarantee a better quality of life and alleviate sickness (Caracciolo et al. 2015; Fernandes et al. 2021). Different drugs have distinct structural features and physicochemical qualities, such as lipophilicity, photosensitivity, and biodegradability (de Oliveira et al. 2020), disrupting ecosystems by bioaccumulating within the food chain at each trophic level and altering local microbial populations. A study found that short-term exposure to a combination of non-steroidal anti-inflammatory drugs (NSAIDs) augmented oxidative stress, thus causing genetic damage in zebra mussels (Parolini and Binelli 2012). Thus, in addition to monitoring and managing CEC emissions, it is essential to eradicate them from the pretentious areas (Fernandes et al. 2021). Due to their shared molecular targets, several medicines administered regularly and concurrently and released into the environment may interact and influence non-target species synergistically (Calisto and Esteves 2009; Fernandes et al. 2021).

Safe elimination of hazardous pharmaceutical contaminants from the environment can be achieved by microbe-mediated degradation (Karishma et al. 2023). Most traditional treatment approaches are time consuming, expensive, and have a narrow scope. Therefore, employing the microbe-mediated degradation method is imperative to reduce pharmaceuticals from the environment (Fang et al. 2012). Certain co-substrates, such as carbon sources, have been shown to help bacteria such as *Rhodococcus rhodochrous*, *Bacillus subtilis*, *Pseudomonas putida*, *Pseudomonas fluorescens*, *Sphingomonas herbicidovorans*, *Arthrobacter spp.*, *Enterobacter spp.*, and *Achromobacter denitrificans* to reduce the noxious effects of pharmaceutical contaminants (Gauthier et al. 2010; Reis et al. 2014).

This chapter emphasizes bacteria-mediated bioremediation and the possible metabolic degradation pathway of xenobiotics, such as NSAIDs, phthalate-based chemical plasticizers, skin-whitening agents, surface-active agents, and sunscreens. This chapter also deals with the omics-based unveiling of specific upregulated genes and gene clusters involved in xenobiotic degradation.

2 Bacteria-Mediated Biodegradation of PCPs

2.1 *Phthalate-Based Plasticizers*

Phthalates such as dibutyl phthalate (DBP), acetyl tributyl citrate (ATBC), bis(2-ethyl hexyl) phthalate (DEHP), and phthalic acid represent plasticizing agents or chemicals that add endurance, elasticity, and durability to plastics. An inclusive multi-OMIC study based on a proteogenomic and metabolomic approach has revealed the inherent ability of marine bacteria to abate plasticizer toxicity. Leaching out of these hazardous chemical plasticizers leads to severe ecotoxicological disruptions. The study revealed strains of *Mycobacterium* and *Halomonas* as potent phthalate-based plasticizer-degrading organisms. Genome sequencing indicated that both strains have genes associated with phthalate degradation and encode proteins and enzymes, including esterases, cutinases, lipases, and enzymes for phthalate metabolism and biodegradation (Wright et al. 2020). Comprehensive proteomic and metabolomic analyses have elucidated the pathways and enzymes behind plasticizers DBP, DEHP, and ATBC degradation (Wright et al. 2020).

Mycobacterium sp. DBP42 is known to convert phthalate into protocatechuate in the presence of enzymes like phthalate dioxygenase and dihydroxy phthalate decarboxylase (Stingley et al. 2004). In contrast, *Halomonas* sp. ATBC28 follows the benzoate degradation pathway to catabolize DBP (Wright et al. 2020). Also, *Aromatoleum aromaticum* EbN1, *Azoarcus evansii* KB740, *Thauera chlorobenzoica* 3CB-1, and *A. toluclasticus* are denitrifying phthalate degraders containing gene clusters that anaerobically degrade phthalate (Boll et al. 2020). The *nph* gene cluster is responsible for the potential biodegradation of 4-nitrophenol (4-NP) unveiled through transcriptome sequencing (Sengupta et al. 2019; Yamamoto et al. 2011). *Rhodococcus* sp. BUPNP1, *Arthrobacter,* and *Bacillus* degrade 4-NP via 4-nitrocatechol (4-NC) and benzene 1,2,4-triol (BT) coded by the *nph* gene group (Sengupta et al. 2019). 4-NP xenodegrons are effectively eradicated by the genera *Burkholderia* and *Moraxella* (Wang et al. 2017). The analysis of the *Rhodococcus* sp. strain BUPNP1 genome depicted the *nph* gene cluster harboring the *cbdABC* genes, encoding the NphA, NphB, NphC, CatA, CatB, and CatD proteins (Sengupta et al. 2019). A distinctive family of gene regulatory proteins, such as AraC and LysR, regulates the 4-NP degradation genes (Sengupta et al. 2019). *Mycobacterium* sp. DBP42 metabolizes DBP via ester side chain removal and generation of mono-butyl phthalate and phthalate. *Halomonas* sp. ATBC28 removes ester side chains and generates citrate during the biodegradation of ATBC (Wright et al. 2020).

2.2 Sunscreens and Personal Care Products

Sunscreens are substances applied to the skin to protect and safeguard it from the detrimental impacts of UV radiation. It is primarily composed of avobenzone or butyl methoxydibenzoylmethane (BM), oxybenzone benzophenone-3 (BP-3), homosalate (HS), octisalate, 2-ethylhexyl salicylate (ES), octocrylene (OC), and ethyl hexyl methoxycinnamate. All these compounds are considered toxic chemicals, and most are detrimental to the environment. Furthermore, although considered safe and comparatively less toxic, zinc and titanium oxides are the chief ingredients occurring in recalcitrant inorganic UV filters, leaving traces of toxicity.

Alteration of the prevailing microbial strains through genetic manipulation technology, genome editing, and biochemical techniques produces genetically modified organisms with significant gene clusters encoding potent enzymes possessing an augmented potential of xenobiotic bioremediation. The degradation and biotransformation of octocrylene (a principal UV-shielding agent in cosmetics and sunscreens) by microorganisms could overcome the grave ecological consequences (Suleiman et al. 2019). LC-MS/MS analysis has revealed *Mycobacterium agri* and *Gordonia cholesterolivorans* as potent candidates for bioremediation, causing a decrease in octocrylene concentration following massive biofilm formation on the octocrylene droplets (Suleiman et al. 2019). Methylparaben (MP) is used as an antibacterial and antifungal agent in personal care products, and it undergoes O-demethylation to transform into 4-hydroxybenzoic acid (Organization 2024; Xiong et al. 2022). O-demethylation is a part of the xenobiotic biotransformation mechanism, wherein polar groups are liberated for further conjugation. Sunscreens and plastics contain a UV light absorber, oxybenzone, converted to 2,4-dihydroxybenzophenone, an O-demethylation transformation product (Porter et al. 2020; Sun et al. 2023). Expectorants are bronchial secretion enhancers that ease mucus clearance through the respiratory tract. Guaifenesin an expectorant, is biodegraded via O-demethylation to 3-(2-hydroxyphenoxy) propane-1,2-diol as the end product (Porter et al. 2020; Wolfson et al. 2019).

It is important to unravel progressing technologies in characterizing xenobiotic-degrading microbial consortia, resolving their dynamically altering genetic constitution. Recalcitrant UV filters pose an enormous environmental threat and require quick remediation. In this context, a synergistic effect, like a microbial association rather than individual taxa, is intended to have more potential. Consequently, a synthetic consortium of microorganisms prepared through a top-down strategy in enrichment cultures has helped efficiently in biodegradation. Besides being a key ingredient in many PCPs, UV filters are an essential constituent of many wastewater treatment plants (Díaz-Cruz et al. 2008). Textile industries, paints, and food also use UV filters to avoid fading and photodegradation. Avobenzone, methylene bis-benzotriazolyl tetramethyl butylphenol (MBBT), octinoxate, diethylamino hydroxybenzoyl hexyl benzoate (DHHB), methoxyphenyl triazine (BEMT), diethylhexyl butamido triazone (DBT), and oxybenzone are examples of such recalcitrant non-biodegradable organic UV filters (Fagervold et al. 2023). Predominant genera exhibiting BM, OC, and HS degradation potency belong to the class Alphaproteobacteria, Betaproteobacteria, Gammaproteobacteria, and Actinobacteria under the phylum Pseudomonadota (Fagervold et al. 2023; Lashani et al. 2023). Some predominant benzophenone-3 (BP-3) degrading taxa include *Sphingomonas wittichii* strain BP14P (Fagervold et al. 2021), *Rhodococcus oxybenzonivorans* sp. nov., and *Methylophilus* sp. strain FP-6 (Jin et al. 2019). *Sphingopyxis*, *Gordonia*, *Pigmentiphaga*, and *Chitinophaga* have been known to diminish the toxicity of octocrylene effectively (Fagervold et al. 2023). Octisalate is degraded by bacterial strains such as *Mycobacterium*, *Pseudomonas*, *Rhodococcus*, *Comamonas*, and *Acidovorax*. However, *Microbacterium* also has HS xenodegron competency (Fagervold et al. 2023).

Characterizing genes and corresponding protein products implicated with xenobiotic metabolism and biodegradation requires an in-depth, multifaceted approach (O'Donnell et al. 2020). The use of omics-based approaches along with bioinformatic tools is essential to unravel the diversity of microbes (Pal et al. 2022). Genome sequencing has unraveled the presence of organophosphate-degrading and atrazine-degrading genera, *Sphingobium fuliginis* ATCC 27552 and *Arthrobacter* sp. ZXY-2, respectively (Zhao et al. 2018) (Mishra et al. 2021). A comprehensive in silico analysis validates the potential of the novel genes and proteins obtained in the microbial consortium. Exploring the bioremediation potential of microorganisms necessitates an interdisciplinary application of multiple genome editing platforms as well. Therefore, focusing on the perspectives of system biology is not possible by engaging any discrete omics technology. Therefore, omics and bioinformatics should work simultaneously to achieve a tailored microbial genome with explicit degradation-function genes.

2.3 Surface-Active Agents

Surfactants have emerged as foremost environmental toxicants and thus entail ready elimination using biodegradative strategies. The prevalent usage of products such as shampoos and shaving foams leads to contaminated waste production. SDS, methylparaben, and hexadecyl trimethyl ammonium bromide (HDTMA) are typical surfactants used in domestic things such as toothpaste, shaving creams, and preservatives in a variety of cosmetics (Onuche et al. 2016). Since surfactants are discharged into the environment, predominantly in wastewater, a cost-effective and environmentally friendly approach is desirable for wastewater treatment (Chaturvedi and Tiwari 2013).

Bacterial bioremediation is an efficient, economical approach to exploring the biodegradative ability of indigenous strains. A continuous 7-day incubation period is essential for *Escherichia coli, Klebsiella pneumoniae,* and *Bacillus cereus* to degrade SDS (Arora et al. 2022; Onuche et al. 2016), while SDS was degraded by *Klebsiella planticola* and *Proteus vulgaris* following 10 days of incubation (Onuche et al. 2016; Sigoillot and Nguyen 1990). *Vibrio cholerae, Pseudomonas beteli,* and *Escherichia coli* exhibit SDS and MP degradation ability (Onuche et al. 2016). Degradation potential for SDS, isolated from the Yamuna River of the Delhi-NCR region, which is heavily contaminated with surfactants, is displayed by two isolates, *P. songnenensis* YRW-05 and *Pseudomonas koreensis* YRW-02 (Arora et al. 2024).

2.4 Skin-Whitening Agents

Niacinamide, soy proteins, octyldodecanol, hydroxy-citronellal, kojic acid, glycolic acid, and ellagic acid are common ingredients of skin-depigmenting and skin-lightening creams and beauty treatment serums available in the commercial market (Juliano 2022). They all have a standard mode of action through tyrosinase inhibition (Boo 2021). Melanin is an amino acid produced from tyrosine as the initial precursor molecule synthesized in melanocytes of the melanosomes (Boo 2021). It is responsible for skin tone, hair, and eye color pigmentation. Melanocytes produce melanin, which is directly proportional to skin darkening. Tyrosinase is the vital enzyme involved in the oxidation of L-tyrosine to DOPAquinone, which either reacts with cysteine and enters the pheomelanin synthesis pathway or is oxidized to produce eumelanin. (Boo 2021). Tyrosinase is the key enzyme involved in both pheomelanin and eumelanin synthesis. Hence, most skin-lightening agents exhibit a generalized mode of action either by inhibiting the enzymatic activity of tyrosinase (TYR), inhibiting melanosome transfer, or inhibiting transcription of tyrosinase coding genes (Cao et al. 2021; Pillaiyar et al. 2017).

Topical skincare products and cosmetics constitute a primary functional ingredient, niacinamide (nicotinamide). Nicotinamide can be used as a carbon source by *Alcaligenes* sp. strain P156 (Hu et al. 2019). Nicotinic acid is transformed into 6-hydroxynicotinic acid, followed by oxidation into 2,5-dihydroxypyridine

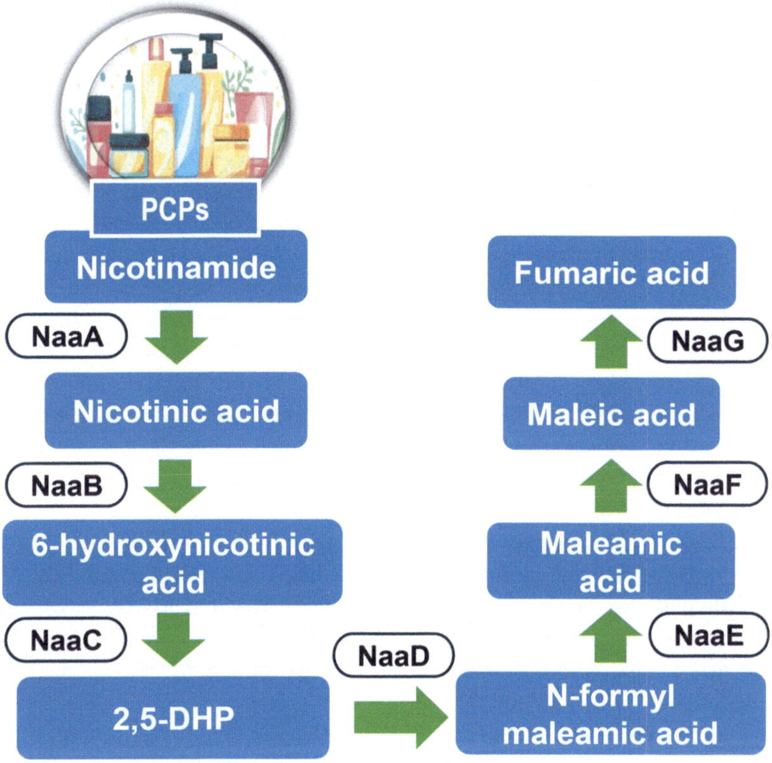

Fig. 1 The catalytic conversion of nicotinamide into fumaric acid is carried out by a cluster of genes, namely, *naaABCDEFG*, encoding for the enzymes NaaA to NaaG in a linear manner

(2,5-DHP) and conversion into N-formylmaleamic acid, which is further converted into fumaric acid via maleamic and maleic acid. The breakdown of nicotinamide is carried out by a sequential series of enzymes, i.e., NaaA to NaaG, encoded by the genes *naaA* to *naaG*, respectively. NaaA converts nicotinamide to nicotinic acid, followed by NaaB, which enzymatically converts nicotinic acid into 6-hydroxynicotinic acid. Further, 6-hydroxynicotinic acid is converted into 2,5-DHP by NaaC, followed by N-formylmaleamic acid into maleamic acid and maleic acid. Finally, the terminal step is catalyzed by the enzyme NaaG, which converts maleic acid into fumaric acid, the end product of nicotinamide biodegradation (Hu et al. 2019). Figure 1 depicts the enzymatic conversion of nicotinamide to fumaric acid.

Arbutin, glycosylated hydroquinone, occurs in face serums, creams, skin-lightening cosmetics, body sprays, and soaps (Boo 2021). *Staphylococcus aureus and S. epidermidis* are prevalent in skin microflora and avert hyperpigmentation by impairing melanogenesis through tyrosinase inhibition and hydrolyzing arbutin to hydroquinone. (Boo 2021).

Parabens (p-hydroxybenzoic acid esters) are the most prevalent additives in cosmetics such as moisturizers, sunscreen, concealers, body lotions, lipsticks, deodorants, and food items such as beer, frozen custard, ice cream, sherbet, and soft drinks. Strains of *Burkholderia cepacia*, *Pseudomonas beteli*, *P. aeruginosa*, *Burkholderia latens*, and *Cladosporium resinae* display endurance to parabens and can degrade MP and PP (Amin et al. 2010). 4-hydroxybenzoic acid is produced from MP via esterase. Hydroxylase acts upon 4-hydroxybenzoic acid to produce protocatechuate, which further undergoes ring cleavage via the β-ketoadipate pathway. The 4-hydroxybenzoic acid has two fates: aerobic or anaerobic decarboxylation into phenol. *Clostridium hydroxybenzoicum* and *Moorella thermoacetica* carry out the anaerobic conversion. However, *Enterobacter cloacae* strain EM is the only strain reported to decarboxylate parabens into phenol [57] aerobically. Hydrolysis, followed by aerobic decarboxylation, converts paraben into phenol via 4-hydroxybenzoic acid. An esterase-hydrolyzing enzyme, PrbA, in *Enterobacter gergoviae* and *E. cloacae* strains is encoded by the *prbA* gene (Valkova et al. 2002). *Enterobacter cloacae* strain EM exhibits significant paraben degradation potency due to the presence of the *pca* operon (Valkova et al. 2001).

Mandelic acid (MA) serves as an anti-aging exfoliant and anti-acne agent. In *Pseudomonas putida*, MA is metabolized via five consecutive steps (Tsou et al. 1990). Mandelate racemase (*mdlA*) tautomerizes S-MA and R-MA. S-MA is oxidized into benzoyl formic acid by S-mandelate dehydrogenase (*mdlB*). Benzoyl formate decarboxylase (*mdlC*) oxidizes benzoyl formic acid into benzaldehyde. NAD^+-dependent benzaldehyde dehydrogenase (gene *mdlD*) catalyzes benzaldehyde to benzoic acid. $NADP^+$-dependent benzaldehyde dehydrogenase (gene *mdlE*) converts benzoic acid into catechol by the β-keto adipate pathway (Tsou et al. 1990). Some other bacteria have also reported the MA degradation pathway gene clusters. *Pseudomonas silensis, Halomonas* sp. KO116, *Marinobacterium georgiense, Burkholderia* sp. Y123, and *Herbaspirillium rubrisubalbicans* also have *mdlCBA* and *mdlXYD* operons (Wang et al. 2022).

3 Bacteria-Mediated Biodegradation of Pharmaceuticals

The biodegradation of polycyclic NSAIDs by certain bacterial strains and the genes involved in the pathway are discussed in the succeeding sections.

3.1 Paracetamol

NSAIDs are among the most common synthetic groups of compounds widely dispersed in our surrounding environment and potable water (Żur et al. 2018a). These pharmaceuticals have pain-relieving, antipyretic, and anti-inflammatory properties. The extensively used, readily accessible medications in this family are monocyclic NSAIDs, which include ibuprofen, acetylsalicylic acid, and paracetamol (Żur et al. 2018a). Among antiphlogistic drugs, paracetamol or acetaminophen (N-acetyl-p-aminophenol or APAP) is the most prevalent. Paracetamol constitutes a benzene core ring replaced by one hydroxyl group and a *para* (1, 4) position substituted amide nitrogen group in its chemical structure (Wu et al. 2012). However, many microbes can utilize ibuprofen and paracetamol as their only carbon and energy source in metabolic pathways. Paracetamol, upon degradation, produces 4-aminophenol and hydroquinone as significant reaction end products, which are toxic xenobiotics necessitating bioremediation (Hu et al. 2013; Zhang et al. 2013). The paracetamol degradation pathway forms other critical intermediates, including 2-hexenoic acid, nitrate, nitrite, succinic acid, malonic acid, oxalic acid, and formic acid. The breakdown of acetaminophen in sludge and soil samples is carried out by many bacterial species belonging to the genera *Pseudomonas, Flavobacterium, Dokdonella, Ensifer,* and *Delftia* (Park and Oh 2020a, b; Rios-Miguel et al. 2022; Rios-Miguel et al. 2021). The APAP biodegradation pathway might involve a deaminase and dioxygenase enzyme (dos S. Grignet et al. 2022; Rios-Miguel et al. 2022; Żur et al. 2018b). It has been reported that five amidases may convert APAP to 4-AP (Chen et al. 2016a; Lee et al. 2015; Rios-Miguel et al. 2022; Yun et al. 2017). An investigation revealed that the primary metabolites in the paracetamol breakdown process in *Rhodococcus* are pyrocatechol, hydroquinone, and AP (Ivshina et al. 2006). Phenolic intermediates have been reported to undergo oxidative deamination during the metabolic breakdown of paracetamol (Wu et al. 2012; Zhang et al. 2013). Only a few reported microbes can break down 4-AP and 4-NP, vital products in the amines and nitrobenzene breakdown pathway (Takenaka et al. 2003; Żur et al. 2018a). The genes encoding ring cleavage enzymes for the biodegradation of 4-AP are plasmid-borne in *Pseudomonas* sp. ST-4 (Khan et al. 2006). In *Pseudomonas putida* DLL-E4, a bacterium that breaks down 4-NP, hydroquinone is degraded by the product of nine genes (Hu et al. 2014; Żur et al. 2018a). Significant genes involved in this pathway are *pnpA, pnpR*, and *pnpC1C2DECX1X2*. The conversion of hydroquinone into the tricarboxylic acid cycle metabolites is regulated through the expression of *pnpC1C2DE* genes (Hu et al. 2014; Żur et al. 2018a). Hydroxyquinol to maleyl acetic acid is converted in the presence of hydroxyquinol 1,2-dioxygenase, encoded by the *pnpC* gene. The product of the *pnpE* gene, maleylacetic acid reductase, transforms maleylacetic acid into beta-ketoadipate. The breakdown of 4-NP is significantly increased when glucose is added, whereas hydroquinone degradation is inhibited. The *pnpR* gene regulates the *pnpC1C2DECX1X2* operon positively, which in turn auto-regulates its own expression as well as the *pnpA* gene (Chen et al. 2016b; Hu et al. 2014; Żur et al. 2018a).

3.2 *Ibuprofen*

Currently, the third-most-used NSAID across the world is ibuprofen, a propionic acid derivative. Ibuprofen (2-[4-(2-methylpropyl)phenyl] propanoic acid) may undergo abiotic transformation, biotransformation, sorption, and desorption in the surrounding environment (Marchlewicz et al. 2016). Ibuprofen is chemically stable owing to the substitutions at the para position of the aromatic ring and its highly branched nature (Balciunas et al. 2020; Lu et al. 2019; Marchlewicz et al. 2017; Rutere et al. 2020; Żur et al. 2018a). The first report of microbial biotransformation of ibuprofen was demonstrated in *Nocardia* sp. strain NRRL 5646 (Chen and Rosazza 1994; Żur et al. 2018a). In aerobic conditions, *Sphingomonas* sp. Ibu-2 can utilize both enantiomers of IBU. A comprehensive metagenomic study revealed that *Sphingomonas* sp. Ibu-2 contains *an ipfABDEF* gene cluster, which converts ibuprofen to isobutyl catechol (Murdoch and Hay 2013; Żur et al. 2018a). CoA ligase IpfF catalyzes ligation of ibuprofen with CoA. IpfABHI catalyzes dihydroxylation of ibuprofen-CoA to 1,2-cis-diol-2-hydroibuprofen. Lastly, the removal of the acyl-CoA is catalyzed by IpfD and IpfE, producing 4-isobutylcatechol (Kagle et al. 2009; Murdoch and Hay 2005, 2013; Żur et al. 2018a). The existence of the *ipfABDEF* gene cluster provides the capacity to degrade ibuprofen and is also found in *Noviherbaspirillum* sp., *Pseudoxanthomonas spadix* BD-a59, and *Rhodospirillales* bacterium 69–11.

3.3 *Diclofenac*

Diclofenac is a non-narcotic analgesic derivative of phenylacetic acid containing a 2,6 dichlorophenyl amino group substituted at the 2-position. The breakdown of polycyclic NSAIDs is significantly challenging and requires potent microbial enzymes for biotransformation. However, there have been many studies on diclofenac's microbial breakdown and transformation in recent years. Bacteria like *Bacillus subtilis*, *Raoultella* sp. DD4, *Rhodococcus ruber*, *Alcaligenes faecalis*, *Labrys portucalensis* F11, *Brevibacillus laterosporus*, *Pseudomonas moorei* KB4, *Staphylococcus aureus*, *Proteus mirabilis*, *Klebsiella* sp. KSC, and *Staphylococcus haemolyticus* are potent diclofenac decomposers (Grandclément et al. 2020; Ivshina et al. 2019; Moreira et al. 2018; Stylianou et al. 2018; Wojcieszyńska et al. 2023; Żur et al. 2020, 2021). Degradation of diclofenac in bacteria was reported for the first time in *Rhodococcus ruber* IEGM 346 and *Raoultella* sp. DD4. The arene ring structure of diclofenac opens due to the breakdown of the C-N bond. Acetoacetic acid, 4, 6, 7-trioxooct-2-enedioic acid, and fumaric acid are end products of the oxidation reaction in diclofenac (Wojcieszyńska et al. 2023). *Labrys portucalensis* F11 undergoes hydroxylation reactions during diclofenac breakdown (Moreira et al. 2018). A significant metabolite, benzoquinone imine, is formed in the above reaction, and further decarboxylation and hydroxylation of the end product occur during diclofenac degradation. The presence of a dihydroxylated structure in the periphery may indicate that dioxygenases are involved in the ring cleavage (Moreira et al. 2018). Based on gene expression studies, upregulation of genes encoding biotransformation enzymes, such as monooxygenase, dihydroxylating and aromatic ring-cleaving dioxygenases, and the cytochrome p450 system, are found to occur in the presence of diclofenac (Wojcieszyńska et al. 2023).

4 Conclusion

The shift from natural cures to synthetic solutions in beauty and healthcare illustrates a broader trend motivated by the need for convenience and precision. The dependence on xenobiotics presents considerable environmental issues due to the accumulation of persistent substances. Investigating and utilizing effective microbial agents that can metabolize these xenobiotics is essential. Investigating their metabolic pathways and related genetic factors is crucial for formulating successful bioremediation techniques. As the industry evolves, it is essential to reconcile the advantages of synthetic products with sustainable methods that safeguard our ecosystems. Focusing on exploring microbial degradation mechanisms can facilitate creative solutions that reduce ecological impacts and foster a more sustainable future.

References

Amin A, Chauhan S, Dare M, Bansal AK (2010) Degradation of parabens by *Pseudomonas* beteli and *Burkholderia latens*. European Journal of Pharmaceutics and Biopharmaceutics 75(2):206–212

Arora J, Ranjan A, Chauhan A, Biswas R, Rajput VD, Sushkova S, Mandzhieva S, Minkina T, Jindal T (2022) Surfactant pollution, an emerging threat to ecosystem: Approaches for effective bacterial degradation. Journal of Applied Microbiology 133(3):1229–1244. https://doi.org/10.1111/jam.15631

Arora J, Ranjan A, Chauhan A, Rajput VD, Sushkova S, Prazdnova EV, Minkina T, Biswas R, Joshi S, Jindal T, Prasad R (2024) A Novel Study on Anionic Surfactant Degradation Potential of Psychrophillic and Psychrotolerant *Pseudomonas* spp. Identified from Surfactant-contaminated River Water. Applied Biochemistry and Biotechnology 196(5):2612–2629. https://doi.org/10.1007/s12010-023-04647-y

Balciunas EM, Kappelmeyer U, Harms H, Heipieper HJ (2020) Increasing ibuprofen degradation in constructed wetlands by bioaugmentation with gravel containing biofilms of an ibuprofen-degrading *Sphingobium yanoikuyae*. Engineering in Life Sciences 20(5–6):160–167

Boll M, Geiger R, Junghare M, Schink B (2020) Microbial degradation of phthalates: biochemistry and environmental implications. Environmental Microbiology Reports 12(1):3–15

Boo YC (2021) Arbutin as a skin depigmenting agent with antimelanogenic and antioxidant properties. Antioxidants 10(7):1129

Bucheli TD, Strobel BW, Hansen HCB (2018) Personal care products are only one of many exposure routes of natural toxic substances to humans and the environment. Cosmetics 5(1):10

Calisto V, Esteves VI (2009) Psychiatric pharmaceuticals in the environment. Chemosphere 77(10):1257–1274

Cao W, Zhou X, McCallum NC, Hu Z, Ni QZ, Kapoor U, Heil CM, Cay KS, Zand T, Mantanona AJ (2021) Unraveling the structure and function of melanin through synthesis. Journal of the American Chemical Society 143(7):2622–2637

Caracciolo AB, Topp E, Grenni P (2015) Pharmaceuticals in the environment: biodegradation and effects on natural microbial communities. A review. Journal of pharmaceutical and biomedical analysis 106:25–36

Chaturvedi AD, Tiwari K (2013) Effect of household detergents (surfactants) degraded through aquatic fungi. Recent Research in Science and Technology 5(5)

Chen Q, Chen K, Ni H, Zhuang W, Wang H, Zhu J, He Q, He J (2016a) A novel amidohydrolase (DmhA) from *Sphingomonas* sp. that can hydrolyze the organophosphorus pesticide dimethoate to dimethoate carboxylic acid and methylamine. Biotechnology letters 38:703–710

Chen Q, Tu H, Huang F, Wang Y, Dong W, Wang W, Li Z, Wang F, Cui Z (2016b) Impact of *pnpR*, a LysR-type regulator-encoding gene, on the cellular processes of *Pseudomonas putida* DLL-E4. FEMS Microbiology Letters 363(12):fnw110

Chen Y, Rosazza JP (1994) Microbial transformation of ibuprofen by a *Nocardia* species. Applied and Environmental Microbiology 60(4):1292–1296

de Oliveira M, Frihling BEF, Velasques J, Magalhães Filho FJC, Cavalheri PS, Migliolo L (2020) Pharmaceuticals residues and xenobiotics contaminants: occurrence, analytical techniques and sustainable alternatives for wastewater treatment. Science of the total environment 705:135568

Díaz-Cruz MS, Llorca M, Barceló D (2008) Organic UV filters and their photodegradates, metabolites and disinfection by-products in the aquatic environment. TrAC Trends in Analytical Chemistry 27(10):873–887

dos S. Grignet R, Barros MG, Panatta AA, Bernal SP, Ottoni JR, Passarini MR, da CS Gonçalves C (2022) Medicines as an emergent contaminant: the review of microbial biodegration potential. Folia Microbiologica 67(2):157–174

Fagervold SK, Rohée C, Lebaron P (2023) Microbial consortia degrade several widely used organic UV filters, but a number of hydrophobic filters remain recalcitrant to biodegradation. Environmental Science and Pollution Research 30(60):125931–125946

Fagervold SK, Rohée C, Rodrigues AM, Stien D, Lebaron P (2021) Efficient degradation of the organic UV filter benzophenone-3 by *Sphingomonas wittichii* strain BP14P isolated from WWTP sludge. Science of the Total Environment 758:143674

Fang Y, Karnjanapiboonwong A, Chase DA, Wang J, Morse AN, Anderson TA (2012) Occurrence, fate, and persistence of gemfibrozil in water and soil. Environmental toxicology and chemistry 31(3):550–555

Fernandes JP, Almeida CMR, Salgado MA, Carvalho MF, Mucha AP (2021) Pharmaceutical compounds in aquatic environments—occurrence, fate and bioremediation prospective. Toxics 9(10):257

Gauthier H, Yargeau V, Cooper DG (2010) Biodegradation of pharmaceuticals by *Rhodococcus rhodochrous* and *Aspergillus niger* by co-metabolism. Science of the Total Environment 408(7):1701–1706

Grandclément C, Piram A, Petit M-E, Seyssiecq I, Laffont-Schwob I, Vanot G, Tiliacos N, Roche N, Doumenq P (2020) Biological Removal and Fate Assessment of Diclofenac Using *Bacillus subtilis* and *Brevibacillus laterosporus* Strains and Ecotoxicological Effects of Diclofenac and 4′-Hydroxy-diclofenac. Journal of Chemistry 2020(1):9789420

Hu C, Zhao S, Li K, Yu H (2019) Microbial degradation of nicotinamide by a strain *Alcaligenes* sp. P156. Scientific Reports 9(1):3647

Hu J, Zhang LL, Chen JM, Liu Y (2013) Degradation of paracetamol by *Pseudomonas aeruginosa* strain HJ1012. Journal of Environmental Science and Health, Part A 48(7):791–799

Hu X, Wang J, Wang F, Chen Q, Huang Y, Cui Z (2014) Complete genome sequence of the p-nitrophenol-degrading bacterium *Pseudomonas putida* DLL-E4. Genome announcements 2(3):10.1128/genomea. 00596–00514

Ivshina I, Rychkova M, Vikhareva E, Chekryshkina L, Mishenina I (2006) Catalysis of the biodegradation of unusable medicines by alkanotrophic rhodococci. Applied Biochemistry and Microbiology 42:392–395

Ivshina IB, Tyumina EA, Kuzmina MV, Vikhareva EV (2019) Features of diclofenac biodegradation by *Rhodococcus ruber* IEGM 346. Scientific reports 9(1):9159

Jin C, Geng Z, Pang X, Zhang Y, Wang G, Ji J, Li X, Guan C (2019) Isolation and characterization of a novel benzophenone-3-degrading bacterium *Methylophilus* sp. strain FP-6. Ecotoxicology and Environmental Safety 186:109780

Juliano CC (2022) Spreading of dangerous skin-lightening products as a result of colourism: a review. Applied Sciences 12(6):3177

Kagle J, Porter AW, Murdoch RW, Rivera-Cancel G, Hay AG (2009) Biodegradation of pharmaceutical and personal care products. Advances in Applied Microbiology 67:65–108

Karishma S, Yaashikaa P, Kumar PS, Kamalesh R, Saravanan A, Rangasamy G (2023) Promising approaches and kinetic prospects of the microbial degradation of pharmaceutical contaminants. Environmental Science: Advances 2(11):1488–1504

Khan SA, Hamayun M, Ahmed S (2006) Degradation of 4-aminophenol by newly isolated *Pseudomonas* sp. strain ST-4. Enzyme and Microbial Technology 38(1–2):10–13

Klaschka U (2015) Naturally toxic: natural substances used in personal care products. Environmental Sciences Europe 27:1–13

Lashani E, Amoozegar MA, Turner RJ, Moghimi H (2023) Use of microbial consortia in bioremediation of metalloid polluted environments. Microorganisms 11(4):891

Lee S, Park E-H, Ko H-J, Bang WG, Kim H-Y, Kim KH, Choi I-G (2015) Crystal structure analysis of a bacterial aryl acylamidase belonging to the amidase signature enzyme family. Biochemical and biophysical research communications 467(2):268–274

Lu Z, Sun W, Li C, Ao X, Yang C, Li S (2019) Bioremoval of non-steroidal anti-inflammatory drugs by *Pseudoxanthomonas* sp. DIN-3 isolated from biological activated carbon process. Water research 161:459–472

Marchlewicz A, Domaradzka D, Guzik U, Wojcieszyńska D (2016) *Bacillus thuringiensis* B1 (2015b) is a gram-positive bacteria able to degrade naproxen and ibuprofen. Water, air, & soil pollution 227:1–8

Marchlewicz A, Guzik U, Smułek W, Wojcieszyńska D (2017) Exploring the degradation of ibuprofen by *Bacillus thuringiensis* B1 (2015b): The new pathway and factors affecting degradation. Molecules 22(10):1676

Marghade DT, Chahande AD, Tiwari MS, Patil PD (2021) Microbial degradation of xenobiotic compounds. Recent advances in microbial degradation: 173–217

Miglani R, Parveen N, Kumar A, Ansari MA, Khanna S, Rawat G, Panda AK, Bisht SS, Upadhyay J, Ansari MN (2022) Degradation of xenobiotic pollutants: an environmentally sustainable approach. Metabolites 12(9):818

Mishra S, Lin Z, Pang S, Zhang W, Bhatt P, Chen S (2021) Recent advanced technologies for the characterization of xenobiotic-degrading microorganisms and microbial communities. Frontiers in Bioengineering and Biotechnology 9:632059

Moreira IS, Bessa VS, Murgolo S, Piccirillo C, Mascolo G, Castro PM (2018) Biodegradation of diclofenac by the bacterial strain *Labrys portucalensis* F11. Ecotoxicology and environmental safety 152:104–113

Murdoch RW, Hay AG (2005) Formation of catechols via removal of acid side chains from ibuprofen and related aromatic acids. Applied and Environmental Microbiology 71(10):6121–6125

Murdoch RW, Hay AG (2013) Genetic and chemical characterization of ibuprofen degradation by *Sphingomonas* Ibu-2. Microbiology 159(Pt_3):621–632

Nguyen PM, Afzal M, Ullah I, Shahid N, Baqar M, Arslan M (2019) Removal of pharmaceuticals and personal care products using constructed wetlands: effective plant-bacteria synergism may enhance degradation efficiency. Environmental science and pollution research 26:21109–21126

Nowak-Lange M, Niedziałkowska K, Lisowska K (2022) Cosmetic preservatives: hazardous micropollutants in need of greater attention? International Journal of Molecular Sciences 23(22):14495

O'Donnell ST, Ross RP, Stanton C (2020) The progress of multi-omics technologies: determining function in lactic acid bacteria using a systems level approach. Frontiers in microbiology 10:3084

Onuche P, Okibe F, Ajibola V (2016) Biodegradation of sodium dodecyl sulphate and methyl paraben in shampoo and hair dressing salon waste by bacteria from sewage treatment sludge. American Chemical Science Journal 14(2):1–9

Organization WH (2024) Evaluation of certain food additives: ninety-sixth report of the Joint FAO/WHO Expert Committee on Food Additives. World Health Organization,

Pal S, Jana A, Mondal KC, Halder SK (2022) Omics Approach to Understanding Microbial Diversity. Biotechnological Advances for Microbiology, Molecular Biology, and Nanotechnology, Apple Academic Press, pp. 25–38

Park S, Oh S (2020a) Activated sludge-degrading analgesic drug acetaminophen: acclimation, microbial community dynamics, degradation characteristics, and bioaugmentation potential. Water research 182:115957

Park S, Oh S (2020b) Detoxification and bioaugmentation potential for acetaminophen and its derivatives using *Ensifer* sp. isolated from activated sludge. Chemosphere 260:127532

Parolini M, Binelli A (2012) Sub-lethal effects induced by a mixture of three non-steroidal anti-inflammatory drugs (NSAIDs) on the freshwater bivalve *Dreissena polymorpha*. Ecotoxicology 21:379–392

Phale PS, Sharma A, Gautam K (2019) Microbial degradation of xenobiotics like aromatic pollutants from the terrestrial environments. Pharmaceuticals and personal care products: waste management and treatment technology, Elsevier, pp 259–278

Pillaiyar T, Manickam M, Namasivayam V (2017) Skin whitening agents: Medicinal chemistry perspective of tyrosinase inhibitors. Journal of enzyme inhibition and medicinal chemistry 32(1):403–425

Piwowarska D, Kiedrzyńska E (2022) Xenobiotics as a contemporary threat to surface waters. Ecohydrology & Hydrobiology 22(2):337–354

Porter AW, Wolfson SJ, Häggblom M, Young LY (2020) Microbial transformation of widely used pharmaceutical and personal care product compounds. F1000Research 9

Reis PJ, Reis AC, Ricken B, Kolvenbach BA, Manaia CM, Corvini PF, Nunes OC (2014) Biodegradation of sulfamethoxazole and other sulfonamides by *Achromobacter denitrificans* PR1. Journal of hazardous materials 280:741–749

Rios-Miguel AB, Smith GJ, Cremers G, van Alen T, Jetten MS, den Camp HJO, Welte CU (2022) Microbial paracetamol degradation involves a high diversity of novel amidase enzyme candidates. Water research X 16:100152

Rios-Miguel AB, Jetten MS, Welte CU (2021) Effect of concentration and hydraulic reaction time on the removal of pharmaceutical compounds in a membrane bioreactor inoculated with activated sludge. Microbial Biotechnology 14(4):1707–1721

Rutere C, Knoop K, Posselt M, Ho A, Horn MA (2020) Ibuprofen degradation and associated bacterial communities in hyporheic zone sediments. Microorganisms 8(8):1245

Sengupta K, Swain MT, Livingstone PG, Whitworth DE, Saha P (2019) Genome sequencing and comparative transcriptomics provide a holistic view of 4-nitrophenol degradation and concurrent fatty acid catabolism by *Rhodococcus* sp. strain BUPNP1. Frontiers in Microbiology 9:3209

Sigoillot J-C, Nguyen M-H (1990) Isolation and characterization of surfactant degrading bacteria in a marine environment. FEMS Microbiology Ecology 6(1):59–67. https://doi.org/10.1111/j.1574-6968.1990.tb03925.x

Stingley RL, Brezna B, Khan AA, Cerniglia CE (2004) Novel organization of genes in a phthalate degradation operon of *Mycobacterium vanbaalenii* PYR-1. Microbiology 150(11):3749–3761

Stylianou K, Hapeshi E, Vasquez MI, Fatta-Kassinos D, Vyrides I (2018) Diclofenac biodegradation by newly isolated *Klebsiella* sp. KSC: microbial intermediates and ecotoxicological assessment. Journal of environmental chemical engineering 6(2):3242–3248

Suleiman M, Schröder C, Kuhn M, Simon A, Stahl A, Frerichs H, Antranikian G (2019) Microbial biofilm formation and degradation of octocrylene, a UV absorber found in sunscreen. Communications biology 2(1):430

Sun J, Rene ER, Tao D, Lu Y, Jin Q, Lam JC-H, Leung KM, He Y (2023) Degradation of organic UV filters in the water environment: A concise review on the mechanism, toxicity, and technologies. Journal of Hazardous Materials: 132822

Takenaka S, Okugawa S, Kadowaki M, Murakami S, Aoki K (2003) The metabolic pathway of 4-aminophenol in *Burkholderia* sp. strain AK-5 differs from that of aniline and aniline with C-4 substituents. Applied and Environmental Microbiology 69(9):5410–5413

Tsou AY, Ransom SC, Gerlt JA, Buechter DD, Babbitt PC, Kenyon GL (1990) Mandelate pathway of *Pseudomonas putida*: sequence relationships involving mandelate racemase,(S)-mandelate dehydrogenase, and benzoylformate decarboxylase and expression of benzoylformate decarboxylase in Escherichia coli. Biochemistry 29(42):9856–9862

Valkova N, Lépine F, Bollet C, Dupont M, Villemur R (2002) *prbA*, a gene coding for an esterase hydrolyzing parabens in *Enterobacter cloacae* and *Enterobacter gergoviae* strains. Journal of bacteriology 184(18):5011–5017

Valkova N, Lépine Fo, Valeanu L, Dupont M, Labrie L, Bisaillon J-G, Beaudet Rj, Shareck Fo, Villemur R (2001) Hydrolysis of 4-hydroxybenzoic acid esters (parabens) and their aerobic transformation into phenol by the resistant *Enterobacter cloacae* strain EM. Applied and environmental microbiology 67(6):2404–2409

Wang Q, Geng S, Wang L, Wen Z, Sun X, Huang H (2022) Bacterial mandelic acid degradation pathway and its application in biotechnology. Journal of Applied Microbiology 133(2):273–286

Wang Y, Hatt JK, Tsementzi D, Rodriguez-R LM, Ruiz-Perez CA, Weigand MR, Kizer H, Maresca G, Krishnan R, Poretsky R (2017) Quantifying the importance of the rare biosphere for microbial community response to organic pollutants in a freshwater ecosystem. Applied and Environmental Microbiology 83(8):e03321–03316

Wojcieszyńska D, Łagoda K, Guzik U (2023) Diclofenac biodegradation by microorganisms and with immobilised systems—a review. Catalysts 13(2):412

Wolfson SJ, Porter AW, Villani TS, Simon JE, Young LY (2019) Pharmaceuticals and personal care products can be transformed by anaerobic microbiomes in the environment and in waste-treatment processes. Environmental toxicology and chemistry 38(7):1585–1593

Wright RJ, Bosch R, Gibson MI, Christie-Oleza JA (2020) Plasticizer degradation by marine bacterial isolates: a proteogenomic and metabolomic characterization. Environmental Science & Technology 54(4):2244–2256

Wu S, Zhang L, Chen J (2012) Paracetamol in the environment and its degradation by microorganisms. Applied microbiology and biotechnology 96:875–884

Xiong Y, Shi Q, Sy ND, Dennis NM, Schlenk D, Gan J (2022) Influence of methylation and demethylation on plant uptake of emerging contaminants. Environment international 170:107612

Yamamoto K, Nishimura M, Kato D-i, Takeo M, Negoro S (2011) Identification and characterization of another 4-nitrophenol degradation gene cluster, nps, in *Rhodococcus* sp. strain PN1. Journal of bioscience and bioengineering 111(6):687–694

Yun H, Liang B, Qiu J, Zhang L, Zhao Y, Jiang J, Wang A (2017) Functional characterization of a novel amidase involved in biotransformation of triclocarban and its dehalogenated congeners in *Ochrobactrum* sp. TCC-2. Environmental Science & Technology 51(1):291–300

Zhang L, Hu J, Zhu R, Zhou Q, Chen J (2013) Degradation of paracetamol by pure bacterial cultures and their microbial consortium. Applied microbiology and biotechnology 97:3687–3698

Zhao X, Wang L, Ma F, Yang J (2018) Characterisation of an efficient atrazine-degrading bacterium, *Arthrobacter* sp. ZXY-2: an attempt to lay the foundation for potential bioaugmentation applications. Biotechnology for biofuels 11:1–10

Żur J, Marchlewicz A, Piński A, Guzik U, Wojcieszyńska D (2021) Degradation of diclofenac by new bacterial strains and its influence on the physiological status of cells. Journal of hazardous materials 403:124000

Żur J, Piński A, Marchlewicz A, Hupert-Kocurek K, Wojcieszyńska D, Guzik U (2018a) Organic micropollutants paracetamol and ibuprofen—toxicity, biodegradation, and genetic background of their utilization by bacteria. Environmental Science and Pollution Research 25:21498–21524

Żur J, Piński A, Wojcieszyńska D, Smułek W, Guzik U (2020) Diclofenac degradation—Enzymes, genetic background and cellular alterations triggered in diclofenac-metabolizing strain *Pseudomonas moorei* KB4. International Journal of Molecular Sciences 21(18):6786

Żur J, Wojcieszyńska D, Hupert-Kocurek K, Marchlewicz A, Guzik U (2018b) Paracetamol–toxicity and microbial utilization. *Pseudomonas moorei* KB4 as a case study for exploring degradation pathway. Chemosphere 206:192–202

Pesticide Bioremediation by Actinobacteria: Current Status and Future Perspectives

Raj Narayan Roy and Sukanta Kumar Sen

Abstract To maintain crop productivity, the use of pesticides becomes a crucial factor in present-day agricultural practices. In India, the Green Revolution indirectly encouraged pesticide use. Population efflux advocates for higher production of secured food. In India, to satisfy the needs of about 140 crore people, it is necessary to use pesticides and provide support. The global usage of pesticides in agriculture has increased steadily in the last couple of decades and reached 3.54 million metric tons in 2021. Though pesticide use becomes a prerogative for the production of target crops, it is reversed to environmental compliance. Though pesticides are applied to target crop production to ensure food security, these are considered hazardous environmental pollutants. It is for their persistence, mobility, long-term effects on the living system, and soil health. **Merely** 2–3% of the pesticides used to mitigate pests are effectively utilized at the target point. The remaining amount is leftovers, which come down to the environment and act as a source of pollutants that cause various health risks. Therefore, their exclusion from the environment or conversion into less hazardous compounds is advised. Various techniques of pesticide bioremediation are recommended depending on the crop type. However, managing the leftover pesticide through the microbial system is highly advisable since it is eco-, environment-, application-, and budget-friendly. Among the environmental microbes, actinobacteria are leading colonizers and thus ubiquitous. They are sufficiently metabolically enriched to harbor many enzymes having great catabolic activities for the conversion/transformation of hazardous chemicals. The strains belonging to the various genera of actinobacteria, including *Arthrobacter*, *Nocardioides*, *Streptomyces*, *Frankia*, and *Rhodococcus,* are rigorously attempted with commonly used pesticides, such as atrazine, chlorpyrifos, carbendazim, deltamethrin, cypermethrin, and lindane, which are noteworthy. An attempt has been

R. N. Roy (✉)
Department of Botany, Dr. Bhupendra Nath Dutta Smriti Mahavidyalaya, Hatgobindapur, Purba-bardhaman, West Bengal, India

S. K. Sen
Department of Botany, Siksha Bhavana, Visva-Bharati, Santiniketan, West Bengal, India

S. Roy, V. Mandal (eds.), *Plant-Microbe Interaction under Xenobiotic Exposure*,
https://doi.org/10.1007/978-981-96-8260-7_20

made to make the next generation of scientists accelerate research on pesticide bioremediation, considering actinobacteria as a model.

Keywords Actinobacteria · Bioremediation · Environment · Pesticides · Pollution

1 Introduction

Pesticides are the compounds (natural or synthetic) applied to protect our commodities from pests. These compounds are classified in different ways based on various criteria. The increased need for agricultural productivity to overcome climatic change has accelerated the utilization of pesticides (Shetty et al. 2008). It is estimated that the annual loss of food productivity due to pest attacks is nearly 45% (Abhilash and Singh 2009). Commonly, 35% of global probable crop yield is destroyed by pre-harvest pests (Oerke 2005). Crop destruction due to pests is linked with crop productivity and its quality. Presently, the application of various pesticides has become inevitable to make sure of the greater yield of pest-free crops, which is indispensable to keep pace with the demand of the increasing population. It is noteworthy that the world's population currently exceeds 8.2 billion, and in India, it is 1.41 billion. Pest-free crops turn human beings away from different diseases (Fenik et al. 2011). Thus, there is a growing demand for pesticides. There has been a steady increase in the global usage of pesticides in agriculture in the last couple of decades. In the year 2021, it reaches 3.54 million metric tons. The **market** dimension of **pesticides in India alone** is expected to rise by **USD 529 million** at a **CAGR of 3.82%** between 2023 and 2028 (technavio.com (2024)https://www.technavio.com/report/pesticide-market-industry-in-india-analysis.

Despite their beneficial activities, pesticides are found to be precarious to the environment. This is due to their stability, mobility, and enduring effects on the living system (Odukkathil and Vasudevan 2016; Lamilla et al. 2021). Merely 2–3% of the pesticides are used to mitigate the effect of pests at their target point. The leftover pesticides may enter soil water through surface runoff, leaching, and percolation. Ultimately, it reaches the biota and human beings through the food chain (World Health Organization, WHO). Accumulating such toxic substances in food and drinking water is a crucial health issue (Mohammed 2009). So, removing these pesticides from the environment or converting them into less hazardous compounds is one of the global burning issues for the research. Various traditional techniques are available to mitigate bioremediation of pesticides. These techniques include excavation, landfilling, aeration, oxidation, and incineration. However, these slow, laborious, less efficient methods are not viable remediation methods, as they lead to several secondary pollutants (Yoshikawa et al. 2017; Bhatt et al. 2021). Thus, bioremediation through the microbial system is the alternative to overcome the issue.

Actinobacteria are ubiquitous Gram-positive filamentous bacteria (Roy and Sen 2002; Bull 2011). The actinobacteria class includes 5 subclasses, 10 orders, 56 families, and 286 genera (Euzeby 2015). They exhibit wide physiological and morphological diversity (Roy and Sen 2011, 2013). Actinobacteria embodies the most renowned assemblage of microorganisms with the capability to synthesize several bioactive compounds (Roy and Sen 2002). Moreover, they are the most prominent source of various enzymes with distinct catabolic activities for pesticide degradation. Actinobacteria can metabolize these noxious compounds through these diverse enzyme systems for growth and development. Thus, pesticide bioremediation employing actinobacteria may be an eco-friendly economic method (Chaturvedi and Khurana 2019). The dominant genus of actinobacteria is *Streptomyces*. Other non-Streptomyces actinomycetes genera include *Actinomyces*, *Kitasatospora*, *Micromonospora*, *Nocardia*, *Micrococcus*, *Arthrobacter*, and *Rhodococcus* (Barka et al. 2015; Ariffin and Rahman 2020*)*.

2 Classification of Pesticides

Pesticides are classified diversely (Fig. 1) based on various criteria.

2.1 According to Toxicity

Two factors, i.e., dose (the quantity of pesticide) and time (period of exposure), are important for the determination of the toxicity of a pesticide. Accordingly, toxicity is of two types: acute and chronic. Acute toxicity of a pesticide denotes toxicity to any life, resulting in a definite short-term exposure to a pesticide. A high acute toxicity is lethal even if only a minute quantity is absorbed into the body. For cataloging pesticides, the WHO has recognized acute toxicity only based on the lethal dosage (LD_{50}) (Table 1).

However, several pesticides have not been classified under the WHO's recommendation to classify pesticides by hazard. These are fumigants (e.g., formaldehyde, hydrogen cyanide, magnesium phosphate) and obsolete pesticides (arsenous oxide, fluoromide, griseofulvin). The fatal impact of pesticide exposure that persists in the long run is known as chronic toxicity. Such toxicity of pesticides is a worry for the common people, and those who work directly with pesticides are prone to pesticide exposure.

2.2 According to the Sources of Origin

Based on their origin, pesticides may be broadly categorized into two groups, i.e., biopesticides and chemical pesticides. The natural sources (bacteria, viruses, fungi, plants, and nematodes) used to control pests are called biopesticides. They are allocated into two categories.

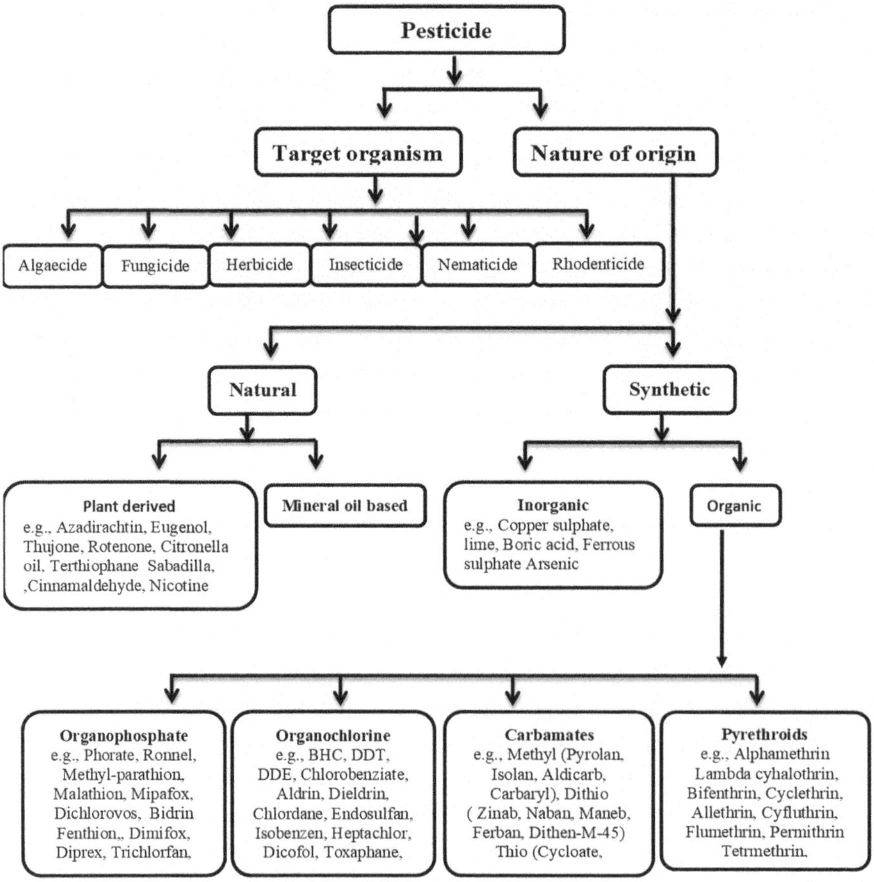

Fig. 1 Pesticides classification

Table 1 Pesticides classification (WHO 2019)

Class		LD_{50} for the rat (mg/kg body weight)	
		Oral	Dermal
Ia	Extremely hazardous	<5	<50
Ib	Highly hazardous	5–50	50–200
II	Moderately hazardous	50–2000	200–2000
III	Slightly hazardous	Over 2000	Over 2000
U	Unlikely to present acute hazard	5000 or	Higher

2.2.1 Bio-pesticide

The natural sources (bacteria, viruses, fungi, plants, and nematodes) used to control pests are called biopesticides. They can be separated into three categories.

2.2.1.1 Microbial Pesticides

Various microorganisms (bacteria, fungi, and protozoa) are used to control pests. These pesticides destroy insects by poisonous substances excreted by microbes or by infecting them.

2.2.1.2 Plant-Incorporated Pesticides

Plants are the natural producers of these pesticides. However, the required gene for the synthesis of desirable pesticides is also inserted within the plant cell through genetic engineering for the production of particular pesticides.

2.2.1.3 Biochemical Pesticides

These are nontoxic natural compounds used in the course of pest control. The sex pheromones of insects (that interfere with the coupling process) and several fra-granced plant extracts (that attract insect pests into traps) are important biochemical insecticides.

2.2.2 Chemical Pesticide

The pesticide whose active ingredient(s) are chemically synthesized is called a chemical pesticide (see Sect. 2.4 for further information).

2.3 According to the Targeted Organism

Pesticides are also categorized based on the targeted organisms (Table 2).

2.4 According to Chemical Structure

Cataloging of pesticides based on their chemical structure/composition is the most appropriate and prevalent method. Based on chemical composition, pesticides may be grouped into four categories (Buchel 1993).

Table 2 Classification of pesticides on the basis of targeted organisms

Type	Effect	Example
Acaricides	Kill or suppress mites and ticks	Diazinon, flumethrin
Algicides	Kill or suppress algae	Stylonema, miizzen
Avicides	Kill or suppress birds	4-aminopyridine
Bactericides	Kill or suppress bacteria	Tetracyclines, monobactams,
Chemosterilants	Make an insect sterile	Triethylenemelamine
Fungicides	Kill or suppress fungi	Nabam, captan
Herbicides	Kill or suppress undesired plants	2,4-D, glyphosate
Insecticides	Kill or suppress insects and other arthropods	Carbamyl, aldrin
Larvicides	Kill or stop the growth of larvae	Methoprene, pyriproxyfen
Molluscicides	Kill or control snails and slugs	Niclosamide, metaldehyde
Nematicides	Plant parasites nematodes	Chlorothalonil
Ovicides	Kill or suppress insect and mite eggs	Imidacloprid, etofenprox
Repellents	Repel bugs	Picaridin, N,N-diethyl-meta-toluamide
Rodenticide	Killing of rodents	Strychnine, chloralose
Termiticides	Kill or suppress termites	Fipronil
Virucides	Inhibit virus	Chlorhexidine, sodium hypochlorite

2.4.1 Organochlorine Pesticides (OCPs)

Organochlorines are very firm structurally; thus, they are stable. The lipophilic and persistent feature of the majority of the OCPs favored their long-term storage in adipose tissue, and they are released into the circulatory system during harsh environmental situations (Waliszewski et al. 2004; Ragab et al. 2016). Structurally, organochlorines can be grouped in five classes based on their structure (Blus 2003).

2.4.1.1 DDT and Its Homologs

The important homologs are dichlorodiphenyldichloroethane (DDD) and dichlorodiphenyldichloroethylene (DDE). The field half-life of several organochlorines (DDD, DDE, and DDT) is approximately 15 years. However, DDT may persist within the human body for more than 50 years (Mrema et al. 2013).

2.4.1.2 Hexachlorocyclohexane (HCH)

HCH is one of the most widely used OCP insecticides for protecting standing crops from various pests. At present, HCH has emerged as a global concern for its persistent bioaccumulative nature and toxicity to non-targeted organisms (Nayyar et al. 2014).

2.4.1.3 Cyclodienes

Cyclodienes are one of the diverse groups of organochlorine insecticides. These groups include several known organochlorine pesticides. Among these, aldrin, chlordane, dieldrin/endrin, endosulfan, and heptachlor are noteworthy.

2.4.1.4 Toxaphene

It is a composite technical mixture of 202 polychlorinated bornanes and other camphenes. Toxaphene was one of the most common globally applied chlorinated pesticides. It is no longer used due to its persistence, bioaccumulative potential, and nontarget toxicity (EC 2005).

2.4.1.5 Mirex and Chlordecone

Mirex is the cyclodiene pesticide used primarily as a stomach poison against ants. Chlordecone is a derivative pesticide of mirex, chlordecone. In vertebrates, mirex exhibits an acute oral toxicity. The persistence and bioaccumulation potential of mirex are very high. Exposure to mirex leads to hyperexcitability, fatty livers, and impairment in the reproductive system, whereas estrogen-controlled processes are affected by chlordecone (Faroon et al. 1995; Dai et al. 1998).

2.4.2 Organophosphate Pesticides

These pesticides consist of a varied group of chemical compounds. Organophosphate pesticides are the esters that result from phosphoric acid. At present, these pesticides are commonly used as pesticides and herbicides. In chemical warfare, these compounds are also used as nerve gas. Many compounds, such as sarin, soman, and tabun, used as nerve gas compounds, are highly toxic agents of this group. These compounds are generally thiol esters or acid anhydride derivatives of phosphorus-containing acids. Globally, more than 100 such pesticides are used, and most of them are compounds of either dimethyl phosphoryl or diethyl phosphoryl (WHO 1986; Taylor 1996).

2.4.3 Carbamate Pesticides

Carbamate pesticides are the dimethyl N-methyl carbamic acid derivative. These are used as fungicides, herbicides, nematicides, and insecticides. Commonly used carbamates are molinate, thiobencarb, propoxur, methiocarb, disulfiram, carbaryl, and pyridostigmine. The toxicity of carbamates varies on the basis of their molecular structure. Acute poisoning by organophosphate or carbamate insecticides is common and often severe.

2.4.4 Pyrethroid Pesticides

These pesticides are natural insecticides obtained from the extract of flowers of chrysanthemum. Structurally, pyrethroid pesticides are the esters of chrysanthemic acid (Soderlund 2012). It causes instabilities in the dynamics of sodium cation channels in the nerve cell membrane, which causes an increase in the opening time of the sodium channels. Such an increase resulted in neuronal hyperexcitation (Narahashi 1996; Garcia et al. 2012; Perry et al. 2013). Efforts for the chemical synthesis of synthetic pyrethroids became successful. Presently, the number of different active pyrethroids is more than 1000. Most pyrethroid insecticides possess several positive features, including low toxicity to fowls and mammals but high toxicity to insects (lethal dose against insects is very low) and quick action, particularly against chewing insects.

3 Pesticide Consumption Scenario in India

In India, the production of pesticides began in 1952. It was initiated with the production of BHC in Calcutta (presently Kolkata). Presently, India is Asia's second-largest pesticide manufacturer, after China, and twelfth internationally (Mathur and Tannan 1999; FAO 2018). In India, technical-grade pesticide production has consistently risen from 5000 metric tons to 102,240 metric tons between 1958 and 1998. The estimated demand for pesticides was to be at Rs. 22 billion (USD 0.5 billion) in 1996–97, about 2% of the overall global market (Kumar 2013). The amount of pesticide use in India has increased a hundredfold over the previous seven decades, from 154 MT in 1953–54 to 57,000 MT in 2016–17. The amount of pesticides India used in 2020 was 61,000 tons. During that time, the total pesticide used in Brazil, China, and Argentina was 377,000 tons, 273,000 tons, and 241,000 tons, respectively. In India during 1994–1995, the use of pesticides was the most in a single year, amounting to 80,000 tons (Chand and Birthal 1997; Agnihotri 2000; Chelliah et al. 2007; FAO 2018). However, pesticide use decreased between 2000 and 2010 due to the prevention or limit on the use of organochlorine pesticides, particularly DDT, BHC, aldrin, etc. One of the key factors for such reduction of pesticide usage was the emergent implementation progression of the program for integrated pesticide management following the Stockholm Convention (Mansouri et al. 2017; Berg et al. 2017). The proper advantage of pesticide application in India is not achieved due to the use of less effective pesticides and the deficiency of knowledge for the proper application of pesticides. Inadequate restrictions for pesticide use have accelerated the presence of pesticide residue in food items in India.

4 Pesticide Contamination of the Environment

Human and industrial activities are responsible for pesticide contamination of the environment. The application and exploitation of chemicals to mitigate crop pests have been enhanced in the agriculture sector. Vast amounts of pesticides and other chemical ingredients are utilized to ensure high productivity and security. Merely 2–3% of the pesticides are used to mitigate the effect of pests at their target point. These leftover pesticides may enter soil water through surface runoff, leaching, and percolation. Ultimately, it reaches the biota and finally human beings through the food chain (Morillo and Villaverde 2017; Villaverde et al. 2017; USEPA).

In agriculture, pesticides are applied to either crops or soil or both. Usually, pesticides are applied as liquid sprays for crops. In the soil, pesticides are used as granules, injected, or sprayed as a fumigant, which are then mixed into the soil's surface. The seeds are occasionally treated with pesticides before seeding (Berg et al. 1999). Thus, pesticides can come into the environment from diverse sources such as runoff from agricultural events, manufacturing and formulation of industries, municipal waste, and soil infiltration.

These pesticide pollution sources are of two types, i.e., point and nonpoint sources. The point sources are distinct pollution sources; on the other hand, nonpoint sources are from a vast area. It is difficult to stress the sources of pollution in an expanded area, such as the transmission of pesticides from several sources to terrestrial and aquatic ecosystems. The familiar point sources for pesticide pollution are agricultural discharge, wastewater emission from agro-industries, and municipal waste. Another primary point source of pesticide contamination in the soil is the washing of spraying containers (Neumann et al. 2002; Spanoghe et al. 2004). On the contrary, nonpoint pollutants do not have any specific discharge point. Its management is challenging because its origin point is not identifiable and needs additional care (Müller et al. 2002).

The physicochemical features of pesticides are responsible for their conveyance and leaching to groundwater. The spreading of pesticides in the soil relies on their solubility in water bodies, their persistent nature, and their adsorption competency of the soil particles. Pesticide solubility is directly linked to the possibility of leaching in the soil. In addition to solubility, the partition coefficient of pesticides has a key role in the pesticide movement. The soil particles absorb several pesticides, and these are not leached out. The withholding of pesticides in the soil and sediments is proportional to the presence of organic matter (Cadkova et al. 2013).

5 Fate of Pesticides in the Environment

The properties of pesticides, soil composition, site condition, and management methods may affect the fate of pesticides (Lourencetti et al. 2008). Their mobility also makes them hazardous to life systems. For that reason, their persistence in the environment may be lethal. Thus, more persistent pesticides cause a more adverse effect on human health. The environmental persistence of pesticides is linked with the physicochemical properties of pesticides and is supported by a lack of pesticide-degrading organisms. The pesticides are also degraded to a lesser extent by other factors such as heat, humidity, light, or vitalization (Sibanda et al. 2011). However, microorganisms are capable of extensive pesticide degradation that persists in the environment. It has been assessed that nearly 99% of the total applied pesticides fail to reach target pests and are coming to the environment, leading to adverse effects on the biotic community of the ecosystem together with humans (Hernández et al. 2013).

6 Threat from Residual Pesticides on the Biosphere

Anthropogenic activities are steadily increasing global environmental contaminants. Among these, pesticides are of great concern owing to their high persistence and toxicity (Ollera et al. 2011). Pesticides in water, soil, and the aerial atmosphere are a significant threat to the environment. Pesticides are the second-largest contaminants in drinking water (Izadifard et al. 2013; Gul et al. 2015). Most pesticides are harmful to living creatures. The intensity may vary with time, quantity, features of the target organism, and environmental situations.

6.1 On Terrestrial Biodiversity

Pesticide exposure is responsible for sublethal effects on terrestrial plants and eliminating nontarget plants. Through the volatilization process, the applied phenoxy herbicides may damage neighboring foliage (Dreistadt et al. 1994). Several herbicides (e.g., glyphosate) make plants disease-prone (Brammall and Higgins 1988), thus hampering their productivity (Locke et al. 1995). Some herbicides, such as imidazolinones, sulfonylureas, and sulfonamides, harshly hamper the productivity of nontarget crops, the natural flora, and wildlife, even at low doses (Fletcher et al. 1993). Pesticides have neurotoxic effects on earthworms. Prolonged exposure to such pesticides leads to their physiological damage (Schreck et al. 2008). The killing of earthworms through residual fungicides in the soil indirectly reduces the bird and mammal population on which they nourish.

Overused pesticides leach into the soil, causing detrimental effects on the microflora. The soil-dwelling microbes help the plant grow appropriately through nutrient uptake, organic matter degradation, and soil fertility escalation. Several fungicides, including chlorothalonil and dinitrophenyl, may interrupt the processes of microbial nitrification and denitrification (Lang and Cai 2009). Triclopyr, a commonly used herbicide, inhibits the microbial transformation of nitrite from ammonia (Pell et al. 1998). Herbicide glyphosate is responsible for impairing the growth and activity of nitrogen-fixing soil bacteria (Santos and Flores 1995). On the other hand, 2,4-D prevents the microbial conversion of ammonia into nitrates in soil (Frankenberger et al. 1991). Triclopyr is toxic to several species of mycorrhizal fungi associated with nutrient uptake (Chakravarty and Sidhu 1987).

6.2 *On Aquatic Biosphere*

The applied pesticides initially contaminate the place of their application, ultimately reaching the aquatic systems. These flowing pesticides are detrimental to nontarget creatures of water bodies. Thus, mistreated pesticides are responsible for a decrease in the population of various aquatic organisms. These pesticides enter the biological system of these aquatic animals through three main routes (Helfrich et al. 2009): through direct skin absorption, through respiratory organs in the course of breathing (e.g., gills), and orally during drinking polluted water. The populations of terrestrial plants, creatures, and birds are also affected by such pesticides. Accretion of pesticides in the food chains through biological magnification is of most conspicuous concern as it directly disturbs the predators and raptors.

Application of pesticides accordingly diminishes the extent of weeds, scrubs, and insects. Such application of insecticides, herbicides, and fungicides also reduces the populations of endangered fauna, including birds. Likewise, prolonged regular utilization of such pesticides leads to bioaccumulation and biomagnification. Nearly 80% of the oxygen in the aquatic ecosystem is in a dissolved condition produced by aquatic plants. This dissolved oxygen is vital for the sustenance of aquatic life. Thus, the destruction of aquatic plants as a side effect of the application of herbicides causes a notable decrease in oxygen. This decreased oxygen level ultimately leads to oxygen deficiency in the fish and affects fish production (Helfrich et al. 2009).

Atrazine indirectly upsets the immune system of several amphibians and some fish species (Forson and Storfer 2006; Rohr et al. 2008). The herbicide glyphosate is highly detrimental to tadpoles and juvenile frogs, leading to a high mortality rate (Relyea 2005). Malathion is very sensitive to plankton and periphytons. Chlorpyrifos and endosulfan also harm amphibians (Sparling and Feller 2009). The accumulation of pesticides in the tissues of various bird species speeds up their death. The bald eagle population in the USA dropped mainly due to exposure to DDT and its derivatives (Liroff 2000).

6.3 *On Human Health*

According to WHO, the cases of pesticide poisoning in developing countries are nearly 3,000,000, with 220,000 deaths yearly (Lourencetti et al. 2008). In these nations, nearly 2.2 million people are at greater risk of exposure (Hicks 2013).

Predominantly, people in rural areas are more susceptible to pesticide poisoning, as there is every possibility of direct daily interaction with pesticides. The newborns, toddlers, and agricultural workers mainly involved with the application of pesticides are more susceptible to the noxious effects of pesticides (Pieper 2005). Pesticides may come into the body of humans in several ways, i.e., ingestion, inhalation, or penetration through the skin (Spear 1991). However, contaminated food is the most important route for the intake of pesticides within the human body. To reach human tissues, toxic compounds must cross several barriers (Hayo and Werf 1996). Despite several defense mechanisms in the human body to remove the toxic compounds, it is absorbed in the circulatory system (Jabbar and Mallick 1994). The toxic compound accumulates within the body system and upsets the human body when it reaches beyond the threshold level (Hayo and Werf 1996). The effect of pesticides on the human body varies significantly based on the nature of the pesticide, dose, age of the victim, etc. Thus, symptoms may manifest within a few hours or come months or years later, i.e., the impacts may be acute or chronic.

6.3.1 Effect of Organochlorine Pesticides

In the human body, these pesticides or their metabolites commonly impair the central nervous system through alternating electrophysiological properties of the nerve cell membranes. As a result, an alteration in the kinetics of the movement of K+ and Na + through these membranes takes place (Narahashi et al. 1992). Ultimately leads to death due to acute poisoning and seizures from apnea (Tordoir and van Sittert 1994). The organochlorine pesticides increase the patient's irritability to various environmental influences such as touch, sound, and light and cause nervousness, nausea, dizziness, vomiting, seizures, and shocks (Lourencetti et al. 2008). Thus, it may cause many neuropsychiatric ailments, including depression, anxiety, and mood alterations.

6.3.2 Effect of Organophosphate Pesticides

In the human body, these pesticides affect the central nervous system by obstructing the activity of the enzyme acetylcholine. The symptoms of toxicity are headache, coma, dizziness, convulsions, nausea, cramps, emesis, excessive salivation, diarrhea, lacrimation, loss of reactions, and urinary inconsistency, leading to fatality (El Nemr and El-Sadaawy 2016; El Nemr et al. 2016). The mode of action of pesticides in the human body varies extensively. The organophosphates and carbamates obstruct the nervous transduction.

6.3.3 Effect of Carbamate Pesticides

The symptoms of carbamate poisoning appear in different organs. In bronchial trees, the symptoms are cough, wheezing, dyspnea, increased secretions, cyanosis, muscarinic bronchoconstriction, and pulmonary edema. In the glandular system, stimulation symptoms are sweating, increased salivation, and lacrimation. In the cardiovascular system, the effects are hypotension and bradycardia. In the eye with blurred vision and miosis. In the urinary bladder, dysfunction and incontinence appear frequently. Gastrointestinal manifestations are abdominal cramps, nausea, diarrhea, vomiting, and incontinence. Motor activity is compromised, and symptoms appear as fasciculation, muscle twitching, weakness, depression of respiratory and circulatory systems, and cramps. The central nervous system is affected, and symptoms appear as hypothermia, restlessness, confusion, emotional lability, slurred speech, drowsiness, Cheyne-Stokes respiration, ataxia, tremor, generalized weakness, convulsion, coma, and areflexia (Garrettson 1993; Reigart and Roberts 2001).

6.3.4 Effect of Pyrethroid Pesticides

In the human body, these pesticides are responsible for autism, neurotoxicity, and reproductive toxicity. In aquatic species, it leads to harsh oxidative stress and histopathological abnormalities (Ahamad and Kumar 2023).

A number of currently used pesticides (CUPs), such as bioethanol, propiconazole, and mancozeb, antagonize androgen receptors. On the other hand, terbuthylazine, propiconazole, bitertanol, malathion, and cypermethrin are inducers of estrogen receptor transactivity and the activity of aromatase (Kjeldsen et al. 2013). Some other CUPs, such as chlorpyrifos, carbendazim, and malathion, impair the development of neurons in youngsters (Hertz-Picciotto et al. 2018), destroy DNA (Wang et al. 2018), and interrupt the cell cycle mechanisms (Sharma et al. 2022), respectively. Exposure to the pesticides of the group pyrethroids may cause skin allergy, aggressiveness, hyperexcitation, and developmental defects, including genital defects and tremors. This exposure may lead to Parkinson's and Alzheimer's diseases (Casida and Durkin 2013).

7 Biodegradation of Pesticides in the Ecosystem

Soil acts as an active filter for pesticides. The process of sorption of pesticides to the soil is strongly linked to the condition of pesticides in the environment. However, the physicochemical features of the pesticides and the properties of the soil and its management are the key factors that determine the pesticide's behavior and movement through the soil compartments (Briceño et al. 2007). After coming in contact with the soil, the pesticides are adsorbed. The adsorption by the soil affects other processes, including leaching, bioavailability, and toxicity toward nontarget organisms (Morillo and Villaverde 2017). Soil organic matter (SOM) content strongly influences the uptake process of organic pesticides (Kah and Brown 2006). This is due to the availability of several binding sites for organic pollutants, particularly for hydrophobic compounds (Alvarez et al. 2017). The involvement of the mineral composition in pesticide adsorption may also be considered, particularly for soils that are poor in organic matter (Kodešová et al. 2011; Cadkova et al. 2013).

Bulk handling of pesticides or utilization in the field or inadvertent release contaminates the soil and water, both surface and ground (Singh and Walker 2006). Various factors influence the biodegradation rate of soils. These are environmental factors (temperature and moisture), physicochemical features of soil, and availability of nutrient sources. These factors influence the microbial population and their activities (Besse-Hoggan et al. 2009). The primary microbial product may be repeatedly degraded and/or transformed by other microorganisms toward entire degradation by the microbial consortium. However, persistent pesticides and dead-end metabolites mount in nature, combined with the soil humus, or move into the food chain, resulting in biomagnification (Fig. 2).

Pesticides are principally applied to the crops and soil of the agricultural field that generally acquires these compounds. However, contamination sites also contain activated sludge, sewage sludge, effluent from insecticide industries, wastewater, sediments, natural waters, adjoining manufacturing areas of insecticide, and even

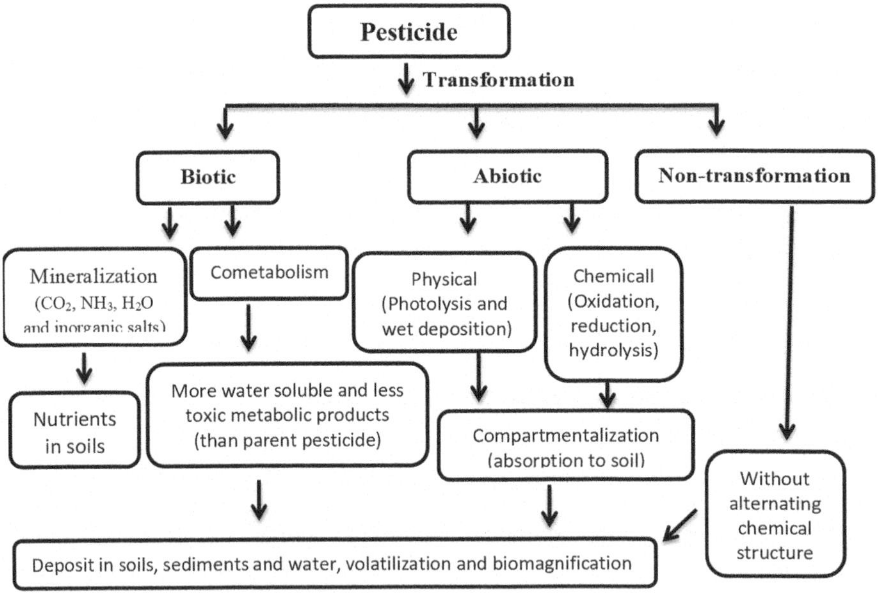

Fig. 2 Fate of Pesticides in environment

certain live organisms. The insecticide-degrading microbes have been screened from these insecticide-contaminated sites (Fuentes et al. 2017). Bioremediating pesticides from polluted soil using microorganisms and phytoremediation is a safe and effective technique. (Zhang et al. 2011).

8 Systems of Bioremediation

8.1 Slurry System

Managing pesticide-contaminated soils under controlled environmental conditions in slurry bioreactors (SBs) is one of the best ex situ bioremediation techniques (Venkata et al. 2006). In this bioreactor, a three-phase (liquid, solid, and gas) collaborating condition is used to create, for escalation, the rate of bioremediation of soil-bound and water-soluble pollutants. The water slurry of the soil and biomass is enabled to degrade the target pollutants (Kensa 2011). By controlling the operative parameters, it is possible to minimize the processing period compared to that of the in situ bioremediation process. The feasible controlling operative parameters include nutrients, mixing, temperature, pH, and supplements, including surfactants (Robles-Gonzalez et al. 2008; Venkata et al. 2006). The SBs particularly upsurge the mass transfer rates, i.e., desorption of contaminants from soil. This transfer translates into faster and accelerated biodegradation rates of contaminants, especially pesticides (Robles-Gonzalez et al. 2008). Native soil microbiota that can metabolize the target contaminants are used for this purpose (Venkata et al. 2006).

8.2 Biobed System

A biobed system is an effective, simple biological system to decrease environmental pesticide contamination. It is adopted mainly to fill the spraying equipment, a typical point source of contamination (Castillo and Tortensson 2007; Castillo et al. 2008). Biobeds are constructed for the adsorption and probable degradation of organic biomixtures consisting of topsoil, peat, and straw (25: 25: 50 vol. %) are filled within a deep pit (60 cm) in the ground. The surface of which is covered with a grass layer (Castillo et al. 2008; Castillo and Tortensson 2007). These absorptions facilitate the biodegradation of the pesticide through the microorganisms of the biomixture, i.e., the core of a biobed. In the biomixture, straw encourages the growth of ligninolytic microorganisms. These ligninolytic microbes produce extracellular ligninolytic enzymes, which include peroxidases and phenoloxidases. The peat ensures the sorption capability and moisture regulation. It also minimizes the pH of the biomixture to make it favorable mainly for fungi and the pesticide-degrading enzymes produced by them (Castillo and Tortensson 2007; Castillo et al. 2008).

9 Actinobacteria

Actinobacteria are Gram-positive filamentous bacteria ubiquitous in nature (Roy and Sen 2002; Bull 2011). The Actinobacteria class includes 5 subclasses, 10 orders, and 56 families (Euzeby 2015). Until 2016, 374 genera of actinobacteria had been documented (Subramani and Sipkema 2019). They exhibit wide physiological and morphological diversity (Roy and Sen 2011, 2013). Moreover, they are the most prominent source of several enzymes with great catabolic activities to degrade pesticides. Actinobacteria can metabolize these noxious compounds through these diverse enzyme systems for growth and development. Thus, pesticide bioremediation employing actinobacteria may be an eco-friendly economic method (Chaturvedi and Khurana 2019). The dominant genus of actinobacteria is *Streptomyces*. Other non-Streptomyces actinomycetes genera include Frankia, *Janibacter*, *Nocardia*, *Kokuria*, *Mycobacterium*, *Nocardia*, *Rhodococcus*, *Arthrobacter*, *Pseudonocardia*, *Arthrobacter*, *Actinomyces*, *Kitasatospora*, *Micromonospora*, *Micrococcus*, and *Rhodococcus* (Barka et al. 2015; Alvarez et al. 2017; Amin et al. 2020; Ariffin and Rahman 2020). The genus *Streptomyces* possesses many species with a vast multiplicity of culture parameters (Shepherd et al. 2010). In contrast, species of non-Streptomyces actinomycetes are reported with low frequency under usual conditions. The majority of them necessitate defined methods for isolation, maintenance, and cultivation (Baltz 2006; Subramani and Aalbersberg 2013).

The phylum Actinobacteria denotes the most documented microbial group capable of producing bioactive compounds (Roy and Sen 2002, 2013; Roy et al. 2006). As a result, actinobacteria have attained much attention in several applications in pharmaceuticals, food industries, biotechnology, agriculture, and the enzyme industry (Mohammadipanah and Wink 2016; Azman et al. 2016; Zhao et al. 2016; Banerjee et al. 2016, 2017). Moreover, there is also attention to exploring Actinobacteria to mitigate the issues related to the clean-up of pollutants. The members of Actinobacteria can be provided as an economical and safe biological tool for eliminating pollutants, as they can metabolize the noxious waste for their growth and development (Chaturvedi and Khurana 2019). The members of the Actinobacteria can utilize pesticides as a carbon source, resulting in the complete degradation of the inorganic compounds entirely and transforming them into non-hazardous to the environment (Jagannathan et al. 2021).

9.1 *Actinobacteria with Pesticide-Degrading Abilities*

The members of the actinobacteria can flourish both in terrestrial and aquatic ecosystems in extreme environmental conditions. They are well known for producing an extensive range of secondary metabolites containing several antibiotics and enzymes. Extensive metabolic diversity is reported within the members of the actinobacteria. Several studies have been carried out on the capability of actinobacteria to degrade pesticides. Several non-Streptomyces actinobacteria have been reported to degrade pesticides from many genera, including *Rhodococcus*, *Gordonia*, *Arthrobacter*, *Frankia*, *Nocardioides*, *Micrococcus*, and *Kocuria*. These actinobacteria can degrade, metabolize, and grow on a good number of chemical groups of pesticides, together with organochlorine (endosulfan, lindane, pentachloronitrobenzene, pentachlorophenol, hexachlorobenzene), organophosphorus (chlorpyrifos, dimethoate), and benzimidazole (carbendazim), triazine (atrazine, simazine, terbuthylazine), and synthetic pyrethroid (deltamethrin, cypermethrin) compounds (Alvarez et al. 2017; Cycoń et al. 2017). Actinobacteria capable of degrading pesticides are listed in Table 3.

9.2 Factors Enhancing Pesticide-Degradation Capability of Actinobacteria

The proficiency of a microorganism is mostly reciprocated to the nutritional and environmental situations. There is no general phenomenon or regulation to ascertain microorganisms' growth and production capability. Under varying cultivation parameters, the production may be stimulated by the fullest expression of the genes (Roy and Sen 2011, 2013). Likewise, it is also important for the members of actinobacteria responsible for bioremediation of pesticides to ensure their optimum utilization.

9.2.1 Fermentation

Pesticides are generally degraded rapidly by various actinobacteria, particularly the members under the genus *Streptomyces* in liquid media under experiment. It takes only days, sometimes merely within a few hours. The duration length may vary depending on the concentration, nature of the pesticide, and the strain involved. Lin et al. (2011) revealed that *Streptomyces* sp. HU-S-01 is capable of completely degrading cypermethrin with a concentration of 30 mg/l within 24 and 30 h; however, at higher concentrations (150 to 250 mg/l), degradation remains incomplete even after 48 h of incubation. Lai et al. (2011) reported that *S. aureus* HP-S-01 can completely degrade deltamethrin (50–300 mg/l) within 7 days. The degradation ability of the microorganisms concerned depends on various factors (Table 4).

9.2.2 Supplementation

Supplement medium is a vital variable for the pesticide-degradation ability in various actinobacteria. In most cases, the pesticide-degradation capability of the members of the actinobacteria is assessed using the compound only in the form of C, N,

Table 3 Sources of pesticide-degrading actinibacteria

Pesticides	Actinobacteria	Reported from	References
α-endosulfan, β-endosulfan	*Arthrobacter* sp.	Soil, India	Kumar et al. (2008)
Acetanilides	*Streptomyces* sp.	Soil sample	Sette et al. (2005)
Aldrin	*Streptomyces*, *Nocardia*, *Mycobacterium*, *Micromonospora* sp.	Soil	Ferguson and Korte (1981)
Aryloxyphenoxy propanoate	*Rhodococcus ruber* JPL-2	Soil, China	Hongming et al. (2015)
Atrazine	*Arthrobacter* sp. DAT1	Soil, China	Wang et al. (2012)
Atrazine	*Rhodococcus* sp. BCH2	Soil, India	Kolekar et al. (2014)
Atrazine	*Rhodococcus* sp. MB-P1	Soil, India	Fazlurrahman et al. (2009)
Atrazine	*Arthrobacter* sp. AKYN10	Soil, India	Sagarkar et al. (2014, 2016)
Atrazine	*Arthrobacter* sp. C3	Soil, China	Wang et al. (2016)
Atrazine	*Frankia alni* ACN14a	Not reported	Rehan et al. (2014)
Atrazine	*Frankia* sp. EuI1c	Not reported	Rehan et al. (2014)
Atrazine	*Rhodococcus* sp.	Not available	Behki et al. (1993)
Carbendazim	*Nocardioides* soli	Soil, China	Sun et al. (2014)
Carbendazim	*Rhodococcus qingshengii* djl-6	Soil, China	Xu et al. (2007)
Carbendazim	*Rhodococcus erythropolis* djl11	Soil, China	Zhang et al. (2013)
Carbendazim	*Nocardioides* sp. SG-4G	Soil, Australia	Pandey et al., (2010a, b)
Carbendazim	*Rhodococcus jialiangiae* djl6–2	Sludge, China	Wang et al. (2010)
Carbaryl	*Micrococcus* sp.	Garden soil	Doddamani and Ninnekar (2001)
Carbofuran	*Streptomyces alanonosinicus*	Saline soil	Jayabarath et al. (2010)
Chlordane	*Streptomyces* sp. A5	Soil, Argentina,	Cuozzo et al. (2012)
Chlorpyrifos	*Kocuria kristinae*	Soil, India	Hamsavathani et al. (2017)
Chlorpyrifos	*Gordonia* sp. JAAS1	Soil,, India	Abraham et al. (2013)
Chlorsulfuron Metsulfuron-methyl	*Streptomyces griseolus*	Soil sample	Zanardini et al. (2002)
Cypermethrin Deltamethrin	*Streptomyces aureus* HP-S-01	Sludge, China	Chen et al., (2011, 2012a, b)
Diazinon	*Streptomyces* spp.	Soil	Briceño et al. (2015)
DDT	*Nocardia* sp., *Streptomyces* sp.	Soil	Chacko et al. (1966)

(continued)

Table 3 (continued)

Pesticides	Actinobacteria	Reported from	References
Dimethoate	*Kocuria turfanensis*	Soil, India	Barot and Chaudhari (2020)
Diuron	*Streptomyces albidoflavus*	Sil, Spain	Castillo et al. (2006)
Dioxin	*Janibacter* sp. YA	River sediment, Japan	Iwai et al. (2005)
Diuron	*Arthrobacter* sp. BS1, BS2	River soil, France	Lamrani et al. (2014)
Endosulfan	*Rhodococcus* MTCC 6716	Gut microflora of an earthworm	Verma et al. (2006)
Glyphosate	*Arthrobacter atrocyaneus* ATCC 13752	Not available	Pipke and Amrhein (1988)
Glyphosate	*Arthrobacter* sp. GLP-1	Not available	Pipke and Amrhein (1988)
Hexachlorobenzene	*Nocardioides* sp. PD653	Soil, Japan	Takagi et al. (2009)
Hexachlorocyclohexane	*Arthrobacter fluorescens* *Arthrobacter giacomelloi*	Soil sample	De Paolis et al. (2013)
Lindane	*Kocuria* sp. DAB-1Y	Soil, Lucknow, India	Kumar et al. (2016)
Lindane	*Streptomyces* consortium (strains A2, A5, A11, M7)	Soil, Argentina	Raimondo et al. (2020)
Lindane (γ-HCH)	*Streptomyces* sp. M7 α-HCH, β-HCH,	Wastewater sediment, Argentina	Benimeli et al. (2003, 2006) and Sineli et al. (2016)
Lindane	*Streptomyces* sp. A2, A5, A11	Soil, Argentina	Fuentes et al. (2011)
Metamitron	*Rhodococcus* spp.	Contaminated soil	Fang et al. (2016)
Methoxychlor	*Streptomyces* sp. A14	Sediments, Argentina	Bourguignon et al. (2014)
Parathion	*Streptomyces venezuelae* ACT1	Marine water	Naveena et al. (2013)
Pentachloronitrobenzene	*Nocardioides* sp. strain PD653	Soil, Japan	Takagi et al. (2009)
Pentachlorophenol	*Kocuria* sp. CL2	Sludge	Karn et al. (2011)
Pentachlorophenol	*Janibacter* sp. AS23	Sediments, Tunisia	Khessairi et al. (2014)
Propazine and Simazine	*Rhodococcus* sp. B-30	Soil, USA	Behki and Khan (1994)
S-ethyl-N,N-dipropylthiocarbamate	*Arthrobacter* sp. strain TE1	Soil isolate	Tam et al. (1987)
Simazine	*Arthrobacter* sp. SD3–25	Soil, China	Guo et al. (2014)
Sulfonylureas	*Streptomyces griseolus*	Soil	Zanardini et al. (2002)

(continued)

Table 3 (continued)

Pesticides	Actinobacteria	Reported from	References
Terbuthylazine	*Arthrobacter aurescens* TC1	Soil, USA	Strong et al. (2002); Silva et al. (2015)
Triazinones	*Arthrobacter* spp. DSM 20389	Sediment, Germany	Wang et al. (2021)

and P as the nutrient sources. On the other hand, in the case of pesticide co-metabolization, an additional nutrient source may increase the degradation of pesticides. Jayabarath et al. (2010) reported a 30% enhanced degradation of pesticide carbofuran in a glycerol asparagine supplemented broth contaminated with 20 mg/l pesticide in comparison to that of pesticide also as a source of carbon inoculated individually with *Streptomyces alanosinicus* and *Streptomyces atratus*. Likewise, one strain of *Streptomyces* sp. exhibited favored growth and enhanced ability due to starch as a co-substrate for eliminating pesticide propoxur (Rahmansyah et al. 2012).

9.2.3 Effect of Monoculture and Consortia

Generally, pure or monoculture is conducted for biotechnological processes for compelling performances. Even so, in many cases, microbial consortia isolated from natural environments have found potential for their productivity in terms of detoxification processes of pesticides. Microbial consortia have been used to eliminate pesticides from wastewater and have performed favorably. However, microbial consortia of natural sources also have some hardships, e.g., complications related to the culture conditions, low competence, time-consuming operation cycles, poor stability, and controllability (Ding et al. 2016). Therefore, increasing attention is being paid to the design and formulation of microbial consortia. For effective design and formulation of microbial consortia, selecting competent strains with suitable combining microorganisms with positive interactions ensures enhanced synergistic pesticide-degrading capabilities (Hu et al. 2017; Zafra et al. 2017).

A microbial consortium can catalyze several processes to stabilize environmental perturbations through different metabolic pathways and enzymatic systems of different organisms. Thus, for the bioremediation of complex structured pesticides, the division of labor among the consortium members ensures several degradation phases (Hays et al. 2015). Research for developing potent strategies for pesticide biodegradation utilizing microbial consortium is the focus area of bioremediation. The elimination of pesticide cypermethrin through a co-culture of Streptomyces *aureus* HP-S-01 and *Bacillus cereus* ZH-3 was found to be an effective co-culture using cypermethrin as the sole source of carbon in liquid media. It also co-metabolized the pesticide in the presence of glucose, particularly with high cypermethrin, even at concentrations of 500 mg/l. At the co-culture, the degradation of cypermethrin was enhanced with a half-life of 13 h (Chen et al. 2012a, b). On the other hand, in the pure cultures, a half-life of around 33–43 h was observed. The

Table 4 Pesticide-degrading actinobacteria used in the bioaugmentation of pesticide-contaminated soils

Actinobacteria	Pesticide degraded	Pesticide dosage	Inoculum size	Pesticide-degrading capability	Reference
Arthrobacter sp.	α-endosulfan β-endosulfan	50 mg/kg	2×10^8 (cfu/g)	Degradation of 73% α-endosulfan and 75% β-endosulfan after 6 weeks	Kumar et al. (2008)
Arthrobacter sp. DAT1	Atrazine	400 mg/kg	1.3×10^7 (cfu/g)	>95% of atrazine removal in 3 days	Wang et al. (2013)
Arthrobacter sp. AK-YN10	Atrazine	100 mg/kg	2.5×10^7 (cfu/g)	Complete removal of atrazine within 12 days	Sagarkar et al. (2016)
Arthrobacter sp. SD3–25	Simazine	200 mg/kg	3.3×10^6 (cfu/g)	>90% of simazine removal within 7 days	Guo et al. (2014)
Arthrobacter aurescens TC1	Terbuthylazine	3.8 mg/kg	5×10^7–2×10^8 (cfu/g)	- 95% of terbuthylazine removal within 3 days	Silva et al. (2015)
Arthrobacter fluorescens	Hexachlorocyclohexane (α-, β-, and γ)	100 mg /l	3.5×10^7 cells/ml	65, 50, and 40% of α-, β-, and γ-HCH have been degraded after 72 h	De Paolis et al. (2013)
Arthrobacter giacomelloi	Hexachlorocyclohexane (α-, β-, and γ)		1.8×10^7 cells/ml	88% degradation of α-, 60% of β-, and 56% of γ-HCH have been degraded after 72 h	De Paolis et al. (2013)
Nocardioides sp. C190	Atrazine	1 mg/kg	1×10^5 cfu/g	The Nocardioides start to mineralize atrazine only after a lag of approximately 7 days	Topp, (2001)

(continued)

Table 4 (continued)

Actinobacteria	Pesticide degraded	Pesticide dosage	Inoculum size	Pesticide-degrading capability	Reference
Rhodococcus jialiangiae djl-6–2	Carbendazim	10 mg/kg	1×10^8 (cfu/g)	Almost 100% carbendazim removal after 20 days	Wang et al. (2010)
Rhodococcus qingshengii djl-6	Carbendazim	2. 8 mg/kg	1×10^8 (cfu/g)	> 93% of carbendazim removal after 14 days of incubation	Chuang et al. (2021)
Streptomyces aureus HP-S-01	Cypermethrin, 3- PBA	50 mg/kg	1×10^6 (cfu/g)	Removal of 81.1% cypermethrin and 73.5% 3-PBA after 10 days	Chen et al. (2011)
Streptomyces consortium (strains A2, A5, A11, M7)	Lindane	35 mg/kg	1×10^7 (cfu/g)	Lindane removal from concentrated slurry (soil: Water ratio of 2:3) within 7 days	Saez et al. (2014)
Streptomyces consortium (strains A2, A5, A11, M7)	Lindane	100 mg/kg	1×10^8 (cfu/g)	81–87% of lindane removal in bioaugmented biomixtures at 60 days of incubation	Saez et al. (2018)
Streptomyces nigra LM01	Atrazine and Nicosulfuron		1×10^8 cfu/g	The highest degradation rates for atrazine and nicosulfuron were 55.76% and 42.29% after 5 d, respectively	Zhao and Wang (2024)

Actinobacteria	Pesticide degraded	Pesticide dosage	Inoculum size	Pesticide-degrading capability	Reference
Streptomyces rochei AJAG7	Fipronil	500 mg/l	1×10^7 spores /ml	The time within which the initial fipronil concentration was reduced by 50% (DT_{50}) was 2.8 d and 4.3 d in liquid medium and in soil, respectively	Abraham and Gajendiran (2019)
Thermobifida cellulosilytica TB100 *T halotolerans* DSM 44931 *Streptomyces colicolor* A3	Lindane	55 µg /ml	1×10^8 cfu/ml	The degradation rates of lindane were 83% by *T. cellulosilytica*, 80% by *T. halotolerans*, and 70% by *S. coelicolor* after 30 d incubation	Usmani et al. (2021)

consortium degraded the cypermethrin into several metabolites through hydrolysis of ester linkages and oxidation of 3-phenoxybenzyl. As the intermediate compounds were metabolized, there was no accumulation in the medium (Chen et al. 2012a, b).

Fuentes et al. (2011) noted that the activity of dechlorinase and removal of lindane was increased up to 12-fold in certain mixed cultures compared to that of the pure cultures. It was also observed that in comparison to consortia of two, three, and four Streptomyces, the latter was the most efficient. Saez et al. (2015) reported a high stability of the members of a consortium of the strains of *Streptomyces* sp. A2, A5, A11, and M7, which is also corroborated by the increased growth and capacity to remove lindane at a range of concentrations from 20 to 50 mg/l. When the consortium was acclimated using the sequential increase of the concentrations of lindane, 40–97% of lindane was removed from the liquid medium. However, the elimination was declined to 33–87% without an acclimation period.

10 Strategies for Enhanced Pesticide Biodegradation by Actinobacteria

10.1 Enzyme-Based Pesticide Degradation

It is an advanced process for excluding pesticides from contaminated environments (Verma et al. 2014). This process has various advantages over direct microbial treatment. Such include high reactivity toward recalcitrant pesticides, lesser sensitivity to the concentration of pesticide, coverage of an extensive range of physicochemical gradients in the environment, and simple field application (Gupta et al. 2015). The microbes under the genus *Streptomyces* are a rich source of a good number of oxidoreductases and hydrolytic enzymes suitable for pesticide degradation. The important associated enzymes include amylase, cellulase, protease, esterase, xylanase, nitrile hydratase, dehydrogenase, laccase, and dehalogenase (Karigar and Rao 2011; Shivlata and Satyanarayana 2015). However, the enzymes of the hydrolase family are mainly responsible for the degradation of the pesticides of the pyrethroid group. However, a novel monooxygenase was identified from *Streptomyces* sp. HU-S-01, capable of degrading beta-cypermethrin (Chen et al. 2013). It also exhibited the ability to completely biodegrade deltamethrin, a pyrethroid pesticide (50–300 mg/l), within 7 days. The strain also degraded 3-phenoxybenzaldehyde, a metabolic product of deltamethrin (Chen et al. 2012a, b).

10.2 Enhanced Bioremediation by Immobilized Cells

Several actinobacteria strains have been explored for bioremediation of various pesticides. The immobilized cells of actinobacteria exhibit their increased tolerance and high degradation ability toward these toxic substances. Further, these cells presented effective degradation through the liquid medium (Briceno et al. 2015). This approach has received significant attention in the field of bioremediation and, particularly, in the management of pesticide-contaminated wastewater (Martins et al. 2013; Yadav et al. 2014; Pradeep and Malavalli 2016). Hence, natural and synthetic carriers, both organic and inorganic, have been used to support cell attachment or entrapment (Dzionek et al. 2016). A suitable carrier for cell immobilization should be nontoxic, non-biodegradable, and manageable. A large surface area for cell mass stacking capability and high mechanical, chemical, and biological stability are also important criteria for a standard carrier. Durability, cost-effectiveness, scale-up amenability, and simple separability from the cells and the media are the criteria for a good carrier. Immobilization is a widely acceptable process in bioremediation as it facilitates better cellular sustainability, better tolerance to elevated pollutant concentrations and pesticide mixtures, and enhanced safeguarding of cells from adverse environments.

Immobilization through entrapment is simple, cost-effective, and non-hazardous for the concerned microorganisms and shields the microorganisms from the toxicity of pollutants (Kurzbaum et al. 2017). Immobilization in alginate beads may be a favorable biotechnological approach for Streptomycetes for the degradation of pesticides from wastewater (Fuentes et al. 2013). Likewise, alginate bead-immobilized *Streptomyces* spp. AC1–6 were found to be 20% more efficient for the elimination of diazinon, an insecticide, than that involving the removal of pesticides by *Streptomyces* strains. This strain's highest removal efficiency for chlorpyrifos was 71.05% in the immobilized state compared to 40.17% in the free state. Thus, immobilized cells can be used to remove pesticides. These studies show that the immobilization of actinobacterial cells may be an operative way for pesticide bioremediation. However, thorough research concerning the proper matrices, durability, and reusability of biomass is vital to escalating the effectiveness of these immobilized cells. Several experimental data have shown more encouraging performance for degradation of pesticides in immobilization than in free cell culture—calcium alginate entrapment of *Streptomyces* sp. AC1–6 exhibited 60% higher diazinon removal than free cells (Briceno et al. 2015). The reusability of immobilized *Streptomyces* strains thrice, 96-h periods for lindane degradation by pure and mixed cultures (Saez et al. 2012). Immobilized cells of a strain of *Streptomyces* were found to be more capable of removing insecticide lindane than the free cells (Saez et al. 2012). However, the low constancy of this support at a limited pH range (Dzionek et al. 2016) influenced alternative supports, including polyurethane sponges, volcanic rocks, polyvinyl alcohol-alginate, biocarbon, cloth sachets, and silicone tubes (Saez et al. 2012; Yadav et al. 2014). When immobilized in silicone tubes, various strains of Streptomyces provide improved performance compared to polyvinyl alcohol-alginate in terms of higher microbial growth. Likewise, cloth sachets exhibited an enhancement of degradation of lindane by nearly 25% for both pure and mixed cultures of *Streptomyces* spp. Chlorpyrifos and pentachlorophenol degrading six strains of *Streptomyces* in consortia presented an upsurge of simultaneous degradation of chlorpyrifos and pentachlorophenol by the immobilization (71% and 14%, respectively), in comparison to free cells (40% and 5%, respectively) (Fuentes et al. 2013).

10.3 Exploiting Synergy Between Actinobacteria and Plants

In nature, the communication between microbial communities and plants plays a vital role in both of their metabolic activities. The microbes residing in the rhizosphere commonly escalate the bioavailability of several plant nutrients. On the other hand, the microbes meet their nutritional demand from the plant root exudates (Alvarez et al. 2017). This interaction between plant and microbe interactions may be exploited to boost the capacity of the plants for phytoremediation, leading to the removal of toxicants from the environment. Several members of actinobacteria can improve the phytoremediation ability of several plants through interacting at the rhizosphere (Kidd et al. 2009).

Several members of actinobacteria, particularly under the genera *Streptomyces, Amycolatopsis*, and *Rhodococcus,* have been commonly used for effective phytoremediation. Polti et al. (2011) observed bioremediation of chromium from contaminated soil through inoculation of *Streptomyces* sp. MC1, along with *Zea mays* plant at soil accelerated. The biomass of the *Z. mays* plants as a result of such interaction was also increased. From another.

A similar experimental report showed the enhanced growth of two strains of *Streptomyces* sp., M7 and A5, respectively (Alvarez et al. 2012). Here, 68% removal of pesticide lindane from the contaminated soil was achieved by inoculating these microbes and the maize plant. This bioremediation process utilizing pesticides is eco-friendly and cost-effective, and it can assist in the elimination of noxious pollutants from the contaminated soil.

11 Enzymes Involved in Pesticide Biodegradation

Bioremediation through microorganisms is a practical, eco-friendly process. Among the microbial community, Actinomycetes are an important community of microbes for the degradation and conversion of pollutants, particularly pesticides, with distinct bioremediation potentiality. It is capable of acting on complex and high concentrations of pesticides. Actinomycetes use pesticides as a carbon source for their growth and development and may synthesize various commercially viable metabolites, including antibiotics, enzymes, and proteins. Several metabolic pathways are involved in the degradation of pesticides. These depend on the pesticide's character, environmental situations, and the microorganism involved. These enzymes also uphold the pesticide degradation in target organisms through intrinsic detoxification processes and provide metabolic resistance. The enzymes laccase, oxidoreductase, and peroxidase are predominantly applied to eliminate polyaromatic hydrocarbon compounds from water (fresh and sea) or the soil (Balaji et al. 2014).

It is found that several species of actinobacteria can consume pesticides as a source of carbon and other molecules (Kandpal 2014; Ghosal et al. 2016). The enzymatic catalyzing metabolic reactions include oxidation, hydrolysis, and dehalogenation. Bioremediation mainly depends on microbial biomass's metabolic capability to transform or remove pollutants. Such ability is associated with the bioavailability of the compounds (Ramakrishnan et al. 2011). The enzymes are preferred due to the scope of their application in harsher environmental conditions, i.e., at elevated concentrations of pollutants, pH, temperature, and salinity, where microbial growth may be hindered. They also do not involve a source of nutrient or necessity acclimatization like that of microbial degradation. The enzymatic bioremediation methods are fast, cost-effective, accessible, and precise (Demarche et al. 2012). Moreover, the production of toxic by-products is significantly minimized in enzyme-aided biotransformation. The difficulties interconnected with microbial bioremediation can be resolved through enzymatic technology, which has improved specificity and execution. Nanotechnology's progress in enzymology has presented a new perception of nanozymes in treating toxicants. This progress makes the enzymatic technique more specific, reproducible, and rapid (Raman and Henning 2013).

12 Role of Catabolic Genes and Genetic Engineering for Pesticide Degradation

Due to genetic plasticity, microorganisms have significant potential for bioremediation. For investigation of the genetic basis of the degradation of pesticides, the involvement of catabolic genes and their implementation in recombinant DNA technology are being studied thoroughly. Several pesticide-degrading genes from various microbes have been documented. The degradative catabolic genes are genomic or extrachromosomal and are also reported in transposons. Advancements in metagenomics and sequencing of the whole genome have opened a new arena for investigating novel genes linked with the degradation of pollutants from culturable and non-culturable microbes. For effective execution of this technology for remediation, the information on the genetic basis of pollutant degradation and the interaction of the microbes and environment is inevitable (Hussain et al. 2009). Microorganisms have the inherent property to respond differently toward various microenvironments and acquire biodegradation (Pieper and Reineke 2000). A new report on the metabolic pathways and bottlenecks of degradation is continuing. These accessible molecular toolboxes are reinforced and used. Introduction of such gene(s) within organisms to modify them has to be done following the regulations and norms (Shimizu 2002; Pieper and Reineke 2000; Cases and De Lorenzo 2005). Shen et al. (2009) revealed the degradation of organophosphate pesticides by the genes encoding hydrolases expressed in *Streptomyces lividans*. On the other hand, Hongming et al. (2015) isolated an aryloxyphenoxypropionate (AOPP)-degrading strain, *Rhodococcus ruber* JPL-2. From this JPL2 strain, gene feh, encoding phenoxaprop-p-ethyl hydrolyzing carboxylesterase (FeH), was isolated and cloned in *E.coli* DH5 (Hongming et al. 2015).

13 Conclusion

Pesticides are a different group of chemical compounds being used worldwide. The increased agricultural production to sustain population growth demands increased consumption of chemical pesticides to control the pests of domestic and farm animals. However, indiscriminate use and misuse of pesticides results in their environmental accumulation, leading to several health issues. Presently, nature is facing ever-increasing pollution problems due to the incessant agricultural extension being implemented to combat hunger. This technology makes comprehensive elimination and management of various environmental contaminants possible. To keep pace with the growing tide of environmental pollution, bioremediation is an efficient, eco-friendly, and cost-effective technology for eliminating pesticides from the environment. The diversified microbial population and their survivability in varied climatic conditions make them potential candidates for the bioremediation process. The microbial biomass can completely metabolize toxic compounds effectively, thus eliminating environmental pollutants.

Among the microbial community, actinobacteria is found to be a potent candidate for the sustainable bioremediation of pesticides. Vast genetic diversity and their ability to thrive in harsh ecological conditions make them a great potential candidate for the putrefaction of pesticides. In addition to biodegradation, this ability ensures nutrient recycling and biodegradation for healing pollution and the ability to survive in polluted conditions. This ability may be enhanced through genetic engineering techniques. Several genetic approaches have been established and applied to optimize the production of degrading enzymes from the organisms pertinent to environments. Research on actinobacteria is going on at the molecular level to know the comprehensive physiological and molecular mechanisms for pesticide bioremediation. The progression of modern technologies like genomics, genetic engineering, metagenomics, and proteomics has provided an important breakthrough in pesticide biodegradation. The biotechnology innovation and recombinant DNA technology have unwrapped a new era for efficient, potential, and clean remedial solutions to safeguard the sustainability and green environment.

References

Abhilash, PC and Singh N (2009) Pesticide use and application: An Indian scenario. J Hazar Mater, 165(13): 1–12

Abraham J, Shanker A, Silambarasan S (2013) Role of *Gordonia* sp. JAAS1 in biodegradation of chlorpyrifos and its hydrolysing metabolite 3,5,6-trichloro-2-pyridinol. Lett Appl Microbiol 57:510–516. https://doi.org/10.1111/lam.12141

Abraham J, Gajendiran A (2019) Biodegradation of fipronil and its metabolite fipronil sulfone by *Streptomyces rochei* strain AJAG7 and its use in bioremediation of contaminated soil. *Pestic Biochem Physiol* 155:90–100

Agnihotri N (2000) Pesticide Consumption in Agriculture in India – an Update. Pestic Res Journa 12(1):150–155

Ahamad A, Kumar J (2023) Pyrethroid pesticides: An overview on classification, toxicological assessment and monitoring. J Hazard Mater Adv **10:** 100284

Alvarez A, Saez JM, Costa JSD, Colin VL, Fuentes MS, Cuozzo SA, Benimeli CS, Polti MA, Amoroso MJ (2017) Actinobacteria: current research and perspectives for bioremediation of pesticides and heavy metals. Chemosphere 166:41–62. https://doi.org/10.1016/j.chemosphere.2016.09.070.

Alvarez A, Yañez ML, Benimeli CS, Amoroso MJ (2012) Maize plants (*Zea mays*) root exudates enhance lindane removal by native *Streptomyces* strains. Int Biodeterior Biodegrad 66 (1): 14–18. https://doi.org/10.1016/j.ibiod.2011.10.001.

Amin DH, Abdallah NA, Abolmaaty A, Tolba S, Wellington EMH (2020) Microbiological and molecular insights on rare actinobacteria harboring bioactive prospective. Bull Natl Res Cent 44: 5, https://doi.org/10.1186/s42269-019-0266-8

Ariffin F, Rahman SA (2020) Biodegradation of Carbofuran; A Review. J Environ Microbiol Toxicol 8:50–57. https://doi.org/10.54987/jemat.v8i1.523

Azman AS, Othman I, Velu SS, Chan KG, Lee LH (2016) Mangrove rare actinobacteria: taxonomy, natural compound, and discovery of bioactivity. Front Microbiol 6: 856

Balaji V, Arulazhagan P, Ebenezer P (2014) Enzymatic bioremediation of polyaromatic hydrocarbons by fungal consortia enriched from petroleum contaminated soil and oil seeds. J Environ Biol 35:521–529

Baltz RH (2006) Marcel faber roundtable: is our antibiotic pipeline unproductive because of starvation, constipation or lack of inspiration? J Ind Microbiol Biotechnol 33: 507–513

Banerjee S, Maity TK, Roy RN (2016) Identification and product optimization of amylolytic *Rhodococcus opacus* GAA 31.1 isolated from gut of *Gryllotalpa africana*. J Genet Eng Biotechnol 10: 490–496

Banerjee S, Maity TK, Roy RN (2017) Protease production by thermo-alkaliphilic novel gut isolate *Kitasatospora cheerisanensis* GAP 12.4 from *Gryllotalpa africana*. Biocatal Biotransform 35 (3): 168–176

Barka EA, Vatsa P, Sanchez L, Vaillant NG, Jacquard C, Klen H-P, Clement C, Ouhdouch Y, Wezel GP (2015) Taxonomy, Physiology, and Natural Products of *Actinobacteria*. Microbiol Mol Biol Rev 80(1):1–43. https://doi.org/10.1128/MMBR.00019-15

Barot J and Chaudhari K (2020) Analysis of dimethoate degradation by *Kocuria turfanensis* using GC-MS. Asian J Microbiol Biotechnol Environ Sci 22:107–110

Behki R, Khan SU (1994) Degradation of atrazine, propazine and simazine by *Rhodococcus* strain B-30. J Agric Food Chem 42(5):1237–1241

Behki R, Topp E, Dick W, Germon P (1993) Metabolism of the herbicide atrazine by *Rhodococcus* strains. App Environ Microbiol 59(6):1955–1959

Benimeli CS, Amoroso MJ, Chaile AP, Castro GR (2003) Isolation of four aquatic streptomycetes strains capable of growth on organochlorine pesticides. Bioresour Technol 89:133–138. https://doi.org/10.1016/s0960-8524(03)00061-0

Benimeli CS, Castro GR, Chaile AP, Amoroso MJ (2006) Lindane removal induction by *Streptomyces* sp. M7. J Basic Microbiol 46:348–357. https://doi.org/10.1002/jobm.200510131

Berg VD F, Kubiak R, Benjey WG, Majewski MS, Yates SR, Reeves GL, Smelt JH (1999) Emission of pesticides into the air. water, air, and soil pollution 115:195–210

Berg VH, Manuweera G, Konradsen F (2017) Global trends in the production and use of DDT for control of malaria and other vector-borne diseases. Malar J **16**: 401 https://doi.org/10.1186/s12936-017-2050-2

Besse-Hoggan P, Alekseeva T, Sancelme M, Delort A, Forano C (2009) Atrazine biodegradation modulated by clays and clay/humic acid complexes. Environmental Pollution 157(10):2837–2844

Bhatt P, Gangola S, Bhandari G, Zhang W, Maithani D, Mishra S, Chen S (2021) New insights into the degradation of systematic pollutants in contaminated environment. Chemosphere 268:128827 https://doi.org/10.1016/j.chemosphere.2020.128827

Blus LJ, (2003) Organochlorine pesticides. In: Hoffman DJ, Rattner BA, Burton GA, Cairns J (Eds), Handbook of Ecotoxicology. 2nd edn. Lewis Publishers, Boca Raton, FL, pp. 313–339.

Bourguignon N, Fuentes MS, Benimeli CS, Cuozzo SA, Amoroso MJ (2014) Aerobic removal of methoxychlor by a native Streptomyces strain: Identification of intermediate metabolites. Int Biodeter Biodegr 96: 80–86

Brammall RA, Higgins VJ (1988) The effect of glyphosate on resistance of tomato to Fusarium crown and root rot disease and on the formation of host structural defensive barriers. Can J Bot 66:1547–1555

Briceno G, Fuentes MS, Rubilar O, Jorquera M, Tortella G, Palma G, Amoroso MJ, Diez MC (2015) Removal of the insecticide diazinon from liquid media by free and immobilized Streptomyces sp. isolated from agricultural soil. J Basic Microbiol 55: 293–302. https://doi.org/10.1002/jobm.201300576

Briceño G, Palma G and Duran N (2007) Influence of organic amendment on the biodegradation and movement of pesticides. Crit Rev Environ Sci Technol 37: 233–271. https://doi.org/10.1080/10643380600987406

Buchel KH (1993) Chemistry of pesticides. John Willey and Sons, New york, USA

Bull AT (2011) Actinobacteria of the extremobiosphere, in Extremophiles Handbook, ed K. Horikoshi (Springer), 1203–1240

Cadkova E, Komarek M, Kaliszova R, Vanek A, Balıkova M (2013) Tebuconazole sorption in contrasting soil types. Soil Sediment Contam, 22:404–414 https://doi.org/10.1080/15320383.2013.733448

Cases I, De Lorenzo V (2005) Promoters in the environment: Transcriptional regulation in its natural context. Nat Rev Microbiol 3:105–118

Casida JE, Durkin KA (2013) Neuroactive insecticides: targets, selectivity, resistance, and secondary effects. Annu Rev Entomol 58:99–117

Castillo MA, Felis N, Aragon P, Cuesta G, Sabater C (2006) Biodegradation of the herbicide diuron by *Streptomyces* isolated from soil. Int Biodeter Biodegr 58: 196–202

Castillo MP, Torstensson L, Stenstrom J (2008) Biobeds for environmental protection from pesticide use: A review. J Agric Food Chem 56: 6206–6219

Castillo MP, Tortensson L (2007) Effect of biobed composition, moisture, and temperature on the degradation of pesticides. J Agric Food Chem 55: 5725–5733

Chacko C I, Lockwood JL, Zabik M (1966) Chlorinated hydrocarbon pesticides: Degradation by microbes. Science 154: 893–895

Chakravarty P, Sidhu SS (1987) Effects of glyphosate, hexazinone and triclopyr on in vitro growth of five species of ectomycorrhizal fungi. Eur J Pathol 17:204–210

Chand R, Birthal PS (1997) Pesticide Use in Indian Agriculture in Relation to Growth in Area and Produciton and Technological Change. Indian J Agric Econ 52: 488–498

Chaturvedi S, Khurana SMP (2019) Importance of actinobacteria for bioremediation. Plant Biotechnology: Progress in Genomic Era, Springer, pp. 277–307

Chelliah RJ, Appasamy PP, Sankar U, Pandey R (2007). Ecotaxes on Polluting Inputs and Outputs. In and others (Ed.), Academic Foundation

Chen S, Geng P, Xiao Y, Hu M (2012a) Bioremediation of β-Cypermethrin and 3-Phenoxybenzaldehyde contaminated soils using *Streptomyces Aureus* HP-S-01. Appl Microbiol Biotechnol 94:505–515 https://doi.org/10.1007/s00253-011-3640-5

Chen S, Lai K, Li Y, Hu M, Zhang Y, Zeng Y (2011) Biodegradation of deltamethrin and its hydrolysis product 3-phenoxybenzaldehyde by a newly isolated *Streptomyces aureus* strain HP-S-01. Appl Microbiol Biotechnol 90: 1471–1483. https://doi.org/10.1007/s00253-011-3136-3

Chen S, Lin Q, Xiao Y, Deng Y, Chang C, Zhong G, Hu M, Zhang LH (2013) Monooxygenase, a novel beta-cypermethrin degrading enzyme from *Streptomyces* sp. Plos One, 8: 1–9

Chen S, Luo J, Hu M, Lai K, Geng P, Huang H (2012b) Enhancement of cypermethrin degradation by a coculture of Bacillus cereus ZH-3 and Streptomyces aureus HPS-01. Bioresour Technol 110: 97–104

Chuang S, Yang H, Wang X, Xue C, Jiang J, Hong Q (2021) Potential effects of *Rhodococcus qingshengii* strain djl-6 on the bioremediation of carbendazim-contaminated soil and the assembly of its microbiome. J Hazard Mater 414:125496. https://doi.org/10.1016/j.jhazmat.2021.125496

Cuozzo SA, Fuentes MS, Bourguignon N, Benimeli CS, Amoroso MJ (2012) Chlordane biodegradation under aerobic conditions by indigenous *Streptomyces* strains. Int Biodeterior Biodegradation, 66:19–24

Cycoń M, Mrozik A, Piotrowska-Seget Z (2017) Bioaugmentation as a strategy for the remediation of pesticide-polluted soil: a review. Chemosphere, 172:52–71

Dai D, Rose RL, Hodgson E (1998) Toxicology of environmentally persistent chemicals: mirex and chlordecone Rev Toxicol 2:477

De Paolis MR, Lippi D, Guerriero E, Polcaro CM, Donati E (2013) Biodegradation of α-, β-, and γ-Hexachlorocyclohexaneby *Arthrobacter fluorescens* and *Arthrobacter giacomelloi* Appl Biochem Biotechnol 170: 514–524 https://doi.org/10.1007/s12010-013-0147-9

Demarche P, Junghanns C, Nair RR, Agathos, SN (2012) Harnessing the power of enzymes for environmental stewardship. Biotechnol Adv 30: 933–953

Ding MZ, Song H, Wang EX, Liu Y, Yuan, YJ (2016) Design and construction of synthetic microbial consortia in China. *Synth Syst Biotechnol* 1: 230–235

Doddamani HP, Ninnekar HZ (2001) Biodegradation of carbaryl by a *Micrococcus* species Curr Microbiol 43: 69–73, https://doi.org/10.1007/s002840010262

Dreistadt SH, Clark JK, Flint ML (1994) Pests of landscape trees and shrubs. An integrated pest management guide. University of California Division of Agriculture and Natural Resources. Publication No. 3359

Dzionek A, Wojcieszy D, Guzik U (2016) Natural carriers in bioremediation: A review. Electron J Biotechnol 23: 28–36

El Nemr A and El-Sadaawy MM (2016) Polychlorinated biphenyl and organochlorine pesticide residues in surface sediments from the Mediterranean Sea (Egypt) Int J Sediment Res 31(1): 144–52

El Nemr A, El-Said GF Khaled A (2016) Risk assessment of organochlorine in Mollusk from Mediterranean Sea Coasts of Egypt. Water Environ Res 88(4): 325–337

European Commission (EC) (2005) Opinion of the scientific panel on contaminants in the food chain on a request from the commission to champlechlor as undesirable in animal feeds. EFSA J. 179:1–32

Euzeby JP (2015) List of Prokaryotic names with standing in nomenclature.: http://www.bacterio.cict.fr

Fang H, Xu T, Cao D, Cheng L, Yu Y (2016) Characterization and genome functional analysis of a novel metamitron-degrading strain *Rhodococcus* spp. MET via both triazinone and phenyl rings cleavage. Sci Rep 6: 32339, https://doi.org/10.1038/srep32339

FAO (Food and Agriculture Organization of the United Nations) (2018) *The future of food and agriculture: Alternative pathways to 2050*. Summary version. Rome. http://www.fao.org/3/CA1553EN/ca1553en.pdf.

Faroon O, Kueberuwa S, Smith L, Derosa C (1995) Toxicol Industr Health 11:1–203

Fazlurrahman, Batra M, Pandey J, Suri CR, Jain RK (2009) Isolation and characterization of an atrazine-degrading *Rhodococcus* sp. strain MB-P1 from contaminated soil. Lett Appl Microbiol 49:721–729. https://doi.org/10.1111/j.1472-765X.2009.02724.x

Fenik J, Tankiewicz M, Biziuk M (2011) Properties and determination of pesticides in fruits and vegetables. Trends Anal Chemi 30(6): 814–826

Ferguson JA, Korte F (1981) Epoxidation of aldrin to exo-dieldrin by soil bacteria. Appl Environ Microbiol 34: 7–15

Fletcher JS, Pfleeger TG, Ratsch HC (1993) Potential environmental risks associated with the new sulfonylurea herbicides. Environ Sci Technol 27:2250–2252

Forson DD, Storfer A (2006) Atrazine increases Ranavirus susceptibility in the tiger salamander (*Ambystoma tigrinum*). Ecol Appl 16:2325–2332

Frankenberger WT, Tabatabai MA Jr, Tabatabai MA (1991) Factors affecting L-asparaginase activity in soils. Biol Fert Soils 11:1–5

Fuentes MS, Briceno GE, Saez JM, Benimeli CS, Diez MC, Amoroso MJ (2013) Enhanced removal of a pesticides mixture by single cultures and consortia of free and immobilized *Streptomyces* strains. BioMed Res Int 2013: 392573 https://doi.org/10.1155/2013/392573.

Fuentes MS, Raimondo EE, Amoroso MJ, Benimeli CS (2017) Removal of a mixture of pesticides by a Streptomyces consortium: Influence of different soil systems. Chemosphere 173:359–367

Fuentes MS, Sáez JM, Benimeli CS, Amoroso MJ (2011) Lindane biodegradation by defined consortia of indigenous *Streptomyces* Strains. Water Air Soil Pollut 222:217–231. https://doi.org/10.1007/s11270-011-0818-5

Garcia FP, Ascencio SYC, Oyarzun JG, Hernandez AC, Alavarado PV (2012) Pesticides: classification, uses and toxicity. Measures of exposure and genotoxic risks. J Res Environ Sci Toxicol 1(11): 279–293

Garrettson LK (1993) Chapter 9-Poisoning. Neurologic Emergencies in Infancy and Childhood (Second Edition) 136–166. https://doi.org/10.1016/B978-0-7506-9419-3.50014-3

Ghosal D, Ghosh S, Dutta TK, Ahn Y (2016) Current state of knowledge in microbial degradation of polycyclic aromatic hydrocarbons (pahs): A review. Front Microbiol 7: 1369

Gul A, Mahmood I, Imadi SR, Shazadi K, Hakeem KR (2015) Effects of pesticides on environment; Chapter 13. In: Hakeem, KR (ed.), Plant, Soil and Microbes (Vol. 1, pp. 253–269). Springer International Publishing, Switzerland.

Guo Q, Zhang J, Wan R, Xie S (2014) Impacts of carbon sources on simazine biodegradation by *Arthrobacter* strain SD3-25 in liquid culture and soil microcosm. Int Biodeterior Biodegrad 89:1–6. https://doi.org/10.1016/j.ibiod.2013.12.018

Gupta S, Pathak B, Fulecar MH (2015) Molecular approaches for biodegradation of polycyclic aromatic hydrocarbon compounds: A review. Rev Environ Sci Biotechnol 14:241–269

Hamsavathani V, Aysha OS, Ajith, AR (2017) Isolation and identification of chlorpyrifos degrading bacteria from agricultural soil. Int J Adv Res 1209–1221, https://doi.org/10.21474/IJAR01/4234

Hayo MG, Werf VD (1996) Assessing the impact of pesticides on the environment. Agric. Ecosyst Environ 60: 81–96

Hays SG, Patrick WG, Ziesack M, Oxman N, Silver PA (2015) Better together: Engineering and application of microbial symbioses. Curr Opin Biotechnol 36: 40–49

Helfrich LA, Weigmann DL, Hipkins P, Stinson ER (2009) Pesticides and aquatic animals: a guide to reducing impacts on aquatic systems. In: Virginia Polytechnic Institute and State University https://pubs.ext.vt.edu/420/420-013/420-013.html.

Hernández AF, Parrón T, Tsatsakis AM, Requena M, Alarcón R, LópezGuarnido O (2013) Toxic effects of pesticide mixtures at a molecular level: Their relevance to human health. Toxicol 307:136–145

Hertz-Picciotto I, Sass JB, Engel S, Bennett DH, Bradman A, Eskenazi B, Lanphear B, Whyatt R (2018) Organophosphate exposures during pregnancy and child neurodevelopment: recommendations for essential policy reforms. PLoS Med 15 https://doi.org/10.1371/journal.pmed.1002671

Hicks B (2013) In: Agricultural Pesticides and Human Health (p. 4). National association of geoscience teachers http://serc.carleton.edu/NAGTWorkshops/health/case_studies/pesticides.html

Hongming L, Xu L, Zhaojian G, Fan Y, Dingbin C, Jianchun Z, Jianhong X, Shunpeng L, Qing H (2015) Isolation of an aryloxyphenoxy propanoate (AOPP) herbicide-degrading strain *Rhodococcus ruber* JPL-2 and the cloning of a novel carboxylesterase gene (feh). Braz J Microbiol 46(2):425–432

Hu J, Xue Y, Guo H, Gao M, Li J, Zhang S, Tsang YF (2017) Design and composition of synthetic fungal-bacterial microbial consortia that improve lignocellulolytic enzyme activity. Bioresour Technol 227:247–255

Hussain S, Siddique T, Saleem M, Arshad M, Khalid A (2009) Impact of pesticides on soil microbial diversity, enzymes, and biochemical reactions. in *Advances in Agronomy*, 102:159–200eds D. L. Sparks and S. H. du Pont (Amsterdam: Elseveir),.

Iwai S, Yamazoe A, Takahashi R, Kurisu F, Yagi O (2005) Degradation of monochlorinated dibeno-p-dioxins by *Janibacter* sp. strain YA isolated from river sediment. Curr Microbiol 51: 353–358

Izadifard M, Achari G, Langford CH (2013) Application of photocatalysts and led light sources in drinking water treatment. Catalysts, 3:726–743 https://doi.org/10.3390/catal3030726

Jabbar A, Mallick S (1994) Pesticides and Environment Situation in Pakistan (Working Paper Series No. 19). Available from Sustainable Development Policy Institute (SDPI)

Jagannathan SV, Manemann EM, Rowe SE, Callender MC, Soto W (2021) Marine Actinomycetes, New Sources of Biotechnological Products. Mar Drugs 19:365. https://doi.org/10.3390/md19070365.

Jayabarath J, Musfira SA, Giridhar R, Shyamsundar S, Arulmurugan R (2010) Biodegradation of carbofuran pesticide by saline soil actinomycetes. Int J Biochem Biotechnol 6: 187–192

Kah M, Brown CD (2006) Adsorption of ionizable pesticides in soils. *Rev Environ Contam Toxicol* 188:149–217

Kandpal V (2014) Biopesticides. Int J Environ Res Dev 4:191–196

Karigar CS, Rao SS (2011) Role of microbial enzymes in the bioremediation of pollutants: A review. Enzyme Res 2011:805187

Karn SK, Chakrabarti SK, Reddy MS (2011) Degradation of pentachlorophenol by *Kocuria* sp. CL2 isolated from secondary sludge of pulp and paper mill. Biodegradation 22:63–69 https://doi.org/10.1007/s10532-010-9376-6

Kensa VM (2011). Bioremediation: An overview. Indian Pollution Control 27: 161–168

Khessairi A, Fhoula I, Jaouani A et al (2014) Pentachlorophenol degradation by *Janibacter* sp., a new actinobacterium isolated from saline sediment of arid land. Biomed Res Int https://doi.org/10.1155/2014/296472

Kidd P, Barcelo J, Bernal MP, Navari-Izzo F, Poschenrieder C, Shilev S, Clemente R, Monterroso C, (2009) Trace element behaviour at the root–soil interface: implications in phytoremediation. Environ Exp Bot 67 (1): 243–259 https://doi.org/10.1016/j.envexpbot.2009.06.013

Kjeldsen LS, Ghisari M, Bonefeld-Jørgensen EC (2013) Currently used pesticides and their mixtures affect the function of sex hormone receptors and aromatase enzyme activity. Toxicol Appl Pharmacol 272(2): 453–64 https://doi.org/10.1016/j.taap.2013.06.028

Kodešová R, Kočárek M, Kodeš V, Drábek O, Kozák J, Hejtmánková K (2011) Pesticide adsorption in relation to soil properties and soil type distribution in regional scale. J Hazard Mater 186: 540–550

Kolekar PD, Phugare SS, Jadhav JP (2014) Biodegradation of atrazine by *Rhodococcus* sp. BCH2 to N-isopropylammelide with subsequent assessment of toxicity of biodegraded metabolites. Environ Sci Pollut Res Int 2: 2334–2345, https://doi.org/10.1007/s11356-013-2151-6

Kumar D, Kumar A, Sharma J (2016) Degradation Study of Lindane by Novel Strains Kocuria Sp. DAB-1Y and *Staphylococcus* sp. DAB-1W. Bioresour Bioprocess 3:53. https://doi.org/10.1186/s40643-016-0130-8

Kumar M, Lakshmi CV, Khanna S (2008) Biodegradation and bioremediation of endosulfan contaminated soil. Bioresour Technol 99: 3116–3122, https://doi.org/10.1016/j.biortech.2007.05.057

Kumar S (2013) Use of pesticides in agriculture and livestock animals and its impact on environment of India. Asian J Environ Sci 8(1): 51–57

Kurzbaum, E, Raizner Y, Cohen O, Suckeveriene RY, Kulikov A, Hakimi B, Menashe O, (2017) Encapsulated *Pseudomonas putida* for phenol biodegradation: Use of a structural membrane for construction of a well-organized confined particle. Water Res 121:37–45

Lai T, Li J, Sua J (2011) Monitoring of beet armyworm Spodoptera exigua (Lepidoptera: Noctuidae) resistance to chlorantraniliprole in China, Pestic Biochem Physiol 101: 198–205 doi:https://doi.org/10.1016/j.pestbp.2011.09.006

Lamilla C, Schalchli H, Briceño, G, Leiva B, Donoso-Piñol P, Barrientos L, Rocha VAL, Freire DMG, Diez MCA (2021) Pesticide biopurification system: A source of biosurfactant-producing bacteria with environmental biotechnology applications, Agronomy 11(4): 624, https://doi.org/10.3390/agronomy11040624

Lamrani MD, Pesce S, Rouard N, Laurent FM (2014) Evidence for cooperative mineralization of diuron by *Arthrobacter* sp. BS2 and *Achromobacter* sp. SP1 isolated from a mixed culture enriched from diuron exposed environments. Chemosphere 117: 208–215

Lang M, Cai Z (2009) Effects of chlorothalonil and carbendazim on nitrification and denitrifi cation in soils. J Environ Sci 21:458–467

Lin QS, Chen SH, Hu MY, Haq MR, Yang L, Li H (2011) Biodegradation of cypermethrin by a newly isolated actinomycetes HU-S-01 from wastewater sludge. Int J Environ Sci Technol 8: 45–56

Liroff RA (2000) Balancing risks of DDT and malaria in the global POPs treaty. Pestic Safety News 4:3–7

Locke D, Landivar JA, Moseley D (1995) The effects of rate and timing of glyphosate applications of defoliation effi ciency, regrowth inhibition, lint yield, fiber quality and seed quality. Proc Beltwide Cotton Conf 2:1088–1090

Lourencetti C, Marchi MRR, Ribeiro ML (2008) Determination of sugar cane herbicides in soil and soil treated with sugar cane vinasse by solid phase extraction and HPLC UV. Talanta 77: 701–709

Mansouri A, Cregut M, Abbes C, Durand MJ, Landoulsi A, Thouand G (2017) The Environmental Issues of DDT Pollution and Bioremediation: a Multidisciplinary Review. Appl Biochem Biotechnol 181(1): 309–339 https://doi.org/10.1007/s12010-016-2214-5

Martins CS, Martins CM, Guedes ML, Santaella ST (2013) Immobilization of microbial cells: A promising tool for treatment of toxic pollutants in industrial wastewater. Afr J Biotechnol 12: 4412–4418

Mathur S, Tannan S (1999) Future of Indian pesticides industry in next millennium. Pestic Inf 24:9–23

Mohammadipanah F, Wink J (2016) Actinobacteria from arid and desert habitats: diversity and biological activity. Front Microbiol 6: 1541 https://doi.org/10.3389/fmicb.2015.01541

Mohammed MS (2009) Degradation of methomyl by the novel bacterial strain *Stenotrophomonas maltophilia* MI. Electronic J Biotechnol 12(4): 1–6

Morillo E, Villaverde J (2017) Advanced technologies for the remediation of pesticide contaminated soils. Sci Total Environ 586: 576–597

Mrema EJ, Rubino FM, Brambilla G, Moretto A, Tsatsakis AM, Colosio C (2013) Persistent Organochlorinated pesticides and mechanisms of their toxicity. Toxicol 307: 74–88 https://doi.org/10.1016/j.tox.2012.11.015

Müller K, Bach M, Hartmann H, Spiteller M, Frede HG (2002) Point-and nonpoint source pesticide contamination in the Zwesterohm catchment, Germany. J Environ Qual 31(1): 309–318

Narahashi T (1996) Neuronal ion channels as the target sites of insecticides. Pharmacol Toxicol 79(1):1–14 https://doi.org/10.1111/j.1600-0773.1996.tb00234.x

Narahashi T, Frey JM, Ginsburg KS, Roy ML (1992) Sodium and GABA-activated channels as the targets of pyrethroids and cyclodienes. Toxicol Lett 64–65:429–436 https://doi.org/10.1016/0378-4274(92)90216-7

Naveena B, Annalakshmi G, Partha N (2013) An efficacious degradation of pesticide by salt tolerant *Streptomyces venezuelae* ACT 1. Bioresour Technol 132: 378–382

Nayyar N, Sangwan N, Kohli P, Verma H, Kumar R, Negi V, Oldach P, Mahato NK, Gupta, V, Lal R (2014) Hexachlorocyclohexane: persistence, toxicity and decontamination. Rev Environ Health 29(1–2): 49–52 https://doi.org/10.1515/reveh-2014-0015

Neumann M, Schulz R, Schäfer K, Müller W, Mannheller W, Liess M (2002) The significance of entry routes as point and nonpoint sources of pesticides in small streams. Water Res 36(4): 835–842

Odukkathil G, Vasudevan N (2016) Residues of endosulfan in surface and subsurface soil and its bioremediation. J Environ Manag 165:72–80

Oerke EC (2005) Crop losses to pests. J Agr Sci 144:31–43

Ollera S, Malatoa J, Sánchez-Pérezb A (2011) Combination of advanced oxidation processes and biological treatments for wastewater decontamination: A review. Sci Total Environ 409(20):4141–4166 https://doi.org/10.1016/j.scitotenv.2010.08.061.15

Pandey G, Dorrian S, Russell R, Brearley C. Kotsonia SE, Oakeshott JG (2010a) Cloning and biochemical characterization of a novel carbendazim (methyl-1H-benzimidazol-2-ylcarbamate)-hydrolyzing esterase from the newly isolated Nocardioides sp. strain SG-4 G and its potential for use in enzymatic bioremediation. Appl Environ Microbiol 76:2940–2945, https://doi.org/10.1128/AEM.02990-09

Pandey G, Dorrian SJ, Russell RJ, Brearley C, Kotsonis S, Oakeshott JG (2010b) Cloning and biochemical characterization of a novel carbendazim (methyl-1H-benzimidazol-2-ylcarbamate)-hydrolyzing esterase from the newly isolated *Nocardioides* sp. strain SG-4G and its potential for use in enzymatic bioremediation. Appl Enviro Microbiol. 76(9):2940–2945

Pell M, Stenberg B, Torstensson L (1998) Potential denitrifi cation and nitrifi cation tests for evaluation of pesticide effects in soil. Ambio 27:24–28

Perry AS, Yamamoto I, Ishaaya I, Perry R (2013) Insecticides in agriculture and environment: Retrospects and prospects. Springer Science and Business Media

Pieper DH (2005) Aerobic degradation of polychlorinated biphenyls. Appl Microbiol Biotechnol 67(2): 170–191

Pieper DH, Reineke W (2000) Engineering bacteria for bioremediation. Curr Opin Biotechnol 11:262–70

Pipke R and Amrhein N (1988) Degradation of the phosphonate herbicide glyphosate by *Arthrobacter atrocyaneus* ATCC 13752. Appl Environ Microbiol 54: 1293–1296

Polti MA, Amoroso MJ, Abate CM, (2011). Intracellular chromium accumulation by *Streptomyces* sp. MC1. Water Air Soil Pollut 214 (1): 49–57. https://doi.org/10.1007/s11270-010-0401-5

Pradeep V, Malavalli U (2016) Use of Ca-alginate immobilized *Pseudomonas aeruginosa* for repeated batch and continuous degradation of Endosulfan. 3 Biotech 6:124

Ragab S, El Sikaily A, El Nemr A (2016) Concentrations and sources of pesticides and PCBs in surficial sediments of the Red Sea coast. Egypt J Aquat Res 42 (4): 365-374

Rahmansyah M, Agustiyani D, Heddy J, Dewi TK (2012) Growth and adaptation of four Streptomyces isolates in the media containing propoxur. ARPN: J Agric Biol Sci, 7: 773–781

Raimondo EE, Saez JM, Aparicio JD, Fuentes MS, Benimeli CS (2020) Bioremediation of lindane-contaminated soils by combining of bioaugmentation and biostimulation: effective scaling-up from microcosms to mesocosms. J Environ Manage 276:111309 https://doi.org/10.1016/j.jenvman.2020.111309

Ramakrishnan B, Megharaj M, Venkateswarlu K, Sethunathan N, Naidu, R (2011) Mixtures of environmental pollutants: Effects on microorganisms and their activities in soils. Rev Environ Contam and Toxicol 211:63–120

Raman K, Henning P (2013) Environmental assessment of enzyme uses in industrial production: A literature review. Clean Prod 42: 228–240

Rehan M, Kluge M, Franzle S, Kellner H, Ullrich R, Hofrichter M (2014) Degradation of atrazine by *Frankia alni* ACN14a: gene regulation, dealkylation, and dechlorination, Appl Microbiol Biotechnol 98: 6125–6135, https://doi.org/10.1007/s00253-014-5665-z

Reigart JR, Roberts JR (2001) Pesticides in children. Pediatr Clin North Am 48(5):1185–98

Relyea RA (2005) The lethal impact of roundup on aquatic and terrestrial amphibians. Ecol Appl 15:1118–1124

Robles-Gonzalez IV, Fava F, Poggi-Varaldo, HM (2008) A review on slurry bioreactors for bioremediation of soils and sediments. Microb Cell Fact 7:5. https://doi.org/10.1186/1475-2859-7-5

Rohr JR, Schotthoefer AM, Raffel TR, Carrick HJ, Halstead N, Hoverman JT, Johnson CM, Johnson LB, Lieske C, Piwoni MD, Schoff PK, Beasley VR (2008) Agrochemicals increase trematode infections in a declining amphibian species. Nature 455:1235–1239

Roy RN, Laskar S, Sen SK (2006) Dibutyl Phthalate, the Bioactive Compound Produced by Streptomyces albidoflavus 321.2. Microbiol Res 161:121–126 https://doi.org/10.1016/j.micres.2005.06.007

Roy RN, Sen SK (2002) Survey of antimicrobial streptomycetes from soils of West Bengal: Characterization and identification of potent broad spectrum antibiotic producing *Streptomyces albiboflavus* 321.2, Hind Antibiot 44: 25–33

Roy RN, Sen SK (2011) Growth and production kinetics of the antimicrobial compound from *Streptomyces albidoflavus* 321.2. Sci Res Essays 6: 2042–2046

Roy RN, SK Sen (2013) Fermentation Studies for the Production of Dibutyl Phthalate, an Ester Bioactive Compound from *Streptomyces albidoflavus* MTCC 3662 Using Low-Priced Substrates. Jordan J Biol Sci 6 (3): 177–181

Saez JM, Alvarez A, Benimeli CS, Amoroso MJ (2014) Enhanced lindane removal from soil slurry by immobilized *Streptomyces* consortium. Int Biodeterior Biodegradation. 93:63–69

Saez JM, Aparicio JD, Amoroso MJ, Benimeli CS (2015) Effect of the acclimation of a *Streptomyces* consortium on lindane biodegradation by free and immobilized cells. Process Biochem 50:1923–1933

Saez JM, Benimeli CS, Amoroso MJ (2012) Lindane removal by pure and mixed cultures of immobilized actinobacteria. Chemosphere 89: 982–987 https://doi.org/10.1016/j.chemosphere.2012.06.057

Saez JM, Bigliardo AL, Raimondo EE, Briceno GE, Polti MA, Benimeli CS (2018) Lindane dissipation in a biomixture: Effect of soil properties and bioaugmentation. Ecotoxicol Environ Saf 156:97–105. https://doi.org/10.1016/j.ecoenv.2018.03.011

Sagarkar S, Bhardwaj P, Storck V, Devers-Lamrani M, Martin- Laurent F, Kapley A (2016) S-triazine degrading bacterial isolate *Arthrobacter* sp. AK-YN10, a candidate for bioaugmentation of atrazine contaminated soil. Appl Microbiol Biotechnol 100:903–913 https://doi.org/10.1007/s00253-015-6975-5

Sagarkar S, Nousiainen A, Shaligram S, Björklöf K, Lindstrom K, Jørgensen, KS. Kapley SA (2014) Soil mesocosm studies on atrazine bioremediation. J Environ Manage.139: 208–216 https://doi.org/10.1016/j.jenvman.2014.02.016

Santos A, Flores M (1995) Effects of glyphosate on nitrogen fixation of free-living heterotrophic bacteria. Lett Appl Microbiol 20:349–352

Schreck E, Geret F, Gontier L, Treilhou M (2008) Neurotoxic effect and metabolic responses induced by a mixture of six pesticides on the earthworm *Aporrectodea caliginosa* nocturna. Chemosphere 71(10):1832–1839

Sette LD, De Oliveira VM, Manfio GP (2005) Isolation and characterization of alachlor-degrading actinomycetes from soil. Antonie Van Leeuwenhoek 87 (2): 81–89. https://doi.org/10.1007/s10482-004-1129-2

Sharma BM Bharat GK Chakraborty P, Martiník J, Audy O, Kukučka P, Přibylová P, Kukreti PK, Sharma A, Kalina J, Steindal EH, Nizzetto L (2022) A comprehensive assessment of endocrine-disrupting chemicals in an Indian food basket: Levels, dietary intakes, and comparison with European data. Environ Pollut: 288:117750

Shen YJ, Lu P, Mei H, Yu HJ, Hong Q, Li SP (2009) Isolation of a methyl parathion-degrading strain *Stenotrophomonas* sp. SMSP-1 and cloning of the ophc2 gene. Biodegradation 21(5):785–792

Shepherd MD, Kharel MK, Bosserman MA, Rohr J (2010) Laboratory maintenance of Streptomyces species Curr Protoc Microbiol 18 (2010):10E.1. https://doi.org/10.1002/9780471729259.mc10e01s18

Shetty PK, Murugan M, Sreeja KG (2008) Crop protection stewardship in India: Wanted or unwanted. Curr Sci 95(4): 457–464

Shimizu H (2002) Metabolic engineering integrating methodologies of molecular breeding and bioprocess systems engineering. J Biosci Bioeng 94: 563–73

Shivlata L, Satyanarayana T (2015) Thermophilic and alkaliphilic Actinobacteria: Biology and potential applications. Front Microbiol 6: 1–29

Sibanda MM, Focke W, Labuschagne FJ, Moyo L, Nhlapo NS, Maity A, Brindley GW, (2011) Degradation of insecticides used for indoor spraying in malaria control and possible solutions. Malar J 10(1):307 Epub: 2011/10/20

Silva VP, Moreira-Santos M, Mateus C, Teixeira T, Ribeiro R, Viegas CA (2015) Evaluation of *Arthrobacter aurescens* strain Tc1 as bioaugmentation bacterium in soils contaminated with the herbicidal substance terbuthylazine. PLoS ONE.10:e0144978. https://doi.org/10.1371/journal.pone.0144978.

Sineli PE, Tortella G, Costa JSD, Benimeli CS, Cuozzo SA (2016) Evidence of α-, β- and γ-HCH mixture aerobic degradation by the native actinobacteria *Streptomyces* sp. M7, World J Microbiol Biotechnol 32:81, https://doi.org/10.1007/s11274-016-2037-0.

Singh BK, Walker A. (2006) Microbial degradation of organophosphorus compounds. FEMS Microbiol. Rev 30(3):428–471

Soderlund DM (2012) Molecular mechanisms of pyrethroid insecticide neurotoxicity: recent advances. Arch Toxicol 86: 165–181

Spanoghe P, Maes A, Steurbaut W (2004) Limitation of point source pesticide pollution: Results of bioremediation system. Commun Agric Appl Biol Sci, 69(4): 719–732

Sparling DW, Feller GM (2009) Toxicity of two insecticides to California, USA, anurans and its relevance to declining amphibian populations. Environ Toxicol Chem 28(8):1696–1703

Spear R (1991) Recognized and possible exposure to pesticides. In: Hayes, WJ, & Laws, ER, (eds.), Handbook of Pesticide Toxicology (pp. 245–274). Academic Press, San Diego, CA, USA

Strong LC, Rosendahl C, Johnson G, Sadowsky MJ, Wackett LP (2002) *Arthrobacter aurescens* TC1 metabolizes diverse s-triazine ring compounds. Appl Environ Microbiol 68:5973–5980 https://doi.org/10.1128/AEM.68.12.5973-5980.2002

Subramani R and Sipkema D (2019) Marine rare actinomycetes: a promising source of structurally diverse and unique novel natural products Mar Drugs 17: 249, https://doi.org/10.3390/md17050249

Subramani R, Aalbersberg W (2013) Culturable rare Actinomycetes: diversity, isolation and marine natural product discovery Appl Microbiol Biotechnol 97:9291–9321

Sun LN, Zhang J, Gong FF, Wang X, Hu G, Li SP, Hong Q (2014) *Nocardioides* soli sp. nov., a carbendazim-degrading bacterium isolated from soil under the longterm application of carbendazim. Int J Syst Evol Microbiol 64: 2047–2052 https://doi.org/10.1099/ijs.0.057935-0

Takagi K, Iwasaki A, Kamei I, Satsuma K, Yoshioka Y, Harada N (2009) Aerobic mineralization of hexachlorobenzene by newly isolated pentachloronitrobenzene-degrading *Nocardioides* sp. strain PD653. Appl Environ Microbiol 75:4452–4458. https://doi.org/10.1128/AEM.02329-08.

Tam ACM, Behki RM, Khan SU (1987) Isolation and characterization of an s-ethyl-N,N-dipropylthiocarbamate- degrading arthrobacter strain and evidence for plasmid-associated s-ethyl-N,Ndipropylthiocarbamate degradation. Appl Environ Microbiol. 53(5): 1088–1093

Taylor P (1996) Anticholinesterase agents. In: Goodman and Gilman's The Pharmacological Basis of Therapeutics. Ed. Hardman JG, Limbird LE, Molinoff PB, Ruddon RW. 9th ed. pp. 161–76

Technavio.com (Country Pesticide Growth Analysis – Size and Forecast 2024–2028)

Topp E (2001) A comparison of three atrazine-degrading bacteria for soil bioremediation. Biol Fertil Soils, 33: 529–534

Tordoir WF, van Sittert NJ (1994) Organochlorines. Toxicol 91(1): 51–57

USEPA (United States of Environmental Protection Agency). About Pesticides. http://www.epa.gov/pesticides/about/index.htm

Usmani Z, Kulp M, Lukk (2021) Bioremediation of lindane contaminated soil: Exploring the potential of actinobacterial strains. Chemosphere 278:130468

Venkata MS, Shailaja S, Rama Krishna M, Reddy KB, Sarma PN (2006) Bioslurry phase degradation of DEP contaminated soil in periodic discontinuous mode operation: Influence of augmentation and substrate partition. Process Biochemistry, 41:644–652

Verma JP, Jaiswal DK, Sagar R (2014) Pesticide relevance and their microbial degradation: A-state-of-art Rev Environ Sci Biotechnol 13: 429–466

Verma K, Agrawal N, Farooq M, Misra RB, Hans R.K (2006) Endosulfan degradation by a *Rhodococcus* strain isolated from earthworm gut. Ecotoxicol Environ Saf 64(3):377–381

Villaverde JJ, Sandín-España P, Sevilla-Morán B, López-Goti C, Alonso-Prados JL (2017) Biopesticides from Natural Products: Current Development, Legislative Framework, and Future Trends. Bioresour 11(2): 5618–5640 https://dx.doi.org/10.15376/biores.11.2.villaverde

Waliszewski SM, Gomez-Arrollo S, Infanzon RM, Carvajal O, Villalobos-Petrini R, Trujillo P, Maxwell M (2004) Persistent organochlorine pesticide levels in bovine fat from Mexico. Food Aditives Contam 21: 774–780

Wang H, Liu Y, Li J, Lin M, Hu X (2016) Biodegradation of atrazine by *Arthrobacter* sp. C3, isolated from the herbicide-contaminated corn field. Int J Environ Sci Technol 13: 257–262

Wang Q, Xie S, Hu R (2013) Bioaugmentation with Arthrobacter sp. strain DAT1 for remediation of heavily atrazine-contaminated soil Int Biodeterior Biodegradation 77: 63–67, https://doi.org/10.1016/j.ibiod.2012.11.003

Wang S, Miltner A, Muskus AM, Nowak KM (2021) Microbial activity and metamitron degrading microbial communities differ between soil and water sediment systems. J Hazard Mater 408: 124293 https://doi.org/10.1016/j.jhazmat.2020.124293.

Wang S, Zhang C, Yan Y (2012) Biodegradation of Methyl Parathion and P-Nitrophenol by a Newly Isolated *Agrobacterium* sp. Strain Yw12. Biodegradation 23:107–116.

Wang Y, Liu C, Wan J, Sun X, Ma W, Ni H (2018) Molecular Cloning and Characterization of a Methyl Parathion Hydrolase from an Organophosphorus-Degrading Bacterium, *Serratia marcescens* MEW06. FEMS Microbiol Lett 365:24 https://doi.org/10.1093/femsle/fny279

Wang Z, Wang Y, Gong F, Zhang J, Hong Q, Li S (2010) Biodegradation of carbendazim by a nove actinobacterium Rhodococcus jialingiae djl-6-2. Chemosphere 81(5): 639–644. https://doi.org/10.1016/j.chemosphere.2010.08.040

WHO (1986) Task Group on Organophosphorus Insectides. Organophosphorus insecticides: A general introduction. WHO, Geneva

WHO (2019) The WHO Recommended Classification of Pesticides by Hazard and Guidelines to Classification 2019

Xu JL, He J, Wang ZC, Wang K, Li WJ, Tang SK, Li SP (2007) *Rhodococcus qingshengii* sp. nov., a carbendazim-degrading bacterium. Int J Syst Evol Microbiol 57: 2754–2757. https://doi.org/10.1099/ijs.0.65095-0

Yadav M, Srivastva N, Sharan R, Nath S, Kumar S (2014) Biodegradation of chlorpyrifos by *Pseudomonas* sp. in a continuous packed bed bioreactor. Bioresour Technol, 165:265–269

Yoshikawa M, Zhang M, Toyota K (2017) Minireview biodegradation of volatile organic compounds and their effects on biodegradability under co-existing conditions. Microbes Environ 32:188–200. https://doi.org/10.1264/jsme2.me16188

Zafra, G, Absalon AE, Anducho-Reyes MA, Fernandez FJ, Cort es-Espinosa DV (2017) Construction of PAH-degrading mixed microbial consortia by induced selection in soil. Chemosphere, 172: 120–126

Zanardini E, Arnoldi A, Boschin G, D'Agostina A, Negri, M, Sorlini C (2002) Microbial degradation of sulfonylurea herbicides: chlorsulfuron and metsulfuron-methyl Microbiology of Composting, Springer, Berlin, Heidelberg pp. 309–319, https://doi.org/10.1007/978-3-662-08724-4_26

Zhang X, Huang Y, Harvey P, Li H, Ren Y, Li J, Wang J, Yang H (2013) Isolation and characterization of carbendazim-degrading *Rhodococcus erythropolis* djl-11. PLoS One 8(10): e74810. https://doi.org/10.1371/journal.pone.0074810

Zhang C, Wang SH, Yan YC (2011) Isomerization and biodegradation of betacypermethrin by *Pseudomonas aeruginosa* CH7 with biosurfactant production. Bioresour Technol 102: 7139–7146

Zhao XQ, Xu XN, Chen LY (2016) Production of enzymes from marine actinobacteria. Adv Food Nutr Res 78:137–151

Zhao S, Wang J (2024) Biodegradation of atrazine and nicosulfuron by *Streptomyces nigra* LM01: Performance, degradative pathway, and possible genes involved. J Hazard Mater 471: 134336 https://doi.org/10.1007/s003740100371

Myco-enzymes as a Potent Tool for the Degradation of Xenobiotics

Somanjana Khatua

Abstract Industrial progress has dramatically increased the release of highly toxic xenobiotic compounds into the environment, many of which are resistant to natural degradation. Addressing this critical issue necessitates innovative and sustainable treatment solutions. Myco-remediation, using fungi for pollutant degradation, has emerged as a promising approach due to fungi's robust metabolic capabilities. White rot fungi, including *Bjerkandera adusta*, *Phanerochaete chrysosporium*, *Pleurotus* sp., and *Trametes versicolor*, are particularly effective in breaking down diverse toxic pollutants such as synthetic dyes, plastics, organophosphate pesticides, and pharmaceutical personal care products. This efficiency is attributed to their production of extracellular enzymes such as laccase, peroxidase, cutinase, and lipase, as well as intracellular cytochrome P450 monooxygenase. However, the activity and durability of these enzymes are influenced by environmental factors, including pH, temperature, availability of minerals, and presence of co-substrates. This chapter highlights the potential of fungal enzymes in xenobiotic degradation, emphasizing the critical role of lignin-modifying enzymes. Future advancements should focus on optimizing operational parameters, developing innovative enzyme immobilization methods, and scaling up biotechnological applications to enhance the effectiveness of fungal enzymes in mitigating environmental contamination.

Keywords Bioremediation · Lignin-modifying enzymes · Pesticides · Plastics · Synthetic dye · White rot fungi

S. Khatua (✉)
Molecular and Applied Mycology Laboratory, Department of Botany, Faculty of Science, University of Allahabad, Prayagraj, Uttar Pradesh, India
e-mail: Somanjana@allduniv.ac.in

S. Roy, V. Mandal (eds.), *Plant-Microbe Interaction under Xenobiotic Exposure*,
https://doi.org/10.1007/978-981-96-8260-7_21

1 Introduction

A safe climate is a fundamental right and crucial for sustaining human life and well-being (Mishra et al. 2021). Nonetheless, the modern era of industrialization and urbanization has led to various human activities such as industrial manufacturing, agricultural operations, and waste disposal that significantly release harmful and toxic substances known as environmental xenobiotics (Štefanac et al. 2021). Such hazardous chemicals include plastics, pesticides, preservatives, plasticizers, dyes, pigments, personal care products, and other synthetic substances. Once released, xenobiotics may undergo several physicochemical or biological changes, and finally, they infiltrate the food chain, leading to detrimental effects. Such transformations can disrupt the microbiota, which is associated with various health issues and, in some cases, increased toxicity (Ortiz et al. 2022). Many xenobiotics are of critical concern due to their carcinogenic, mutagenic and teratogenic properties, which can negatively affect the development and reproduction in humans, aquatic organisms, birds, and other animals (Baker et al. 2019). So far, various strategies have been created to tackle xenobiotic pollution (Khan et al. 2023). However, these approaches often involve high costs and can produce harmful byproducts. As a more sustainable alternative, biological methods such as bioremediation have gained attention. Bioremediation leverages the metabolic activities of living organisms to degrade or transform toxic environmental contaminants into safer compounds (Ellouze and Sayadi 2016).

In this context, fungi have gained recognition as promising candidates for bioremediation as evidenced by growing number of publications highlighting their unique capabilities (Fig. 1). Indeed, fungi offer a distinct advantage over other biological remediation agents, such as algae, archaea, and bacteria, due to their higher tolerance for elevated contamination levels, because of differences in the cellular metabolism (Abdel-Fattah Mostafa et al. 2022). Fungi possess an extensive arsenal of enzymes, such as laccase, esterase, transferase, cytochrome P450 (CYP) monooxygenase, dioxygenase, hydrolase, dehalogenase, peroxidases including manganese peroxidase (MnP), lignin peroxidase (LiP), versatile peroxidase (VP), and dye-decolorizing peroxidases (DyP) which play a vital role in decomposing organic matter (Khan et al. 2023). Among various fungal members, white rot fungi (WRF) are particularly significant for their capacity to produce a range of lignin-modifying enzymes during metabolic processes. Furthermore, species such as *Bjerkandera adusta*, *Irpex lacteus*, *Phanerochaete chrysosporium*, and *Trametes versicolor* have been extensively studied for their non-specific degradation of phenolic compounds, chlorinated herbicides, synthetic dyes, plastics, and petroleum hydrocarbons (Choi et al. 2013; Vaksmaa et al. 2023).

This chapter focuses on significant advancements in fungal xenobiotic metabolism over the last decade, focusing on the critical enzymes driving these processes and their applications in bioremediation and biocatalysis. It delves into recent examples of fungal degradation of various xenobiotics and highlights the pivotal

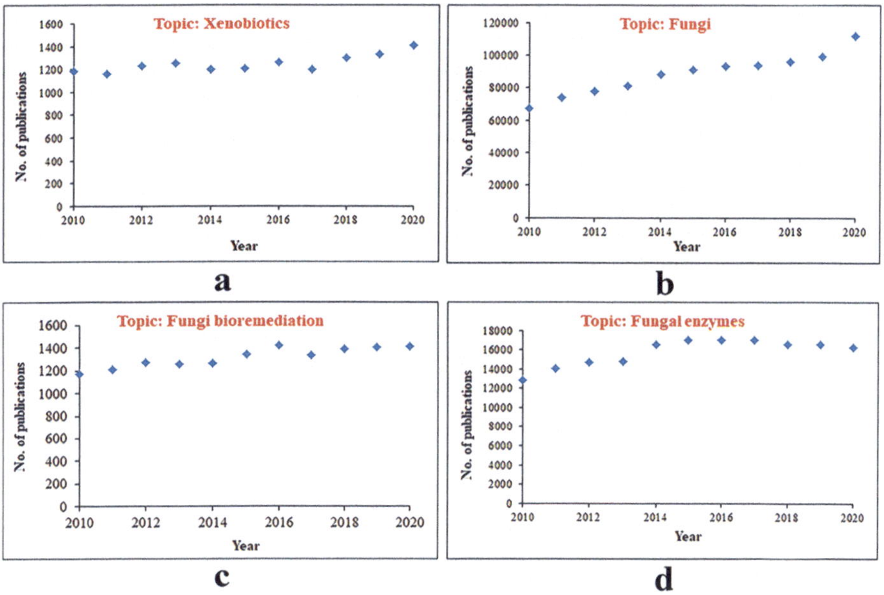

Fig. 1 The number of publications indexed in PubMed (https://pubmed.ncbi.nlm.nih.gov/) between 2010 and 2020 focusing on topics such as (**a**) Xenobiotics, (**b**) Fungi, (**c**) Fungi bioremediation, and (**d**) Fungal enzymes

roles of enzymes like CYPs, laccases, and peroxidases in facilitating these biotransformations.

2 Xenobiotics and Its Impact on the Environment

The term "xenobiotics" is derived from the Greek words xenos (foreign) and bios (life), referring to substances foreign to living organisms (Štefanac et al. 2021). These artificially synthesized toxic chemical compounds are extensively used in industrial, agrochemical, and domestic applications (Rathore et al. 2022). Metabolically, these are defined as chemicals that are extraneous to the normal metabolic processes of living organisms (Abdelsalam et al. 2020). Chemically, most xenobiotics are aromatic compounds, often featuring one or more substituted phenyl functional groups (Baker et al. 2019). These environmental pollutants are widely employed in producing paints, explosives, and several other industrial elements (Singh et al. 2023). Key sources of xenobiotics include the textile industry (synthetic dye), pharmaceutical personal care products (PPCPs), effluents from paper and pulp, plastic, polycyclic aromatic hydrocarbons (PAHs), food production, petroleum industry, and pesticides (Torres-Farradá et al. 2024). Such hazardous substances, released through inadequate waste management and human-induced activities, can have both direct and indirect adverse effects on plants, animals, and humans. Numerous comprehensive reviews have examined the far-reaching environmental impacts of xenobiotics, highlighting their detrimental effects on natural habitats, biological diversity, and human well-being (Embrandiri et al. 2016; Miglani et al. 2022; Piwowarska and Kiedrzyńska 2022; Sharma and Kumar 2024). Researchers have noted that these chemicals contribute to eutrophication, disrupting faunal diversity in aquatic ecosystems, and harm soil organisms by reducing microbial activity and altering soil properties. Continuous exposure to these compounds can damage the immune, reproductive, and nervous systems in living organisms and, in some cases, lead to various forms of cancer. The persistence, high toxicity, and limited biodegradability of xenobiotic compounds further amplify their impact, posing significant threats to various ecosystems including flora, fauna, and human health (Mishra et al. 2021; Rathore et al. 2022; Miglani et al. 2022). These findings underscore the urgency of developing strategies to reduce the negative impacts of xenobiotics and preserve ecological balance.

3 Mitigation Strategies for Xenobiotics

Several physicochemical techniques have been employed to mitigate xenobiotic pollution, including adsorption, coagulation, filtration, precipitation, electrolysis, chemical oxidation, ozonation, and membrane separation. While effective in specific contexts, these methods are often economically unfeasible and can generate hazardous intermediates or byproducts that may pose even more significant environmental risks than the original pollutants (Miglani et al. 2022). To overcome these challenges, biological approaches leveraging plants and microorganisms are gaining prominence as sustainable and eco-friendly alternatives. These methods have spurred significant research focused on the effective degradation of xenobiotics through natural processes (Khan et al. 2023). The core concept of bioremediation involves utilizing microbes to eliminate xenobiotics from the natural environment by mineralizing them into harmless end products, such as CO_2 and H_2O (Mishra et al. 2021).

4 Mycoremediation of Xenobiotics

Fungi-mediated remediation is now acknowledged as an effective approach, leveraging the organism's metabolic capabilities to reduce or eliminate xenobiotic contaminants. This potential stems from their ability to produce a wide array of enzymes that can aid in the decomposition of xenobiotics into safer constituents (Khan et al. 2023; Torres-Farradá et al. 2024). Fungi, especially, are widely researched microorganisms for plastic decomposition because of their capacity to break down synthetic plastic polymers through hydrolysis. They are often considered more effective than bacteria in degrading hydrocarbon polymers due to their ability to attach to and colonize the hydrophobic surfaces of plastics. Additionally, they can infiltrate and decompose persistent hydrocarbons by producing potent enzymes secreted externally for degradation (Khatua et al. 2024). Among the fungi capable of detoxification, WRF is particularly notable for their distinctive capacity to break down lignin, a complex aromatic biopolymer, resulting in the characteristic "white rot" appearance in the substrates they affect. These fungi, primarily classified under Basidiomycetes, with some contributions from Ascomycetes, are exceptionally well-suited for xenobiotic degradation due to the structural similarities between lignin and many xenobiotic compounds (Asgher et al. 2008). Genera such as *Trametes*, *Pleurotus*, and *Phanerochaete* are frequently employed in bioremediation processes, as their enzymatic systems can transform toxic xenobiotics into less harmful or inert compounds (Baker et al. 2019). Advancements in biotechnology indicate that directly utilizing microbial enzymes or engineering microbes to produce these enzymes under controlled conditions could significantly augment the efficacy and speed of xenobiotic degradation (Rathore et al. 2022).

5 Enzyme Systems of WRF

The primary enzymes involved in xenobiotic bioremediation belong to the oxidoreductase family, which includes laccases, oxygenases, and heme peroxidases such as LiP, MnP, VP, and DyP (Fig. 2) (Baker et al. 2019). Additionally, fungi produce a range of other oxidoreductases, including glyoxal oxidase, aryl alcohol oxidase (also known as veratryl alcohol oxidase), pyranose 2-oxidase (glucose 1-oxidase), cellobiose/quinone oxidoreductase, and cellobiose dehydrogenase. These proteins play a complementary role in lignin degradation by regulating the activity of key enzymes such as LiP, MnP, and laccases through the reduction of methoxy radicals produced during the process. Furthermore, enzymes such as glyoxal oxidase, glucose oxidase, and veratryl alcohol oxidase contribute to the generation of H_2O_2, a critical cofactor required by peroxidases to carry out lignin degradation effectively (Asgher et al. 2008; Abdel-Hamid et al. 2013; Falade et al. 2017). These fungal enzymes are critical in breaking down various materials, including wood, plastic, paint, and jet fuel, into nutrients. Some of these enzymes are already used in industries like pulp and paper processing and fine chemical synthesis. Recent research has even suggested that lignin-degrading fungi, such as *P. chrysosporium* and *Trametes versicolor*, could replace specific chemical processes traditionally used in paper production (Viswanath et al. 2014).

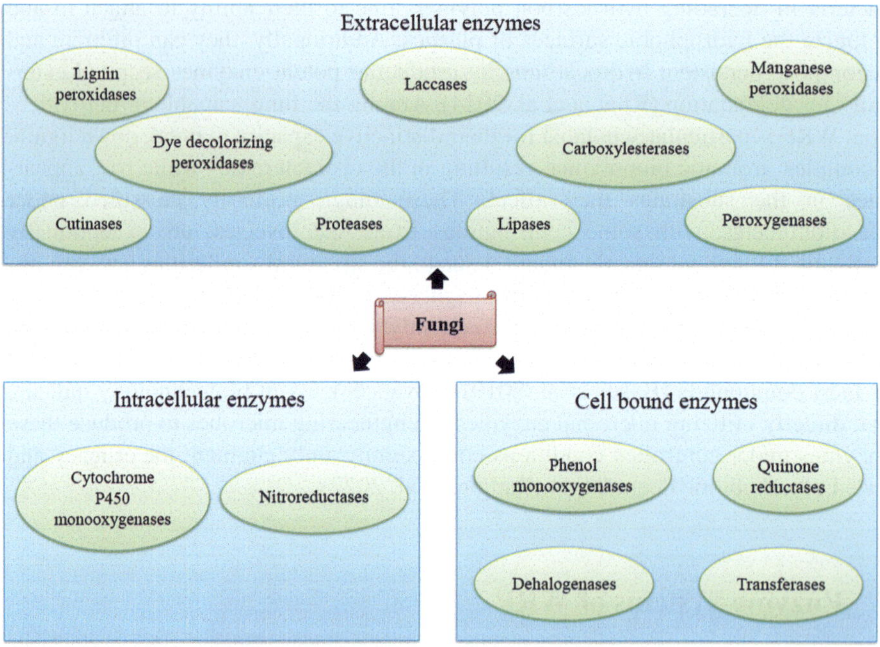

Fig. 2 The figure summarizes several representative enzymes involved in xenobiotic degradation, highlighting their roles in the degradation and transformation of harmful pollutants

5.1 Laccase (EC 1.10.3.2; Benzenediol, Oxygen Oxidoreductase, and p*-Diphenol Oxidase)*

Laccases, a class of well-studied glycosylated blue multi-copper oxidases, are abundantly produced by microorganisms, particularly WRF, where they play a significant role in lignin degradation. Beyond their involvement in the breakdown of organic bio-polymers, laccases also contribute to various physiological and ecological processes, such as sporulation, pigment synthesis, fruiting body development, stress response, and plant pathogenesis (Baldrian 2004; Arregui et al. 2019). Some fungi, such as *Pycnoporus cinnabarinus*, lack LiP and MnP and rely solely on laccase as their primary ligninolytic enzyme for efficient lignin decomposition (Baldrian 2004).

In 1896, the first fungal laccase was identified by Bertrand and Laborde (Viswanath et al. 2014). The first three-dimensional structure of laccase was determined in 1998 from *Coprinus cinereus* (Loi et al. 2021). More than a hundred laccases from fungi have been isolated and studied in detail (Torres-Farradá et al. 2024). Laccases from WRF are particularly notable for their stability under extreme pH and temperature conditions. They also exhibit higher redox potential than bacterial laccases, enabling fungi to oxidize a broader range of substrates. These enzymes can efficiently oxidize various molecules, including phenols, ortho- and para-diphenols, polyphenols, aminophenols, aromatic amines, aliphatic amines, and some non-phenolic substrates, through one-electron oxidation mechanisms (Michniewicz et al. 2008; Ellouze and Sayadi 2016; Balcázar-López et al. 2016). This enzymatic versatility has made laccases invaluable for a wide array of industrial applications, such as the bleaching and decolorization of textile dyes, detoxification of effluents, bioremediation, and greenfuel production (Ellouze and Sayadi 2016). For example, *T. versicolor* produces laccases with a high redox potential, making them effective for degrading complex pollutants and synthetic dyes such as triphenylmethane, anthraquinone, indigo, and azo dyes (Legerská et al. 2018). Laccases from other fungal species, including *Aspergillus* spp., *Coriolus versicolor*, *Funalia trogii*, *P. chrysosporium*, *Pleurotus ostreatus*, and *Trametes trogii* have also been extensively studied for their ability to decolorize dyes (Isanapong and Mataraj 2018). These results highlight the promising role of laccases in tackling environmental challenges and their application across diverse industries.

A typical fungal laccase has a molecular weight (Mw) of approximately 60–70 kDa and an isoelectric point near pH 4.0 (Arregui et al. 2019). These enzymes generally consist of about 550 amino acids and exhibit stability generally at temperatures below 40 °C (Torres-Farradá et al. 2024). However, Baldrian (2004) purified laccase from *Daedalea quercina*, which was identified as a monomeric protein with Mw of 69 kDa. This enzyme displayed thermal stability up to 45 °C and demonstrated enhanced stability at alkaline pH levels. Its activity was enhanced in the presence of Cu^{2+} ions but was inhibited by Mn^{2+}, sodium azide, dithiothreitol, and cysteine.

Laccases contain four copper atoms distributed across three active sites: T1, T2, and T3. During the catalytic process, the T1 copper accepts electrons from the substrate and transfers them to the trinuclear copper cluster (T2 and T3). This cluster is key in reducing O_2 to H_2O, with a transient peroxo or peroxide intermediate forming during the reaction. Unlike peroxidases, which require H_2O_2 as an electron acceptor, laccases use O_2, avoiding the complications associated with H_2O_2-induced enzyme inactivation (Khan et al. 2023). While laccases generally have a lower oxidation potential than peroxidases, they can oxidize non-phenolic substrates in the presence of appropriate redox mediators. These mediators include fungal metabolites and natural compounds like veratryl alcohol (VA), *p*-coumaric acid, and gallic acid, which extend the range of substrates that laccases can act upon. Additionally, chemical artificial substrates such as 2,2-azino-bis(3-ethylbenzothiazoline-6-sulfonate) (ABTS) and n-hydroxybenzotriazole (HBT) are also excellent laccase mediators (Torres-Farradá et al. 2024). A laccase produced in the yeast *Pichia pastoris* was engineered through site-directed mutagenesis to enhance the electron transfer rate. This suggests that laccases can be engineered to improve the efficiency of specific bioremediation processes (Viswanath et al. 2014). Ongoing research is dedicated to discovering laccases with improved efficiency from natural sources or through protein modification techniques to enhance their catalytic properties, thus broadening their industrial applications (Chen et al. 2012).

5.2 *Peroxidase (EC 1.11.1.7; Hydrogen Peroxide Oxidoreductase)*

Peroxidase catalyzes the oxidation of a wide variety of substrates (denoted by AH) using H_2O_2 as the oxidizing agent (Nakayama and Amachi 1999). The reaction proceeds as follows:

$$AH + H_2O_2 \rightarrow A + 2H_2O$$

These complex proteins, typically glycosylated, contain numerous disulfide linkages and incorporate multiple calcium ions and a heme cofactor. This structural complexity may make it intrinsically challenging for certain organisms to produce them (Singh et al. 2024). In 1986, a novel peroxidase was identified in the extracellular culture filtrate of the Hyphomycete *Arthromyces ramosus*, which sparked significant interest due to its potential applications in clinical diagnostics. Following this discovery, a similar peroxidase was detected in the culture filtrate of *Coprinus cinereus* (Ink Cap Basidiomycete) (Nakayama and Amachi 1999). Since then, studying these enzymes has become a growing area of interest. Fungal peroxidases are generally categorized into two major classes (Conesa et al. 2002):

Class I: Intracellular peroxidases, such as yeast cytochrome c peroxidase
Class II: Extracellular fungal peroxidases, including LiP and MnP

5.3 *LiP (EC 1.11.1.14; Diarylpropane: Oxygen, H_2O_2, Oxidoreductase)*

LiPs are monomeric, glycosylated enzymes with Mw ranging from 37 to 50 kDa, composed of 343 amino acid residues, 370 water molecules, a heme group, four carbohydrate moieties, and two calcium ions (Kumar and Chandra 2020). The pH and temperature optima of LiPs generally range between pH 2–5 and 35–55 °C (Asgher et al. 2008). Structurally, LiPs are predominantly helicoidal, consisting of eight major helices, eight minor helices, and two antiparallel beta sheets. These enzymes are organized into two domains, with heminic groups located on either side. The heme groups are embedded within the protein structure but remain accessible to solvents via two small channels, facilitating their catalytic activity (Kumar and Chandra 2020).

The catalytic cycle of LiP involves three key reactions. First, H_2O_2 acts as the electron acceptor, oxidizing the resting ferric enzyme (Fe^{3+}) to form Compound I ($Fe^{4+} = O$). In the second step, Compound I oxidizes a substrate (S) via an one-electron transfer, generating Compound II ($Fe^{4+} - O$) and an oxidized substrate (S^+). In the final step, Compound II reacts with another substrate molecule, regenerating the resting enzyme and releasing an additional oxidized substrate along with H_2O

(Martínez 2002). The immobilization of LiP has been found to improve its pH and temperature optima, thermostability, and catalytic efficiency. Certain natural secondary metabolites, including VA and 2-chloro-1,4-dimethoxybenzene, can function as redox mediators. However, LiP activity could be inhibited by certain compounds. For instance, LiP from *P. chrysosporium* was inhibited by ethylenediaminetetraacetic acid (EDTA) and N,N,N′,N′-tetramethylethylenediamine (TEMED). Interestingly, it could be reversed with an excess of zinc ions and various other metallic ions. Additionally, the oxidation efficiency of LiP is influenced by the molar ratio of H_2O_2 to the target pollutant (Asgher et al. 2008).

LiP was first identified as an extracellular enzyme in 1983 from ligninolytic cultures of *P. chrysosporium*. Since then, LiP has been detected in various microorganisms, particularly white-rot Basidiomycetes (Hussaan et al. 2021). Reported LiP-producing fungi to include *Arthromyces ramosus*, *B. adusta*, *Cerena maxima*, *Ceriporiopsis subvermispora*, *Coriolopsis polizona*, *Cylindrobasidium evolvens*, *Daedaleopsis septentrionalis*, *Lentinus tigrinus*, *Pleurotus eryngii*, *Phlebia radiata*, and multiple species of *Trametes*, such as *T. versicolor*, *T. ochracea*, *T. hirsuta*, *T. gallica*, and *T. villosa* (Hussaan et al. 2021; Sarker et al. 2021). However, not all WRF produce LiP (Singh et al. 2024). LiP induces H_2O_2-dependent oxidative depolymerization of lignin, effectively cleaving ether (C–O–C) and carbon–carbon (Cα–Cβ) bonds within the polymers (Chowdhary et al. 2019a; Singh et al. 2024). With a high redox potential ranging from 700 to 1400 mV, LiP surpasses laccases and MnP in catalytic efficiency. This high redox potential enables LiP to oxidize phenolic and non-phenolic components of lignin without needing mediators, demonstrating exceptional oxidative capabilities (Singh et al. 2024). LiP can mineralize recalcitrant aromatic compounds, including veratryl alcohol (3,4-dimethoxybenzyl), methoxybenzenes, and other complex substrates. It catalyzes a variety of reactions, such as Cα–Cβ bond cleavage, aryl–Cα bond cleavage, aromatic ring opening, phenolic oxidation, and demethoxylation (Asgher et al. 2008; Chowdhary et al. 2019a). Its broad substrate specificity, high redox potential, and ability to function efficiently with specific mediators make LiP a powerful enzyme for industrial applications. These include the degradation of complex aromatic pollutants, lignin valorization, environmental remediation, and bio-based material production (Christian et al. 2005).

5.4 *MnP (EC 1.11.1.13; Mn^{2+}: H_2O_2 Oxidoreductases)*

MnPs are glycoproteins with Mw ranging from 32 to 62.5 kDa, containing approximately 350 amino acid residues and an iron protoporphyrin IX (heme) prosthetic group. These enzymes exhibit optimal activity at pH levels ranging from 4 to 7 and temperature between 40 and 60 °C (Asgher et al. 2008; Chowdhary et al. 2019b). MnPs are secreted in multiple isoforms under carbon- and nitrogen-limited conditions when supplemented with Mn^{2+} and VA. Mn^{2+} acts as a mediator in the catalysis process. The catalytic cycle begins with the ferric MnP enzyme reacting with H_2O_2 to form compound I, a Fe^{4+} oxo-porphyrin radical complex. Mn^{2+} donates one electron to compound I, generating compound II and forming Mn^{3+}. The chelated Mn^{3+} acts as charge transfer mediators, enabling the oxidation of a range of phenolic substrates (Chowdhary et al. 2019b). Besides H_2O_2, MnP can utilize organic hydroperoxides as sources of oxidizing equivalents. However, MnP activity can be inhibited by sodium azide, ascorbic acid, β-mercaptoethanol, and dithiothreitol. The activity of MnP can be enhanced by cooxidants such as glutathione, unsaturated fatty acids, and Tween 80. Immobilization techniques, such as using carriers like sodium alginate, gelatin, or chitosan combined with glutaraldehyde as a cross-linking agent, have proven effective in improving the stability of MnP (Asgher et al. 2008).

MnP was first identified in *P. chrysosporium* (Martínez 2002). In addition to this well-known fungus, MnP is predominantly produced by various WRF, including *Agaricus bisporus*, *Agrocybe praecox*, *Ceriporiopsis subvermispora*, *Dichomitus squalens*, *Echinodontium taxodii* 2538, *Fusarium* sp. strain ZH-H2, *Ganoderma lucidum*, *I. lacteus*, *Lentinula edodes*, *Penicillium* sp., *Phlebia radiata*, *P. ostreatus*, *Pleurotus pulmonarius*, *P. eryngii*, *Pseudolagarobasidium* sp. PP17–33, *Schizophyllum* sp., *Stereum ostrea*, and *T. versicolor* (Qin et al. 2014; Kumar and Arora 2022). Research has shown that purified MnPs from some of these fungi can effectively degrade azo and anthraquinone dyes, polycyclic aromatic hydrocarbons, and 2,4,6-trinitrotoluene. Additionally, MnP is capable of degrading persistent aromatic contaminants, including industrial dye, polycyclic aromatic hydrocarbon, chlorophenol, nitroaromatic compounds, and antibiotics (Qin et al. 2014). Recently, MnP has gained attention for its potential in wastewater treatment and a range of industrial applications, including dye decolorization, biopulping, biobleaching, juice clarification, bioethanol production, and the degradation of xenobiotic compounds (Kumar and Arora 2022). However, the use of MnP is constrained by various challenges like the slow growth and low yield of native enzyme-producing fungi, as well as the absence of efficient methods for recombinant enzyme production (Järvinen et al. 2012).

5.5 Versatile Peroxidases (VPs, EC 1.11. 1.16; Reactive Black 5:H_2O_2 Oxidoreductase)

VPs are heme-containing enzymes that serve as a functional hybrid between MnPs and LiPs. These enzymes can oxidize Mn^{2+} and a broad spectrum of substrates, including VA, phenolic and non-phenolic compounds, large aromatic molecules, and dyes, even without manganese. VPs exhibit catalytic activity similar to both MnPs and LiPs by oxidizing Mn^{2+} to Mn^{3+}, degrading lignin model compounds, and oxidizing aromatic compounds such as VA and *p*-dimethoxybenzene (Torres-Farradá et al. 2024). Their optimal pH for Mn^{2+} oxidation is around 5; while for aromatic compound oxidation, it is closer to pH 3, reflecting their dual functionality (Hernández-Bueno et al. 2021).

First discovered in *P. eryngii*, VPs have since been studied extensively in several WRF, including *Bjerkandera* sp., *P. ostreatus*, and *P. pulmonarius* (Hernández-Bueno et al. 2021; Torres-Farradá et al. 2024). The activity of these enzymes is influenced by medium polarity, with organic solvents such as acetonitrile, dimethyl sulfoxide, and ethanol inhibiting the enzyme's activity, while oxidizing agents such as acetosyringone and 2,2,6,6-tetramethyl-1-piperidinyloxy (TEMPO) enhance the enzyme's ability to degrade dyes. VPs also exhibit strong affinities for H_2O_2, ferulic acid, and various hydroquinones, adding to their potential in biotechnological applications, especially in dye degradation and other industrial processes (Asgher et al. 2008).

5.6 DyP-Type Peroxidases (EC 1.11.1.19)

DyPs are a unique class of heme peroxidases that exhibit significant differences in both sequence and structure compared to other peroxidase families. These enzymes are widely distributed and can be found both intra- and extracellularly (Sugano and Yoshida 2021). The first DyP-type peroxidase was identified in *B. adusta*, and since then, these enzymes have been discovered in a variety of fungi, including *Auricularia auricula-judae*, *Exidia glandulosa*, *I. lacteus*, *P. ostreatus*, *Mycena epipterygia*, *Mycetinis scorodonius*, and *Termitomyces albuminosus* (Colpa et al. 2014; Cordas et al. 2022).

DyPs are particularly known for their ability to oxidize synthetic anthraquinone dyes in the presence of H_2O_2 (Cordas et al. 2022). However, synthetic dyes are not their natural substrates. Instead, DyPs are believed to target naturally occurring anthraquinone compounds like alizarin, which plants produce. In WRF that parasitize trees, DyPs may break down antifungal anthraquinones, facilitating the parasitic process. One of the key features of DyPs is their remarkable stability, which makes them highly suitable for industrial applications. For example, the purified dimeric DyP-type peroxidase MsP1 from *M. scorodonius* shows exceptional stability even under extreme conditions, including high pressure (up to 250 MPa) and temperature (up to 70 °C) (Torres-Farradá et al. 2024). This robust stability makes DyPs promising candidates for industrial processes, including dye degradation and other environmental and biotechnological applications.

5.7 CYP Monooxygenase (EC 1.14.x.x; Mixed-Function Oxidases)

CYP enzymes are membrane-bound, heme-containing monooxygenases recognized for their characteristic absorbance peak at 450 nm. These enzymes are highly versatile and catalyze the cleavage of molecular oxygen, incorporating one oxygen atom into a substrate molecule and reducing the other to H_2O (Khan et al. 2023). The general stoichiometry of the reaction (where S-H represents the substrate) is as follows (Zanger 2021):

$$\mathrm{NADPH} + \mathrm{H}^{+} + \mathrm{O}_2 + \mathrm{S} - \mathrm{H} \rightarrow \mathrm{NADP} + + \mathrm{H}_2\mathrm{O} + \mathrm{S} - OH$$

A recent high-throughput sequencing study estimates that approximately 5.1 million fungal species exist on Earth, roughly six times the total number of plant species. Fungal adaptations to diverse ecological niches are often facilitated by CYP monooxygenases. These enzymes are essential for various fungal functions, including metabolite production related to pathogenesis, decomposition, substrate utilization, and the detoxification of xenobiotics (Durairaj et al. 2016). For instance, in the model white-rot fungus, *P. chrysosporium*, CYP monooxygenases are critical for breaking down lignin and other environmental toxicants (Syed and Yadav 2012). These enzymes exhibit a broad catalytic spectrum, performing various reactions such as hydroxylation, epoxidation, dehalogenation, and so forth on diverse substrates. Beyond their ligninolytic role, intracellular CYPs also contribute to non-ligninolytic processes, such as pesticide transformation, enabling fungi to metabolize a wide range of substances, including pesticides, hydrocarbons, drugs, and steroids (Khan et al. 2023). The industrial application of CYP enzymes is hindered by several challenges, including instability, low catalytic efficiency, inadequate expression yield, poor solvent tolerance, and high cofactor cost. However, innovative approaches, such as immobilizing CYPs, have shown promise in addressing these limitations and improving performance in industrial settings (Durairaj et al. 2016).

5.8 Cutinases (E.C. 3.1.1.74)

Cutinases are extracellular serine esterases with an α/β fold structure comprising a central β-sheet of five parallel strands surrounded by helices. Their active site, usually exposed (except in *Trichoderma reesei*), contains a catalytic triad of Ser-His-Asp/Glu (Temporiti et al. 2022). These enzymes are naturally produced by fungi such as *Aspergillus oryzae*, *Cryptococcus* sp. strain S-2, *Fusarium solani*, *Humicola insolens*, *Penicillium citrinum*, and *P. pastoris* (Srikanth et al. 2022; Khatua et al. 2024). Functionally, cutinases play an essential role in fungal pathogenesis by breaking down ester bonds. Their versatility also makes them valuable in industrial applications, including the hydrolysis of high Mw polyesters, esterification, and transesterification, as well as the modification of synthetic fibers (Temporiti et al. 2022).

5.9 Lipase (E.C. 3.1.1.3; Triacylglycerolacyl Hydrolases)

Fungal lipases are extracellular enzymes responsible for hydrolyzing ester bonds in tri-, di-, and mono-glycerides, converting these into free fatty acids and glycerol. These enzymes are key players in lipid metabolism, facilitating digestion, absorption, and reconstitution (Temporiti et al. 2022). Lipases also contribute to the degradation of xenobiotic compounds and are produced by various fungi, including *Aspergillus tubingensis*, *Candida antarctica*, *Microsphaeropsis arundinis*, *Penicillium simplicissimum*, *Pseudozyma antarctica*, *Rhizopus delemar*, *Thermomyces lanuginosus*, and *Yarrowia lipolytica* (Srikanth et al. 2022; Khatua et al. 2024). These enzymes' robust catalytic potential underscores their significance in environmental remediation, biocatalysis, and the development of sustainable industrial processes.

6 Decolorization of Synthetic Dyes by Purified Fungal Enzymes

Artificial dyes are extensively used in numerous trades such as textiles, plastics, paper, cosmetics, and food (Bhattacharya et al. 2011). However, improper disposal of dye-polluted wastewater, estimated at nearly 11 million tons annually, poses significant environmental challenges, as many dyes are toxic, recalcitrant, and even carcinogenic (Gao et al. 2023). Among these, azo dyes dominate the textile industry for dyeing cotton fabrics. At the same time, anthracene derivatives such as Remazol Brilliant Blue R (RBBR) are noteworthy due to their structural resemblance to polycyclic aromatic compounds, which are substrates of ligninolytic peroxidases (Christian et al. 2005). In this context, using fungi to degrade or sequester pollutants offers a promising solution.

The ability of WRF to decolorize dyes was first reported by Glenn and Gold (1983), who observed sulfonated polymeric dye decolorization as a measure of ligninolytic activity in *P. chrysosporium*. Subsequent studies demonstrated the decolorization potential of fungi such as *B. adusta* and *T. versicolor*; although the enzymes responsible remained unidentified at the time (Heinfling et al. 1997). Later, WRF were utilized to decolorize industrial effluents containing nitrated stilbene sulfonic acids, leading to further research on lignin-modifying enzymes under dye-decolorizing conditions (Knapp and Newby 1999).

As demonstrated in Table 1, various fungi, including *Coriolopsis* spp. *Irpex lacteus*, *Ganoderma* spp., *Pleurotus* spp., *Phanerochaete* spp., *Trametes* spp. have demonstrated significant potential in dye decolorization, with laccase being one of the most commonly studied enzymes due to its oxidative activity and broad substrate specificity on aromatic rings (Ben Younes et al. 2016). Biological decolorization can occur through dyes' adsorption (biosorption) within the microbial biomass or through enzymatic biodegradation (Pandey et al. 2018). Among the mediators tested, synthetic mediators like ABTS have been found to be most effective, owing to their compatibility with laccase activity, while natural mediators such as vanillin and its derivatives achieve 40–65% decolorization of synthetic dyes (Chhabra et al. 2008). However, the application of soluble laccase is limited by its instability and non-reusability, making the process costly. To address this, immobilization techniques have been explored to enhance laccase stability and reusability. For example, *F. trogii* cultures immobilized on a *Luffa cylindrica* sponge effectively decolorized Reactive Black 5 (RB5) (Mazmanci and Ünyayar 2005). Similarly, partially purified acidic laccase from *Coriolopsis gallica*, entrapped in calcium alginate beads, exhibited high efficiency in decolorizing the anthraquinone dye RBBR, with immobilization improving the enzyme's stability under varying pH and temperature conditions. These findings emphasize the promising potential of immobilized fungal laccase systems in industrial dye treatment processes (Daâssi et al. 2014). Furthermore, *Trametes* laccase, immobilized as cross-linked enzyme aggregates (CLEAs), exhibited enhanced stability and functionality. Both free and CLEA-immobilized laccase efficiently removed 11 trace organic contaminants from simulated wastewater, with CLEA laccase achieving removal through a combination of biocatalytic activity and adsorption (George et al. 2022).

As outlined in Table 1, recalcitrant dyes can effectively be decolorized by peroxidases, particularly MnP, when used with suitable redox mediators. Immobilized peroxidases have demonstrated superior performance to their free counterparts, particularly in the continuous decolorization of dyes in wastewater treatment reactors. For instance, *I. lacteus* immobilized on polyurethane foam (PUF) cubes produced high and sustained levels of MnP (Kasinath et al. 2003), while *B. adusta* immobilized on PUF was reported to produce six different MnP isoenzymes (Nakamura et al. 1999). MnP immobilized in a sol-gel matrix from *G. lucidum* effectively decolorized various dye effluents (Iqbal and Asgher 2013). These results emphasize the potency of immobilized enzyme systems in sustainable wastewater remediation.

Recent research has increasingly focused on harnessing fungi for wastewater treatment applications. However, challenges remain with the use of free fungal organisms, which often clog pipelines and are difficult to reuse. Similarly, employing fungal enzyme extracts for pollutant degradation is hindered by high production costs, susceptibility to environmental stress, and limited catalyst recovery. To address these limitations, fermenting solid substrates with live fungi has emerged as promising to generate effective and environmentally friendly biocatalysts (Gao et al. 2023). Despite these advancements, fungal decolorization processes are often time-intensive, necessitating the development of innovative strategies to expedite treatment and improve the efficiency of fungal applications in textile effluent remediation (Moreira-Neto et al. 2013).

Table 1 Ability of various fungal species and their associated enzymes to degrade synthetic dyes

Fungal species	Enzyme(s) involved	Synthetic dyes degraded	Remark	References
Aspergillus flavus	Laccase	CR	98.86% degradation of the dye was achieved within 96 h through biosorption and biodegradation	Bhattacharya et al. (2011)
Bjerkandera adusta strain CX-9	LiP (45.22 kDa), MnP (30.12 kDa)	Acid blue 158, cibacet brilliant blue BG, polymeric dye R, RBBR, remazol brilliant violet 5R, indigo carmine, MG	MnP achieved up to 91–12% decolorization of the dyes within 12 h, while LiP decolorized the dyes in the 15–89% range after 12 h of incubation.	Bouacem et al. (2018)
Cerrena unicolor	Laccase	Azo and anthraquinone dyes	Converted the dyes at pH 3.5 in aqueous solution	Michniewicz et al. (2008)
Cerrena sp.	Laccase	RBBR	90% decoloration efficiency was achieved by incubating 8 U/mL of the enzyme in 0.01% dye solution within 20 min	Chen et al. (2012)
Clitopilus scyphoides, Ganoderma resinaceum	MnP	BPB, Bromothymol blue, RBBR	Complete degradation was noted	Rao et al. (2019)
Coprinus comatus	Laccase	Reactive Dark Blue KR, Reactive orange 1, Reactive Red X-3B, CR, RBBR, Reactive Brilliant Blue X-BR, Reactive Brilliant Blue K-GR, Reactive Brilliant Blue K-3R	The enzyme independently decolorized >90% of Reactive Dark Blue KR within 12 h, with even higher efficiencies observed for the rest of the dyes in the presence of HOBT	Bao et al. (2012)
Coriolopsis gallica	Laccase entrapped into calcium alginate beads	RBBR, RB5, Bismark Brown R	Immobilization improved the thermal and storage stabilities of the enzyme	Daâssi et al. (2014)
Coriolopsis byrsina, Lentinus strigellus, Lentinus sp., *Pycnoporus sanguineus, Phellinus rimosus*	Laccase	Orange II, RBBR, Cibacron Blue 3GA, Methylene blue, crystal violet	Efficient decolorization was observed within 48 h.	Gomes et al. (2009)

(continued)

Table 1 (continued)

Fungal species	Enzyme(s) involved	Synthetic dyes degraded	Remark	References
Coriolopsis polyzona, Hypholoma fasciculare, Stropharia rugosoannulata, Trametes versicolor	Laccase	Acid Red 299, Direct Blue 1, Direct Red 28, Disperse Red 1, Disperse Yellow 3, RB5, Reactive Red 4, Reactive Yellow 81, NY3, Disperse Blue 1, RBBR	NY3, Direct blue 1 and RBBR were degraded more efficiently	Junghanns et al. (2008)
Coriolus versicolor f. *antarcticus*	Laccase	Malachite Green, Azure B, Poly R-478, Anthraquinone Blue, CR, Xylidine	Whole cultures decolorized 28–100% of five different dyes within an hour	Levin et al. (2004)
Cyathus bulleri	Laccase	Reactive Orange 1, Reactive Red 198	ABTS was effective even at low mediator/dye ratios, achieving 80–95% decolorization	Chhabra et al. (2008)
Daedalea dickinsii	Laccase, MnP, LiP	Disperse violet-63, Disperse orange-30	Dyes were converted into simpler secondary amines, such as diphenylamine and 3-methyldiphenylamine	Mehmood et al. (2020)
Daedalea quercina	Laccase	Chicago sky blue, poly B-411, RBBR, trypan blue and reactive blue 2	Decolorized the dyes in the absence of mediators	Baldrian (2004)
Datronia caperata, Polyporus tenuiculus, Pycnoporus sanguineus	Laccase	Orange II	In the presence of ABTS, 96–100% decolorization of the dye was achieved within 1–48 h	Abrahão et al. (2008)
Dichomitus squalens	Laccase, MnP	Orange G and RBBR	On Kirk medium containing 500 mg/L of synthetic dyes, static cultivation significantly increased decolorization after the 14th day	Eichlerová et al. (2005)

(continued)

Table 1 (continued)

Fungal species	Enzyme(s) involved	Synthetic dyes degraded	Remark	References
Fomes fomentarius	Laccase	RBBR	80% decolorization was achieved at an initial concentration of 150 mg/L RBBR with 0.2 U/mL of laccase	Neifar et al. (2010)
Funalia trogii	Laccase	RB5	Cultures immobilized on *Luffa cylindrica* sponge decolorized the dye	Mazmanci and Ünyayar (2005)
	Laccase	RBBR and Drimaren Blue CL-BR	Removed color entirely at the end of 48 h of incubation	Erkurt et al. (2007)
Ganoderma boninense, Ganoderma miniatocinctum, Ganoderma zonatum, Ganoderma tornatum	Laccase, MnP	RBBR	*G. boninense* exhibited the best decolorization effect, followed by *G. miniatocinctum*	Siddiqui et al. (2019)
Ganoderma lucidum	Laccase	RBBR	Mediator (vanillin and HBT) and metal ions (Na^+ and Cu^{2+}) enhanced ability of laccase to decolorize RBBR	Qin et al. (2019)
	Laccase, MnP	RBBR	Peach-Palm sheaths showed considerable potential in the process of dye removal	Chicatto et al. (2018)
Peniophora cinerea, Pleurotus ostreatus, Trametes villosa	Laccase	Cibacron Brilliant Blue H-GR and Cibacron Red FN-2BL	Demonstrated remarkable efficiency in decolorizing synthetic textile effluent, achieving 74–77% in vivo and 60–74% in vitro.	Moreira-Neto et al. (2013)
Ischnoderma resinosum	Laccase, MnP	RB5, Reactive Blue 19, Reactive Red 22, and Reactive Yellow 15	Mn-P effectively decolorized all tested dyes, while laccase showed no activity with RB 5 and Reactive Red 22	Kokol et al. (2007)
Ischnoderma resinosum, Pleurotus calyptratus	Laccase	Orange G	Shaken cultivation significantly hindered the decolorization	Eichlerová et al. (2005)

(continued)

Table 1 (continued)

Fungal species	Enzyme(s) involved	Synthetic dyes degraded	Remark	References
Irpex lacteus	MnP	RBBR	Cultures immobilized on polyurethane foam and pine wood exhibited rapid decolorization	Kasinath et al. (2003)
	Laccase, MnP	RBBR, RB5	The dye was degraded faster than RB5	Máximo and Costa-Ferreira (2004)
	Laccase, MnP	RO16	90% decolorization efficiency in 3 days	Tavčar et al. (2006)
	MnP	RO16, RBBR, BPB	79%, 90%, and 94% decolorization of R016, RBBR, and BPB was noted within 1 day, with an initial dye concentration of 25 mg/L	Šušla and Svobodová (2007)
	Laccase	RO16, RBBR	The highest activity was noted with ABTS	Svobodová et al. (2008)
	MnP	RBBR, Remazol Brilliant Violet 5R, Direct Red 5B, Indigo Carmine and Methyl Green	All the tested dyes and simulated textile wastewater were efficiently decolorized	Qin et al. (2014)
Lentinula (*Lentinus*) *edodes*	MnP	RBBR	The absence of manganic ions and H_2O_2 markedly reduced dye decolorization	Boer et al. (2004)
Lentinus polychrous	Laccase	RBBR	87% decolorization efficiency within 6 h	Isanapong and Mataraj (2018)
Lenzites elegans	Laccase	Brilliant green CR	Maximum decolorization achieved in broth culture was 92.77% and 98.8% for brilliant green and CR, respectively	Pandey et al. (2018)
Marasmius cladophyllus	Laccase	RBBR	99% decolorization was achieved after 12 days of incubation in Glucose Minimal liquid medium (pH 5.5), with a recorded enzyme activity of 285 U/L	Muradi et al. (2017)

(continued)

Table 1 (continued)

Fungal species	Enzyme(s) involved	Synthetic dyes degraded	Remark	References
Microporus vernicipes, Peniophora incarnata, Perenniporia subacida, Phanerochaete sordida, Phlebia acerina, Phlebia radiata	Laccase, MnP	RBBR	*P. sordida* efficiently decolorized RBBR within 5 days, whereas other fungi required 6–10 days to achieve similar results	Lee et al. (2020)
Mycena purpureofusca	Laccase	RBBR	Efficient decolorization was achieved using crude laccase at low concentrations without the need for redox mediators	Sun (2012)
Oudemansiella canarii	Laccase (41 kDa)	CR	5 U enzyme decolorized 80% of 50 mg/L the dye within 1 day at 30 °C and pH 5.5	Iark et al. (2019)
Panus strigellus	Laccase	RBBR, RB 220, MG	The crude enzyme showed >90% decolorization of RB 220 within 24 h. In comparison, 60% and 68% of RBBR and MG decolorization were achieved within 72 h, respectively	Cardoso et al. (2018)
Perenniporia tephropora	Laccase	RBBR, neolane blue and neolane pink	>80% decolorization efficiency was noted	Ben Younes et al. (2007)

(continued)

Table 1 (continued)

Fungal species	Enzyme(s) involved	Synthetic dyes degraded	Remark	References
Phanerochaete chrysosporium	MnP	Direct Blue 15	Immobilized culture on $ZrOCl_2$-activated pumice demonstrated a decolorizing efficiency of 95–100% as a biocatalyst	Pazarlioglu et al. (2005)
	MnP	Direct Blue 15, and Direct Green 6	Maximum MnP activity was obtained at 37 °C and pH 4.5 after 6 days of incubation	Ürek and Pazarlioglu (2007)
	MnP	Methylene Blue	Kissiris appeared as a suitable support material for immobilizing the culture, enhancing its decolorizing capacity	Karimi et al. (2006)
	LiP, MnP	Direct Red 80, Mordant Blue-9	Maximum enzyme activities were found with Direct Red 80 than Mordant-9 Blue	Pakshirajan et al. (2011)
	MnP	Methylene blue	Under optimal conditions (temperature 35 °C, pH 5), a maximum decolorization of 84.8% was achieved	Zeng et al. (2015)
	MnP	CR	The dye was adsorbed onto biomass and subsequently decolorized, achieving 85% decolorization in MnP cultures	Bosco et al. (2017)
Phanerochaete sordida	LiP, laccase	RB5	Complete decoloration was achieved within 3 days, and the resulting degradation products were non-phytotoxic	Permpornsakul et al. (2016)

(continued)

Table 1 (continued)

Fungal species	Enzyme(s) involved	Synthetic dyes degraded	Remark	References
Pleurotus ostreatus	MnP, MiP	Phenol red, orthocresol red, meta-cresol purple, bromophenol red, bromocresol purple, BPB, bromocresol green	The presence of an additional bromine group alongside the methyl group on sulfonphthalein chromophores reduced the decolorization rate	Shrivastava et al. (2005)
	Laccase, MnP	Drimarene blue K2RL	97.86% decolorization was achieved within 3 days at pH 3 and 30 °C by supplementing the medium with 1% glucose and 1% ammonium sulfate	Kanwal et al. (2010)
	Laccase, MnP	Drimarene brilliant red K-4BL, drimarene orange K-GL	73% Drimarene brilliant red K-4BL removal was noted after 10 days, accompanied by laccase and MnP production of 64 and 42 IU/mL, respectively	Kamal et al. (2011)
	Laccase	RBBR	Two purified laccases, POXC and POXA3, were able to degrade RBBR in vitro without the need for redox mediators	Palmieri et al. (2005)
	Laccase	RBBR, acetyl yellow G	The dyes highly induced levels of *pox3* and *pox4* genes	Garrido-Bazán et al. (2016)
	Laccase	RBBR, BPB, MG, methyl orange	Within 24 h, removal rates were 73.1%, 79.1%, 84.9%, and 91.5% for methyl orange, BPB, RBBR, and MG, respectively at the concentration of 100 mg/L	Zhuo et al. (2019)

(continued)

Table 1 (continued)

Fungal species	Enzyme(s) involved	Synthetic dyes degraded	Remark	References
Pleurotus pulmonarius	Laccase	Amido black, CR, trypan blue, methyl green, RBBR, methyl violet, ethyl violet, brilliant cresyl blue, methylene blue, poly R-478	All dyes were completely decolorized after 6 days of cultivation, except for methylene blue and Poly R-478, which were only partially decolorized	Tychanowicz et al. (2004)
Pleurotus sajor-caju	MnP	CR	MnP extracted from banana peel culture showed 95% discoloration within 1 h (optimum temperature 35 °C, pH 4.0, and enzyme concentration 1.4 U/mL)	Yehia and Rodriguez-Couto (2017)
	Laccase	Orange G	Enzyme activity significantly increased to 279.27 IU/mL in the presence of *Punica granatum* peel extract as a redox mediator, enhancing dye decolorization to 85%	Bibi et al. (2019)
Pycnoporus sanguineus	Laccase	Orange G, amaranth, BPB and MG	Partial and complete decolorization of azo and triphenylmethane dyes, respectively	Pointing and Vrijmoed (2000)
	Laccase (61.4 kDa)	RBBR	Efficient decolorization was achieved without the addition of redox mediators	Lu et al. (2007)
Schizophyllum commune	MnP, Laccase	Solar golden yellow R	Complete decolorization was achieved within 2 days of incubation with 1% glucose under optimal conditions	Asgher et al. (2008)
Scytalidium thermophilum	Laccase	CR, MG, bromo-cresol green, and phenol red	99% decoloration efficiency was achieved at a dye concentration of 50 mg/L using 0.1 U/mL of laccase	Ben Younes et al. (2012)
Thelephora sp.	Laccase	Orange G, CR, and amido black 10B	Amido black 10B was degraded more effectively	Selvam et al. (2003)

(continued)

Table 1 (continued)

Fungal species	Enzyme(s) involved	Synthetic dyes degraded	Remark	References
Trametes hirsuta	Laccase	Triarylmethane, indigoid, azo, and anthraquinone dyes	Immobilization of the enzyme on alumina increased thermal stability and tolerance against inhibitor	Abadulla et al. (2000)
	Laccase, MnP	RBBR, acid blue 129	A ten-fold increase in dye decolorization rate was noted after adding $MnSO_4$, H_2O_2, and violuric acid	Anita et al. (2019)
Trametes hirsuta, Phanerochaete chrysosporium, Pleurotus ostreatus	Laccase, MnP	RBBR	When supplemented with 0.1 mM glucose, the tested fungi could remove phenolic and PAH compounds	Batista-García et al. (2017)
Trametes hirsuta, Pycnoporus sp., *Irpex* sp.	Laccase	RBBR	*T. hirsuta* showed maximum biodecolorization potential under static conditions after 5 days of incubation	Bibi and Bhatti (2012)
Trametes polyzona KU-RNW027	MnP, laccase	RBBR, Reactive Blue 120, Reactive Yellow 160, Reactive Orange 107, Reactive Red 198, Reactive Red 180	Purified enzymes (1 U/mL) achieved 33–100% decolorization of 25 mg/L synthetic dyes	Lueangjaroenkit et al. (2019)
Trametes versicolor	Mnp, Laccase	Remazol Brilliant Blue RR, Remazol Red RR, Remazol Yellow RR	A notable increase in decolorization rate was observed on 5th day	Toh et al. (2003)
	LiP	RBBR	Veratryl alcohol enhanced and stabilized decolorization rates of RBBR by LiP	Christian et al. (2005)
	Laccase	Orange 2, Acid Orange 6	Orange 2 was degraded more rapidly	Legerská et al. (2018)
	Laccase	RBBR	One of the RBBR degradation byproducts exhibited an orange color	Pype et al. (2019)

(continued)

Table 1 (continued)

Fungal species	Enzyme(s) involved	Synthetic dyes degraded	Remark	References
Trametes trogii	Laccase, MnP, GOx	MG, Xylidine, Ponceau 2R, Anthraquinone Blue	Laccase and MnP played significant roles in xylidine degradation and MG decolorization	Levin et al. (2005)
	Laccase	MG, BCG	Complete degradation was achieved within 6 h at a 50 mg/L initial dye concentration	Ben Younes et al. (2016)

ABTS 2,2′-azino-bis(3-ethylbenzthiazoline)-6-Sulfonate, *BCG* bromo-cresol green, *BPB* bromophenol blue, *CR* congo red, *HOBT* 1-hydroxybenzotriazol, *MG* malachite green, *MnP* manganese peroxidase, *MIP* manganese-independent peroxidase, *NY3* acid blue 62, *PAH* polycyclic aromatic hydrocarbons, *RBBR* remazol brilliant blue R/reactive blue 19, *RB5* reactive black 5, *RB220* reactive blue 220, *RO16* reactive orange 16

7 Biodegradation of Pesticides Through Fungal Enzymes

The application of pesticides is essential to maintain crop health, protect against pests, and optimize agricultural yields. In Canada, approximately 35 million kg of pesticides are applied annually, while India uses about 56 million kg each year (Vaithyanathan et al. 2022). A significant portion comprises organochlorine or chlorinated compounds, which consist of carbon, hydrogen, and oxygen atoms with chlorine-substituted aliphatic or aromatic rings. These non-polar compounds tend to bioaccumulate in animal tissues, leading to their transfer into the food chain (Magnoli et al. 2023). For instance, the pyrethroid insecticide cyfluthrin, commonly used on vegetables, leaves toxic residues that can exceed maximum residue limits, posing risks to both human health and the environment (Swathy et al. 2024). Addressing these challenges necessitates effective solutions to mitigate the harmful impacts of such compounds. Fungi, with their remarkable biochemical versatility, have shown immense potential in the biodegradation of pesticides, offering promising avenues for reducing environmental contamination and associated risks (Bhandari 2017).

In this context, several fungal strains and their associated enzymes have demonstrated promising potential for pesticide degradation (Fig. 3). For example, *T. versicolor* laccase, coupled with an HBT mediator, efficiently degraded the herbicide isoproturon and the insecticide imiprothrin, showcasing rapid degradation capabilities (Mir-Tutusaus et al. 2014; Sarker et al. 2021). Similarly, *Fusarium proliferatum* CF2, isolated from pesticide-contaminated agricultural fields, exhibited robust degradation kinetics, utilizing 50 mg/L allethrin as its sole carbon source, tolerating up to 1000 mg/L, and achieving complete degradation within 144 hours under optimal conditions (Bhatt et al. 2020). In another study, laccase from *Pleurotus dryinus*, produced via biosolid bioaugmentation, exhibited broad-spectrum pesticide degradation without requiring additional mediators, emphasizing its potential as a cost-effective bioremediation agent (Vaithyanathan et al. 2022). Moreover,

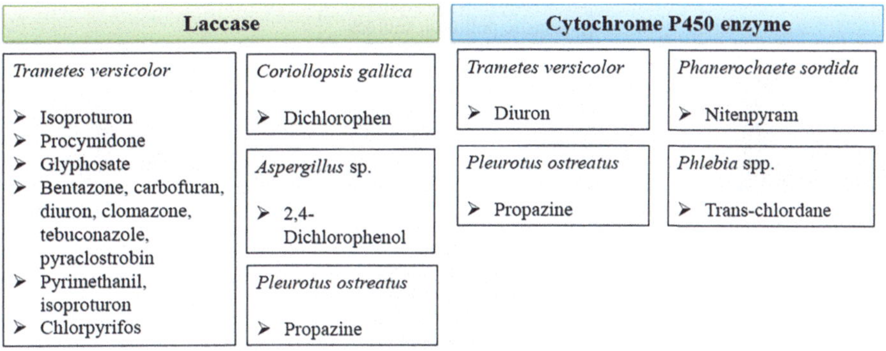

Fig. 3 Fungal source and concomitant enzymes related to pesticide degradation. (Adapted from Sarker et al. (2021) and Khan et al. (2023))

P. chrysosporium effectively degraded neonicotinoids such as acetamiprid, clothianidin, imidacloprid, and thiacloprid in nutrient-rich media. This activity was linked to two CYP enzymes, CYP5037B3 and CYP5147A3, identified as key players in neonicotinoid metabolism (Mori et al. 2021).

Basidiomycota fungi have been suggested to possess the ability to sense xenobiotics in their environment. In response, they synthesize extracellular oxidative enzymes that modify and degrade compounds such as 2,4-dichlorophenoxyacetic acid, facilitating their adsorption and subsequent entry into fungal cells (Magnoli et al. 2023). The fungal degradation of pesticides typically thus proceeds through a three-step process: (1) transformation of the parent pesticide into non-toxic, soluble compounds via reduction, hydrolysis, or oxidation, mediated by fungal reductive and oxidative enzymes; (2) conjugation to enhance water solubility; and (3) conversion of intermediates into non-toxic end products by enzymes like peroxidases and oxygenases, completing the detoxification cycle (Swathy et al. 2024). These mechanisms underscore the potential of fungal bioremediation as a sustainable approach to mitigate pesticide pollution.

8 Biodegradation of Plastics Through Fungal Enzymes

The global production of plastic has grown dramatically, from 1.5 to 2 million metric tons (MMT) in 1950 to 380 MMT by 2015, and it is projected to double by 2050 (Gross and Enck 2021). In 2019 alone, approximately 400 MMT of plastic waste was generated worldwide, with estimates suggesting the amount may have doubled in 2020 due to the COVID-19 pandemic (Yuan et al. 2021). As a result, plastic pollution has become a significant concern for researchers globally, evidenced by the substantial increase in related publications. Once introduced into the environment, plastics can enter aquatic, terrestrial, and atmospheric systems, either directly or through various pathways (Fig. 4). The situation worsens as plastics fragment

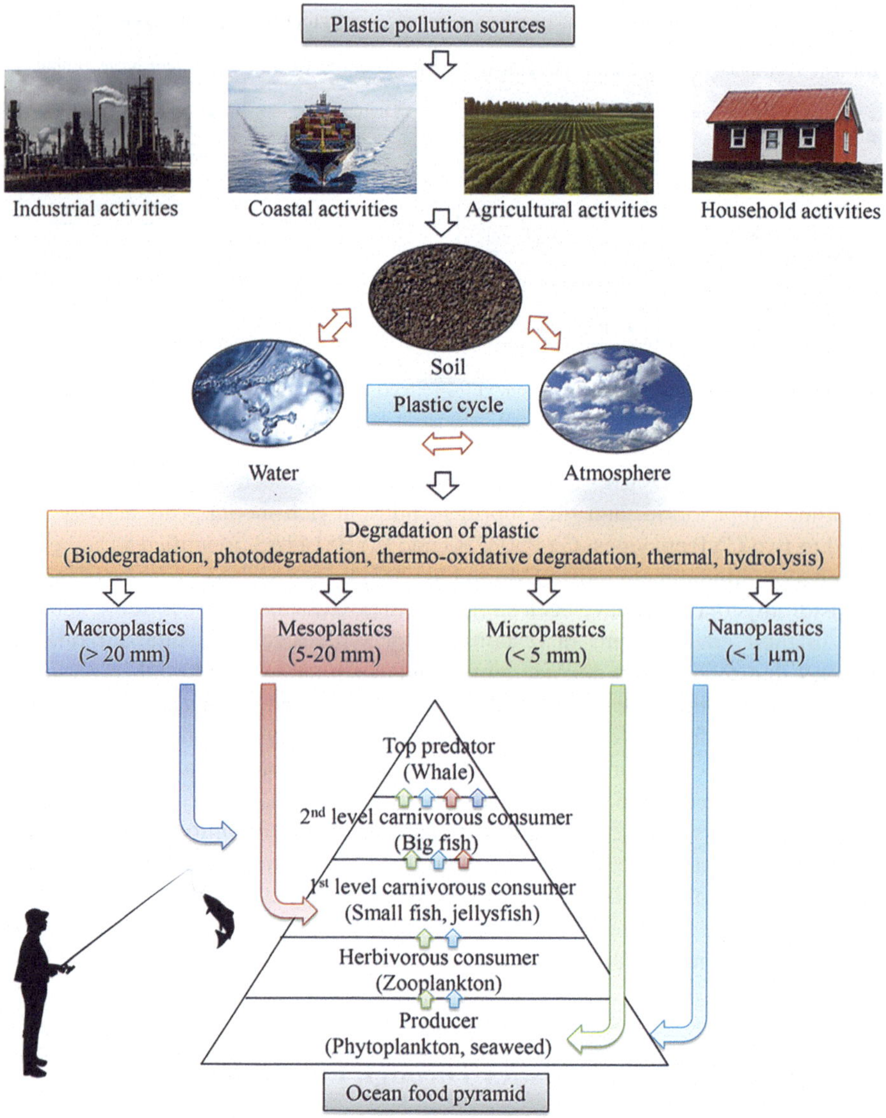

Fig. 4 A summary of sources, transports and routes of plastics. (Adapted from Urbanek et al. (2018) and Oliveira et al. (2020))

progressively into smaller particles, known as microplastics (MPs), which have been detected in the atmosphere, soil, and aquatic environments. MPs pose direct threats to aquatic organisms, causing issues such as gastrointestinal problems, hormonal imbalances, microbiome alterations, reduced growth rates, reproductive complications, disruptions in lipid metabolism, and feeding obstructions (Fig. 5). In response to these challenges, there has been growing interest in biodegradation as a

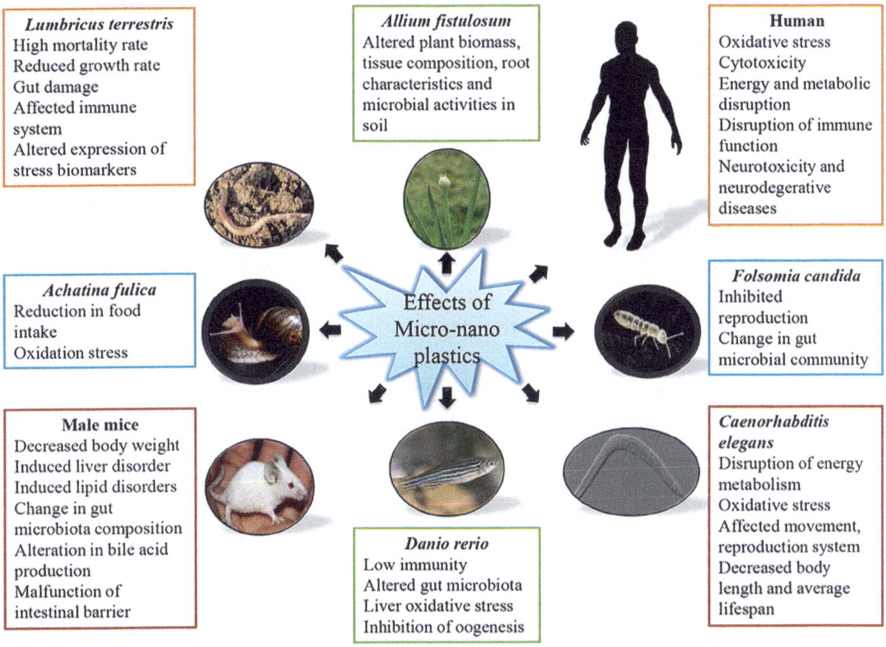

Fig. 5 Biological effects of micro-nano plastics on different organisms. (Adapted from Barría et al. (2020), Sana et al. (2020), and Allouzi et al. (2021))

more cost-effective and efficient solution compared to traditional waste disposal methods (Zahra et al. 2010). Fungi, in particular, have been identified as effective degraders of polymeric materials, with their ability to produce and secrete proteins significantly surpassing that of bacteria (Khatua et al. 2024).

The breakdown of plastic polymers typically involves oxidative enzymes like peroxidases, laccases, and monooxygenases, as well as hydrolytic enzymes such as esterases, cutinases, lipases, proteases, and ureases. These enzymes, which are secreted by fungal hyphae, serve a dual purpose: they initiate the degradation of the polymers and transform the resulting fragments into forms that can be assimilated and mineralized by the fungal mycelium (Okal et al. 2023). Oxidative enzymes such as fungal laccases and peroxidases, which are primarily involved in lignin degradation, have effectively broken down plastics like polyethylene (PE) and polyvinyl chloride (PVC). In contrast, hydrolytic enzymes, including cutinases and lipases, are particularly efficient in degrading plastics like polyethylene terephthalate (PET) and polyurethane (PUR). These enzymes are crucial in facilitating the breakdown of various synthetic polymers (Temporiti et al. 2022). For example, fungi such as *Fusarium falciforme*, *Humboldtia brunonis*, *Lasiodiplodia theobromae*, *Paecilomyces lilacinum*, *Psychotria flavida*, *Purpureocillum lilacinum*, and *Zelererion maritimum* secrete oxidoreductase enzymes that effectively depolymerize polyolefins, including PE, polystyrene (PS), polypropylene (PP), and PVC. These enzymes contribute to the breakdown of these persistent synthetic plastics, demonstrating the potential of fungi in plastic biodegradation (Okal et al. 2023). In another study, *Alternaria alternata* utilized laccase (59.49 kDa) and peroxidase (36.7 kDa) to degrade PE, while *Aspergillus flavus* employed laccase-like multicopper oxidases for the breakdown of high-density polyethylene (HDPE) MPs. Khruengsai et al. (2021) reported that *Aspergillus niger*, *Diaporthe italiana*, *Stagonosporopsis citrulli*, and *Thyrostroma jaczewskii* were able to degrade low-density polyethylene (LDPE) through the production of oxidative enzymes, including laccase, MnP, and LiP. Similarly, *P. ostreatus* and *T. versicolor* used laccase enzymes to degrade oxo-biodegradable plastics and nylon. These findings highlight the diverse enzymatic mechanisms that fungi employ in the degradation of various plastic materials. In addition, *Fusarium solani* produces cutinases and esterases that target polyethylene terephthalate (PET) and polycaprolactone (PCL) (Khatua et al. 2024). These fungal enzymes demonstrate versatility in depolymerizing synthetic polymers, highlighting fungi as a sustainable solution for addressing the growing environmental challenge of plastic waste.

9 Deactivation of PPCPs by Purified Enzymes

PPCPs are emerging persistent pollutants commonly found in aquatic environments. Despite their typically low concentrations, even trace amounts of PPCPs can significantly impact aquatic ecosystems, which may affect human health (Kang et al. 2021). The biodegradation of these contaminants through WRF and their oxidoreductase enzymes offers a promising, cost-effective, and eco-friendly approach for water treatment. The efficacy of fungal cultures in removing PPCPs is influenced by various factors, including the fungal species involved, the enzymes secreted, the molecular structure of the target compounds, and the composition of the culture medium (Naghdi et al. 2018). Fungi, especially WRF, utilize four main mechanisms for PPCP degradation, which can work independently or synergistically: (1) bioabsorption and immobilization of PPCPs by fungal hyphae, (2) the action of extracellular enzymes like peroxidases and phenolases, (3) intracellular enzyme complexes such as CYP, and (4) the production of reactive oxygen species (Malik et al. 2023).

In a study on *Trametes polyzona* KURNW027, purified ligninolytic enzymes were shown to effectively deactivate selected pharmaceutical products across various chemical classes, including tetracycline, β-lactam antibiotics, and quinolones. Of the enzymes tested, MnP2 exhibited the highest efficiency in deactivating all four antibiotics. The enzymes MnP1, MnP2, and Lac1 each demonstrated the ability to entirely deactivate tetracycline and doxycycline within 1–3 days at a 25 mg/L concentration. MnP2 achieved complete deactivation of amoxicillin, while MnP1 and Lac1 showed lower deactivation rates of 45% and 25%, respectively. MnP2 was also effective against ciprofloxacin (a quinolone), although with a moderate deactivation rate of 73%. Remarkably, these reactions were carried out without any redox mediator (Lueangjaroenkit et al. 2019). The use of redox mediators has also been explored to enhance enzyme activity. For instance, HBT improved the activity of laccase from *T. versicolor* IFO-6482 in degrading tetracycline (Suda et al. 2012), while VA boosted the LiP activity of *P. chrysosporium* in the degradation of tetracycline and oxytetracycline (Wen et al. 2009). These findings underscore the potential of fungal enzymes, particularly ligninolytic enzymes, as an effective and sustainable method for bioremediation of environmental and pharmaceutical contaminants.

10 Conclusion and Future Directions

Anthropogenic activities have significantly increased the release of pollutants into the environment, threatening ecological balance and human health. Fungal enzymes, particularly those derived from WRF, have shown immense potential as tools for environmental remediation. Enzymes such as laccases, peroxidases, and CYP enzymes are particularly promising due to their ability to degrade persistent pollutants in an eco-friendly and cost-effective manner. Their adaptability to diverse environmental conditions and capability to target a broad spectrum of xenobiotics make them indispensable in addressing global pollution challenges. Despite their promise, several challenges remain, such as optimizing enzyme production, stability, and efficiency under industrial-scale conditions. Future research should focus on enhancing fungal enzyme scalability, identifying and utilizing synergistic enzyme combinations, and developing innovative strategies to improve enzyme stability and activity. Advancements in enzyme immobilization could further enhance enzymatic efficiency, though current limitations of this approach require resolution. By addressing these challenges, fungal enzymes can play a pivotal role in advancing sustainable and effective solutions for environmental remediation.

References

Abadulla E, Tzanov T, Costa S, Robra KH, Cavaco-Paulo A, Gübitz GM (2000) Decolorization and detoxification of textile dyes with a laccase from *Trametes hirsuta*. Appl Environ Microbiol 66(8):3357–3362

Abdel-Fattah Mostafa A, Yassin MT, Dawoud TM, Al-Otibi FO, Sayed SR (2022) Mycodegradation of diazinon pesticide utilizing fungal strains isolated from polluted soil. Environ Res 212(Pt C):113421

Abdel-Hamid AM, Solbiati JO, Cann IK (2013) Insights into lignin degradation and its potential industrial applications. Adv Appl Microbiol 82:1–28

Abdelsalam NA, Ramadan AT, ElRakaiby MT, Aziz RK (2020) Toxicomicrobiomics: The human microbiome vs. pharmaceutical, dietary, and environmental xenobiotics. Front Pharmacol 11:390

Abrahão MC, Gugliotta ADM, Da Silva R, Fujieda RJY, Boscolo M, Gomes E (2008) Ligninolytic activity from newly isolated basidiomycete strains and effect of these enzymes on the azo dye orange II decolourization. Ann Microbiol 58:427–432

Allouzi MMA, Tang DYY, Chew KW, Rinklebe J, Bolan N, Allouzi SMA, Show PL (2021) Micro (nano) plastic pollution: the ecological influence on soil-plant system and human health. SciTotal Environ 788:147815

Anita SH, Sari FP, Yanto DHY (2019) Decolorization of synthetic dyes by ligninolytic enzymes from *Trametes hirsuta* D7. Makara J Sci 23:44–50

Arregui L, Ayala M, Gómez-Gil X, Gutiérrez-Soto G, Hernández-Luna CE, Herrera de Los Santos M, Levin L, Rojo-Domínguez A, Romero-Martínez D, Saparrat MCN, Trujillo-Roldán MA, Valdez-Cruz NA (2019) Laccases: structure, function, and potential application in water bioremediation. Microb Cell Fact 18(1):200

Asgher M, Bhatti HN, Ashraf M, Legge RL (2008) Recent developments in biodegradation of industrial pollutants by white rot fungi and their enzyme system. Biodegradation 19(6):771–783

Baker P, Tiroumalechetty A, Mohan R (2019) Fungal enzymes for bioremediation of xenobiotic compounds. In: Yadav A, Singh S, Mishra S, Gupta A (eds) Recent Advancement in White Biotechnology Through Fungi. Fungal Biology, Cham, pp. 463–489

Balcázar-López E, Méndez-Lorenzo LH, Batista-García RA, Esquivel-Naranjo U, Ayala M, Kumar VV, Savary O, Cabana H, Herrera-Estrella A, Folch-Mallol JL (2016) Xenobiotic compounds degradation by heterologous expression of a *Trametes sanguineus* laccase in *Trichoderma atroviride*. PLOS One 11(2):e0147997

Baldrian P (2004) Purification and characterization of laccase from the white-rot fungus *Daedalea quercina* and decolorization of synthetic dyes by the enzyme. Appl Microbiol Biotechnol 63:560–563

Bao S, Teng Z, Ding S (2012) Heterologous expression and characterization of a novel laccase isoenzyme with dyes decolorization potential from *Coprinus comatus*. Mol Biol Rep 40:1927–1936

Barría C, Brandts I, Tort L, Oliveira M, Teles M (2020) Effect of nanoplastics on fish health and performance: A review. Mar Pollut Bull 151:110791

Batista-García RA, Kumar VV, Ariste A, Tovar-Herrera OE, Savary O, Peidro-Guzmán H, González-Abradelo D, Jackson SA, Dobson AD, Sánchez-Carbente MDR, Folch-Mallol JL, Leduc R, Cabana H (2017) Simple screening protocol for identification of potential mycoremediation tools for the elimination of polycyclic aromatic hydrocarbons and phenols from hyperalkalophile industrial effluents. J Environ Manag 198:1–11

Ben Younes S, Bouallagui Z, Sayadi S (2012) Catalytic behavior and detoxifying ability of an atypical homotrimeric laccase from the thermophilic strain *Scytalidium thermophilum* on selected azo and triarylmethane dyes. J Mol Catal B Enzym 79;41–48

Ben Younes S, Cherif I, Dhouib A, Sayadi S (2016) *Trametes trogii*: A biologic powerful tool for dyes decolorization and detoxification. Catal Lett 146:204–211

Ben Younes S, Mechichi T, Sayadi S (2007) Purification and characterization of the laccase secreted by the white rot fungus *Perenniporia tephropora* and its role in the decolourization of synthetic dyes. J Appl Microbiol 102(4):1033–1042

Bhandari G (2017) Mycoremediation: An eco-friendly approach for degradation of pesticides. In: Prasad R (eds) Mycoremediation and Environmental Sustainability. Fungal Biology. Springer, Cham, pp. 119–131

Bhatt P, Zhang W, Lin Z, Pang S, Huang Y, Chen S (2020) Biodegradation of Allethrin by a novel fungus *Fusarium proliferatum* strain CF2, isolated from contaminated soils. Microorganisms 8(4):593

Bhattacharya S, Das A, Mangai G, Vignesh K, Sangeetha J (2011) Mycoremediation of Congo red dye by filamentous fungi. Braz J Microbiol 42:1526–1536

Bibi I, Bhatti HN (2012) Enhanced biodecolorization of reactive dyes by Basidiomycetes under static conditions. Appl Biochem Biotechnol 166:2078–2090

Bibi I, Javed S, Ata S, Majid F, Kamal S, Sultan M, Jilani K, Umair M, Khan MI, Iqbal M (2019) Biodegradation of synthetic orange G dye by *Plearotus sojar-caju* with *Punica granatum* peal as natural mediator. Biocatal Agric Biotechnol 22:101420

Boer CG, Obici L, de Souza CG, Peralta RM (2004) Decolorization of synthetic dyes by solid state cultures of *Lentinula* (*Lentinus*) *edodes* producing manganese peroxidase as the main ligninolytic enzyme. Bioresour Technol 94(2):107–112

Bosco F, Mollea C, Ruggeri B (2017) Decolorization of Congo Red by *Phanerochaete chrysosporium*: The role of biosorption and biodegradation. Environ Technol 38:2581–2588

Bouacem K, Rekik H, Jaouadi NZ, Zenati B, Kourdali S, El Hattab M, Badis A, Annane R, Bejar S, Hacene H, Bouanane-Darenfed A, Jaouadi B (2018) Purification and characterization of two novel peroxidases from the dye-decolorizing fungus *Bjerkandera adusta* strain CX-9. Int J Biol Macromol 106:636–646

Cardoso BK, Linde GA, Colauto NB, Valle JSD (2018) *Panus strigellus* laccase decolorizes anthraquinone, azo, and triphenylmethane dyes. Biocatal Agric Biotechnol 16:558–563

Chen S-C, Wu P-H, Su Y-C, Wen T-N, Wei Y-S, Wang N-C, Hsu C-A, Wang AH-J, Shyur L-F (2012) Biochemical characterization of a novel laccase from the basidiomycete fungus *Cerrena* sp. WR1. Protein Eng Des Sel 25:761–769

Chhabra M, Mishra S, Sreekrishnan TR (2008) Mediator-assisted decolorization and detoxification of textile dyes/dye mixture by *Cyathus bulleri* laccase. Appl Biochem Biotechnol 151:587–598

Chicatto JA, Rainert KT, Gonçalves MJ, Helm CV, Altmajer-Vaz D, Tavares LBB (2018) Decolorization of textile industry wastewater in solid state fermentation with Peach-Palm (*Bactris gasipaes*) residue. Braz J Biol 78:718–727

Choi YS, Long Y, Kim MJ, Kim JJ, Kim GH (2013) Decolorization and degradation of synthetic dyes by *Irpex lacteus* KUC8958. J Environ Sci Health A Tox Hazard Subst Environ Eng 48(5):501–508

Chowdhary P, More N, Yadav A, Bharagava RN (2019a) Ligninolytic enzymes: An introduction and applications in the food industry. In: Kuddus M (ed) Enzymes in Food Biotechnology: Production, Applications, and Future Prospects. Academic Press Inc, pp. 181–195

Chowdhary P, Shukla G, Raj G, Ferreira LZR, Bharagava RN (2019b) Microbial manganese peroxidase: a ligninolytic enzyme and its ample opportunities in research. SN Appl Sci 1:45

Christian V, Shrivastava R, Shukla D, Modi H, Rajiv B, Vyas M (2005) Mediator role of veratryl alcohol in the lignin peroxidase-catalyzed oxidative decolorization of Remazol Brilliant Blue R. Enzyme Microb Technol 36:426–431

Colpa DI, Fraaije MW, van Bloois E (2014) DyP-type peroxidases: a promising and versatile class of enzymes. J Ind Microbiol Biotechnol 41(1):1–7

Conesa A, Punt PJ, van den Hondel CAMJJ (2002) Fungal peroxidases: molecular aspects and applications. J Biotechnol 93:143–158

Cordas CM, Nguyen GS, Valério GN, Jønsson M, Söllner K, Aune IH, Wentzel A, Moura JJG (2022) Discovery and characterization of a novel Dyp-type peroxidase from a marine actinobacterium isolated from Trondheim fjord, Norway. J Inorg Biochem 226:111651

Daâssi D, Rodriguez-Couto S, Nasri M, Mechichi T (2014) Biodegradation of textile dyes by immobilized laccase form *Coriolopsis galica* into Ca-alginate beads. Int Biodeter Biodegr 90:71–78

Durairaj P, Hur JS, Yun H (2016) Versatile biocatalysis of fungal cytochrome P450 monooxygenases. Microb Cell Fact 15:125

Eichlerová I, Homolka L, Lisá L, Nerud F (2005) Orange G and Remazol Brilliant Blue R decolorization by white rot fungi *Dichomitus squalens*, *Ischnoderma resinosum* and *Pleurotus calyptratus*, Chemosphere 60(3):398–404

Ellouze M, Sayadi S (2016) White-rot fungi and their enzymes as a biotechnological tool for xenobiotic bioremediation. In: Hosam El-Din Saleh M, Rehab Abdel Rahman O (eds) Management of Hazardous Wastes. Intech Open Limited, Gate Court, London, pp. 103–120

Embrandiri A, Kiyasudeen SK, Rupani PF, Ibrahim MH (2016). Environmental xenobiotics and its effects on natural ecosystem. In: Singh A, Prasad S, Singh R (eds) Plant Responses to Xenobiotics. Springer, Singapore, pp. 1–18

Erkurt EA, Ünyayar A, Kumbur H (2007) Decolorization of synthetic dyes by white rot fungi involving laccase enzyme in the process. Proc Biochem 42:1429–1435

Falade AO, Nwodo UU, Iweriebor BC, Green E, Mabinya LV, Okoh AI (2017) Lignin peroxidase functionalities and prospective applications. Microbiologyopen 6(1):e00394

Gao Y, Croze B, Birch QT, Nadagouda MN, Mahendra S (2023) Sorghum-grown fungal biocatalysts for synthetic dye degradation. Water Res X 19:100181

Garrido-Bazán V, Téllez-Téllez M, Herrera-Estrella A, Díaz-Godínez G, Nava-Galicia S, Villalobos-López MÁ, Arroyo-Becerra A, Bibbins-Martínez M (2016) Effect of textile dyes on activity and differential regulation of laccase genes from *Pleurotus ostreatus* grown in submerged fermentation. AMB Express 6:1–9

George J, Rajendran DS, Venkataraman S, Rathankumar AK, Saikia K, Muthusamy S, Singh I, Singh I, Sinha S, Ramkumar S, Cabana H, Vaidyanathan VK (2022) Insolubilization of Tramates versicolor laccase as cross-linked enzyme aggregates for the remediation of trace organic contaminants from municipal wastewater. Environ Res 209:112882

Glenn JK, Gold MH (1983) Decolorization of several polymeric dyes by the lignin–degrading basidiomycete *Phanerochaete chrysosporium*. Appl Environ Microbiol 45:1741–1747

Gomes E, Aguiar AP, Carvalho CC, Bonfá MRB, Silva RD, Boscolo M (2009) Ligninases production by Basidiomycetes strains on lignocellulosic agricultural residues and their application in the decolorization of synthetic dyes. Braz J Microbiol 40:31–39

Gross L, Enck J (2021) Confronting plastic pollution to protect environmental and public health. PLoS Biol 19(3):e3001131

Heinfling A, Bergbauer M, Szewzyk U (1997) Biodegradation of azo and phthalocyanine dyes by *Trametes versicolor* and *Bjerkandera adusta*. Appl Microbiol Biotechnol 48:261–266

Hernández-Bueno NS, Suárez-Rodríguez R, Balcázar-López E, Folch-Mallol JL, Ramírez-Trujillo JA, Iturriaga G (2021) A versatile peroxidase from the fungus *Bjerkandera adusta* confers abiotic stress tolerance in transgenic tobacco plants. Plants (Basel) 10(5):859

Hussaan M, Amna, Javed MT, Akram MS, Ali S (2021) Physiological and molecular basis of bioremediation of micropollutants. In: Hasanuzzaman M, Prasad MNV (eds.) Handbook of Bioremediation: Physiological, Molecular and Biotechnological Interventions. Academic Press Inc, pp. 447–464

Iark D, Buzzo AJDR, Garcia JAA, Côrrea VG, Helm CV, Corrêa RCG, Peralta RA, Moreira RDFPM, Bracht A, Peralta RM (2019) Enzymatic degradation and detoxification of azo dye Congo red by a new laccase from *Oudemansiella canarii*. Bioresour Technol 289:121655

Iqbal HMN, Asgher M (2013) Decolorization applicability of sol-gel matrix immobilized manganese peroxidase produced from an indigenous white rot fungal strain *Ganoderma lucidum*. BMC Biotechnol 13:56

Isanapong J, Mataraj S (2018) Application of ligninolytic enzyme of *Lentinus polychrous* on synthetic dye decolorization. IOP Conf Ser Earth Environ Sci 185:012004

Järvinen J, Taskila S, Isomäki R, Ojamo H (2012) Screening of white-rot fungi manganese peroxidases: a comparison between the specific activities of the enzyme from different native producers. MB Express 2:62

Junghanns C Krauss G, Schlosser D (2008) Potential of aquatic fungi derived from diverse freshwater environments to decolourize synthetic azo and anthraquinone dyes. Bioresour Technol 99:1225–1235

Kamal S, Asgher M, Khalil-ur-Rehman Q, Zahir ZA (2011) Hyperproduction of laccase by *Pleurotus ostreatus* IBL-02 during decolorization of Drimarene Brulliant Red K-4BL. Fresen Environ Bull 20:1478–1486

Kang BR, Kim SY, Kang M, Lee TK (2021) Removal of pharmaceuticals and personal care products using native fungal enzymes extracted during the ligninolytic process. Environ Res 195:110878

Kanwal N, Asgher M, Bhatti HN, Sheikh MA (2010) Ligninase synthesis and decolorization of drimarine Blue K2RL by *Pleurotus ostreatus* IBL-02 in carbon and nitrogen sufficient shake flask medium. Fresen Environ Bull 19:63–68

Karimi A, Vahabzadeh F, Bonakdarpour B (2006) Use of *Phanerochaete chrysosporium* immobilized on kissiris for synthetic dye decolourization: Involvement of manganese peroxidase. World J Microbiol Biotechnol 22:1251–1257

Kasinath A, Novotný Č, Svobodová K, Patel KC, Šašek V (2003) Decolorization of synthetic dyes by *Irpex lacteus* in liquid cultures and packed-bed bioreactor. Enzyme Microbiol Technol 32:167–173

Khan MF, Hof C, Niemcová P, Murphy CD (2023) Recent advances in fungal xenobiotic metabolism: enzymes and applications. World J Microbiol Biotechnol 39:296

Khatua S, Simal-Gandara J, Acharya K (2024) Myco-remediation of plastic pollution: current knowledge and future prospects. Biodegradation 35(3):249–279

Khruengsai S, Sripahco T, Pripdeevech P (2021) Low-density polyethylene film biodegradation potential by fungal species from Thailand. J Fungi 7(8):594

Knapp JS, Newby PS (1999) The decolourization of a chemical industry effluent by white rot fungi. Water Res 33(2):575–577

Kokol V, Doliška A, Eichlerová I, Baldrian P, Nerud F (2007) Decolorization of textile dyes by whole cultures of *Ischnoderma resinosum* and by purified laccase and Mn-peroxidase. Enzym Microb Technol 40:1673–1677

Kumar A, Arora PK (2022) Biotechnological applications of manganese peroxidases for sustainable management. Front Environ Sci 10:875157

Kumar A, Chandra R (2020) Ligninolytic enzymes and its mechanisms for degradation of lignocellulosic waste in environment. Heliyon 6:e03170

Lee AH, Lee H, Heo YM, Lim YW, Kim C-M, Kim G-H, Chang W, Kim J-J (2020) A proposed stepwise screening framework for the selection of polycyclic aromatic hydrocarbon (PAH)-degrading white rot fungi. Bioprocess Biosyst Eng 43:767–783

Legerská B, Chmelová D, Ondrejovič M (2018) Decolourization and detoxification of monoazo dyes by laccase from the white–rot fungus *Trametes versicolor*. J Biotechnol 285:84–90

Levin L, Forchiassin FF, Viale A (2005) Ligninolytic enzyme production and dye decolorization by *Trametes trogii*: application of the Plackett-Burman experimental design to evaluate nutritional requirements. Proc Biochem 40:1381–1387

Levin L, Papinutti L, Forchiassin F (2004) Evaluation of four Argentinean white rot fungi; their ability to produces lignin-modifyng enzymes and decolorize industrial dyes. Biores Technol 2:169–176

Loi M, Glazunova O, Fedorova T, Logrieco AF, Mulè G (2021) Fungal laccases: The forefront of enzymes for sustainability. J Fungi 7(12):1048

Lu L, Zhao M, Zhang B-B, Yu S-Y, Bian X-J, Wang W, Wang Y (2007) Purification and characterization of laccase from *Pycnoporus sanguineus* and decolorization of an anthraquinone dye by the enzyme. Appl Microbiol Biotechnol 74:1232–1239

Lueangjaroenkit P, Teerapatsakul C, Sakka K, Sakka M, Kimura T, Kunitake E, Chitradon L (2019). Two manganese peroxidases and a laccase of *Trametes polyzona* KU-RNW027 with novel properties for dye and pharmaceutical product degradation in redox mediator-free system. Mycobiology 47(2):217–229

Magnoli K, Carranza C, Aluffi M, Magnoli C, Barberis C (2023) Fungal biodegradation of chlorinated herbicides: an overview with an emphasis on 2,4-D in Argentina. Biodegradation 34:199–214

Malik S, Bora J, Nag S, Sinha S, Mondal S, Rustagi S, Hazra R, Kumar H, Rajput VD, Minkina T, Sadier NS, Almutary AG (2023) Fungal-based remediation in the treatment of anthropogenic activities and pharmaceutical-pollutant-contaminated wastewater. Water 15(12):2262

Martínez AT (2002) Molecular biology and structure-function of lignin-degrading heme peroxidases. Enzyme and Microbial Technology 30(4):425–444

Máximo C, Costa-Ferreira M (2004) Decolorization of reactive textile dyes by *Irpex lacteus* and lignin modifying enzymes. Proc Biochem 39:1475–1479

Mazmanci MA, Ünyayar A (2005) Decolourisation of Reactive Black 5by *Funalia trogii* immobilized on Luffa cylindrica sponge. Process Biochem 40(1):337–342

Mehmood RT, Asad MJ, Asgher M, Hadri SH, Gulfraz M, Wu JD, Bhatti MI, Zaman N, Ahmed D (2020) First report of using response surface methodology for the biodegradation of single azo disperse dyes by indigenous *Daedalea dickinsii*-IEBL-2. Romanian Biotechnol Lett 25:1236–1245

Michniewicz A, Ledakowicz S, Ullrich R, Hofrichter M (2008) Kinetics of the enzymatic decolorization of textile dyes by laccase from *Cerrena unicolor*. Dyes Pigm 77:295–302

Miglani R, Parveen N, Kumar A, Ansari MA, Khanna S, Rawat G, Panda AK, Bisht SS, Upadhyay J, Ansari MN (2022) Degradation of xenobiotic pollutants: an environmentally sustainable approach. Metabolites 12(9):818

Mir-Tutusaus JA, Masís-Mora M, Corcellas C, Eljarrat E, Barceló D, Sarrà M, Caminal G, Vicent T, Rodríguez-Rodríguez CE (2014) Degradation of selected agrochemicals by the white rot fungus *Trametes versicolor*. Sci Total Environ 500:235–242

Mishra S, Lin Z, Pang S, Zhang W, Bhatt P, Chen S (2021) Recent advanced technologies for the characterization of xenobiotic-degrading microorganisms and microbial communities. Front Bioeng Biotechnol 9:632059

Moreira-Neto SL, Mussatto SI, Machado KMG, Milagres AMF (2013) Decolorization of salt-alkaline effluent with industrial reactive dyes by laccase-producing basidiomycetes strains. Lett Appl Microbiol 56:283–290

Mori T, Ohno H, Ichinose H, Kawagishi H, Hirai H (2021) White-rot fungus *Phanerochaete chrysosporium* metabolizes chloropyridinyl-type neonicotinoid insecticides by an N-dealkylation reaction catalyzed by two cytochrome P450s. J Hazard Mater 402:123831

Muradi NA, Husaini A, Zulkharnain A, Roslan HZ, Guan TM (2017) Optimization of remazol brilliant blue r dye decolourization and laccase enzyme production by *Marasmius cladophyllus* using response surface methodology. Malays Appl Biol 46(1):1–8

Naghdi M, Taheran M, Brar SK, Kermanshahi-Pour A, Verma M, Surampalli RY (2018) Removal of pharmaceutical compounds in water and wastewater using fungal oxidoreductase enzymes. Environ Pollut 234:190–213

Nakamura Y, Sungusia MG, Sawada T, Kuwahara M (1999) Lignindegrading enzyme production by *Bjerkandera adusta* immobilized on polyurethane foam. J Biosci Bioeng 88:41–47

Nakayama T, Amachi T (1999) Fungal peroxidase: its structure, function, and application. J Mol Catal B Enzym 6:185–198

Neifar M, Jaouani A, Ellouze-Ghorbel R, Ellouze-Chaabouni S (2010) Purification, characterization and decolourization ability of *Fomes fomentarius* laccase produced in solid medium. J Mol Catal B Enzym 64:68–74

Okal EJ, Heng G, Magige EA, Khan S, Wu S, Ge Z, Zhang T, Mortimer PE, Xu J (2023) Insights into the mechanisms involved in the fungal degradation of plastics. Ecotoxicol Environ Saf 262:115202

Oliveira J, Belchior A, da Silva VD, Rotter A, Petrovski Ž, Almeida PL, Lourenço ND, Gaudêncio SP (2020) Marine environmental plastic pollution: Mitigation by microorganism degradation and recycling valorization. Front Mar Sci 7:567126

Ortiz P, Torres-Sánchez A, López-Moreno A, Cerk K, Ruiz-Moreno Á, Monteoliva-Sánchez M, Ampatzoglou A, Aguilera M, Gruszecka-Kosowska A (2022) Impact of cumulative environmental and dietary xenobiotics on human microbiota: Risk assessment for one health. J Xenobiot 12(1):56–63

Pakshirajan K, Jaiswal S, Das RK (2011) Biodecolourization of azo dyes using *Phanerochaete chrysosporium*: Effect of culture conditions and enzyme activities. J Sci Ind Res 70:987–991

Palmieri G, Cennamo G, Sannia G (2005) Remazol Brilliant Blue R decolourization by the fungus *Pleurotus ostreatus* and its oxidative enzymatic system. Enzym Microb Technol 36:17–24

Pandey RK, Tewari S, Tewari L (2018) Lignolytic mushroom Lenzites elegans WDP2: Laccase production, characterization, and bioremediation of synthetic dyes. Ecotoxicol Environ Saf 158:50–58

Pazarlioglu N K, Urek R O, Ergun F (2005) Biodecolourization of Direct Blue 15 by immobilized *Phanerochaete chrysosporium*. Process Biochem 40:1923–1929

Permpornsakul P, Prasongsuk S, Lotrakul P, Eveleigh D, Kobayashi D, Imai T, Punnapayak H (2016) Biological treatment of reactive black 5 by resupinate white rot fungus *Phanerochaete sordida* PBU 0057. Pol J Environ Stud 25:1167–1176

Piwowarska D, Kiedrzyńska E (2022) Xenobiotics as a contemporary threat to surface waters. Ecohydrology and Hydrobiology 22(2):337–354

Pointing S, Vrijmoed L (2000) Decolorization of azo and triphenylmethane dyes by *Pycnoporus sanguineus* producing laccase as the sole phenoloxidase. World J Microbiol Biotechnol 16:317–318

Pype R, Flahaut S, Debaste F (2019) On the importance of mechanisms analysis in the degradation of micropollutants by laccases: The case of Remazol Brilliant Blue R. Environ Technol Innov 14:100324

Qin P, Wu Y, Adil B, Wang J, Gu Y, Yu X, Zhao K, Zhang X, Ma M, Chen Q, Chen X, Zhang Z, Xiang Q (2019) Optimization of laccase from *Ganoderma lucidum* decolorizing Remazol Brilliant Blue R and Glac1 as main laccase-contributing gene. Molecules 24:3914

Qin X, Zhang J, Zhang X, Yang Y (2014) Induction, purification and characterization of a novel manganese peroxidase from *Irpex lacteus* CD2 and its application in the decolorization of different types of dye. PLOS One 9:e113282

Rao RG, Ravichandran A, Kandalam G, Kumar SA, Swaraj S, Sridhar M (2019) Screening of wild basidiomycetes and evaluation of the biodegradation potential of dyes and lignin by manganese peroxidases. BioResources 14:6558–6576

Rathore S, Varshney A, Mohan S, Dahiya P (2022) An innovative approach of bioremediation in enzymatic degradation of xenobiotics. Biotechnol Genet Eng Rev 38(1):1–32

Sana SS, Dogiparthi LK, Gangadhar L, Chakravorty A, Abhishek N (2020) Effects of microplastics and nanoplastics on marine environment and human health. Environ Sci Pollut Res Int 27(36):44743–44756

Sarker A, Nandi R, Kim J-E, Islam T (2021) Remediation of chemical pesticides from contaminated sites through potential microorganisms and their functional enzymes: Prospects and challenges. Environ Technol Innov 2021:101777

Selvam K, Swaminathan K, Chae KS (2003) Decolourization of azo dyes and a dye industry effluent by a white rot fungus *Thelephora* sp. Bioresour Technol 88(2):115–119

Sharma K, Kumar P (2024) Environmental threats posed by xenobiotics. In: Kumar P, Chaudhary V, van Hullebusch ED, Busquets R, Srivastav AL (eds.) Bioremediation of Emerging Contaminants from Soils. Elsevier: Amsterdam, The Netherlands, pp. 183–201

Shrivastava R, Christian V, Vyas BRM (2005) Enzymatic decolorization of sulfonphthalein dyes. Enzyme Microb Technol 36:333–337

Siddiqui Y, Surendran A, Fishal EMM (2019) Inhibition of lignin degrading enzymes of *Ganoderma* spp.: An alternative control of basal stem rot disease of oil palm. Int J Agric Biol 3:523–530

Singh AK, Fernandez-Lafuente R, Schmidt JE, Boczkaj G, Bilal M (2024) Biocatalytic functionalities of lignin peroxidase-based systems in lignin depolymerization and pollutants removal from environmental matrices. Curr Pollution Rep 10:345–361

Singh VK, Gunasekaran P, Kumari M, Krishnan D, Ramachandran VK (2023) Animal sourced biopolymer for mitigating xenobiotics and hazardous materials. Phys Sci Rev 9(4):1731–1746

Srikanth M, Sandeep TSRS, Sucharitha K, Godi S (2022) Biodegradation of plastic polymers by fungi: a brief review. Bioresour Bioprocess 9:42

Štefanac T, Grgas D, Landeka Dragičević T (2021) Xenobiotics-division and methods of detection: A review. J Xenobiot 11(4):130–141

Suda T, Hata T, Kawai S, Okamura H, Nishida T (2012) Treatment of tetracycline antibiotics by laccase in the presence of 1-hydroxybenzotriazole. Bioresour Technol 103:498–501

Sugano Y, Yoshida T (2021) DyP-type peroxidases: Recent advances and perspectives. Int J Mol Sci 22(11):5556

Sun S (2012) Decolourization characterizations of crude enzymes from a novel basidiomycete *Mycena purpureofusca*. Afr J Microbiol Res 6:3501–3509

Šušla M, Svobodová K (2007) Effect of various synthetic dyes on the production of manganese-dependent peroxidase isoenzymes by immobilized *Irpex lacteus*. World J Microbiol Biotechnol 24:225–230

Svobodová K, Majcherczyk A, Novotný Č, Kües U (2008) Implication of mycelium-associated laccase from *Irpex lacteus* in the decolorization of synthetic dyes. Biores Technol 99:463–471

Swathy K, Vivekanandhan P, Yuvaraj A, Sarayut P, Kim JS, Krutmuang P (2024) Biodegradation of pesticide in agricultural soil employing entomopathogenic fungi: Current state of the art and future perspectives. Heliyon 10(1):e23406

Syed K, Yadav JS (2012) P450 monooxygenases (P450ome) of the model white rot fungus *Phanerochaete chrysosporium*. Crit Rev Microbiol 38:339–363

Tavčar M, Svobodová K, Kuplenk J, Novotný C, Pavko A (2006) Biodegradation of azo dye RO16 in different reactors by immobilized *Irpex lacteus*. Acta Chim Slovenica 53:338–343

Temporiti MEE, Nicola L, Nielsen E, Tosi S (2022) Fungal enzymes involved in plastics biodegradation. Microorganisms 10(6):1180

Toh Y, Jia J, Yen L, Obbard JP, Ting Y (2003) Decolorization of azo dyes by white-rot fungi (WRF) isolated in Singapore. Enzyme Microb Technol 33:569–575

Torres-Farradá G, Thijs S, Rineau F, Guerra G, Vangronsveld J (2024) White rot fungi as tools for the bioremediation of xenobiotics: A review. J Fungi (Basel) 10(3):167

Tychanowicz GK, Zilly A, Souza CGM, Peralta RM (2004) Decolourization of industrial dyes by solid-state cultures of *Pleurotus pulmonarius*. Proc Biochem 39:855–859

Urbanek AK, Rymowicz W, Mirończuk AM (2018) Degradation of plastics and plastic-degrading bacteria in cold marine habitats. Appl Microbiol Biotechnol 102:7669–7678

Ürek RO, Pazarlioglu NK (2007) Enhanced production of manganese peroxidase by *Phanerochaete chrysosporium*. Braz Arch Biol Technol 50(6):913–920

Vaithyanathan VK, Vaidyanathan VK, Cabana H (2022) Laccase-driven transformation of high priority pesticides without redox mediators: towards bioremediation of contaminated wastewaters. Front Bioeng Biotechnol 9:770435

Vaksmaa A, Guerrero-Cruz S, Ghosh P, Zeghal E, Hernando-Morales V, Niemann H (2023) Role of fungi in bioremediation of emerging pollutants. Front Mar Sci 10:1070905

Viswanath B, Rajesh B, Janardhan A, Kumar AP, Narasimha G (2014) Fungal laccases and their applications in bioremediation. Enzyme Res 2014:163242

Wen X, Jia Y, Li J (2009) Degradation of tetracycline and oxytetracycline by crude lignin peroxidase prepared from *Phanerochaete chrysosporium*–A white rot fungus. Chemosphere 75:1003–1007

Yehia RS, Rodriguez-Couto S (2017) Discoloration of the azo dye Congo Red by manganese-dependent peroxidase from *Pleurotus sajor caju*. Appl Biochem Microbiol 53:222–229

Yuan X, Wang X, Sarkar B, Ok YS (2021) The COVID-19 pandemic necessitates a shift to a plastic circular economy. Nat Rev Earth Environ 2:659–660

Zahra S, Abbas SS, Mahsa MT, Mohsen N (2010) Biodegradation of low-density polyethylene (LDPE) by isolated fungi in solid waste medium. Waste Manag 30(3):396–401

Zanger UM (2021) P450 monooxygenase system. In: Offermanns S, Rosenthal W (eds). Encyclopedia of Molecular Pharmacology. Springer, Cham, pp. 1211–1219

Zeng G, Cheng M, Huang D, Lai C, Xu P, Wei Z, Li N, Zhang C, He X, He Y (2015) Study of the degradation of methylene blue by semi-solid-state fermentation of agricultural residues with *Phanerochaete chrysosporium* and reutilization of fermented residues. Waste Manag 38:424–430

Zhuo R, Zhang J, Yu H, Ma F, Zhang X (2019) The roles of *Pleurotus ostreatus* HAUCC 162 laccase isoenzymes in decolorization of synthetic dyes and the transformation pathways. Chemosphere 234:733–745

Exploring the Role of Earthworm Gut Bacteria in Minimizing the Noxious Effects of Xenobiotic Compounds on Plant Growth

Moumita Parveen, Sarmistha Paul, Kunal Kumar Saha, and Narayan Chandra Mandal

Abstract Xenobiotic compounds are released into the environment by several anthropogenic activities. In the past few decades, the use of xenobiotic compounds has increased exponentially due to rapid industrialization, globalization, and population explosion. The recalcitrant nature of these pollutants jeopardized the livelihood of all living beings. Microbe-assisted bioremediation i.e., the use of taxonomically diverse groups of bacteria and fungi is one of the most cost-effective and environmentally friendly techniques for reducing the noxious effects of such toxic pollutants. Interestingly, earthworms play a vital role in maintaining soil structure and function in terrestrial ecosystems. Yet very little information is available pertaining to the role of the gut microbes of these annelids in shaping soil chemistry. In this chapter, we will highlight the role of earthworm gut bacteria (EGB) in promoting soil health and plant growth while facilitating the degradation of xenobiotic pollutants such as pesticides, polycyclic aromatic hydrocarbons (PAHs), polychlorinated biphenyls (PCBs), and pharmaceuticals, which are ingested from the soil. Additionally, the involvement of xenobiotic compounds in shaping the EGB community and plant-microbe interaction in such polluted soil will be discussed.

Keywords Agroecosystem · Earthworm gut bacteria · Plant growth promotion · Soil health · Xenobiotics

---

Moumita Parveen and Sarmistha Paul contributed equally with all other contributors.

M. Parveen · S. Paul · K. K. Saha · N. C. Mandal (✉)
Mycology and Plant Pathology Laboratory, Department of Botany, Siksha Bhavana (Institute of Science), Visva-Bharati, Santiniketan, West Bengal, India
e-mail: narayanchandra.mandal@visva-bharati.ac.in

S. Roy, V. Mandal (eds.), *Plant-Microbe Interaction under Xenobiotic Exposure*,
https://doi.org/10.1007/978-981-96-8260-7_22

1 Introduction

The word "xenobiotic" traces its origin to two Greek words: "xenos", meaning foreign, and "bios," meaning life. Xenobiotics are synthetic compounds that organisms can encounter but are not naturally produced, making them foreign to life and ecosystems (Stefanac et al. 2021). Due to population growth, industrialization, urbanization, and globalization, numerous xenobiotic compounds are utilized for agricultural, industrial, and domestic purposes, resulting in their release into environmental matrices (Malla et al. 2018). Once released, these compounds act as pollutants because they cannot be degraded naturally, or the degradation is very slow. They get incorporated into the food chain and cause harmful effects at each trophic level through biomagnification. Xenobiotics may be present as both micro and macro-environmental pollutants (mg/L to μg/L) (Miglani et al. 2022). The effect of xenobiotics on plants, animals, and the environment is alarming. Long-term exposure to low-level xenobiotics may cause toxic, mutagenic, carcinogenic, or teratogenic effects (Elekwachi et al. 2014; Liu et al. 2019). Considering the harmful impact of xenobiotic compounds, several laws and regulations have been adopted worldwide. Xenobiotics can be categorized based on their nature, uses, physical state, and pathophysiological effects. Polycyclic aromatic hydrocarbons (PAHs), phenolics, halogenated compounds, personal care products (PCPs), azodyes, pharmaceutical active compounds (PHACs), pesticides, nitroaromatic compounds, chlorinated compounds, triazines, and heavy metals are some of the xenobiotic compounds that adversely affect the environment by their long-term persistence and slow or no biodegradation (Stefanac et al. 2021; Godheja et al. 2016). The primary route through which xenobiotics enter the food chain is the soil. Plants absorb the pollutants from soil and are fed by humans and animals; thus, the non-biodegradable pollutants biomagnify along the food chain (Capozzi and Spagnuolo 2024). This chapter will discuss the major types of xenobiotic compounds and their effects on agricultural fields, earthworm activity, and plants. It is well known that earthworms play a significant role in soil health. They improve nutrient content, increase the soil's ability to retain water, and enhance microbial activity within the soil. Due to these facets, earthworms are considered a good indicator of soil health. Earthworms enhance plant growth by harboring beneficial plant growth-promoting (PGP) bacteria in their guts. Therefore, we will highlight the role of earthworm gut bacteria (EGB) in promoting soil health and plant growth. We will discuss in detail the impact of xenobiotics on the interaction between plant and earthworm gut microbes and potential measures that can be taken to mitigate the harmful effects of xenobiotics by utilizing earthworm gut bacteria.

2 Types of Xenobiotics and Their Harmful Effect on Agricultural Fields, Earthworms, and Plants

2.1 Major Types of Xenobiotics and Their Effect

Xenobiotic compounds rarely exist as natural products and contain structural elements that cannot be synthesized biochemically. Microorganisms were not exposed to these structures during the evolution of catabolic enzymes and pathways and have not developed the capability to use these compounds as their sole carbon and energy source (Rieger et al. 2002). Synthetic compounds are considered xenobiotics when certain groups' numbers and sterical orientation differ from naturally existing compounds. For example, aromatic or aliphatic compounds carrying several methyl (CH_3) groups are commonly found. However, ortho-dimethyl or isopropyl-trimethyl substituted aromatic compounds and branched aliphatic compounds like tert-butanol or tert-butylmethyl ether have foreign structural elements that limit the biodegradability of these molecules (Rieger et al. 2002). Table 1 lists some significant types of xenobiotic compounds and their harmful effects. The degree of hazard caused depends on the chemical nature of the compound, as well as the extent and duration of exposure. Xenobiotics such as pesticides are not degraded naturally and undergo biomagnification; thus, they accumulate along the trophic levels and are known to cause endocrinal anomalies and toxicity in aquatic animals (Wang et al. 2022). They contaminate air, water, and soil health and can lead to biodiversity loss. Similarly, heavy metals released due to numerous anthropogenic activities cause several negative impacts, such as soil and water pollution. They also cause nephrotoxicity and hepatotoxicity in humans and animals (Sall et al. 2022). Polycyclic aromatic hydrocarbons (PAHs) cause toxicity to aquatic and avian ecosystems (Honda and Suzuki 2020). Meanwhile, polychlorinated biphenyls (PCBs) such as PCB-156,180,19 are responsible for neurological disorders (Yadav 2020). Pharmaceuticals, however, harm the reproductive potential of aquatic, terrestrial, and arboreal animals. It is known that scavengers play an important role in the food web by keeping the ecosystem free from dead matter. Pharmaceuticals have a lethal effect on scavengers, thus disrupting the food web (Patneedi and Prasadu 2015). While halocarbons such as CFCs and H(C) FCS root global warming and climate change (Piers and Joshi 2005). Thus, different types of xenobiotics have various effects. In the upcoming section, we will briefly discuss the harmful effects of xenobiotics on agricultural fields, earthworms, and plants.

2.2 Effect of Xenobiotics on Agricultural Fields

Population explosion has led to increased demand for food. Modern agriculture is excessively dependent on chemical fertilizers and pesticides to meet the growing demand. We have reached a stage where agriculture greatly relies on high-yielding

Table 1 Different types of xenobiotics and their harmful effects

Types of xenobiotics	Harmful effect on plants	Harmful effect on animals	Reference
Pesticides (Insecticides, herbicides, fungicides)	Biomagnification and bioaccumulation hazard Disruption of plant growth and development Soil contamination Death of essential microorganisms Toxicity to non-target plants Development of pesticide resistance	Toxicity in aquatic animals Toxicity to insects and pollinators Endocrine disruption and reproductive effects in mammals Acute and chronic toxicity in birds Bioaccumulation and trophic cascade effects Pesticide resistance and ecological imbalance	Wang et al. (2022), Parween et al. (2016), and Hashimi et al. (2020)
Heavy metals	Toxicity to plant growth and development Pb is one of the most toxic heavy metals to plants; it inhibits seed germination, root elongation, and overall plant growth Cd interferes with the absorption of essential nutrients such as Ca, K, and Mg, leading to nutrient imbalances Disruption of nutrient uptake and transport (Fe deficiency, Ca and Mg imbalance) Oxidative stress and cellular damage Disruption of photosynthesis Root damage and impaired water uptake Translocation and accumulation in edible parts of plants	Toxicity and bioaccumulation Neurological effects (Pb poisoning in mammals, Hg affects the central nervous system) Carcinogenic effect Cellular and organ damage Metabolic disorder Nephrotoxicity and hepatotoxicity Reproductive and developmental effects Cardiovascular and respiratory effects Immune system suppression Ecological and behavioral effects	Saha et al. (2025) Sall et al. (2022), Govind and Madhuri (2014), and Vajargah (2021)
Polycyclic aromatic hydrocarbons (PAHs)	PAH uptake by plants results in deposition of contaminants on the plant surface Effects on plant growth and development Biochemical changes (plants under PAH stress show increased levels of antioxidants such as SOD, catalase, and peroxidase, which play a role in detoxifying ROS and mitigating oxidative damage) Prolonged or high-level exposure to PAHs may overwhelm the plant's antioxidant defenses, leading to cellular damage	Toxicological effects on animals (carcinogenicity, genotoxicity, reproductive and developmental toxicity, neurotoxicity, immune system disruption, oxidative stress, and cellular damage) Immunosuppression and hormonal disorders Adverse effects on ecology and wildlife health	Honda and Suzuki (2020), Sun et al. (2021), and Jajoo (2017)

Polychlorinated biphenyls (PCBs) PCB-156,180,19	Toxicological effects on plants (growth inhibition, chlorophyll degradation, and photosynthetic disruption) Oxidative stress and cellular damage Reproductive and developmental impacts (PCB exposure can disrupt flower and fruit development, leading to reduced yields and reproductive failure) Altered nutrient uptake	Toxicological effects on animals Neurological disorder (wide range of neural abnormalities, i.e., abnormal reflexes and neural tissue damage) Carcinogenicity Endocrine disruption Reproductive toxicity Immune system impairment	Faroon and Ruiz (2016) and Yadav (2020)
Polybrominated biphenyls (PBBs)	Effects of PBBs on plants include growth inhibition, chlorophyll degradation, and photosynthetic disruption Oxidative Stress (ROS accumulation can damage cellular structures, including lipids, proteins, and DNA, and can cause lipid peroxidation, which further damages cell membranes) Reproductive and developmental effects	Endocrine disruption Immune system suppression Disorders of the thyroid gland and related hormones (adverse effects on hormone T3 and T4 secretion) Reproductive and developmental toxicity	Curtis et al. (2019), Zhu et al. (2014), and Sharma et al. (2019a, b)
Pharmaceuticals Analgesics, antibiotics, antiepileptics, antiseptics, beta-blockers, estrogenic drugs	Effects on plant physiology (growth inhibition, toxicity, morphological changes, photosynthetic efficiency changes, altered water and nutrient uptake) Act as endocrine disruptors (synthetic estrogens and anti-androgens can alter plant hormone levels, resulting in abnormal growth and reproduction)	Cellular and tissue damage Endocrine disruption Toxicity Behavioral changes and antibiotic resistance. Adverse effect on the reproductive potential of aquatic, terrestrial, and arboreal animals Lethal effect on scavengers	Patneedi and Prasadu (2015)
Halocarbons CFCs, H(C)FCS, CH_3CCl_3, CCl_4, CFC-12, HFC-134a	Global warming and climate change Loss of biodiversity and habitat destruction. Interference with photosynthesis. Disruption of plant growth and development. Oxidative stress and cellular damage Disruption of nutrient uptake	Neurotoxicity Reproductive and developmental toxicity Immunotoxicity	Kroeze and Reijnders (1992) and Piers and Joshi (2005)
Synthetic polymers	Physical interference with plant growth. Impediments to soil structure and health	Alteration in the food chain and food webs Aquatic and soil pollution Physical harm from ingestion and entanglement. Toxicity of additives and degradation products	Griffin (1980) and Moore (2008)

varieties grown by applying fertilizers and pesticides. Pesticides are xenobiotic compounds used to produce cheap crops, thus making food available to all. Pesticides include insecticides, fungicides, herbicides, weedicides, nematicides, and rodenticides, of which insecticides constitute 77% of the total pesticides used (Chauhan and Singhal 2006). Although pesticides have several advantages when used in moderation. However, indiscriminate and disproportionate use may lead to their residue in the food chain, which may, in turn, exert harmful effects on humans and animals. Pesticides, such as chlordane, dieldrin, hexachlorobenzene, and endrin, are persistent organic pollutants (POPs) and remain in the environment for extended periods.

Research indicates that pesticide residues can bioaccumulate and may exceed a concentration of over 70,000 times the original level (Kim et al. 2017a, b; Zaka et al. 2019).

Heavy metal contamination is another major problem in agricultural fields. Heavy metals are considered a part of soil; however, they are considered toxicants at concentrations above permissible limits (Osmani et al. 2015). WHO has set the permissible limits/MRL of metal concentrations (mg Kg^{-1}) in soil and plants. Metals with high permissible limits are generally considered safe. The permissible limit of Zn is highest in soil, followed by Cu and Pb, while Cd has the least permissible limit. Hence, even a low concentration of Cd can cause significant toxicity. The major sources of heavy metals in the agricultural sector are fertilization, pesticides, livestock manure, and wastewater. Soil acidity and low macro-nutrients are among the main problems associated with heavy metal toxicity in agriculture. Some other xenobiotic compounds that threaten the agroecosystem are chlordane, dioxins, PCBs, PAHs, DDT, and nitroaromatics.

2.3 *Effect of Xenobiotics on Earthworms*

Earthworms play an important role in maintaining soil quality by regulating soil processes such as fertilization and organic matter decomposition, so they are also called "soil engineers." They are considered important biomonitoring organisms in soil ecosystems and are widely used to estimate soil quality. In soil, earthworms are exposed to different types of xenobiotic compounds. Litskas et al. (2019) studied the effect of doxycycline (DOX) on earthworms and found that the drug harmed juvenile count but had no significant impact on adult mortality and body weight. Similarly, in a study, De Bernardi et al. (2022) found that earthworms can tolerate copper contamination in soil at low concentrations. However, they observed adverse effects on overall reproductive output at high concentrations. Reports also state that these annelids recover from genotoxic damage at intermediate concentrations of pollutants and do not bioaccumulate them. Nevertheless, irreversible damage is observed beyond a certain threshold. The in-depth mechanism by which earthworms deal with xenobiotic stress must be studied in detail. Analysis of the changes in the intestinal microbiome of earthworms under the influence of xenobiotics, which has not been extensively studied to date and has a vast scope.

2.4 *Effect of Xenobiotics on Plants*

Xenobiotics adversely affect a plant's morphological and physiological characteristics in several ways. Xenobiotics with diverse structures and designs cause gene expression, regulation, and signal transduction changes in higher plants (Ramel et al. 2012). They also cause DNA damage in plants due to oxidative stress and ROS production (Pande et al. 2022). Pesticide residues and their degradation by-products are taken up through the roots via xylem to the entire plant, causing deleterious effects on the plant. These effects include overproduction of ROS, oxidative stress, DNA damage, photosynthetic blockage, necrosis, chlorosis, leaves twisting, and ultimately plant death (Sharma et al. 2018, 2019a, b). Few metals such as Zn, Cu, Mo, and Fe are needed in trace amounts for plant growth but are toxic at high concentrations (Kumar and Aery 2016). Few nonessential metals such as Pd, Cd, As, and Hg are toxic even at low concentrations.

3 Role of Earthworms in Soil Health and Plant Growth

The imperative role of earthworms in boosting soil fertility and restoring soil health is well-acknowledged among the masses. Even scientists like Charles Darwin endorsed the role of earthworms in maintaining soil ecosystems (Reed 1982). They are called "ecosystem engineers" and are essential for improving soil quality and encouraging plant development through improved soil aeration, porosity, and water retention brought about by their burrowing activity (Eisenhauer 2010). Their activity creates the ideal conditions for root development. Earthworms consume organic materials and produce casts that contain nutrients like nitrogen, phosphate, and potassium that plants may use for their development (Kim et al. 2017a, b). Additionally, earthworms increase microbial activity, which supports helpful microorganisms such as mycorrhizal fungi, nitrogen-fixing bacteria, and many other plant growth-promoting bacteria. These microorganisms produce phytohormones that promote plant development and improve the cycling of nutrients. By encouraging microbial variety, earthworms also inhibit soil-borne pathogens, lowering the prevalence of diseases. Earthworms are essential for sustainable agriculture and soil restoration because of the cumulative impacts of these activities, which include enhanced soil fertility, better root systems, and higher crop productivity (Wu et al. 2012). Their impact on soil health and plant development is discussed below.

3.1 Role of Earthworms in Soil Health

3.1.1 Soil Porosity and Structure

Earthworms dig into the soil and create pathways that enhance water infiltration and aeration. Their digging habits improve soil aggregation, which lessens compaction and encourages root penetration (Lavelle et al. 2004). Additionally, these pathways let nutrients and water travel to deeper soil layers, which is good for plant roots for nutrient absorption (Bottinelli et al. 2010).

3.1.2 Cycling of Nutrients

Earthworms break organic matter and turn it into casts rich in nutrients. Nitrogen, phosphorus, potassium, and other micronutrients are abundant in these casts and are easily absorbed by plants. Earthworm activity speeds up decomposition, increasing organic matter turnover and encouraging nutrient cycling (Edwards and Arancon 2022).

3.1.3 Microbial Activity in the Soil

In the soil, earthworm activity promotes microbiological diversity and abundance. Earthworms facilitate microbial colonization by breaking up organic matter and combining it with mineral soil. These microorganisms inhibit soil-borne diseases and aid in the mineralization of nutrients. Earthworms increase biochemical processes like nitrogen fixing and organic matter breakdown by fostering microbial growth and activity in their guts and casts. The fertility of the soil and the health of the plants depend on these microbial activities (Chaoui et al. 2003).

3.1.4 Soil pH

Earthworms' biological processes of burrowing, eating, and excreting organic debris can substantially affect adequate soil pH. Through both direct and indirect methods, earthworms are essential in controlling the pH of the soil. They are important contributors to soil fertility and health because of their capacity to improve nutrient cycling and mix soil layers. Earthworms assist soils in achieving a more balanced pH, which generally improves soil functionality. However, depending on the soil type and the environmental conditions present, their effects may vary (Desie et al. 2020).

3.1.5 Soil Temperature

Soil can be insulated by earthworm-processed organic matter, which reduces temperature changes and makes the environment more stable for plant and microbial activity (Edwards 2004).

3.1.6 Carbon Sequestration

Earthworms help stabilize organic carbon in deeper soil layers by burying and combining organic wastes (Don et al. 2008). Earthworms help form Particulate Organic Matter (POM) by adding POM to aggregates, which reduces its decomposition (Bossuyt et al. 2006).

3.1.7 Pollutant Detoxification

By encasing heavy metals in their castings, earthworms can lower the soil's toxicity and bioavailability. By accelerating the microbial breakdown of pesticide residues and other pollutants, earthworms help to create cleaner soils (Wang et al. 2018).

3.1.8 Bioindicator of Soil Health

Earthworms are known as indicator species. They can be used as ecosystem stability and soil fertility markers because of their diversity, abundance, and activity. Since earthworms are extremely sensitive to pollutants, including heavy metals, herbicides, and industrial toxins, they are reliable markers of soil quality (Fründ et al. 2010).

3.2 Role of Earthworms in Plant Growth

3.2.1 Availability of Nutrient

The vital nutrients (nitrate, ammonium, and soluble phosphorus) that are incorporated into earthworm casts can be easily absorbed by plants. These nutrients support robust plant growth and increase agricultural yields (Dobson et al. 2017).

3.2.2 Root Growth

Earthworm activity improves soil aeration and structure; as a result, roots can grow deeper and have more access to water and nutrients. Research has indicated that plants cultivated in soils containing earthworms display higher branching and root biomass (Edwards and Lofty 1980).

3.2.3 Plant-Microbe Interactions

Earthworm activity creates hotspots of microbial activity that are advantageous to plants by increasing microbial biomass and variety in soil, especially in their burrows and casts (Brown and Doube 2004). By promoting the growth of nitrogen-fixing bacteria, earthworms increase the amount of nitrogen available for plant absorption. According to Lavelle and Spain (2001), earthworms stimulate mycorrhizal colonization of plant roots, which increases nutrient and water absorption via changing soil structure and nutrient dynamics. Favorable circumstances are created for plant growth-promoting rhizobacteria (PGPR) by earthworm casts and burrowing, which suppress plant diseases and produce phytohormones (Medina-Sauza et al. 2019). According to Blouin et al. (2005), interactions among earthworms, soil microorganisms, and plant roots can trigger defensive (induced systemic resistance) mechanisms in plants, increasing their resistance to pests and diseases.

3.2.4 Hormones for Plant Growth

In earthworm casts or vermicompost, hormones such as auxins and cytokinins promote root elongation and shoot development, which aids in plant growth. Earthworm activity encourages root branching and nutrient uptake by indirectly encouraging soil bacteria to produce indole acetic acid (IAA) (Arancon et al. 2006).

3.2.5 Plant Disease Suppression

The influence of earthworms, soil properties, and microbial populations reduces plant diseases. By strengthening plant resilience and reducing the incidence of soil-borne disease, their activity encourages healthier plant growth. Antimicrobial substances produced by beneficial bacteria associated with earthworm-modified soils inhibit the growth of diseases such as *Pythium* sp. and *Fusarium* sp. (Akca et al. 2014).

4 Utility of Earthworm Gut Bacteria in Reducing the Harmful Effect of Xenobiotic Compounds on Plant Growth

Xenobiotics can negatively impact plant growth by altering soil health, disrupting nutrient cycling, and leading to toxicity through bioaccumulation. A novel approach to mitigate these effects involves using beneficial bacteria found in the gut of earthworms (*Eisenia fetida* and others), which play a crucial role in bioremediation processes. The earthworm's gut hosts a diverse community of bacteria that can break down organic matter, including complex pollutants and xenobiotics. These bacteria have evolved mechanisms to detoxify harmful substances, transforming them into less toxic or non-toxic forms through enzymatic processes such as biotransformation, mineralization, and biodegradation (Sokolova et al. 2012). Some beneficial bacteria found in the earthworm gut include *Bacteroides*, *Firmicutes*, and *Proteobacteria* (Desai et al. 2019). These bacteria have shown significant capacity for degrading xenobiotic compounds, thus enhancing the earthworm's ability to survive in polluted environments.

Earthworm gut bacteria facilitate the degradation of xenobiotic pollutants, such as pesticides, PAHs, PCBs, pharmaceuticals, and heavy metals, which are ingested from the soil. Thus, it plays a significant role in reducing the harmful effect of xenobiotic compounds on plant growth. For instance, research has shown that earthworms exposed to soils contaminated with cadmium or lead have higher survival rates than other soil organisms, mainly due to the detoxification processes facilitated by their gut bacteria (García et al. 2020). The detoxification of heavy metals by earthworm gut bacteria (EGB) benefits both the worms and plants by reducing the bioavailability of these metals, thus mitigating their toxic effects on plant growth. Recent studies have highlighted the role of gut bacteria in earthworms and their potential to alleviate the harmful effects of xenobiotics. This could provide valuable insights into sustainable agriculture and bioremediation strategies. This chapter explores how gut bacteria in earthworms assist in reducing the toxicity of xenobiotic compounds while promoting plant growth.

4.1 Strategies by Which Earthworm Gut Bacteria Mitigate Xenobiotics

Xenobiotics can enter the environment through agricultural runoff, industrial waste, or even atmospheric deposition. Several studies have demonstrated that EGB can mitigate the effects of xenobiotic compounds such as heavy metals, pesticides, and pharmaceutical contaminants. Through their unique gut microbiota, earthworms contribute to the bioremediation of contaminated soils by utilizing various strategies to detoxify these harmful substances. Some of the major detoxification mechanisms are as follows:

4.1.1 Microbial Metabolism and Enzyme Production

The specific microbial communities in the gut help earthworms metabolize and transform harmful compounds into less toxic forms, which are then excreted by the earthworm or absorbed by surrounding soil microbes. These detoxified compounds are less detrimental to plants growing in contaminated soils (Sun et al. 2020). The gut microbes also generate various enzymes, including cytochrome p450s, glucuronidases, and esterases, which catalyze detoxification processes (Liu et al. 2020a). Reports suggest that EGB can break down PCBs and PAHs, common industrial pollutants, into simpler, less toxic metabolites (Krauss et al. 2000). Further, Jiang et al. (2021) reported that cytochrome P450 (CYP450) and glutathione S-transferase (GST) were secreted by EGB in response to decabromodiphenyl ethane (DBDPE) exposure, which resulted in biotransformation of DBDPE.

4.1.2 Biotransformation of Xenobiotics

Earthworm gut bacteria also contribute to the biotransformation of xenobiotics. This process involves the conversion of these chemicals into forms that are easier for the organism to excrete. Some bacteria in the gut use reductive dehalogenation to remove halogen atoms from organic pollutants, such as organochlorine pesticides (Mohn and Tiedje 1992). This transformation can either neutralize the toxicity of the compounds or make them more biodegradable, enhancing the detoxification process. The primary mechanism of bacteria xenobiotic transformation involves using metabolic pathways like aerobic and anaerobic respiration, fermentation, and sulfate reduction to break down harmful compounds.

4.1.3 Heavy Metal Sequestration and Detoxification

Many earthworm species, especially those living in polluted soils, are exposed to heavy metals, such as lead (Pb), mercury (Hg), and cadmium (Cd). The EGB communities are more tolerant to heavy metal contamination than soil microbiota (Liu et al. 2020b). Gut microbiota is key in detoxifying these metals through various mechanisms, including biosorption, bioaccumulation, metal efflux pumps, and chelation by siderophores and metallothionein (Mandal et al. 2021). Certain bacteria possess metal-binding proteins (metallothionein) or excrete extracellular polymers that trap metals in the gut, preventing them from being absorbed by the worm (Sun et al. 2020). Some microbial species can also convert these metals into less toxic forms, further aiding in detoxification.

4.1.4 Microbial Biofilm Formation

The gut microbiota of earthworms often forms biofilms, clusters of bacteria that adhere to surfaces within the gut. These biofilms play a protective role by preventing the absorption of xenobiotics into the worm's body. Moreover, biofilms act as a local barrier to the toxicity of pollutants, reducing their harmful effects on both the bacteria and the earthworm. The biofilms in the earthworm gut create an environment that facilitates the efficient breakdown of xenobiotics by the microbiota, promoting detoxification (Sun et al. 2020).

4.1.5 Symbiotic Relationship Between Earthworms and Microorganisms

The relationship between earthworms and their gut bacteria is symbiotic. Earthworms provide a stable environment for microbes to thrive, offering nutrients from the decomposed organic material in their gut. In return, the bacteria help break down and detoxify harmful xenobiotics that the earthworms ingest through contaminated soil. This mutualistic relationship enhances the survival and fitness of both the earthworm and its microbiota (Edwards and Arancon 2022).

Thus, earthworm gut bacteria play a crucial role in detoxifying xenobiotics through various mechanisms, including enzymatic degradation, biotransformation, heavy metal sequestration, and biofilm formation. Their symbiotic relationship with earthworms enhances the resilience of both the worm and its microbial community against environmental pollutants. Further research into these microbial processes holds promise for improving bioremediation strategies and managing soil health in contaminated areas.

4.2 Role of Earthworm Gut Bacteria on Plant Growth Under Xenobiotic Stress

The growing xenobiotic contamination of soil ecosystems seriously threatens the health and production of plants. Earthworms are now important promoters of plant growth and soil detoxification because of the bacteria they carry in their guts. The symbiotic relationships between earthworms and their gut microbiota are examined in this article, focusing on how they help plants develop and reduce xenobiotic stress (Dawood et al. 2017). To illustrate methods and applications, specific case studies and examples are provided. The gut of an earthworm is a unique ecosystem inhabited by various microorganisms that can break down or change xenobiotics. Pseudomonas, Bacillus, Streptomyces, and Flavobacterium are important bacterial genera that use a variety of enzymatic pathways, including reductases, dioxygenases, and monooxygenases, to break down harmful substances. For example, research has demonstrated that *Pseudomonas putida*, isolated from the gut system of *Eisenia fetida*, can break down PAHs such as pyrene and phenanthrene. By changing metals into less accessible forms, *Bacillus* sp. also supports phytoremediation and shows tolerance to heavy metals (Banerjee et al. 2019).

Under xenobiotic stress, earthworm-associated bacteria facilitate plant growth through different approaches. Earthworms break down organic matter and xenobiotics by gut bacteria and release essential nutrients like nitrogen, phosphorus, and potassium, making them available to plants. Plant growth-promoting hormones such as gibberellins, cytokinin, and indole-3-acetic acid (IAA) are produced by gut bacteria and improve root elongation, cell division, and overall plant health. By introducing gut bacteria into the soil, earthworms help to strengthen the microbial community's ability to outcompete diseases and improve soil health.

The basis of the world's food production is the soil ecosystem, but contaminants and unsustainable farming methods are destroying it. Because of their capacity to enhance soil fertility and structure, earthworms are highly acknowledged as ecosystem engineers. The bacteria that live in earthworm digestive tracts substantially impact soil biochemistry and plant growth in addition to their physical impacts. These gut microbes create bioactive molecules that aid plant development, improve nutrient cycling, and break down toxic components. This article examines these interactions and focuses on their significance in agriculture. A diverse bacterial population comprising species like *Pseudomonas* sp., *Bacillus* sp., *Azospirillum* sp., *Flavobacterium* sp., and many others is found in the gut of earthworms. These microbes enhance plant growth in several ways (Chang et al. 2021).

4.2.1 Nutrient Mineralization

Gut bacteria break down complex organic matter into bioavailable nutrients, such as nitrogen (N), phosphorus (P), and potassium (K).

4.2.1.1 Phosphate Solubilization

Globally, phosphate deficiency is a significant barrier to agricultural output. Although phosphate is abundant in soils, it is mostly found in forms that plants cannot use. Phosphate-solubilizing microorganisms and other biological processes provide environmentally beneficial substitutes for chemical fertilizers (Saha et al. 2024). Earthworms, also called "ecosystem engineers," impact soil fertility due to their burrowing and feeding habits. The role of gut bacteria in phosphate solubilization is an important but less studied component of their function. Through a variety of processes, earthworm gut microorganisms dissolve insoluble phosphate minerals like (a) Organic acid production: Numerous gut bacteria generate organic acids, including lactic acid, citric acid, and gluconic acid, which chelate phosphate-bound cations and generate soluble phosphate. (b) Enzymatic activity: The gut bacteria can produce enzymes that break down phosphate into bioaccessible forms. (c)

Acidification of microenvironments: Phosphate dissolves more easily when the metabolic processes of gut microflora lower the pH (Biswas et al. 2018).

It has been observed that several bacterial genera actively participate in phosphate solubilization and are isolated from the earthworm gut system (Table 2). These consist of *Bacillus* sp. (Andleeb et al. 2022), *Pseudomonas* sp. (Bhakta et al. 2022), and many more. These bacteria can be applied in agriculture to improve phosphate using efficiency in crop plants by some strategies under xenobiotic stress conditions. (a) Vermicomposting: Composting by earthworms increases the bioavailability of phosphate in organic fertilizers. (b) Development of inoculants: Phosphate-solubilizing bacteria from earthworm guts are isolated and can be used as biofertilizers. The gut bacteria of earthworms are essential for the solubilization of phosphate and provide a sustainable means of increasing phosphate availability in soils. Reducing reliance on chemical fertilizers by incorporating this biological approach into agricultural practices can promote effective and environmentally friendly nutrient management.

4.2.1.2 Potassium Solubilization

Plants' physiological processes, such as activation of enzymes, maintenance of water potential, and stress tolerance, require potassium in large amounts. Although it is abundant in soils, much of it is locked in mineral forms, such as mica and feldspar, that plants cannot use directly. Soil microbial community, including earthworm gut bacteria, is significant in the solubilization of potassium and its bioavailability for plant uptake. Earthworms are considered ecosystem engineers, supporting this process by increasing habitats and promoting the conditions for microbial activity in their guts and casts.

Potassium solubilization by earthworm gut bacteria uses various mechanisms to solubilize potassium: (a) Production of organic acids: Earthworm gut bacteria secrete organic acids (citric acid, oxalic acid, and acetic acid), which chelate cations associated with potassium-bearing minerals to make soluble potassium ions. (b) Acidification: The metabolism of gut microbes decreases pH, which increases the solubility of insoluble potassium compounds. (c) Use of enzymes: Some enhance the breakdown of more complex minerals rich in potassium by the secretion of enzymes. (d) Chelation and ion exchange: Microbial metabolites can exchange ions, replacing potassium from mineral surfaces with hydrogen or other cations.

Several microbial taxa are associated with potassium solubilization in the earthworm gut microbiome. The important genera of potassium solubilizers are *Pseudomonas* sp. and *Bacillus* sp. (Table 2). Using gut bacteria from earthworms to solubilize potassium provides an earth-friendly alternative to increasing soil fertility and decreasing dependence on harmful chemical fertilizers by using them as (a) Biofertilizers: Isolation and development of potassium-solubilizing bacteria from earthworm guts as microbial inoculants when soil is deficient in potassium, the activity of earthworms increases the nutrients in the crop. (b) Vermicomposting:

Table 2 Different types of earthworm gut bacteria and their role in plant growth promotion

Bacterial genus	Earthworm	Role in plant growth promotion	Reference
***Bacillus* sp.** (Firmicutes)	*E. fetida* *Pheretima guillelmi* *Aporrectodea molleri* *Metaphire posthuma*	Phosphate solubilization, potassium solubilization, nitrogen fixation, and IAA production	Andleeb et al. (2022), Wu et al. (2012), Yakkou et al. (2022), Hussain et al. (2016), and Biswas et al. (2018)
***Pseudomonas* sp.** (Proteobacteria)	*M. posthuman* *A. molleri*	Phosphate solubilization, potassium solubilization, and nitrogen fixation	Bhakta et al. (2022) and Yakkou et al. (2022)
***Aeromonas* sp.** (Proteobacteria)	*A. molleri*	Phosphate solubilization	Yakkou et al. (2022)
***Staphylococcus* sp.** (Firmicutes)	*M. posthuma*	Phosphate solubilization	Biswas et al. (2018)
***Serratia* sp.** (Proteobacteria)	*E. fetida*	Phosphate solubilization	Hussain et al. (2016)
***Azospirillum* sp.** (Proteobacteria)	*Eudrilus* sp.	Phosphate solubilization and nitrogen fixation	Gopal et al. (2009) and Pathma and Sakthivel (2012)
***Streptomyces* sp.** (Actinobacteria)	*E. fetida*	Potassium solubilization	Liu et al. (2016)
***Aeromonas* sp.** (Proteobacteria)	*A. molleri*	Potassium solubilization	Yakkou et al. (2022)
***Micrococcus* sp.** (Actinobacteria)	*Enantiodrilus borelli* *Pontoscolex corethrurus*	Potassium solubilization	Picón and Teisaire (2012)
***Rhizobium* sp.** (Proteobacteria)	*E. fetida*	Nitrogen fixation	Jiang et al. (2023)
***Clostridium* sp.** (Firmicutes)	*E. fetida*	Nitrogen fixation	Khyade (2018)
***Azotobacter* sp.** (Proteobacteria)	*Eudrilus* sp.	Nitrogen fixation	Gopal et al. (2009) and Pathma and Sakthivel (2012)
***Nitrosomonas* sp.** (Proteobacteria)	*Eudrilus* sp.	Nitrogen fixation	Gopal et al. (2009) and Pathma and Sakthivel (2012)
***Staphylococcus* sp.** (Firmicutes)	*M. posthuma*	IAA production	Biswas et al. (2018)
***Rhodococcus* sp.** (Actinobacteria)	*E. fetida*	IAA production	Singh et al. (2016)
***Enterobacter* sp.** (Proteobacteria)	*M. posthuma*	IAA production	Arraktham et al. (2016)

Bioavailable potassium enrichment in organic fertilizers mediated by composting with earthworms (Liu et al. 2016).

4.2.1.3 Nitrogen Fixation

In many agricultural systems, nitrogen availability is a limiting factor. Biological nitrogen fixation—the conversion of atmospheric nitrogen (N_2) into ammonia (NH_3) by microorganisms—occurs naturally. While soil-dwelling and plant-associated nitrogen fixers have been the focus of much attention, earthworms also host nitrogen-fixing bacteria in their guts. They can increase nitrogen availability in soil, thus lowering reliance on chemical nitrogen fertilizers.

Earthworm gut bacteria are important contributors to nitrogen fixation through the following: (a) Activity of nitrogenase: It has been reported that some gut bacteria have the *nif* genes encoding nitrogenase, the enzyme complex needed to reduce atmospheric nitrogen; (b) Anaerobic microenvironments: The gut of earthworms supplies micro-anaerobic habitats that are suitable for the function of the bacteria. (c) Synergistic interactions: The digestion of organic matter in the gut supplies the energy and carbon sources required for gut bacteria to fix nitrogen. The earthworm gut contains several bacteria that may fix nitrogen, including *Azospirillum* sp., *Klebsiella* sp., *Rhizobium* sp., and *Clostridium* sp. The nitrogen-fixing ability of earthworm gut bacteria has promising applications in sustainable agriculture: (a) Soil nitrogen enrichment: The activity of earthworms causes an increase in the number of nitrogen-fixing bacteria in the soil. (b) Biofertilizers: Isolating nitrogen-fixing bacteria from earthworms' guts can lead to microbial inoculants' development. Integrating beneficial bacteria into composting systems can further enhance nitrogen fixation. (c) Crop productivity: The activity of earthworms in soils promotes plant growth and overall yield by creating improved nitrogen dynamics. The nitrogen fixation process carried out by earthworm gut bacteria is very crucial for soil fertility and ecosystem health by enhancing nitrogen management and lessening the negative environmental effects of synthetic fertilizers (Hussain et al. 2016).

4.2.2 Hormone Regulation

Phytohormones are naturally occurring substances that control the physicochemical processes in plants and their activity and yield. Auxins, gibberellins, cytokinins, ethylene, and abscisic acid are phytohormones that promote plant growth and stress tolerance. These hormones can also be produced endogenously by plants; however, soil bacteria, such as those found in the gut of the earthworms, will affect the hormone's availability and function. Earthworms placed in soil encourage greater microbial diversity and activity and create their gut microbiome, which is involved in hormone secretion and regulation. Regulation of hormones by earthworm gut bacteria occurs via multiple mechanisms. Various bacterial taxa are involved in hormone production (IAA), i.e., *Staphylococcus* sp., *Bacillus* sp., and *Enterobacter* sp. (Table 2) (Arraktham et al. 2016).

4.3 Prospect of Meta-Omics to Facilitate the Discovery and Eco-physiological Alteration of Earthworm Gut Bacterial Bioresource Under Xenobiotic Stress

Opposite to limitations associated with conventional culture-based methods, meta-omics strategies (including metagenomics, metatranscriptomics, metaproteomics, and metabolomics) often provide a robust alternative in exploring microbial community dynamics, metabolic profiles, and various eco-physiological adaptations among microorganisms that aid in their survival in rapidly altering environmental conditions. However, it is noteworthy that despite all its benefits, the employment of meta-omics strategies is still in its inception. If employed strategically, meta-omics-based tools can curtail various knowledge gaps associated with conventional approaches and provide deep insight into the uncharted part of the microbial world. For instance, metagenomic analysis (both functional as well as sequence driven) following high-throughput next-generation sequencing technologies, including Ion Torrent, Illumina, and Nanopore, enabled microbiologists to rapidly determine the entire microbial community (including both culturable and non-culturable microbiota), an assemblage of novel genes conferring resistance to xenobiotics or altering environmental conditions in various habitats (Mishra et al. 2021). For example, Xia et al. (2022) reported several interesting findings following whole genome shotgun sequencing on the Illumina platform of earthworm gut and soil microbiota under vanadium stress. These include the alteration of earthworm gut microbiota under vanadium stress in a concentration-dependent manner. Where bacterial taxa like *Aeromicrobium*, *Clostridium*, and *Klebsiella were* found to be increasing, on the other hand, genera like *Gemmata*, *Mycobacterium*, and *Marmoricola* significantly decreased with increased concentration of vanadium. Contrastingly, soil genera like *Acinetobacter*, *Comamonas*, and *Sphingobacterium* significantly increased, whereas genera like *Luteitalea*, *Haloferula*, and *Shewanella* significantly decreased with increasing vanadium concentration. Parallelly, one of the most intriguing findings is the distribution of vanadium-degrading bacteria and vanadium-degrading genes. During metagenomic analysis, it was observed that vanadium-degrading bacteria were more diverse but less abundant in soil compared to earthworm gut; concurrently, vanadium-degrading genes are more abundant in soil.

Similarly, following 16S rRNA gene sequencing on the Illumina platform, Wang et al. (2019) reported varying assemblages of microbiota in the gut of earthworm *Metaphire sieboldi* and soil under arsenic stress, where the concentration of arsenite was higher in the gut and the concentration of arsenate was higher in the soil, where the gut microbiota was dominated by members of Actinobacteria (57.5%), Firmicutes (20.7%), and Proteobacteria (13.5%). Contrarily, soil microbiota was dominated by members of Bacteroidetes (26.1%), Proteobacteria (27%), and Actinobacteria (24.6%) under arsenic stress. Following high-throughput quantitative real-time PCR analysis, they have also reported the distribution of 19 arsenic biotransformation genes in the gut and soil environment. During this analysis, it was observed that genes related to the biotransformation of As(V) to As(III) and arsenic efflux pumps were predominant in the earthworm gut environment compared to soil. On the other hand, genes related to arsenic methylation were significantly less in the gut environment than in soil. These findings suggest the employment of differential and niche-specific vanadium and arsenic resistance strategies by soil and earthworm gut bacteria; in another fascinating yet elusive finding, Xia et al. (2023) reported the involvement of earthworm gut virome in shaping the gut bacterial community under benzo[a]pyrene stress following metagenomics and metatranscriptomics analysis. It was observed that under control conditions, taxa, including Siphoviridae, Microviridae, Myoviridae, and Podoviridae, dominated the earthworm gut virome. Species diversity of the phage community was lowest at lower levels of benzo[a]pyrene and gradually increased with increased concentration of benzo[a]pyrene. It was also observed that the abundance of lysogenic phages is 56%, 88%, and 43% under control, with low and high benzo[a]pyrene concentrations, respectively. The sudden drop in lysogenic phages at high concentrations of benzo[a]pyrene indicates activation of prophages and release of virions. In parallel, this finding was further corroborated by metatranscriptomic analysis, which revealed significant upregulation of lysogenic marker genes (related to transposase, integrase, excisionase, resolvase, and recombinase) under low concentrations of benzo[a]pyrene. This implies that under low benzo[a]pyrene concentrations, phages in earthworm guts integrated into bacterial genomes and survived as lysogenic phages; on the other hand, phages tend to become lytic at high concentrations of benzo[a]pyrene. Such altered virome assemblage significantly influences the community dynamics of gut bacteria in earthworms under stress.

In addition to metagenomics and metatranscriptomics, metabolomic studies often provide a functional readout of cellular response against various stresses. This aids in a better understanding of organisms' eco-physiological adaptations under stressful environments. Techniques including LC-MS, GC-MS, GC/LC-TOF/QTOF-MS, UPLC-HRMS, and NMR are routinely used to determine the metabolomic profile of various microorganisms. Earthworms, especially *Eisenia fetida*, which is often considered as an excellent model organism to evaluate soil health, may also function for studying environmental metabolomics studies in contaminated environments. Coelomic fluid from earthworms is considered to be ideal for metabolomic analysis (Zhang et al. 2021). Following metagenomics and metabolomic studies, Fu et al. (2024) reported that co-inoculation with polyethylene microplastics and pesticide imidacloprid to *Eisenia fetida* causes massive disruption of amino acid, fatty acid, and carbohydrate metabolism pathways. Additionally, it also caused a significantly altered microbial community with an increased prevalence of pathogens in the earthworm gut.

5 Conclusion

The obnoxious effect of xenobiotics, along with record alteration of global climate change, cannot be averted overnight in the agriculture sector. Hastening any drastic, unscientific measure will only exacerbate the situation. Therefore, an integrated management practice can be employed, which will involve a reduction in the use/release of xenobiotics in industries along with restoration of soil health over time. In this context, it is noteworthy to mention that the use of earthworms to restore and maintain soil health in agricultural fields is an ancient practice, which recently gained mass recognition again following worldwide renewed interest in organic farming. Earthworms are also reported to survive in various stressful environments, especially soil contaminated with various xenobiotics. Where the physiological response of earthworms to cope with this xenobiotic stress is a well-researched aspect, the crucial involvement of earthworm gut bacteria is a relatively less discussed phenomenon. Such gut bacteria are reported to not only tolerate various xenobiotic stresses to aid the survival of earthworms in contaminated environments but also promote the growth of various crops in the same contaminated soil. This chapter provides a comprehensive idea of the detailed mechanisms employed by earthworm gut bacteria for xenobiotic stress mitigation, restoration of soil health, and simultaneous growth promotion of various crop plants. Additionally, the importance of multi-omics strategies to decipher community dynamics and eco-physiological adaptations of various earthworm gut microbes under various xenobiotic stresses was briefly discussed. Hope these interdisciplinary discussions will map new avenues for developing new eco-friendly strategies for mitigating xenobiotic stress with earthworms and earthworm gut bacteria at central.

References

Akca, İ., Yilmaz, N.D.K. and Kızılkaya, R., 2014. Evaluation of suppression of rhizomania disease by earthworm (*Lumbricus terrestris* L.) and its effects on soil microbial activity in different sugar beet cultivars. *Archives of Agronomy and Soil Science*, *60*(11), pp. 1565–1575. https://doi.org/10.1080/03650340.2014.901502

Andleeb, S., Shafique, I., Naseer, A., Abbasi, W.A., Ejaz, S., Liaqat, I., Ali, S., Khan, M.F., Ahmed, F. and Ali, N.M., 2022. Molecular characterization of plant growth-promoting vermi-bacteria associated with *Eisenia fetida* gastrointestinal tract. *Plos One*, *17*(6), p. e0269946. https://doi.org/10.1371/journal.pone.0269946

Arancon, N.Q., Edwards, C.A. and Bierman, P., 2006. Influences of vermicomposts on field strawberries: Part 2. Effects on soil microbiological and chemical properties. *Bioresource Technology*, *97*(6), pp. 831–840. https://doi.org/10.1016/j.biortech.2005.04.016

Arraktham, S., Tancho, A., Niamsup, P. and Rattanawaree, P., 2016. The potential of bacteria isolated from earthworm intestines, Vermicompost and liquid vermicompost to produce indole-3-acetic acid (IAA).

Banerjee, A., Biswas, J.K., Pant, D., Sarkar, B., Chaudhuri, P., Rai, M. and Meers, E., 2019. Enteric bacteria from the earthworm (*Metaphire posthuma*) promote plant growth and remediate toxic trace elements. *Journal of Environmental Management, 250*, p. 109530. https://doi.org/10.1016/j.jenvman.2019.109530

Bhakta, J.N., Sarkar, B. and Brahma, P., 2022. Isolation and characterization of potential phosphate solubilizing bacteria from earthworm (*Metaphire posthuma*) for applying as biofertilizer. *Organic Agriculture, 12*(3), pp. 431–444. https://doi.org/10.1007/s13165-022-00400-7

Biswas, J.K., Banerjee, A., Rai, M., Naidu, R., Biswas, B., Vithanage, M., Dash, M.C., Sarkar, S.K. and Meers, E., 2018. Potential application of selected metal resistant phosphate solubilizing bacteria isolated from the gut of earthworm (*Metaphire posthuma*) in plant growth promotion. *Geoderma, 330*, pp. 117–124. https://doi.org/10.1016/j.geoderma.2018.05.034

Blouin, M., Zuily-Fodil, Y., Pham-Thi, A.T., Laffray, D., Reversat, G., Pando, A., Tondoh, J. and Lavelle, P., 2005. Belowground organism activities affect plant aboveground phenotype, inducing plant tolerance to parasites. *Ecology Letters, 8*(2), pp. 202–208. https://doi.org/10.1111/j.1461-0248.2004.00711.x

Bossuyt, H., Six, J. and Hendrix, P.F., 2006. Interactive effects of functionally different earthworm species on aggregation and incorporation and decomposition of newly added residue carbon. *Geoderma, 130*(1–2), pp. 14–25. https://doi.org/10.1016/j.geoderma.2005.01.005

Bottinelli, N., Henry-des-Tureaux, T., Hallaire, V., Mathieu, J., Benard, Y., Tran, T.D. and Jouquet, P., 2010. Earthworms accelerate soil porosity turnover under watering conditions. *Geoderma, 156*(1–2), pp. 43–47. https://doi.org/10.1016/j.geoderma.2010.01.006

Brown, G.G. and Doube, B.M., 2004. Functional interactions between earthworms, microorganisms, organic matter, and plants. In *Earthworm ecology* (pp. 213–239). CRC Press.

Capozzi, F., Spagnuolo, V., 2024. The Potential of Plants to Absorb Xenobiotics. *Plants*, 13(7), 922. https://doi.org/10.3390/plants13070922

Chang, X., Sun, Y., Zhao, L., Li, X., Yang, S., Weng, L. and Li, Y., 2021. Exposure to fomesafen alters the gut microbiota and the physiology of the earthworm *Pheretima guillelmi. Chemosphere, 284*, p. 131290. https://doi.org/10.1016/j.chemosphere.2021.131290

Chaoui, H.I., Zibilske, L.M. and Ohno, T., 2003. Effects of earthworm casts and compost on soil microbial activity and plant nutrient availability. *Soil Biology and Biochemistry, 35*(2), pp. 295–302. https://doi.org/10.1016/S0038-0717(02)00279-1

Chauhan, R.S., Singhal, L., 2006 Harmful effects of pesticides and their control through cowpathy. *International Journal of Cow Science*, 2(1), 61–70.

Curtis, S.W., Terrell, M.L., Jacobson, M.H., Cobb, D.O., Jiang, V.S., Neblett, M.F., Gerkowicz, S.A., Spencer, J.B., Marder, M.E., Barr, D.B., 2019. Thyroid hormone levels associate with exposure to polychlorinated biphenyls and polybrominated biphenyls in adults exposed as children. *Journal of Environmental Health*, 18(1). https://doi.org/10.1186/s12940-019-0509-z

Dawood, M., Wahid, A., Hashmi, M.Z., Mukhtar, S. and Malik, Z., 2017. Use of earthworms in biomonitoring of soil xenobiotics. *Xenobiotics in the soil environment: monitoring, toxicity and management*, pp. 73–88. https://doi.org/10.1007/978-3-319-47744-2_6

De Bernardi, A., Marini, E., Casucci, C., Tiano, L., Marcheggiani, F., Vischetti, C., 2022. Copper Monitoring in Vineyard Soils of Central Italy Subjected to Three Antifungal Treatments, and Effects of Sub-Lethal Copper Doses on the Earthworm *Eisenia fetida*. Toxics, *10*, 310. https://doi.org/10.3390/toxics10060310

Desai, M., Haigh, M., Walkington, H., 2019. Phytoremediation: Metal decontamination of soils after the sequential forestation of former opencast coal land. *Science of The Total Environment*, 656, 670–680. https://doi.org/10.1016/j.scitotenv.2018.11.327

Desie, E., Van Meerbeek, K., De Wandeler, H., Bruelheide, H., Domisch, T., Jaroszewicz, B., Joly, F.X., Vancampenhout, K., Vesterdal, L. and Muys, B., 2020. Positive feedback loop between earthworms, humus form and soil pH reinforces earthworm abundance in European forests. *Functional Ecology, 34*(12), pp. 2598–2610. https://doi.org/10.1111/1365-2435.13668

Dobson, A.M., Blossey, B. and Richardson, J.B., 2017. Invasive earthworms change nutrient availability and uptake by forest understory plants. *Plant and Soil, 421*, pp. 175–190. https://doi.org/10.1007/s11104-017-3412-9

Don, A., Steinberg, B., Schöning, I., Pritsch, K., Joschko, M., Gleixner, G. and Schulze, E.D., 2008. Organic carbon sequestration in earthworm burrows. *Soil Biology and Biochemistry*, *40*(7), pp. 1803–1812. https://doi.org/10.1016/j.soilbio.2008.03.003

Edwards, C.A. and Lofty, J.R., 1980. Effects of earthworm inoculation upon the root growth of direct drilled cereals. *Journal of Applied Ecology*, pp. 533–543. https://doi.org/10.2307/2402635

Edwards, C.A., 2004. The importance of earthworms as key representatives of the soil fauna. In *Earthworm ecology* (pp. 3–11). CRC Press.

Edwards, C.A., Arancon, N.Q., 2022. Interactions Between Earthworms, Microorganisms, and Other Invertebrates. In: Biology and Ecology of Earthworms. Springer, New York, NY. https://doi.org/10.1007/978-0-387-74943-3_9

Eisenhauer, N., 2010. The action of an animal ecosystem engineer: Identification of the main mechanisms of earthworm impacts on soil microarthropods. *Pedobiologia*, *53*(6), pp. 343–352. https://doi.org/10.1016/j.pedobi.2010.04.003

Elekwachi, O., Andresen, J., Hodgman, T.C., 2014. Global use of bioremediation technologies for decontamination of ecosystems. *Journal of Bioremediation & Biodegradation*, 5, 1–30. https://doi.org/10.4172/2155-6199.1000225

Faroon, O., Ruiz, P., 2016. Polychlorinated biphenyls: New evidence from the last decade. *Toxicology and Industrial Health*, 32(11), 1825–1847. https://doi.org/10.1177/0748233715587849

Fründ, H.C., Graefe, U. and Tischer, S., 2010. Earthworms as bioindicators of soil quality. In *Biology of earthworms* (pp. 261–278). Berlin, Heidelberg: Springer Berlin Heidelberg. https://doi.org/10.1007/978-3-642-14636-7_16

Fu, H., Zhu, L., Chen, L., Zhang, L., Mao, L., Wu, C., Chang, Y., Jiang, J., Jiang, H. and Liu, X., 2024. Metabolomics and microbiomics revealed the combined effects of different-sized polystyrene microplastics and imidacloprid on earthworm intestinal health and function. *Environmental Pollution*, *361*, p. 124799. https://doi.org/10.1016/j.envpol.2024.124799

García, C., Gutiérrez, S., Obrador, A., Farnandez, M.D., 2020. Study of Zn availability, uptake, and effects on earthworms of zinc oxide nanoparticle *versus* bulk applied to two agricultural soils: Acidic and calcareous. *Chemosphere*, 239, 124814. https://doi.org/10.1016/j.chemosphere.2019.124814

Godheja, J., Shekhar, S.K., Siddiqui, S.A., Modi, D.R., 2016. Xenobiotic compounds present in soil and water: A review on remediation strategies. *Journal of Environmental & Analytical Toxicology,* 6, 392. https://doi.org/10.4172/2161-0525.1000392

Gopal, M., Gupta, A., Sunil, E. and Thomas, G.V., 2009. Amplification of plant beneficial microbial communities during conversion of coconut leaf substrate to vermicompost by *Eudrilus* sp. *Current Microbiology*, *59*, pp. 15–20. https://doi.org/10.1007/s00284-009-9388-9

Govind, P., Madhuri, S., 2014. Heavy Metals Causing Toxicity in Animals and Fishes. *Research Journal of Animal, Veterinary and Fishery Sciences*, 2(2), 17–23. ISSN 2320–6535

Griffin, G.J.L., 1980. Synthetic polymers and the living environment. *Pure and Applied Chemistry*, *52*, 399–407.

Hashimi, M.H., Hashimi, R., Qasi, Ryan., 2020. Toxic Effects of Pesticides on Humans, Plants, Animals, Pollinators and Beneficial Organisms. *Asian Plant Research Journal,* 5(4), 37–47. https://doi.org/10.9734/aprj/2020/v5i430114

Honda, M., Suzuki, N., 2020. Toxicities of polycyclic aromatic hydrocarbons for aquatic animals. *International Journal of Environmental Research and Public Health*, 17(4), 1363. https://doi.org/10.3390/ijerph17041363

Hussain, N., Singh, A., Saha, S., Kumar, M.V.S., Bhattacharyya, P. and Bhattacharya, S.S., 2016. Excellent N-fixing and P-solubilizing traits in earthworm gut-isolated bacteria: a vermicompost based assessment with vegetable market waste and rice straw feed mixtures. *Bioresource Technology*, *222*, pp. 165–174. https://doi.org/10.1016/j.biortech.2016.09.115

Jajoo, A., 2017. Effects of Environmental Pollutants Polycyclic Aromatic Hydrocarbons (PAH) on Photosynthetic Processes. *Photosynthesis: Structures, Mechanisms, and Applications,* 249–259. https://doi.org/10.1007/978-3-319-48873-8_11

Jiang, L., Ling, S., Fu, M., Peng, C. Zhang, W., Lin, K., Zhou, B., 2021.Bioaccumulation, elimination and metabolism in earthworms and microbial indices responses after exposure to decabromodiphenyl ethane in a soil-earthworm-microbe system. *Environmental Pollution*, 289, 117965. https://doi.org/10.1016/j.envpol.2021.117965

Jiang, X., Lu, C., Hu, R., Shi, W., Zhou, L., Wen, P., Jiang, Y. and Lo, Y.M., 2023. Nutritional and microbiological effects of vermicompost tea in hydroponic cultivation of maple peas (Pisum sativum var. arvense L.). *Food Science & Nutrition, 11*(6), pp. 3184–3202. https://doi.org/10.1002/fsn3.3299

Khyade, V.B., 2018. Bacterial diversity in the alimentary canal of earthworms. *J Bacteriol Mycol Open Access, 6*(3), pp. 183–185. https://doi.org/10.15406/jbmoa.2018.06.00200

Kim, K.H., Kabir, E., Jahan, S.A., 2017a. Exposure to pesticides and the associated human health effects. *Science of Total Environment*, 575, 525–535. https://doi.org/10.1016/j.scitotenv.2016.09.009

Kim, Y.N., Robinson, B., Lee, K.A., Boyer, S. and Dickinson, N., 2017b. Interactions between earthworm burrowing, growth of a leguminous shrub and nitrogen cycling in a former agricultural soil. *Applied Soil Ecology, 110*, pp. 79–87. https://doi.org/10.1016/j.apsoil.2016.10.011

Krauss, M., Wilcke, W., Zech, W., 2000. Availability of Polycyclic Aromatic Hydrocarbons (PAHs) and Polychlorinated Biphenyls (PCBs) to Earthworms in Urban Soils. *Environmental Science and Technology*, 34(20), 4335–4340. https://doi.org/10.1021/es001137s

Kroeze, C., Reijnders, L., 1992. Halocarbons and global warming. *Science of Total Environment*, 111, 1–24. https://doi.org/10.1016/0048-9697(92)90042-Q

Kumar, A., Aery, N.C., 2016. Impact, Metabolism, and Toxicity of Heavy Metals in Plants. In *Plant Responses to Xenobiotics*; Singh, A., Prasad, S., Singh, R., Eds.; Springer: Singapore. https://doi.org/10.1007/978-981-10-2860-1_7

Lavelle, P. and Spain, A.V., 2001. Soil organisms. *Soil Ecology*, pp. 201–356. https://doi.org/10.1007/0-306-48162-6_3

Lavelle, P., Charpentier, F., Villenave, C., Rossi, J.P., Derouard, L., Pashanasi, B., André, J., Ponge, J.F. and Bernier, N., 2004. Effects of earthworms on soil organic matter and nutrient dynamics at a landscape scale over decades. In *Earthworm ecology* (pp. 145–160). CRC Press.

Litskas, V. D., Karamanlis, X.N., Prousali, S.P., Koveos, D.S., 2019. The xenobiotic doxycycline affects nitrogen transformations in soil and impacts earthworms and cultivated plants. *Journal of Environmental Science and Health, Part A*, 54(14), 441–1447. https://doi.org/10.1080/10934529.2019.1655368.

Liu, D., Lian, B. and Wang, B., 2016. Solubilization of potassium containing minerals by high temperature resistant *Streptomyces* sp. isolated from earthworm's gut. *Acta Geochimica, 35*, pp. 262–270. https://doi.org/10.1007/s11631-016-0106-6

Liu, J., Xie, X., & Li, S.M., 2020a. Increasing cytochrome P450 enzyme diversity by identification of two distinct cyclodipeptide dimerases. *Chemical Communications*, 56(75), 11042–11045. https://doi.org/10.1039/d0cc04772d

Liu, L., Bilal, M., Duan, X.H., Iqbal, H.M.N., Iqbal, N., 2019. Mitigation of environmental pollution by genetically engineered bacteria current challenges and future perspectives. *Science of Total Environment,* 667, 444–454. https://doi.org/10.1016/j.scitotenv.2019.02.390

Liu, P., Yang, Y., Li, M., 2020b. Responses of soil and earthworm gut bacterial communities to heavy metal contamination. *Environmental Pollution*, 265, 114921 https://doi.org/10.1016/j.envpol.2020.114921.

Malla, M.A., Dubey, A., Yadav, S., Kumar, A., Hashem, A., Abd-Allah, E.F., 2018. Understanding and designing the strategies for the microbe-mediated remediation of environmental contaminants using omics approaches. *Frontiers in Microbiology,* 9, 1132. https://doi.org/10.3389/fmicb.2018.01132

Mandal, S., Saha, K.K., Mandal, N.C., 2021. Molecular Insight Into Key Eco-Physiological Process in Bioremediating and Plant-Growth-Promoting Bacteria. *Frontiers in Agronomy,* 3, 664126. https://doi.org/10.3389/fagro.2021.664126.

Medina-Sauza, R.M., Álvarez-Jiménez, M., Delhal, A., Reverchon, F., Blouin, M., Guerrero-Analco, J.A., Cerdán, C.R., Guevara, R., Villain, L. and Barois, I., 2019. Earthworms building up soil microbiota, a review. *Frontiers in Environmental Science*, *7*, p. 81. https://doi.org/10.3389/fenvs.2019.00081

Miglani, R., Parveen, N., Kumar, A., Ansari, M.A., Khanna, S., Rawat, G., Panda, A.K., Bisht, S.S., Upadhyay, J., Ansari, M.N., 2022. Degradation of Xenobiotic Pollutants: An Environmentally Sustainable Approach. *Metabolites*, 12(9), 818. https://doi.org/10.3390/metabo12090818

Mishra, S., Lin, Z., Pang, S., Zhang, W., Bhatt, P. and Chen, S., 2021. Recent advanced technologies for the characterization of xenobiotic-degrading microorganisms and microbial communities. *Frontiers in Bioengineering and Biotechnology*, *9*, p. 632059. https://doi.org/10.3389/fbioe.2021.632059

Mohn, W.W., Tiedje, J.M., 1992 Microbial reductive dehalogenation. *Microbiological Reviews*, 56(3). https://doi.org/10.1128/mr.56.3.482-507.1992

Moore, C.J., 2008. Synthetic polymers in the marine environment: A rapidly increasing, long-term threat. *Environmental Research,* 108(2), 131–139. https://doi.org/10.1016/j.envres.2008.07.025

Osmani, M., Bani, A., Hoxha, B., 2015. Heavy Metals and Ni Phytoextractionin in the Metallurgical Area Soils in Elbasan. *Albanian Journal of Agricultural Sciences,* 14(4), 414–419.

Pande, V., Pandey, S.C., Sati, D., Bhatt, P., Samant, M., 2022. Microbial interventions in bioremediation of heavy metal contaminants in agroecosystem. *Frontiers in Microbiology,* 13, 824084. https://doi.org/10.3389/fmicb.2022.824084

Parween, T., Jan, S., Mahmood, U., Fatma, T., Siddiqui, Z.H., 2016. Selective effect of pesticides on plant – a review. *Critical Reviews in Food Science and Nutrition*, 56(1), 160–179. https://doi.org/10.1080/10408398.2013.787969

Pathma, J. and Sakthivel, N., 2012. Microbial diversity of vermicompost bacteria that exhibit useful agricultural traits and waste management potential. *SpringerPlus*, *1*, pp. 1–19. https://doi.org/10.1186/2193-1801-1-26

Patneedi, C.B., Prasadu, K.D., 2015. Impact of Pharmaceutical Wastes on Human Life and Environment. *Rasayan Journal of Chemistry*, 8, 67–70.

Picón, M.C. and Teisaire, E.S., 2012. Identification of the intestinal microbial community of Glossoscolecidae earthworms (Annelida: Oligochaeta). *Munis Entomol Zool*, *7*, pp. 1035–1043.

Piers, M.D.F., Joshi, M., 2005. The role of halocarbons in the climate change of the troposphere and stratosphere. *Climate Change*, 71(1–2), 249–266. https://doi.org/10.1007/s10584-005-5955-7

Ramel, F., Sulmon, C., Serra, A.A., Gouesbet, G., Couée, I., 2012. Xenobiotic sensing and signalling in higher plants. *Journal of Experimental Botany*, *63*, 3999–4014. https://doi.org/10.1093/jxb/ers102

Reed, E.S., 1982. Darwin's earthworms: A case study in evolutionary psychology. *Behaviorism*, *10*(2), pp. 165–185. https://www.jstor.org/stable/27759005

Rieger, P.G., Meier, H.M., Gerle, M., Vogt, U., Groth, T., Knackmuss, H.G., 2002. Xenobiotics in the environment: present and future strategies to obviate the problem of biological persistence. *Journal of Biotechnology,* 9(1),101–123. https://doi.org/10.1016/S0168-1656(01)00422-9

Saha, K.K., Mandal, S., Barman, A., Mondal, S., Chatterjee, S. and Mandal, N.C., 2024. Genomic insight of phosphate solubilization and plant growth promotion of two taxonomically distinct winter crops by *Enterobacter* sp. DRP3. *Journal of Applied Microbiology*, *135*(6), lxae146. https://doi.org/10.1093/jambio/lxae146

Saha, K.K., Mandal, S., Barman, A., Chatterjee, S. and Mandal, N.C., 2025. Deciphering the genomic and physiological basis of pH dependent siderophore production in Enterobacter sp. DRP3 and mitigation of lead stress in rice seedlings. *Journal of Hazardous Materials*, p. 137587.

Sall, M.L., Diaw, A.K.D., Gningue-Sall, D., Efremova Aaron, S., Aaron, J.J., 2022. Toxic heavy metals: Impact on the environment and human health, and treatment with conducting organic polymers, a review. *Environmental Science and Pollution Research,* 27, 29927–29942. https://doi.org/10.1007/s11356-020-09354-3

Sharma, A., Kumar, V., Kumar, R., Shahzad, B., Thukral, A.K., Bhardwaj, R., Tejada Moral, M., 2018. Brassinosteroid-mediated pesticide detoxification in plants: A mini-review. *Cogent Food and Agriculture*, *4*, 1436212. https://doi.org/10.1080/23311932.2018.1436212

Sharma, A., Kumar, V., Thukral, A.K., Bhardwaj, R., 2019a. Responses of plants to pesticide toxicity: An overview. *Planta Daninha*, 37, e019184291. https://doi.org/10.1590/S0100-83582019370100065

Sharma, S., Nagpal, A.V., Kaur, I., 2019b. Biphenyls: health impacts and toxicity evaluation. *Environmental contaminants and natural products*, 116–137.

Singh, A., Tiwari, R., Sharma, A., Adak, A., Singh, S., Prasanna, R., Saxena, A.K., Nain, L. and Singh, R.V., 2016. Taxonomic and functional diversity of the culturable microbiomes of epigeic earthworms and their prospects in agriculture. *Journal of Basic Microbiology*, *56*(9), pp. 1009–1020. https://doi.org/10.1002/jobm.201500779

Sokolova, I.M., Frederich, M., Bagwe, R., Lannig, G., Sukhotin, A.A., 2012. Energy homeostasis as an integrative tool for assessing limits of environmental stress tolerance in aquatic invertebrates. *Marine Environmental Research,* 79, 1–15. https://doi.org/10.1016/j.marenvres.2012.04.003

Stefanac, T., Grgas, D., Dragecevic, T.L., 2021. Xenobiotics—Division and Methods of Detection: A Review. *Journal of Xenobiotics*, *11*(4), 130–141. https://doi.org/10.3390/jox11040009

Sun, K., Song, Y., He, F., Jing, M., Tang, J., Liu, R., 2021. A review of human and animal exposure to polycyclic aromatic hydrocarbons: Health risk and adverse effects, photo-induced toxicity and microplastic regulating effect. *Science of the Total Environment* 773, 145403. https://doi.org/10.1016/j.scitotenv.2021.145403

Sun, M., Chao, H., Zheng, X., Deng, S., Ye, M., Hu, F., 2020. Ecological role of earthworm intestinal bacteria in terrestrial environments: A review. *Science of the Total Environment,* 740, 140008. https://doi.org/10.1016/j.scitotenv.2020.140008.

Vajargah, M.F., 2021. A Review on the Effects of Heavy Metals on Aquatic Animals. *Journal of Biomedical Research and Environmental Science*, 30, 2(9): 865–869. https://doi.org/10.37871/jbres1324

Wang, C., Rong, H., Liu, H., Wang, X., Gao, Y., Deng, R., Liu, R., Liu, Y. and Zhang, D., 2018. Detoxification mechanisms, defense responses, and toxicity threshold in the earthworm *Eisenia foetida* exposed to ciprofloxacin-polluted soils. *Science of the Total Environment*, *612*, pp. 442–449. https://doi.org/10.1016/j.scitotenv.2017.08.120

Wang, D., Saleh, N.B., Byro, A., Zepp, R., Demessie, E.S., Luxton, T.P., Ho, K.T., Burgess, R.M., Flury, M, White, J.C., Su, C., 2022. Nano-enabled pesticides for sustainable agriculture and global food security. Nature Nanotechnology, 17, 347–360. https://doi.org/10.1038/s41565-022-01082-8

Wang, H.T., Zhu, D., Li, G., Zheng, F., Ding, J., O'Connor, P.J., Zhu, Y.G. and Xue, X.M., 2019. Effects of arsenic on gut microbiota and its biotransformation genes in earthworm *Metaphire sieboldi*. *Environmental science & technology*, *53*(7), pp. 3841–3849. https://doi.org/10.1016/j.jes.2022.07.033

Wu, F., Wan, J.H.C., Wu, S. and Wong, M., 2012. Effects of earthworms and plant growth–promoting rhizobacteria (PGPR) on Availability of nitrogen, phosphorus, and potassium in soil. *Journal of Plant Nutrition and Soil Science*, *175*(3), pp. 423–433. https://doi.org/10.1002/jpln.201100022

Xia, R., Shi, Y., Wang, X., Wu, Y., Sun, M. and Hu, F., 2022. Metagenomic sequencing reveals that the assembly of functional genes and taxa varied highly and lacked redundancy in the earthworm gut compared with soil under vanadium stress. *Msystems*, *7*(1), pp. e01253-21. https://doi.org/10.1128/mSystems.01253-21

Xia, R., Sun, M., Balcázar, J.L., Yu, P., Hu, F. and Alvarez, P.J., 2023. Benzo [a] pyrene stress impacts adaptive strategies and ecological functions of earthworm intestinal viromes. *The ISME Journal*, *17*(7), pp. 1004–1014. https://doi.org/10.1038/s41396-023-01408-x

Yadav, A. (2020). Potential impact of Polychlorinated biphenyls on plant and environmental health. *Int. J. Plant. Soil. Sci*, *32*, 51–57.

Yakkou, L., Houida, S., Bilen, S., Kaya, L.O., Raouane, M., Amghar, S. and El Harti, A., 2022. Assessment of earthworm (*Aporrectodea molleri*)'s coelomic fluid-associated bacteria on different plant growth-promoting traits and maize germination and seedling growth. *Biocatalysis and Agricultural Biotechnology*, *42*, 102341. https://doi.org/10.1016/j.bcab.2022.102341

Zaka, S.M., Iqbal, N., Saeed, Q., Akrem, A., Batool, M., Khan, A.A., Anwar, A., Bibi, M., Azeem, S., Rizvi, D.E.N., 2019. Toxic effects of some insecticides, herbicides, and plant essential oils against Tribolium confusum Jacquelin du val (Insecta: Coleoptera: Tenebrionidae). *Saudi Journal of Biological Science,* 26(7), 1767–1771. https://doi.org/10.1016/j.sjbs.2018.05.012

Zhang, L.J., Qian, L., Ding, L.Y., Wang, L., Wong, M.H. and Tao, H.C., 2021. Ecological and toxicological assessments of anthropogenic contaminants based on environmental metabolomics. *Environmental Science and Ecotechnology*, *5*, p. 100081. https://doi.org/10.1016/j.ese.2021.100081

Zhu, Z.C., Chen, S.J., Zheng, J., Tian, M., Feng, A.H., Luo, X.J., Mai, B.X., 2014. Occurrence of brominated flame retardants (BFRs), organochlorine pesticides (OCPs), and polychlorinated biphenyls (PCBs) in agricultural soils in a BFR-manufacturing region of North China. *Science of the Total Environment*, 481, 47–54. https://doi.org/10.1016/j.scitotenv.2014.02.023

Nanobiosensors in the Prediction of Environmental Pollutants: Recent Advances and Challenges

Hemanga Kumar Das, Simran Choudhury, and Usha Sabharwal

Abstract The progressing and exponentially growing world has witnessed major technological advancements to their benefit. However, these advancements have their own share of cons in this world, especially concerning environmental contamination. These contaminants or pollutants have been known to enter biological bodies and ecosystems and cause severe damage. This is evident by many facts; for example, according to WHO, in 2019, 99% of our population were living in areas where air quality was not at par with the prescribed safety levels, and pollution has led to over 6.5 million premature deaths globally. Apart from this, we also come across the previously presumed inert pollutant, plastics, entering living systems and the food chain. The world has recently been very active in the eradication of pollution as a whole to ensure a secure planet for the generations to come. Thus, various novel technologies are being researched and analysed. One of such is nanotechnology. With efficient sensing and monitoring being the crucial and foremost step in combating pollution, highly efficient nanobiosensors are in the talks. Nanobiosensors are built of materials that have unique and tailored properties to make them precise and highly efficient. With such devices already in place for diagnostic, medicine, and food safety purposes, nanobiosensors are seen as a resort in bringing about efficient environmental pollutant monitoring, and hence create enhanced alertness amongst the population. With varying types of components and functionalities, nanobiosensors might prove to be a way to a cleaner earth; yet, there do reside challenges in their success.

H. K. Das
Department of Biotechnology and Biomedical Engineering, U.V. Patel College of Engineering, Ganpat University, Mehsana, Gujarat, India

S. Choudhury
Department of Life Sciences, Parul Institute of Applied Sciences, Parul University, Vadodara, Gujarat, India

U. Sabharwal (✉)
Department of Life Science, Parul Institute of Applied Sciences, Parul University, Waghodia, Vadodara, Gujarat, India

S. Roy, V. Mandal (eds.), *Plant-Microbe Interaction under Xenobiotic Exposure*,
https://doi.org/10.1007/978-981-96-8260-7_23

Keywords Bioreceptors · Environment · Nanobiosensors · Remediation · Transducers · Xenobiotics

1 Introduction

The utilization of modern technologies and equipment has indeed made our lives easier and increased productivity in various fields, but has simultaneously rendered our environment with vast amounts of various pollutants. These environmental pollutants are a result of various anthropogenic activities like agriculture, mining, waste disposal, and industrial activities, and include various elements from plastics and pesticides to heavy metals. All these pollutants degrade the quality of our environment and make it unsafe for the survival of humans and all other species, thus endangering various organisms. These xenobiotics have been found to enter biological systems, implicating their entrance in the food chain and web, and their significance in bioaccumulation and biomagnification, which would have long-standing effects. Environmental pollutants like phenols, plastics, and pesticides are known to have widespread toxicological impacts (Pironti et al. 2021).

The exponential utilization of various equipment, fuel, and chemicals is indeed indispensable, but tracking the pollutants, thus produced, is also the need of the hour. And in order to act upon the same, '*nanobiosensors*' are being developed. It might pose a question as to why nanobiosensors are employed while various other techniques are available; this is due to the iconic features possessed by nanobiosensors as depicted in Fig. 1. Nanobiosensors are often known for their real-time tracking, and also their precautionary nature. While most sensing techniques are known to detect the results of certain processes, nanosensors have the ability to detect in a precautionary manner, thus facilitating smarter resource allocation. Apart from this, nanobiosensors have high surface-to-volume ratio, and unique physical, chemical, biological, and mechanical properties, and are often made of robust nanoparticles (NPs) of silver, platinum, carbon-based nanomaterials, and palladium (Javaid et al. 2021; Huang et al. 2021; Gözde et al. 2022).

Xenobiotics can be of various types depending on their source, such as industrial chemicals, pharmaceuticals, agricultural chemicals, manufacturing by-products, and petroleum products, and can illicit acute health impacts to carcinogenic impacts. These substances can biotransform within biological systems to induce immunogenic responses and cause diseases and impairments. Apart from this, they can also interfere with drug delivery in patients exposed to the same (Wasi et al. 2013; Bhaskar Mahanayak 2024). Our body is certainly equipped with processes to combat pollutants as well. For example, the aryl hydrocarbon receptor (AHR) of our body can sense xenobiotics and activate detoxification enzymes to combat the same, but their prolonged activation can result in immunotoxicity, and also polycyclic aromatic hydrocarbons (PAHs) released due to incomplete combustion and use of

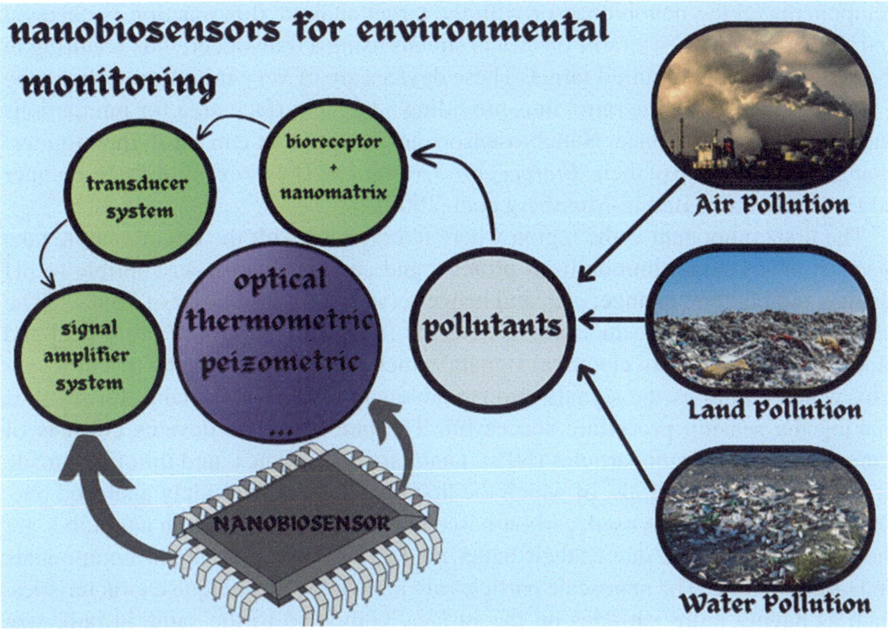

Fig. 1 Working pathway for nanobiosensor

petroleum-based products in vehicles, industries, and cooking can perturb the immune system (Suzuki et al. 2020). However, pollutants can be classified as inert, less reactive, highly reactive, and specifically reactive, based on their interaction with receptors within biological systems (Uddin et al. 2021).

All the pollutants that are being released into the environment are injurious to the health of not just humans, but also animals and plants, of which many are a part of the agricultural ecosystem. Thus, a regular exposure to such pollutants, even at a minute dose, can cause accumulation of these toxic substances, and hence, hasty yet productive actions are of utmost importance. Sensing of environmental pollutants and measuring their presence is one of the foremost steps in combating pollution and achieving a cleaner environment. Although much has been done towards this approach, there are various significant challenges. In the light of such issues, this chapter seeks to highlight certain recent advancements and challenges in the success of this strategy.

2 Insights on Nanobiosensors and Their Basic Types

Nanobiosensors are nanoscale devices that are used for the recognition and measurement of certain biological analytes/elements, by employing certain biological reactions on their surface. These reactions are possible upon the interaction of the components on the nanobiosensor with the target analyte; this reaction is detected and converted into measurable electrical signals using a transducer, thus resulting in the detection of the required target. These devices are of very minute size, but have a high surface-to-volume ratio, thus providing a larger surface area for interactions and reactions to take place. Nanobiosensors are devices that can sense the minutest changes with the help of their '*Bioreceptor-Transducer-Detector*' system (Baeumner 2003) (Ron 2007) (Badihi-Mossberg et al. 2007).

The first component is the region where interactions with the target analyte take place; it may contain immobilized proteins and enzymes and is susceptible to pH change, temperature change, etc., and hence serves as the first point of contact. The second component (transducer) is responsible for the conversion of biological and biochemical signals into electrical signals, which are passed on to the detector system, which amplifies the signals into readable and visually detectable forms, thus making the sensing procedure successful. The built of these devices consists of nanomaterials like nanoparticles (NPs), nanowires, nanotubes, and thin films made of noncrystalline materials, of which the first one is the most widely analysed one. Of all the nanomaterials used, carbon-based nanoparticles and carbon nanotubes are the most widely used, due to their better interaction with bioreceptor components and target analyte. The nanoscale particles are known to have unique characteristics, such as having more particles on the surface, compared to the same in bulk size (Malik et al. 2013). Therefore, these materials become a preferred material for such specialized devices.

Nanobiosensors can be classified into various types depending on the processes involved in their sensing mechanisms. It can be categorized based on the bioreceptor component, the transducer component (Ambrin et al. 2020), or overall system as well. Hence, here we list some of the significant variations of nanobiosensors. Although this table (Table 1) consists of a variety of the same, it is not exhaustive.

3 Working Mechanism of Nanobiosensors

Nanobiosensors work by various mechanisms to identify and quantify biological responses and transform them into signals that can be analysed. The essential components are a bioreceptor, a transducer, and a detector system (Barreiros Dos Santos et al. 2022).

Bioreceptor identifies the target analyte of interest, for example, antibodies or nucleic acids (Saadati et al. 2019). The transducer transforms the bioanalyte-bioreceptor interaction into a signal that can be measured, which is usually electrical (Pablos et al. 2024). The detector system captures the signal from the transducer and amplifies it for correct analysis (Pablos et al. 2024). The bioreceptor identifies and binds to the analyte (Wang 2016). The analyte-bioreceptor interaction creates a change. The transducer translates this alteration into an electric signal. The detector system processes, amplifies, and shows the signal (Bag and Mandal 2023).

Table 1 Variations in nanobiosensors based on the bioreceptor type, transducer type, nanomaterial type, and overall processes

Variation	Details	References
Enzyme-based nanobiosensor	As the name suggests, these devices employ immobilized enzymes as bioreceptors to interact with target analyte. In order to build this type, the reactions of the target analyte must be known, as this would determine the enzyme to be used, along with determining the feasibility of enzyme immobilization, required for performing the redox reactions with the analyte on its surface	Ambrin et al. (2020)
Aptamer-based nanobiosensor	For this type, the bioreceptors used are nucleic acids, which are known to have high affinity binding. This is similar to antibody-based nanobiosensors, but has the advantage of sensing high toxicity and low immunogenicity molecules as well (Ambrin et al. 2020).	Ambrin et al. (2020)
Antibody-based nanobiosensor	As the name suggests, these consist of antibodies as bioreceptors, and undergo high affinity binding with specific analytes, hence guaranteeing specificity in detection and lower levels of false alarms, with ease of immobilization. However, low immunogenic substances may go undetected	Ambrin et al. (2020)
Electrochemical nanobiosensor	In this type of sensor, the potential difference detected by transducer is produced by the reaction of ions and electrons producing an electrochemical energy. They are mostly based on metallic NPs and utilize the conversion of biochemical energy to electrical energy	Malik et al. (2013) and Ambrin et al. (2020)
Amperometric nanobiosensor	This type is related to enzyme-based sensors, as they sense the electrical energy produced by the redox reactions between analyte and bioreceptor, as is seen in enzyme-based reactions	Okon and Ronkainen (2017), Ambrin et al. (2020)
Potentiometric nanobiosensor	As the name suggests, these devices have two pH electrodes, whose potential difference is measured upon difference in ion concentration	Ambrin et al. (2020)
Optical nanobiosensor	These are one of the most common types, as they enable real-time, direct, and tag-free sensing. This is done by analysing the reaction of analytes with enzymes, which produces luminescence. It is used in monitoring of environmental pollutants	Damborský et al. (2016) and Ambrin et al. (2020)
Piezoelectric nanobiosensor	These sensors employ materials that generate specific electrical waves upon experiencing mechanical stress (piezoelectric materials, like tantalite and quartz), and are paired with the biocomponent to sense and generate electric signals	Ambrin et al. (2020)

(continued)

Table 1 (continued)

Variation	Details	References
Thermometric nanobiosensor	These sensors utilize the fundamental property of biocomponents and reactions to release or absorb energy	Ramanathan and Danielsson (2001) and Damborský et al. (2016)
NP-based acoustic wave nanobiosensor	The acoustic wave properties of biological layers (such as attenuation, delay time, resonating frequency, and velocity) are well defined. Upon the interaction of analytes with biological layers of the bioreceptor, the changes are detected	Voiculescu and Nordin (2012) and Malik et al. (2013)
NP-based magnetic nanobiosensor	These sensors utilize the superparamagnetic effect of nanoscale materials, mostly ferrite particles, to create a magnetic bioassay for detection	Malik et al. (2013) and Nabaei et al. (2018)
Nanotube-based nanobiosensor	Nanotubes are hollow, cylindrical nanostructures that have unique mechanical, chemical, and electrical properties, which are used in electrical and optical sensing. The most prominent material is carbon nanotubes	Lei and Ju (2010), Malik et al. (2013)
Nanowire-based nanobiosensor	These biosensors are made of unique one-dimensional nanoscale wire structures, for example, silicon nanowires, and are employed due to facilitating improved charge carrier movement within them; although very less in number, they are under analysis	Wanekaya et al. (2006) and Malik et al. (2013), and Ambhorkar et al. (2018)
Nanofibre-based nanobiosensor	Nanofibres have improved immobilization site for bioreceptors and better sensitivity, especially in nanofibre mats and arrays, that bestow upon them unique properties for improved sensing, thus facilitating environment monitoring and food safety-related sensing	Matlock-Colangelo and Baeumner (2012)
Nanoparticle nanobiosensor	These are the most widely used nanobiosensor material, as they provide high surface-to-volume ratio, yet are easier to fabricate	Bellan et al. (2011) and Malik et al. (2013)
Microbial nanobiosensor	Microbial nanobiosensors are those that employ a system composed of microbial cells, like bacteria, yeast, and electrodes to perform sensing. For example, biosensor is composed of *Acetobacter xylinum* cell and oxygen electrodes. These cells can be engineered microbes as well. This innovation is facilitated by the various enzymes produced by the microbe, which act as bioreceptors	De Silva et al. (2010)
Hybrid nanobiosensor	These nanobiosensors contain different biochemical components as bioreceptor, or even different strains of microbes, that catalyse a coupled reaction to facilitate sensing	De Silva et al. (2010)
Cantilever nanobiosensor	Cantilever nanobiosensors utilize minute mechanical displacements of nanoscale cantilevers, due to interaction with target analytes	Basu et al. (2020)

(continued)

Table 1 (continued)

Variation	Details	References
Quantum dot–based nanobiosensors	These sensors employ nanocrystals as the nanomaterial for immobilization of bioreceptor component, for example, quantum dots attached to DNA probes	Zhang et al. (2005)

3.1 Methods of Signal Transduction

Electrochemical nanosensors recognize modifications in the nanomaterial resistance upon analyte binding due to changes in charge carrier scattering, depletion, or accumulation. Carbon nanotubes, conductive polymers, or metal oxide nanowires can serve as gates in field-effect transistors (Syedmoradi et al. 2019). Electromagnetic or plasmonic nanosensors may function through electromagnetic or plasmonic principles. Refractive-index-dependent localized surface plasmon resonance increases detection of very weak interactions. Spectroscopic nanosensors, for example, employing surface-enhanced Raman spectroscopy, are also employed (Vasilantonakis et al. 2020). Mechanical nanosensors are a second class of nanosensors. Biological nanosensors frequently employ fluorescence as the transduction mechanism because of its sensitivity and ease of quantification. This may involve protein conjugation of active nanoparticles, site-directed mutagenesis to develop indicator proteins, or designing nanomaterials for attachment sites with bioreceptors (Hang et al. 2021) (Fig. 2).

Selection and optimization of nanomaterials are paramount for biosensing purposes, depending on their physical and chemical attributes, and on their energy-sensitive and energy-selective reactions (Farzin et al. 2018). Nanofabrication methods are used to tailor nanomaterials for performance. Nanomaterials such as nanoparticles, nanocables, nanofilms, and nanofibres provide properties such as large surface area and effective electron transfer, enhancing the sensitivity and stability of biosensors. Carbon nanotubes, for instance, improve electrical detection because of their electron transfer properties and have been utilized to enhance enzyme performance in catalytic biosensors (Yang et al. 2020).

3.2 Sensitivity of Nanobiosensors

Nanomaterials increase the sensitivity of nanobiosensors through a number of critical properties and mechanisms. The surface-to-volume ratio of nanomaterials is higher, enabling increased immobilization of biomolecules. Increased surface area enables more interaction between the bioreceptor and the analyte (Shandilya et al. 2019). Electrical conductivity of some nanomaterials, like graphene and carbon nanotubes, has greater electrical conductivity, enhancing the sensitivity and response rate of electrochemical biosensors. Graphene, for example, has the lowest

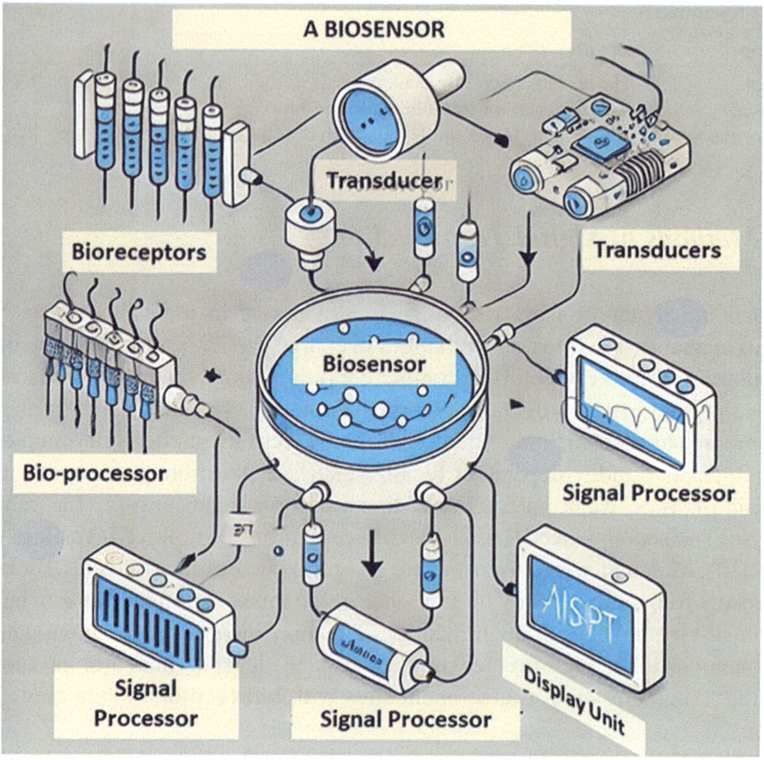

Fig. 2 Working mechanism and different components of biosensors

resistivity, allowing for fast and sensitive measurement (Gan and Hu 2011). Signal amplification nanomaterials are capable of amplifying desired signals, resulting in enhanced biosensor performance, such as greater sensitivities and lower detection limits. They serve as mediators, electroactive species, and detection nanoprobes, enhancing signal transduction. Biocompatibility of nanomaterials can be designed to be biocompatible, which is crucial for investigating biomolecules such as DNA, viruses, and bacteria. Special properties nanomaterials possess special properties such as high thermal and electrical conductivity, which are responsible for the creation of sensitive and selective nanosensors (Wu et al. 2014). Improved electron transfer materials such as carbon nanotubes facilitate electron transfer, which is especially beneficial in catalytic biosensors to enhance enzyme performance.

Depending on the mechanism of transduction employed in biosensors (electrochemical, optical, thermoelectric, etc.), nanomaterials are applied for enhancing performance. Property variations such as conductance, resonant frequency, or absorption/fluorescence upon binding to target analytes are utilized for sensing (Bollella and Katz 2020). Compatibility-for-modification nanomaterials such as graphitic carbon nitride can be adapted to improve their electrochemical biosensing performance. Such adaptations may involve changing the surface to enhance biocompatibility, decrease the cost of production, and enhance the ratio of surface area to volume.

3.3 Utilization of Sensing in Monitoring of Environmental Pollutants: Current Strategies, Requirements, and Benefits

As is known, pollutants are not static in nature, but spread across the environment by anthropogenic and microbial activities. They enter natural cycles, food web, and living systems, and accumulate over time. This causes unpredictable toxic impacts and is especially life-threatening when they accumulate in tissues within living beings causing dysfunctions (Gözde et al. 2022).

The detection of pollution levels in the environment is a critical first step. Although various conventional strategies are in place, such as the chemical test methods, they are unable to keep up with the passing time, and often require expert personnel for handling and operations. Also, not many real-time strategies are available. And in case of lack of expertise, precision and accuracy of results are compromised.

Nanobiosensors are a promise towards resolving these grievances, and hence are known as the best solutions in environmental monitoring. A detailed analysis of environment requires examination of physical, chemical, organic, and biological contaminants, with sensitivity towards each, even at low concentrations. Air and water contaminants were the most analysed, as well as easily polluted by contaminants like phenolic compounds, pesticides, pathogens, gases, and heavy metals (Gözde et al. 2022).

Heavy metals and pesticides are indeed among the most highly researched ones, for which sensors are sought to be developed. Heavy metals are the most concerning contaminants, as they are toxic even at low concentrations. The major sources of heavy metal pollutants are pesticide wastes. Nanobiosensors have bioreceptors as one of the components. This component renders sensitivity and selectivity in detection process. An example of nanobiosensor is a cantilever nanobiosensor having alkaline phosphatase that has been used in river water pollutant detection at micro- and nanoscales. Heavy metals like zinc, mercury, and lead were detected. Such detections can also prove to be a method to deduce the source of the pollutant; for example, abundant concentration of cadmium in the ecosystem suggests exponential agricultural and industrial activity. Monitoring of soil physiology and pollutants is an important aspect of ensuring a healthy agricultural ecosystem (Gözde et al. 2022). This is due to the intertwining of food industry, agricultural ecosystem, and land, air, and water system, which can lead to the entry of contaminants in the food. When these contaminants are heavy metals, upon consumption (by inhalation and ingestion) these metals can replace the essential metals of the body from their chelates, and cause toxicity and dysfunction in the body. Also, there are elements that are essential micronutrients but in excess quantities are toxic to the body (copper and zinc). All this leads to risks of cardiovascular diseases, hypertension, kidney damage, and growth inhibition to name a few. As has been mentioned in Table 1, there are a variety of nanobiosensors that can be used to determine heavy metal contamination.

Another prominent environmental pollutant is agricultural chemicals, like pesticides and herbicides. As is evident, the use of agricultural chemicals is responsible for bioleaching, eutrophication, land quality degradation, etc., while impacting the ecosystem and living organisms. The overuse of nitrogen and phosphorus from agriculture is known to account for over 55% of eutrophication in freshwater bodies in regions of EU. Also, there are estimates that over 4 million tonnes of pesticides are annually used, which are mostly sprayed and may not reach target sites and organisms (95% herbicides and 98% insecticides), which leads to ammonia and nitrogen oxides that pollute water bodies and environment as a whole (Buckley and Carney 2013; Rad et al. 2022; Motasim et al. 2024). These chemicals can enter the environment and even food, as was detected by acetylcholinesterase nanobiosensor (Singh 2021). Likewise, various other nanobiosensors are used in the detection of pesticide pollutants in the environment; for example, the molecularly imprinted polymer is one such material for efficiency (Ambrin et al. 2020). The utilization of optical fluorescent nanobiosensors for soil health monitoring is also on the rise (Thakur et al. 2022).

Water pollution is another prominent issue, especially today, when the world is experiencing an intense shortage of potable water. Also, often there is existence of pathogens and contaminants in the water bodies, which degrade their quality, and make them unsuitable for human use. Industrial activities, sewage discharge, and other human activities render freshwater bodies with loads of waste. It is also no surprise to find agricultural chemical residues in such water bodies. Therefore, fabrication of sensors for water quality monitoring is of utmost importance. DNA-based nanobiosensors are one of those devices used for the same. Conformational changes in the biological component of the DNA-based nanobiosensors are the key in the detection process (Soukarié et al. 2020).

Air pollution is an aspect that needs efficient monitoring, with optical, electrical, and mass sensors being the most appropriate ones. These sensors detect radiations and generate electrical signals to provide information on the presence of pollutants. The use of miniaturized sensors is a cost-effective and efficient way of sensing air pollutants, both indoor and outdoor. The bioreceptor component interacts with the pollutants and generates electrical signals with the transducer system (Costa et al. 2020). One of the most impactful strategies in air pollutant monitoring is continuous monitoring of the same. However, conventional techniques, such as gas chromatography-mass spectroscopy (GC-MS), require extremely sophisticated machines, expensive systems, and expertise, which is not feasible enough. But this is not so in the case of nanobiosensors, such as gas sensors, where the bioreceptors are specialized nanomaterials. These nanomaterials have high surface-to-volume ratio, catalytic properties, and enhanced specificity. Apart from this, advanced nanobiotechnology can also contribute to the elimination of air pollutants (Mohamed and Awad 2022). Thus, advanced nanomaterials can aid in efficient sensing and pollutant elimination of almost all pollutants, from air, water, and soil (Aguilar-Pérez et al. 2020).

The use of nanobiosensors in environmental monitoring is not just about contaminant detection but also includes sensing humidity, temperature, and other atmospheric conditions for a comprehensive overview. And unlike conventional strategies, these versatile devices often require very less sample sizes, as in cantilever nanobiosensors, and also multiplexing the various sensing functionalities can lead to a reduction in time and energy consumption. Also, the utilization of nanomaterials renders the ability to intensify the electrical signals, and better carbon compatibility for conjugation with biomolecules, and has been reported to enhance the performance of biosensors (increase sensitivity and specificity, and decrease detection limit) (Kulkarni et al. 2022).

4 Challenges and Future Prospects

Although nanobiotechnology and nanosciences have witnessed rapid growth, the implementation of such novel strategies requires intense analysis and studies, and may not be feasible everywhere and every time due to constraints of time, technology, and expense. One of the significant hindrances is the immobilization of the biological components used in bioreceptors, without losing their bioactivity. However, immobilization of biomolecules on a nanomatrix has shown feasibility. Yet, other challenges like automation, integration, and multiplexing still do exist. Another challenge can be the inconsistency in nanomaterial synthesis as there are different methods for achieving the same. There are also inconsistencies in biosensor technology itself, especially at the micro level, and also often, sensors are developed for a single purpose, and multiplexing becomes a hefty task for the developer. In addition, commercial-level fabrication and market acceptance also prove to be an obstacle. Thus, all such challenges make an emphasis on strength, weakness, opportunity, and threat (SWOT) analysis for nanobiosensors (Kulkarni et al. 2022). Despite all this, still there can exist performance limits in nanobiosensors that must be determined mathematically (Nair and Alam 2006).

Nanobiosensors are indeed eco-friendly and better remote access devices, along with better chemical reactivity and electrocatalytic properties, which are also being tested to not just detect air pollutants, but also eliminate them. The microchip industry is inspired by similar principles. The enhanced catalytic properties of nanomaterials can prove to be a milestone in air pollutant detection-cum-elimination systems; this can be integrated with technologies like Internet of Things (IoT) and artificial intelligence (AI) for better user interface and automation of the process, leading to concepts like Internet of Bio-Nano Things (IoBNT) and advancing the field of synthetic biology (Rai and Duran 2011; Pramanik et al. 2020; Mohamed and Awad 2022; Kulkarni et al. 2022). The worldwide biosensor market is estimated to grow to around \$47.54 billion by 2030 at a compound annual growth rate (CAGR) of 6.7%. The cost of production is high owing to intricate technology and materials, which can hamper adoption. Regulatory policies are different across the world, with strict standards from regulatory bodies such as the Food and Drug Administration (FDA) and Europe, the Middle East, and Africa (EMEA), requiring intensive testing for efficacy and safety prior to market introduction (https://www.polarismarket-research.com).

The global sensor market is expected to cross the trillion dollar market by 2025, and it is expected that wearable sensors will prove to be the largest players in the same, due to the increasing consumer psychology of desiring to monitor their immediate environment for precautionary measures and health concerns, especially regarding noise and air pollution (Gözde et al. 2022). This shows the potential of the sensor industry, in which nanobiosensors can become a key player, in the near future. Investing in nanobiosensors would prove to be a key to micropollutant detection (Kuswandi 2019) and elimination in the near future.

5 Conclusion

While bringing any technology into the market, it is essential to previously undergo a SWOT test for the product, as the ultimate goal of any technology is commercialization and benefiting mankind. Nanobiosensors, being a multidisciplinary technology, have definitely acquired their own share of strengths and weaknesses, which require detailed analysis. These strengths can be furthered by expanding research on the same. However, it is necessary to test the biocompatibility of any technology before releasing it into the environment.

References

Aguilar-Pérez KM, Heya MS, Parra-Saldívar R, Iqbal HMN (2020) Nano-biomaterials in-focus as sensing/detection cues for environmental pollutants. Case Stud Chem Environ Eng 2:100055. https://doi.org/10.1016/j.cscee.2020.100055

Ambhorkar P, Wang Z, Ko H, et al (2018) Nanowire-Based Biosensors: From Growth to Applications. Micromachines 9:679. https://doi.org/10.3390/mi9120679

Ambrin G, Kausar H, Naz R, Ahmad A (2020) Current Status of Nanosensors in Biological Sciences. In: Mohsin Mohd, Naz R, Ahmad A (eds) Nanobiosensors for Agricultural, Medical and Environmental Applications. Springer Singapore, Singapore, pp. 15–41

Badihi-Mossberg M, Buchner V, Rishpon J (2007) Electrochemical Biosensors for Pollutants in the Environment. Electroanalysis 19:2015–2028. https://doi.org/10.1002/elan.200703946

Baeumner AJ (2003) Biosensors for environmental pollutants and food contaminants. Anal Bioanal Chem 377:434–445. https://doi.org/10.1007/s00216-003-2158-9

Bag S, Mandal D (2023) Overview of Biosensors and Its Application in Health Care. In: Dutta G, Biswas A (eds) Next Generation Smart Nano-Bio-Devices. Springer Nature Singapore, Singapore, pp. 29–60

Barreiros Dos Santos M, Rodriguez-Lorenzo L, Queirós R, Espiña B (2022) Fundamentals of Biosensors and Detection Methods. In: Caballero D, Kundu SC, Reis RL (eds) Microfluidics and Biosensors in Cancer Research. Springer International Publishing, Cham, pp. 3–29

Basu AK, Basu A, Bhattacharya S (2020) Micro/Nano fabricated cantilever based biosensor platform: A review and recent progress. Enzyme Microb Technol 139:109558. https://doi.org/10.1016/j.enzmictec.2020.109558

Bellan LM, Wu D, Langer RS (2011) Current trends in nanobiosensor technology. WIREs Nanomedicine Nanobiotechnology 3:229–246. https://doi.org/10.1002/wnan.136

Bhaskar Mahanayak (2024) Biotransformation reactions of xenobiotics: Mechanisms and implications for environmental and human health. World J Biol Pharm Health Sci 19:158–164. https://doi.org/10.30574/wjbphs.2024.19.1.0406

Bollella P, Katz E (2020) Enzyme-Based Biosensors: Tackling Electron Transfer Issues. Sensors 20:3517. https://doi.org/10.3390/s20123517

Buckley C, Carney P (2013) The potential to reduce the risk of diffuse pollution from agriculture while improving economic performance at farm level. Environ Sci Policy 25:118–126. https://doi.org/10.1016/j.envsci.2012.10.002

Costa D, Sebastián J, Romero CM, et al (2020) Nanoparticles for New Pharmaceuticals: Metabolites from Actinobacteria. In: Dasgupta N, Ranjan S, Lichtfouse E (eds) Environmental Nanotechnology Volume 4. Springer International Publishing, Cham, pp. 195–213

Damborský P, Švitel J, Katrlík J (2016) Optical biosensors. Essays Biochem 60:91–100. https://doi.org/10.1042/EBC20150010

Farzin L, Shamsipur M, Samandari L, Sheibani S (2018) Advances in the design of nanomaterial-based electrochemical affinity and enzymatic biosensors for metabolic biomarkers: A review. Microchim Acta 185:276. https://doi.org/10.1007/s00604-018-2820-8

Gan T, Hu S (2011) Electrochemical sensors based on graphene materials. Microchim Acta 175:1–19. https://doi.org/10.1007/s00604-011-0639-7

Gözde KA, Dulta K, Chauhan P, Chauhan PK (2022) Biomimetic Material-Based Biosensor for Environmental Monitoring. In: Singh RP, Ukhurebor KE, Singh J, et al. (eds) Nanobiosensors for Environmental Monitoring. Springer International Publishing, Cham, pp. 191–202

Hang X, He S, Dong Z, et al (2021) Nanosensors for single cell mechanical interrogation. Biosens Bioelectron 179:113086. https://doi.org/10.1016/j.bios.2021.113086

https://www.polarismarketresearch.com. PMR Biosensors Market Size, Trends and Revenue Forecast, 2034. In: Polaris. https://www.polarismarketresearch.com/industry-analysis/biosensors-market. Accessed 20 Feb 2025

Huang X, Zhu Y, Kianfar E (2021) Nano Biosensors: Properties, applications and electrochemical techniques. J Mater Res Technol 12:1649–1672. https://doi.org/10.1016/j.jmrt.2021.03.048

Javaid M, Haleem A, Singh RP, et al (2021) Exploring the potential of nanosensors: A brief overview. Sens Int 2:100130. https://doi.org/10.1016/j.sintl.2021.100130

Kulkarni MB, Ayachit NH, Aminabhavi TM (2022) Recent Advancements in Nanobiosensors: Current Trends, Challenges, Applications, and Future Scope. Biosensors 12:892. https://doi.org/10.3390/bios12100892

Kuswandi B (2019) Nanobiosensor approaches for pollutant monitoring. Environ Chem Lett 17:975–990. https://doi.org/10.1007/s10311-018-00853-x

Lei J, Ju H (2010) Nanotubes in biosensing. WIREs Nanomedicine Nanobiotechnology 2:496–509. https://doi.org/10.1002/wnan.94

Malik P, Katyal V, Malik V, et al (2013) Nanobiosensors: Concepts and Variations. ISRN Nanomater 2013:1–9. https://doi.org/10.1155/2013/327435

Matlock-Colangelo L, Baeumner AJ (2012) Recent progress in the design of nanofiber-based biosensing devices. Lab Chip 12:2612. https://doi.org/10.1039/c2lc21240d

Mohamed EF, Awad G (2022) Development of nano-sensor and biosensor as an air pollution detection technique for the foreseeable future. In: Comprehensive Analytical Chemistry. Elsevier, pp. 163–188

Motasim AM, Samsuri Abd W, Nabayi A, et al (2024) Urea application in soil: processes, losses, and alternatives—a review. Discov Agric 2:42. https://doi.org/10.1007/s44279-024-00060-z

Nabaei V, Chandrawati R, Heidari H (2018) Magnetic biosensors: Modelling and simulation. Biosens Bioelectron 103:69–86. https://doi.org/10.1016/j.bios.2017.12.023

Nair PR, Alam MA (2006) Performance limits of nanobiosensors. Appl Phys Lett 88:233120. https://doi.org/10.1063/1.2211310
Okon SL, Ronkainen NJ (2017) Enzyme-Based Electrochemical Glutamate Biosensors. In: Rahman MM, Asiri AM (eds) Electrochemical Sensors Technology. InTech
P. Ravindra S. De Silva, Ajith P. Madurapperuma, Masatake Higashi, et al (2010) The Extraction of Symbolic Postures to Transfer Social Cues into Robot. INTECH Open Access Publisher
Pablos JL, Manzano M, Vallet-Regí M (2024) Biosensors. In: Sensory Polymers. Elsevier, pp. 271–307
Pironti C, Ricciardi M, Proto A, et al (2021) Endocrine-Disrupting Compounds: An Overview on Their Occurrence in the Aquatic Environment and Human Exposure. Water 13:1347. https://doi.org/10.3390/w13101347
Pramanik PKD, Solanki A, Debnath A, et al (2020) Advancing Modern Healthcare With Nanotechnology, Nanobiosensors, and Internet of Nano Things: Taxonomies, Applications, Architecture, and Challenges. IEEE Access 8:65230–65266. https://doi.org/10.1109/ACCESS.2020.2984269
Rad SM, Ray AK, Barghi S (2022) Water Pollution and Agriculture Pesticide. Clean Technol 4:1088–1102. https://doi.org/10.3390/cleantechnol4040066
Rai M, Duran N (eds) (2011) Metal Nanoparticles in Microbiology. Springer Berlin Heidelberg, Berlin, Heidelberg
Ramanathan K, Danielsson B (2001) Principles and applications of thermal biosensors. Biosens Bioelectron 16:417–423. https://doi.org/10.1016/S0956-5663(01)00124-5
Ron EZ (2007) Biosensing environmental pollution. Curr Opin Biotechnol 18:252–256. https://doi.org/10.1016/j.copbio.2007.05.005
Saadati A, Hassanpour S, Guardia MDL, et al (2019) Recent advances on application of peptide nucleic acids as a bioreceptor in biosensors development. TrAC Trends Anal Chem 114:56–68. https://doi.org/10.1016/j.trac.2019.02.030
Shandilya R, Bhargava A, Bunkar N, et al (2019) Nanobiosensors: Point-of-care approaches for cancer diagnostics. Biosens Bioelectron 130:147–165. https://doi.org/10.1016/j.bios.2019.01.034
Singh RP (2021) Recent Trends, Prospects, and Challenges of Nanobiosensors in Agriculture. In: Pudake RN, Jain U, Kole C (eds) Biosensors in Agriculture: Recent Trends and Future Perspectives. Springer International Publishing, Cham, pp. 3–13
Soukarié D, Ecochard V, Salomé L (2020) DNA-based nanobiosensors for monitoring of water quality. Int J Hyg Environ Health 226:113485. https://doi.org/10.1016/j.ijheh.2020.113485
Suzuki T, Hidaka T, Kumagai Y, Yamamoto M (2020) Environmental pollutants and the immune response. Nat Immunol 21:1486–1495. https://doi.org/10.1038/s41590-020-0802-6
Syedmoradi L, Ahmadi A, Norton ML, Omidfar K (2019) A review on nanomaterial-based field effect transistor technology for biomarker detection. Microchim Acta 186:739. https://doi.org/10.1007/s00604-019-3850-6
Thakur M, Wang B, Verma ML (2022) Development and applications of nanobiosensors for sustainable agricultural and food industries: Recent developments, challenges and perspectives. Environ Technol Innov 26:102371. https://doi.org/10.1016/j.eti.2022.102371
Uddin MM, Zakeel MCM, Zavahir JS, et al (2021) Heavy Metal Accumulation in Rice and Aquatic Plants Used as Human Food: A General Review. Toxics 9:360. https://doi.org/10.3390/toxics9120360
Vasilantonakis N, Wurtz GA, Zayats AV (2020) Refractive Index Sensing with Anisotropic Hyperbolic Metamaterials. In: De La Rue R, Herzig HP, Gerken M (eds) Biomedical Optical Sensors. Springer International Publishing, Cham, pp. 81–107
Voiculescu I, Nordin AN (2012) Acoustic wave based MEMS devices for biosensing applications. Biosens Bioelectron 33:1–9. https://doi.org/10.1016/j.bios.2011.12.041
Wanekaya AK, Chen W, Myung NV, Mulchandani A (2006) Nanowire-Based Electrochemical Biosensors. Electroanalysis 18:533–550. https://doi.org/10.1002/elan.200503449
Wang J (2016) Self-propelled affinity biosensors: Moving the receptor around the sample. Biosens Bioelectron 76:234–242. https://doi.org/10.1016/j.bios.2015.04.095

Wasi S, Tabrez S, Ahmad M (2013) Toxicological effects of major environmental pollutants: an overview. Environ Monit Assess 185:2585–2593. https://doi.org/10.1007/s10661-012-2732-8

Wu L, Xiong E, Zhang X, et al (2014) Nanomaterials as signal amplification elements in DNA-based electrochemical sensing. Nano Today 9:197–211. https://doi.org/10.1016/j.nantod.2014.04.002

Yang T, Zhan L, Huang CZ (2020) Recent insights into functionalized electrospun nanofibrous films for chemo-/bio-sensors. TrAC Trends Anal Chem 124:115813. https://doi.org/10.1016/j.trac.2020.115813

Zhang C-Y, Yeh H-C, Kuroki MT, Wang T-H (2005) Single-quantum-dot-based DNA nanosensor. Nat Mater 4:826–831. https://doi.org/10.1038/nmat1508

Nanobiosensors in Environmental Pollution Control: Innovations and Future Prospects

Chandra Mukherjee, Debabrata Misra, and Vivekananda Mandal

Abstract In recent years, there has been tremendous progress in the development of nanobiosensors, which are a critical need of the hour and could be integrated with point-of-care testing for various applications, including healthcare, the environment, energy harvesting, electronics, and the food industry. In order to create biosensors with high selectivity, uniqueness, and sensitivity, nanomaterials play a crucial role in effectively sensing bioreceptors like cells, enzymes, and antibodies. Without nanomaterials, it is nearly impossible to carry out any application in science and technology. Nanomaterials' unique electrical, thermal, chemical, optical, mechanical, and physical properties set them apart from tiny particles employed in a wide range of applications. There is a growing global demand for a secure workplace as people become more health conscious. This chapter briefly reviewed environmental pollution issues, such as the classification of pollutants and the specific nanosensors used to detect them. This will also encompass the recent breakthroughs in advanced nanotechnology and nanobiotechnology applications in environmental pollutant detection using nanobiosensors for air pollutants and sensing environmental pathogenic microbes that could help control environmental pollution and could reduce costs and save time. However, for large-scale applications, the nanotechnology industries should promote safety regulations, including developing biosources related to nanostructures, to make this technology commercially viable and to develop advanced technologies, such as environmentally friendly materials and green systems, that are compatible with environmental requirements. In order to address the demand for vast quantities of highly pure, structurally well-identified, and precisely functionalized nanomaterials, green nanofabrication also needs to be aggressively pursued.

C. Mukherjee
Department of Basic Science (Chemistry), The Neotia University, Sarisa, West Bengal, India
e-mail: chandra.mukherjee@tnu.in

D. Misra (✉) · V. Mandal
Plant and Microbial Physiology and Biochemistry Laboratory, Department of Botany, University of Gour Banga, Mokdumpur, Malda, West Bengal, India
e-mail: vivek.bot@ugb.ac.in

S. Roy, V. Mandal (eds.), *Plant-Microbe Interaction under Xenobiotic Exposure*,
https://doi.org/10.1007/978-981-96-8260-7_24

Keywords Environment · Nanomaterials · Nanosensors · Xenobiotics

1 Introduction

Nanobiosensor technology is the broad term for integrating biosensors and nanostructured materials. Furthermore, the advancement of nano/biotechnology and its impact on the design of portable ultrasensitive devices is likely one of the most effective means of resolving the aforementioned issues, as well as the cumulative need for the development of a quick, inexpensive, and extremely sensitive instrument for analyzing environmental pollutants, particularly xenobiotics. The environmentally friendly synthesis of nanosensors entails using green chemistry concepts to reduce environmental impact, limit the use of hazardous ingredients, and increase sustainability. Figures 1 and 2 show the typical methodologies and strategies for producing eco-friendly nanosensors. To improve biocompatibility and sustainability, reactions are generally conducted in solvent-free conditions where feasible or by replacing harmful organic solvents with water or ionic liquids to reduce toxicity using renewable precursors such as bio-based polymers (e.g., chitosan and cellulose) or agricultural waste for nanoparticle synthesis and stabilization (Fuster et al. 2023). Synthetic pathways are designed to ensure minimal production of hazardous by-products, and cytotoxicity is reduced by coating or by incorporating benign additives. Processes are employed to allow recovery of materials used in synthesis

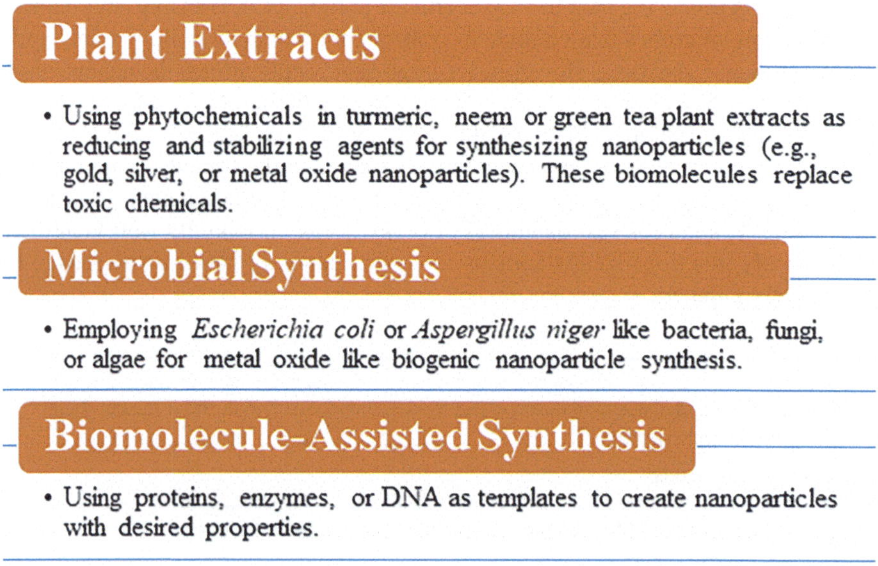

Fig. 1 Green synthesis of nanosensors

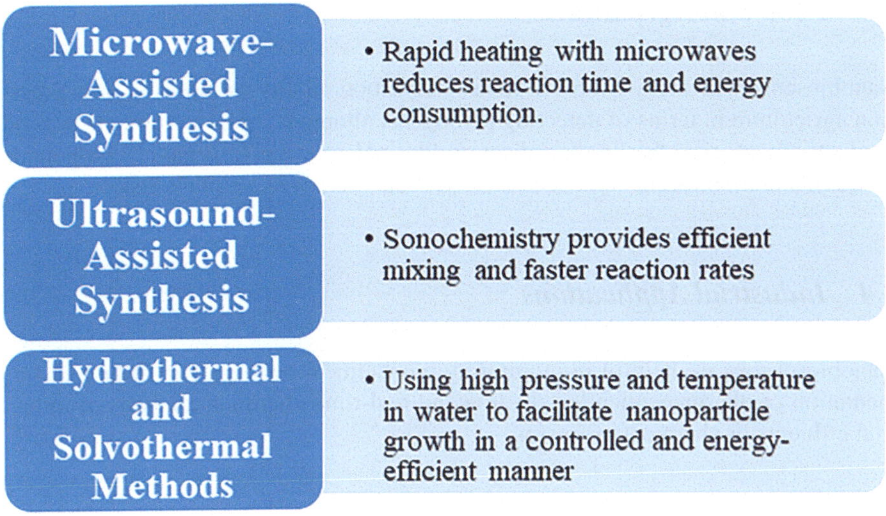

Fig. 2 Energy-efficient techniques of nanosensor synthesis

(e.g., solvents and reducing agents), and catalysts and nanosensors are developed that can be recovered and reused to minimize waste generation (Nyabadza et al. 2023).

Eco-friendly synthesized biocompatible nanosensors have diverse applications across multiple fields due to their high sensitivity, specificity, and rapid response. The following are some typical applications of nanobiosensors:

1.1 Medical Diagnostics

Nanobiosensors are used in detecting biomarkers for diseases, such as cancer, diabetes, or cardiovascular conditions. For example, glucose nanosensors use graphene oxide for diabetes monitoring. Nanosensors can also identify pathogenic bacteria or viruses, including real-time detection of COVID-19. Point-of-care devices use portable biosensors for onsite diagnostics.

1.2 Environmental Monitoring

Nanobiosensors can sense pollutants like heavy metals, pesticides, or toxic gases in water, air, and soil and biological contaminants like microbes in environmental samples. For example, nanobiosensors based on TiO_2 nanoparticles can detect arsenic and mercury ions in water through colorimetric changes.

1.3 Food and Agriculture

Nanobiosensors are very useful in food safety, food quality monitoring, and precision agriculture in terms of detecting pathogens, allergens, or contaminants in food products; measuring freshness indicators like pH or volatile organic compounds (VOCs); and monitoring soil conditions, pesticide residues, or plant health.

1.4 Industrial Applications

Nanobiosensors are helpful in monitoring biochemical processes in industrial fermentation or pharmaceutical production and real-time chemical analysis of industrial effluents or chemical processes.

1.5 Defense and Security

Nanobiosensors can identify bio-weapons, or toxic agents and explosives in trace amount.

2 Key Environmental Applications and Advantages of Nanobiosensors

Nanobiosensors can accurately detect pollutants by combining tailored recognition elements with efficient transduction mechanisms, making them an effective and long-term tool for addressing environmental issues, ensuring clean water, air, and soil, and contributing to global sustainability goals (Sharma et al. 2025).

2.1 Water Quality Monitoring

Nanobiosensors are used in the detection of heavy metals (e.g., arsenic, mercury, and cadmium), monitoring pesticide residues, nitrates, and phosphates, and detection of pathogens like *E. coli* or viruses in water.

2.2 Air Quality Monitoring

Nanosensors can detect toxic gases (e.g., CO, NOx, SO_2) and monitor volatile organic compounds (VOCs).

2.3 Soil and Agricultural Applications

Recently they have been used in monitoring pesticide residues and heavy metals in soil and also in assessing microbial activity and soil quality.

2.4 Industrial Effluent Monitoring

Industrial applications of nanosensors include the detection of dyes, pharmaceutical residues, and toxic chemicals in wastewater and the monitoring of thermal pollution and organic contaminants.

2.5 Ecosystem Health Assessment

Nanobiosensors can assess ecosystem health status by detecting harmful microalgal blooms (HABs) and toxins in aquatic environments and by monitoring endocrine-disrupting compounds affecting wildlife.

The main advantages of nanobiosensors in the prediction of environmental pollutants are as follows (Javaid et al. 2021):

(i) Selectivity: Sensors are functionalized with specific elements (e.g., enzymes, antibodies, and nanomaterials) to interact only with the target pollutant.
(ii) High Sensitivity: Nanobiosensors can detect pollutants even at ultra-low concentrations, are critical for trace analysis, and enhance detection capabilities by amplifying the signal.
(iii) Real-time Monitoring: Advanced sensors offer rapid detection and immediate feedback capabilities.
(iv) Portability: Many environmental sensors are compact and easy to deploy in field conditions.
(v) Versatility: Can detect a wide range of environmental pollutants.

3 Detection of Diverse Types of Pollutants: Chemical as Well as Pathogenic

Indeed, nanosensors can efficiently detect contaminants. Nanosensors detect even minute amounts of chemical pollutants (like heavy metals, gases, and organic compounds) and biological pollutants (like pathogens and toxins). The effectiveness of nanosensors in detecting pollutants is based on the synthesis and characterization techniques methodology and how their functionalization and sensing mechanisms are optimized for various types of pollutants. In order to identify particular biological or chemical substances with high sensitivity and selectivity, nanobiosensors are analytical tools that integrate biological recognition elements (such as enzymes, antibodies, and DNA) with nanomaterials. In comparison to traditional biosensors, their nanoscale characteristics offer improved surface-to-volume ratios, distinct optical, electrical, and catalytic properties, special qualities like excellent conductivity, enhanced sensitivity, and the capacity to interact with both chemical and biological substances, and quicker reaction times (Huang et al. 2021).

Sensors detect pollutants via precise interactions between the pollutant (analyte) and a sensing device, resulting in a detectable signal (Huang et al. 2023). The method typically consists of three major steps:

3.1 Recognition

The sensor's active element detects the pollutant by physical, chemical, or biological interactions.

3.2 Transduction

This is the conversion of an interaction into a detectable signal, such as electrical, optical, or thermal.

3.3 Signal Processing

The signal is evaluated and quantified to identify pollutant concentrations.

4 Mechanisms of Pollutant Detection

Nanobiosensors are rapidly being used in environmental applications due to their high sensitivity, specificity, and capacity to detect a wide range of pollutants and toxins at extremely low concentrations. Different sensing mechanisms are used depending on the type of sensor and pollutant:

4.1 Chemical Sensing Mechanisms

Sensors can function differently to various chemicals as follows.

4.1.1 Electrochemical Sensors

This type of nanobiosensor uses electrodes to measure changes in electrical properties (current, voltage, resistance) caused by the pollutant. For example, heavy metals (lead, mercury) are detected by using electrodes modified with nanomaterials.

4.1.2 Gas Sensors

They can measure conductivity, capacitance, or resistance changes when gases like CO_2, NOx, or SO_2 interact with the sensor surface. For example, metal oxide sensors (e.g., ZnO or TiO_2) can detect toxic gases through surface reactions (Moulahoum et al. 2023; Mukherjee et al. 2017).

4.2 Optical Sensing Mechanisms

4.2.1 Absorbance or Fluorescence

Pollutants change the optical properties of a sensing element, such as fluorescence intensity or absorbance. Quantum dots fluoresce differently in the presence of pesticides.

4.2.2 Colorimetric Sensors

Visual color changes occur due to chemical reactions with the pollutant. For example, gold nanoparticles change color when detecting arsenic in water.

4.2.3 Surface Plasmon Resonance (SPR)

This type of optical nanosensor measures changes in refractive index caused by the binding of pollutants on a metallic surface. Organic pollutants like phenols can be detected by this sensing method.

4.3 *Biological Sensing Mechanisms*

Following are some of the nanobio sensing mechanisms.

4.3.1 Enzymatic Biosensors

Enzyme-based sensors are used for detecting pesticides like organophosphates. Enzymes catalyze reactions with pollutants, producing measurable by-products (e.g., protons or electrons).

4.3.2 Immunosensors

Antibodies bind specifically to pollutants (e.g., pathogens like *E. coli* in water, toxins), generating measurable signals.

4.3.3 DNA Sensors

Detect genetic material from pollutants, such as microbial contamination in water. DNA aptamers are used for detecting harmful algal blooms.

4.4 *Physical Sensing Mechanisms*

Physical sensing mechanisms include:

4.4.1 Mass-Based Sensors

This type of nanobiosensor measures changes in mass when pollutants bind to the sensor surface. For example, mixed nanoferrites (MNF) (Mukherjee et al. 2024; Srivastava et al. 2020) and quartz crystal microbalance (QCM) sensors for detecting volatile organic compounds (VOCs).

4.4.2 Thermal Sensors

Thermal sensors detect heat changes caused by reactions between the pollutant and the sensor. Greenhouse gases are detected based on such exothermic reactions.

4.4.3 Magnetic Sensors

Magnetic sensors use magnetic nanoparticles that bind to specific pollutants and detect changes in magnetic properties such as heavy metals are detected by using superparamagnetic nanoparticles (Skládal 2024).

The various types of nanobiosensors having environmental applications are classified according to their working principles, nanomaterials, and environmental applications, as shown in Table 1.

5 Designing and Creating Compatible Nanosensors to Detect Diverse Pollutants

Nanosensors can detect pollutants with high efficiency, detecting even trace amounts of chemical pollutants (such as heavy metals, gases, and organic compounds) and biological pollutants (such as pathogens and toxins) due to their unique properties, which include a high surface-to-volume ratio, sensitivity, and the ability to interact with both chemical and biological substances. The efficiency of nanosensors in detecting pollutants is determined by the approach used in their synthesis and characterization, as well as how their functionalization and sensing mechanisms are tuned for different types of pollutants (Javaid et al. 2021). In order to obtain high specificity and sensitivity for various pollutants, nanomaterials with customized recognition elements and transduction processes must be included in the design and development of nanosensors that can detect a wide range of pollutants.

The following techniques can be used to create nanosensors that can identify a variety of contaminants.

5.1 Selection of Nanomaterials

The synthesis of nanobiosensors involves integrating nanomaterials with biological components. The synthesis can be categorized based on the type of nanomaterial and its functionalization. Different nanomaterials offer distinct properties that make them suitable for detecting different pollutants, so the choice of nanomaterial plays a crucial role in the sensor's ability to detect various pollutants (Asghar et al. 2024). By incorporating nanomaterials, sensors can achieve enhanced capabilities in terms of superior performance, including lower detection limits, faster response times, and improved selectivity through the following mechanisms:

(i) Signal Amplification: High surface area increases interaction with pollutants, amplifying detectable signals (e.g., electrical and optical).
(ii) Catalytic Activity: Some nanomaterials (e.g., metal oxides) catalyze reactions with pollutants, making them detectable.

Table 1 Different types of nanobiosensors in environmental applications

Nanobiosensors type	Working principle	Nanomaterials used	Applications	References
Electrochemical nanobiosensors	Detect environmental analytes based on changes in electrical signals (e.g., current, voltage, impedance) when a target interacts with the biosensor.	Carbon nanotubes (CNTs), graphene, gold nanoparticles, and metal oxides (ZnO, TiO_2).	Heavy metal detection (e.g., lead, cadmium), monitoring pesticide residues in water and soil, and detecting phenolic compounds and nitrates in water bodies.	Khanna (2012)
Optical nanobiosensors	Changes in optical properties such as fluorescence, surface plasmon resonance (SPR), or absorbance can be used for detection.	Quantum dots, silver and gold nanoparticles, and plasmonic nanostructures.	Monitoring pollutants like pesticides and herbicides, detecting organic pollutants such as dyes or pharmaceuticals in wastewater, and real-time detection of pathogenic microorganisms in water.	Zhang et al. (2023)
Piezoelectric nanobiosensors	Measure mass or resonance frequency changes on the sensor's surface when the analyte binds to the recognition element.	Zinc oxide nanowires, gold-coated quartz crystals.	Detection of volatile organic compounds (VOCs) in air; monitoring microbial contamination in soil and water.	Alberti et al. (2021)
Thermal nanobiosensors	Detect analytes by measuring heat changes during biological interactions, such as enzyme-substrate reactions.	Nanostructured polymers, gold nanoparticles.	Detection of toxic gases and greenhouse gases (e.g., CO_2, CH_2) and monitoring industrial effluents for thermal pollution.	Srivastava et al. (2024)
Magnetic nanobiosensors	Use magnetic nanoparticles to detect analytes based on changes in magnetic fields or relaxation times.	Superparamagnetic iron oxide nanoparticles (SPIONs).	Monitoring pollutants in water via magnetic separation techniques and detecting algal toxins in water systems.	Aguilar-Pérez et al. (2020)
Colorimetric nanobiosensors	Change color visually or spectroscopically upon interaction with the target analyte.	Gold nanoparticles, silver nanoparticles, and carbon dots.	Detection of arsenic, mercury, or lead in drinking water; monitoring pesticides and herbicides in agricultural runoff; and identifying microbial contamination in water.	Beni and Jabbari (2022)
Surface-enhanced Raman scattering (SERS) nanobiosensors	Enhance Raman signals of target molecules using plasmonic nanostructures.	Gold and silver nanoparticles, nanostars, and nanorods.	Trace detection of pollutants like dyes, drugs, or pesticides in water and detection of nanoplastics and microplastics in aquatic environments.	Zarejousheghani et al. (2021)

(iii) Selective Binding: Functionalized nanomaterials (e.g., with antibodies, enzymes, or aptamers) ensure selective interaction with specific pollutants.

Some common nanomaterials and their roles in pollutant detection include (Kumar et al. 2024; Sargazi et al. 2022):

5.1.1 Nanomaterials in Detecting Chemical Pollutants

Nanomaterials that are functionalized to interact specifically with chemical pollutants, enabling their detection through various sensing mechanisms, are:

5.1.1.1 Metal Nanoparticles (e.g., Gold, Silver, and Copper)

Properties High surface area, plasmonic effects.

Applications

(i) Used in colorimetric sensors to detect heavy metals like mercury, arsenic, or lead. These metals cause changes in the optical properties of gold or silver nanoparticles, resulting in visible color changes.
(ii) Enhance the sensitivity of surface-enhanced Raman scattering (SERS) sensors for detecting trace organic pollutants (e.g., pesticides and dyes).
(iii) Used in optical sensors due to their plasmonic properties (e.g., surface plasmon resonance or SPR).
(iv) It can be functionalized with chemical ligands to detect pollutants like heavy metals.

5.1.1.2 Metal Oxide Nanoparticles (e.g., ZnO, TiO_2, and SnO_2)

Properties High sensitivity to gases, photocatalytic properties.

Applications

(i) Used in gas sensors to detect pollutants like NOx, NH_2, CO, and SO_2. These nanoparticles interact with gas molecules, causing measurable electrical conductivity or resistance changes.
(ii) Detect organic pollutants (e.g., phenols) through photocatalytic degradation and optical signal changes.

5.1.1.3 Carbon-Based Nanomaterials, e.g., Graphene, Carbon Nanotubes (CNTs)

Properties Excellent electrical conductivity and high mechanical strength.

Applications

(i) Used in electrochemical sensors to detect heavy metals, pesticides, and VOCs by measuring changes in current or voltage during pollutant binding.
(ii) Functionalized graphene sheets or carbon nanotubes are highly effective in detecting nitrates or phosphates in water systems.
(iii) They are known for their high electrical conductivity and large surface area.
(iv) It can detect gases like carbon monoxide, ammonia, and nitrogen dioxide through changes in their electrical conductivity.

5.1.1.4 Quantum Dots (QDs)

Properties High photoluminescence size-tunable optical properties.

Applications

(i) Used in fluorescent sensors to detect pesticides and herbicides in water and soil by observing quenching or enhancement of fluorescence.
(ii) Fluorescent nanoparticles can be engineered to bind to specific molecules.

Examples of Sensor Applications for Chemical Pollutants

(i) Gold nanoparticle-based colorimetric sensors for arsenic detection in water.
(ii) Graphene-based electrochemical sensors for VOCs in industrial emissions.

5.1.2 Nanomaterials in Detecting Pathogenic Pollutants

Nanomaterials enhance the interaction between the sensor's recognition element and pathogens, allowing for precise and rapid detection.

5.1.2.1 Metal Nanoparticles

Properties Enhanced catalytic activity, antimicrobial effects, and tunable optical properties

Applications

(i) Gold nanoparticles are functionalized with antibodies or DNA probes to detect pathogens like *E. coli*, *Salmonella*, or viruses (e.g., COVID-19) using colorimetric or electrochemical sensors.
(ii) Silver nanoparticles exhibit antimicrobial properties and are used in biosensors to detect bacterial contamination.
(iii) Metal nanoparticles can be functionalized with specific bioreceptors (e.g., antibodies or aptamers) for pathogen detection.

5.1.2.2 Magnetic Nanoparticles (e.g., F_2O_3)

Properties Superparamagnetic behavior, biocompatibility.

Applications
(i) Used in immunosensors for isolating and detecting pathogens through magnetic separation and signal generation.
(ii) Facilitate rapid pathogen concentration and detection in water or food samples.

5.1.2.3 Carbon-Based Nanomaterials

Applications
(i) Graphene oxide is used in DNA biosensors to detect microbial contamination through hybridization with pathogen-specific DNA sequences.
(ii) CNTs and graphene can be functionalized with biochemicals or biological receptors to enhance selectivity.
(iii) Carbon nanotubes enhance the performance of electrochemical sensors for detecting bacteria like *E. coli* or *Staphylococcus* species.

5.1.2.4 Nanostructured Polymers

Properties Tunable porosity, biofunctionalization.

Applications
(i) Nanostructured polymers are integrated with biosensors to capture and detect pathogens like viruses and bacteria based on size exclusion or affinity interactions.
(ii) Polymers at the nanoscale can be engineered to interact with specific chemicals or biological agents and change properties such as color or optical absorbance.

5.1.2.5 Quantum Dots (QDs)

Applications Used to detect toxins and pathogens by observing fluorescence intensity shifts.

Examples of Sensor Applications for Pathogenic Pollutants
(i) Magnetic nanoparticle-enhanced immunosensors for detecting *E. coli* in drinking water.
(ii) Quantum dot-based fluorescent sensors for rapid detection of viruses in medical diagnostics.

5.2 *Functionalization with Specific Recognition Elements*

Nanosensors are frequently functionalized with particular receptors that engage selectively with the target pollutant to detect various contaminants (Sargazi et al. 2022). These components of recognition could consist of:

5.2.1 Biorecognition Elements

These are antibodies and enzymes that specifically bind to pathogens or toxic substances. Example: Antibodies can be used to detect bacteria or viruses, while enzymes can be used to detect pesticides or pollutants in water.

5.2.2 Molecular Imprinting

In order to enable the sensor to identify and attach to the pollutant precisely, a "template" of the target molecule is created using molecular imprinting technology. For instance, when a particular toxin is present, a polymer matrix imprinted with its structure can specifically collect and detect it.

5.2.3 Chemical Receptors

Chemical groups, such as thiols, amines, or carboxyl groups, can be affixed to the surface of nanomaterials to preferentially bond with particular molecules, thereby aiding in the detection of chemical contaminants, such as gases and heavy metals. For instance, thiol groups on gold nanoparticles have the ability to bind to mercury ions selectively, changing the sensor in a way that can be measured. Multi-walled carbon nanotubes (MWCNTs) have 90% removal efficiency for toluene removal with π-π interaction.

5.3 *Integration of Transduction Mechanisms*

Transduction transforms a pollutant's interaction with the identification element into a detectable signal. Various transduction techniques are employed based on the kind of pollutant:

5.3.1 Electrochemical Transduction

Current, voltage, or impedance changes can be measured upon interaction with the pollutant. Example: Carbon nanotube-based electrochemical sensors can detect heavy metals like lead, cadmium, or arsenic by measuring changes in electrical resistance.

5.3.2 Optical Transduction

Nanomaterials can exhibit fluorescence, colorimetric changes, or surface plasmon resonance (SPR) in response to pollutants. Example: Gold nanoparticles (AuNPs) can be used in colorimetric sensors to detect heavy metals. A color change occurs when the metal ions bind to the nanoparticle surface.

5.3.3 Mass-Based Transduction

Changes in frequency or resonance due to the adsorption of pollutants on the sensor surface can be detected. Example: Quartz crystal microbalance (QCM) sensors detect VOCs or gases by measuring the shift in resonance frequency when the pollutant adsorbs on the sensor surface.

5.3.4 Thermal Transduction

Changes in temperature or heat generated during a chemical reaction with the pollutant can be measured. Example: Metal oxide semiconductors can detect gases like methane by catalyzing a reaction that generates heat.

5.4 Fabrication

Methods of nanosensor fabrication are of two types:

5.4.1 Top-Down Approach

Techniques like lithography and etching are used to create nanoscale structures.

5.4.2 Bottom-Up Approach

Nanomaterials are synthesized atom by atom or molecule by molecule, such as in self-assembly or templated growth.

5.5 Multiplexing (Multifunctional Sensors)

It is possible to create multifunctional nanosensors that can detect several contaminants simultaneously. These sensors combine nanomaterials, recognition components, and transduction mechanisms to detect various contaminants simultaneously. Example: A multi-sensor array using graphene to detect gases, gold nanoparticles to detect heavy metals, and quantum dots (QDs) to detect pathogens can be employed to monitor air, water, and food for various contaminants.

6 Sensing Mechanisms of Nanobiosensors to Detect Pollutants

Nanosensors can detect pollutants through various mechanisms, which depend on the nature of the pollutant and the sensing approach. Here are the main ways in which nanosensors can detect pollutants:

6.1 Chemical Interaction

Nanosensors can detect pollutants like gases, heavy metals, and organic chemicals by using nanomaterials (e.g., nanoparticles, nanotubes, or nanowires) that interact with the pollutants at the surface. These interactions result in measurable changes in the sensor's properties, such as:

(i) Electrical conductivity
(ii) Optical properties (e.g., color change, fluorescence)
(iii) Mass (e.g., change in frequency in mass-sensitive sensors)

Example: Metal oxide nanoparticles (e.g., ZnO or SnO_2) can detect VOCs by altering their conductivity upon gas interaction in varied proportions.

6.2 *Biological Detection*

For pathogen detection, nanosensors can utilize specific bioreceptors (e.g., antibodies, aptamers, or enzymes) that bind to pathogens or toxins, triggering a measurable signal change which may be affected by the change in pH.

Example: Gold nanoparticles (AuNPs) conjugated with specific antibodies can bind to *E. coli*, causing a color change that can be detected visually.

6.3 *Optical Changes*

Many nanosensors use optical transduction, where a pollutant interaction causes a change in the optical properties (e.g., fluorescence, surface plasmon resonance) of the nanomaterial.

Example: Quantum dots (QDs) are semiconductor nanoparticles that can fluoresce in response to specific chemicals or biological agents by shifting fluorescence intensity or wavelength.

6.4 *Mass-Based Detection*

Nanosensors like Quartz Crystal Microbalance (QCM) and Surface Acoustic Wave (SAW) sensors detect pollutants by measuring shifts in mass or frequency when pollutants interact with the sensor surface.

Example: Carbon nanotubes (CNTs)-based sensors can detect volatile organic compounds (VOCs) by measuring frequency shifts caused by the adsorption of the VOCs on the CNTs.

7 Nanobiosensor as an Air Pollution Detection Technique: Oxides of Carbon Sensing, Oxides of Nitrogen Sensing, VOCs Sensing, Airborne Pathogen Detection

Nanobiosensors provide an innovative and highly effective approach for detecting various air pollutants, including carbon oxides, nitrogen oxides, VOCs, and airborne pathogens. They are perfect for onsite and real-time pollution monitoring because of their special qualities, which include large surface area, bio capabilities, and quick reaction times. Environmental monitoring, public health, and safety depend on the sensitive, selective, and quick detection of contaminants made possible by combining nanomaterials like metal oxides, carbon nanotubes, and graphene with biorecognition components.

Table 2 Nanobiosensor applications for air pollution detection

Pollutant	Nanomaterials used	Detection mechanism	Advantages	References
Carbon oxides (CO & CO_2)	Tin oxide (SnO), graphene oxide (GO), carbon nanotubes (CNTs)	Chemiresistive, optical, electrochemical	High sensitivity to low concentrations, rapid response	Reshma et al. (2024), Najafi and Ghaemi (2024)
Nitrogen oxides (NO & NO_2)	Metal oxide nanoparticles (ZnO, TiO, SnO), graphene	Chemiresistive, surface plasmon resonance	High selectivity and sensitivity, real-time detection	Zhang et al. (2024)
VOCs chemiresistive, capacitive (benzene, toluene, formaldehyde)	Carbon nanotubes (CNTs), graphene oxide (GO), metal oxide nanostructures, metal ferrites	Co-precipitation, sol-gel,	Detect a wide range of VOCs, sensitive to low concentrations	Feng et al. (2024) and Kulkarni et al. (2022)
Airborne pathogens	Gold nanoparticles (AuNPs), carbon nanotubes (CNTs), quantum dots (QDs), titanium dioxide (TiO_2)	Biorecognition (antibodies, aptamers), optical, electrical	High specificity, real-time detection, low-cost portable sensors	Akgönüllü and Denizli (2022) and Malik et al. (2023)

Because of their great sensitivity, selectivity, and real-time operation, nanobiosensors present a viable method for detecting air pollution. To detect different pollutants, such as carbon oxides (COx), nitrogen oxides (NOx), volatile organic compounds (VOCs), and airborne pathogens, they use nanomaterials like nanoparticles, nanotubes, and nanowires. Here, we examine the use of nanobiosensors in the detection of these contaminants as well as the function of various nanomaterials in each sensing application (Table 2).

Monitoring air pollution, especially in metropolitan areas and industrial emissions, depends heavily on detecting low amounts of gaseous pollutants and airborne pathogens, which can be effectively detected using nanobiosensors.

8 Nanobiosensor in Water Quality Monitoring: Sensing Metal Ions, Chemicals, Drugs, and Pathogens

The capacity of nanobiosensors to identify a variety of contaminants at low concentrations has made them highly effective instruments for monitoring water quality. These sensors provide real time, onsite, and extremely sensitive monitoring by detecting metal ions, chemicals, medicines, and pathogens in water using nanomaterials and biorecognition components. Below, we examine how metal ions, chemicals, medications, and infections may be detected using nanobiosensors in water quality monitoring (Table 3).

Table 3 Nanobiosensor applications in water quality monitoring

Pollutant	Nanomaterials used	Detection mechanism	Advantages	References
Metal ions (Pb^{2+}, Hg^{2+}, Cd^{2+}, As^{3+}, Cu^{2+})	Gold nanoparticles (AuNPs), graphene, CNTs, ZnO, SnO	Chemiresistive, optical, electrochemical	High sensitivity to low concentrations, real-time detection	Kumalasari et al. (2024)
Chemicals and drugs (pesticides, antibiotics)	Quantum dots (QDs), gold nanoparticles (AuNPs), graphene, CNTs	Fluorescence, surface plasmon resonance (SPR), electrochemical	Detect a wide range of chemicals and drugs, sensitive and selective	Abdellatif et al. (2022)
Pathogens (*E. coli*, *Salmonella*, *Cholera*)	Gold nanoparticles (AuNPs), CNTs, quantum dots (QDs), silver nanoparticles (AgNPs)	Biorecognition (antibodies, aptamers), electrochemical, SERS	High specificity and sensitivity, multiplexing potential	Zhu et al. (2025)

Water quality monitoring is being revolutionized by nanobiosensors, which offer real time, extremely sensitive, and precise detection of various pollutants, such as metal ions, chemicals, pharmaceuticals, and viruses. Their ability to operate with nanomaterials such as gold nanoparticles, graphene, carbon nanotubes, andquantum dots enables detection at low concentrations, making them valuable tools for environmental monitoring and public health protection. The continued development of nanobiosensors will enhance our ability to monitor water quality more efficiently, ensuring safer water supplies and better environmental health management.

9 Nanobiosensors in Sustainable Agriculture and Soil Health

Nanobiosensors are becoming increasingly crucial for managing soil health and sustainable agriculture. These sensors make it possible to precisely monitor environmental elements, plant health, and soil conditions, which promotes more sustainable agricultural methods, efficient resource usage, and less environmental impact. Using nanomaterials and biorecognition components, nanobiosensors can identify various environmental factors, including pesticide residues, soil pathogens, plant diseases, and soil nutrients and moisture levels. This part examines how nanobiosensors might improve soil health and agricultural sustainability.

9.1 Nanobiosensors for Soil Nutrient Monitoring

To maintain soil fertility and maximize agricultural output, it is crucial to track soil nutrients, including nitrogen (N), phosphorus (P), potassium (K), and micronutrients (such as iron, zinc, and copper). Farmers may better monitor soil health by using nanobiosensors, which offer a way to assess nutrient levels in real time and with high precision.

9.1.1 Mechanisms for Nutrient Detection

Nutrients may be detected by the nanobiosensors through any of the following sensing mechanisms.

9.1.1.1 Electrochemical Sensing

Electrochemical sensors frequently employ nanomaterials such as graphene, gold nanoparticles (AuNPs), and carbon nanotubes (CNTs). By monitoring changes in current or voltage upon interaction with the target nutrient, these sensors are able to identify nutrient ions in the soil, including phosphate (PO_4^{3-}), ammonium (NH_4^+), and nitrate (NO_4^-).

9.1.1.2 Optical Sensing

Gold nanoparticles (AuNPs) and quantum dots (QDs) can be functionalized with specific ligands that attach to nutrient ions, producing colorimetric or fluorescent changes that are simple to identify and measure.

9.1.1.3 Field-Effect Transistor (FET) Sensors

FET-based sensors can detect nutritional ions by using nanomaterials such as graphene oxide (GO) and carbon nanotubes (CNTs). These sensors provide real-time information on the availability of nutrients by altering their electrical characteristics in response to particular ions.

Nanomaterials Used

(i) Graphene oxide (GO): Used in FET-based sensors to detect nitrogen, phosphorus, and other important nutrients.
(ii) Gold nanoparticles (AuNPs): Functionalized with ligands to detect specific nutrients in the soil, providing colorimetric or fluorescent signals.
(iii) Carbon nanotubes (CNTs): High surface area and electrical conductivity make CNTs suitable for detecting soil nutrients via electrochemical methods.

Advantages

(i) Real time, onsite monitoring of soil nutrients.
(ii) High sensitivity and specificity, allowing detection of low concentrations.
(iii) Reduced need for chemical fertilizers by providing precise nutrient measurements.

9.2 *Nanobiosensors for Soil Pathogen Detection*

Rhizoctonia, Pythium, and Fusarium are a few examples of soilborne diseases that can majorly affect soil health and crop output. By enabling prompt response, early identification of these infections lessens the need for chemical pesticides and protects the environment in the following manner.

9.2.1 Biorecognition Elements

Nanobiosensors can be functionalized with antibodies, aptamers, or enzymes that specifically bind to pathogens. When the pathogens interact with these recognition elements, it triggers a measurable change, such as a change in electrical signal, optical properties, or mass.

9.2.2 Surface-Enhanced Raman Spectroscopy (SERS)

Gold nanoparticles (AuNPs) and silver nanoparticles (AgNPs) are often used for SERS-based detection of pathogens. When pathogens bind to these nanoparticles, they enhance the Raman signal, allowing for highly sensitive pathogen detection in soil samples.

9.2.3 Electrochemical Sensors

CNTs, graphene, and AuNPs can be used in electrochemical sensors to detect specific bacterial pathogens or fungi by measuring changes in electrical signals caused by pathogen binding.

Nanomaterials Used

(i) Gold nanoparticles (AuNPs): Functionalized with antibodies or aptamers to specifically detect soil pathogens, such as *Fusarium* and *Pythium.*
(ii) Carbon nanotubes (CNTs): CNTs can be functionalized to detect specific soilborne bacteria or fungi through changes in electrical conductivity.
(iii) Quantum dots (QDs): Used for fluorescent detection of pathogens in soil, providing highly sensitive and specific results.

Advantages

(i) High sensitivity to low concentrations of pathogens.
(ii) Ability to detect a wide range of soilborne pathogens.
(iii) Real time, onsite detection reduces the need for laboratory analysis.

9.3 *Nanobiosensors for Pesticide Residue Detection*

In order to prevent agricultural practices from causing long-term environmental damage, it is essential to identify pesticide residues in soil. To ensure food safety and environmental health, nanobiosensors provide an effective and affordable way to monitor pesticide levels in the soil by the following mechanisms:

9.3.1 Biorecognition Elements

Nanobiosensors can be designed to detect specific pesticide molecules by functionalizing nanomaterials with antibodies, aptamers, or enzymes that bind to the pesticide molecules. The interaction leads to a measurable change, such as a change in optical signal (e.g., colorimetric), electrochemical signal, or mass.

9.3.2 Surface Plasmon Resonance (SPR)

Gold nanoparticles (AuNPs) or silver nanoparticles (AgNPs) can be used in SPR sensors to detect pesticide residues by measuring changes in the refractive index when pesticide molecules bind to the surface of the nanoparticles.

9.3.3 Electrochemical Detection

Nanomaterials like graphene or carbon nanotubes (CNTs) are often used in electrochemical sensors to detect pesticides by measuring changes in voltage or current caused by the interaction of pesticide molecules with the nanomaterial surface.

Nanomaterials Used

(i) Gold nanoparticles (AuNPs): Used in SPR-based sensors for the detection of pesticide residues in soil.
(ii) Carbon nanotubes (CNTs): CNTs can be functionalized with specific receptors for pesticide molecules, providing sensitive electrochemical detection.
(iii) Graphene oxide (GO): Used in electrochemical sensors to detect pesticide residues in the soil through changes in electrical conductivity.

Advantages

(i) Real time, onsite detection of pesticide residues.
(ii) High sensitivity and specificity for a wide range of pesticides.
(iii) Helps reduce pesticide overuse and contamination of the environment.

9.4 *Nanobiosensors for Soil Moisture and Temperature Monitoring*

Soil moisture and temperature are critical factors for plant growth and crop productivity. Monitoring these factors with the help of sensing mechanisms can optimize irrigation practices and improve the management of water resources in agriculture in the following mechanisms.

9.4.1 Humidity Sensors

Nanomaterials such as carbon nanotubes (CNTs) or graphene oxide (GO) can be used to create humidity sensors that detect changes in water content in the soil. These sensors change their electrical properties as the moisture content of the soil changes.

9.4.2 Thermal Sensors

Nanowires made from materials like silicon or tin oxide (SnO) can be used to detect changes in soil temperature. These sensors measure changes in resistance or voltage as the temperature of the soil fluctuates.

Nanomaterials Used

(i) Carbon nanotubes (CNTs): Used to detect soil moisture due to their high surface area and conductivity properties.
(ii) Graphene oxide (GO) is employed in moisture sensing due to its hydrophilic nature, allowing it to absorb water molecules and change its electrical properties.

Advantages

(i) Real-time monitoring of soil moisture and temperature.
(ii) Helps optimize irrigation practices, reducing water usage and improving crop yield.
(iii) Easy integration into smart farming systems for efficient resource management.

9.5 *Nanobiosensors for Plant Disease Monitoring*

Early detection of plant diseases caused by pathogens like bacteria, fungi, and viruses is crucial for effective disease management and ensuring crop health. Nanobiosensors provide a rapid and cost-effective way to detect plant diseases by the following mechanisms.

9.5.1 Biorecognition Elements

Nanobiosensors can be functionalized with antibodies or aptamers to bind to plant pathogens specifically. When the pathogen interacts with the sensor, it causes a change in optical, electrical, or mass properties.

9.5.2 Optical and Fluorescent Sensing

Gold nanoparticles (AuNPs) and quantum dots (QDs) can be used to detect plant pathogens. These materials emit a fluorescent signal when they bind to pathogen-specific molecules, enabling detection.

Nanomaterials Used

(i) Gold nanoparticles (AuNPs): Functionalized with antibodies to detect pathogens that cause plant diseases.
(ii) Carbon nanotubes (CNTs): CNTs are used in electrochemical sensors to detect plant pathogens through changes in conductivity.
(iii) Quantum dots (QDs): Employed in fluorescent sensing for the detection of plant diseases.

10 Major Challenges in the Production and Application of Nanobiosensors and Its Solutions

Applications for nanobiosensors include industrial operations, healthcare, agriculture, and environmental monitoring. Nevertheless, a number of obstacles stand in the way of the development and use of nanobiosensors, notwithstanding their promise. The production of nanomaterials, their effects on the environment, and regulatory frameworks are only a few of these difficulties. However, advances in technology, better manufacturing techniques, and the creation of transparent regulatory frameworks are all necessary to overcome the difficulties in developing and using nanobiosensors. The following are some of the main obstacles to developing and using nanobiosensors and their solutions.

10.1 Synthesis of Nanomaterials

The synthesis of nanomaterials is crucial for developing functional nanobiosensors, but there are several challenges involved in this process (Table 4).

Table 4 Major challenges with the synthesis of nanomaterials and solutions

Sl No.	Parameter	Challenges	Solutions	References
1.	Scalability	It is challenging to produce nanomaterials on a wide scale while preserving their homogeneity and superior qualities. Scalability is a problem for many synthesis methods, both top-down (like lithography) and bottom-up (like chemical vapor deposition).	Adopting continuous flow synthesis techniques (e.g., microreactors) can help produce large quantities of uniform nanomaterials, overcoming scalability issues.	Nagime and Chandak (2024)
2.	Reproducibility	When synthesizing nanomaterials, it might be difficult to get consistent results. Sensor performance might vary depending on the size, shape, and surface characteristics of the nanomaterials. For the commercial manufacturing of nanobiosensors, batch-to-batch repeatability is essential.	Reproducibility may be guaranteed by standardizing the synthesis processes and enhancing quality control systems. Nanomaterial property variability can be decreased by using automated procedures that provide fine control over reaction conditions.	Nagime and Chandak (2024)
3.	Cost	The production of superior nanomaterials may be prohibitively expensive. For instance, costly raw ingredients, intricate manufacturing processes, and specialized equipment might raise production costs and restrict the broad application of nanobiosensors.	It is essential to develop scalable and affordable synthesis techniques. Strategies, such as green synthesis methods, which employ natural resources like plants or microbes to produce nanomaterials, are being investigated by researchers. This is more ecologically friendly and lowers the cost of electricity and raw materials.	Das and Siddique (2024)
4.	Toxicity of nanomaterials	Certain nanomaterials, especially metals (such as copper, zinc oxide, and silver), may be harmful to the environment and living things. This makes it extremely difficult for them to be used safely and sustainably in biosensing applications, especially in the medical or environmental domains.	Give top priority to the usage of non-toxic and biocompatible nanomaterials, such as silica, carbon-based materials (including graphene and carbon nanotubes), and biodegradable nanoparticles. The ecotoxicology of nanoparticles is being extensively studied in order to find safer substitutes for environmental applications.	Ray et al. (2009) and Mathew et al. (2024)

10.2 Functionalization of Nanomaterials

In order to facilitate selective sensing, functionalization is the process of affixing specific chemical or biological recognition components (such as enzymes, aptamers, or antibodies) to the surface of nanomaterials. However, there are several functionalization-related obstacles, which are tabulated in Table 5.

Table 5 Major challenges with the functionalization of nanomaterials and solutions

Sl No.	Parameter	Challenges	Solution	References
1.	Stability of functional groups	It might be difficult to guarantee the functionalized nanomaterials' stability and activity over time. The sensor's sensitivity or selectivity may be lost due to the degradation of the chemical bonds that hold the recognition components and the nanomaterial surface together.	It is essential to increase the stability of surface-functionalized nanomaterials. Scholars can investigate using more resilient functional groups and coatings that withstand deterioration, such as covalent bonds and self-assembling monolayers that provide longer term stability.	Negahdary and Mabbott (2025)
2.	Non-specific binding	The biosensor may produce false positives or negatives due to non-specific interactions between the recognition element and other environmental chemicals.	Selective coatings or steric shielding agents, such as polyethylene glycol (PEG), can be added to the surface of the nanomaterial to reduce non-specific binding and stop unwanted interactions. Additionally, interference from non-target molecules can be minimized by adjusting the density and orientation of functional groups.	Singh et al. (2024)
3.	Optimal functionalization	Optimal functionalization is frequently difficult to achieve. Low sensitivity can result from having too few functional groups on the surface, whereas steric hindrance can be caused by having too many. It takes great skill to optimize the functionalization procedure to optimize the sensor's performance.	The surface modification can not only allow improving the mechanical, chemical and thermal properties of polymer, silica, inorganic and hybrid nanomaterials, but also provide those new unique features, as well as their biological activity.	Kassie et al. (2024)

10.3 Sensor Performance Issues

Ensuring the optimal performance of nanobiosensors in real-world conditions is essential for their practical application. The performance-related common challenges are mentioned in Table 6.

Table 6 Major challenges with the performance of nanomaterials and solutions

Sl No.	Parameter	Challenges	Solution	References
1.	Sensitivity and detection limits	The purpose of nanobiosensors is to identify target analytes, such as pollutants, infections, or biomarkers, at low concentrations. To increase sensitivity and decrease detection thresholds, further advancements are required, as detection limitations could not always satisfy the requirements of certain applications.	High surface area nanomaterials, such as quantum dots, nanowires, or nanoparticles, can increase a sensor's sensitivity. Additionally, detection can be greatly increased by integrating nanomaterials with microfluidic systems or optical enhancement methods such as plasmonic nanostructures or surface-enhanced Raman spectroscopy (SERS).	Devi et al. (2025)
2.	Specificity	One significant problem is ensuring selective detection. False results might result from the influence of other compounds in complicated biological or environmental samples. It is essential to increase the selectivity of nanobiosensors for the target analytes while preventing interference from other substances.	To increase sensor specificity, functionalize nanobiosensors with highly specific recognition components like aptamers, antibodies, or receptors made for specific target molecules. Selectivity can also be increased, and cross-reactivity decreased, by employing several recognition layers.	Nagime and Chandak (2024)
3.	Long-term stability	Nanobiosensors must continue to perform steadily over time in real-world settings. Sensor performance may be impacted by environmental variables such as pH, humidity, and temperature. Stability over an extended period without regular recalibration or replacement is crucial for practical application.	Nanobiosensors should be made with self-healing mechanisms to prevent sensor deterioration or enclosed in protective materials to maintain long-term stability. The longevity of the sensor will be increased by ensuring that its active parts are shielded from adverse environmental circumstances.	Nagime and Chandak (2024)

10.4 Integration into Practical Systems

Integrating nanobiosensors into practical systems—whether for healthcare diagnostics, agriculture, or environmental monitoring—presents numerous challenges, as shown in Table 7.

Table 7 Major challenges with practical integration of nanomaterials and solutions

Sl No.	Parameter	Challenges	Solution	References
1.	Miniaturization and portability	Sensors must be small and portable to be effective in field applications. It is challenging to reduce the size of sensors without sacrificing their functionality, especially when integrating nanomaterials with electronics or microfluidic systems.	Advances in microelectronics and microfluidics can facilitate the miniaturization of nanobiosensors. Utilizing flexible substrates and low-power designs will enable the development of portable, wearable, or field-deployable sensors for real-time monitoring.	Shukla et al. (2024)
2.	Interfacing with detection instruments	In order to understand the signals produced by the sensor (such as optical or electrical signals), nanobiosensors frequently need sophisticated detection tools or readout systems. It is difficult to create integrated, affordable, and user-friendly devices that can instantly analyze data from nanobiosensors.	To facilitate easier data interpretation, wireless communicationandcloud-based platforms can be integrated with nanobiosensors. This allows users to access and analyze sensor data remotely via smartphones or other portable devices, enabling real-time monitoring.	Chauhan and Saxena (2021)
3.	Power supply and energy efficiency	Many nanosensors need an energy source; therefore, it is critical to make sure they are sustainable and energy-efficient, particularly for long-term outdoor applications. Using energy-harvesting techniques or incorporating low-power designs can increase the total usefulness.	Incorporating energy-efficient electronics and integrating energy-harvesting techniques (e.g., solar cells, piezoelectric devices) can help provide sustainable power for nanobiosensors, making them more practical for long-term field applications without requiring frequent battery replacements.	Tejashwini et al. (2024)

10.5 Environmental and Health Concerns

Significant health- and environment-related issues in the use of nanomaterials in biosensors are mentioned in Table 8.

Table 8 Major challenges with environmental and health concerns regarding the use of nanomaterial and solutions

Sl No.	Parameter	Challenges	Solution	References
1.	Environmental impact	Human health and ecosystems may be negatively impacted by nanomaterials. A thorough assessment of their toxicity and environmental bioaccumulation is required. Environmental impact analyses and safe disposal techniques are crucial to ensuring nanobiosensors do not provide long-term hazards.	Green chemistry and biodegradable nanomaterials research can assist in lowering the environmental dangers related to nanobiosensor disposal. Nanobiosensors will also be employed in an ecologically responsible way provided life cycle assessments (LCAs) and safe disposal procedures are established.	Pervez and Yadav (2023)
2.	Regulatory issues	The regulatory framework for nanobiosensors is still under development. There is uncertainty around standards, certifications, and safety regulations for their widespread use, especially in sensitive applications such as food safety, healthcare, and environmental monitoring.	Governments and regulatory agencies must establish clear guidelines and standards for developing, testing, and commercializing nanobiosensors. This includes establishing protocols for toxicity testing, data security, and quality assurance for nanobiosensors in medical, environmental, and industrial applications.	Mehta et al. (2023)
3.	Ethical concerns	Integrating nanobiosensors in *medicine* and *environmental monitoring* raises ethical questions, particularly regarding privacy, *data security*, and the potential for misuse in surveillance or other unintended applications.	It should be developed to address privacy, data security, and misuse concerns. Public discussions and transparent regulatory frameworks can help manage concerns about the impact of nanobiosensors on society, ensuring that these technologies are used responsibly and ethically.	Mehta et al. (2023)

Table 9 Major challenges with cost and commercialization concerns and solutions

Sl No.	Parameter	Challenges	Solution	References
1.	High manufacturing costs	The *complexity* and *costliness* of nanomaterial synthesis and sensor fabrication may make them too expensive for large-scale deployment, particularly in resource-constrained settings.	Research into *low-cost manufacturing techniques*, such as *roll-to-roll processing* and *3D printing*, can help reduce production costs. As nanobiosensors become more widespread, economies of scale will also lead to cost reductions over time.	Pandey and Jain (2020), Sunaina et al. (2024)
2.	Commercialization barriers	Moving from *laboratory-scale research* to *commercial products* requires substantial *investment*, *infrastructure*, and *regulatory approval*. The path to commercialization can be lengthy, and there are challenges in developing *affordable* and *scalable manufacturing techniques*.	Collaboration between *academia*, *industry*, and *government* can facilitate the transition from research to commercial products. Additionally, supporting *start-ups* and *small businesses* focused on nanobiosensors can foster innovation and accelerate the development of market-ready products.	Pandey and Jain (2020), Sunaina et al. (2024)

10.6 Cost and Commercialization

While the production of nanobiosensors holds great promise, scaling up their production and making them commercially viable for widespread use presents several challenges, as shown in Table 9.

11 Conclusions and Future Prospects

Nanobiosensors have much potential, but several problems, including synthesis scalability, reproducibility, sensor performance, integration into real-world systems, and regulatory and environmental impact concerns, hinder their development and use. To overcome these obstacles and guarantee the safe and efficient use of nanobiosensors in various applications, further research and development, interdisciplinary cooperation, and precise regulatory frameworks are needed. As these obstacles are addressed, nanobiosensors will probably become more significant in various industries, including environmental monitoring, agriculture, and healthcare.

Nanobiosensors have much potential for use in food safety, agriculture, healthcare, and environmental monitoring. However, their development, use, and commercialization present several difficulties, including scalability, cost, performance, integration, and environmental effects.

The development of cost-effective, scalable, and sustainable manufacturing processes and improvements in functionalization techniques and synthesis procedures are all necessary to overcome these obstacles. Furthermore, resolving toxicological, regulatory, and ethical issues is crucial to the broad acceptance and use of nanobiosensors.

Nanobiosensors have the potential to transform many sectors by overcoming these obstacles and offering more effective, affordable, and sustainable solutions for pressing global issues, including agricultural production, healthcare diagnostics, and environmental contamination. Nanobiosensors are expected to become essential to contemporary scientific and industrial systems as research and technology progress, helping to create a healthier and more sustainable future.

Nanobiosensors hold immense potential for improving human health, protecting the environment, and advancing technology in various sectors. Their continued development focuses on enhancing sensitivity, scalability, and affordability with the following advancements:

- Integration with Internet of Things (IoT): Smart nanobiosensors connected to cloud networks for real-time monitoring.
- Wearable Biosensors: Nanotechnology-enabled wearable devices for continuous health monitoring.
- Multiplexed Sensing: Detecting multiple analytes simultaneously using advanced nanostructures.

References

Abdellatif AAH, Younis MA, Alsharidah M et al. (2022) Biomedical applications of quantum dots: overview, challenges, and clinical potential. Int J Nanomed 17:1951–1970.

Aguilar-Pérez KM, Heya MS, Parra-Saldívar R et al (2020) Nano-biomaterials in-focus as sensing/detection cues for environmental pollutants. Case Studies in Chem Environ Eng 2:100055. ISSN 2666-0164. https://doi.org/10.1016/j.cscee.2020.100055.

Akgönüllü S, Denizli A (2022) Recent advances in optical biosensing approaches for biomarkers detection. Biosensors and Bioelectronics: X 12:100269.

Alberti G, Zanoni C, Magnaghi LR et al (2021) Gold and silver nanoparticle-based colorimetric sensors: new trends and applications. Chemosensors *9*(11):305. https://doi.org/10.3390/chemosensors9110305.

Asghar N, Hussain A, Nguyen DA et al (2024) Advancement in nanomaterials for environmental pollutants remediation: a systematic review on bibliometrics analysis, material types, synthesis pathways, and related mechanisms. J Nanobiotechnol 22:26. https://doi.org/10.1186/s12951-023-02151-3

Beni AA, Jabbari H (2022) Nanomaterials for environmental applications. Res Engineer 15:100467. ISSN 2590-1230. https://doi.org/10.1016/j.rineng.2022.100467.

Chauhan N, Saxena K (2021) Chapter 4 – Nanoscale interface techniques for standardized integration of nanosensors in current devices. In: Nanosensors for smart manufacturing, Micro and Nano Technologies, pp 91–114.

Das S, Siddique IM (2024) Navigating the nanoscale: challenges and opportunities in nanomaterial science. Euro J Adv Engin Tech 11(4):32–40.

Devi, S., Parambath, A.B. and Arora, K., 2025. Challenges, significance, and prospects of nanomaterials-based next-generation biosensors. In: Smart and Intelligent Nanostructured Materials for Next-Generation Biosensors (pp. 337–423). Elsevier. https://doi.org/10.1016/B978-0-443-19146-6.00001-0

Feng P, Yang F, Shi X et al. (2024) Carbon nanotubes perpendicularly grown on graphene oxide nanosheets derived from metal-organic frameworks: synergistic reinforcement of poly (l-lactic acid) scaffold. J Mat Res Tech 33:892–905

Fuster MG, Moulefera I, Muñoz MN et al (2023) Synthesis of cellulose nanoparticles from ionic liquid solutions for biomedical applications. Polymers (Basel);15(2):382. https://doi.org/10.3390/polym15020382. PMID: 36679262; PMCID: PMC9867531.

Huang CW, Lin C, Nguyen MK et al (2023) A review of biosensor for environmental monitoring: principle, application, and corresponding achievement of sustainable development goals. Bioengineering 14(1):58–80. https://doi.org/10.1080/21655979.2022.2095089. PMID: 37377408; PMCID: PMC10308875.

Huang X, Zhu Y, Kianfar E (2021) Nano biosensors: properties, applications and electrochemical techniques. J Mat Res Tech 12:1649–1672. ISSN 2238-7854, https://doi.org/10.1016/j.jmrt.2021.03.048.

Javaid M, Haleem A, Singh RP et al (2021) Exploring the potential of nanosensors: a brief overview. Sensor Int 2:100130. ISSN 2666-3511, https://doi.org/10.1016/j.sintl.2021.100130.

Kassie BB, Daget TM, Tassew DF (2024) Synthesis, functionalization, and commercial application of cellulose-based nanomaterials. Int J Biol Macromol 278(4):134990.

Khanna VK (2012) Thermal nanosensors. In: Nanosensors 1st edn, CRC Press, Boca Raton, pp. 46, eBook ISBN9780429093951. https://doi.org/10.1201/b11289.

Kulkarni DD, Joshi VR, Shetty SN et al (2022) Comparative study between co-precipitation and sol gel method for the preparation of ferrites. Int J Adv Res Sci Comm Tech (IJARSCT) 2(4).

Kumalasari MR, Alfanaar R, Andreani AS (2024) Gold nanoparticles (AuNPs): a versatile material for biosensor application. Talanta Open 9:100327.

Kumar A, Shaikh WA, Mukherjee C et al (2024) Biochar-based nanocomposites: a novel and sustainable solution to (waste) water contamination: In: Nanotechnology for environmental management, Routledge, Taylor and Francis.

Malik S, Singh J, Goyat R et al (2023) Nanomaterials-based biosensor and their applications: A review. Heliyon 9(9), e19929.

Mathew J, Shivaji PM, Pratihar S (2024) Chapter 20 – Nanomaterials: challenges and environmental toxicity. In: Nanomaterials in environmental analysis, pp 471–481.

Mehta N, Shetty S, Prajapati BG et al (2023) Regulatory and ethical concerns in the use of nanomaterials. In: Alzheimer's disease and advanced drug delivery strategies, Academic Press.

Moulahoum H, Ghorbanizamani F, Celik EG et al (2023) Chapter 6 – Nanoconjugated materials as sensors in point-of-care diagnostic tools: detection of small molecules and viruses. In: Verma SK, Das AK (ed) Comprehensive analytical chemistry, Elsevier 102:203–231. ISSN 0166-526X, ISBN 9780443131998, https://doi.org/10.1016/bs.coac.2023.02.002.

Mukherjee C, Choudhury S, Baildya N et al (2024) Green synthesis and DFT study of Nickel Zinc Ferrite nanoparticles: a highly sensitive room temperature VOC sensing material. Int J Env Agri Biotech 9(6).

Mukherjee C, Mondal R, Dey S et al (2017) Nanocrystalline Copper Nickel Zinc Ferrite: efficient sensing materials for ethanol and acetone at room temperature. IEEE Sensors J 17(9):2662–2669.

Nagime PV, Chandak VS (2024) A comprehensive review of nanomaterials synthesis: physical, chemical, and biological approaches and emerging challenges. Biocat Agri Biotech 62:103458.

Najafi P, Ghaemi A (2024) Chemiresistor gas sensors: design, challenges, and strategies: a comprehensive review. Chem Engr J 498:154999.

Negahdary M, Mabbott S (2025) Automated synthesis and processing of functional nanomaterials: advances and perspectives. Coord Chem Rev 523(1): 216249.

Nyabadza A, McCarthy É, Makhesana M et al (2023) A review of physical, chemical and biological synthesis methods of bimetallic nanoparticles and applications in sensing, water treatment, biomedicine, catalysis and hydrogen storage. Adv Col Int Sci 321:103010. ISSN 0001-8686, https://doi.org/10.1016/j.cis.2023.103010.

Pandey G, Jain P (2020) Assessing the nanotechnology on the grounds of costs, benefits, and risks. J Basic Appl Sci 9 (63).

Pervez A, Yadav M (2023) Chapter 15 – Nanomaterials in the environment: impacts and challenges. In: Nanoparticles and plant-microbe interactions- an environmental perspective nanomaterial- plant interactions, pp 389–414.

Ray PC, Yu H, Fu PP (2009) Toxicity and environmental risks of nanomaterials: challenges and future needs. J Environ Sci Health C Environ Carcinog Ecotoxicol Rev 27(1):1–35.

Reshma RP, Abishek NS, Gopalakrishna KN (2024) Synthesis and characterization of graphene oxide, tin oxide, and reduced graphene oxide-tin oxide nanocomposites. Inorg Chem Commun 165:112451.

Sargazi A, Fatima I, Kiani MH et al (2022) Fluorescent-based nanosensors for selective detection of a wide range of biological macromolecules: a comprehensive review. Int J Biol Macromol 206:115–147. ISSN 0141-8130, https://doi.org/10.1016/j.ijbiomac.2022.02.137. (https://www.sciencedirect.com/science/article/pii/S0141813022003890)

Sharma M, Mahajan P, Alsubaie AS et al (2025) Next-generation nanomaterials-based biosensors: real-time biosensing devices for detecting emerging environmental pollutants. Mat Today Sust 29:101068. ISSN 2589-2347, https://doi.org/10.1016/j.mtsust.2024.101068.

Shukla SK, Mangla B, Hussain CM et al (2024) Nanotechnology in miniaturization: an emerging trend to fabricate future devices. (n.p.): Springer.

Singh R, Srinivas SP, Kumawat M, Daima HK (2024) Ligand-based surface engineering of nanomaterials: trends, challenges, and biomedical perspectives. Open Nano 15:100194.

Skládal P (2024) Piezoelectric biosensors: shedding light on principles and applications. Mikrochim Acta 191(4):184. https://doi.org/10.1007/s00604-024-06257-9. PMID: 38451295; PMCID: PMC10920441.

Srivastava KR, Awasthi S, Mishra PK, Srivastava PK (2020) Chapter 13 – Biosensors/molecular tools for detection of waterborne pathogens In: VaraPrasad MN, Grobelak A (ed) Waterborne pathogens, Butterworth-Heinemann:237–277. ISBN 9780128187838, https://doi.org/10.1016/B978-0-12-818783-8.00013-X.

Srivastava S, Wang W, Zhou W et al (2024) Machine learning-assisted surface-enhanced Raman spectroscopy detection for environmental applications: a review. Environ Sci Technol 58(47):20830–20848. https://doi.org/10.1021/acs.est.4c06737.

Sunaina, Yadav KK, Jha M (2024) Chapter-7: Challenges for commercialization of nanoemulsions. In: Industrial applications of nanoemulsion. Micro and Nano Technologies, pp 179–198.

Tejashwini DM, Deepa HM, Deshmukh VV et al (2024) Challenges and opportunities in nanomaterials for energy storage. In: Advanced nanomaterials for energy storage devices.

Zarejousheghani M, Rahimi P, Borsdorf H (2021) Molecularly imprinted polymer-based sensors for priority pollutants. Sensors 21(7): 2406. https://doi.org/10.3390/s21072406.

Zhang K, Song X, Liu M et al (2023) Review on the use of magnetic nanoparticles in the detection of environmental pollutants. Water 15(17):3077. https://doi.org/10.3390/w15173077.

Zhang L, Li F, Yang Y et al (2024) Polyoxometalates/metal–organic frameworks-derived ZnO/ZnWO4 nanoparticles for highly sensitive and selective ppb-level NO2 detection. Chem Engr J 499:156604.

Zhu R, Martínez-Roque MA, Figueroa-Miranda G et al (2025) Plasmon-enhanced fluorescence and electrochemical aptasensor for SARS-CoV-2 Spike protein detection. Talanta 281:126760.

Integration of Nanobiosensors with IoT for Real-Time Environmental Monitoring and Data Analytics

Bhaskar Mallick, Zisan Ahamed, and Prashanta Kumar Mitra

Abstract The integration of nanotechnology with the Internet of Things (IoT) has revolutionised environmental surveillance by enabling real-time, high-precision monitoring of contaminants. Nanobiosensors, owing to their nanoscale size and unique properties, provide high sensitivity and specificity for detecting environmental pollutants. These sensors, consisting of a recognition element and a transducer, convert biological reactions into measurable signals. When combined with IoT networks, nanobiosensors can continuously collect and transmit environmental data, allowing for immediate detection of hazardous substances and facilitating remote monitoring in inaccessible areas. Their miniaturisation enhances portability, affordability, and adaptability for diverse applications, including air and water quality assessment, soil health monitoring, and food safety. Additionally, IoT-enabled nanobiosensors support smart agriculture by optimising crop conditions and pest management while minimising environmental impact. Smart wearable applications also benefit from this technology by providing real-time health metrics and proactive wellness tracking. However, challenges such as interoperability, data security, and power supply constraints must be addressed to ensure efficient deployment. The integration of artificial intelligence (AI) with IoT further enhances data analysis capabilities, enabling predictive modelling and real-time decision-making. As IoT networks expand, they facilitate broader adoption of nanobiosensors, significantly improving environmental monitoring, public health protection, and sustainability efforts. This fusion of IoT and nanotechnology represents a promising advancement in smart environmental management, offering innovative solutions to global ecological challenges.

Keywords Artificial intelligence · Environment · Environmental monitoring · Nanoparticles · Nanotechnology

---

B. Mallick · Z. Ahamed
Department of Ecological Studies & International Centre for Ecological Engineering (ICEE), University of Kalyani, Kalyani, Nadia, West Bengal, India

P. K. Mitra (✉)
Department of Lifescience, Shri Rawatpura Sarkar University, Raipur, Chattishgarh, India

S. Roy, V. Mandal (eds.), *Plant-Microbe Interaction under Xenobiotic Exposure*,
https://doi.org/10.1007/978-981-96-8260-7_25

1 Introduction

Including nanotechnology with the Internet of Things (IoT) marks an impact on advancement in environmental surveillance (Akhtar and Perwej 2020; Benitez 2023; Twentyman 2024; Chen et al. 2017; Dumé 2018; Vikesland 2018; MIT Engineers Build Advanced Microprocessor Out of Carbon Nanotubes 2019; Dash et al. 2021; Pajooh et al. 2021; Soliman et al. 2020; Tang et al. 2019). Nanobiosensors provide high sensitivity and specificity for identifying a range of environmental contaminants because of their distinctive characteristics and nanoscale size (Huang et al. 2021a, b; Ramesh et al. 2022). When integrated with IoT technologies, these sensors facilitate real-time monitoring and data management, enhancing our ability to safeguard environmental health (Huang et al. 2021a, b; Ramesh et al. 2022; Verma et al. 2022). Highly sensitive instruments that can identify pollutants at extremely low concentrations have been developed due to recent advancements in nanobiosensor technology (Kulkarni et al. 2022; Thakur et al. 2022). Cutting-edge innovations in nanomaterials have improved biosensors' performance, allowing for rapid and accurate detection of pollutants (Malik et al. 2021; Fu et al. 2024). Tailoring the surface of nanobiosensors with specific receptors enhances their ability to selectively bind to target analytes, increasing specificity for pollutants like heavy metals and pesticides (Naresh and Lee 2021; Fu et al. 2024). Because of their small dimensions, nanobiosensors are used to create portable instruments that can be used in various environments, from industrial sites to remote natural areas (Javaid et al. 2021; Kulkarni et al. 2022).

Combining IoT technologies with nanobiosensors has produced many creative uses, including real-time monitoring, in which IoT-enabled nanobiosensors can continuously collect and transmit data on environmental conditions, allowing for the immediate detection of hazardous pollutants. Data analytics with advanced algorithms can analyse the information gathered from these sensors to spot patterns, predict pollution events, and inform decision-making processes. Lastly, remote sensing sensors can be deployed in hard-to-reach areas, providing valuable data without manual sampling and analysis (Huang et al. 2021a, b; Ramesh et al. 2022). Despite the promising potential of nanobiosensors and IoT, several challenges remain in data security, including transmitting sensitive environmental data; it brings up data integrity and privacy issues, emphasising the necessity for robust security measures. Furthermore, the lack of established procedures for sensor calibration and data reporting may impede the comparability of results across various studies and applications. Also, navigating the regulatory landscape for deploying these technologies can be complex, particularly concerning environmental safety and public health (Dressler and Kargl 2012; Alabdulatif et al. 2023; Gaobotse et al. 2022). Emerging developments and trends are anticipated to impact nanobiosensors' use in monitoring the environment. The ongoing advancement of IoT networks will enhance connectivity, enabling more effective and comprehensive monitoring solutions. Data sharing among sensors leads to more comprehensive environmental monitoring systems. Incorporating artificial intelligence (AI) into data analysis will enable more sophisticated interpretations of sensor data, improving predictive capabilities and response strategies. The growing need for sustainable monitoring systems will prompt more research into eco-friendly materials and energy-efficient sensor designs (Tovar-Lopez 2023; Alabdulatif et al. 2023; Darwish et al. 2024; Kulkarni et al. 2022; Malik et al. 2023).

In summary, combining nanobiosensors and IoT technologies offers a revolutionary environmental monitoring approach (Fig. 1). These systems improve our ability to accumulate and analyse data in real time to detect and manage environmental pollutants effectively. To achieve all of the potential of this synergy, addressing the existing constraints will be crucial to providing opportunities for creative approaches to environmental health management (Fig. 1).

2 Understanding Nanobiosensors

2.1 Definition and Functionality

Nanobiosensors are analytical devices that utilise nanomaterials to determine chemicals and biological substances at the nanoscale. Usually, these sensors consist of a recognition element and a transducer that converts the biological reaction into a measurable signal (Fig. 1). These sensors use biological molecules to connect to and specifically recognise the target analyte, such as nucleic acids, enzymes, or antibodies. The transducer subsequently transforms the biological response into a measurable signal, like an electrical, visual, or electrochemical one. This unique characteristic of nanomaterials, specifically their larger surface area, improves these sensors' sensitivity and effectiveness and enhanced reactivity, making nanobiosensors particularly effective for detecting low concentrations of pollutants. The ability to manipulate the size and form of nanoparticles enables the optimisation of their characteristics for specific applications, enhancing selectivity and performance. At the nanoscale, materials may exhibit quantum effects that can be harnessed for improved detection capabilities, such as fluorescence enhancement or changes in electrical conductivity. Nanomaterial-based nanobiosensors that utilise the unique features of nanomaterials like nanotubes, nanoparticles, and nanowires are used to improve their practicality. These materials enormously increase the sensitivity and specificity of the sensors, making them particularly effective in detecting biological and chemical substances (Banigo et al. 2020; Kulkarni et al. 2022; Naresh and Lee 2021; Ramesh et al. 2022).

Nanoparticles are particles with measurements in the range of nanometers (1–100 nm). They can be made from different materials, including metals (like gold and silver), oxides, and polymers (Altammar 2023; Khan et al. 2019); as an example, gold nanoparticles are often used for their optical properties in surface-enhanced Raman scattering (SERS) applications (López-Lorente 2021; Tian et al. 2014). Like carbon nanotubes (CNTs), nanotubes are cylinder-shaped nanostructures with

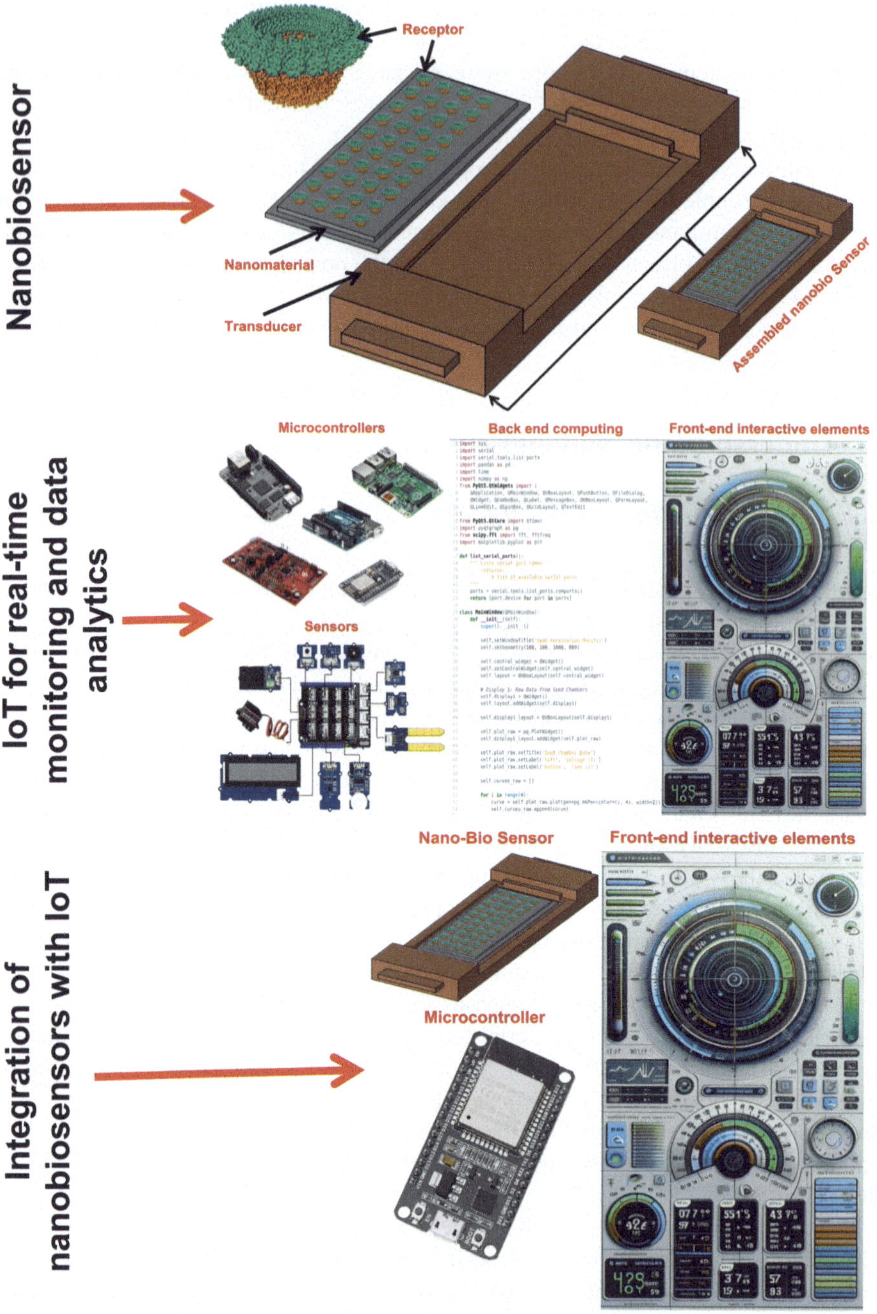

Fig. 1 Components of nanobiosensor, IoT, and their integration for developing monitoring systems

exceptional electrical, thermal, and mechanical properties. They are suitable for electrochemical sensing applications because of their resilient electrical conductivity and huge surface area. CNTs can facilitate rapid electron transfer, enhancing the biosensor's sensitivity (Anzar et al. 2020; Norizan et al. 2020; Patel et al. 2020). Nanowires are one-dimensional nanostructures with diameters in the nanometer range and can be considerably longer. They can function as transducers in biosensors, transforming biological interactions into detectable electrical signals (Li et al. 2011; Verardo et al. 2018).

In summary, because of the high surface-to-volume ratio and enhanced reactivity of nanomaterials, nanobiosensors can detect extremely low concentrations of analytes. The small size and efficient mass transfer properties of nanomaterials enable rapid detection and analysis of target analytes. Nanobiosensors can be miniaturised, allowing for the development of devices that are easy to use, adaptable, and affordable for point-of-care and in-situ monitoring applications.

2.2 *Types of Nanobiosensors*

2.2.1 Electrochemical Nanobiosensors

Electrochemical nanobiosensors are specialised nanobiosensors that are used for electrochemical transduction methods to detect and quantify target analytes. Functionalisation of the electrode surface with specific aspects of recognition like DNA, antibodies, and enzymes ensures selectivity for the target molecules. These sensors measure changes in current, voltage, or impedance in response to the way in which the recognition element and analyte interact, providing real-time and sensitive detection capabilities. Changes in current can be monitored, resulting in redox reactions taking place at the surface of the electrode, while the target analyte binds to the recognition element. For example, in the devices used for glucose monitoring in diabetes management and detection of neurotransmitters, commonly, the current is proportional to the concentration of the target analyte. Voltage changes are measured using potentiometric sensors that can measure the voltage change in response to the binding of the analyte. Potentiometric sensors can indicate the presence of specific analytes, often involving ion-selective electrodes, used for measuring pH levels, ion concentrations, and particular biomolecules. Impedimetric sensors measure changes in the sensor's impedance due to the binding of analytes, providing information about the charge transfer resistance and indicating the binding events and the concentration of analytes. This kind of method is beneficial for detecting biomolecular interactions and monitoring cell behaviour capabilities (Baranwal et al. 2022; Cho et al. 2020; Hammond et al. 2016; Irkham et al. 2023; Raziyan et al. 2024; Wang et al. 2021; Zhu et al. 2014).

2.2.2 Optical Nanobiosensors

These advanced analytical devices utilise light-based techniques to detect and quantify biological and chemical substances at the nanoscale. These sensors capitalise on the interaction between light and nanomaterials that can measure variations in light characteristics such as scattering, absorbance, or fluorescence, which occur due to the binding of target analytes (Naresh and Lee 2021; Munawar et al. 2019; Ramesh et al. 2022). Standard methods include fluorescence, absorbance spectroscopy, and surface plasmon resonance (SPR). The unique optical properties of nanomaterials, such as enhanced light absorption and scattering, allow for the finding of extremely low analyte concentrations. Fluorescence-based sensors utilise fluorescent labels or dyes that emit light upon excitation. When the target analyte binds to the recognition element, the presence of the analyte is indicated by a change in the wavelength or intensity of the fluorescence.

It is commonly used in biological assays for detecting specific proteins, nucleic acids, or pathogens (Li et al. 2015; Semeniak et al. 2022). Surface plasmon resonance (SPR) sensors exploit the phenomenon of surface plasmons—coherent oscillations of free electrons at the interface between a dielectric and a metal. When the metal surface is illuminated at a particular angle, it can excite these plasmons, leading to a change in the reflected light intensity or angle, which correlates with the binding of analytes. It is commonly used for real-time biomolecular association monitoring, including drug-target and antigen-antibody interactions (Howe et al. 2019; Kumar and Seo 2023; Miyazaki et al. 2017). Raman spectroscopy measures the inelastic scattering of light (Raman scattering) to provide information about molecular vibrations, utilising that surface-enhanced Raman scattering (SERS) can amplify the signal using nanostructured surfaces, enabling the detection of an analyte in low concentrations. It is helpful in identifying chemical compounds and detecting pollutants in environmental samples (Balasubramani et al. 2024; Serebrennikova et al. 2021).

2.2.3 Mass-Sensitive Nanobiosensors

Mass-sensitive nanobiosensors are analytical devices that detect changes in mass or frequency to ascertain the target analytes' concentration and existence. These sensors provide extremely sensitive and precise detection capabilities by utilising resonance frequency changes and mass measurement principles.

These sensors work because a material's resonance frequency can be influenced by its mass. A detectable shift in frequency results from the additional mass when a target analyte attaches to the sensor surface. Low concentrations of analytes can be detected using mass-sensitive sensors because they can distinguish even minute changes in mass. One mass-sensitive nanobiosensor is the quartz crystal microbalance (QCM), which is made out of a thin quartz crystal that oscillates at a particular frequency. The mass of the crystal shifts when an analyte attaches to its surface, changing the oscillation frequency of the crystal. The mass of the bound analyte directly relates to the frequency change. QCM sensors are widely used for detecting biomolecules, such as proteins and DNA, as well as environmental pollutants and toxins (Heydari and Haghayegh 2014; Jandas et al. 2021; Lim et al. 2020; Thies et al. 2017). Surface acoustic waves that move across a piezoelectric material's surface are used in surface acoustic wave (SAW) sensors. The mass and mechanical characteristics of the sensor surface are modified when an analyte binds to it, changing the acoustic waves' amplitude and velocity. These sensors are employed in gas sensing, detecting pathogens, and monitoring biochemical interactions (Go et al. 2017; Gouda et al. 2023; Huang et al. 2021b). Microcantilever sensors are tiny beams that can bend or deflect in response to mass changes on their surface. The mass and mechanical characteristics of the sensor surface are modified when an analyte binds to it, changing the acoustic waves' amplitude and velocity causing a measurable deflection, which can be monitored using optical or electrical methods. Microcantilever sensors are used to detect biomolecules, environmental pollutants, and chemical agents (Chaudhary and Gupta 2009; Singamaneni et al. 2008; Lang et al. 2017).

3 The Internet of Things (IoT)

3.1 Definition and Components

The Internet of Things (IoT) refers to a network of linked devices that exchange information while communicating online (Fig. 1). These devices might be anything from primary industrial tools to everyday home commodities, all equipped with software and sensors that allow them to collect and exchange data. The Internet of Things facilitates the seamless integration of the digital and physical worlds, hence improving system automation, control, and monitoring. Sensors are critical components of Internet of Things (IoT) devices that collect environmental data. They are capable of measuring a wide range of variables, including light, motion, temperature, and humidity (Mishra and Mishra 2023; Kumar et al. 2019). Communication protocols make data transmission between devices and the cloud or other devices possible. They guarantee the efficient and safe transmission of data. Wi-Fi, Zigbee, Bluetooth, cellular networks (e.g. 5G, 4G), and LoRaWAN are standard communication protocols. Sensor data is analysed and processed by data processing units. They may be found in the cloud or on the device (edge computing). User interfaces enable interaction with IoT systems, allowing users to visualise data and control devices. These interfaces can be web-based applications, mobile apps, or dashboards. Examples include mobile apps for smart home devices, web dashboards for industrial IoT, and voice-activated assistants (Srivastava and Srivastava 2023; IoT Technologies and Protocols | Microsoft Azure n.d.). Actuators are devices that perform actions depending on commands from the data processing unit. They can control physical systems or devices in response to analysed data. Examples are motors, valves, and relays that can turn on lights, adjust thermostats, or open/close doors.

IoT devices can continually monitor and gather data, giving real-time insights into environmental conditions and system performance. IoT systems can enhance operational efficiency and reduce resource consumption by automating processes and enabling remote monitoring. The data generated by IoT devices can be analysed to inform better decision-making, leading to improved outcomes in various applications. IoT systems can also predict equipment failures by monitoring performance metrics, allowing for proactive maintenance and reducing downtime (Redekar et al. 2022).

3.2 *Role of IoT in Environmental Monitoring*

The IoT plays an important role in environmental monitoring because it allows the integration of many sensors and devices to gather and analyse data from multiple sources. This connectivity facilitates real-time insights into environmental conditions, such as pollution levels, weather patterns, and biodiversity, allowing for timely interventions and informed decision-making. IoT devices with sensors can constantly monitor environmental factors such as water quality, air quality, temperature, and soil moisture. This real-time data collection enables immediate awareness of changing conditions, facilitating rapid responses to environmental issues. IoT systems can acquire data from various sources, such as satellite imaging, weather stations, and ground-based sensors. This thorough data integration gives a complete picture of environmental conditions, improving the accuracy and reliability of monitoring activities. IoT technologies enable the deployment of air and water quality sensors that can detect pollutants and hazardous substances. Authorities can promptly mitigate health risks and environmental damage by monitoring pollution levels in real time. An example is smart air quality monitoring stations that can track levels of ozone (O3), particulate matter (PM), and nitrogen dioxide (NO2), providing critical data for public health advisories. IoT devices, like camera traps and acoustic sensors, can monitor wildlife populations and their habitats. This data helps researchers and conservationists understand species behaviour, track migration patterns, and assess ecosystem health. An example is smart tracking collars on endangered species, which can provide real-time location data, aiding in conservation efforts and habitat protection. IoT technologies can optimise agricultural practices by monitoring crop health, soil conditions, and water usage. This precision farming approach enhances resource efficiency and reduces environmental impact. Soil moisture sensors can trigger irrigation systems only when needed, conserving water and improving crop yields. Real-time data on environmental conditions by IoT systems can be crucial for managing natural disasters like wildfires, earthquakes, and floods. IoT technologies can reduce the costs associated with traditional monitoring methods by automating data collection and analysis, leading to more efficient resource allocation. IoT systems can provide accessible data to the public, raising awareness of environmental issues and encouraging community involvement in conservation efforts (Ullo and Sinha 2020; Katie 2024; Narayana et al. 2024; Popescu et al. 2024).

4 Synergy Between Nanobiosensors and IoT

IoT connectivity enables nanobiosensors to transmit collected data in real time to central databases or cloud platforms (Fig. 1). This immediate data transfer allows instant information access, enabling stakeholders to respond quickly to environmental changes. The data collected from multiple nanobiosensors can be aggregated in a central database, where advanced data analysis and machine-learning techniques can be used. This centralised analysis helps identify trends, detect anomalies, and predict potential pollution events. When integrated with IoT, their capabilities are amplified, allowing for comprehensive surveillance. Receiving real-time data enables authorities and environmental agencies to implement rapid response measures. For example, if a spike in pollutant levels is detected, immediate actions can be taken to mitigate the impact, such as issuing health advisories or deploying cleanup crews. The continuous flow of data allows policymakers and environmental managers to make defensible choices in light of existing circumstances. This data-driven approach enhances the effectiveness of environmental regulations and resource management strategies. IoT-enabled nanobiosensors can be easily scaled and deployed in various locations, adapting to different monitoring needs (Alabdulatif et al. 2023; Kulkarni et al. 2022; Mutunga et al. 2024; Tovar-Lopez 2023).

Integrating IoT networks with nanobiosensors facilitates remote monitoring and control of environmental conditions, allowing stakeholders to access real-time data from virtually anywhere. This capability is especially valuable for monitoring remote or hard-to-reach areas where traditional environmental assessment methods may be impractical, costly, or time-consuming. IoT networks allow stakeholders to access data nanobiosensors collect in real time through web-based platforms or mobile applications. This immediate access to information enables informed decision-making and timely interventions. This flexibility expands the scope of environmental monitoring efforts. IoT systems can be configured to send automated alerts and notifications when specific thresholds are exceeded, such as pollutant levels or environmental parameters. This feature ensures that stakeholders are promptly informed of critical changes, allowing for rapid response. In addition to monitoring, IoT networks enable remote control of nanobiosensor devices. Stakeholders can adjust settings, recalibrate sensors, or initiate data collection remotely, enhancing operational efficiency (Alabdulatif et al. 2023; Mutunga et al. 2024; Tovar-Lopez 2023; Huang et al. 2021a; Yadav et al. 2023).

Remote access to data reduces the need for physical site visits, saving time and resources. This efficiency is particularly beneficial for organisations managing multiple monitoring sites. Continuous remote monitoring ensures data is collected consistently, providing a comprehensive view of environmental conditions over time. This long-term data collection is essential for trend analysis and environmental assessments. Personnel could be under threat when monitoring potentially hazardous circumstances, such as industrial sites or places impacted by natural catastrophes. Remote monitoring minimises the need for human presence in potentially dangerous situations. Remote monitoring can lower operational costs by reducing the need for manual monitoring and site visits. Organisations may more efficiently deploy resources rather than routine checks by concentrating on data analysis and reaction. Accessing real-time data from remote locations supports decision-making procedures based on data. Stakeholders can respond to environmental changes based on current conditions rather than relying on outdated information (Kumar et al. 2019).

Integrating IoT platforms with nanobiosensors allows for the compilation of data from numerous sources, giving stakeholders a thorough understanding of the state of the environment. This capability is essential for effective pollution management and environmental monitoring, as it enables the analysis of complex datasets to identify trends, assess risks, and support informed decision-making. IoT platforms can collect and consolidate data from numerous nanobiosensors deployed across various locations. A comprehensive understanding of environmental factors, including soil conditions, water quality, and air quality, is made possible by this unified data management. Integrating advanced analytics tools enables stakeholders to process and analyse large volumes of data. Anomalies, patterns, and correlations in the data can be found using statistical techniques and machine-learning algorithms, enhancing the understanding of environmental dynamics. Visualisation tools transform complex data into intuitive graphical representations like charts, maps, and dashboards. These visualisations make it easier for stakeholders to interpret data, identify trends, and communicate findings effectively. Integrated IoT platforms provide real-time data visualisation, allowing stakeholders to monitor environmental conditions as they change. This capability is crucial for timely decision-making and rapid response to pollution events. By aggregating data from multiple nanobiosensors, stakeholders gain a more complete understanding of environmental conditions. Data integration allows for assessing environmental risks by correlating pollutant levels with health outcomes or ecological impacts. Access to integrated data and visualisations supports data-driven decision-making. Policymakers, environmental managers, and community stakeholders can make informed choices regarding pollution management and resource allocation. Visualisation tools facilitate effective communication of complex data to diverse audiences, including policymakers, the public, and environmental organisations (Bharti et al. 2024; Mutunga et al. 2024; Prabhu et al. 2024; Tovar-Lopez 2023).

5 Applications of Integrated Nanobiosensors and IoT

These sensors help monitor the environment since they can identify contaminants and diseases in the air and water, enabling timely responses to environmental hazards and ensuring public safety. Nanobiosensors can monitor soil health, crop conditions, and pest presence, optimising agricultural practices and improving yield while minimising environmental impact (Mohamed and Awad 2022; Salouti and Derakhshan 2020). Nanobiosensors can be used to detect contaminants and pathogens in food products, ensuring food quality and safety throughout the supply chain (Guruprasath et al. 2024; Muthukrishnan 2022). Integration with IoT allows for the development of smart wearables that can track health metrics and provide user feedback, promoting wellness and proactive health management. The body's drug levels can be tracked with these sensors, and medications can be released accordingly, enhancing the efficacy of treatments. Combining nanobiosensors and IoT facilitates remote health diagnostics and consultations, making healthcare more accessible, especially in underserved areas. Additionally, integrated nanobiosensors can continuously monitor vital signs and biochemical markers in real time, providing valuable data for personalised healthcare and early disease detection. Integrated nanobiosensors can be deployed for the rapid detection of biological threats, aiding in national security and public health responses (Bharti et al. 2024; Mahmoud and Fawzy 2021; Muthukrishnan 2022; Tovar-Lopez 2023; Yadav et al. 2023).

6 Challenges in Integration

6.1 Technical Challenges

While integrating IoT with nanobiosensors offers significant advantages for environmental monitoring, a number of technological issues need to be resolved to promise efficient deployment and functioning. Key challenges include interoperability, data security, and power supply issues. Interoperability pertains to the ability of different sensors, devices, and communication protocols to work together seamlessly within an IoT ecosystem. The variety of manufacturers and technologies means that sensors may operate on different communication protocols (e.g. Zigbee, LoRaWAN, Wi-Fi), making integrating them into a unified system difficult. Sensors may output data in different formats, complicating data aggregation and analysis. Ensuring that various components of an IoT system (sensors, gateways, cloud platforms) communicate effectively requires careful planning and implementation. Utilising widely accepted communication standards can enhance compatibility among devices. Implementing middleware can facilitate communication between different systems and data formats, enabling smoother integration (IoT Technologies and Protocols | Microsoft Azure n.d.; Aledhari et al. 2022; Dada et al. 2024; Javaid et al. 2022; Rajak et al. 2023; Srivastava and Srivastava 2023).

Data security involves protecting sensitive environmental data collected by nanobiosensors from unauthorised access, breaches, and cyberthreats. IoT devices, including nanobiosensors, can be targets for cyberattacks, which may compromise data integrity and privacy. In order to safeguard against illegal access and interception, data must be encrypted during transmission and storage. Strong access control measures must be implemented to ensure that only individuals with permission may access private information. Using advanced encryption standards for data transmission and storage can enhance security. Conducting regular security assessments and updates can help identify vulnerabilities and mitigate risks. Employing multifactor authentication and role-based access control can strengthen data protection measures (Atlam and Wills 2019; Alabdulatif et al. 2023; Jaime et al. 2023).

Many nanobiosensors require a reliable power source, which can be particularly challenging in remote or off-grid locations. Remote monitoring sites may lack access to traditional power sources, making it difficult to maintain continuous sensor operation. The need for frequent battery replacements can lead to increased maintenance costs and operational downtime. High energy consumption by sensors can limit their deployment in energy-constrained environments. Utilising energy harvesting methods (e.g. solar, wind, or kinetic energy) can provide sustainable power solutions for remote sensors. Developing low-power nanobiosensors that can operate efficiently with minimal energy requirements can extend battery life and reduce maintenance needs. Implementing intelligent power management systems that optimise energy usage based on operational needs can enhance the longevity of power sources (Naresh and Lee 2021).

6.2 Regulatory and Ethical Considerations

The implementation of IoT-enabled nanobiosensors in order to ensure responsible and efficient implementation and environmental monitoring brings up significant ethical and regulatory issues that need to be addressed, including data privacy, environmental impact, and the need for standardised protocols (Atlam and Wills 2019; Alabdulatif et al. 2023; Tovar-Lopez 2023).

The collection and storage of environmental data by nanobiosensors raise concerns about individual privacy, particularly in cases where the data may be linked to specific locations or individuals. It is necessary to ensure that personal information, such as home addresses or contact details, is not inadvertently collected or shared, determine the appropriate level of consent required for data collection, especially in public spaces, implement measures to anonymise data and prevent the identification of individuals based on sensor data, establish guidelines that emphasise privacy protection first for data collection, storage, and dissemination, use strategies to safeguard individual identities, such as data aggregation, pseudonymisation, and differential privacy, provide clear information about data collection, and obtain consent from individuals or communities where appropriate (Egon and Temiloluwa 2023; Wang et al. 2023; Okorie et al. 2024; Tu and Gao 2021; Thakur et al. 2022; Wasti et al. 2023).

The deployment of nanobiosensors raises concerns about their potential environmental impact, particularly regarding the materials used in their construction and the disposal of used sensors. Some nanomaterials used in sensors may have unknown or unintended environmental consequences. Ensuring the safe and responsible disposal of used sensors is necessary to prevent environmental contamination. It is also necessary to reduce sensor deployments' physical and ecological footprint, especially in sensitive ecosystems, evaluate the potential environmental risks associated with nanobiosensor deployment and developing mitigation strategies, select nanomaterials with low toxicity and high biodegradability for sensor construction, and establish programs for collecting and properly disposing or recycling used sensors (Ray et al. 2009; Wahab et al. 2024).

The lack of established sensor deployment and data reporting procedures can lead to inconsistencies and challenges in data interpretation and comparison. Sensor placement and density differences can affect data quality and comparability across monitoring sites. Compiling and analysing data from many sources may be challenging if there are no established reporting guidelines or data formats. To address these issues, it is crucial to establish quality control measures and calibration procedures to ensure data accuracy and reliability. Collaboration with regulatory agencies, industry groups, and scientific organisations is essential to develop standardised protocols for sensor deployment, data collection, and reporting. Implementing Quality Assurance and Quality Control (QA/QC) measures and establishing guidelines for sensor calibration, maintenance, and data validation are key steps to maintaining data quality and consistency. Furthermore, promoting open data standards and data-sharing platforms will facilitate data exchange and collaboration among stakeholders (Ketshabetswe et al. 2019; Ma et al. 2024; Skålvik et al. 2023; Teh et al. 2020).

7 Future Directions

7.1 Advancements in Nanotechnology

Significant progress in the creation of nanobiosensors is being driven by ongoing nanotechnology research and innovation, particularly in enhancing their sensitivity and selectivity for environmental monitoring applications. These advancements can potentially revolutionise how we detect and analyse pollutants, contributing to more effective environmental management and protection (Kulkarni et al. 2022).

Researchers are developing nanoparticles with tailored surface properties to enhance their interaction with specific target analytes. This functionalisation can improve the selectivity and sensitivity of nanobiosensors. Combining different nanomaterials (e.g. metal nanoparticles with carbon-based materials) can create hybrid sensors that leverage the unique properties of each component, resulting in enhanced performance. Utilising materials that exhibit surface plasmon resonance (SPR) can amplify the signal detected by nanobiosensors, enabling the identification of contaminants at lower concentrations. Advances in electrochemical sensing techniques, such as the use of nanostructured electrodes, can increase the sensitivity and response time of nanobiosensors, enabling rapid detection of contaminants. Integrating multiple sensing functions onto a single chip allows for the simultaneous detection of multiple analytes. This miniaturisation reduces costs and enhances the portability and usability of nanobiosensors in field applications. Incorporating microfluidic systems with nanobiosensors enables precise control over fluid handling and sample processing, enhancing the precision and effectiveness of environmental surveillance. Integrating machine-learning algorithms with nanobiosensor data can enhance data analysis, enabling better pattern recognition, anomaly detection, and predictive modelling of environmental conditions. Enhanced data processing capabilities allow for real-time analysis of sensor data, providing immediate insights into environmental quality and enabling timely interventions. Research is focusing on developing environmentally friendly nanomaterials that minimise ecological impact. Biodegradable nanobiosensors can reduce waste and environmental burden while maintaining performance. Utilising natural materials, such as cellulose or chitosan, in the fabrication of nanobiosensors can enhance biocompatibility and reduce toxicity concerns (Baruah et al. 2024; Khazaei et al. 2023; Malik et al. 2023; Tovar-Lopez 2023; Virk et al. 2024).

Several pollutants, such as pesticides, heavy metals, and chemical compounds, can be detected using sophisticated nanobiosensors at lower concentrations than traditional methods. This capability is crucial for early detection and response to environmental contamination. Enhanced sensitivity and selectivity allow for real-time monitoring of water quality parameters, such as microbial contamination and chemical pollutants, ensuring safe drinking water and protecting aquatic ecosystems. Improved detection methods for nanobiosensors allow them to track airborne contaminants such as particulate matter and volatile organic compounds (VOCs), providing critical data for public health and regulatory compliance. Advanced nanobiosensors can assess soil contamination and nutrient levels, supporting sustainable agricultural practices and land management strategies. Nanobiosensors can be used to monitor environmental changes affecting biodiversity, such as habitat degradation and pollution, aiding conservation efforts and ecosystem management (Kulkarni et al. 2022; Mohamed and Awad 2022; Ramesh et al. 2022; Thakur et al. 2022).

7.2 Integration with Artificial Intelligence

Incorporating artificial intelligence (AI) into IoT systems, mainly those utilising nanobiosensors, significantly enhances data analysis capabilities. This integration enables predictive modelling, facilitates real-time decision-making, and improves responses to environmental changes, ultimately leading to more effective environmental management and protection (Alahi et al. 2023; Gouiza et al. 2024; Marengo 2024).

AI algorithms can analyse large datasets generated by nanobiosensors to identify patterns and correlations that may not be immediately apparent. This capability allows for deeper insights into environmental conditions and pollutant behaviour. AI can automatically detect anomalies in sensor data, such as sudden spikes in pollutant levels. Early identification of these anomalies enables timely interventions to mitigate potential environmental hazards. AI can utilise historical data and real-time sensor inputs to create predictive models that forecast environmental changes, such as pollution events or shifts in ecosystem health. This predictive capability allows stakeholders to prepare for and respond to potential issues proactively. AI can simulate various scenarios based on different environmental conditions and human activities, helping stakeholders understand the potential impacts of their decisions and actions. AI can facilitate automated responses to specific environmental conditions. For example, if pollutant levels exceed a specific threshold, the system can automatically trigger alerts or initiate remediation actions. AI can analyse data to identify areas that require immediate attention, allowing more efficient allocation of resources, such as personnel and equipment, to address environmental issues by analysing data on soil health, water quality, and air quality. AI can recommend sustainable agricultural practices and pollution reduction strategies that optimise resource use while minimising environmental impact (Alahi et al. 2023; Gouiza et al. 2024; Kulkarni et al. 2022; Marengo 2024; Mohamed and Awad 2022; Ramesh et al. 2022; Thakur et al. 2022).

AI algorithms can analyse real-time data from nanobiosensors to monitor water quality, predict contamination events, optimise water treatment processes, and manage water resources more effectively. AI can use data from air quality sensors to predict pollution levels based on weather patterns, traffic conditions, and industrial activities. This information can inform public health advisories and regulatory actions. Farmers may make better decisions regarding irrigation, fertilisation, and pest management using AI-driven soil and crop health data analysis, which will increase yields and lessen their effect on the environment. Also, AI can enhance disaster response efforts by predicting the impact of environmental disasters, such as floods or wildfires, based on real-time sensor data. This capability allows for better preparedness and resource allocation during emergencies. AI can analyse data from environmental sensors to track changes in biodiversity and ecosystem health, providing valuable insights for conservation efforts and habitat management (Alahi et al. 2023; Gouiza et al. 2024; Kulkarni et al. 2022; Mohamed and Awad 2022; Marengo 2024; Ramesh et al. 2022; Thakur et al. 2022).

The effectiveness of AI algorithms relies heavily on the quality and accuracy of the data collected by nanobiosensors. Ensuring data integrity through robust calibration and quality control measures is essential. AI can sometimes lead to 'black box' scenarios, where the decision-making process is not transparent. Developing explainable AI models that provide insights into how decisions are made is important. Integrating AI with existing IoT systems can be complex and requires significant infrastructure and expertise investment. Stakeholders must carefully plan and execute integration strategies to maximise benefits. Using AI in environmental monitoring raises ethical questions regarding data privacy, consent, and the potential for biased decision-making. Addressing these concerns is crucial for building trust among stakeholders (Alahi et al. 2023; Gouiza et al. 2024; Ketshabetswe et al. 2019; Kulkarni et al. 2022; Mohamed and Awad 2022; Ma et al. 2024; Marengo 2024; Skålvik et al. 2023; Teh et al. 2020; Thakur et al. 2022; Ramesh et al. 2022; Marengo 2024).

7.3 *Expansion of IoT Networks*

The expansion of IoT networks is paving the way for the widespread deployment of nanobiosensors across various environmental settings. As these networks become more sophisticated, they enable more comprehensive monitoring and management strategies, significantly enhancing our ability to address environmental challenges (Alabdulatif et al. 2023; Javaid et al. 2021; Kulkarni et al. 2022; Tovar-Lopez 2023).

The introduction of 5G technology promises faster data transmission rates, less latency, and more capacity to connect many devices at once (Pons et al. 2023). This advancement facilitates real-time data transmission from nanobiosensors deployed across vast areas. Deploying sensors in remote or difficult-to-reach areas is made possible by technologies like Sigfox and LoRaWAN, which enable long-range communication with minimal power consumption. The decreasing cost of nanobiosensors and IoT devices makes deploying a more significant number of sensors economically viable across diverse environments. Widespread adoption in both urban and rural areas is encouraged by this cost. The processing of massive amounts of data produced by nanobiosensors is made possible by combining artificial intelligence and advanced data analytics with IoT technologies. This capability enhances the value of the data collected, driving further interest in deploying these technologies. The demand for efficient monitoring systems is being driven by heightened public awareness of environmental challenges, including pollution, biodiversity loss, and climate change. IoT networks equipped with nanobiosensors provide a proactive approach to environmental management (Bharti et al. 2024; De Camargo et al. 2021; Guruprasath et al. 2024; Haxhibeqiri et al. 2018; Mahmoud and Fawzy 2021; Mohamed and Awad 2022; Muthukrishnan 2022; Salouti and Derakhshan 2020; Tovar-Lopez 2023; Yadav et al. 2023) (Fig. 2).

8 Conclusion

Integrating nanobiosensors with IoT technologies marks a significant advancement in environmental monitoring. This powerful synergy enables remote monitoring, real-time data collection, and comprehensive data analysis, fundamentally transforming how we detect and manage environmental pollutants. By using the capabilities of nanobiosensors, stakeholders can gain immediate information about air, water, and soil quality, facilitating timely interventions and informed decision-making. However, the full realisation of this potential hinges on addressing several critical challenges. Technical issues such as interoperability, data security, and power supply must be resolved to ensure seamless operation and reliability of these systems. Additionally, regulatory and ethical considerations such as data privacy, environmental impact, and the creation of standardised protocols are crucial for building trust and ensuring the responsible deployment of technologies. As research and technology continue to evolve, the future of environmental monitoring will

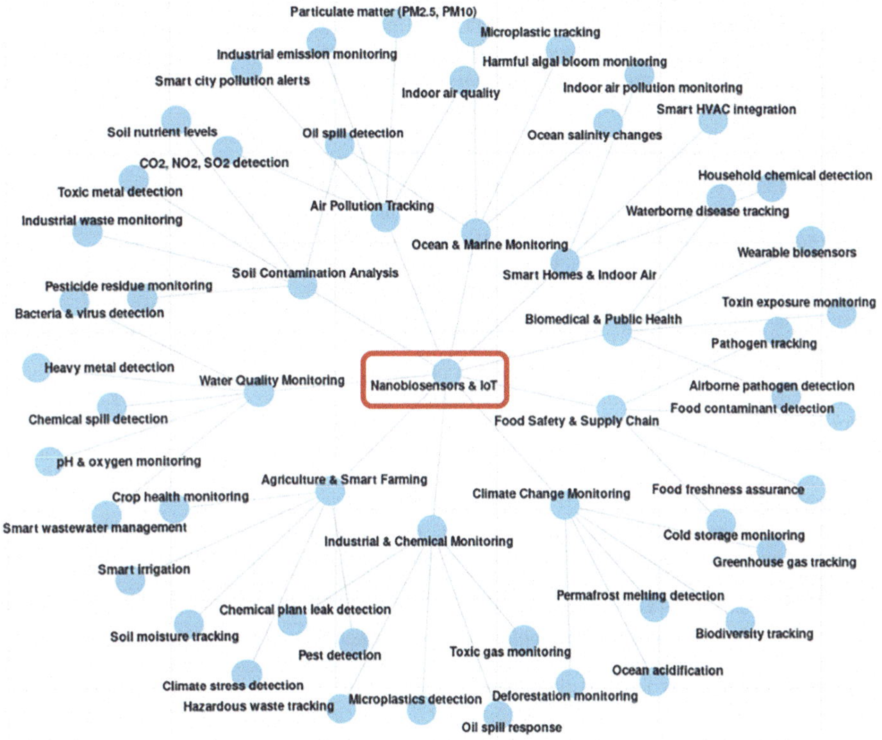

Fig. 2 Vast and complex network of potential applications of integrated nanobiosensor & IoT

undoubtedly be shaped by the innovative combination of nanobiosensors and IoT. This integration encourages a more sustainable approach to environmental stewardship and promises improved pollution detection and control capabilities. We can clear the path by embracing these advancements and collaboratively addressing the associated challenges for a healthier and more robust planet, safeguarding our natural resources for future generations.

References

Akhtar, N. and Perwej, Y., (2020) The internet of nano things (IoNT) existing state and future Prospects. GSC Advanced Research and Reviews, 5(2), pp. 131–150. https://doi.org/10.30574/gscarr.2020.5.2.0110

Alabdulatif, A., Thilakarathne, N. N., Lawal, Z. K., Fahim, K. E., & Zakari, R. Y. (2023). Internet of Nano-Things (IoNT): A Comprehensive Review from Architecture to Security and Privacy Challenges. *Sensors*, *23*(5), 2807. https://doi.org/10.3390/s23052807

Alahi, M. E. E., Sukkuea, A., Tina, F. W., Nag, A., Kurdthongmee, W., Suwannarat, K., & Mukhopadhyay, S. C. (2023). Integration of IoT-Enabled Technologies and Artificial intelligence (AI) for Smart City Scenario: recent advancements and future trends. Sensors, 23(11), 5206. https://doi.org/10.3390/s23115206

Aledhari, M., Razzak, R., Qolomany, B., Al-Fuqaha, A., & Saeed, F. (2022). Biomedical IOT: enabling technologies, architectural elements, challenges, and future directions. IEEE Access, 10, 31306–31339. https://doi.org/10.1109/access.2022.3159235

Altammar, K. A. (2023). A review on nanoparticles: characteristics, synthesis, applications, and challenges. *Frontiers in Microbiology*, *14*. https://doi.org/10.3389/fmicb.2023.1155622

Anzar, N., Hasan, R., Tyagi, M., Yadav, N., & Narang, J. (2020). Carbon nanotube – A review on Synthesis, Properties and plethora of applications in the field of biomedical science. *Sensors International*, *1*, 100003. https://doi.org/10.1016/j.sintl.2020.100003

Atlam, H. F., & Wills, G. B. (2019). IoT Security, Privacy, Safety and Ethics. In Internet of things (pp. 123–149). https://doi.org/10.1007/978-3-030-18732-3_8

Balasubramani, S. R., Nagappan, K., Ewe, L. S., Viswanathan, K., Yew, W. K., Thiraviam, S. G., Sadana, S., Samuel, T. K. D., & Kamalakannan, K. (2024). Recent developments in surface-enhanced Raman spectroscopy in the field of chemical, biochemical and clinical application–a critical review. *International Journal of Environmental & Analytical Chemistry*, 1–27. https://doi.org/10.1080/03067319.2024.2339453

Banigo, A., Azeez, T., Ejeta, K., Lateef, A., & Ajuogu, E. (2020). Nanobiosensors: applications in biomedical technology. *IOP Conference Series Materials Science and Engineering*, *805*(1), 012028. https://doi.org/10.1088/1757-899x/805/1/012028

Baranwal, J., Barse, B., Gatto, G., Broncova, G., & Kumar, A. (2022). Electrochemical Sensors and their applications: a review. *Chemosensors*, *10*(9), 363. https://doi.org/10.3390/chemosensors10090363

Baruah, A., Newar, R., Das, S., Kalita, N., Nath, M., Ghosh, P., Chinnam, S., Sarma, H., & Narayan, M. (2024). Biomedical applications of graphene-based nanomaterials: recent progress, challenges, and prospects in highly sensitive biosensors. Discover Nano, 19(1). https://doi.org/10.1186/s11671-024-04032-6

Benitez, C. (2023, February 18). *The ultimate list of internet of things statistics for 2025*. Findstack. https://findstack.com/internet-of-things-statistics/

Bharti, A., Jain, U., & Chauhan, N. (2024). From lab to field: Nano-Biosensors for real-time plant nutrient tracking. Plant Nano Biology, 9, 100079. https://doi.org/10.1016/j.plana.2024.100079

Chaudhary, M., & Gupta, A. (2009). Microcantilever-based sensors. Defence Science Journal, 59(6), 634–641. https://doi.org/10.14429/dsj.59.1569

Chen, Y., Lu, S., Zhang, S., Li, Y., Qu, Z., Chen, Y., Lu, B., Wang, X. and Feng, X., (2017) Skin-like biosensor system via electrochemical channels for non-invasive blood glucose monitoring. Science Advances, 3(12). https://www.science.org/doi/10.1126/sciadv.1701629

Cho, I., Kim, D. H., & Park, S. (2020). Electrochemical biosensors: perspective on functional nanomaterials for on-site analysis. *Biomaterials Research*, *24*(1). https://doi.org/10.1186/s40824-019-0181-y

Dada, N. M. A., Majemite, N. M. T., Obaigbena, N. A., Daraojimba, N. O. H., Oliha, N. J. S., & Nwokediegwu, N. Z. Q. S. (2024). Review of smart water management: IoT and AI in water and wastewater treatment. World Journal of Advanced Research and Reviews, 21(1), 1373–1382. https://doi.org/10.30574/wjarr.2024.21.1.0171

Darwish, M. A., Abd-Elaziem, W., Elsheikh, A., & Zayed, A. A. (2024). Advancements in Nanomaterials for Nanosensors: A Comprehensive review. *Nanoscale Advances*. https://doi.org/10.1039/d4na00214h

Dash, S., Soni, G., Patnaik, A., Liaskos, C., Pitsillides, A., & Akyildiz, I. F. (2021). Switched-Beam Graphene plasmonic nanoantenna in the Terahertz Wave region. *Plasmonics*, *16*(5), 1855–1864. https://doi.org/10.1007/s11468-021-01449-y

De Camargo, E. T., Spanhol, F. A., & Souza, Á. R. C. E. (2021). Deployment of a LoRaWAN network and evaluation of tracking devices in the context of smart cities. Journal of Internet Services and Applications, 12(1). https://doi.org/10.1186/s13174-021-00138-7

Dressler, F., & Kargl, F. (2012). Towards security in nano-communication: Challenges and opportunities. *Nano Communication Networks*, *3*(3), 151–160. https://doi.org/10.1016/j.nancom.2012.08.001

Dumé, I. (2018, October 9). *Spray-on antennas for the Internet of Things – Physics World*. Physics World. https://physicsworld.com/a/spray-on-antennas-for-the-internet-of-things/

Egon, A., & Temiloluwa, F. (2023). Privacy and Ethical Implications of IoT Data Collection and Usage. Journal of Computer Science.

Fu, Y., Liu, T., Wang, H., Wang, Z., Hou, L., Jiang, J., & Xu, T. (2024). Applications of nanomaterial technology in biosensing. *Journal of Science Advanced Materials and Devices*, *9*(2), 100694. https://doi.org/10.1016/j.jsamd.2024.100694

Gaobotse, G., Mbunge, E., Batani, J., & Muchemwa, B. (2022). Non-invasive smart implants in healthcare: Redefining healthcare services delivery through sensors and emerging digital health technologies. *Sensors International*, *3*, 100156. https://doi.org/10.1016/j.sintl.2022.100156

Go, D. B., Atashbar, M. Z., Ramshani, Z., & Chang, H. (2017). Surface acoustic wave devices for chemical sensing and microfluidics: a review and perspective. *Analytical Methods*, *9*(28), 4112–4134. https://doi.org/10.1039/c7ay00690j

Gouda, M., Ghazzawy, H. S., Alqahtani, N., & Li, X. (2023). The recent development of acoustic sensors as effective chemical detecting tools for biological cells and their bioactivities. *Molecules*, *28*(12), 4855. https://doi.org/10.3390/molecules28124855

Gouiza, N., Jebari, H., & Reklaoui, K. (2024). INTEGRATIONOFIOT-ENABLEDTECHNOLOGIES AND ARTIFICIAL INTELLIGENCE IN DIVERSE DOMAINS: RECENT... ResearchGate. https://www.researchgate.net/publication/379034933_INTEGRATION_OF_IOT-ENABLED_TECHNOLOGIES_AND_ARTIFICIAL_INTELLIGENCE_IN_DIVERSE_DOMAINS_RECENT_ADVANCEMENTS_AND_FUTURE_TRENDS

Guruprasath, N., Sankarganesh, P., Adeyeye, S. A. O., Babu, A. S., & Parthasarathy, V. (2024). Review on emerging applications of nanobiosensor in food safety. Journal of Food Science, 89(7), 3950–3972. https://doi.org/10.1111/1750-3841.17149

Hammond, J. L., Formisano, N., Estrela, P., Carrara, S., & Tkac, J. (2016). Electrochemical biosensors and nanobiosensors. *Essays in Biochemistry*, *60*(1), 69–80. https://doi.org/10.1042/ebc20150008

Haxhibeqiri, J., De Poorter, E., Moerman, I., & Hoebeke, J. (2018). A survey of LORAWAN for IoT: From Technology to Application. Sensors, 18(11), 3995. https://doi.org/10.3390/s18113995

Heydari, S., & Haghayegh, G. H. (2014). Application of nanoparticles in quartz crystal microbalance biosensors. *Journal of Sensor Technology*, *04*(02), 81–100. https://doi.org/10.4236/jst.2014.42009

Howe, C. L., Webb, K. F., Abayzeed, S. A., Anderson, D. J., Denning, C., & Russell, N. A. (2019). Surface plasmon resonance imaging of excitable cells. *Journal of Physics D Applied Physics*, *52*(10), 104001. https://doi.org/10.1088/1361-6463/aaf849

Huang, X., Zhu, Y., & Kianfar, E. (2021a). Nano Biosensors: Properties, applications and electrochemical techniques. *Journal of Materials Research and Technology*, *12*, 1649–1672. https://doi.org/10.1016/j.jmrt.2021.03.048

Huang, Y., Das, P. K., & Bhethanabotla, V. R. (2021b). Surface acoustic waves in biosensing applications. *Sensors and Actuators Reports*, *3*, 100041. https://doi.org/10.1016/j.snr.2021.100041

IoT Technologies and Protocols | Microsoft Azure. (n.d.). https://azure.microsoft.com/en-in/solutions/iot/iot-technology-protocols

Irkham, I., Ibrahim, A. U., Pwavodi, P. C., Al-Turjman, F., & Hartati, Y. W. (2023). Smart Graphene-Based Electrochemical Nanobiosensor for Clinical Diagnosis: review. *Sensors*, *23*(4), 2240. https://doi.org/10.3390/s23042240

P J Jandas, Prabakaran, K., Luo, J., & Derry Holaday M G (2021). Effective utilisation of quartz crystal microbalance as a tool for biosensing applications. *Sensors and Actuators a Physical*, *331*, 113020. https://doi.org/10.1016/j.sna.2021.113020

Jaime, F. J., Muñoz, A., Rodríguez-Gómez, F., & Jerez-Calero, A. (2023). Strengthening privacy and data security in biomedical microelectromechanical systems by IoT communication security and protection in smart healthcare. Sensors, 23(21), 8944. https://doi.org/10.3390/s23218944

Javaid, M., Haleem, A., Singh, R. P., & Suman, R. (2022). Enhancing smart farming through the applications of Agriculture 4.0 technologies. International Journal of Intelligent Networks, 3, 150–164. https://doi.org/10.1016/j.ijin.2022.09.004

Javaid, M., Haleem, A., Singh, R. P., Rab, S., & Suman, R. (2021). Exploring the potential of nanosensors: A brief overview. *Sensors International*, *2*, 100130. https://doi.org/10.1016/j.sintl.2021.100130

Katie, B. (2024). Internet of things (IoT) for environmental monitoring. International Journal of Computing and Engineering, 6(3), 29–42. https://doi.org/10.47941/ijce.2139

Ketshabetswe, L. K., Zungeru, A. M., Mangwala, M., Chuma, J. M., & Sigweni, B. (2019). Communication protocols for wireless sensor networks: A survey and comparison. Heliyon, 5(5), e01591. https://doi.org/10.1016/j.heliyon.2019.e01591

Khan, I., Saeed, K., & Khan, I. (2019). Nanoparticles: Properties, applications and toxicities. *Arabian Journal of Chemistry*, *12*(7), 908–931. https://doi.org/10.1016/j.arabjc.2017.05.011

Khazaei, M., Hosseini, M. S., Haghighi, A. M., & Misaghi, M. (2023). Nanosensors and their applications in early diagnosis of cancer. Sensing and Bio-Sensing Research, 41, 100569. https://doi.org/10.1016/j.sbsr.2023.100569

Kulkarni, M. B., Ayachit, N. H., & Aminabhavi, T. M. (2022). Recent advancements in nanobiosensors: current trends, challenges, applications, and future scope. *Biosensors*, *12*(10), 892. https://doi.org/10.3390/bios12100892

Kumar, S., & Seo, S. (2023). Plasmonic sensors: a new frontier in nanotechnology. *Biosensors*, *13*(3), 385. https://doi.org/10.3390/bios13030385

Kumar, S., Tiwari, P., & Zymbler, M. L. (2019). Internet of Things is a revolutionary approach for future technology enhancement: a review. Journal of Big Data, 6(1). https://doi.org/10.1186/s40537-019-0268-2

Lang, H. P., Hegner, M., & Gerber, C. (2017). Nanomechanical Cantilever array sensors. In Springer handbooks (pp. 457–485). https://doi.org/10.1007/978-3-662-54357-3_15

Li, J., Zhang, Y., To, S., You, L., & Sun, Y. (2011). Effect of nanowire number, diameter, and doping density on Nano-FET biosensor sensitivity. *ACS Nano*, *5*(8), 6661–6668. https://doi.org/10.1021/nn202182p

Li, M., Cushing, S. K., & Wu, N. (2015). Plasmon-enhanced optical sensors: a review. *The Analyst*, *140*(2), 386–406. https://doi.org/10.1039/c4an01079e

Lim, H. J., Saha, T., Tey, B. T., Tan, W. S., & Ooi, C. W. (2020). Quartz crystal microbalance-based biosensors as rapid diagnostic devices for infectious diseases. *Biosensors and Bioelectronics*, *168*, 112513. https://doi.org/10.1016/j.bios.2020.112513

López-Lorente, Á. I. (2021). Recent developments on gold nanostructures for surface enhanced Raman spectroscopy: Particle shape, substrates and analytical applications. A review. *Analytica Chimica Acta*, *1168*, 338474. https://doi.org/10.1016/j.aca.2021.338474

Ma, Z., Jørgensen, B. N., & Grace, Z., MA. (2024). A systematic data characteristic understanding framework towards physical-sensor big data challenges. Journal of Big Data, 11(1). https://doi.org/10.1186/s40537-024-00942-5

Mahmoud, A. E. D., & Fawzy, M. (2021). Nanosensors and nanobiosensors for monitoring the environmental pollutants. In Topics in mining, metallurgy and materials engineering (pp. 229–246). https://doi.org/10.1007/978-3-030-68031-2_9

Malik, P., Gupta, R., Malik, V., & Ameta, R. K. (2021). Emerging nanomaterials for improved biosensing. *Measurement Sensors*, *16*, 100050. https://doi.org/10.1016/j.measen.2021.100050

Malik, S., Singh, J., Goyat, R., Saharan, Y., Chaudhry, V., Umar, A., Ibrahim, A. A., Akbar, S., Ameen, S., & Baskoutas, S. (2023). Nanomaterials-based biosensor and their applications: A review. *Heliyon*, *9*(9), e19929. https://doi.org/10.1016/j.heliyon.2023.e19929

Marengo, A. (2024). Navigating the nexus of AI and IoT: A comprehensive review of data analytics and privacy paradigms. Internet of Things, 27, 101318. https://doi.org/10.1016/j.iot.2024.101318

MIT engineers build advanced microprocessor out of carbon nanotubes. (2019, August 28). MIT News | Massachusetts Institute of Technology. https://news.mit.edu/2019/carbon-nanotubes-microprocessor-0828

Mishra, V., & Mishra, V. (2023, September 4). Introduction to the Internet of Things: Concepts, Technologies, and Applications – IFACET. IFACET – IITK Foundation for Advanced Continuing Education & Training. https://ifacet.iitk.ac.in/knowledge-hub/internet-of-things/introduction-to-the-internet-of-things-concepts-technologies-and-applications/

Miyazaki, C. M., Shimizu, F. M., & Ferreira, M. (2017). Surface plasmon resonance (SPR) for sensors and biosensors. In *Elsevier eBooks* (pp. 183–200). https://doi.org/10.1016/b978-0-323-49778-7.00006-0

Mohamed, E. F., & Awad, G. (2022). Development of nano-sensor and biosensor as an air pollution detection technique for the foreseeable future. In Comprehensive analytical chemistry (pp. 163–188). https://doi.org/10.1016/bs.coac.2021.11.003

Munawar, A., Ong, Y., Schirhagl, R., Tahir, M. A., Khan, W. S., & Bajwa, S. Z. (2019). Nanosensors for diagnosis with optical, electric and mechanical transducers. *RSC Advances*, *9*(12), 6793–6803. https://doi.org/10.1039/c8ra10144b

Muthukrishnan, L. (2022). An overview on the nanotechnological expansion, toxicity assessment and remediating approaches in Agriculture and Food industry. Environmental Technology & Innovation, 25, 102136. https://doi.org/10.1016/j.eti.2021.102136

Mutunga, T., Sinanovic, S., & Harrison, C. S. (2024). Integrating wireless remote sensing and sensors for monitoring pesticide pollution in surface and groundwater. Sensors, 24(10), 3191. https://doi.org/10.3390/s24103191

Narayana, T. L., Venkatesh, C., Kiran, A., Chinna Babu J, Kumar, A., Khan, S. B., Almusharraf, A., & Quasim, M. T. (2024). Advances in real time smart monitoring of environmental parameters using IoT and sensors. Heliyon, 10(7), e28195. https://doi.org/10.1016/j.heliyon.2024.e28195

Naresh, V., & Lee, N. (2021). A review on biosensors and recent development of Nanostructured Materials-Enabled biosensors. *Sensors*, *21*(4), 1109. https://doi.org/10.3390/s21041109

Norizan, M. N., Moklis, M. H., Demon, S. Z. N., Halim, N. A., Samsuri, A., Mohamad, I. S., Knight, V. F., & Abdullah, N. (2020). Carbon nanotubes: functionalisation and their application in chemical sensors. *RSC Advances*, *10*(71), 43704–43732. https://doi.org/10.1039/d0ra09438b

Okorie, N. G. N., Udeh, N. C. A., Adaga, N. E. M., DaraOjimba, N. O. D., & Oriekhoe, N. O. I. (2024). Ethical considerations in data collection and analysis: a review: investigating ethical practices and challenges in modern data collection and analysis. International Journal of Applied Research in Social Sciences, 6(1), 1–22. https://doi.org/10.51594/ijarss.v6i1.688

Pajooh, H. H., Rashid, M., Alam, F., & Demidenko, S. (2021). Multi-Layer Blockchain-Based security architecture for internet of things. *Sensors*, *21*(3), 772. https://doi.org/10.3390/s21030772

Patel, D. K., Kim, H., Dutta, S. D., Ganguly, K., & Lim, K. (2020). Carbon Nanotubes-Based Nanomaterials and their agricultural and biotechnological applications. *Materials*, *13*(7), 1679. https://doi.org/10.3390/ma13071679

Pons, M., Valenzuela, E., Rodríguez, B., Nolazco-Flores, J. A., & Del-Valle-Soto, C. (2023). Utilisation of 5G Technologies in IoT Applications: Current Limitations by Interference and Network Optimisation Difficulties—A Review. Sensors, 23(8), 3876. https://doi.org/10.3390/s23083876

Popescu, S. M., Mansoor, S., Wani, O. A., Kumar, S. S., Sharma, V., Sharma, A., Arya, V. M., Kirkham, M. B., Hou, D., Bolan, N., & Chung, Y. S. (2024). Artificial intelligence and IoT driven technologies for environmental pollution monitoring and management. Frontiers in Environmental Science, 12. https://doi.org/10.3389/fenvs.2024.1336088

Prabhu, S., Nithiya, M., Venkatesan, B., Mohanapriya, R., Raja G. M., & Kumar, M. S. (2024). Smart Biosensors and IoT with Machine Learning Applications in Environmental Monitoring.

In African Journal of Bio-Sciences (pp. 2927–2950). https://www.afjbs.com/uploads/paper/1822c13148d14464121f89db92718385.pdf

Rajak, P., Ganguly, A., Adhikary, S., & Bhattacharya, S. (2023). Internet of Things and smart sensors in agriculture: Scopes and challenges. Journal of Agriculture and Food Research, 14, 100776. https://doi.org/10.1016/j.jafr.2023.100776

Ramesh, M., Janani, R., Deepa, C., & Rajeshkumar, L. (2022). Nanotechnology-Enabled Biosensors: A review of fundamentals, design principles, materials, and applications. *Biosensors, 13*(1), 40. https://doi.org/10.3390/bios13010040

Ray, P. C., Yu, H., & Fu, P. P. (2009). Toxicity and environmental risks of nanomaterials: challenges and future needs. Journal of Environmental Science and Health Part C, 27(1), 1–35. https://doi.org/10.1080/10590500802708267

Raziyan, M. S., Palevicius, A., & Janusas, G. (2024). RevIew—Electrochemical biosensors for Alpha-Fetoprotein Detection: Recent advances and Future Perspectives. *Journal of the Electrochemical Society*. https://doi.org/10.1149/1945-7111/ad586f

Redekar, A., Deb, D., & Ozana, S. (2022). Functionality Analysis of Electric Actuators in Renewable Energy Systems—A review. Sensors, 22(11), 4273. https://doi.org/10.3390/s22114273

Salouti, M., & Derakhshan, F. K. (2020). Biosensors and nanobiosensors in environmental applications. In Springer eBooks (pp. 515–591). https://doi.org/10.1007/978-981-15-2985-6_26

Semeniak, D., Cruz, D. F., Chilkoti, A., & Mikkelsen, M. H. (2022). Plasmonic fluorescence enhancement in diagnostics for clinical tests at Point-of-Care: A review of Recent technologies. *Advanced Materials, 35*(34). https://doi.org/10.1002/adma.202107986

Serebrennikova, K. V., Berlina, A. N., Sotnikov, D. V., Zherdev, A. V., & Dzantiev, B. B. (2021). Raman Scattering-Based Biosensing: New Prospects and Opportunities. *Biosensors, 11*(12), 512. https://doi.org/10.3390/bios11120512

Singamaneni, S., LeMieux, M. C., Lang, H. P., Gerber, C., Lam, Y., Zauscher, S., Datskos, P. G., Lavrik, N. V., Jiang, H., Naik, R. R., Bunning, T. J., & Tsukruk, V. V. (2008). Bimaterial microcantilevers as a hybrid sensing platform. Advanced Materials, 20(4), 653–680. https://doi.org/10.1002/adma.200701667

Skålvik, A. M., Saetre, C., Frøysa, K., Bjørk, R. N., & Tengberg, A. (2023). Challenges, limitations, and measurement strategies to ensure data quality in deep-sea sensors. Frontiers in Marine Science, 10. https://doi.org/10.3389/fmars.2023.1152236

Soliman, W. G., Swathi, C., Yasasvi, T., Priya, B. K., & Reddy, D. A. (2020). Review on poly(ethylene oxide)-based electrolyte and anode nanomaterials for the internet of things node-level lithium-ion batteries. *Materials Today Proceedings, 42*, 429–435. https://doi.org/10.1016/j.matpr.2020.09.806

Srivastava, S., & Srivastava, S. (2023, September 4). IoT networks: communication protocols, security, and Infrastructure – IFACET. IFACET – IITK Foundation for Advanced Continuing Education & Training. https://ifacet.iitk.ac.in/knowledge-hub/internet-of-things/iot-networks-communication-protocols-security-and-infrastructure/

Tang, N., Zhou, C., Xu, L., Jiang, Y., Qu, H., & Duan, X. (2019). A fully integrated wireless flexible ammonia sensor fabricated by Soft Nano-Lithography. *ACS Sensors, 4*(3), 726–732. https://doi.org/10.1021/acssensors.8b01690

Teh, H. Y., Kempa-Liehr, A. W., & Wang, K. I. (2020). Sensor Data Quality: A Systematic Review. Journal of Big Data, 7(1). https://doi.org/10.1186/s40537-020-0285-1

Thakur, M., Wang, B., & Verma, M. L. (2022). Development and applications of nanobiosensors for sustainable agricultural and food industries: Recent developments, challenges and perspectives. *Environmental Technology & Innovation, 26*, 102371. https://doi.org/10.1016/j.eti.2022.102371

Thies, J., Kuhn, P., Thürmann, B., Dübel, S., & Dietzel, A. (2017). Microfluidic quartz-crystal-microbalance (QCM) sensors with specialised immunoassays for extended measurement range and improved reusability. *Microelectronic Engineering, 179*, 25–30. https://doi.org/10.1016/j.mee.2017.04.023

Tian, F., Bonnier, F., Casey, A., Shanahan, A. E., & Byrne, H. J. (2014). Surface enhanced Raman scattering with gold nanoparticles: effect of particle shape. *Analytical Methods*, *6*(22), 9116–9123. https://doi.org/10.1039/c4ay02112f

Tovar-Lopez, F. J. (2023). Recent progress in micro- and Nanotechnology-Enabled sensors for biomedical and environmental challenges. *Sensors*, *23*(12), 5406. https://doi.org/10.3390/s23125406

Tu, J., & Gao, W. (2021). Ethical considerations of wearable technologies in human research. Advanced Healthcare Materials, 10(17). https://doi.org/10.1002/adhm.202100127

Twentyman, J. (2024, February 18). Internet of things sparks race to replace the battery. *Financial Times*. https://www.ft.com/content/3ba7fc12-8205-11e9-a7f0-77d3101896ec

Ullo, S. L., & Sinha, G. R. (2020). Advances in smart environment monitoring systems using IoT and sensors. Sensors, 20(11), 3113. https://doi.org/10.3390/s20113113

Verardo, D., Lindberg, F. W., Anttu, N., Niman, C. S., Lard, M., Dabkowska, A. P., Nylander, T., Månsson, A., Prinz, C. N., & Linke, H. (2018). Nanowires for biosensing: lightguiding of fluorescence as a function of diameter and wavelength. *Nano Letters*, *18*(8), 4796–4802. https://doi.org/10.1021/acs.nanolett.8b01360

Verma, D., Singh, K. R., Yadav, A. K., Nayak, V., Singh, J., Solanki, P. R., & Singh, R. P. (2022). Internet of things (IoT) in nano-integrated wearable biosensor devices for healthcare applications. *Biosensors and Bioelectronics X*, *11*, 100153. https://doi.org/10.1016/j.biosx.2022.100153

Vikesland, P. J. (2018). Nanosensors for water quality monitoring. Nature Nanotechnology, 13(8), 651–660. https://doi.org/10.1038/s41565-018-0209-9

Virk, V., Deepak, H., Taneja, K., Srivastava, R., & Giri, S. (2024). Amelioration in nanobiosensors for the control of plant diseases: current status and future challenges. Frontiers in Nanotechnology, 6. https://doi.org/10.3389/fnano.2024.1310165

Wahab, A., Muhammad, M., Ullah, S., Abdi, G., Shah, G. M., Zaman, W., & Ayaz, A. (2024). Agriculture and environmental management through nanotechnology: Eco-friendly nanomaterial synthesis for soil-plant systems, food safety, and sustainability. The Science of the Total Environment, 926, 171862. https://doi.org/10.1016/j.scitotenv.2024.171862

Wang, C., Liu, M., Wang, Z., Li, S., Deng, Y., & He, N. (2021). Point-of-care diagnostics for infectious diseases: From methods to devices. *Nano Today*, *37*, 101092. https://doi.org/10.1016/j.nantod.2021.101092

Wang, M., Qin, Y., Liu, J., & Li, W. (2023). Identifying personal physiological data risks to the Internet of Everything: the case of facial data breach risks. Humanities and Social Sciences Communications, 10(1). https://doi.org/10.1057/s41599-023-01673-3

Wasti, S., Lee, I. H., Kim, S., Lee, J., & Kim, H. (2023). Ethical and legal challenges in nanomedical innovations: a scoping review. Frontiers in Genetics, 14. https://doi.org/10.3389/fgene.2023.1163392

Yadav, A., Yadav, K., Ahmad, R., & Abd-Elsalam, K. A. (2023). Emerging frontiers in nanotechnology for precision agriculture: Advancements, hurdles and prospects. Agrochemicals, 2(2), 220–256. https://doi.org/10.3390/agrochemicals2020016

Zhu, C., Yang, G., Li, H., Du, D., & Lin, Y. (2014). Electrochemical sensors and biosensors based on nanomaterials and nanostructures. *Analytical Chemistry*, *87*(1), 230–249. https://doi.org/10.1021/ac5039863

The manufacturer's authorised representative in the EU is Springer Nature Customer Service Centre GmbH, Europaplatz 3, 69115 Heidelberg, Germany. If you have any concerns regarding our products, please contact ProductSafety@springernature.com

Printed and bound by CPI Group (UK) Ltd, Croydon, CR0 4YY
01/10/2026
02225922-0018